Advanced High Strength Steel and Press Hardening

Advanced High Strength Steel and Press Hardening

Proceedings of the 4th International Conference on
Advanced High Strength Steel and Press Hardening (ICHSU2018)

Hefei, China, 20 – 22 August 2018

Editors

Yisheng Zhang
Huazhong University of Science and Technology, China

Mingtu Ma
China Automotive Engineering Research Institute, China

World Scientific

NEW JERSEY • LONDON • SINGAPORE • BEIJING • SHANGHAI • HONG KONG • TAIPEI • CHENNAI • TOKYO

Published by

World Scientific Publishing Co. Pte. Ltd.
5 Toh Tuck Link, Singapore 596224
USA office: 27 Warren Street, Suite 401-402, Hackensack, NJ 07601
UK office: 57 Shelton Street, Covent Garden, London WC2H 9HE

Library of Congress Cataloging-in-Publication Data
Names: International Conference on Advanced High Strength Steel and Press Hardening
(4th : 2018 : Hefei, China) | Zhang, Yisheng (1951), editor. | Ma, Mingtu, editor.
Title: Advanced high strength steel and press hardening : proceedings of the 4th International Conference on Advanced High Strength Steel and Press Hardening (ICHSU2018), Hefei, China, 20-22 August 2018 / edited by Yisheng Zhang, Huazhong University of Science and Technology, China, Mingtu Ma, China Automotive Engineering Research Institute, China.
Description: New Jersey : World Scientific, [2018]
Identifiers: LCCN 2018052949 | ISBN 9789813277977 (hardcover)
Subjects: LCSH: Steel--Heat treatment--Congresses. | Hard-facing--Congresses. |
Steel, High strength--Congresses.
Classification: LCC TN751 .I33 2018 | DDC 669/.142--dc23
LC record available at https://lccn.loc.gov/2018052949

British Library Cataloguing-in-Publication Data
A catalogue record for this book is available from the British Library.

For any available supplementary material, please visit
https://www.worldscientific.com/worldscibooks/10.1142/11203#t=suppl

Desk Editor: Tay Yu Shan

Typeset by Stallion Press
Email: enquiries@stallionpress.com

Preface

With continuous development and successful industrial applications of the press hardening technology, there are urgent needs to expand the influence of industrial and academic community within this field by international exchange platforms. In the rapidly growing market of advanced high strength steel and press hardening, the availability of comprehensive knowledge for the exploration of economic potential of this technology is extremely important to sustain its competitiveness. We have realized that research and development, at both academic and industrial application levels, is one of the most important prerequisites for continuing innovation.

In the light of these developments, the 4th International Conference on Advanced High Strength Steel and Press Hardening (ICHSU2018) was held in Hefei, China, to bring together experts all over the world to share their experiences and results of their research in advanced press high strength steel (AHSS), and press hardening technology. After review, we have selected 87 articles to be included in this review volume to cover hot stamping of press hardened steel and high strength aluminum alloy sheet, as well as new hot stamping equipment and design tools for automation.

The organizing committee will like to take this opportunity to express our thanks to the contributors, editors, reviewers and publisher in making this review volume possible. Finally, our thanks also go to all the authors, speakers and participants for their great contributions to the success of ICHSU2018. We will be looking forward to seeing you again in our next international conference on hot stamping.

Yisheng Zhang and Mingtu Ma

ICHSU2018 Committees

Organized by:

- Imperial College London
- State Key Laboratory of Materials Processing and Die & Mould Technology, Huazhong University of Science and Technology
- State Key Laboratory of Vehicle NVH and Safety Technology, China Automotive Engineering Research Institute Co., Ltd.
- National Engineering Laboratory of Clean Energy Vehicle & Power Train Systems, Tongji University
- Key Laboratory of Hubei Province for Modern Vehicle Component, Wuhan University of Technology

Co-organized by:

Ma'anshan Iron and Steel Company
Materials Institution, Chinese Mechanical Engineering Society
Materials Committee of SAE-China
Shanghai Society of Automotive Engineers

Conference Organization

Chairmen

Jianguo Lin, Imperial College London
Mingtu Ma, China Automotive Engineering Research Institute

Vice-Chairmen

Han Dong, Shanghai University
Yisheng Zhang, Huazhong University of Science and Technology
Jianpin Lin, Tongji University
Xuejun Jin, Shanghai Jiaotong University
Lin Hua, Wuhan University of Technology
Fuh-Kuo Chen, National Taiwan University
Jian Zhang, Ma'anshan Iron and Steel Group Co., Ltd.

Organizing Committee

Yisheng Zhang, Huazhong University of Science and Technology

Editorial Committee

Yisheng Zhang, Huazhong University of Science and Technology
Mingtu Ma, China Automotive Engineering Research Institute

Contents

Materials & Testing

Alloying Design and Process Strategy for High Performance 1800 MPa Press Hardening Steel

J. Bian[1,†], H. Z. Lu[2], W. J. Wang[2] and A. M. Guo[2]
[1]Niobium Tech Asia, 068898 Singapore
[2]CITIC Metal, 100004 Beijing, China
[†]E-mail: jian.bian@niobiumtech.com

Press hardening steel (PHS) of 1500MPa grade has been widely used in the car body structure of modern vehicles today. The global trend for light-weighting and safety will drive OEMs to further increase the strength of car body structure in the future. This has given steel industry momentum to develop new advanced high strength steels (AHSS) including PH1800 or even PH2000. The conventional steel grade of PH1500 is based on the Mn-B concept (22MnB5). For the new grade of PH1800, basically only carbon content needs to be increased from 0.22 (PH1500) to 0.32% to reach 1800MPa strength after press hardening. However, due to higher strength related to higher carbon content, PH1800 has lower fracture resistance under impact load, especially at lower temperature or under the influence of hydrogen embrittlement. The delayed cracking due to hydrogen embrittlement is the major concern for OEMs to use high strength PHS currently. Therefore, the development of PH1800 should be focused on the improvement of resistance to hydrogen embrittlement which requires the fundamental changes in alloying design and process optimization compared to conventional PH1500.This paper will highlight the alloying design and process strategy for the new steel grade of PH1800 with the target to improve the crash performance by increasing the fracture resistance under impact load or during bending deformation, especially the resistance to hydrogen induced delayed cracking. The key solution is to achieve the optimum steel microstructure characterized by clean steel matrix with less inclusions and large particles, refined martensite substructure with high share of high angle grain boundary, high homogeneity with less or no banding structure and finely-dispersed precipitates of NbC as hydrogen traps. Based on the experiences gained from development of Nb alloyed PH1500, the new alloying design and process strategy for PH1800 will be discussed.

Keywords: High strength PHS; Bendability; Hydrogen embrittlement; Alloy design; Process strategy; Microstructure optimization; Crash performance.

1. Introduction

The global trend for light-weighting and safety will drive OEMs to further increase the strength of car body structure in the future. In this aspect press hardening provides state-of art light-weighting technology involving high strength, lightweight design and modern manufacturing process. The advantage to increase steel strength through press hardening is that the steel formability during forming process will not deteriorate with increasing strength, as cold forming high strength steels generally do, due to high temperature process. Due to this reason many OEMs are planning to use PH1800 or even PH2000 to replace current PH1500 in the next 2 or 3 years. Basically, in order to reach 1800 MPa strength, only carbon content of the steel needs to be increased from 0.22 % (PH1500) to 0.32 % without changing other alloys and production processes. However, the strength increase beyond 1500MPa will bring about critical problems in performance of PH1800 when used in the car body structure because of reduced resistance to high speed loading (bending behavior and impact energy during crash), especially to hydrogen embrittlement compared to PH1500. The increase of carbon content in martensitic steels like PH1800 will make the steel more sensitive to hydrogen embrittlement which is the major concern of global OEMs to use PH1800 today. Consequently, whether PH1800 can be used by global OEMs in the future depends on if we can develop high performance PH1800 successfully. In order to achieve the target, we need to make fundamental changes in alloying design and appropriate process strategy for steelmaking which will be highlighted in this paper.

2. Development Challenges and Requirements on High Performance PH1800

The crash performance of PH1800 depends on its resistance to high speed impact without premature cracking, especially at lower temperature or under hydrogen influence. Therefore, the development of PH1800 should be focused on improvement of crash relevant steel properties like impact energy, ductile to brittle transition temperature, bending performance and sensitivity to hydrogen embrittlement. In the previous developments and research work [1-3] it was found that these crash relevant properties can be significantly improved on PH1500 only by Nb microalloying for grain refinement and precipitation of nano-sized NbC. However, for PH1800 based on the similar alloying concept like 22MnB5 these properties cannot be sufficiently improved despite 0.05%Nb microalloying and resulted significant grain refinement. Fig. 1 and Fig. 2 show the comparison of performance in bending test (VDA-238-100) and hydrogen embrittlement test between PH1500 and PH1800. These results indicate that PH1800 cannot perform as safely as PH1500 in the crash event if the steel design keeps the traditional Mn-B-Ti concept. The reason is that PH1800 has lower resistance against cracking under heavy impact due to higher strength related to higher C content. For such strength level the impurity elements like S, P and N, the microstructural defects like inclusions and large particles of precipitate and finally the grain size and substructure become much more critical to martensitic steels. Since the traditional Mn-B concepts were not designed for automotive application involving important safety issues, it is necessary to modify the alloying concept for PH1800 which will be introduced in the next paragraph.

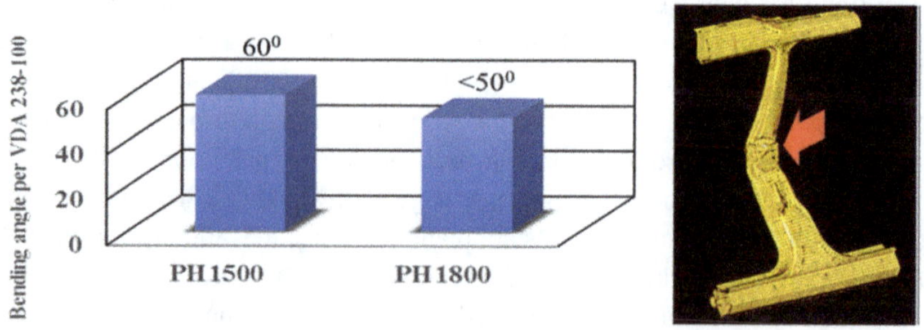

Fig. 1. Comparison of bending angle between PH1500 (22MnB5) and PH1800 (0.32C-1.2Mn-B-Ti-0.05%Nb).

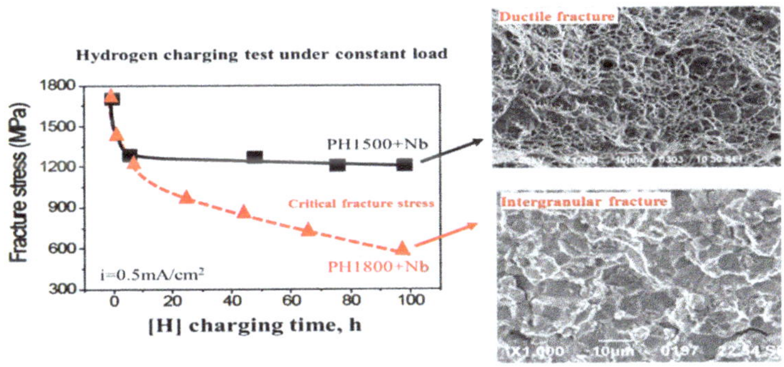

Fig. 2. Comparison of resistance to hydrogen embrittlement behavior between PH1500 (22MnB5-0.05%Nb) and PH1800 (0.32C-1.2Mn-B-Ti-0.05%Nb).

3. Alloying Design for High Performance PH1800

Table 1 shows the alloying design of high performance PH1800 (steel 4) in comparison with other press hardening steels.

Table 1. Alloying design of high performance PH1800 compared to other PHS.

	Wet. %	C	Si	S	P	Mn	Cr	Ti	B	Nb	Mo
1	PH1500	0.24	0.25	0.015	0.025	1.2	0.15	0.03	25	0	0
2	PH1500+Nb	0.24	0.25	0.015	0.025	1.2	0.15	0.03	25	0.05	0
3	PH1800+Nb	0.32	0.25	0.015	0.025	1.2	0.15	0.03	25	0.05	0
4	New design PH1800	0.32	0.6	0.002	0.01	0.8	0.2	x	x	0.05	0.2

Steel 1 is the conventional PH1500 based on 22MnB5. Steel 2 is Nb alloyed PH1500 which was developed in cooperation between several steelmakers and CBMM Brazil. Due to significant grain refinement and precipitation of nano-sized NbC, Nb alloyed PH1500 shows higher performance in bending test (VDA-238-100) and hydrogen embrittlement test compared to conventional PH1500 (steel 1) [1-3]. Today several OEMs in China are using Nb alloyed PH1500 developed by WISCO China. In the further development, same amount of Nb (0.05 %) was added in PH1800 (steel 3) without changing other alloying elements of conventional Mn-B concept. Also, the production processes, especially the hot rolling process, follow the traditional way as PH1500. As a result, the performance of the steel is insufficient in bending test and hydrogen embrittlement test compared to PH1500 shown in Fig. 1-2. Based on these results the new alloying design was proposed as steel 4 in Table 1. Compared to PH1800 of traditional alloying design (steel 3) following alloying changes were made:

- Si: increased from 0.25 to 0.6%
- S: decreased from 0.015 to 0.002%
- P: decreased from 0.025 to 0.01%
- Mn: decreased from 1.2 to 0.8%
- Cr: increased from 0.15 to 0.2%

- Ti: recommended to remove from the concept, if possible
- B: recommended to remove from the concept, if possible
- Nb: added with 0.05%
- Mo: added with 0.2%

The functionality of each alloying element in PH1800 will be explained as follows:

- Si: increases carbon diffusivity and facilitates carbon homogenization in austenite during reheating process of press hardening if ferrite-perlite banding structure exists. Si also reduces hydrogen diffusivity, thus reduces the sensitivity to hydrogen embrittlement of high strength PHS [4]. Si content beyond 0.6% will deteriorate the surface quality after hot rolling and require longer pickling time.

- **S**: S forms MnS during steelmaking which reduces impact energy and bending performance of high strength PHS, thus S should be kept below 20ppm.

- **P**: reduces the grain boundary cohesion and facilitates intergranular fracture under hydrogen influence, thus P should be kept below 0.01%.

- **Mn**: intends to segregate during casting and phase transformation resulting in ferrite-perlite banding structure which is detrimental to mechanical properties, especially the bending performance. Mn also reduces the grain boundary cohesion and facilitates brittle fracture under hydrogen influence, thus Mn should be limited to 0.8%. However, Mn is one of the strongest elements to facilitate hardability. The reduced hardability due to Mn reduction should be compensated by Cr and Mo. Also, higher carbon content contributes to hardability effectively. Nb in solid solution increases hardability as well.

- **Cr**: is the strong element for hardability and does not segregate as much as Mn due to lower addition. Cr is increased to 0.2% to compensate the hardability due to Mn reduction.

- **Ti**: forms TIN at high temperature. TiN precipitates with cubic morphology usually serve as crack initiators during rolling and forming process on one hand, on the other hand TIN will trigger early precipitation of Nb (co-precipitation with Ti) at high temperature leaving less Nb to precipitate at lower temperature for nano-sized NbC. If Hardability is sufficient for PH1800 without B, both Ti and B should be removed from the concept of PH1800.

- **B**: retards ferrite transformation and increases hardability enabling longer transfer time during press hardening. However, if B is added, Ti must be added to protect B from forming BN losing hardability effect. B itself has negative effect on impact toughness, furthermore requires Ti to stabilize N causing more negative effects. In fact, PH1800 has higher C than PH1500, therefore higher hardability is expected. It is beneficial to remove both B and Ti from PH1800 if the hardability is sufficient. However, this should be verified in the lab process.

- **Mo**: is found to have synergy with Nb which further refine the microstructure after press hardening. Mo also reduces the carbon diffusivity, thus reduces the particle size of NbC, especially after annealing and press hardening processes. Mo increases grain boundary cohesion like Nb. These positive results were verified in PH1500 with alloying of 0.2% Mo and 0.05%Nb.

- **Nb:** is essential for grain refinement of martensite substructure leading to high share of high angle grain boundary in the steel matrix to increase the critical fracture stress. The nano-sized particles of NbC precipitated during hot rolling, annealing and press hardening help to pin the grain boundary to keep grain small on one hand and on the other hand to serve as deep hydrogen traps reducing hydrogen diffusivity. With reduced hydrogen diffusivity the sensitivity to hydrogen embrittlement of PH1800 decreases as well, because hydrogen cannot easily segregate to the areas with stress concentration like voids, microcracks, inclusions, large precipitates causing damages. Consequently, Nb microalloying is the key to develop high performance PH1800. However, it was found in TEM investigations that precipitation behavior of Nb in PH1800 is different compared to PH1500 (comparison between steel 2 and 3 in Table 1). Due to higher carbon content in PH1800 the solubility of Nb decreases leading to more Nb precipitation with Ti at high temperature during rolling and subsequent cooling on the run-out table. These particles of (TiNb)(NC) precipitated at high temperature have generally the size range between 50-250nm and precipitation density 0.458 number/µm which is 2.8 times higher than that in PH1500. Consequently, NbC particles in the nano range between 5-10 nm in the final microstructure after press hardening are much less than in PH1500 as shown in Fig. 3. As above mentioned these nano-sized particles of NbC can effectively pin the grain boundary for grain refinement on one hand and on the other hand to serve as deep hydrogen traps to increase the resistance to hydrogen embrittlement. That is the major reason why in the hydrogen charging tests under constant load the critical fracture stress of PH1800 (steel 3) was reduced to less than 600MPa while PH1500 (steel 2) with same amount of Nb microalloying and production conditions achieved more than 1200MPa critical fracture stress after 100h hydrogen charging (Fig. 2). Due to the carbon limitation of PH1800 more than 0.05% Nb cannot be completely dissolved at the slab reheating temperature of 1250 °C. Therefore, we have to modify the process conditions fundamentally, especially implementation of TMCP for hot rolling, in order to strengthen the effect of grain refinement through Nb and to control the precipitation of Nb for sufficient nano-sized particles of NbC in the final microstructure after press hardening which will be explained in the next paragraph.

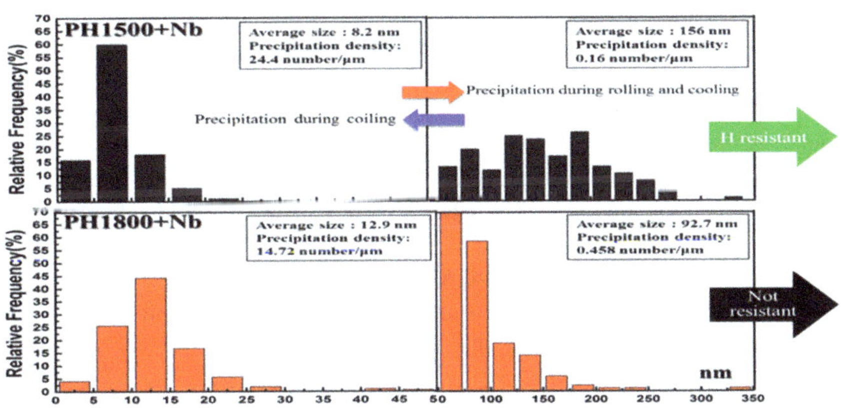

Fig. 3. Comparison of Nb precipitation behavior in PH1500 (steel 2) and PH1800 (steel 3) with 0.05% Nb microalloying (coiling temperature: 620 °C and press hardening after 930 °C reheating).

4. Process Strategy for High Performance PH1800

It is well known that steel properties depend on microstructure and the microstructure will be decisively controlled through the entire metallurgical processes like hot rolling, cold rolling, annealing and other post heat treatments in addition to alloying design. Especially, thermo-mechanically controlled process (TMCP) is one of microstructural control techniques, combining controlled rolling and controlled cooling, to obtain excellent properties, such as high strength, excellent toughness and weldability. Over decades TMCP has been widely applied to produce high quality Nb containing steels both in hot strip and plate for different applications such as oil/gas transportation, structure and engineering. However, for the cold rolled automotive steels, TMCP is seldom practiced in hot rolling mills till today and the final microstructure will be usually adjusted through annealing process. This conventional approach for cold rolled automotive steels has taken one important aspect not into account that the microstructure of hot strip has important influence on the final microstructure after annealing regarding, phase constitutes, grain size, homogeneity and precipitation of microalloying elements. Therefore, to produce high strength steels containing high content of Nb the microstructure control should start with hot rolling process via TMCP in which rolling schedule, cooling strategy and coiling temperature should be modified accordingly in order to explore the maximum benefits of Nb on grain refinement and precipitation hardening.

As for PH1500 based on 22MnB5, conventional hot rolling process is commonly practiced and the coiling temperature is usually between 620 and 650 °C to precipitate most of Nb during coiling process. The high coiling temperature is generally applied to reduce the strength for cold rolling. Recent investigations [5] found that as rolled microstructure obtained from high and low coiling temperature has significant influence on the microstructure transformation during inter-critical annealing, thus on the final mechanical properties of high strength DP steels (0.08C-2.2Mn-0.04Nb). The major results are summarized as follows:

- Nb microalloying (in solid solution) increases hardability and promotes bainite transformation in the hot strip.
- Nb microalloying in combination with low coiling temperature (500 °C) can suppress ferrite-perlite microstructure which is often characterized by banding structure due to high transformation temperature. Perlite is the cluster of high carbon containing phase which requires longer time for carbon homogenization during austenitization.
- If the hot strip is coiled at 620 °C, 50% of Nb is precipitated during coiling. If coiled at 500 °C, only 10% of Nb is precipitated. Nb in solution after hot rolling will precipitate in-situ during annealing leading to very small particle size and high particle density.
- Nb alloyed steels promote austenite formation and accelerate alloying homogenization due to grain refinement during annealing process [6].
- Nb microalloying in combination with lower coiling temperature leads to a

better balance of strength and ductility due to finer, more homogeneous and recrystallized microstructure after inter-critical annealing.

Based on these results the new process strategy for PH1800 was designed and showed in Fig. 4. The target is to strengthen the grain refinement and explore the maximum benefits of Nb to PH1800. The key point is to implement TMCP in hot rolling and subsequently low temperature coiling. Due to Nb addition in 0.05% the non-recrystallization temperature is increased which makes it possible to do the finishing rolling in the non-recrystallization temperature region leading to austenite pan-caking. After rolling finished, fast cooling (>100 °C/s) to avoid ferrite transformation during cooling process. Coiling temperature should be set between 550-600 °C, preferentially 550 °C. Pan-caked austenite and low coiling temperature will make significant grain refinement even without pining of NbC on the grain boundary. At such low coiling temperature most of Nb will remain in solid solution and Nb in solid solution will increase the hardability leading to transformation of bainite, perlite and ferrite after hot rolling. The typical banding structure of ferrite-perlite can be avoided which is detrimental to the bending behavior in rolling direction. The fine-grained microstructure of hot strip will accelerate austenite transformation and homogenization processes during press hardening and refine the final microstructure in the component. Of course, due to lower coiling temperature the strength of hot strip will increase by more than 100MPa and requires higher rolling force for cold rolling. During annealing and subsequent press hardening, Nb, which remains in solid solution after hot rolling, will precipitate in-situ and pin the grain boundary to preserve fine-grained microstructure even at higher temperature (>950 °C) for press hardening. Due to pining effect of NbC, annealing temperature should be set 20-30 °C higher to achieve complete recrystallization compared to non Nb steels. The in-situ precipitated particles of NbC during annealing and press hardening are much smaller in size and higher in density which plays an important role to trap hydrogen to increase the resistance to hydrogen embrittlement. The particles of NbC precipitated during hot rolling and coiling will become coarser through annealing and press hardening which will not be as effective as hydrogen traps as in-situ precipitated during annealing and press hardening. Finally, due to higher carbon content of PH1800 and related lower Ac3 temperature, the reheating temperature for press hardening can also be lowered to 900 °C, which is beneficial to prevent coarsening of microstructure and Nb precipitates during reheating process.

The benefits and role of Nb in such process strategy for high performance PH1800 will be summarized as follows:

- Fine-grained and homogeneous microstructure without banding structure through TMCP rolling and low temperature coiling. Nb increases non-recrystallization temperature and makes it possible to do finish rolling in non-recrystallization temperature region. Low temperature coiling keeps most of Nb in solid solution.
- Nb in-situ precipitation can pin grain boundary more effectively during annealing and press hardening to preserve fine-grained microstructure even at high temperature.

▪ Particles of in-situ precipitated NbC have smaller size, higher density and more homogeneously distributed in the steel matrix as more effective hydrogen traps to make PH1800 less sensitive to hydrogen embrittlement.

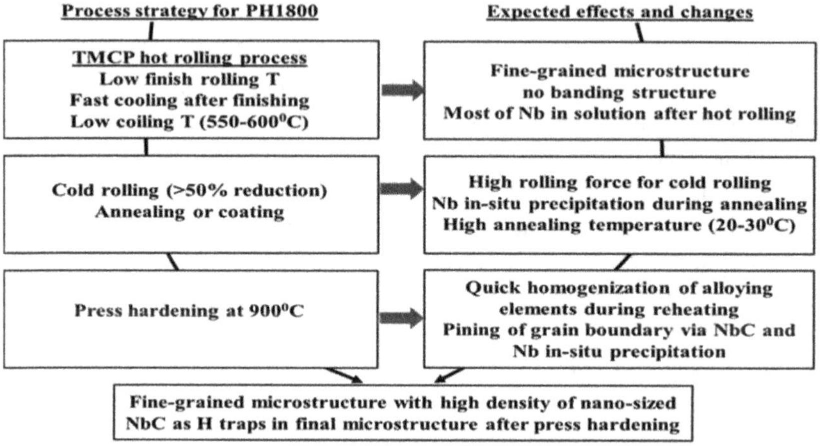

Fig. 4. Process strategy and its impact on the Nb precipitation behavior and the performance of PH1800.

5. Discussion

Crash performance of press hardening steel with martensitic microstructure is dependent on its fracture resistance, especially at high impact speed; lower temperature and in the worst-case scenario under hydrogen influence at the same time. Compared to PH1500, PH1800 has lower fracture resistance due to higher strength because higher carbon content in martensitic steel which is necessary for the required strength reduces ductility, impact toughness, most dramatically the resistance to hydrogen embrittlement. Under these circumstances any premature cracking or fracture in the press hardened components made from PH1800 will lead to uncontrolled intrusion of crash front into the passenger cell. Consequently, the development of PH1800 should be focused on compensation of fracture resistance reduction caused by high carbon content through microstructure engineering in the entire metallurgical processes. The target is to achieve the optimum steel microstructure with following features:

▪ Clean steel matrix with less impurity, inclusions and large particles of precipitates like TiN.

▪ Refined martensite substructure with high share of high angle grain boundary.

▪ High homogeneity with less or no banding structure of ferrite and perlite.

▪ High density of finely dispersed precipitates of NbC as hydrogen traps.

In order to achieve such microstructure for PH1800 it is essential to make fundamental changes to the traditional alloying design of Mn-B steels and simultaneously to modify the current metallurgical processes for PH1500 based on 22MnB5, especially the hot rolling process.

The alloying design of steel 4 in table 1 shows one example of innovative alloying designs for high performance PH1800. The key of the concept is Nb microalloying for making grain refinement and producing nano-sized precipitates of NbC. Both play the important roles to increase the resistance to hydrogen embrittlement which is the most critical problem for PH1800 to be used safely in the car body structure today. Furthermore, grain refinement improves impact energy at lower temperature and bending behaviour in 3-point-bending test contributing to high fracture resistance. To achieve high homogeneity of martensitic microstructure with less second phases and banding structure it is beneficial to keep S, P and N as low as possible in the steelmaking process. It was found that the increase of S content from 30 to 60ppm in PH1500 could reduce impact energy in the stacked Charpy tests by half and the bending angle was less than 40 degrees due to elongated MnS inclusions at the bending corner. Today some leading European Steelmakers produce PH1500 with S less than 20ppm. Since PH1800 is more critical to such inclusions it is recommended to keep S at least below 20ppm. P tends to segregate on the grain boundary, especially during tempering process, and reduces grain boundary cohesion making PH1800 more vulnerable for intergranular fracture under hydrogen attack. Therefore, it is desirable to keep P below 0.01% in PH1800. N is detrimental to impact and bending performance of PH1800 due to formation of TIN. However, to keep B in solution for hardability, Ti must be added to stabilize N. The more N in PH1800, the more Ti is required leading to more TIN formation at high temperature. TIN has cubic morphology and usually serves as stress concentrator. During cold rolling and forming process microcracks can easily occur around TIN particles. The more TIN particles formed in PH1800, the more detrimental effects can be expected. Furthermore, if Nb is added in the steel containing Ti, a part of Nb will co-precipitate with Ti to form (TiNb)(NC) making the particle size even bigger. It is estimated that 10 to 20% of Nb will precipitate with Ti in PH1800 resulting in less Nb for precipitation at lower temperature to form nano-sized NbC particles as hydrogen traps. Fig. 2 shows clearly that PH1800 containing same amount of Nb (0.05%) processed under same conditions has much lower critical fracture stress in the hydrogen charging test under constant load compared to PH1500 because the density of nano-sized Nb precipitates in the range 5-10nm is much lower than in PH1500 (Fig. 3) which indicates much less effective hydrogen traps in PH1800. Now the question is how to increase the density of nano-sized NbC precipitates in PH1800. Due to the higher carbon content (0.32%), the maximum Nb content which can be dissolved at slab reheating temperature of 1250 °C is 0.05%. Higher Nb content is impossible for PH1800. Nevertheless, it is possible to make Nb more effective in PH1800 through alloying and process optimization. From alloying side, it would be beneficial to remove Ti from the concept, if the hardability of PH1800 is sufficient even without B. If B is not added, Ti is not necessary either. B can increase hardability only by suppressing ferrite formation during the transfer time in press hardening. By removing B, the hardability can be compensated by C, Cr, Mo and Nb in the alloying design of steel 4. However, this should be verified in the lab process. From process side, the target is to achieve the maximum grain refinement with effective use of

Nb. The strategy is to make the maximum grain refinement through TMCP in hot rolling and subsequent low temperature coiling (controlled rolling and controlled cooling). During rolling process most of Nb remains in solid solution and increases non-recrystallization temperature to such that finish rolling can be done in the non-recrystallization temperature region with the result of pan-caked austenite. Pan-caked austenite with high density of nucleation sites will transform into very fine and homogeneous microstructure of bainite, ferrite and perlite without banding structure if the coiling temperature is selected between 550 and 600 °C. The advantage of such process strategy is to achieve the maximum grain refinement while keeping most of Nb in solid solution after hot rolling process. During subsequent annealing and press hardening process Nb will precipitate in-situ pining grain boundary to preserve the fine-grained microstructure from hot rolling process. Due to high precipitation potential and short process time in annealing and press hardening (compared to coiling time) a high density of nano-sized particles of NbC is achievable for PH1800 which will play a key role in the resistance to hydrogen embrittlement.

Currently, CBMM is cooperating with several global steelmakers to develop high performance PH1800 based on the alloying designs and process strategy introduced in this paper.

6. Conclusion

PH1800 can offer additional 10-15% weight reduction for the same components made from PH1500. However, PH1800 has lower fracture resistance under heavy impact load, especially at lower temperature or/and hydrogen influence due to higher strength related to higher carbon content in the steel. Investigations show clearly that PH1800 based on the traditional Mn-B alloying concept and produced through conventional processes cannot be safely used in the body structure as PH1500. Consequently, high performance PH1800 should be produced through innovative alloying design and optimized process conditions. Nb microalloying is the key of the new alloying concept for grain refinement and TMCP hot rolling process in combination with low coiling temperature enables Nb to be more effective in PH1800 containing higher carbon content. Pan-caked austenite and subsequent low temperature coiling makes grain refinement and creates ideal conditions for press hardening process with high microstructural homogeneity. Simultaneously, low coiling temperature keeps most of Nb in solid solution after hot rolling which will precipitate in-situ during annealing and press hardening and result in high density of nano-sized NbC which is responsible for high resistance to hydrogen embrittlement of PH1800.

References

1. J. Bian and H. Mohrbacher, Novel alloying design for press hardening steels with better crash performance, in *Proc. of AIST International Symposium on New Developments in Advanced High Strength Sheet Steels*, (Colorado, USA, 2013).

2. J. Bian, H. Z. Liu, W. J. Wang and A. M. Guo, Metallurgical solutions to improve bending and crash performance of press hardening steels. *Proceedings of Hot Sheet Metal Forming of High-performance Steel* 601, (2017).
3. J. Bian, H. Mohrbacher, H. Z. Lu and W. J. Wang, Development of press hardening steel with high resistance to hydrogen embrittlement, in *HSLA Steels 2015, Microalloying 2015 & OES 2015,* (Hangzhou, China, 2015).
4. R. J. Song, N. Fonstein, N. Pottore, H. J. Jun, D. Bhattacharya and S. Jansto, Effect of Nb on delayed fracture resistance of ultrahigh strength martensitic steels, in *HSLA Steels 2015, Microalloying 2015 & OES 2015,* (Hangzhou, China, 2015).
5. R. J. Song, N. Fonstein, H. J. Jun, N. Pottore, D. Bhattacharya and S. Jansto, Effects of Nb on Microstructural Evolution and Mechanical Properties of Low-Carbon Cold-Rolled Dual-Phase, *Metallog. Microstruct. Anal.* **3**, 174 (2014).
6. O. A. Girina, N. Fonstein and D. Bhattacharya, Effect of Nb on the Phase Transformation and Mechanical Properties of Advanced High Strength Dual-Phase Steels, *New Developments on Metallurgy and Applications of High Strength Steels,* (SAE, Buenos Aires, 2008).

Development of Press Hardening Steel and Application Technology in Ma Steel

Y. G. Liu[a], H. Zhan[b], H. R. Gu[c], L. Cui[d], W. Zhang[e] and J. C. Jin[f,†]

Technology Center of Ma'anshan Iron and Steel Company, Ma'anshan, China
E-mail: [a]lyghotmail@hotmail.com, [b]40588205@qq.com, [c]gohero1234@163.com
[d]cuilei1219@163.com, [e]zwniat@126.com, [f,†]jinjiachun1988@163.com

The boron alloyed steel can get ultra-high strength after hot forming process, which can improve the safety while realizing automobile lightweight. So HFS has been used more and more widely. Ma steel has developed HFS since 2013. At present, the HFS produced by Ma steel covers 1300MPa~1800MPa, we can supply the HFS by different product category, such as HR, PO, CR and coated product. In the frontier research of HFS, Ma steel has developed zinc based and high bendability HFS, which have been recognized by customers. On the other hand, in the research of hot stamping technology, TRB and TWB products developed by Ma steel and its partner were used in hot forming parts, which provide a more extensive way for vehicle lightweight selection.

Keywords: Hot forming; Light weight; Press hardening steel.

1. Introduction

With the increasingly severe energy crisis and the urgent need to control greenhouse gases, countries around the world are drawing up new regulations on CO2 emissions, and the lightweight of automobile plays a vital role in realizing energy saving and emission reduction. According to the survey, if the body weight is averagely reduced by 10 %, the fuel consumption will be reduced by 6 % to 8 %, and the emission will be reduced by 10 % [1]. There are three main ways to achieve the body lightweight: 1) the material; 2) the process; 3) the structure. In the selection of car body material, due to the high cost performance and good performance, ultra-high strength steel is widely used, especially in hot stamping steel. Especially hot stamping steel, which has Ultra-high strength after forming, and can realize lightweight car body and improve the safety performance at the same time, has become a necessary material for many automobile manufacturers to design new models.

According to the authoritative forecast, the global demand for hot-stamping parts in 2018 is about 547 million pieces. The hot-stamping forming production line will reach 580 pieces, an increase of about 30 pieces compared with 2017, and it will reach more than 180 pieces only in China [2].

As a super large steel joint venture, Ma steel now defines automotive steel as its strategic product and spends a lot of money and energy to improve the automobile plate brand every year.

2. Development of Press Hardening Steel and Application Technology in Ma Steel

Ma steel has been developing hot forming steel products since 2011. So far, it has accumulated technologies for about 7 years and plays a role in the field of hot forming. According to statistics, in 2017 alone, Ma steel produced and sold 50,000 tons of hot-forming steel.

At present, Ma steel is one of the few domestic steel mills that can provide the whole series of hot forming steel products. According to the delivery state, the products can be divided into four categories: hot rolling, acid washing, cold rolling and coating products; the strength level is covered with 600 MPa~1800 MPa. The progress of hot forming technology of Ma steel is described in detail below.

2.1. *Zinc-based coating hot forming steel products*

Compared with AlSi hot-forming steel, zinc-based hot forming steel has the advantages of sacrificial anode protection, better corrosion resistance, and lower friction coefficient at high temperature, which is conducive to the forming of complex parts, and can improve the service life of mold. Therefore, it has become a research hotspot. Due to the low melting point of zinc, it is easy to melt and even volatilize in the process of hot forming, and there are LME risks in the process of use (shown in Fig. 1 below [3]), thus restricting the promotion and application of hot-forming steel based on zinc coating. Ma steel completed the first round of zinc-based coating hot forming steel product production in July 2017 (shown in Fig. 2), and dozens of technological experiments have been carried out on the industrial production line (shown in Fig. 2). In this paper, the coating morphology, microstructure, corrosion resistance and welding performance of the parts obtained under different technological conditions are studied and analyzed, and the best forming process suitable for zinc-based hot forming steel is found out, which lay a foundation for mass production.

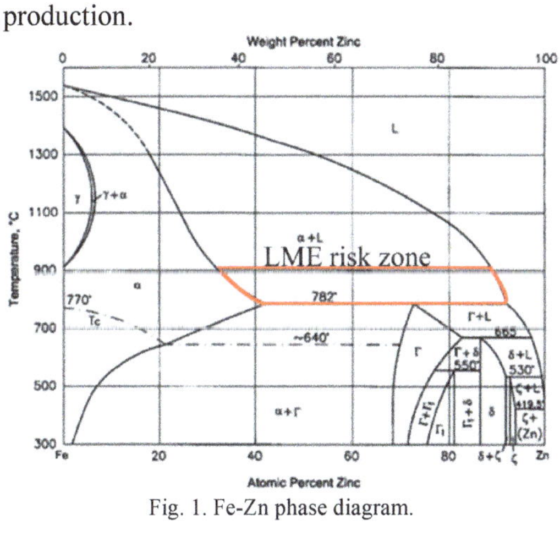

Fig. 1. Fe-Zn phase diagram.

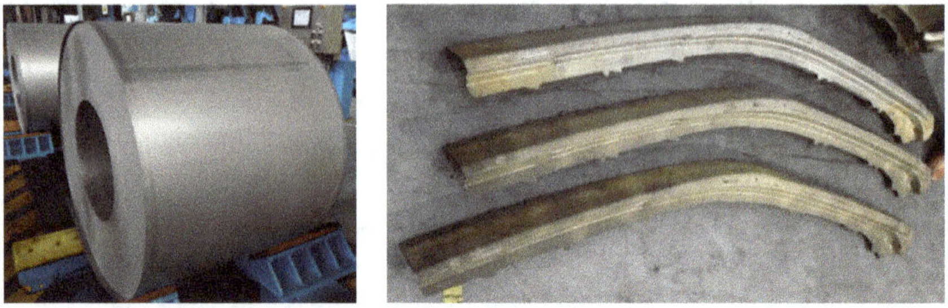

Fig. 2. The first roll of galvanized sheet hot forming steel and finished parts.

2.2. *High cold bending - resistance to delay cracking of hot forming steel products*

At present, the hot forming steel 22MnB5 is widely used, which has high strength after hot forming; however, it also has the risk of poor toughness and delayed cracking. The performance is that the cold bending angle cannot meet the requirements of the high-end mainframe factory [4]. On the other hand, there is a risk of hydrogen to delay cracking [5]. For the insufficient cold bending performance, the main improvement measure is the refinement of grain. For the lack of hydrogen brittleness, the main measure is increasing hydrogen trap to catch hydrogen.

Through subsequent comparative studies and analysis, the result shows that the size of original austenite grain decreased from 12 quartile m before optimized to 7.5 quartile m after optimized (shown in Fig. 3 below), and the hydrogen brittleness, cold bending performance, and the impact resistance performance are significantly improved (shown in Fig. 4 below).

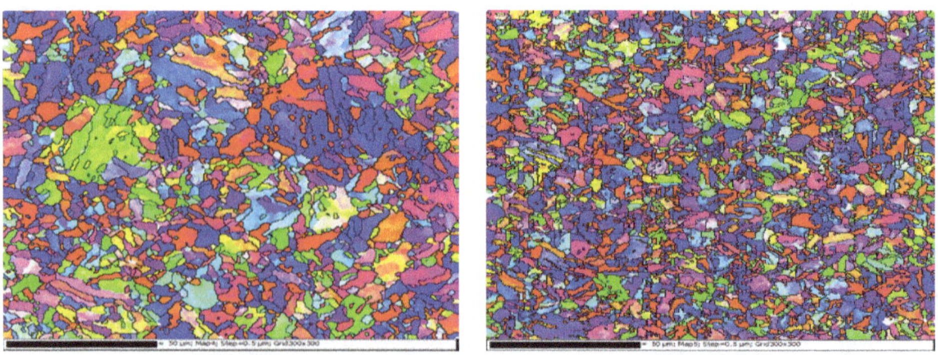

(a) 22MnB5 (b) M1500LW

Fig. 3. Comparison of microstructures.

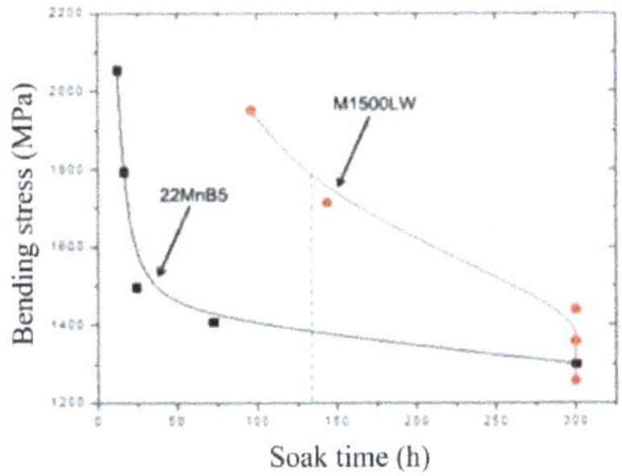

(a) The hydrogen brittleness

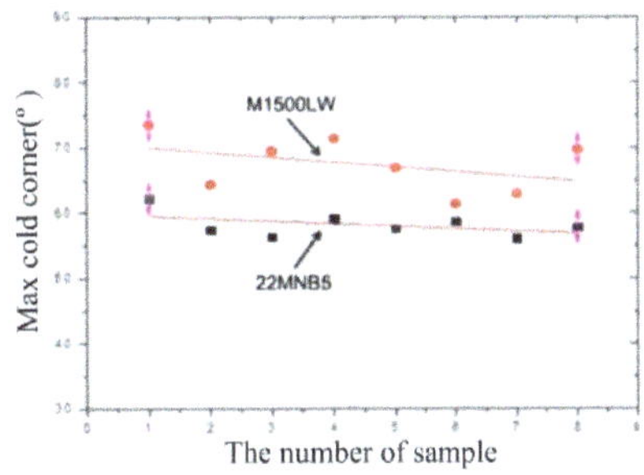

(b) Cold bending performance

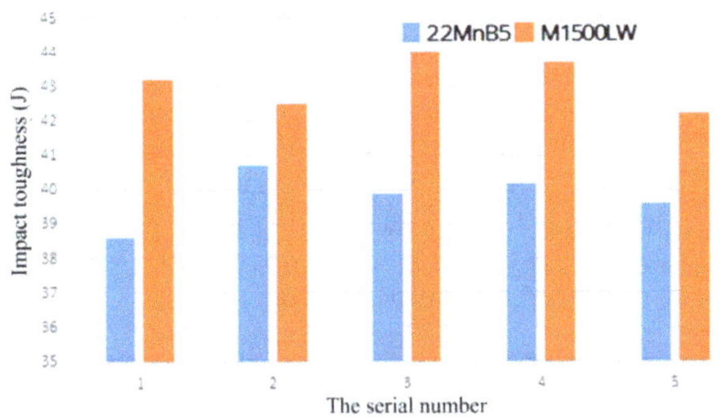

(c) Resistance performance

Fig. 4. Comparison of comprehensive performance.

2.3. *Whole developmental process and comprehensive performance evaluation of hot forming parts*

At present, Ma steel has a professional CAE analysis team of thermal forming, which can use professional analysis software to simulate and analyze the forming process of parts, especially for the cooling process and microstructure analysis, so as to ensure that the parts actually produced meet the requirements.

On the other hand, through the cooperation with partners, Ma steel can complete the whole process of mathematical model analysis die development, part production, part comprehensive performance evaluation and other work, so as to meet customers' requirements for thermal forming design and performance evaluation.

3. Conclusion

(1) Based on traditional Mn-B series hot forming steel, Ma steel has successfully developed a series of products: such as hot forming steel with high cold-bending and anti-delay cracking performance, hot forming steel with galvanized coating, soft zone material with laser welding, and bare plate with high corrosion resistance;

(2) Combined with TRB and TWB technology, variable thickness and strength of hot-forming parts are realized, which provide various schemes for lightweight material selection;
(3) Combined with CAE simulation analysis of hot forming, development of hot forming mold as well as design and production of hot forming parts, we have already possessed a package of technical service capabilities, and preliminarily established evaluation standards for hot forming parts.

References

1. T. Senuma, Physical metallurgy of modern high strength steel sheets, *Isij Int.* **6**, 520 (2001).
2. M. Eriksson and M. Oldenburg, Testing and evaluation of material data for analysis of forming and hardening of boron steel components, *Modelling Simulation of Mater. Sci. Eng.* **10**, 277 (2002).
3. L. Dosdat, J. Petitjean and T. Vietoris, Corrosion resistance of different metallic coatings on press-hardened steels for automotive, *Steel Res. Int.* **82**, 726 (2011).
4. VDA238-100, Plate bending test for metallic materials, in Germany standard 2010.
5. H. Z. Lu, Industry Forum. **15**, 11 (2015).

Effects of Al on Microstructure and Mechanical Properties of Ultrahigh Carbon Steel (UHCS)

W. L. Zhang, H. Wang, H. F. Xu, C. Y. Wang and W. Q. Cao†

Special steel department of Central Iron and Steel Research Institute, Beijing, 100081, China

†E-mail: caowenquan@nercast.com

The effects of aluminum addition on the microstructure and mechanical properties of ultrahigh carbon steel (UHCS) after quenching and tempering were studied in this paper. The results showed that the addition of aluminum significantly enhanced the hardness and temperature resistance. The main reason was attributed to that the addition of aluminum inhibited the transformation of ε-carbide to cementite during tempering and thus resulted in the very high hardness during higher temperature tempering. And with increasing the aluminum content up to 6 %, the carbide network was completely disappeared because increasing addition of aluminum increased the eutectic carbon content significantly. This study not only demonstrated the effects of aluminum on the carbide network but revealed the extra high hardness of the ultrahigh carbon steel alloyed with aluminum, which provided a useful way to develop low density and high hardness bearing steel materials.

Keywords: Aluminum alloying; Ultrahigh carbon steel; Temperature resistance; Bearing steel.

1. Introduction

Ultrahigh carbon steels (UHCSs), which are hypereutectoid steels (1 to 2.1 pct C), have been studied extensively for their unique mechanical properties [1-9]. UHCSs have high strength, high hardness and good abrasion resistance, but are prone to form network carbides due to carbon segregation, which leads to poor plasticity and high brittleness at room temperature [10, 11]. It was demonstrated that addition of aluminum into the UHCS could inhibit the precipitation of network carbides, but tends to form the graphite in UHCSs because aluminum is one of the graphitized elements [1]. Some researchers have shown that the hardness of UHCS alloyed with aluminum after quenching and tempering is over 65 HRC, indicating an extra high hardness could be obtained in high carbon steel [12]. Due to the advantages of suppression of the carbide network by aluminum alloying and the extra high hardness of the UHCS, it could be imagined that addition of aluminum into the UHCS could result in a very promising bearing steel without carbide network but with extra high hardness, which would could enhance the contact fatigue life and wear resistance comparing with that of the conventional bearing steel of GCr15 [13-15].

In this work, the effects of aluminum content on the microstructure and mechanical properties of ultrahigh carbon steel of UHCS-1.4C were studied. The microstructure and hardness of the UHCS with different aluminum content were examined and the effects of aluminum on the hardness and microstructure were discussed, which would provide a fundamental understanding on the aluminum addition effects on the microstructure and mechanical properties of the ultrahigh carbon steels and thus the possibility to design low density bearing steel with extra high hardness.

2. Experimental Procedure

Three kinds of 1.4%C-UHCSs with different aluminum content were prepared by induction melting and ingot casting, and finally forged into bars (60 mm in diameter). And the GCr15 bearing steel for comparison was obtained from the special steel company. The chemical compositions of these steels were given in Table 1. The specimens cut from the bars were heated to different temperatures from 700-1000 °C for 30 minutes and quenched into oil. Then these samples were then cryogenic processed at -73 °C for 2 hours and finally tempered at 160 °C for 2 hours. In order to examine the temperature resistance, the specimens of both UHCSs and the GCr15 after quenching and cryogenic treated at -73 °C for 2 hours were tempered at different temperatures form 160 to 600 °C for 2 hours. The microstructure examinations were examined by scanning electron microscopy (SEM) and transmission electron microscopy (TEM). The hardness was measured by Rockwell hardness tester. In addition, the high temperature hardness of the samples tempered at 160 °C was tested after heating 30 min at 200~500 °C by high temperature hardness tester.

Table 1. Compositions of Ultra-high Carbon steels and GCr15 bearing steel (Wt. Pct.).

Materials	C	Cr	Si	Mn	Al	Fe
UHCS-2Al	1.33	1.54	0.27	0.45	2.05	Balance
UHCS-4Al	1.40	1.68	0.33	0.47	4.7	Balance
UHCS-6Al	1.39	1.62	0.31	0.45	5.98	Balance
GCr15	1.0	1.5	0.25	0.35	-	Balance

3. Results and Discussion

3.1. *Microstructures after forged and determination of phase transition temperatures*

Microstructures of our designed UHCSs after forging were examined by SEM and were shown in Fig. 1(a), (b) and (c). It can be seen very clearly that the number and size of the carbides in grain boundary decreased significantly with the increase of Al content. Fig. 1(c) indicates that the carbide network completely disappeared in the UHCS with 6 % aluminum addition, which is the big problem in in the conventional GCr15 [16]. Thus the addition of Al element obviously inhibited the network carbides of UHCSs. The UHCSs with Al containing were austenized at different temperature with 30minutes and then quenched into oil. The harness evolution is regarded as a function of austenization temperature is plotted in Fig. 1(d). It can be seen from Fig. 1(d) that the A_{c1} temperatures of UHCS-2Al, UHCS-4Al and UHCS-6Al were about 775 °C, 800 °C and 825 °C respectively, and the hardened temperatures were about 825 °C, 850 °C and 875 °C respectively.

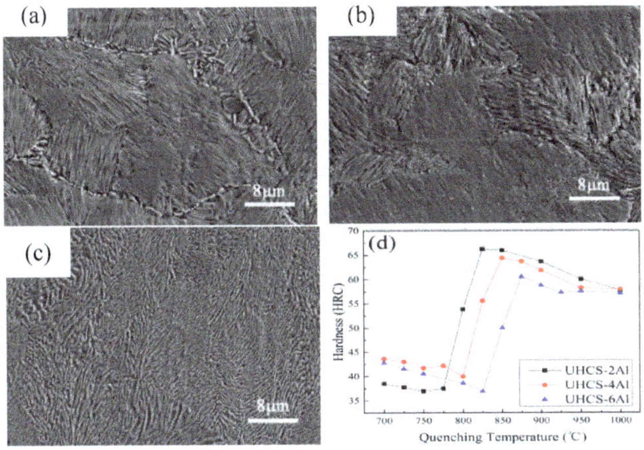

Fig. 1. Microstructures and hardness of UHCSs with Al containing after forging and the hardness of UHCSs with Al containing after quenching at different temperatures, (a) UHCS-2Al, (b) UHCS-4Al, (c) UHCS-6Al and (d) Hardness as a function of austenization temperature.

3.2. *Hardness evolution after different heat treatments*

The heat treatment effects on the microhardness were examined by the hardness evolution as a function of austenized temperature. It can be seen from Fig. 2(a) that after quenching the microhardness remains constant firstly, then increases sharply with increasing of the austenization temperature, and lastly slightly decreases with further increasing of the austenization temperature. The relationship between hardness and austenization temperature could be described by the Sigmoidal trend for all the three UHCSs. After cryogenic treatment, the Sigmoidal trend is still remained, but the hardness is slightly increased comparing with that before cryogenic treatment as shown in Fig. 2(b), especially in the high temperature range. After tempering at 160 °C with 2hours, the hardness is still a Sigmoidal trend with temperature as shown in Fig. 2(c).

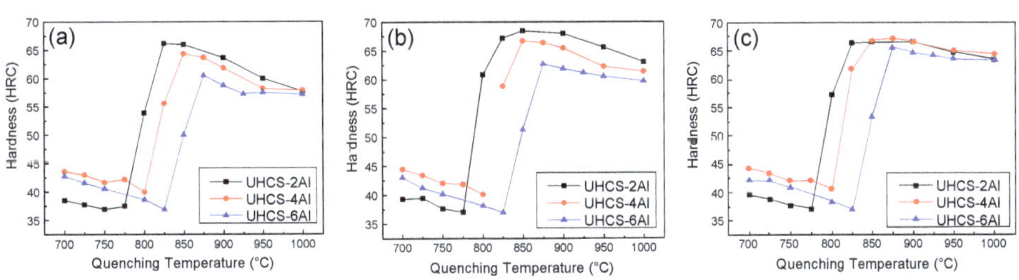

Fig. 2. Hardness of UHCSs quenched at different temperatures, (a) quenching hardness;(b) cryogenic treatment hardness; (c) tempering hardness.

It can be seen from Fig. 2, the hardness of UHCSs with Al containing could reached 65~68 HRC after quenching, cryogenic treatment and tempering at low temperature, indicating a very high hardness in the designed steels.

3.3. *Tempering stability and temperature resistance*

In order to explore the effect of Al content on the tempering stability, GCr15, UHCS-2Al, UHCS-4Al and UHCS-6Al were firstly quenched at 840 °C, 850 °C, 875 °C, 900 °C, and cryogenic treated at -73 °C for 2 hours, and the tempering processes were at 160~600 °C for 2 hours. After heat treatment, the hardness of the sample was measured, as shown in Fig. 3(a).

The hardness of GCr15 steel was 62 HRC after tempered at 200 °C. When the tempering temperature exceeded 200 °C, the tempering hardness decreased continuously with the increase of tempering temperature. The hardness of UHCSs with Al containing was stable at higher level (63~65 HRC) within certain temperature range, which is 200~360 °C for UHCS-2Al, 200~400 °C for UHCS-4Al and UHCS-6Al. The hardness began to decrease as the tempering temperature rose further. This temperature resistance during temperature indicates that the addition of Al inhibited the transition from ε-carbide to cementite during tempering.

The high temperature hardness was tested for UHCS-2Al, UHCS-4Al and UHCS-6Al, which were firstly quenched at 850 °C, 875 °C and 900 °C, then cryogenic treated and lastly tempered at 160 °C for 2 hours. The result was shown in Fig. 3(b). It was shown that 3 kinds of UHCSs with Al containing had good temperature resistance. The hardness is still maintained above 58HRC under 400 °C for all three UHCSs. This greatly enhanced the potential of ultra-high carbon steel with Al containing as high temperature bearing steel.

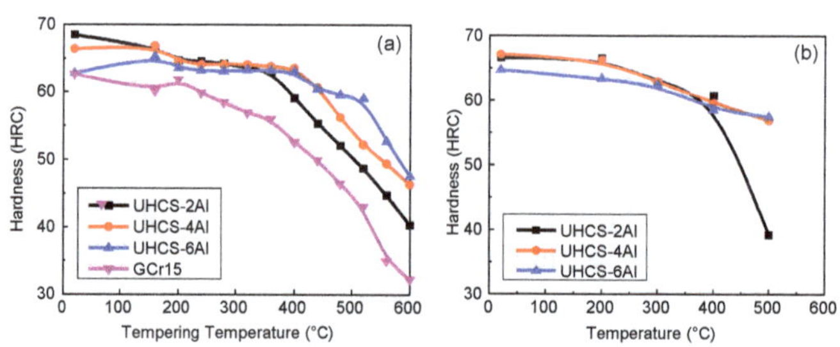

Fig. 3. Hardness of UHCSs and GCr15 tempered at different temperatures in (a) and the hardness at high temperatures in (b).

3.4. *Microstructure stability with temperature*

Figure 4 shows the morphology and diffraction pattern observed by transmission electron microscope of UHCSs with aluminum containing after the tempered at different temperature from 160-560 °C with 2 hours. It was clear that when UHCS-6Al steel was tempered at 200 °C~520 °C, the nanoscale ε-carbides did not decompose, which confirmed that the high hardness and the high temperature resistance is attributed to the stability of the ε-carbides.

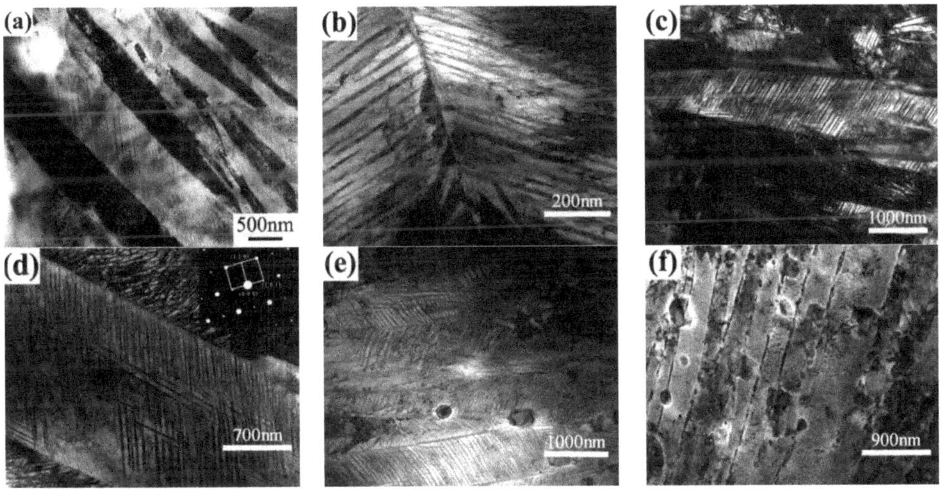

Fig. 4. Microstructures of UHCS-6Al after tempering, (a) 160 °C, (b) 360 °C, (c) 440 °C, (d) 480 °C, (e) 520 °C and (f) 560 °C.

4. Conclusion

Based on the studied on the effects of aluminum addition into the UHCS of 1.4C-1.5Cr steel, it was indicated that the addition of aluminum is benefited to the increasing of the hardness, temperature resistance and the reduction of the density. The main conclusion could be made as following,

(1) The addition of Al inhibited the precipitation of carbide networks and significantly improved of phase transition temperature of ultrahigh carbon steel.

(2) UHCSs with aluminum containing had high hardness of up to 65~68 HRC after quenching and tempering, which would benefit to the wear-resistance and fatigue properties.

(3) The addition of Al significantly improved the tempering stability and temperature resistance of ultrahigh carbon steel, due to inhibition of ε-carbide transforming into granular cemetite in ultrahigh carbon steel alloyed with aluminum.

Acknowledgments

This research is supported by the National Key R&D Program of China (2016YFB0300101 and 2017YFB0304401) and the National Natural Science Foundation of China (NSFC, NO. 51371057).

References

1. O. D. Sherby and T. Oyama, Ultrahigh Carbon Steels, *JOM* **37**, 50 (1985).

2. B. Walser and O. D. Sherby, Mechanical behavior of superplastic ultrahigh carbon steels at elevated temperature, *Metall. Trans. A* **10**, 1461 (1979).
3. T. Oyama and O. D. Sherby, Application of the divorced eutectoid transformation to the development of fine-grained, spheroidized structures in ultrahigh carbon steels, *Scr. Mater.* **18**, 799 (1984).
4. C. K. Syn and D. R. Lesuer, Influence of microstructure on tensile properties of spheroidized ultrahigh-carbon (1.8 Pct C) steel, *Metall. Mater. Trans. A* **25**, 1481 (1994).
5. M. L. Young and J. D. Almer, Load partitioning between ferrite and cementite during elasto-plastic deformation of an ultrahigh-carbon steel, *Acta Mater.* **55**, 1999 (2007).
6. A. Fernández-Vicente and M. Carsí, Toughness dependence on the microstructural parameters for an ultrahigh carbon steel (1.3 wt.% C), *Mater. Sci. Eng. A* **335**, 175 (2002).
7. H. Li and B. Wang, New Spheroidizing Technique of Ultra-High Carbon Steel with Aluminum Addition, *J. Iron Steel Res. Int.* **13**, 9 (2006).
8. Z. L. Zhang and Y. N. Liu, Processing and properties of ultrahigh-carbon (1.6%C) steel, *Mater. Sci. Eng. A* **483**, 64 (2008).
9. E. M. Taleff and C. K. Syn, Pearlite in ultrahigh carbon steels: Heat treatments and mechanical properties, *Metall. Mater. Trans. A* **27**, 111 (1996).
10. E. M. Taleff and M. Nagao, High-strain-rate super plasticity in ultrahigh-carbon steel containing 10 wt.% Al (UHCS-10Al), *Scr. Mater.* **34**, 1919 (1996).
11. M. D. Hecht and B. A. Webler, Digital image analysis to quantify carbide networks in ultrahigh carbon steels, *Mater. Charact.* **117**, 134 (2016).
12. H. Sunada and J. Wadsworth, Mechanical properties and microstructure of heat-terated ultrahigh carbon steels, *Mater. Sci. Eng.* **38**, 35 (1979).
13. X. M. Chen and Y. N. Liu, Tribological Behavior of 1.41 wt% C Ultrahigh-carbon Steel With Quenching and Low-temperature Tempering Treatment, *Tribol. Lett.* **38**, 79 (2010).
14. H. J. Liu and J. J. Sun, Rolling contact fatigue behavior of an ultrahigh carbon steel, *Acta Metall.* **50**, 1446 (2014).
15. H. J. Liu and J. J. Sun, Improved rolling contact fatigue life for an ultrahigh-carbon steel with nanobainitic microstructure, *Scr. Mater.* **91-92**, 17 (2014).
16. N. Miao, Technological Practice of Refining Carbide Network of GCr15 Bearing Steel, *Bearing* (1998).

Microstructures and Properties of a Press Hardening Steel Subjected to Twice Partitioning Treatment

L. J. Wang[a,†] and C. M. Liu[b]

Key Laboratory for Anisotropy and Texture of Materials, Northeastern University, Shenyang, Liaoning, China

E-mail: [a,†]lijunwang@mail.neu.edu.cn, [b]cmliu@mail.neu.edu.cn

As more and more automobile components are manufactured using low alloy steels by press hardening technology, attempts for improved ductility have been made by designing new chemistries and processing paths. Traditional quenching and partitioning processing has been investigated intensively in laboratory scale to achieve excellent ductility through introducing finely and uniformly austenite into martensite, whereas its successful application in industrial hot stamping lines is restricted due to complicated process and enormous equipment investment. In present research, a low carbon medium manganese chemistry is designed. A novel twice partitioning process that combines intercritical pretreatment annealing with slack quenching was explored accordingly to integrate with the new chemistry design to obtain adequate volume fraction of retain austenite at room temperature. The austenite was adequately stabilized by Mn and C enrichment in twice partitioning process. A fine mixed microstructure of martensite and austenite (about 4vol.%) was obtained which offers an excellent combination of ultimate tensile strengths of 1700MPa and significantly improved elongation of 12 %.

Keywords: Press hardening steel; Twice partitioning; Retained austenite.

1. Introduction

More and more structural and safety automobile components are manufactured using press hardening (PH) technology because of its advantages of excellent shape accuracy and possibility of producing ultra-high strength parts without any springback [1]. Though the strength is raised greatly, the car safety performance is only improved slightly, because the strength increment of martensite is achieved is at the cost of ductility [2, 3]. Therefore, improving the strength and elongation simultaneously became challenging in developing new generation of press hardening steels (PHSs) [4].

Recently a novel heat treatment of quenching and partitioning (Q&P) proposed by Speer et al has been applied for developing advanced high strength steels composed of mixed microstructure of austenite and martensite, which grants increased strength without compromising ductility [5, 6]. Inspired by this idea, effects of Q&P processing on microstructures and properties of classical and newly designed PHSs have been studied extensively [7-11]. Since the quenching stop temperature involved in Q&P processing must be controlled strictly between martensite start (M_s) and finish (M_f) temperatures, and enormous equipment investment is demanded for online reheating or isothermal holding, which makes the industrialization of PH-Q&P processing very difficult. As an alternative process, quenching and dynamical partitioning (Q&DP) treatment in which slack quenching was invented to replace isothermal holding or heating in Q&P, failed to obtain adequate volume fraction of retained austenite due to the weak austenite stabilization effect caused by inadequate partitioning conditions [9].

To compensate the weak stabilization effect of carbon enrichment in Q&DP treatment, certain amount of austenite forming element Mn is adopted together with non-carbide forming element Si into low carbon steels. Additionally, proposed is a novel heat-treatment process named twice partitioning (TP) that combines inter-critical annealing with slack quenching. While the former pretreatment is to accelerate the kinetics of substitutional element like Mn partitioning from reheated ferrite to reverted austenite, the slack quenching is to promote carbon dynamic partitioning after a rapid austenization treatment of the resultant microstructure from pretreatment. Due to the large gradient distribution of alloying elements in matrix and resultantly gradual martensitic phase transformation caused by TP treatment, carbon partitioning is promoted further by extended temperature range from M_s to M_f. In present research, the microstructures evolution and mechanical properties of a cold rolled low carbon medium manganese steel sheet subjected to TP treatment were studied.

2. Experimental Methods

The investigated steel with chemical composition shown in Table 1, was smelted using a 150kg vacuum induction furnace. After the ingot was hot rolled into plate with 4 mm thickness, annealing at 620 °C for 4 h was performed. The annealed plate was then cold rolled into sheet with 1.8 mm thickness followed by inter-critical annealing at 680 °C for 5 h. The Q&DP treatment was conducted in following procedure: after a rapid austenization at a temperature in the range of 900~980 °C for 5 min, the specimens were air cooled to room temperature. Tensile test on sheet specimen with dimensions shown in Fig. 1 was performed at room temperature. The microstructure characterizations were carried out using optical microscope (OM) and transmission electron microscope (TEM). The volume fraction of austenite at room temperature was determined through measuring the integrated intensities of $(111)_\gamma$ and $(110)_\alpha$ peaks of X-ray diffraction (XRD) spectrum conducted on TP treated specimens using Cu-K_α radiation.

Table 1. Chemical composition of the test steel (wt.%).

C	Mn	Si	P	S	Ti
0.24	3.0	1.8	0.009	0.002	0.0014

3. Experimental Results and Discussions

The OM microstructures of the test steel hot rolled plate and cold rolled sheet before and after intercritical annealing are shown in Fig. 2, which exhibit martensite, tempered sorbite and $(\alpha+\gamma)$ two-phase microstructures respectively. As a medium manganese steel, the test steel has adequate hardenability to obtain full martensite upon air cooling after hot rolling. To regain formability before cold rolling, the hot rolled plate needs to be annealed at temperatures above 600 °C, which may lie in the intercritical region of (α+Fe_3C)/(α+γ+Fe_3C)/(α+γ). Regardless of the primary annealing temperature, the cold rolled sheet ought to be secondary annealed within temperature range amid of (α+γ) region, making Mn element partitioning to austenite and Si element to ferrite respectively. Through controlling annealing temperature and duration, the elements contents in austenite may be several times higher than that in ferrite.

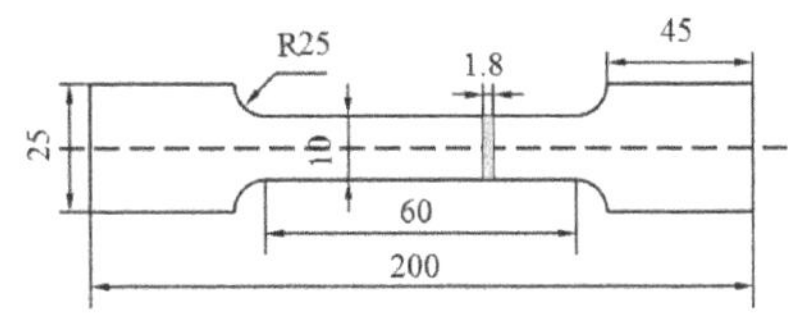

Fig. 1. Specifications of tensile test sample dimensions (mm).

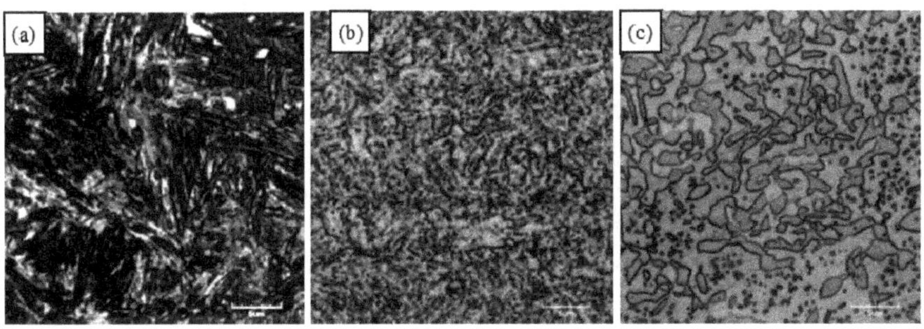

Fig. 2. Microstructures of hot rolled plate and cold rolled sheet before and after intercritical annealing. (a) hot rolled, martensitic structure, (b) cold rolled, sorbite structure, (c) annealed, ferrite + austenite structure.

Subsequent to secondary annealing, the sheet samples were subjected to air quenching after soaking at different temperature for 5min The optical microstructures of the samples are shown in Fig. 3. The metallographes show that, crypto-crystal martensitic microstructures are obtained through TP treatment due to the ultra-fine grain size of prior austenite. The intrinsic and extrinsic factors that hinder austenization and austenite grain growth are Si-enriched and Mn-depleted ferrite and transient soaking time respectively. As the soaking temperature is elevated, the austenite gain coarsens resulting in oriented lath microstructure after slack quenching. As shown in Fig. 4, TEM observation confirms that the microstructures are mainly composed of lath martensite and small fraction of twin martensite, indicating a large carbon concentration in local areas. Different from traditional quenching treatment, the heterogeneous distribution of alloying elements and carbon in local areas featured by TP process enlarges the opportunities of getting twin martensite and retained austenite.

Though the retained austenite is difficult to be detected in both OM and TEM micrographes, the FCC peaks in XRD spectrums confirm the existence of austenite in the microstructure, as shown in Fig. 5. Both the heterogeneous distribution of substitutional alloying elements brought about by prolonged secondary annealing plus transient austenization and carbon enrichment due to dynamic partitioning during slack quenching contribute to stabilize retained austenite at room temperature. The results of quantitative analysis on volume fraction of retained austenite based on X-ray measurement of the samples with different austenization temperatures are listed in Table 2 along with the mechanical properties. It is found that, as the heating temperature rises from 900°C to 980°C, the volume fraction of retained austenite increases slightly till it reach maximum at 940°C, then decreases gradually. Because 2~4vol.% retained austenite exists in martensite microstructure, the yield strength decreases to about 900MPa. Though Ultra-high ultimate tensile strengths about 1700MPa are achieved, the elongations above 10% are obtained, which is much higher than that of traditional boron steels(6%~7%). The elongation reaches the maximum of 12.1% at the heating temperature of 920°C, a bit lower than 940°C at which the maximum fraction of retained austenite is obtained, this should be attributed to the contribution of grain refinement obtained at 920°C.

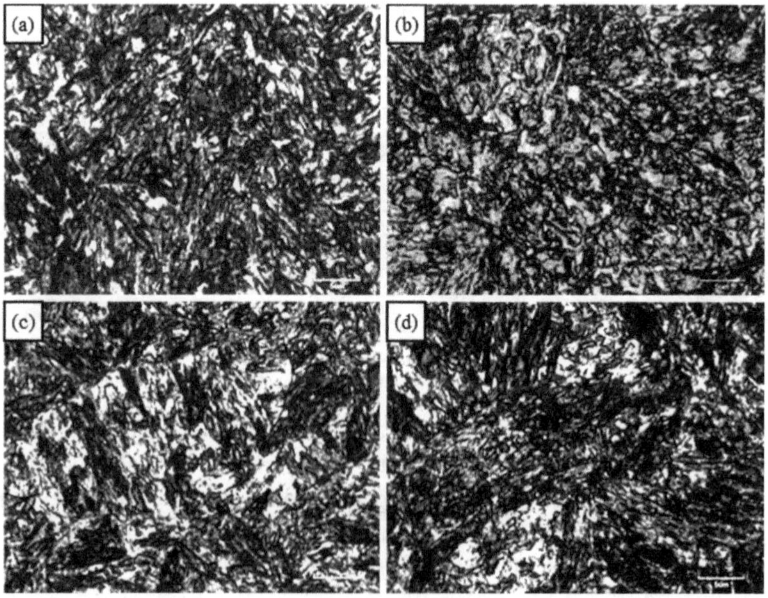

Fig. 3. Microstructures of test steel sheet air quenched from different heating temperature: (a) 900°C, (b) 920°C, (c) 940°C, (d) 960°C.

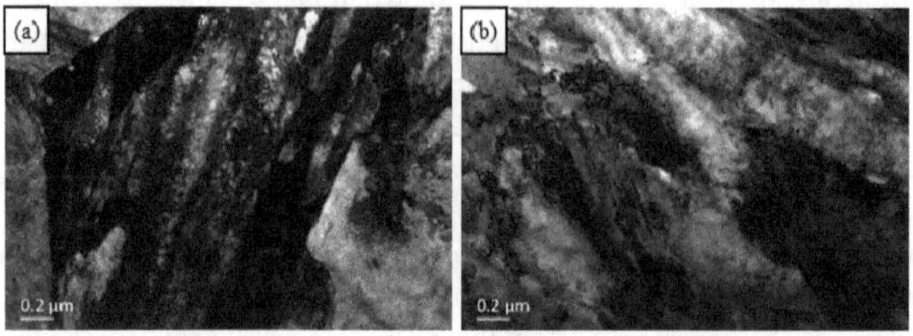

Fig. 4. TEM micrographs of test steel sheet air quenched from 920°C (a) lath martensite, (b) twin martensite.

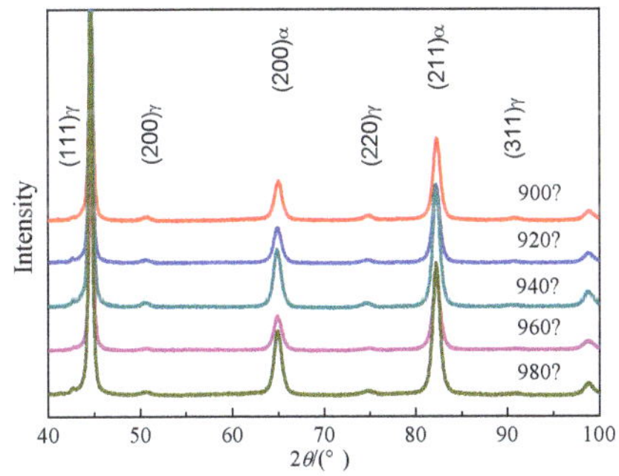

Fig. 5. XRD spectrums of testing steel sheets air quenched from different heating temperatures.

Table 2. Volume fraction of retained austenite and mechanical properties of testing steel sheets air quenched from different heating temperatures.

AT °C	Ψγ %	$Rp_{0.2}$ MPa	Rm MPa	A %
900	2.2	912	1710	10.6
920	2.5	948	1776	12.1
940	4.3	957	1763	11.3
960	3.4	949	1788	10.6
980	2.8	943	1805	10.0

4. Summary

A microstructure of equiaxial austenite islands distributed in the matrix of ferrite is achieved in medium manganese test steel by prolonged secondary inter-critical annealing of cold rolled sheet after cold rolling after hot rolling and primary annealing. The mixed microstructures of small amount of retained austenite within martensite matrix are obtained by subsequent rapid heating and slack quenching. The best properties obtained are Rel=950 MPa, Rm=1700 MPa, and A=12.1%.

Acknowledgments

This work is supported by the National Natural Science Foundation of China (NSFC No. 51571053).

References

1. J. Zhou, B. Y. Wang, M. D. Huang and D. Cui, Effect of hot stamping parameters on the mechanical properties and microstructure of cold-rolled 22MnB5 steel strips, *Int. J. Miner. Metall. Mater.* **21**, 544 (2014).
2. H. Karbasian and A. E. Tekkaya, A review on hot stamping, *J. Mater. Proc. Tech.* **210**, 2103 (2010).

3. T. Taylor, G. Fourlaris, P. Evans and G. Bright, New generation ultrahigh strength boron steel for automotive hot stamping technologies, *Mater. Sci. Tech.* **30**, 818 (2014).
4. M. Naderi, M. Abbasi and A. Saeed-Akbari, Enhanced Mechanical Properties of a Hot-Stamped Advanced High-Strength Steel via Tempering Treatment, *Metall. Mater. Trans. A* **44A**, 1852 (2013).
5. J. G. Speer, D. K. Matlock, B. C. De Cooman and J. G. Schroth, Carbon partitioning into austenite after martensite transformation, *Acta Mater.* **51**, 2611 (2003).
6. D.V. Edmondsa, K. He F. C. Rizzo, B. C. De Cooman, D. K. Matlock and J. G. Speer, Quenching and partitioning martensite-A novel steel heat treatment, *Mater. Sci. Eng. A* **25**, 438 (2006).
7. E. J. Seo, L. Cho, and B. C. De Cooman, Application of quenching and partitioning (Q&P) processing to press hardening steel, *Metall. Mater. Trans. A* **45A**, 4022 (2014).
8. H. P. Liu, X. W. Lu, X. J. Jin, H. Dong and J. Shi, Enhanced mechanical properties of a hot stamped advanced high-strength steel treated by quenching and partitioning process, *Scr. Mater.* **64**, 749 (2011).
9. X. H. Han, Y. Y. Zhong, S. L. Tan, Y. N. Ding and J. Chen, Microstructure and performance evaluations on Q&P hot stamping parts of several UHSS sheet metals, *Sci. China Techn. Sci.* **60**, 1692 (2017).
10. L. O. Wolf, F. Nürnberger, D. Rodman and H. J. Maier, 1-step "quenching and partitioning" of the press-hardening steel 22MnB5, *Steel Res. Int.* **88**, 1600307 (2017).
11. G. Li, M. T. Ma, X. Mao and C. Zhao, New process of hot stamping in combination with Q-P-T treatment for higher dtrength-ductility auto-parts, *Adv. Mater. Res.* **1063**, 223 (2015).
12. W. J. Liu, An Introduction to advanced hot-formed steel for automobile, *Acta Metall. Sin. (Engl. Lett.)* **27**, 373 (2014).

Microstructural and Mechanical Properties Development During Hot Stamping Process in a New Environmental Friendly Steel

K. H. Hu[†], X. P. Mao, R. Ge, S. Z. Wang, Y. Yang, Z. F. Wang and T. Gong

Wuhan Branch of Baosteel Central Research Institute (R&D Center of Wuhan Iron & Steel Co., Ltd.), Wuhan, Hubei, 430080, China

[†]E-mail: hukh@baosteel.com

This presents a detailed characterization of the microstructral and mechanical properties development of a new environmentally friendly steel during the hot stamping process. Hot stamping components with a fully martensitic structure can improve the crashworthiness safety of vehicle. And now the steel product type is cold rolled products or Al-Si coating products, but there is higher cost due to a long process for manufacturing the two products. This study presents an innovative approach of developing a low cost and carbon emission environmentally friendly steel. The steel sheet is manufactured by thin slab rolled directly, so the process is shortened substantially. And the thinnest thickness of the steel sheet can be produced to 1.2 mm. After hot stamping process, the ultimate tensile strength (UTS) of the investigated steel was increased by 243%. There were no obvious work hardening properties. Compared to the cold rolled products for hot stamping, there are comparative mechanical properties. The current window of the test steel spot welding is 4.5 KA. The developed steel with ultra high strength, excellent spot weld capability, and good dimension precision of parts has been applied to the vehicle body for A-pillars, B-pillars, frontal longitudinal beam and so on.

Keywords: Hot stamping process; Microstructure; Automobile parts.

1. Introduction

The weight reduction of vehicle body has become a driving force to decrease fuel consumption and emission by using ultra-high strength steel [1]. However, Parts manufacture for ultra-high strength steel is difficult due to the lower formability and severe springback [2-3]. Hot stamping is an innovative approach to perform the deformation of complex parts and have good dimensional accuracy [4]. Now cold-rolled 22MnB5 steel grade and Al-Si coating steel is widely used in hot stamping processes. The processes to produce the hot stamping steel are very long. And compare to cold forming parts, the cost is high for hot stamping parts. Therefore, it is necessary to find out a solution for reducing the cost of hot stamping parts.

The purpose of this study was to develop newly green ecological cost-efficient hot stamping steel. And hot stamping test of the developed steel was carried out in order to evaluate microstructures, mechanical properties, dynamic strain performance and spot weldability after hot stamping process.

2. Materials and Methods

The new steel in the present work is WHF1500H press hardening steel with a chemical composition (wt%) of 0.21 C, 1.15 Mn, 0.24 Si, 0.008 P, 0.002 S, 0.19 Cr, 0.002 B and balance Fe. The developed steel was produced by thin slab casting and direct rolling (TSCR) line of Wuhan Iron & Steel Co., Ltd.

Hot stamping B-pillar parts were performed using a trial production line for hot stamping. The steel sheet was firstly austenitized for 280 s at 930 °C, then formed and quenched by a die with water cooling system. The microstructure of the specimens from B-pillar was observed using optical microscopy (OM) and transmission electron microscope (TEM). And hardness and tensile test were used to examine mechanical properties of B-pillar.

Specimens with dimensions of 1.5 ×210 ×190 (mm) were cut from the developed steel sheets. Then they were austenitized at 930 °C for 300 s, subsequently, quench hardened by a flat dies. The high speed tensile test and welding test were used to evaluate dynamic strain performance and weldability of the developed steel.

3. Results and Discussion

3.1. *Microstructures evolution*

As-received microstructure of developed steel is shown in Fig. 1. The microstructure of the specimen before hot stamping process appears to contain polygonal ferrite and 29 % blocky pearlite. The white phase is ferrite while the black phase with long block morphology is pearlite. Optical micrographs of the developed steel are austenitized for 280 s at 930 °C is presented in Fig. 2. The grain size of prior austenite is 7.8 μm. The austenitized steel sheets are deformed into B-pillar parts in a die with water cooling system. Base on Fig. 3 observed morphologies, the specimens from the B-pillar are fully martensite, the observed packet size is fine and homogeneously dispersed throughout the microstructure, and the depth of decarburized layer on the surface of the B-pillar part is 0.02 mm, which would ensure an ultra-high strength for the B-pillar parts. A typical TEM image of hot stamping parts with martensitic microstructure is presented in Fig. 4. Obviously, auto-tempered martensite is formed, and acicular-shaped ε-carbide appeared and dispersed on the martensitic lath, which lead to a very good combination strength and ductility.

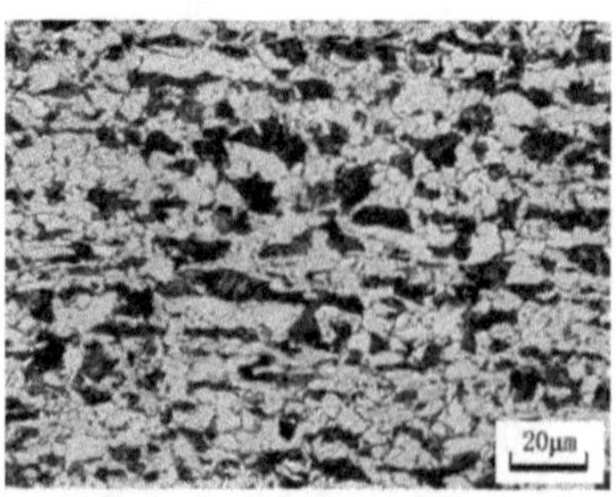

Fig. 1. As-received microstructure of the developed steel.

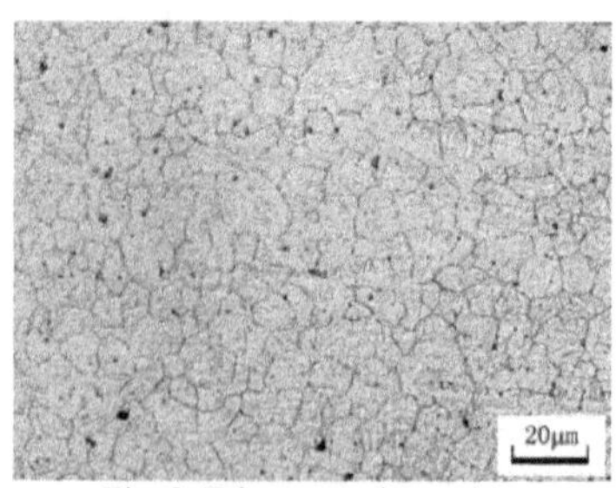

Fig. 2. Prior austenite grains.

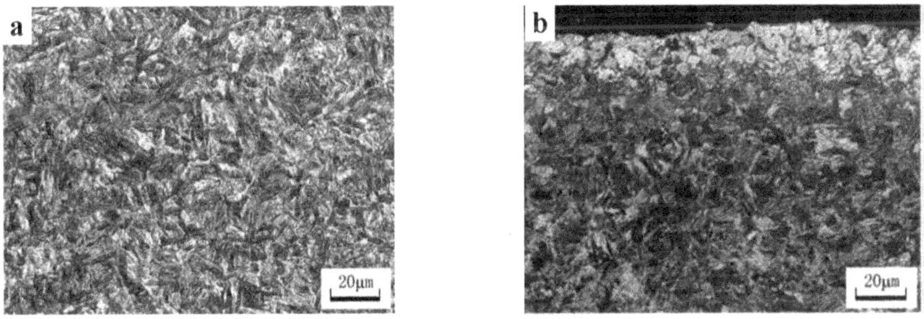

Fig. 3. Microstructure and decarburized layer of the B-pillar parts.

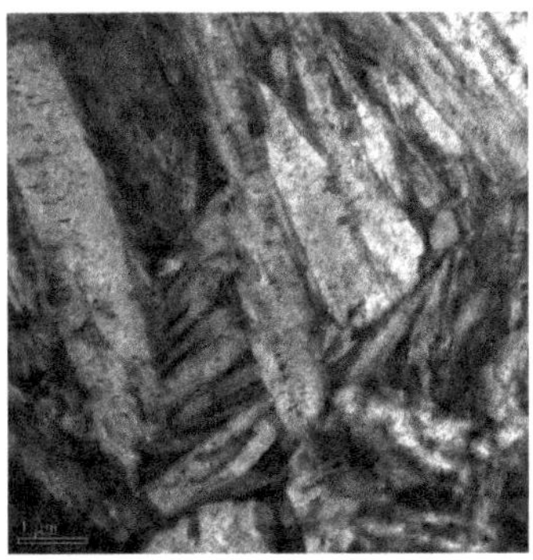

Fig. 4. TEM image of the microstructure of the B-pillar parts after hot stamping.

3.2. *Mechanical properties*

Fig. 5a shows the engineering stress-strain curves of the developed steel before and after hot stamping process. Before hot stamping process, the yield strength of the developed steel is from 461 MPa to 471 MPa, Tensile strength is from 594 MPa to 605 MPa, and the total elongation is above 18 %. During hot stamping process, the austenite-martensite phase transformation leads to a sharply increase of the tensile strength by 243 %, of the yield strength by 138 %.

Fig. 5b shows the dynamic tensile curves of the developed steel for quench hardened specimens by a flat dies. For each engineering stress-strain curve with different strain rate, the developed steels show continuous yielding and considerable strength value with the increase of strain rate. Compared with the high work hardening properties of the conventional steel, less work hardening rate is found for the developed steel. That is due to the fine martensitic packet and block hinder dislocations spreading into grain boundaries, which results in no significant accumulation of dislocations inside martensitic lath, consequently, less work hardening appears for the developed steel.

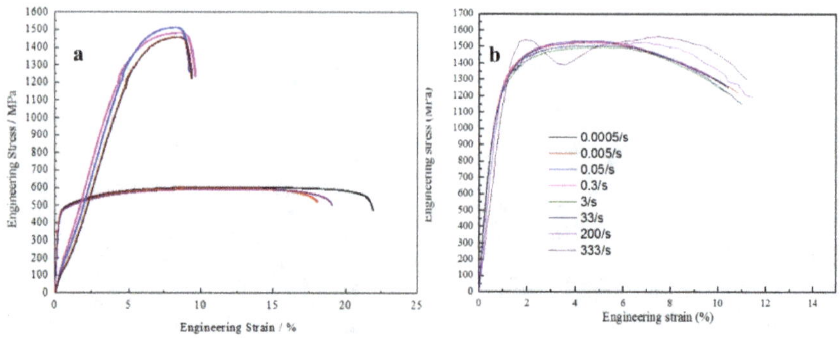

Fig. 5. The engineering stress-strain curves of the developed steel. (a) the static tensile curves before and after hot stamping process; (b) the dynamic tensile curves of the developed steel.

3.3. *Dimensional tolerance of B-pillar*

The dimensional tolerance of the B-pillar is detected by ATOS 3D scanner. Fig. 6 shows the results of the measurement dimensional tolerance. It can be seen that the measured size tolerance of pilot hole is 0.11 mm, which is 0.25 mm for the assembly surface, which is minus 0.52 for the non-assembly surface. The B-pillar parts by the developed steel produced have good dimension precision to compare with the standard requirements, which the size tolerance of the assembly hole and assembly surface are less than ±0.50mm, the non-assembly surface is less than ±1.0 mm.

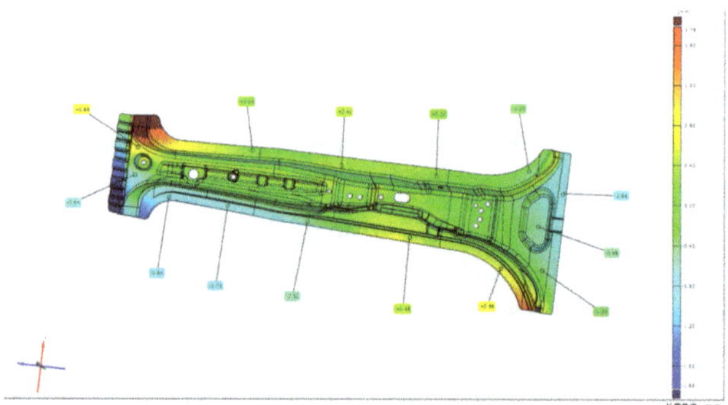

Fig. 6. The dimensional tolerance of B-pillar.

3.4. *Weldability*

It is very important for hot stamping parts that there is good weldability. Fig. 7 shows the nugget diameter as a function of welding current. The diameter of nugget is from 5.0 mm to 8.5 mm with the increase of welding current from 7.8 KA to 12.3 KA. The current window of the developed steel spot welding is 4.5 KA. Fig. 8 shows that the microstructure of welding spot zone is martensite. So the developed steel has excellent welding performance and can fully meet the requirements of automobile manufacturing.

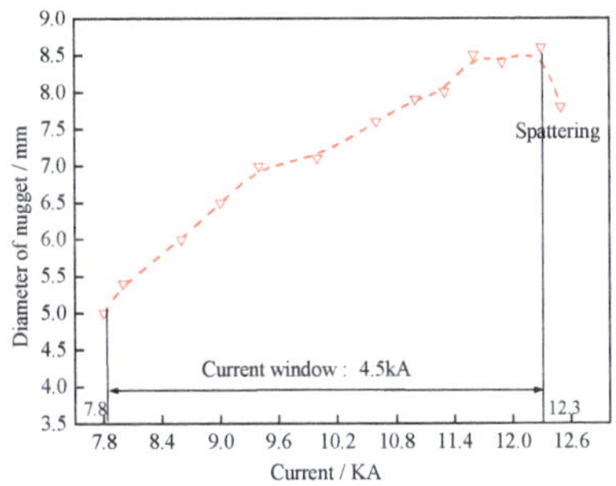

Fig. 7. Nugget diameter as a function of welding current.

Fig. 8. The microstructure of welding spot.

4. Conclusion

(1) A new steel with short manufacturing process show a sharply increase of the tensile strength by 243 % after hot stamping process as compared to the UTS of steel before hot stamping process. And less work hardening appears for the developed steel.

(2) The microstructures of the developed steel before hot stamping process appears to contain polygonal ferrite and 29 % blocky pearlite, the grain size of prior austenite is 7.8 μm. And the microstructure is all homogeneous lath martensite after hot stamping,

(3) For the developed steel, the current window of the test steel spot welding is 4.5kA, and the B-pillar parts by the developed steel produced have good dimension precision.

References

1. N. Li, J. Lin, D. S. Balint and T. A. Dean, Experimental characterization of the effects of thermal conditions on austenite formation for hot stamping of boron steel, *J. Mater. Process. Tech.* **231,** 254 (2016).
2. M. Naderi, M. Ketabchi, M. Abbasi and W. Bleck, Analysis of microstructure and mechanical properties of different boron and non-boron alloyed steels after being hot stamped, *Procedia Eng.* **10**, 460 (2011).
3. M. Naderi, M. Ketabchi, M. Abbasi and W. Bleck, Analysis of microstructure and mechanical properties of different high strength carbon steels after hot stamping, *J. Mater. Process. Tech.* **211**, 1117 (2011).
4. M. Abbasi, A. Saeed-Akbari and M. Naderi, The effect of strain rate and deformation temperature on the chanracteristics of isothermally hot compressed boron-alloyed steel, *Mater. Sci. Eng. A* **538**, 356 (2012).

Warm Rolling and Warm Stamping Performance of the Novel Nb-Mo Microalloyed Medium Mn Steels

H. J. Pan[1], H. Ding[1,a,†], Y. F. Jiang[1], M. H. Cai[1,b] and Y. S. Zhang[2]

[1] *School of Materials and Metallurgy, Northeastern University, Shenyang, 110819, China*

[2] *State Key Laboratory of Material Processing and Die & Mould Technology, Huazhong University of Science and Technology, Wuhan, 430074, China*

E-mail: [a,†]dinghuaneu@163.com, [b]cmhing@126.com

Compared with the hot/cold working of the medium Mn steels, the warm working has received more and more attention for the lower processing temperature and excellent mechanical properties, but the studies about the warm working of the medium Mn steels have just been started. Hence, the warm rolling and warm stamping performance of the Nb-Mo micro alloyed medium Mn steels will be investigated in this work. The results showed that the warm rolled medium Mn steels after 650 °C for 30 min exhibited 950 MPa yield strength (YS), 1340 MPa ultimate tensile strength (UTS), 41 % total elongation (EI) and the UTS×EI>54.9 GPa%, which were better than the hot/cold rolled medium Mn steels. The warm-stamped U-shaped parts of the warm rolled medium Mn steels presented more than YS of 1100 MPa, UTS of 1300 MPa and EI of 33 %, the UTS×EI>42.9 GPa%, which are more than triple times than that of the hot stamped 22MnB5 parts. Moreover, the warm stamped parts presented better surface quality for the smaller yield point extension. Therefore, the warm working may be one of the promising techniques to implement the application of the medium Mn steels.

Keywords: Medium Mn steel; Warm rolling; Warm stamping; Mechanical properties.

1. Introduction

The increasing demand to save fuel consumption by decreasing vehicle weight and maintaining or increasing crash performance has led to the use of advanced high strength steels (AHSS) in car bodies. Among the various candidate steels of the 3rd generation AHSS, the medium Mn steels containing 3-12 wt.% Mn have received significant attention for the large fraction of ultrafine/nanostructured austenite, and a good tradeoff between the mechanical properties and the costs of materials [1]. It is well known that the better mechanical properties of the medium Mn steels is primarily due to the large volume fraction of the meta-stable retained austenite (V_γ). Hence many studies have been done to achieve the large V_γ and improve the stability of retained austenite by controlling rolling process and intercritical annealing (IA) conditions [2]. Recently, the studies about the medium Mn steels mainly focus on the hot/cold rolled blanks, and many works has been well reviewed by Suh et al [3].

Warm rolling normally in the temperature region of about 500-800 °C offers considerable cost savings due to the lower rolling temperature and roll wear [4]. Moreover, the retained austenite exhibited lath and equiaxed mixed grain morphology, which will benefit the retained austenite distributing in different stability levels. Hence, the warm rolled medium Mn steels can exhibit better ductility upon loading for the continuous transformation induced plasticity (TRIP) effect resulted by the mixed grain morphology [1]. Zhang et al. also pointed out that warm rolling could significantly increase YS, UTS and the El of the medium Mn steels by the refined lath thickness and the remained large fractioned austenite [4]. But the studies about warm rolled medium Mn steels were rather limited.

Hot/warm stamping was the novel ways to overcome the springback of the AHSS during the cold working resulted by the relatively high strength and stiffness. The boron steels are the main stamped material, but its further development was restricted for the lower elongation (< 8 %). Recently, many researchers attempted to apply the medium Mn steels to replace the boron steels, and many works have been well reviewed in our previous works [5]. Until now, the studies about the stamped medium Mn steels mainly focus on the hot/cold rolled medium Mn steels. The studies about the warm rolled medium Mn steels were also limited.

Therefore, in the present work, the medium Mn steels will adopt warm rolling and warm stamping processing and forming. Moreover, the microstructures and the mechanical properties will be detailed discussed. The aim of the work is expected to yield ultra-high strength and ductility medium Mn stamped parts from the view of the rolling and stamping approach.

2. Experimental

2.1. *Material and experimental design*

The composition of the medium Mn steel used in the study is Fe-0.17C-5.56Mn-1.21Al-0.05Nb-0.2Mo (wt. %) The fully austenitic temperature of the experimental steel was approximately 750 °C (calculated by TCFE5). The ingots were cast using a 50 Kg vacuum furnace, heated at 1200 °C for 1 h and hot forged to rods of section size ~100 mm×35 mm, followed by air cooling to room temperature. Small billets of 100 mm × 100 mm × 35 mm were cut from the as-received slab. The phase fraction of experimental steels based on an equilibrium thermodynamic analysis predicted by Thermo-Calc is presented in Fig. 1(a). Equilibrium austenite fractions were used as input to the model. A pronounced peak is observed at 650 °C, resulting in maximum austenite fraction (~ 31 %) retained at room temperature for the experimental steels in Fig. 1(b).

Small billets were reheated to 650 °C for 1 h for austenitization. Warm rolling was performed using a four-high rolling machine, in the temperature range from 500 to 650 °C in seven passes with approximately 95 % total reduction. The final sheet thickness was about 1.8 to 2.3 mm. The warm stamping experimental line and the U-shaped tool, designed for this study is same to the equipment in our previous report [6]. The as-received warm-rolled blanks with dimension of 180 mm×260 mm were reheated to 650 °C for 5 min in this study to ensure the continuity of stamping process in practice.

2.2. *Characterization methods and tensile testing*

Tensile specimens were prepared from the warm-stamped parts, parallel to the rolling direction with a gauge section of 25 mm × 6 mm, according to the ASTM E8/E8M standard. Uniaxial tensile tests were carried out on a SANSCMT-5000 machine at a constant strain rate of 1×10^{-4} s^{-1}. Microstructural examinations were performed using field-emission Scanning Electron Microscope (FE-SEM, Supra, SSX-550) and field-emission Transmission Electron Microscope (FE-TEM, FEI, Tecnai G2 F20, operated at 200 kV). SEM specimens were first etched in a 2 % nital solution and then a mixed solution of $Na_2S_2O_5$ water solution and alcoholic solution of picric acid. TEM specimens were mechanically ground to 60~80 μm in thickness and then punched to make round disks of 3 mm in diameter. The disks were electro-polished in a solution of 95% glacial acetic acid (CH3COOH) and 5 % perchloric acid ($HClO_4$) at 16 °C using a twin-jet polisher (Struers, Tenupol). The volume fraction of austenite was determined by X-ray Diffraction (XRD, Rigaku, D/Max2250/PC) using a Cu-Kα radiation source (λ = 1.5405 Å). The γ fraction was calculated using the integrated intensities of all diffraction peaks.

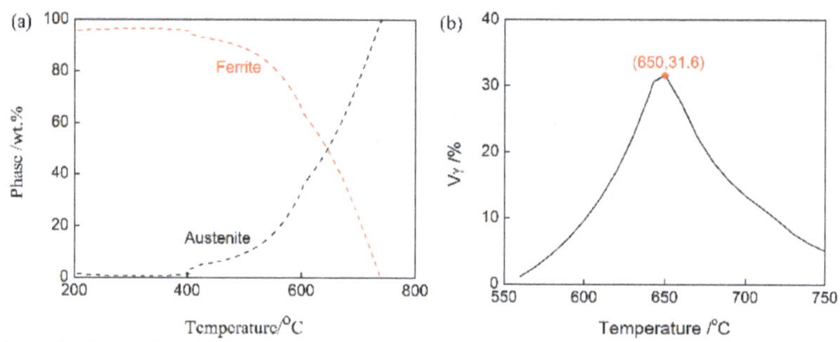

Fig. 1. Schematic illustration of the predictive model development for austenite stabilization as a function of temperature for the experimental steels. (a) phase fractions, (b) calculated retained austenite fraction.

3. Results and Discussion

3.1. *Microstructure and the tensile behaviors of the warm rolled medium Mn steels*

Fig. 2 presented the SEM and TEM results of the warm rolled medium Mn steels after intercritical annealing (IA). The microstructure mainly consisted of retained austenite and ferrite, shown in Fig. 2 (a). Generally, the hot rolled medium Mn steels after IA usually exhibited lamellar microstructure. While cold rolled medium Mn steels often presented equiaxed microstructures after IA resulted by the higher stored energy [1]. In the warm rolling case, owning to the relatively lower rolling temperature and a certain amount of deformation stored energy, the warm rolled medium Mn steels exhibited lath and equiaxed mixed microstructure. So as to further characterize the fine microstructure of the warm rolled samples, the representative TEM morphologies were shown in Fig. 2(b-c). The retained austenite can be divided into two types based on the microstrcture morphology: film-like austenite and blocky austenite. Different morphology of austenite will benefit the better mechanical properties for the continuous TRIP effect caused by the different stability levels [1-3].

The engineering stress-strain curve of the WRed medium Mn steels after IA is displayed in Fig. 3. The stress-strain curves exhibit clear yielding plateaus, followed by significant serrated flow before fracture. The yield strength (YS) and ultimate tensile strength (UTS) respectively reaches 942 MPa and 1341 MPa, and the total elongation (EI) reaches 41 %. The tensile deformation leads to the transformation of about 25 % retained austenite to martensite. Comparted with the HR/CRed medium Mn steels, the relatively higher UTS×EI (PSE) values of the medium Mn steels can be achieved by warm rolling process.

Fig. 2. The SEM and TEM results of the warm rolled medium Mn steels after IA.

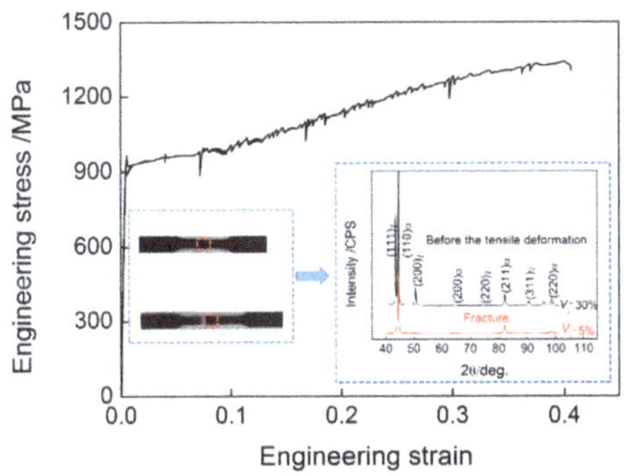

Fig. 3. The engineering stress-strain curve of the WRed medium Mn steels after IA and the volume fraction of retained austenite before or after the tensile deformation.

3.2 *Microstructures and the mechanical properties of the warm stamped medium Mn U-shaped parts*

The macrographs of the stamped parts were shown in Fig. 4. Compared with the relatively thicker oxidation layer (about 300 μm) observed in the 22MnB5 (Fig. 4(c)), the oxidation phenomenon of the Medium Mn steels (Fig. 4(b)) almost can be ignored for both stamping temperatures. In addition, the relatively higher accuracy of size and shape was also obtained for the springback angle ≤ 1°(Fig. 4(a)).

The SEM microstructure of the medium Mn steels stamped parts mainly consisted of retained austenite, ferrite and martensite, shown in Fig. 5(a). The morphology of austenite was similar to the warm rolled medium Mn steels. The initial V_γ (Fig.5(b)) was lower than the warm rolled planks, which may related to the relatively shorter annealing time and the stamping deformation. After the tensile deformation, the V_γ was only 4%, which indicated that about 23% V_γ made the contribution to the mechanical properties.

The engineering stress-strain curves of the WRed medium Mn steels after warm stamping are displayed in Fig. 6. The stress-strain curves exhibit clear yielding plateaus, followed by significant serrated flow before fracture. Compared with the warm rolled samples, the warm stamped parts exhibit relatively higher YS without compromise of UTS. While the EI decrease up to about 8%. Moreover, the Lüders band propagation (LBP) strain is also lower compared with the warm rolled samples.

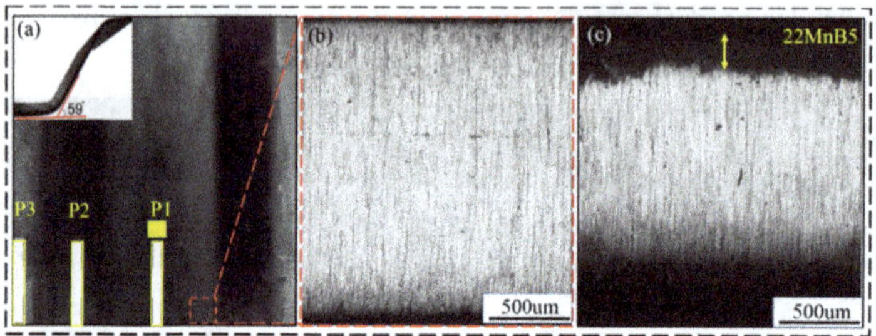

Fig. 4. The macrographs and oxidation behaviors of the experimental steels: (a) The macrographs, the tensile position and the characterization position of U-shaped medium Mn steels stamped parts; (b) The oxidation behaviors of U-shaped medium Mn steels stamped parts; (c) The oxidation behaviors of 22MnB5 at 950 °C for 5min.

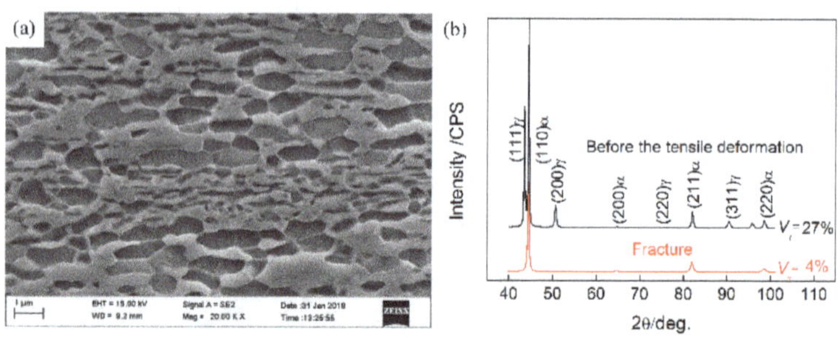

Fig. 5. (a) The SEM microstructure; (b) The volume fraction of retained austenite before or after the tensile deformation.

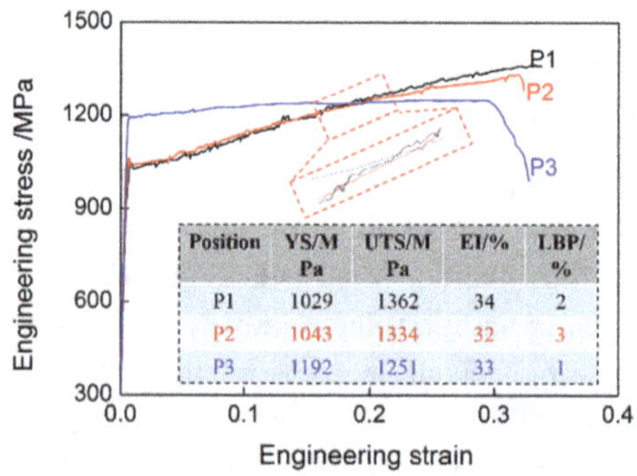

Fig. 6. The engineering stress-strain curves of the WRed medium Mn steels after warm stamping at position P1, P2, P3.

Table 1 present the results of previous work about the stamping of medium Mn steels and present date. The warm stamped U-shaped parts of the warm rolled medium Mn steels exhibited 1300 MPa UTS and 32 %EI, the PSE values was triple times higher than that of the stamped 22MnB5 parts. In addition, the warm rolled medium Mn steels usually presented relatively large yield point extension, but the tensile curves of the warm stamped parts almost presented smaller yield point elongation, which may be related to the higher dislocation density of the stamped parts caused by the stamping deformation [5]. The warm working not only can enhance the mechanical properties, but also can further reduce the energy consumption, when compared with the hot or cold working of the medium Mn steels. Hence, the warm working may be one of the promising techniques to implement the application of the medium Mn steels.

Table 1. A comparison of tensile properties of the present warm-stamped Nb-Mo microalloyed medium Mn alloy with others' experimental data.

Mn (wt.%)	AT(°C)	UTS（MPa）	TE(%)	PSE(GPa·%)	Ref.
4~7	790~840	1420	12	17	[7]
5	800	1520	11.3	17.2	[8]
5~8	760	1880	16	30	[7]
5.56	700	1424	22	31.3	[5]
	760	1717	16	27.5	
5.56	650	1300	33	42.9	Present

4. Conclusion

In this study, warm rolling and stamping performance of Fe-0.17C-5.56Mn-1.21Al-0.05Nb-0.2Mo (wt. %) steels were investigated and the following conclusions can be drawn:

(1) The warm rolled medium Mn steels after IA exhibited lath and equiaxed mixed microstructure for the relatively lower rolling temperature and a certain amount of deformation energy. It presented 1340 MPa UTS, 41% EI and the PSE>54.9 GPa·%, which were better than the hot/cold rolled medium Mn steels.

(2) The warm stamped U-shaped parts of the warm rolled medium Mn steels showed 1300 MPa UTS and 33 %EI, the PSE values >42.9 GPa%, which was triple times higher than that of the stamped 22MnB5 parts. The stamped parts presented better surface quality for the smaller yield point extension.

Acknowledgments

The present work was financially supported by the National Natural Science Foundation, China (No. U1760205) and Henan Provincial Science and Technology Cooperation Project, China (No.182106000016).

References

1. B. Hu, H. W. Luo, F. Yang and H. Dong, Recent progress in medium-Mn steels made with new designing strategies, *J. Mater. Sci. Technol.* **33**, 1457 (2017).
2. J. Han, S. H. Kang, S. J. Lee and Y. K. Lee, Fabrication of bimodal-grained Al-free medium Mn steel by double intercritical annealing and its tensile properties, *J. Alloys Compd.* **681**, 580 (2016).
3. D. Suh, W. Kim and S. J. Medium, Mn transformation-induced plasticity steels: Recent progress and challenges, *Scr. Mater.* **126**, 63 (2016).
4. R. Zhang, W. Q. Cao, Z. J. Peng, J. Shi, H. Dong and C. X. Huang, Intercritical rolling induced ultrafine microstructure and excellent mechanical properties of the medium-Mn steel, *Mater. Sci. Eng. A* **583**, 84 (2013).
5. H. J. Pan, M. H. Cai, H. Ding, H. S. Huang, B. Zhu, Y. L. Wang and Y. S. Zhang, Microstructure evolution and enhanced performance of a novel Nb-Mo micro alloyed medium Mn alloy fabricated by low-temperature rolling and warm stamping, *Mater. Des.* **134**, 352 (2017).
6. C. Y. Chen, C. C. Chen and J. R. Yang, Microstructure characterization of nanometer carbides heterogeneous precipitation in Ti–Nb and Ti–Nb–Mo steel, *Mater. Charact.* **88**, 69 (2014).
7. Y. Chang, C. Y. Wang, K. M. Zhao, H. Dong and J. W. Yan, An introduction to medium-Mn steel: metallurgy, mechanical properties and warm stamping process, *Mater. Des.* **94**, 424 (2016).
8. X. D. Li, Y. Chang, C. Y. Wang, P. Hu and H. Dong, Comparison of the hot-stamped boron alloyed steel and the warm-stamped medium-Mn steel on microstructure and mechanical properties, *Mater. Sci. Eng. A* **679**, 240 (2017).

Effect of Micro Alloying on Hardenability and Mechanical Properties of the 22MnB5 Hot Stamped Steel

A. M. Liu†, Y. Feng, L. F. Song, Y. Zhao and M. T. Ma
China Automotive Engineering Research Institute Co., Ltd, Chongqing, 401122, China
†E-mail: Liuanmin@caeri.com.cn

J. Li
Chery Automobile Co., Ltd., Wuhu, 241006, China
E-mail: lijun6@mychery.com

Two kinds of hot stamped steel 0.04%Nb and 0.04%Nb+0.04%V had been designed and developed on the basis of traditional 22MnB5 hot stamped steel. The hardenability test of three steels was carried out by end quenching method. The results show that the hardenability of 22MnB5 has been improved after adding trace elements, in which the hardenability of 0.04%Nb+0.04%V hot stamped steel is the strongest, and the hardenability of 0.04%Nb steel is not as good as 22MnB5. From the micro level analysis, it is considered that V has a better effect on hardenability than Nb. A comparative test on mechanical properties and bending properties of three kinds of steels had been carried out. The results show that the addition of Nb element alone has the best effect in improving bending properties. The strength and elongation of the newly designed two kinds of hot stamped steels have not been obviously improved. It is believed that it is related to the different influence of the hot stamping process and trace elements on the structure and properties of martensite.

Keywords: Micro alloying; Hot stamped steel; Hardenability; Martensite.

1. Introduction

At present, 1.5 GPa grade hot stamped steel has been widely used in our country, and the hot stamped steel with the strength of 1.8 GPa and 2.0 GPa has been developed successfully used in our country [1-2]. The toughness of the hot stamped steel will decrease gradually with the increase of strength level, and brittle fracture will occur easily during the collision process which is due to the low toughness level. At present, the domestic and foreign automobile enterprises have put forward higher strength and toughness matching requirements for hot stamped steel. A large number of studies have shown that adding certain micro alloying elements in conventional hot stamped steel can improve the strength and toughness of the matrix such as Nb and V etc. By adding Nb element, the grain structure can be refined and fine grain strengthening effect will be achieved, which is conducive to improving the strength, plasticity and toughness of steel at the same time. The addition of V can enhance the strength and toughness of steel by precipitation strengthening during cooling process. In recent years, the industry at home and abroad tends to improve the strength and toughness of steel by adding Nb and V at the same time [3]. However, the effect of composite Micro alloying on the properties of hot stamping steel has not been clearly reported. Some research results have shown that the addition of multiple elements in steel is not necessarily a superposition of the effect of single element addition. Based on the traditional 22MnB5, two new composite micro alloyed steels have been designed and developed. Three kinds of steels were tested for hardenability, mechanical and bending properties, and the influence mechanism and degree of the Nb and V elements on the comprehensive properties of 22MnB5 were analyzed from the micro layer, which will provide valuable guidance for the design and development of more advanced hot stamped steel.

2. Materials

The composite micro alloying idea was used to design new materials. Different composition schemes are as shown in Table 1. The quenching temperature is 930 °C, the holding time is 5 min, and the mold temperature is about 150 °C.

Table 1. Design schemes for different components (wt%).

No.	Chemical elements								
	C	Si	Mn	P	S	Cr	B	Nb	V
1#	0.22	0.26	1.44	0.008	0.004	0.17	0.002	–	–
2#	0.22	0.25	1.45	0.009	0.005	0.16	0.002	0.04	–
3#	0.23	0.26	1.44	0.009	0.006	0.17	0.002	0.04	0.04

3. Test Process

3.1. *Hardenability test*

According to the standard of GB/T 225-2006, the end quenching test of 3 kinds of steels was carried out. Before the test, the material was normalised at 900 °C, and then the end quenching test was carried out after the heat preservation of 25min. The specimen size is 25 mm 100 mm (Figure 1), and the quenching method after austenitic heat treatment is water quenching. 450 HV is considered as the basis for testing the depth of hardened layer.

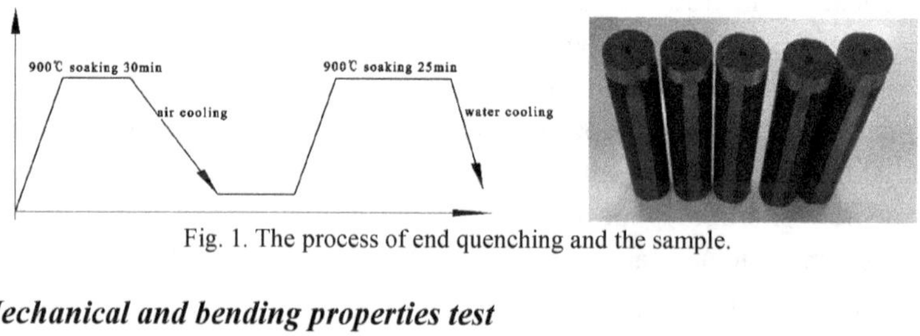

Fig. 1. The process of end quenching and the sample.

3.2. *Mechanical and bending properties test*

Referring to the VDA 238-100 standard of GSA, the comparative tests on the extreme tip cold bending performance of different steels were carried out. The sample size is 60 mm*60 mm, the roller diameter is 30 mm, the minimum spacing between the left and right roller is 3.5 mm, the loading speed is 20 mm/min, the pre bending force is 30 N (Figure 2). The test equipment is the normal material performance testing machine. The extreme tip cold bending angle of each sample was calculated referring to VDA 238-100 standard. In addition, there were 5 times repeated stretching and bending tests carried out under each component condition.

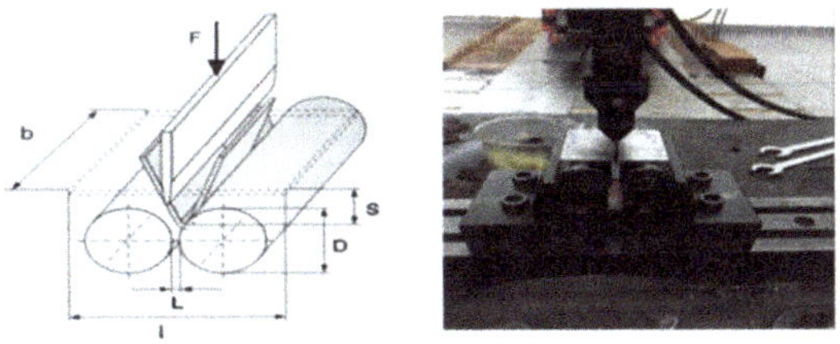

Fig. 2. The bending test process.

4. Test Results

4.1. *Hardenability*

The end quenching curves of the three steels is shown in Figure 3. The distance between the quenched surface to the surface which is not less than the 42 HRC hardness is as the basis for evaluating the hardenability of the material. The results show that the hardened layer depth of 0.04% Nb hot stamped steel is smaller than that of the 22MnB5 steel. The hardened depth of 0.04%Nb+0.04%V hot stamped steel has a significant increase compared with the other two materials and has the highest hardenability. As shown in Figure 4, the micro hardness of the specimen is gradually reduced after the end quenching process, and the microstructure gradually evolves from martensite (M) to bainite (B), pearlite (P) and ferrite (F).

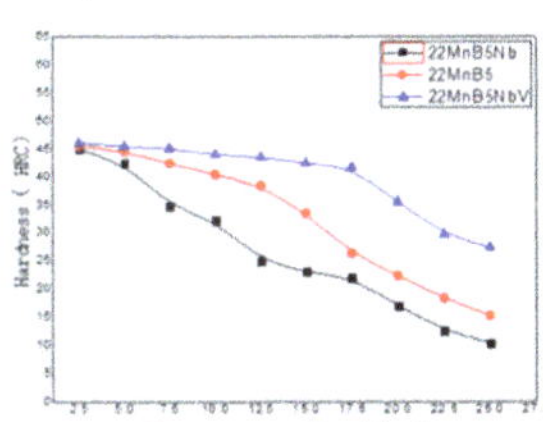

Fig. 3. End quenching curves.

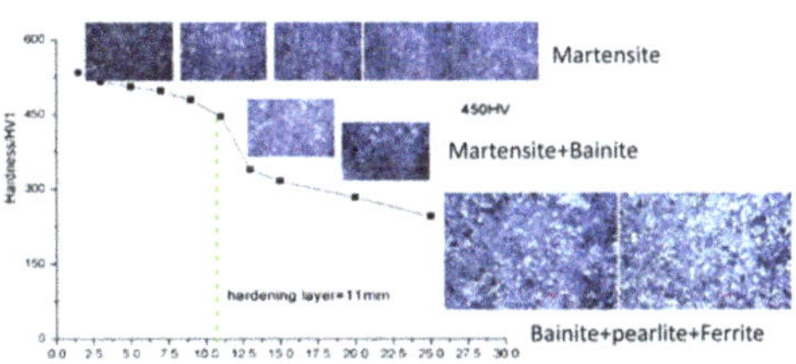

Fig. 4. Hardness and microstructure changes.

The results show that the addition of Nb and V elements has the best effect on the hardenability of 22MnB5 steel while the addition of Nb alone was not conducive to the improvement of hardenability. For steel, except cobalt elements, the other elements can promote the critical cooling transition curve to move right and enhance the hardenability of the material as long as it can be fused to the high temperature austenite before quenching. First of all, Nb elements are usually combined with C elements in steel to form interstitial phases such as $NbC \sim NbC_{0.87}$. The V element is also easy to combine with C elements to form interstitial phases such as $VC \sim V_3C_4$. At room temperature, Nb and V exist in the matrix in the form of precipitated particles. But there are significant differences between the two states at high temperature. The solid fusion state of Nb and V elements under high temperature condition can be described by the solid fusion product equation. The formula (1) ~ formula (2) are the solid fusion product equations of Nb and V. [X] represents the solid melting amount of elements in steel (%). For the three kinds of steels, [C] is totally unified to 0.22. For the sample adding 0.04% Nb element ([Nb]:0.04), assuming that the Nb element is completely fused, the corresponding fixed melting temperature of 1130 °C can be obtained by the formula (1) calculation. For the steel with 0.04% V added, by formula (2), it can be concluded that the temperature of V is 810 °C. It can be seen that the solid melting temperature of V element is obviously lower than that of Nb. The general austenitizing temperature of hot stamping is from 890 to 950 °C, and it is obvious that the Nb element can only exist in the form of precipitated phase in this temperature range, while the V element can be fully fused in the matrix of steel. So the V element is beneficial to improve the hardenability of the material, while the Nb element has no effect on the hardenability of the material. Secondly, under high temperature, the grain growth of austenite is hindered by NbC, the matrix microstructure is refined, and the critical cooling transition curve will be moved to the left (the hardenability of coarse grain steel is stronger), and the hardenability of the steel will be reduced. In addition, the combination of Nb elements and C elements to form carbides will also lead to the decrease of C element concentration in the matrix, which will further reduce the hardenability for medium and low carbon steel [4- 6].

$$\lg[Nb]\cdot[C]=3.42-\frac{7900}{T(K)} \tag{1}$$

$$\lg[V]\cdot[C]=6.72-\frac{9500}{T(K)} \tag{2}$$

4.2. *Mechanical and bending properties*

The tensile and mechanical properties of three kinds of steels are shown in Table 2. At first, the tensile strength and elongation of the three steels are almost the same. The maximum bending angle of the samples added with Nb is the highest, while the samples with Nb and V are the second ones, both of which are higher than those without the addition of trace elements. The experimental results show that the effect of micro alloying on tensile mechanical properties of 22MnB5 hot stamped steel is not so significant, but the effect of micro alloying on bending properties is obvious. The martensitic morphology of three kinds of steels is shown in Figure 5. It can be seen that the martensite microstructure of Nb is the smallest, while the samples containing Nb and V are slightly larger, and the martensite in 22MnB5 is the largest.

Table 2. Comparison of mechanical and bending properties of three kinds of steels.

No.	Rp0.2 (MPa)	Rm (MPa)	δ (%)	Θm (°)
22MnB5	1188	1588	6.9	58
22MnB5Nb	1191	1586	7.1	73
22MnB5NbV	1194	1595	7.0	69

The micro alloying treatment has no obvious improvement on the strength and plasticity of 22MnB5. It is considered that the increase of non martensitic structure in the matrix will lead to the decrease of dislocation density in the matrix after quenching process, which will decrease the dislocation strengthening effect and soften the material for the single addition of Nb. This softening effect should be counteracted with the fine grain strengthening effect of martensite. In the end, there was no obvious change in the strength and plasticity of the material. For the sample adding Nb and V simultaneously, it can be considered that under the condition that Nb has no significant influence on the strength of the material, V can improve the strength of the material by the precipitation strengthening effect. However the precipitation of V has to meet thermodynamic and kinetic conditions. Although the quenching process is beneficial to the precipitation of VC, but the precipitation process of VC is restricted because the cooling rate during the quenching process is too fast. Therefore, the effective precipitation of VC is very small and the enhancement effect is very weak. For the bending properties, the increase of grain boundary area and the large angle grain boundary ratio in the matrix which is due to the grain refinement effect of Nb and the precipitation effect of V can effectively prevent the dislocation movement during the bending process, which can effectively improve the flexural toughness of the material.

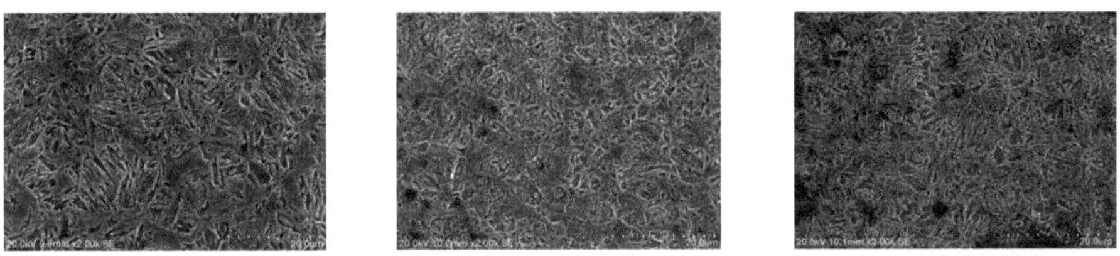

(a) 22MnB5 (b) 22MnB5Nb (c)22MnB5NbV

Fig. 5. Comparison of quenched martensite microstructure of three kinds of steels.

5. Conclusion

Two kinds of micro alloyed steels had been designed and developed on the basis of traditional 22MnB5 hot stamped steel. The hardenability test of three kinds of steels was carried out by end quenching method. The results show that after adding trace elements such as Nb and V, the hardenability of 22MnB5 has been improved. The hardenability of 0.04%Nb+0.04%V hot stamped steel is the strongest, and the hardenability of 0.04%Nb hot stamped steel decreases. The effect of Nb and V on the hardenability of hot stamped steel is opposite, which is related to the difference of their solid melting characteristics in the matrix. There is no obvious change in the strength of 22MnB5 steel through the micro alloying treatment. It is considered that the effect of Nb and V on the microstructure and properties of the matrix is different under the condition of hot stamping. The bending properties of 22MnB5 steel have been improved after micro alloying treatment, which is related to the refine of matrix and precipitation phase, which hinders the movement of dislocation during the bending process.

Acknowledgments

Thanks to the capital and technical support provided by the national key R & D project "The key technology for the forming and application of ultra high strength steel auto parts" (NO: 2016YFB0101605).

References

1. M. T. Ma, H. L. Yi, H. Z. Lu. On The Lightweight of Automobile, *Enginerring Science* **11**, 20 (2009).
2. H. Karbasian and A. E. Tekkaya, A review on hot stamping, *J. Mater. Process. Tech.* **210**, 2103 (2010).
3. P. Chen and K. Muammer, Simulation of springback variation in stamping of advanced high strength steels, *J. Mater. Process. Tech.* **190**, 189 (2007).
4. P. Rozenak and I. M. Robertson, HVEM studies of the effects of hydrogen on the deformation and fracture of AISI type 316 austenitic stainless steel, *Acta Metall. Mater.* **38**, 31 (1990).
5. J. Bian, H. Z. Lu and W. J. Wang, Impact of alloying design on the crash relevant material properties of press hardening steel based on Mn-B concept, in *The 2nd International Conference on Advanced High Strength Steel and Press Hardening (ICHSU)*, (ChangSha, China, 2015).
6. W. S. Choi and B. C. D. Cooman, Characterization of the Bendability of Press-hardened 22MnB5 Steel, *Steel Res. Int.* **5**, 824 (2014).

Study on Extreme Tip Cold Bending Performance of Micro Alloyed Hot-stamped Steel

Y. Feng†, M. T. Ma, Y. Zhao, L. F. Song and J. P. Zhang
China Automotive Engineering Research Institute Co., Ltd., Chongqing, 401122, China
†E-mail: fengyi@caeri.com.cn

X. J. Jin
Shanghai Jiao Tong University, Shanghai, 200240, China
E-mail: jin@sjtu.edu.cn

Based on the composition of traditional 22MnB5 hot stamping steel, two new types of hot-stamped steels with Nb/V composite micro alloying was designed and developed. According to the requirements of the VDA 238-100 standard, comparative tests on the extreme tip cold-bending performance of the 22MnB5, 22MnB5Nb and 22MnB5NbV steels and their corresponding microstructure inspection tests were carried out. The results show that by adding micro elements such as Nb and V, with the refinement of grains and precipitation of carbon-nitrogen compound particles of the second phase, the extreme tip cold bending performance of 22MnB5 hot-stamped steel is improved.

Keywords: Hot-stamped steel; Micro alloying; Extreme tip cold bending performance; Energy absorption.

1. Summary

Lightweight is the most important technical means to achieve a good match between automobile power, safety and fuel saving [1]. The hot stamping technology combines the high temperature stamping and the quenching strengthening which can obtain the 1500 MPa strength grade parts, and perfectly solves the contradiction between the super high strength and the good formability [2-4]. As a safety component, hot stamped parts often bear the angular bending stress along the surface of the vehicle during the collision process. The hot stamped part should have enough strength and excellent toughness to meet the requirements of the collision and energy absorption under the stress state of the plane from the safety considerations. Improving the toughness of matrix by micro alloying has become a hot spot in the field of hot stamped at home and abroad [1]. In this paper, two kinds of Nb/V micro alloyed hot stamped steels had been designed on the basis of traditional 22MnB5 hot stamped steel. According to the VDA 238-100 standard of the German Steel Association, the comparison test and microstructure test of the extreme tip cold bending properties of different components of hot stamped steel were carried out. The mechanism of improving the flexural toughness of 22MnB5 by refining the grain structure and the second phase precipitation strengthening was discussed from the micro level.

2. Material

The composite micro alloying idea was used to design new materials. Different composition schemes are as shown in Table 1. The quenching temperature is 930 °C, the holding time is 5min, and the mold temperature is about 150 °C.

Table 1. Design schemes for different components (wt%).

No.	Chemical element								
	C	Si	Mn	P	S	Cr	B	Nb	V
1#	0.24	0.26	1.44	0.008	0.004	0.17	0.002	_	
2#	0.25	0.25	1.42	0.009	0.005	0.16	0.002	0.04	_
3#	0.24	0.26	1.44	0.009	0.006	0.19	0.002	0.04	0.04

3. Test Process

Referring to the VDA 238-100 standard of German Steel Association, the comparative tests on the Extreme tip cold bending performance of different steels were carried out. The sample size is 60 mm*60 mm, the roller diameter is 30 mm, the minimum spacing between the left and right roller is 3.5 mm, the loading speed is 20 mm/min, the pre bending force is 30 N (Figure 1). The test equipment is the normal material performance testing machine. The bending load displacement curve is recorded in real time in the bending process. The extreme tip cold bending angle of each sample was calculated referring to VDA 238-100 standard. The area below the curve corresponding to the peak load section is defined as the energy absorption value of the sample. Three times tests were repeated under each component.

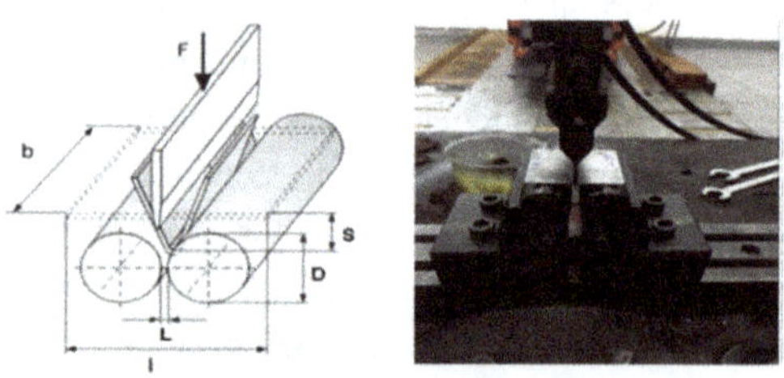

Fig. 1. The bending test process.

4. Test Result

As can be seen in Figure 2, the extreme tip cold bending angle of 22MnB5+NbV (3#) is higher than that of 22MnB5 (1#), but slightly lower than 22MnB5+Nb (2#). It shows that the increasing effect of Nb on the flexural toughness of the steel is better than that of Nb/V compound micro alloying method. As is shown in Figure 3, we can see that the limit bending energy absorption of 3# is also higher than that of 1#, but it is still slightly lower than 2#. In addition, as shown in Figure 3, it also can be seen that the peak bending load of 3# samples is higher than that of 1# and 2# samples which indicates the most favorable for improving the strength of the material.

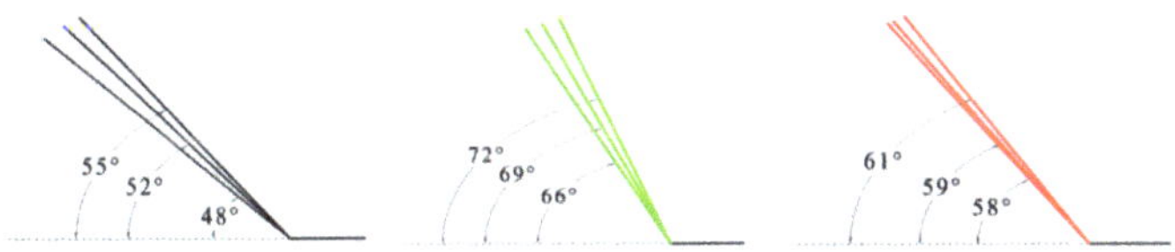

(1) 1# (2) 2# (3) 3#

Fig. 2. Comparison of extreme tip cold bending angle.

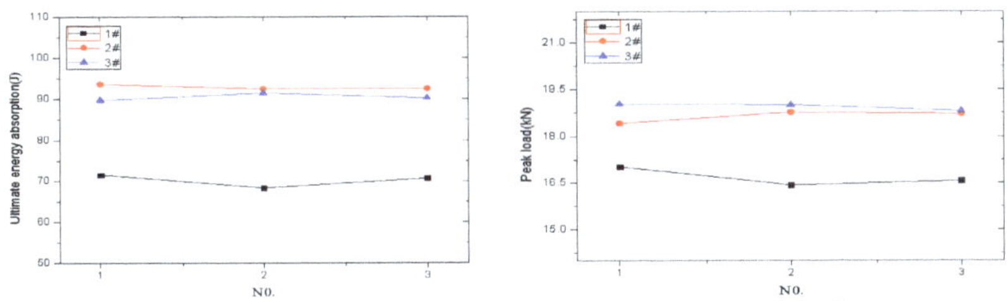

(a) Energy absorption (b) Peak load

Fig. 3. Comparison of energy absorption value and peak load force of extreme sharp cold bending.

5. Discussion

First of all, adding Nb element played a role in improving the extreme tip cold bending property of 22MnB5 hot formed steel. The analysis shows that Nb is a typical carbon and nitrogen compound forming element. The solid solution temperature of Nb (C/N) is about 1150 °C, and the general austenitizing temperature for 22MnB5 is 880 °C -950 °C. So during the austenitizing process, Nb (C, N) can precipitate along the dislocation line or grain boundary, thus refining the austenite grain. As is shown in Figure 4, the original austenite size of 1# is bigger than that of 2#. The analysis shows that the proportion of large angle grain boundary (> 15 degrees) in the matrix of 2# sample is higher than that of the 1# sample. In the process of the bending deformation, these large angle grain boundaries can impede the motion of dislocation due to the bending deformation, and can significantly increase the deformation transfer path distance and consume part of the deformation energy which hinders the deformation expansion in the matrix [5]. The thinner the matrix is, the larger the number of large angle grain boundaries is, and the more significant the effect of deformation prevention is. Secondly, when the external load is certain, the degree of dislocation plug is weaker than that of coarse grain, and the stress concentration effect is relatively lower which is beneficial to enhance the toughness of the matrix, so that there is a certain promotion of extreme tip cold bending angle (up to about 10%). In addition, by adding a certain amount of Nb or V, the amount of the retained austenite will be increased through the promotion of hardenability of 22MnB5 hot stamped steel. These retained austenite structures will produce TRIP effect due to deformation in the process of bending, so as to achieve higher processing hardening and deformation quantity, and further play an important role of improving the extreme tip cold bending angle of the material [6].

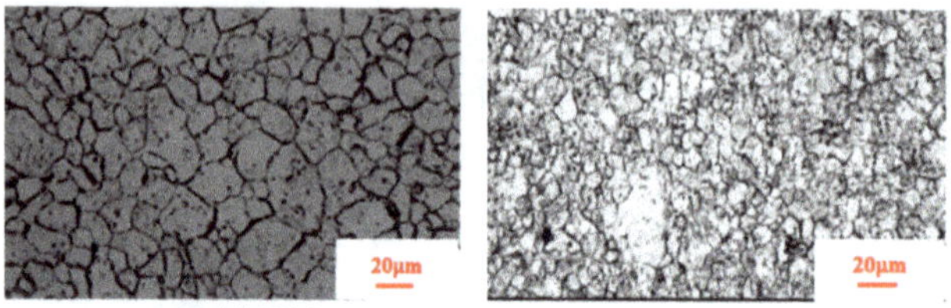

(a) 1# (OM) (b) 2# (OM)

Fig. 4. Comparison of original austenite grain structure between 1# and 2# samples.

Secondly, the extreme tip cold bending angle of 22MnB5 hot stamped steel has also been significantly enhanced by adding the Nb and V elements at the same time, but the effect is not as obvious as that of adding Nb alone, but the limit bending load of the sample is increased. The analysis indicates that V is also a typical carbon and nitrogen compound forming element. But there is a significant difference between the characteristics of V and Nb. The solid solution temperature of V (C, N) is at 700 °C～800 °C which does not play an important role in the refinement of austenite grain. But in the process of quenching, V (C, N) can precipitate in a large amount along the grain boundary, dislocation line and interior of grain. In addition, V can also improve the hardenability of matrix and improve the mechanical properties of quenched martensite structure. Therefore, the extreme tip cold bending limit load of 22MnB5 can be enhanced by adding trace amounts of V. However, it also can be seen that compared with 2# samples, the ultimate bending load capacity of 3# samples added with Nb and V elements did not increase significantly. Because the optimum precipitation temperature of the V element is from 550 °C to 650 °C, and the cooling rate of quenching process is rapid, so it is not conducive to a large number of V (C, N) precipitation from the kinetic point of view, so the strengthening effect is weak. In addition, according to the existing research results, it is shown that in the condition of composite micro alloying, each trace element is not completely independent. One element tends to attach to another element to form a multi-element paragenetic second phase and this kind of secondary phase particles are often large in size (Figure 5). According to the basic material theory, these coarsening particles can not improve the strength or toughness of the steel substrate, and this symbiotic precipitation process will weaken the strengthening effect of Nb and V element. Therefore, the bending limit load of 3# samples is slightly higher than that of the 2# sample, but the effect of the bending energy absorption is slightly reduced.

As in Figure 6, as the bending process continues, the size of martensite decreases gradually which is resulted from the fragmentation of the matrix structure during the bending process. The proportions of the large angle grain boundary in the 1# and 3# samples were reduced after bending process through calculation. This is due to the deformation or rotation of the internal grain in the matrix, which leads to the tendency of the internal crystal orientation to coincide with the loading direction which leads to the reducing of proportion of the large angle grain boundary in the matrix [7-8]. However, it also can be calculated that the proportion changing of the large angle grain boundary before and after bending process of 3# sample is less than that of 1# sample. The possible reasons for the analysis are as follows: on the one hand, the number of grain in the unit volume of the 3# sample is more. When the total deformation energy consumption is certain, the deformation energy consumed by the single grain or the adjacent grain in the 3# sample is lower than that of the 1# sample, which leads to the decrease of the grain boundary angle between some grains. On the other hand, the precipitation of the second phase particles in the 3# sample will impede the grain deformation or rotation, which inhibits the decrease of the grain boundary ratio in the large angle.

(a) VC (b) NbC (c) (Nb+V) C

Fig. 5. Typical precipitates in microalloyed 22MnB5 matrix.

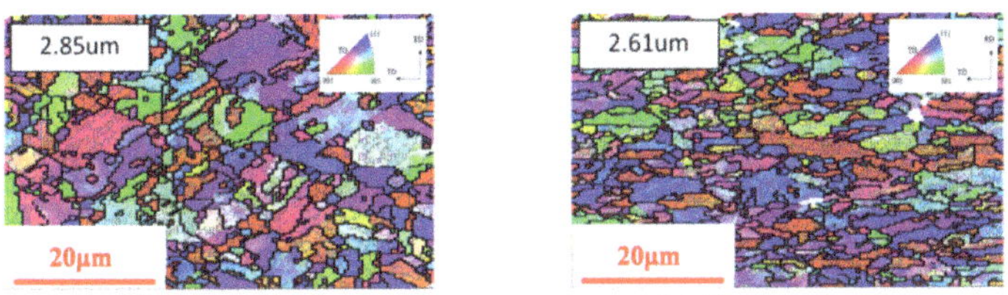

(a) 3# -Before bending (b) 3# -After bending

Fig. 6. Grain orientation changes before and after bending process (3#).

6. Conclusion

Two kinds of micro alloyed hot stamped steel were designed on the basis of 22MnB5. According to the VDA-238 standard, the cold bending properties of three kinds of hot stamped steel were compared. The extreme tip cold bending angle and bending energy absorption value of 22MnB5+NbV are higher than those of 22MnB5, but slightly lower than 22MnB5+Nb. The refinement of austenite grain from Nb (C, N) and the increase of the retained austenite content which dues to adding V element are two key factors to improve the cold bending performance of 22MnB5.

Acknowledgments

This work is supported by the National Natural Science Foundation of China: the joint fund for the innovation and development of the China automotive industry, "the mechanism and regulation of the strength and toughness of the ultra high intensity hot stamped organization for automotive use" (No: U1564203).

References

1. H. Karbasian and A. E. Tekava, A review on hot stamped, *J. Mater. Process. Tech.* **210**, 2103 (2010).
2. P. Chen and K. Muammer, Simulation of springback variation in forming of advanced high strength steels, *J. Mater. Process. Tech.* **190**, 189 (2007).
3. P. Rozenak and I. M. Robertson, HVEM studies of the effects of hydrogen on the deformation and fracture of AISI type 316 austenitic stainless steel, *Acta Metall. Mater.* **38**, 31 (1990).
4. I. M. Robertson and H. K. Birnbaum, An HVEM study of hydrogen effects on the deformation and fracture of nickel, *Acta Metall. Mater.* **34**, 353 (1986).
5. J. Bian et al. , Impact of alloying design on the crash relevant material properties of press hardening steel based on Mn-B concept, In *Advanced High Strength Steel And Press Hardening:Proceedings of the 2nd International Conference (ICHSU2015)*, (ChangSha, China, 2015).
6. J. Bian, H. Z. Lu and W. J. Wang, Development of high performance press hardening steel through microstructure engineering, in *Advanced High Strength Steel And Press Hardening:Proceedings of the 3rd International Conference (ICHSU2016)*, (Xi'an, China, 2016).
7. VDA238-100, Plate bending test for metallic materials, In Germany standard 2010.
8. W. S. Choi and B. C. D. Cooman, Characterization of the Bendability of Press-Hardened 22MnB5 Steel, *Steel Res. Int.* **85**, 824 (2014).

Overview on Coatings Developed on Press Hardened Steels for Automotive Applications

P. Drillet
ArcelorMittal Global R&D, Automotive Products
Voie Romaine BP 30320, 57283, France
E-mail: pascal.drillet@arcelormittal.com

The Press Hardened Steels (PHS) are more and more competitive in the lightening race engaged by the automotive manufacturers, due to the unique advantage to perform parts with very high mechanical properties at high temperatures enabling the production of complex geometries without spring-back. In this frame the pre-coating solution is the key of PHS success. In fact, the final in-use properties like corrosion resistance and spot welding are fully dependent on the coating and its microstructure after the austenitization step. The behavior of the Al-Si and Zn based coatings during austenitization and Hot Forming is described. The Al-Si, which is the mainstream PHS solution on the market, shows the widest process window and the best perforating corrosion resistance: it exhibits a high efficient barrier effect created by its own corrosion products, leading to a very limited attack of the steel substrate. Nevertheless, the Al-Si remains sensitive to red rusting appearance from the surface, especially in areas with a too thin E-coat or areas where E-coat is locally eliminated (like stone chipping). Zn based coatings could give an answer on that point, but the process-window is too narrow. To keep a large process window a new Al-Si-Zn-Mg bath solution has been developed in lab and validated through several industrial trials. This new hot dip coating, is described and the improvements on the red rusting are illustrated by lab results, in a wide process-window. The challenge in the development of coatings dedicated to PHS is to satisfy the main requests of the final customers by keeping a solution working in a large industrial range of production for the Hot Stamper.

Keywords: AlSi and Zn coatings; PHS process window; Al-Si-Zn-Mg coatings; Corrosion behaviour.

1. Introduction

Hot stamping steels obtained after austenization and die quenching of boron steels, have been developed to satisfy requirements in terms of reduction of fuel consumption and increase of safety performances mainly for body in white applications [1]. In this frame ArcelorMittal has developed an aluminized boron steel (Usibor® 1500 AlSi), for direct hot stamping applications [2]. This AlSi pre-coated solution initially proposed to prevent scaling and decarburization during the austenitization step gives also a very good protection against corrosion, particularly in terms of perforating corrosion [3]. During austenitization, the Al-Si coating transforms completely by inter-diffusion and solidification reactions, which define the final microstructure and particularly the top layer responsible for the in-use properties. This fully martensitic grade coated with AlSi, shows a very rapid growth in anti-intrusion applications of automotive structures, thanks to its superior mechanical properties after hot stamping and its very good anti-corrosion performance.

For some specific cases, a few customers ask for edge galvanic protection. To satisfy this requirement Zn-based coatings were developed for Hot-Forming [4,5]. During austenitization the Zn-based coatings are also fully transformed by inter-diffusion with a high kinetic of phases transformation. The width of the process window of each pre-coated solution is compared in terms of final In Use properties.

To improve the AlSi coating efficiency in terms of rest rust pitting, especially in areas with a too thin E-coat or areas where E-coat is locally eliminated by stone chipping, a new coating based on a dedicated bath chemistry, Usibor® 1500 Ultraprotect, have been developed. Some first results are presented in this paper.

2. Process Window Description

2.1. *Hot stamping parameters*

The control of the hot stamping including the inter-stage quenching operation, generally performed by independent hot stampers (more & more car makers have their own stamping lines), is essential to achieve the good in-use properties. For this reason, it is very important to work closely with the hot stampers to determine the best process windows. The full phase transformations occurring on 22MnB5 with an AlSi coating of 150 g/m^2 are illustrated by Fig. 1. The final coating microstructure, particularly the top surface layer determines the in-use properties like spot weldability and paint adhesion.
The coating microstructures obtained after an austenitization at 900 °C for a total time inside the oven of 5 minutes on Zn based coatings like 22MnB5 GI and 22MnB5 GA are reported in Fig. 2. The final coatings microstructures are completely different between the two delivery materials for a similar heat treatment: the remaining Fe-Zn gamma phases are much higher on the GI coating (phases involved in the galvanic coupling protection of the final part). A stay for a longer time of this GI coating inside the furnace could lead in a similar coating microstructure as observed on GA. So, in that case the level of alloying (process window), is also a key point to determine the final in-use properties.

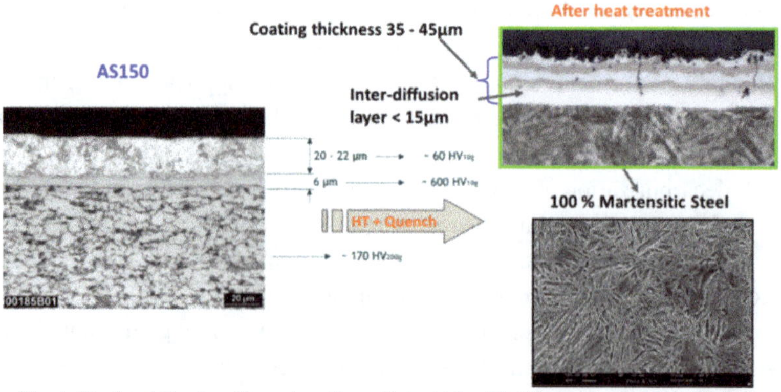

Fig. 1. Steel and Coating Phase transformations during Hot Stamping on 22MnB5 AlSi.

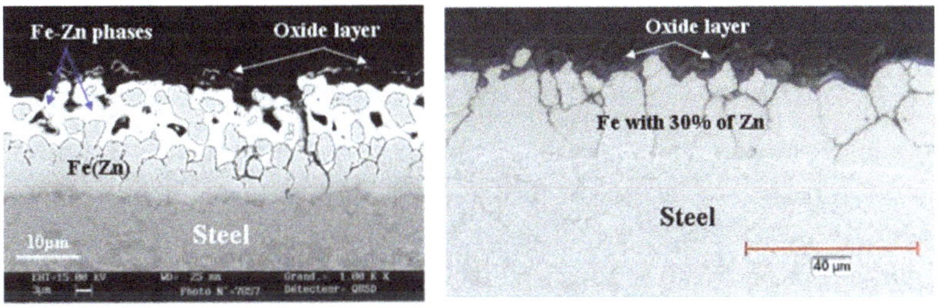

GI-coating microstructure obtained after hot stamping

GA-coating microstructure obtained after hot stamping

Fig. 2. Final coating microstructure obtained on 22MnB5 GI and GA after HT at 900°C - 5minutes.

2.2. *Definition of the process window – example of the AlSi coating*

The process window for the Hot Stamper is defined by the couples Time/Temperature acceptable to ensure the final in-use properties. As discussed previously, this window is mainly determined by the coating surface microstructure.

During austenitization of the AlSi coating the coating fully transformed by interdiffusion at high temperature, the iron diffuses in the coating when Al and Si diffuse in the iron. These reactions lead to the formation of various components, as illustrated by Fig. 1. For shorter times, there is some remaining Al liquid phase not fully alloyed which conduct to the formation of a smooth surface after solidification. For longer dwell times the coating is over-alloyed, with risks of bad homogeneous spot weldability and high hydrogen pick-up which could lead to delayed fracture if there are some deviations in the mounting of these parts on the body-in white in terms of stress level [6].

The coating microstructure is also function of the kinetics of the phase transformations: for very high heating speed the ternary tau1 phase is not able to form a continuous sub-layer as illustrated in the picture 1 (5-10 min) [7]. But the resulting coating microstructure (Fig. 3) shows that it is not necessary to have a continuous ternary layer to get a good surface in terms of micro-roughness. The key point to ensure a suitable micro-roughness for the paint-ability is to preserve the Fe_2Al_5 as a continuous layer at the top coating surface.

In the case of too long HT the interfacial layer thickness is the best parameter to predict the maximum iron content acceptable at the coating surface to ensure a good range of welding, a maximum of 15 micrometers is usually taken as a reference on that point (Fig. 1). This is typically the kind of process window necessary to establish with each hot stamper.

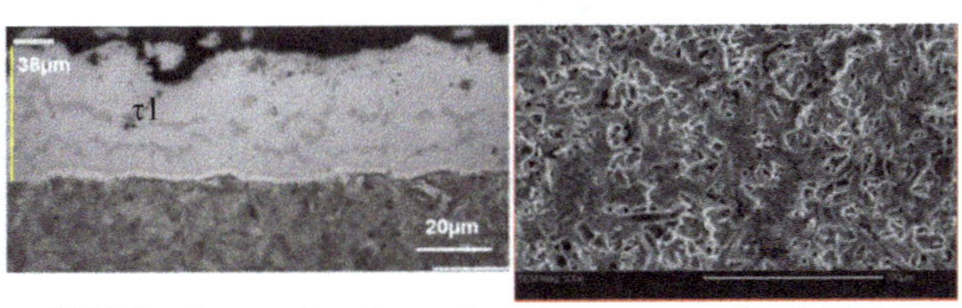

Fig. 3. Effect of high heating speed on the coating microstructure – Steel of 1.5mm of thickness.

2.3. *Differences in terms of process window between AlSi and Zn based coatings*

2.3.1. *Width of the process window on AlSi vs perforating corrosion performance [8]*

The perforating corrosion resistance of aluminized press hardened steel 22MnB5 AlSi in cyclic corrosion tests is compared with the corrosion resistance of bare 22MnB5 and conventional GA coated steel for cold stamping. The coating is observed in the external radius of an omega part with a punch radius of 4 mm, a cyclic VDA621-415 test has been performed on non-painted samples. The results after 10 weeks are reported on the figure 4: half of the uncoated PHS has been lost and there is no more residual GA coating on the CR reference while only a very limited steel has been attacked on the PHS AlSi (<30 μm).

An SEM examination of the cross section after 2 weeks of a new VDA corrosion test shows some local AlSi coating dissolution (Fig. 5). All corroded areas are filled with corrosion products. SEM-EDS analyses performed on the corrosion products show the presence of Fe, Al and Si in significant amounts. In situation of corrosion, AlSi and exposed steel (in cracks for example) corrode simultaneously, allowing the oxidation of Al, Si and Fe elements. Due to the pH increase (oxygen reduction), Fe, Al and Si corrosion products precipitate in oxides. This corrosion products layer strongly inhibits the steel corrosion rate in cavities due to a blocking effect on oxygen diffusion towards the steel. It explains the very limited corrosion propagation rate in steel cavities: similar depth of attack between 6 and 12 weeks of test (70 μm). Although the AlSi coating is not sacrificial for steel, it is an active coating acting like a barrier for the corrosion propagation. This resulting very large process window is one key of the success of this pre-coated AlSi solution.

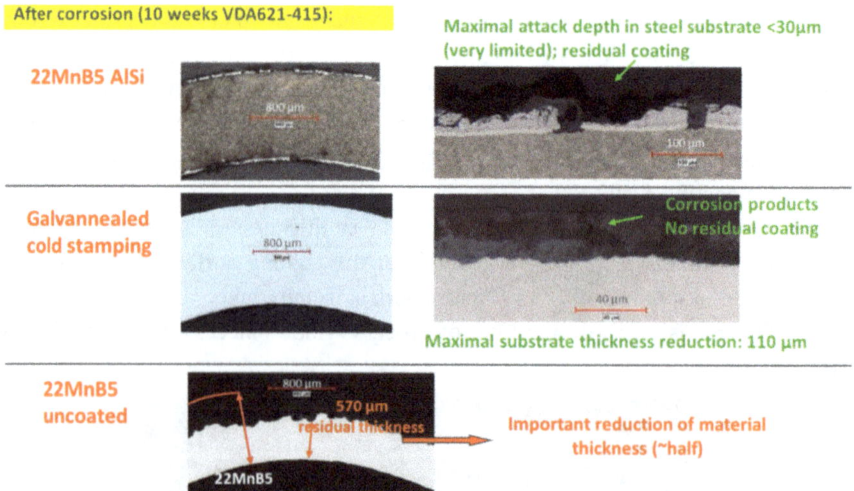

Fig. 4. Comparison of the perforated corrosion attack depth for non-painted samples after 10 weeks of VDA621-415 test between PHS 22MnB5 AlSi, CR GA reference material and PHS uncoated 22MnB5.

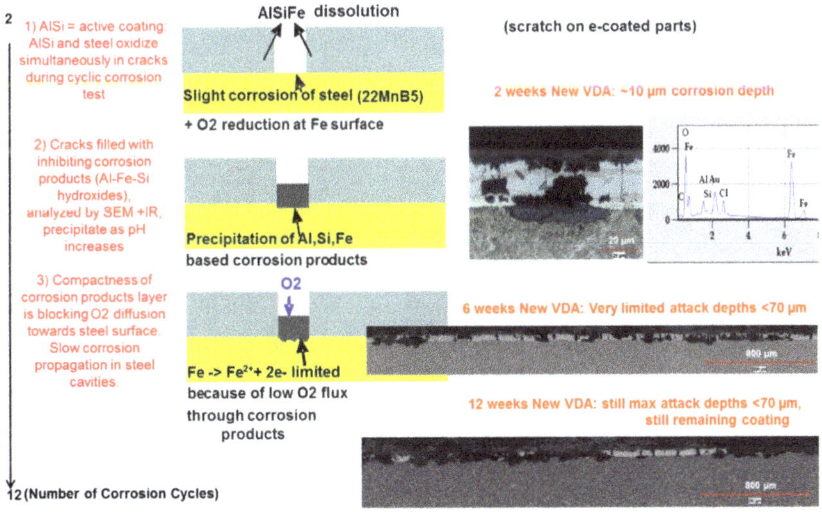

Fig. 5. Mechanisms on the high resistance of 22MnB5 AlSi PHS solutions against perforating corrosion.

2.3.2. *Width of the process window between 22MnB5 AlSi & 22MnB5 GA [9]*

Studies conducted always on the perforated corrosion resistance on non-painted samples have been performed on 22MnB5 GA (GA of 12 µm). The comparative results with AlSi shown a progressive deterioration of perforating corrosion resistance on the GA coating with the increasing dwell time which corresponds to the progressive loss of phases which are responsible in the sacrificial power of the GA coating[9)]. The process window of this GA coating is then much narrower than exhibited by the AlSi coating. The only way to increase the width of the process window is to increase the Zn base coating thickness.

3. New Generation of Coatings for PHS

Nevertheless, the AlSi coating remains sensitive to the red rusting, especially in areas with a too thin E-coat or areas where E-coat is locally eliminated by stone chipping for instance. On this purpose, we have developed in lab and validated through several industrial trials a new coating solution obtained from an Al bath including Silicon, Zinc, Magnesium and Iron, called Usibor® Ultraprotect. Due the growth of complex oxide layers at the top surface during austenitzation this new coating becomes phosphatable by keeping the strong points of the AS coatings (direct hot-stamping, excellent perforating corrosion resistance, very large process window). This new solution reduces strongly the sensitivity to red rust pitting. After 3 cycles in VDA 233-102 corrosion test it is possible to see that materials coated with AlSi and GI coating present 100% red rust on their surface whereas for Usibor® Ultraprotect, surface is slightly corroded (Fig. 6a). From phosphate panels after 12 cycles of VDA 233-102 corrosion test it is possible to see (Fig. 6b) that the Usibor® Ultraprotect presents even lower red rust than cold stamping materials (GI and GA).

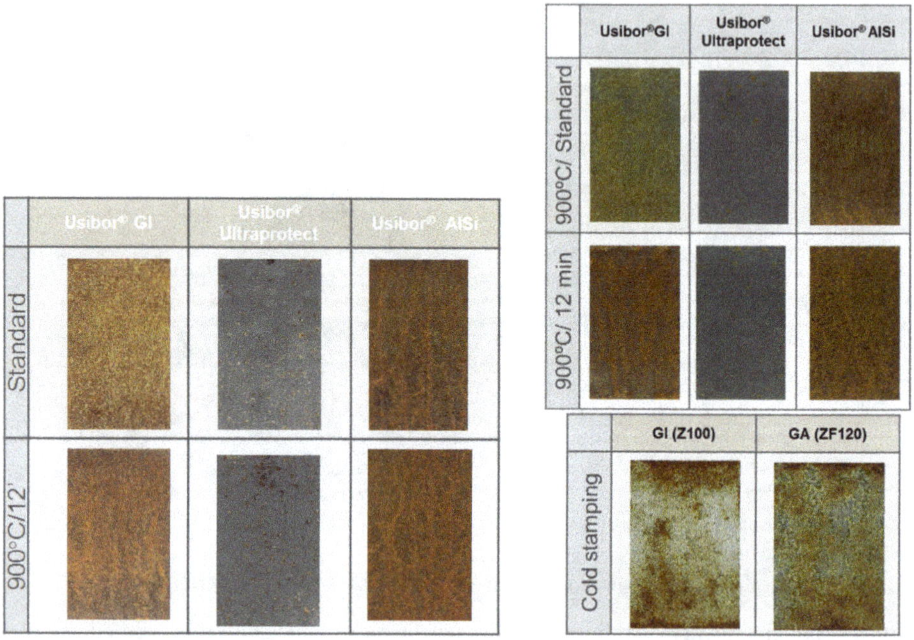

a) Unpainted panels b) Unpainted & phosphate panels

Fig. 6. Red rust pitting: a) After 3 Cycles of VDA; b) After 12 cycles of VDA (phosphate panels).

4. Conclusion

The performances of GA & GI are strongly depending on the austenitizing process window, which is much narrower than for AlSi. Less sensitive to process deviation, the AlSi coating appears to be the most cost-efficient solution for main applications, but for some cases a lower red rust pitting appearance is requested. The new development of the Usibor® Ultraprotect shows huge improvements on this topic by keeping the advantage of AlSi. This New coating still under development, obtained in an Al bath with addition of Si, Fe, Zn and Mg, exhibits a very promising potential.

References

1. H. Guyon, Bad Nauheim, Material in Car Body Engineering, (2013), http://www.automotive-circle.com/content/download/176371/3535838/file/.
2. L. Vaissière, J. P. Laurent and A. Reinhardt, Development of Pre-coated Boron Steel for Applications on PSA Peugeot Citroën and RENAULT Bodies in White, *SAE T. J. of Mater. Manuf.* **111**, 909 (2003).
3. C. Allély, L. Dosdat, O. Clauzeau and P. Volovitch, Anticorrosion mechanisms of aluminized steel for hot stamping, *Surf. Coat. Tech*, **238**, 188 (2014).
4. T. Kurz, H. Schwinghammer, G. Luckeneder, T. Manzenreiter and A. Sommer, Zinc Coated Press-Hardening Steel-Challenges and Solutions, in *5th Int. CHS2 Conference,* (Toronto, Canada, 2015).

5. T. Setck, J. Gerdenitsch, A. Tomandl, W.Achleitner, J. Faderl, T. Lavric, T. Boiadjieva-Scherzer and H. Kronberger, Zinc-chromium coated steel sheet: Properties and production, in *8th International Galvatech Conference*, (Genova, Italy, 2011).
6. C. Georges, T. Sturel, P. Drillet and J. M. Mataigne, Absorption/Desorption of Diffusible Hydrogen in Aluminized Boron Steel, *ISIJ International* **53**, 1295 (2013).
7. P. Drillet, R. Grigorieva, G. Leuillier, and T. Vietoris, Study of cracks propagation inside the steel on press hardened steel zinc based coatings, in *8th International Galvatech Conference*, (Genova, Italy, 2011).
8. C. Allely, J. Petitjean and T. Vietoris, Corrosion Resistance of Zinc Based and Aluminized Coatings on Press-Hardened Steels for Automotive, in *3rd International CHS2 Conference,* (Kassel, Germany, 2011).
9. L. Dosdat, C. Allely, KIMAB corrosion Seminar, Stockholm (2013). http://www.stockholmresilience.org/download/18.3186f824143d05551ad3e11/1459560367804/SRC_arsrapport2013_sida.pdf

High Temperature Oxidation Behavior and Morphology Evolution of Galvannealed Coating in Hot Stamping

K. Wang[1], Z. J. Zhu[1], B. Zhu[1], Y. S. Zhang[1,†], S. Q. Li[2], H. Q. Li[3], T. Wu[3], J. C. Jin[4] and L. Cui[4]

[1]State Key Laboratory of Materials Processing and Die and Mould Technology, Huazhong University of Science & Technology, Wuhan, 430074, China

[2]School of Mechanical Science and Engineering, Huazhong University of Science & Technology, Wuhan, 430074, China

[3]Vision Hot Stamping Technology Co. Ltd., Dongguan, 523460, China

[4]Technology Center of Ma'anshan Iron and Steel Company, Ma'anshan, China

[†]E-mail: zhangys@mail.hust.edu.cn

In this paper, high temperature oxidation behavior and surface morphology evolution of galvannealed (GA) coating during heating and dwell stages were studied. The variations of chemical compositions and thickness of oxide layer with time were analyzed. In addition, the influence of austenization temperature on integrity of oxide layer was discussed. The results showed that the oxide layer on GA coating was mainly composed of ZnO. During austenization treatment, the morphology of surface oxides underwent a transition procedure from particles to clusters and finally to a compact layer. In addition, the thickness of oxide layer did not increase significantly during dwell stage. Furthermore, the austenization temperature should not exceed 900 °C to avoid rupture of the oxide layer and suppress zinc loss in the coating.

Keywords: Hot stamping; Galvannealed coating; Oxidation behavior; Morphology evolution.

1. Introduction

The application of hot stamping steel parts is one of the effective ways to achieve energy conservation and emission reduction of automobile, with the advantages including weight reduction and improvement of passive safety [1, 2]. In general, the austenization temperature used in hot stamping process is above 850 °C. The direct contact of steel with air at elevated temperature could result in severe surface oxidation and decarburization, which degrades part performance and tool life. To prevent oxidation and decarburization of hot stamping steel at elevated temperature and provide certain protection for hot stamping parts in service, coating technology is widely used for surface protection of hot stamping steel [3-5].

Galvannealed (GA) coating could provide both barrier protection and cathodic protection for hot stamping steel [6]. However, liquid zinc exists in the coating for a certain period during heating stage of GA coated steel, due to the fact that most Fe-Zn alloys has a relatively lower melting points than austenization temperature. To suppress zinc loss resulting from zinc evaporation at high temperatures, GA coated hot stamping steel should be heated in oxidizing atmosphere. The generated oxide layer on GA coating is necessary to decrease zinc loss. Meanwhile, the surface appearance of GA coating could be changed by oxidation. At present, there are few researches about high temperature oxidation process and surface morphology evolution of GA coating during hot stamping.

In this study, morphology evolution due to high temperature oxidation of GA coating was firstly discussed. The compositions and thickness variation of surface oxide layer with time was then analyzed. Furthermore, the influence of autenization temperature on integrity of oxide layer was investigated.

2. Materials and Experiments

2.1. *Materials*

The tested material was GA coated boron steel sheet with a thickness of 1.5 mm. The chemical compositions of the 22MnB5 steel substrate are shown in Table 1. The average coating weight in as-delivered condition is 80g/m^2 per side and the coating thickness is about 12 μm with iron content of about 10wt%.

Table 1. Chemical compositions (wt.%) of GA coated steel substrate.

C	Mn	Si	Cr	Ti	Ni	Nb	B
0.23	1.02	0.24	0.16	0.023	0.029	0.012	0.001

2.2. *Experiments*

The heat treatment of GA coated steel was carried out using Gleeble thermo-mechanical simulator. The specimen dimensions and test procedures are illustrated in Fig. 1. The uniform temperature region of the specimen is about 8 mm. Three heating rates were adopted for different temperature ranges to simulate the heating stage of the blank in furnace. A heating rate of 15 °C/s was used below 600 °C. For the temperature range from 600 to 750 °C, the heating rate was 6 °C/s. From 750 °C to the target temperature, a lower heating rate of 3 °C/s was adopted. The specimens were divided into two groups and the heat treatment conditions are shown in Table 2. Specimens in group 1 were used to investigate the progressive oxidation process of GA coating with temperature increasing. The influence of different austenization temperature on coating morphology was evaluated using specimens in group 2. After heat treatment, the specimens were cooled to room temperature by compressed air with a cooling rate greater than 30 °C/s. The temperature region was cut from the specimen for coating detection.

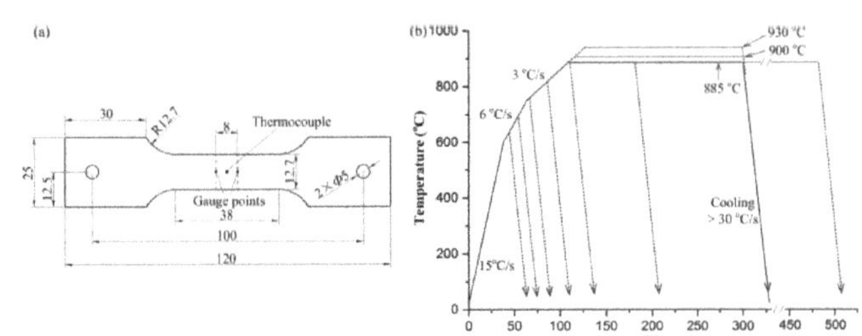

Fig. 1. Specimen dimensions (mm) and test procedures of high temperature oxidation experiments.

Table 2. Heat treatment conditions of the GA coated specimens.

Group 1		Group 2	
Time (s)	Temperature (°C)	Time (s)	Temperature (°C)
47	642	300	885
54	684	300	900
69	762	300	910
85	810		
110	885		
180	885		
300	885		
480	885		

The morphologies of GA coating were observed using an optical microscopy (KEYENCE VHX-1000C) and a scanning electron microscopy (ESEM, FEI/QUANYA 200). Chemical compositions on coating surface were detected by energy disperse spectroscopy (EDS) to identify the surface oxides.

3. Results and Discussion

3.1. *Surface morphology evolution of the GA coating*

The progressive evolution process of surface morphologies of GA coating during heating and dwell are illustrated in Fig.2. Two types of surface morphologies could be observed on the coating under the initial condition, as shown in Fig.2a. Region A showed a flat appearance while region B showed a granular morphology, which resulted from phase change during galvannealed treatment. During heating stage from the initial condition to 642 °C, the coating surface morphology had no significant change and there was no obvious oxides being observed (Fig.2b). When the temperature increased to 684 °C, the coating surface showed an up-and-down appearance due to the formation of some oxide particles (Fig.2c). At 762 °C, large-sized oxide clusters could be observed and more areas of coating surface were covered by the oxides, as shown in Fig.2d. Almost the entire coating surface was covered by oxides with temperature increasing to 810°C (Fig.2e). A majority of the oxides were in the shape of clusters and a small number of dispersive oxide particles could also be observed. The oxide particles disappeared at the target temperature 885 °C (Fig.2f). As a result, the coating surface was entirely covered by oxide clusters. The oxide clusters showed a scraggly appearance and the adjacent clusters were not continuous. In the following dwell stage, the oxide clusters gradually merged with each other. At the heating time of 180 s, the scraggly appearance disappeared and the coating surface was covered by a compact oxide layer with a continuous and ravined morphology. With time increasing to 300 s, the ravine size became larger dramatically compared with 180 s (Fig.2h). However, as the heating time increased from 300 s to 480 s, the morphology of oxide layer had no significant change, which indicates retard of the oxidation process.

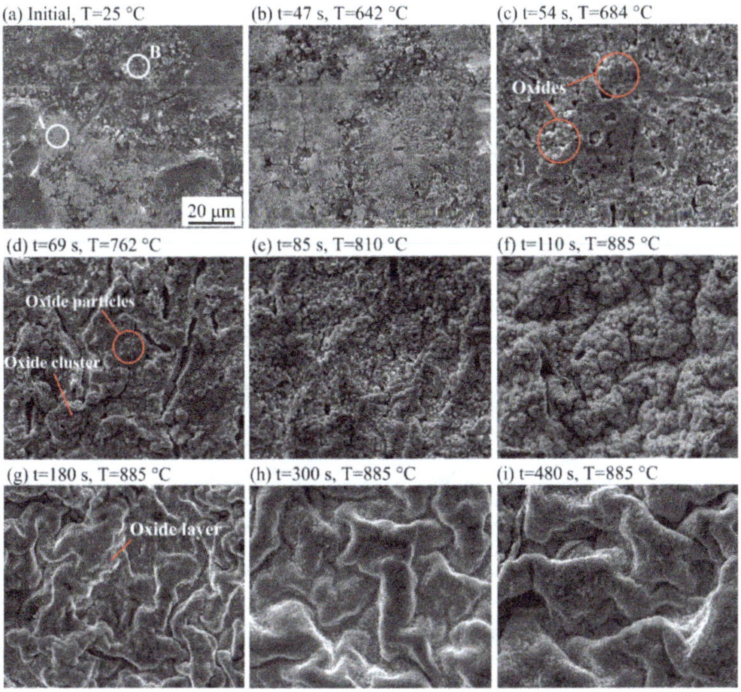

Fig. 2. Surface morphologies evolution of GA coating during heating and dwell: (a) initial condition; (b-f) heating-up stage; (g-i) dwell stage.

3.2. *Variations of chemical compositions and thickness of oxide layer*

The surface chemical compositions of GA coating under different heat treatment conditions are shown in Table 3. The degree of surface oxidation after heating to 642 °C was unapparent with only an O content of 3.58 wt%. When the temperature increased to 684 °C, the O content increased dramatically to 9.56 wt% with the formation of visible surface oxides as shown in Fig.2c. Finally, the surface O content of the coating was higher than 15 wt% as time progressed and heating temperature increased. According to the atom ratio of Zn/O, it could be estimated that the oxide layer on GA coating was mainly composed of ZnO. In addition, a small amount of manganese (Mn) in substrate diffused to coating surface during heating of GA coated steel. The coating surface after dwell for several minutes was enriched with about 4 wt% Mn.

Figure 3 shows the cross-sectional morphologies and average thickness of the oxide layer on GA coating after heating at 885 °C for 300 s and 480 s. It could be observed that the oxide layer was compact and continuous, which would be beneficial to suppress zinc loss during heating and dwell. Besides, the thickness of oxide layer did not increased significantly. The average thickness only increased from 2.02 μm at 300 s to 2.32 μm at 480 s. This is due to the fact that the ion diffusion coefficients of Zn and O in ZnO respectively have a small magnitude of 10^{-6} cm^2s^{-1} and 10^{-15} cm^2s^{-1}, which are far less than the magnitude of 10 cm^2s^{-1} for Fe in Fe_3O_4 [8].

Table 3. Surface EDS analysis results of GA coating under different conditions.

t/s-T/°C	Zn		Fe		Mn		O	
	wt. %	at. %	wt. %	at. %	wt. %	at. %	wt. %	at. %
47-642	87.60	77.95	8.81	9.10	—	—	3.58	12.95
54-684	80.51	61.50	9.33	8.27	0.60	0.54	9.56	29.68
85-810	73.92	50.06	6.99	5.49	4.14	3.32	14.95	41.13
180-885	70.98	47.65	7.06	5.50	6.75	5.35	15.22	41.50
300-885	75.13	48.52	4.31	3.22	2.74	2.09	17.64	46.17
480-885	73.12	46.73	4.58	3.40	4.36	3.29	17.94	46.58

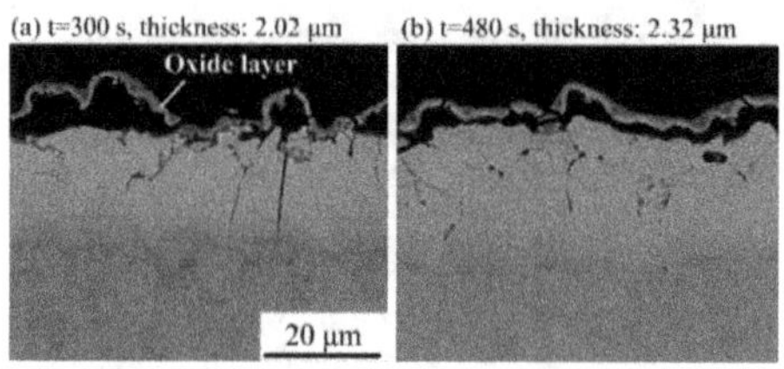

Fig. 3. Cross-sectional morphologies of the oxide layer on GA coating after heating for 300 s and 480 s.

3.3. *Influence of austenization temperature on surface morphology of GA coating*

Both the surface and cross-sectional morphologies of the GA coating after heating at different temperatures for 5 min are shown in Fig. 4. The 8 mm uniform temperature region is marked by two dashed lines. The oxide layer maintained integrated and continuous at the heating temperature of 885 °C (Fig.4a). A few local rupture of oxide layer occurred in the specimen heated at 900 °C, resulting in the formation of white ZnO powder due to oxidation of liquid zinc beneath the layer (Fig.4b). As the heating temperature increased to 930 °C, most of the oxide layer was ruptured because of excessive boiling of liquid zinc. As for the coating thickness, the GA coating heated at 885 °C and 900 °C had a thickness around 20 μm. And the thickness of coating heated at 930 °C was only 14 μm, which indicates that zinc loss could not be suppressed due to rupture of oxide layer.

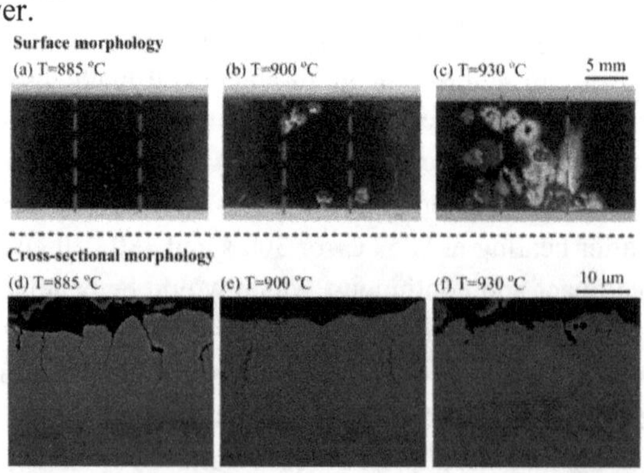

Fig. 4. Surface morphologies and cross-sectional morphologies of GA coating after heating at different temperatures for 5 min.

4. Conclusion

During austenization treatment of GA coated hot stamping steel, the oxide morphology on coating surface underwent a transformation process from particles to clusters and finally to a compact continuous oxide layer. As a result, the coating surface showed a ravined morphology. The oxide layer were mainly composed of ZnO and there was around 4.0 wt.% manganese enriched on coating surface.

The thickness of oxide layer on GA coating did not changed significantly with time increasing. The average thickness of oxide layer increased from 2.02 μm at 300 s to 2.32 μm at 480 s. To avoid the rupture of oxide layer and then suppress zinc loss in the coating, the heating temperature should not exceed 900 °C.

Acknowledgments

This research work was financially supported by the National Natural Science Foundation of China (grant no. U1564203 and U1760205) and National Major Science Technology Project of China (grant no. 2018zx04023001).

References

1. W. S. Choi and B. C. De. Cooman, Characterization of the bendability of press-hardened 22MnB5 steel, *Steel Res. Int.* **85**, 824 (2014).
2. K. Mori, P. F. Bariani, B. A. Behrens, A. Brosius, S. Bruschi, T. Maeno, M. Merklein and J. Yanagimoto, Hot stamping of ultra-high strength steel parts, *CIRP Annals* **66**, 755 (2017).
3. C. Allély, L. Dosdat, O. Clauzeau, K. Ogle and P. Volovitch, Anticorrosion mechanisms of aluminized steel for hot stamping, *Surf. Coat. Tech.* **238**, 188 (2014).
4. M. Windmann, A. Röttger and W. Theisen, Phase formation at the interface between a boron alloyed steel substrate and an Al-rich coating, *Surf. Coat. Tech.* **226**, 130 (2013).
5. K. Wang, Y. M. Jin, B. Zhu and Y. S. Zhang, Investigation on cracking characteristics of Al–Si coating on hot stamping boron steel parts based on surface strain analysis, *Surf. Coat. Tech.* **282**, 309 (2017).
6. D. W. Fan and B. C. De. Cooman, State-of-the-knowledge on coating systems for hot stamped parts, *Steel Research Int.* **83**, 412 (2012).
7. G. Vourlias, N. Pistofidis, K. Chrissafis and G. Stergioudis, Zinc coatings for oxidation protection of ferrous substrates, *Journal of Thermal Analysis and Calorimetry* **90**, 769 (2007).

Corrosion Resistance of Zinc-based Coating for Hot Forming and Its Comparison with Aluminum-silicon Coating

Z. T. Li[1,a], J. C. Jin[1,b,†], K. Wang[2], Z. J. Wang[2], C. Y. Sun[1,c] and W. Zhang[1,d]

[1]Technology Center of Ma'anshan Iron and Steel Company, Ma'anshan, China
[2]State Key Laboratory of Materials Processing and Die & Mould Technology, Huazhong University of Science and Technology, Wuhan 430074, China
E-mail: [a]lizitao2011@163.com, [b,†]jinjiachun1988@163.com
[c]308025796@qq.com, [d]zwniat@126.com

Hot formed part is widely used in the automotive industry to meet increasing requirements of safety and light weight. However, the corrosion protection is also a major aspect for automotive materials. For this reason, aluminized, galvanized and galvannealed hot forming steels were developed. Here the corrosion resistance of hot formed zinc-based coating was evaluated by cycle corrosion test. And it was compared with aluminum-silicon coated steel.

Keywords: Corrosion resistance; Hot forming; Zinc-based coating; Aluminum-silicon coating.

1. Introduction

Increasingly stringent environmental laws and regulations have promoted the automotive industry's demand for lightweight technology. The hot-formed parts have advantages of high strength and no springback problem, compared with the cold-formed high-strength steel [1]. And the weight of the vehicle body can be greatly reduced while the safety of the vehicle body structure is also ensured by application of hot forming steel [2]. So during the past decade, it is widely used for automobile safety parts and structural parts[3].

Use of uncoated hot forming steel is limited by the problem of surface decarburization and oxidation peeling in the process of hot stamping, which should be removed by shot blasting. Hot forming steel with coatings has better surface quality after hot forming and the shot blasting process can be avoided [4,5]. On the other hand, corrosion protection can also be enhanced by the coatings [6-9]. For these reasons, the hot forming steel with aluminum-silicon coating is widely used. However, the lack of cathodic corrosion protection of aluminum-silicon coating lead to the development of zinc and zinc-iron alloy coating, which is expected to provide cathodic corrosion protection for the base steel [10,11].

In this paper, the corrosion resistance of hot formed zinc-based coating and aluminum-silicon coated steel was evaluated by cycle corrosion test.

2. Experimental

Two steel substrate was common boron alloyed carbon manganese steel 22MnB5 with a thickness of 1.5 mm. zinc-iron coating weight was 70 g/m^2 on one side, its iron content is about 10 wt. %. And aluminum-silicon coating weight was 70 g/m^2. The both materials were hot formed by tablet quenching.

(1) The corrosion test was carried out according to JASO M610-1992.One cycle of the test is consist of 2 hours salt spray, 4 hours dry and 2 hours moist. It was performed in Ascott CC1000IP cyclic corrosion test chamber.

(2) Electrochemical experiments were performed using a CHI 660 Potentiostat/Galvanostat. The electrolyte is 5 wt% NaCl solution. The temperature is ambient temperature. The reference electrode used was saturated calomel electrode.

(3) Cross-sections were analyzed by scanning electron microscopy (SEM) and energy dispersive X-ray spectroscopy (EDX).

3. Results and Discussion

3.1. *Coating microstructure after hot forming*

Coating microstructure of GA after hot forming is showed as Figure 1. The coating was etched by inhibited HNO_3 solution. From the figure, no obvious Γ-ZnFe Phase was detected. The EDS shows that Zn content is 14.12 wt%~37.12 wt%, from the top of coatings to the coating next to the base steel, which indicates that the α-iron is the major phase.

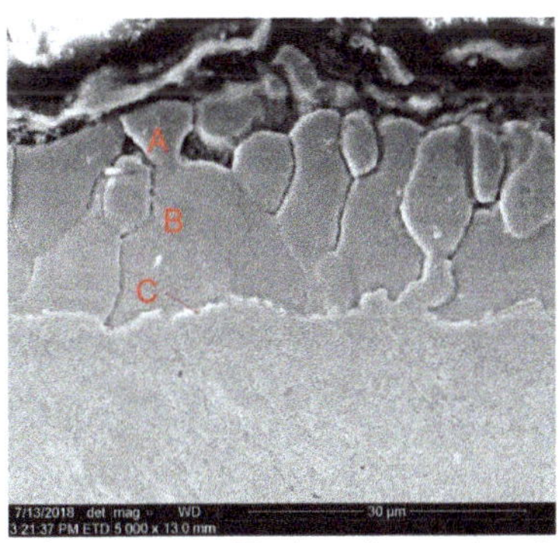

Fig. 1. Coating microstructure of Zn-iron alloy coating after hot forming.

Table 1. Composition distribution of Zn-iron alloy coating after hot forming

No.	Element	Wt. %	At%
A	Fe	62.88	66.47
	Zn	37.12	33.53
B	Fe	67.35	70.71
	Zn	32.65	29.29
C	Fe	85.88	87.68
	Zn	14.12	12.32

3.2. *Electrochemical behavior*

The results of the potentio dynamic polarization scans are shown in Fig. 2. The AlSi coating zero current potential is found to-0.562 V, which is very close to -0.591V of the base 22MnB5.While, the Zn-iron alloy coating is -0.734V. So it can be concluded that the AlSi coating cannot be an sacrificial anode protection coating for the base steel. The Zn-iron alloy coating zero current potential is 143 mV negative than the base steel. With this, a sacrificial anode protection coating for the base steel maybe expected.

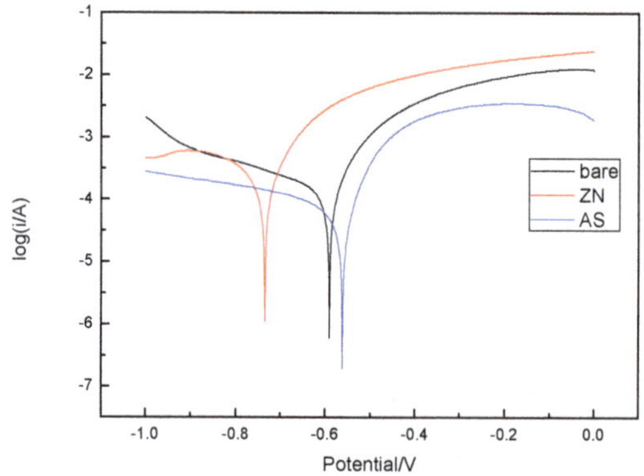

Fig. 2. Potentio dynamic polarization curve after hot stamping (bare-22MnB5, ZN-Zn-iron alloy, AS-AlSi).

3.3. *Corroded sample investigation*

The surface after 1 cycle test was recorded as the Fig. 3. It shows that the yellow rust was appeared on both samples after 1 cycle test. This is caused by the diffused iron in the metallic coating during the austenitizing heat treatment, does not mean the coating is breached [7].

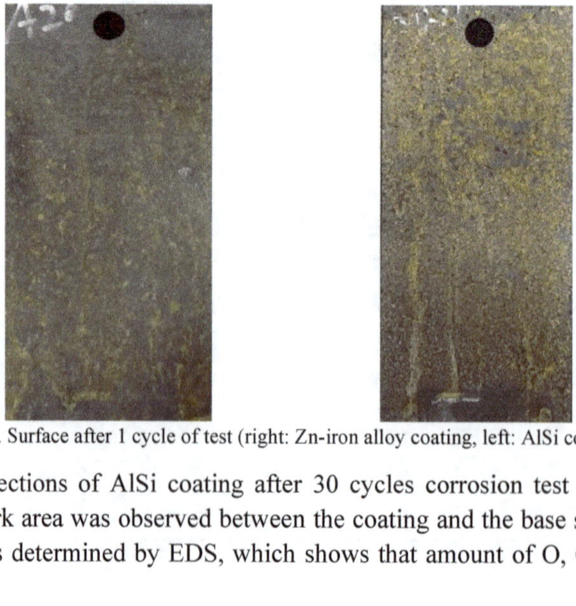

Fig. 3. Surface after 1 cycle of test (right: Zn-iron alloy coating, left: AlSi coating).

The cross-sections of AlSi coating after 30 cycles corrosion test is showed in the Fig. 4. Some dark area was observed between the coating and the base steel. The content of dark area was determined by EDS, which shows that amount of O, Cl elements were remained in the area. So it could be sure that the area is the corrosion product. As the considerable quantity coating was observed above the corrosion product, it can also be concluded that simultaneous corrosion of the AlSi coating and of the steel substrate, which demonstrates a lack of sacrificial anode protection of AlSi coating for the base steel.

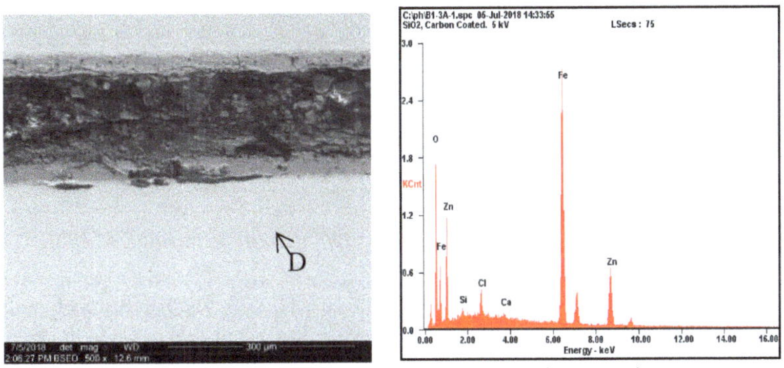

Fig. 4. The cross-sections of AlSi coating after 30 cycle's corrosion test.

The cross-sections of Zn-iron alloy coating after 30 cycles corrosion test is showed in the Fig. 5. It is observed that the dark corrosion product is almost above the coating and no obvious corrosion product can be found beneath the coating. Compared with AlSi coating, the Zn-iron alloy coating shows an advantage of sacrificial anode protection for the base steel, whose zero current potential is 143 mV negative than the base steel. However, O and Cl elements are found in microcrack, which means that corrosion can occur preferentially in the microcrack area as the corrosion solution immersed. So it can be concluded that the controlled microcracks is benefit to the improvement of the corrosion resistance.

Caitlin Dever et al. [12] have assessed the electrochemical properties of Zn-coated PHS annealed for varying times at 900 °C to determine if these coatings can provide cathodic protection to the underlying substrate. The work shows that it is improbable that α-Fe (Zn) will be able to provide robust cathodic protection to the bare substrate. Here in this work, cathodic protection is detected although the primary microstructure of the Zn-iron alloy coating is α-Fe (Zn) phase, especially compared with AlSi coating.

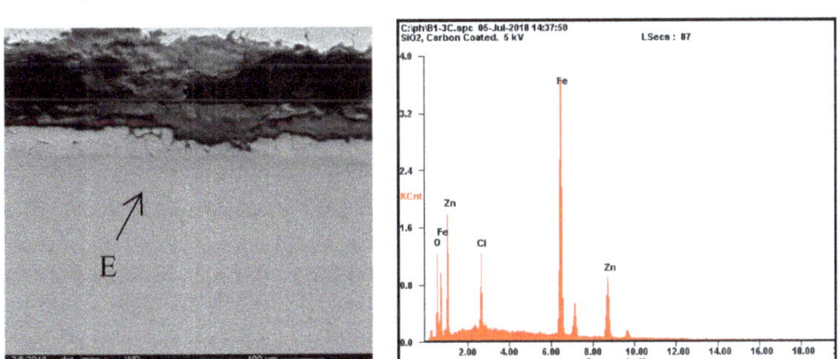

Fig. 5. The cross-sections of Zn-iron alloy coating after 30 cycle's corrosion test.

4. Summary

(1) Electrochemical test shows that the AlSi coating cannot be an sacrificial anode protection coating for the base steel and the Zn-iron alloy coating zero current potential was 143 mV negative than the base steel.
(2) The corrosion product observed beneath AlSi coating demonstrates a lack of sacrificial anode protection of the coating for the base steel.
(3) Cycle corrosion test showed that Zn-iron alloy coating has cathodic protection although the microstructure is primary α-Fe (Zn) phase.

References

1. H. Karbasian and A. E. Tekkaya, A review on hot stamping, *J. Mater. Process. Tech.* **210**, 2103 (2010).
2. H. Sun, H. U. Ping, M. A. Ning, G. Shen and B. Liu, Application of hot forming high strength steel parts on car body in side impact, *Chin. J. Mech. Eng.* **2**, 252 (2010).
3. Z. Gui, W. Liang and Y. Zhang, Formability of aluminum-silicon coated boron steel in hot stamping process, *T. Nonferr. Metal. Soc. China* **6**, 1750 (2014).
4. Z. X. Gui, W. K. Liang, Y. Liu and Y. S. Zhang, Thermo-mechanical behavior of the Al - Si alloy coated hot stamping boron steel, *Mater. Des.* **60**, 26 (2014).
5. D. W. Fan and B. C. D. Cooman, State of the knowledge on coating systems for hot stamped parts, *Steel Res. Int.* **5**, 412 (2012).
6. M. Suehiro, J. Maki, K. Kusumi, M. Ohgami and T. Miyakoshi, Properties of aluminized steels for hot-forming, in *Int. Body Eng. Conference and Exposition*, (2003).
7. L. Dosdat, J. Petitjean, T. Vietoris and O. Clauzeau, Corrosion resistance of different metallic coatings on press-hardened steels for automotive, *Steel Res. Int.* **6**, 726 (2011).
8. H. S. Hwang, J. C. Park, J. S. Kim and K. T. Min, Investigations on corrosion performance of galvanized 22MnB5 for hot press hardening under salt spray test, in *GALVATECH APGALVA 2013: proceedings of 9th International Conference on Zinc and Zinc Alloy Coated Steel Sheet, 2nd Asia-Pacific Galvanizing Conference*, (Beijing, China, 2013).
9. C. Allély, L. Dosdat, O. Clauzeau, K. Ogle and P. Volovitch, Anticorrosion mechanisms of aluminized steel for hot stamping, *Surface and Coatings Tech.* **2**, 188 (2014).
10. R. Autengruber, G. Luckeneder and A. W. Hassel, Corrosion of press-hardened galvanized steel, *Corrosion Sci.* 12 (2012).
11. T. Kurz, G. Luckeneder, T. Manzenreiter, H. Schwinghammer and A. Sommer, Zinc coated press-hardening steel-challenges and solutions, *Sae Tech. Paper*. 2015.
12. C. Dever, J. Kish and J. McDermid, Corrosion Properties of Hot Hip Zinc Galvanized Coatings on 22MnB5 Press Hardened Steels, in *GALVATECH 2017: 11th International Conference on Zinc and Zinc Alloy Coated Steel Sheet*, (Tokyo, Japan, 2017).

Comparative Study of Corrosion Resistance of Different Types of Press-hardened Steels

H. Peng[†], X. P. Mao, X. Q. Huang, H. Wang, J. H. Song, T. Pang, Y. Ma, Z. Peng and K. H. Hu
Wuhan Branch of Baosteel Central Research Institute (R & D Center of Wuhan Iron & Steel Co., Ltd.), Wuhan, 430080, China
[†]E-mail: penghao1986@baosteel.com

Properties and corrosion resistance of the electrophoretic paint films on two types of uncoated press-hardened 22MnB5 steels (cold-rolled and CSP hot-rolled) and an Al-Si steel based on 22MnB5 were compared in the present contribution. The physicochemical properties of the electrophoretic paint films on these three types of press-hardened steels were found to be proximal, and all meet the requirements of related standards of automobile manufacturers. The corrosion resistance of unpainted press-hardened samples of the three kinds of materials was found to be poor. The poor corrosion resistance of the Al-Si sample is related to the high iron concentration and the presence of microcracks in the coating after hot-stamping process.

Keywords: Corrosion resistance; Press-hardened steel; Al-Si coating; CSP.

1. Introduction

The hot stamping (or hot-press forming, press-hardening) process is a new technology widely used in producing advanced high strength steel parts for automotive applications in recent years. After heating the steel blanks of press-hardening steel in a furnace for a few minutes to the austenitizing temperature (≈ 880-930 °C), it is quickly transferred to a press in which it is immediately formed and simultaneously hardened by die quenching. The final press-hardened automotive component has a fully martensitic structure and exhibits a tensile strength up to 1500 MPa. The parts produced by this technology are mainly used in the anti-collision components of automobiles, such as bumper beams, door reinforcements, A and B-pillars reinforcements, and so on [1-4].

During the hot stamping process of uncoated 22MnB5 steel (cold-rolled, hot rolled, etc.), exposure to air at high temperature will cause surface oxidation (scaling). In order not to affect the paintability, the press-hardened parts need to be shot blasted or pickled to remove the surface oxidized layer (scale). However, the Al-Si press-hardening steel which is widely used at present can prevent the surface oxidation of the steel blanks during the hot stamping process, and the final parts can be directly painted without shot blasting. However, it exists some problems in Al-Si press-hardening steels, such as high cost, presence of microcracks and high iron concentration in the coating after hot stamping, no cathodic protection ability and so on [1, 5-7].

In order to fully evaluate the corrosion resistance of different types of press-hardened steels, the present work compares the properties and corrosion resistance of electrophoretic paint films on two types of uncoated (cold-rolled and CSP hot-rolled) and an Al-Si press-hardened steel based on 22MnB5. From the point of view of automobile manufacturers, as the press-hardened parts are mostly structural parts, while the painting of body-in-white is limited by structure design and electrophoretic painting process, some areas are poorly electro-painted. Therefore, the corrosion resistance evaluation of press-hardened parts should be carried out before and after painting separately.

2. Experimental

2.1. *Sample preparation*

All press-hardening steel materials used in this work (uncoated cold-rolled and CSP hot-rolled steel, Al-Si steel) were sampled in the form of blanks from coils produced by automobile steel lines.

In order to simulate the actual hot stamping process of automobile parts, three kinds of steel blanks were hot stamped by plate quenching with the same process parameters. Three kinds of plate samples were heated in a furnace to 930 °C and kept for 300 s, then transferred to the plate quenching die, stamped and rapidly cooled to room temperature by die quenching. Two kinds of uncoated steel samples (cold rolled and CSP hot-rolled) were covered with scales on surface after hot stamping and need to be shot blasted, while no shot blasting is performed for Al-Si samples. Three kinds of samples processed by plate quenching were finally cut by laser into electrophoretic and salt spray test samples.

The specifications of three kinds of press-hardened samples, the main process parameters of plate quenching and the final sample sizes are shown in Table 1.

Table 1. Test materials and main process parameters of plate quenching.

Types of press hardening steels	Thickness (mm)	Heating temperature (°C)	Heating duration (s)	Press pressure (t)	Holding time (s)
Cold rolled	1.50				
CSP hot-rolled	1.50	930	300	230	15
Al-Si	1.50				

2.2. *Preparation and performance test of electrophoretic paint film*

Three kinds of press-hardened samples for electrophoretic painting processed by the steps mentioned in 2.1 (size 70 mm × 150 mm), are entrusted to a automobile manufacturer and electro-painted in a paint production line together with the automobile components. The complete preparation of the electrophoretic samples consists of the following steps: Washing→ Degreasing→ Washing→ Surface adjusting→ Phosphating→ Electrophoretic painting→ Washing→ Bake hardening.

The physicochemical properties and the resistance to neutral salt spray test of the electrophoretic films of the three kinds of samples were determined respectively. The test standards involved and instruments used are shown in Table 2.

Table 2. Test standards and instruments for measuring physicochemical properties and resistance to neutral salt spray of electrophoretic paint films of three press-hardened samples.

No.	Test items	Test standards	Test instruments
1	Film thickness	GB/T 13452.2-2008 (ISO 2808 : 2007, IDT)	Leica DM6000M Optical microscope
2	Paint adhesion	GB/T 9286-1998 (ISO 2409 : 1992, IDT)	Model 295 Cross cut tester
3	Film hardness (by pencil test)	GB/T 6739-2006 (ISO 15184 : 1998, IDT)	Zhonghua 101 Drawing pencil
4	Resistance to liquids	GB/T 9274-1988 (ISO 2812 : 1974, IDT)	
5	Resistance to neutral salt spray	GB/T 1771-2007 (ISO 7253 : 1996, IDT)	Q-FOG CCT 600 Cyclic corrosion chamber

2.3. *Salt spray test of unpainted samples*

Three kinds of press-hardened samples processed by the steps mentioned in 2.1 (size 50 mm × 100 mm) were cleaned by using absolute ethyl alcohol and ultra-pure water and dried. The salt spray test was carried out in a Q-FOG CCT 600 cyclic corrosion chamber. The test was based on the standard GB/T 1771-2007 and carried out with a continuous spraying method, the chamber temperature was 35 °C and the salt solution was 5% NaCl (wt%). In addition, the surface and cross-section morphology of the Al-Si sample surface and the composition determination of the Al-Si coating after plate quenching were measured by QUANTA FEG450 field emission scanning electron microscope and EDAX energy dispersive spectrometer.

3. Results and Discussions

3.1. *Physicochemical properties of the electrophoretic paint films*

The physicochemical properties of the electrophoretic paint films of the three kinds of samples are shown in Table 3.

Table 3. Physicochemical properties of the electrophoretic paint films of the samples.

	Film thickness (μm)	Paint adhesion	Film hardness	Resistance to liquids			
				acid	alkali	petrol	engine oil
Cold rolled	26.5	0 scale	3H	No apparent change of the paint film were found on sample surface			
CSP hot-rolled	25.7	0 scale	3H				
Al-Si	24.0	0 scale	4H				

It can be seen from Table 3 that the electrophoretic paint film thicknesses of the three samples are close, and the paint adhesion of the films is excellent (0 scale). In addition, after immersion in acid (0.1N H_2SO_4), alkali (0.1N HaOH), petrol (92#) and engine oil for a certain time, no apparent change of the paint film were found on sample surface. The physicochemical properties of the films meet the requirements of related standards of the main automobile manufacturers, and the paintability of the three kinds of materials was good.

3.2. *Resistance to neutral salt spray of the electrophoretic paint films*

After neutral salt spray test for 1000 hours, red corrosion products appeared on the scribed lines of the three kinds of electro-painted samples (see Figure 1), in which the corrosion products on the electro-painted Al-Si sample are less. This is due to the physical barrier protection of the Al-Si coating, which hinders the diffusion of corrosion media into the steel base [5].

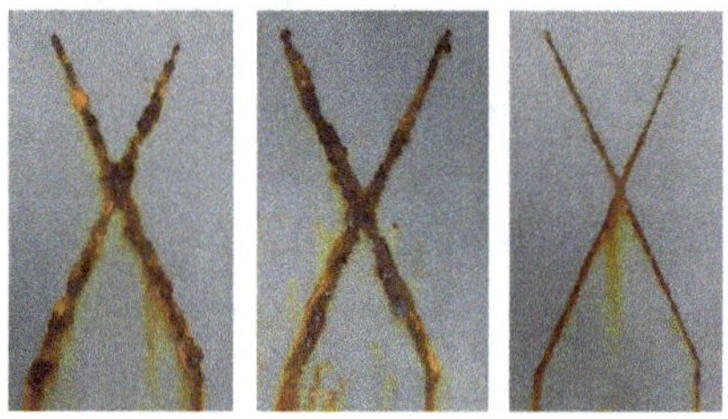

Cold rolled CSP hot-rolled Al-Si

Fig. 1. Electro-painted samples of press-hardened steels after neutral salt spray test for 1000 h.

After removing the corrosion products on the electro-painted surfaces of the three kinds of samples according to the standard method (see Figure 2), the maximum width of the single side at the scratches of all the samples is less than 0.5 mm, which met the requirements of most automobile manufacturers. This shows that the adhesion of the three kinds of samples was good. A homogeneous and continuous phosphating film is formed between the uncoated steels (cold rolled and CSP hot-rolled) and the electrophoretic paint film, based on the microscopic inspection of the cross section of the electro-painted samples (not shown here), which greatly improves paint adhesion. The Al-Si sample has no phosphating film between Al-Si coating and electrophoretic paint film, because of the "anchoring effect" of coating and film, the electrophoretic paint film has also good adhesion [1].

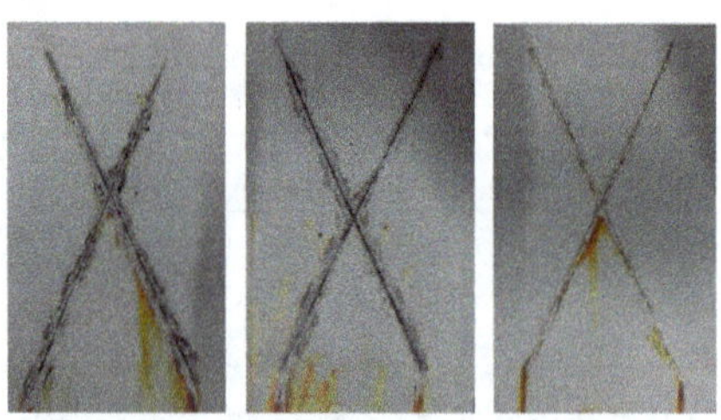

Cold rolled CSP hot-rolled Al-Si

Fig. 2. Electro-painted samples of press-hardened steels after neutral salt spray test for 1000 h (after scraping the scribe lines).

3.3. *Resistance to neutral salt spray of unpainted samples*

Figure 3 shows the unpainted samples of the three kinds of materials after neutral salt spray test for 9 hours.

Cold rolled CSP hot-rolled Al-Si

Fig. 3. Unpainted samples of press-hardened steels after neutral salt spray test for 9 h.

It can be seen from Figure 3 that after neutral salt spray test for 9 h, the surfaces of cold rolled and CSP hot-rolled samples are completely covered with red rust, while the surface of Al-Si sample has a relatively small area of red rust, which is due to the physical barrier protection of the Al-Si coating, hinders the diffusion of corrosion media into the steel base [5]. According to the microscopic inspection of the cross section of the unpainted samples (not shown here) after salt spray test, the red rust on the surface of Al-Si sample was mainly located on the surface of the coating.

However, red rust appears on the surface of Al-Si sample after salt spray test for a short time. This indicates that the corrosion resistance of press-hardened Al-Si steel is poor, which is related not only to the presence of microcracks, but also to the high iron concentration in the coating (*cf.* 3.4). Moreover, the corrosion potential of Al-Si coating is so close to the steel base and does not have cathodic protection ability [7].

3.4. *Microstructure and composition measurements of Al-Si coatings*

The cross section microstructure of the Al-Si coating after plate quenching is shown in Figure 4. It can be seen that the microcracks in the coating are distributed in a three-dimensional network, and the microcracks extend longitudinally until the inter-diffusion layer between the steel substrate and the coating. The formation of microcracks is related to the different coefficients of thermal expansion of coating and steel substrate.

The energy dispersive analysis of different regions of the coating is shown in Figure 4. The measured contents of Al, Si and Fe corresponding to the regions indicated in Figure 4 are shown in Table 4.

It can be seen from Table 4 that the iron concentration in the Al-Si coating after plate quenching is as high as 44～62%. This is related to the enrichment of iron in the coating due to the diffusion of the coating and the steel substrate during the austenitizing heat treatment. The high iron content in the coating and the presence of microcracks lead to the appearance of red rust on the surface of the uncoated sample after a short time of salt spray test.

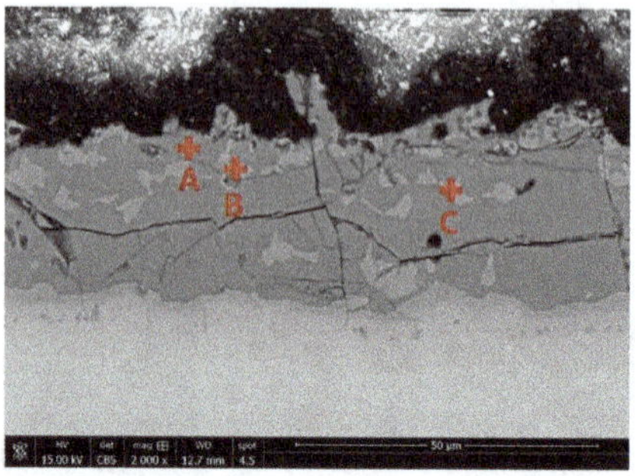

Fig. 4. Cross section microstructure of Al-Si sample after plate quenching.

Table 4. Energy dispersive analysis of Al-Si coating after plate quenching.

Regions	Al (wt%)	Si (wt%)	Fe (wt%)
A	52.68	3.13	44.19
B	28.31	10.95	60.74
C	27.96	9.92	62.12

It can be seen from Table 4 that the iron concentration in the Al-Si coating after plate quenching is as high as 44～62%. This is related to the enrichment of iron in the coating due to the diffusion of the coating and the steel substrate during the austenitizing heat treatment. The high iron content in the coating and the presence of microcracks lead to the appearance of red rust on the surface of the uncoated sample after a short time of salt spray test.

4. Conclusion

(1) The physicochemical properties (Film thickness, Paint adhesion, Film hardness, Resistance to liquids) of the electrophoretic paint films on press-hardened uncoated 22MnB5 steels (cold-rolled and CSP hot-rolled) and Al-Si steel were found to be proximal,

(2) The resistance to neutral salt spray of the electrophoretic paint films of the three kinds of samples all meets the requirements of automobile manufacturers. After neutral salt spray test for 1000 hours, the maximum width of the single side at the scratches of all the samples is less than 0.5 mm,

(3) After plate quenching, the corrosion resistance of unpainted samples of the three materials was poor. The poor corrosion resistance of Al-Si sample is related to the presence of microcracks and the high iron concentration in the coating.

References

1. D. W. Fan and B. C. D. Cooman, State-of-the-Knowledge on Coating Systems for Hot Stamped Parts, *Steel Res. Int.* **83**, 412 (2012).
2. R. R. Xing, S. L. Zeng, W. J. Jin and Q. Liu, Application of hot-forming technology for automobile, *P. Soc. Auto. Eng. China* 951 (2010).

3. J. Zhang, S. M. Jiang and Q. F. Zhang, Research progress on hot formed steel coating, *Heat treatment of Metals* **40**, 169 (2015).
4. H. Chen and C. N. Jing, Application and development of thermoforming technology in automobile lightweighting, *Heat treatment of Metals* **41**, 61 (2016).
5. M. Suehiro, J. Maki, K. Kusumi, M. Ohgami and T. Miyakoshi, Properties of Aluminized Steels for Hot-Forming, *Nippon Steel Techn. Rep.* **88**, 16 (2003).
6. L. Dosdat, J. Petitjean, T. Vietoris, and O. Clauzeau, Corrosion Resistance of Different Metallic Coatings on Press-Hardened Steels for Automotive, *Steel Res. Int.* **82**, 726 (2011).
7. C. Allely, L. Dosdat, O. Clauzeau, K. Ogle and P. Volovitch, Anticorrosion mechanisms of aluminized steel for hot stamping, *Surf. Coat. Tech.* **238**, 188 (2014).

The Effect of Dew Point Temperatures at Heating Section on the Inhibition Layer Morphology of Galvanized High Strength Steel

E. Q. Ma[a,†], L. Cui[b], H. Zhan and L. Chen
Technology Center of Ma'anshan Iron and Steel Company, Ma'anshan, Anhui, China
E-mail: [a,†]erqingma@163.com, [b]cuilei1219@163.com

The effect of different dew point temperatures at heating section on the inhibition layer morphology of galvanized high strength steel was studied. 600 MPa grade hot-dip galvanized dual phase steel was annealed on the continuous hot galvanizing line at different dew point temperatures of heating section. The surface morphology of inhibition layer was scanned by SEM, and the chemical compositions of the inhibition layer was analyzed by EDS. The results show that Fe-Al phase particles in inhibition layer of Galvanizing DP600 Steel are dense and complete when the dew point temperature of heating section is from -30 °C to +15 °C. When the dew point temperature is from 0 °C to +15 °C, the size of particles is about 0.1 micron. The smaller and more compact of Fe-Al phase particles, the stronger of the adhesion of zinc coating.

Keywords: Dew point; Inhibition layer; Galvanized; High strength steel.

1. Introduction

With the strengthening of automobile lightweight and safety regulations [1], dual-phase steel is widely used in automobile structural parts, reinforcing parts and anti-collision parts due to its advantages of high strength, good forming and welding performance. In order to improve the corrosion resistance of car body, galvanized dual-phase steel plate is widely used [2-3]. Due to the addition of a quantity of surface oxidation elements such as Si, Mn, etc., in the dual-phase steel, has certain influence to the zinc layer adhesion, causing skip plating or dezincification in stamping process. Many scholars studied under laboratory conditions of hot galvanizing technology parameters on the Si and Mn element surface oxidation behavior research [4-7]. However, relatively few studies have been conducted on the influence of relative process parameters on the zinc coating. In this paper, the influence of the dew point temperature of the heating section on the Fe-Al phase inhibition layer has been studied.

2. Test Materials and Methods

The test material was taken from 600 MPa class hot-dip galvanized dual-phase steel produced by a continuous hot-dip galvanizing production line. The thickness of the double-sided plating layer was 100 g/m^2, the thickness of the sample was 2.0 mm, and the sample size was 18 mm × 20 mm. The chemical composition of experimental steel is shown in Table 1.

Table. 1 Chemical composition of experimental steel

Element	C	Si	Mn	S	P	Al	Cr+Mo
Wt%	0.05~0.10	<0.30	1.0~2.5	<0.010	<0.020	0.03~0.08	0.20~0.50

The dew point temperature of heating section is adjusted to -30 °C, -20 °C, -10 °C, 0 °C and +15 °C, and the zinc layer surface morphology and inhibition layer morphology was compared and analyzed at the five different dew point temperatures. The annealing furnace atmosphere is 5 $\%H_2$+95 $\%N_2$, and the continuous annealing temperature process diagram is shown in Figure 1.

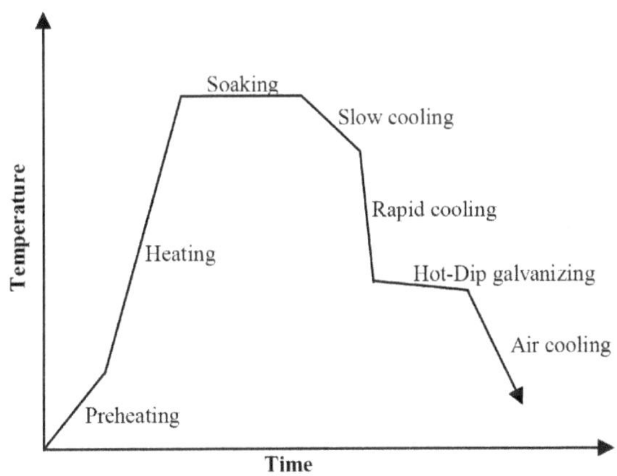

Fig. 1. The continuous annealing temperature process diagram.

The zinc layer of steel sheet is corroded by 5 % HCl solvent containing corrosion inhibitor until the zinc layer disappears, and the zinc layer morphology and inhibition layer surface are analyzed by scanning electron microscope (SEM) by means of FEI QUANT 450 after using alcohol electromagnetic concussion. The components of the inhibition layer were qualitatively analyzed by the attached EDAX spectrometer, with an acceleration voltage of about 5 kV, a high vacuum mode and a working distance of about 10 mm.

3. Test Results

3.1. *Surface morphology of zinc layer*

Figure 1 is surface SEM morphology of zinc layer at the five kinds of dew point temperature. Finishing processing traces can be found in Figure 2(a)~(e), and the zinc layer is complete and no leakage point.

Figure 2(a)~(c) can still distinguish the grain boundary of zinc layer (as shown in the dotted line). In Figure 2(d)~(e), the grain boundary of zinc layer is not obvious, and the homogeneity of zinc layer is better.

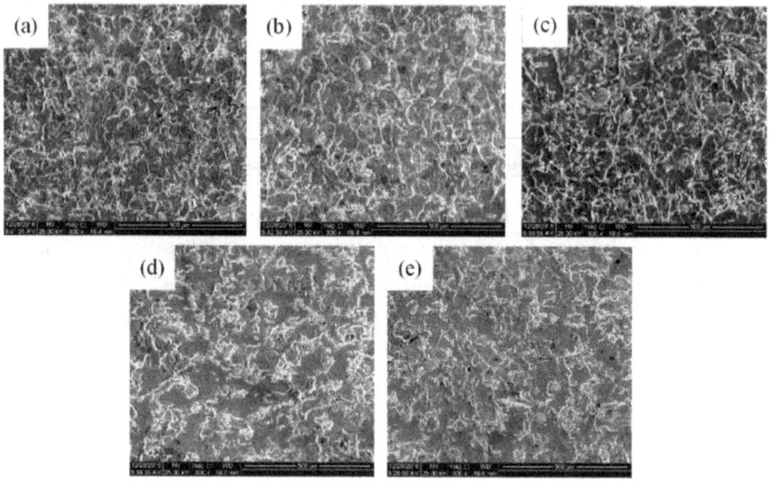

Fig. 2. Surface morphology of zinc layer at different dew point temperatures: (a) -30°C; (b) -20°C; (c) -10°C; (d) 0°C; (e) + 15°C.

3.2. *Morphology of inhibition layer*

Figure 3 shows the morphology of inhibition layer at different dew point temperatures of heating sections under the 20,000 times magnification. In Figure 3(a)~(e), the Fe-Al phase particles of inhibiting layer are uniformly distributed on the iron matrix, and the inhibition layer is complete and dense. In Figure 3(a)~(c), the inhibition layer particles in Figure 3 are uniformly distributed. Fe-Al phase particles are round and uniform in size. The average particle size of Fe-Al phase is about 0.3 μm. Figure 3(d)~(e) the Fe-Al phase is much smaller. The average particle size of Fe-Al phase is about 0.1μm. Compared with Figure 3(d)~(e), the Fe-Al phase particles of inhibition layer in Figure 3(a)~(c) are smaller, and more tightly arranged, which has better adhesion performance of zinc layer.

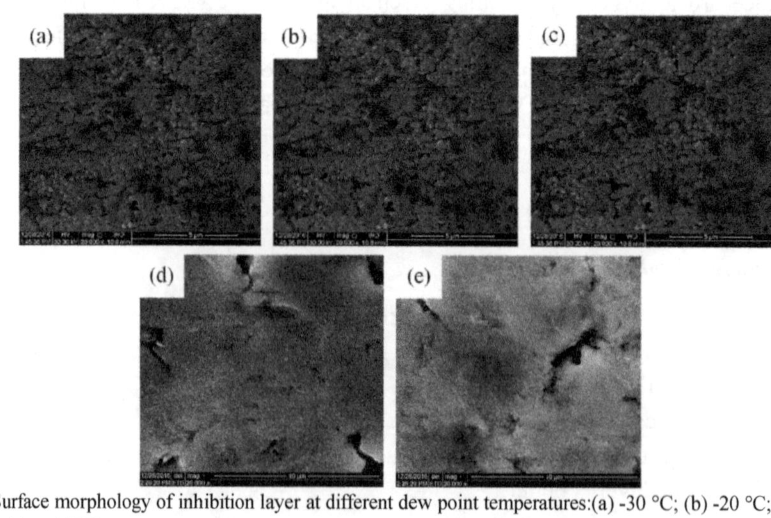

Fig. 3. Surface morphology of inhibition layer at different dew point temperatures:(a) -30 °C; (b) -20 °C; (c) -10 °C; (d) 0 °C; (e) + 15 °C.

3.3. *Formation of inhibition layer*

Figure 4 shows the surface scanning energy spectrum at different dew point temperatures of heating section. Figure 5 shows the corresponding detection results of chemical composition energy spectrum. The surface detection components are Fe, Al and Mn. When the dew point temperature is -20 °C， -30 °C and -10 °C, energy spectrum analysis of Al content is 4.4 %, 4.9 % and 5.4 %. When the dew point of 0 °C and + 15 °C, Al content is 1.1 %, 1.5 %. It is not hard to find that when the dew point temperature is negative and positive, there is a significant difference in Al content.

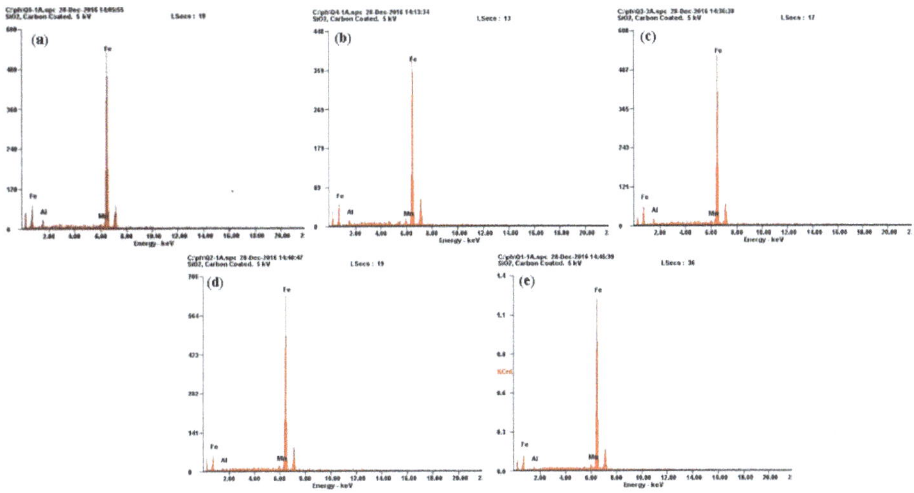

Fig. 4. EDS of zinc layer at different dew point temperatures: (a) -30 °C; (b) -20 °C; (c) -10 °C; (d) 0 °C; (e) +15 °C.

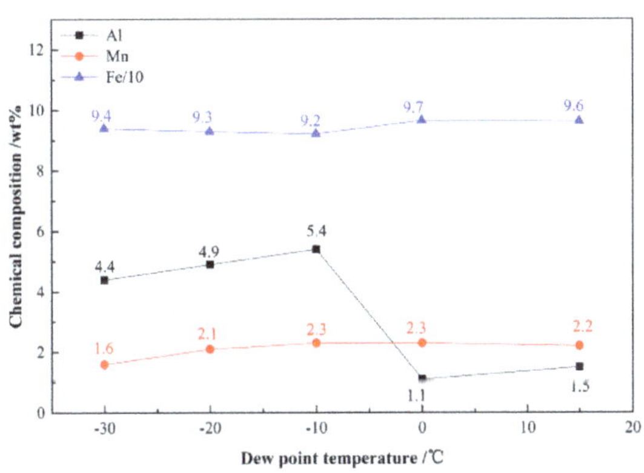

Fig. 5. Chemical compositions of EDS at different dew point temperatures.

4. Analysis and Discussion

Because the heat of formation of Fe-Al phase is higher than that of Fe-Zn phase [8], when the steel sheet enters the zinc liquid, the Fe-Al compound is preferentially formed and adheres to the surface of the steel sheet, which can hinder the diffusion of iron into zinc liquid. This effectively inhibits the formation of brittle Fe-Zn metal parts and improves the adhesion of the zinc layer. In general, the adhesive Fe_2Al_5 (or $FeAl_3$) films formed in the early stage of this type of plating are called inhibiting layer [9]. In addition, the adhesion work of Fe/Fe_2Al_5 interface is the largest and the adhesion performance is the strongest among all the interface adhesion work [10]. Therefore, the denser and more complete the intermediate film of Fe-Al phase is, the better adhesion performance of Zn layer is.

As the detection depth of scanning electron microscope energy spectrum is much larger than that of the inhibition layer, the Fe content in the matrix is 97.5 %. Assuming that the EDS spectrum detection depth is constant, the suppression layer is composed of a Fe_2Al_5 phase, and the Al content in the matrix is 0.2%. Assuming that the thickness of the inhibition layer is x_1, the thickness of the substrate detected by the energy spectrum is x_2, and Fe and Al are respectively a_1, b_1, c_1 and a_2, b_2 and c_2 in the inhibition layer, substrate and total amount, and the following relations can be obtained by Eq. 1 and Eq. 2.

$$a_1x_1 + b_1x_2 - c_1 = 0 \tag{1}$$

$$a_2x_1 + b_2x_2 - c_2 = 0 \tag{2}$$

Table 2 shows the calculated ratio of the thickness of the inhibition layer. When the dew point temperature is positive, the thickness of the inhibition layer is significantly reduced, which is about 1/4 of the thickness of the inhibition layer when the dew point temperature is negative. High dew point temperature can reduce the thickness of surface oxidation layer, which is helpful to improve the coating performance [11].

Table 2. The calculated ratio of the thickness of the inhibition layer, %.

DP/°C	a_1	b_1	c_1	a_2	b_2	c_2	x_1	x_2
-30	0.453	0.975	0.940	0.547	0.002	0.044	8 %	92 %
-20	0.453	0.975	0.930	0.547	0.002	0.049	9 %	91 %
-10	0.453	0.975	0.922	0.547	0.002	0.054	10 %	90 %
0	0.453	0.975	0.966	0.547	0.002	0.011	2 %	98 %
+15	0.453	0.975	0.963	0.547	0.002	0.015	2 %	98 %

5. Conclusion

The particles of Fe-Al phase inhibition layer in 600 MPa galvanized dual-phase steel are uniformly and continuously dense, and the zinc coating has good adhesion performance. Compared with -30 °C ~-10 °C, the dew point temperature is 0~ +15 °C, the particle size of the inhibition layer is finer and denser, and the thickness of the inhibition layer is obviously thinner, which helps to improve the adhesion of the zinc layer.

References

1. G. Y. Li and M. T. Ma, Auto steels production in China-status and prospect, *Steel Rolling* **4**, 23 (2014).
2. W. Q. Yang and H. Z. Liu, Study on anti-Corrosion property of different kinds of galvanized steel sheet, *Total Corrosion Control* **5**, 28 (2015).
3. C. C. Feng, G. L. Feng and Q. Y. Song, Corrosion resistance experiment of a vehicle body, *Corrosion & Protection* **4**, 366 (2015).
4. K. Shinoda, T. Yamamoto and S. Suzuki, Characterization of selective oxidation of manganese in surface layers of Fe–Mn alloys by different analytical methods, *ISIJ Int.* **53**, 2000 (2013).
5. H. H. Wang, C. Y. Qi, Y. P. Li, S. M. Jiang, Q. F. Zhang and G. Q. Yu, Effects of dew point on the galvanizability of DP590, *T. Mater. Heat Treat.* **36**, 177 (2015).
6. W. Mao and W. G. Sloof, Relation between oxygen activity gradient in the internal oxidation zone of Mn alloyed steel and the composition of oxide precipitates, Scripta Mater. **135**, 29 (2017).
7. Y. P. Li, Q. F. Zhang and S. M. Jiang, Effect of hydrogen content on surface selective oxidation of the Mn-Si dual phase steel during continuous annealing and galvanizing process, *Heat Treat. Met.* **37**, 95 (2012).
8. G. M. Song and W. G. Sloof, Effect of alloying element segregation on the work of adhesion of metallic coating on metallic substrate: Application to zinc coatings on steel substrates, *Surf. Coat. Tech.* **205**, 4634 (2011).
9. L. Zhu, *Hot galvanizing of steel*, (Chemical Industry Press, 2006).
10. C. D. Hu, S. M. Jiang and Y. P. Li, Effect of inhibition layer and ζ-FeZn13 on adhesion of dual phase steel coating, *J. Iron Steel Res.* **25**, 37 (2013).
11. L. Cho, S. J. Lee, M. S. Kim, Y. H. Kim and B. C. D. Cooman, Influence of the dew point on the coating quality of GA TRIP steel, in *Proceedings Abstracts of Asia Steel International Conference 2012 (Asia Steel 2012)*, (Beijing, China, 2012).

Applications of the Digital Image Correlation (DIC) Technique for High-temperature Strain Measurement: A Review

R. Zhang, Z. Shao† and J. Lin
Department of Mechanical Engineering, Imperial College London, London, SW7 2AZ, UK
†E-mail: z.shao12@imperial.ac.uk

Displacement and strain are critical data to obtain by materials mechanical tests. The digital image correlation (DIC) technique has been increasingly employed for contactless measurements of displacement and strain applied in experimental mechanics. DIC is relatively simple to use at ambient condition, but it becomes challenging for testing at high temperature due to surface condition change, sample color change, or coating peeling off. In order to develop the DIC method of strain measurement for boron steel at elevated temperature or under hot stamping conditions, the high-temperature applications of the DIC technique are reviewed and compared for different materials and various testing conditions. The potential problems for DIC testing at high temperature are identified and summarized. Solutions for boron steel tests under hot stamping conditions are proposed in this paper.

Keywords: Digital image correlation; High temperature; Thermo-mechanical test; Strain measurement; Hot stamping.

1. Introduction to the DIC Technique

The digital image correlation (DIC) technique is a non-contact full-field strain measurement method achieved by comparing the grey intensity changes of a series of images captured continuously at different instants. Fig. 1 shows the schematic for a typical two-dimensional (2D) DIC tests for sheet metal forming applications. Deformation history can be recorded for samples with stochastic coating pattern and strain field can be calculated through the DIC method.

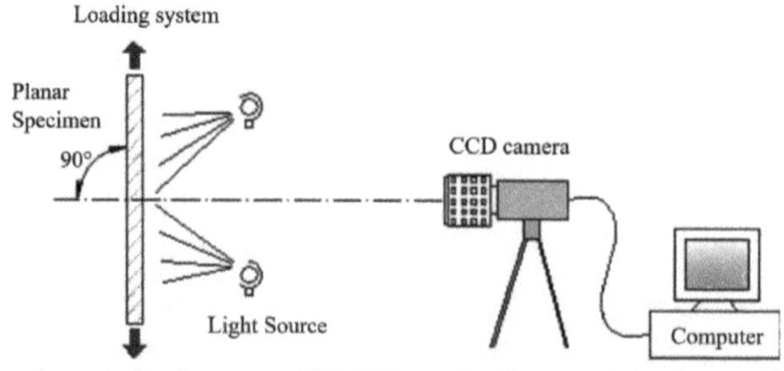

Fig. 1. The schematic for the set-up of 2D DIC test for sheet metal forming application [1].

The DIC technique was firstly developed by a research group at the University of South Carolina in 1980 [2]. In comparison with using conventional methods, such as strain gauge and extensometer, DIC has been increasingly adopted for strain measurement given that its advantages [3]: (1) Contactless measurement; (2) High flexibility in terms of both spatial resolution and dynamical performance; (3) Capability of wide applications for samples deformed under complex conditions. Verhulp et al. [4] measured the strain field in open-cell structures using a three-dimensional DIC technique, in which high-resolution computed tomography images were applied. Bueno and Lambros [5] investigated the behaviour of the crack growth resistance of functionally graded materials using the DIC technique at room temperature.

The accuracy of strain measurement by DIC has been intensively investigated and improved. Hild and Roux [6] reviewed the process used to obtain the strain field from images recorded at different instants. Pan et al. [1] reviewed the methodologies of the 2D DIC technique for strain field measurement and extended the application of DIC for nanoscale deformation measurement by using high-spatial-resolution microscopes. Mazzoleni [3] identified the problems of uncertainty in DIC measurement and proposed solutions in order to reduce the expected data dispersion. Khoo et al. [7] reviewed systematically the development of the DIC algorithms, and presented the improvement with respect to computation efficiency and measurement accuracy.

However, applications for high-temperature testing are challenging [8, 9]. In this paper, a review on the applications of DIC for stain measurement at high temperatures is presented. The potential problems and solutions are summarized. This will provide a guideline to design thermo-mechanical test programme for boron steel by using DIC to measure deformation under hot stamping conditions.

2. DIC for High-Temperature Applications

2.1. *Applied patterns or coating*

Novak and Zok [10] applied alumina or zirconia paints on the specimen surface to generate high contrast speckle patterns in order to measure strain field for SiC-bearing composite and Ni-based superalloy at temperatures up to 1500 °C. The alumina paint was also used by Wang [11] for full-field strain measurement up to 2000 °C. The paint was sprayed onto specimen surface by using an airbrush, followed by a drying at room temperature and then at 200 °C to increase the bonding strength. The agreement of Young's modulus for Inconel 718 Ni-based superalloy and graphite at different temperatures has validated the effectiveness of this method. Chen [9] investigated different coating materials and found that a coating material (VHT Flameproof), consisting of amorphous precipitated silica and titanium dioxide with a maximum resistant temperature of 1100 °C, provides high ductility performance for deformation tests at high temperature. However, at high temperature, the stability of the coating material is greatly affected by the chemical reaction with the base material, including aluminum, steel and nickel alloy. A stochastic speckle pattern composing of a black background and white dots was employed for thermo-mechanical tensile tests [12] and formability tests [13] (as shown in Fig. 2) of aluminum alloy 6082 under hot stamping conditions (the maximum required temperature of 535 °C). Rooyen and Becker [8] also employed the same coating to characterize material mechanical properties, including elastic moduli and Poisson's ratio, for 12% Cr steels at temperatures up to 900 °C. A vacuum condition of 44 Pa was set to minimize the oxidation of the steel samples. Guo et al. [14] used tungsten for applying a speckle pattern by the plasma spraying method. This speckle pattern was demonstrated to be capable of strain measurement using DIC for carbon fiber specimens up to 2600 °C. Mao et al. [15] sprayed a coating material, consisting of high-temperature glue and ZrO2 powder with the size of 10 to 20 μm, onto the surface of C/SiC composite specimen by a high-pressure airbrush gun, and measured displacement distribution up to 1600 °C using DIC. Valeri et al. [16] created a speckle pattern by applying a high-temperature silica-based ceramic black paint with an average size of 50 μm, against a thin layer of ultra-high-temperature resistant yttrium oxide white spray paint, to measure the flow stress response for 304 stainless steel from room temperature to 900 °C. The speckle pattern was left for 24 h to dry the pattern fully.

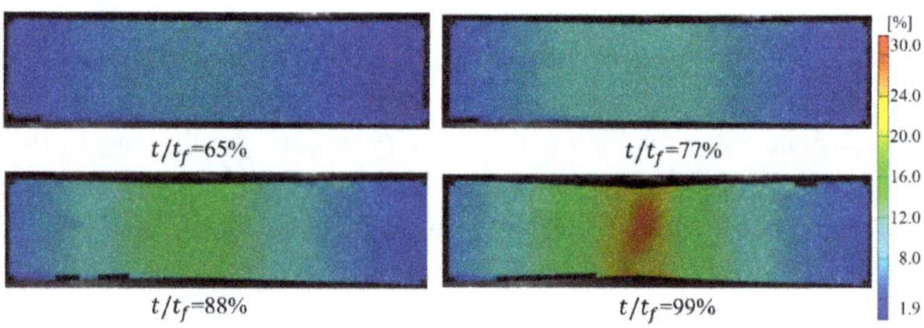

Fig. 2. The speckle pattern of a black background and white dots, and the calculated strain field at the different stages (t_f is the final fracture time).

Speckle patterns can be generated by other methods instead of spraying paint. Grant et al. [17] used 1200 grit silicon carbide papers to abrade the specimen surface for a random distribution of scratches. The capability of the method was demonstrated by characterizing the mechanical properties, including Young's modulus and the coefficient of thermal expansion, of a Ni-based superalloy, at room temperature and up to 1000 °C. Meyer and Waas [18] exploited the wavelength difference between the rough sample surface and illumination to develop a new speckle pattern because of destructive interference. This technique was validated by determining the strain field for ceramic matrix composite monotonically loaded at temperatures up to 1315 °C. Lee et al. [19] investigated the formability of 22MnB5 at high temperatures up to 930 °C. In order to obtain limit strains under different strain states using DIC, square grids were engraved on the sample surface using a laser machine. Argon was purged into the test chamber in order to eliminate the oxidation of the samples.

The speckle patterns and other patterns applied in the DIC tests for displacement and strain measurement at elevated temperatures are summarized in Table 1 and Table 2, respectively. The pattern quality is also affected by the deformation speed and base material.

Table 1. Speckle patterns employed for strain measurement for different materials at elevated temperatures using the DIC technique.

Base material	Speckle patterns	Temperature/°C	Vacuum condition
SiC-bearing composite and Ni-based superalloy	Alumina or zirconia paints. (White spray dots only)	1500	Not required
Inconel 718 Ni-based superalloy and graphite	Alumina paint. (White spray dots only)	2000	Not required
Aluminum, stainless steel and nickel alloy	VHT Flameproof	1100	Required
Aluminum alloy	VHT Flameproof. (Black spray background and white spray dots)	550	Not required
12% Cr steel	VHT Flameproof. (Black spray background and white spray dots)	900	Required
Carbon fiber	Tungsten sprayed by a plasma spraying method	2600	Not required
C/SiC composite	High-temperature glue and ZrO2 power	1600	Not required
304 Stainless steel	White yttrium oxide spray	900	Required

Table 2. Other patterns employed for strain measurement for different materials at high temperatures using DIC technique.

Base material	Other patterns	Temperature/°C	Vacuum condition
Ni-based superalloy	Sample surface was abraded using 1200 grit silicon carbide papers	1000	Not required
Ceramic matrix composite	Relying on wavelength difference between a rough sample surface and illumination	1315	Not required
Boron steel (22MnB5)	Square grids were engraved on the surface of samples using a laser machine	930	Required

2.2. *Applied imaging systems to record deformation process*

To eliminate the influence of thermal radiation from high-temperature objects on the grayscale intensity of images captured during deformation, a blue bandpass filter (BP470) with a blue illumination with mid-wavelength of 450 nm was employed in [9-11, 16-18, 20], as shown in Fig. 3. The filter aims to eliminate the light of wavelengths which have higher radiation energy at elevated temperatures, while the illumination with the associated bandpass wavelength increases the intensity of the captured images. Zhang et al. [21] adopted an ultraviolet narrow band filter and a ultraviolet sensitive charged coupled device (CCD) camera for the imaging system, in which the sample was illuminated by a ultraviolet light. Mao et al. [15] employed an optical bandpass filter (490 nm) and two white lights for illumination to decrease the influence of high-temperature thermal radiation and to compensate for the light intensity, respectively.

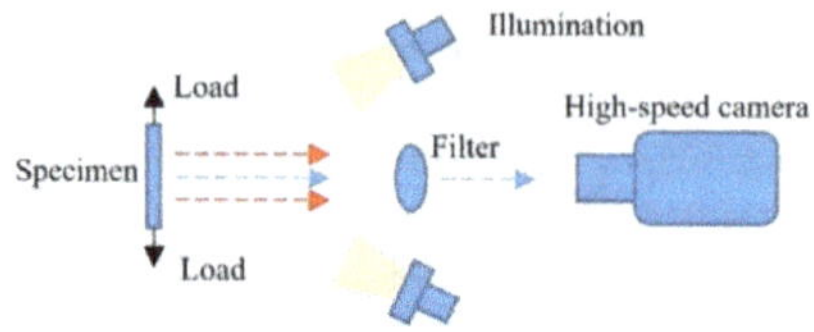

Fig. 3. A schematic of an imaging system with a bandpass filter and an associated wavelength illumination.

3. Conclusion

The difficulties which limit the DIC applications at elevated temperature are: (1) Few coating materials are both high-temperature resistant and being ductile at high temperature. (2) Coating materials usually have low adhesion and cannot attach firmly to base material at high temperature; (3) Chemical reactions between coating material and base material could cause speckle pattern color change and worse adhesion; (4) Oxidization of the base material, such as boron steel and titanium alloy, at high temperature would deteriorate the homogeneous surface; (5) Excessive surface radiation for high temperature tests leads to image saturation and decorrelation; (6) Convection current in the test environment optically distorts the images captured, resulting in spurious measurement.

DIC has been applied for thermo-mechanical tests for aluminum alloy under hot stamping conditions in the previous study, but no application has been proposed for boron steel testing under hot stamping conditions. Using high-temperature resistance coating for samples tested in vacuum conditions can be a solution. Future work will be conducted to investigate the feasibility and stability of the proposed method.

Acknowledgments

Financial support from EPSRC Impact Acceleration Account Funding (grant number: EP/R511547/1) is gratefully acknowledged by the authors. The author R Zhang appreciates financial support from the CSC-Imperial Scholarship.

References

1. B. Pan, K. Qian, H. Xie and A. Asundi, Two-dimensional digital image correlation for in-plane displacement and strain measurement: a review, *Meas. Sci. Technol.* **6**, 152 (2009).
2. W. H. Peters and W. F. Ranson, Digital imaging techniques in experimental stress analysis, *Opt. Eng.* **3**,427 (1982).
3. P. Mazzoleni, Uncertainty estimation and reduction in digital image correlation measurements, PhD thesis, Polytechnic University of Milan (2013).
4. E. Verhulp, B. van Rietbergen and R. Huiskes, A three-dimensional digital image correlation technique for strain measurements in microstructures, *J. Biomech.* **9**,1313 (2004).

5. J. Abanto-Bueno and J. Lambros, Investigation of crack growth in functionally graded materials using digital image correlation, *Eng. Fract. Mech.* **14**, 1695 (2002).
6. F. Hild and S. Roux, Digital image correlation: from displacement measurement to identification of elastic properties–a review, *Strain* **2**, 69 (2010).
7. S. W. Khoo, S. Karuppanan and C. S. Tan, A review of surface deformation and strain measurement using two-dimensional digital image correlation, *Metrol. Meas. Syst.* **3**, 461 (2016).
8. M. V. Rooyen and T. H. Becker, High-temperature tensile property measurements using digital image correlation over a non-uniform temperature field, *J. Srtain Anal. Eng.* 7, 117 (2018).
9. X. Chen, N. Xu, L. Yang and D. Xiang, High temperature displacement and strain measurement using a monochromatic light illuminated stereo digital image correlation system, *Meas. Sci. Technol.* **12**, 125 (2012).
10. M. D. Novak and F. W. Zok, High-temperature materials testing with full-field strain measurement: experimental design and practice, *Rev. Sci. Instrum.* **11**, 303 (2011).
11. W. Wang, C. Xu, H. Jin, S. Meng, Y. Zhang and W. Xie, Measurement of high temperature full-field strain up to 2000°C using digital image correlation, *Meas. Sci. Technol.* **3**, 35 (2017).
12. Z. Shao, N. Li, J. Lin and T. A. Dean, Strain measurement and error analysis in thermo-mechanical tensile tests of sheet metals for hot stamping applications, *P. I. Mech. Eng. C-J Mec.* **203-210**, (2017).
13. Z. Shao, N. Li, J. Lin and T. Dean, Formability evaluation for sheet metals under hot stamping conditions by a novel biaxial testing system and a new materials model, *Int. J. Mech. Sci.* 149 (2017).
14. X. Guo, J. Liang, Z. Tang, B. Cao and M. Yu, High-temperature digital image correlation method for full-field deformation measurement captured with filters at 2600 °C using spraying to form speckle patterns, *Opt. Eng.* **6**, 063101 (2014).
15. W. Mao, J. Chen, M. Si, R. Zhang, Q. Ma, D. Fang and X. Chen, High temperature digital image correlation evaluation of in-situ failure mechanism: An experimental framework with application to C/SiC composites, *Mater. Sci. Eng. A*. 26 (2016).
16. G. Valeri, B. Koohbor, A. Kidane and M.A. Sutton, Determining the tensile response of materials at high temperature using DIC and the Virtual Fields Method, *Opt. Laser. Eng.* 53 (2017).
17. B. Grant, H. Stone, P. Withers and M. Preuss, High-temperature strain field measurement using digital image correlation, *J. Strain Anal. Eng.* **4**, 263 (2009).
18. P. Meyer and A. Waas, Measurement of in situ-full-field strain maps on ceramic matrix composites at elevated temperature using digital image correlation, *Exp. Mech.* **5**,795 (2015).
19. R. S. Lee, Y. K. Lin, T. W. Chien, Experimental and theoretical studies on formability of 22MnB5 at elevated temperatures by Gleeble simulator, *Procedia. Eng.* 1682 (2014).
20. B. Pan, D. Wu, Z. Wang and Y. Xia, High-temperature digital image correlation method for full-field deformation measurement at 1200 °C, *Meas. Sci. Technol.* **1**, 015701 (2010).
21. H. Zhang, Y. Wang, L. Chen, J. Sun and L. Yang, High temperature deformation measurement based on ultraviolet DIC, in *Int. Conference on Experimental Mechanics (ICEM)*, (2015).

Investigation of the Constitutive Model for Warm Stamping of 7075-T6 Aluminum Alloy Sheet

Z. J. Zhu, B. Zhu, K. Wang, Y. L. Wang[†] and Y. S. Zhang
State Key Laboratory of Materials Processing and Die & Mold Technology, Huazhong University of Science & Technology, Wuhan 430074, China
[†]E-mail: wangyilin@hust.edu.cn

Warm stamping has potential application in processing 7075-T6 aluminum alloy for using the better formability of the alloy at elevated temperature. The constitutive model of 7075-T6 aluminum alloy at elevated temperature was investigated for the simulation of warm stamping. The uniaxial tensile tests on 7075-T6 aluminum alloy were conducted at elevated temperatures (ranging from 373 to 523 K) and different strain rates (ranging from 0.01 to 1 s^{-1}) by using a Gleeble-3500 simulator. Serials of continuous images of the specimens were taken by a general camera and Digital Image Correlation (DIC) method was adopted to obtain the strain fields during tensile tests. With the strain field data, the constitutive model 7075-T6 aluminum alloy was built, then the simulation works were done for the warm stamping with the constitutive model. Both the result of simulation and experimental works showed that the warm stamped part exhibited comprehensive property on smaller springback and higher final hardness when the optimal parameters are as following: the sheet was heated and holed for 10 minutes in a furnace preheated to 273 K, and then rapidly transferred onto a press where forming and quenching take place. Finally, the experimental result for warm stamped part showed that the springback angle was 4.61° and hardness was 180.6 HV.

Keywords: Warm forming; 7075-T6 aluminum alloy; Digital Image Correlation; Constitutive model.

1. Introduction

The high strength aluminum alloys such as 7075 aluminum alloys used in automotive parts have potential of emissions reduction [1]. It limits the applications of 7075 aluminum alloys that inefficient and costly for HFQ-artificial aging process and poor formability for cold stamping. One proper solution is warm stamping. The suitable warm stamping condition sufficiently improves formability of the alloy while retaining the high strength of as-received T6 temper. Hu et al conducted warm tensile test to investigate optimal forming temperature. It indicated that the intrinsic high strength of 7075-T6 is perfectly preserved after 473 K warm forming due to the combined effect of the decreased precipitation strengthening and increased dislocation strengthening. Further raising temperature to 523 K led to severely decreased hardness due to sharply coarsening of matrix precipitates [2]. Wang et al. investigated the deep drawing and stretch formability of 7075-T6 aluminum alloy through limiting drawing ratio tests at different forming temperatures. And the optimal drawing temperatures are ranging from 433 - 493 K [3-4].

Constitutive models of materials at elevated temperatures are commonly used to simulate and predict the formability of materials. The ability of measuring strain accurately and precisely is the key to the development of reliable constitutive models. A method known as Digital Image Correlation is applied to acquire strain of specimen surface during deformation stage by comparing sequential digital images (a reference image and deformed images). Comparing with traditional strain gauges, the DIC method has advantages of high accuracy and efficiency, low cost and adaptive to non-contact measurement [5]. Hoult et al. implemented DIC method to general tensile tests by using common CCD cameras to obtain digital images. The out-of-plane motion error was amended by properly increasing the effective distance between the camera and the specimen. The solution achieves mean strain errors of less than 5 $\mu\varepsilon$ when comparing DIC to foil strain gage measurements [6].

In this paper, the uniaxial tensile tests of 7075 aluminum alloy sheets were conducted by using Gleeble 3500 simulator with DIC method at elevated temperature. The Johnson cook constitutive model was developed with flow curves acquired by DIC method. Simulation and experiment were implemented to explore the feasibility of warm stamping of 7075 aluminum alloy.

2. Experiment

The Chemical compositions of as-received 7075-T6 aluminum alloy used in experiment are shown in Table 1. The original Vickers hardness was 192.3 HV and the ultimate tensile strength (UTS) was 592 MPa. To evaluate the flow behaviors of the AA7075-T6 at elevated temperatures, tensile tests were performed in a Gleeble-3500 thermo-mechanical process simulator combined with DIC method. Specimens were cut to the dimensions in rolling direction. Heat resistant paint is used to spray the random gray speckle patterns on the surface of specimen, as shown in Figure 1. The procedure of uniaxial tensile tests is shown in Figure 2, thickness of sheet is 1.5 mm. The specimens were heated to 373, 473 and 523 K at a heating rate of 2 K/s and kept for 5 min to obtain temperature homogeneity, then specimens were isothermally deformed to fracture at strain rates of 0.01 and 0.1 s^{-1}, then air cooling to room temperature. The camera simultaneously recorded the sequential digital images at forming stage. A software GOM-correlate was used to compare deformed image and reference image to determine how much movement had occurred, as shown in Figure 3. Then the strain at specific time was calculated.

Table 1. Chemical compositions of the as received 7075-T6 aluminum alloy (wt.%).

Elements	Al	Zn	Mg	Cu	Cr	Fe	Si	Al
Contents	90.70	5.37	2.18	1.25	0.24	0.13	0.07	Balance

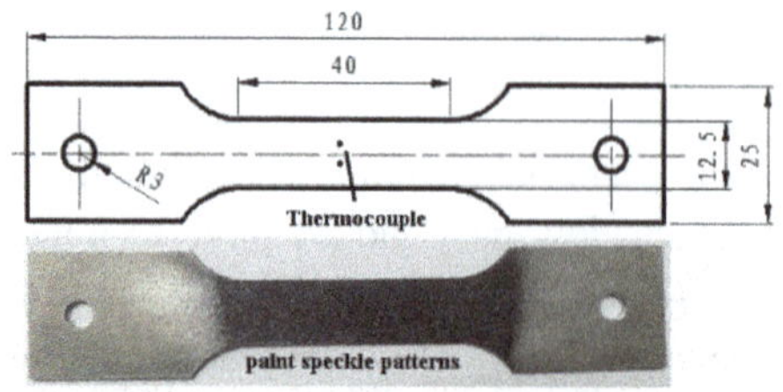

Fig. 1. Dimensions and speckle patterns of the tensile specimens.

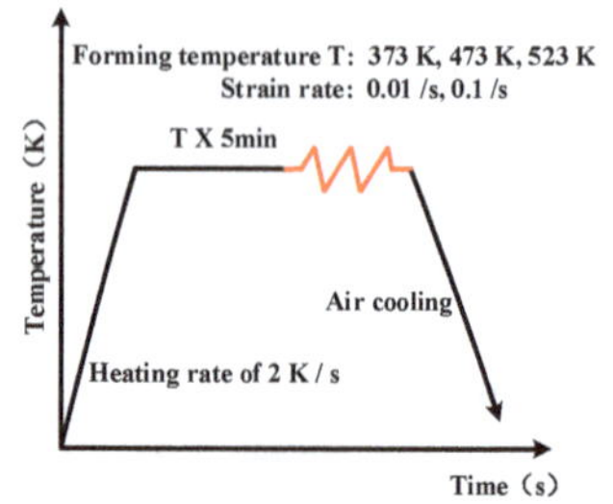

Fig. 2. Procedures for uniaxial tensile tests.

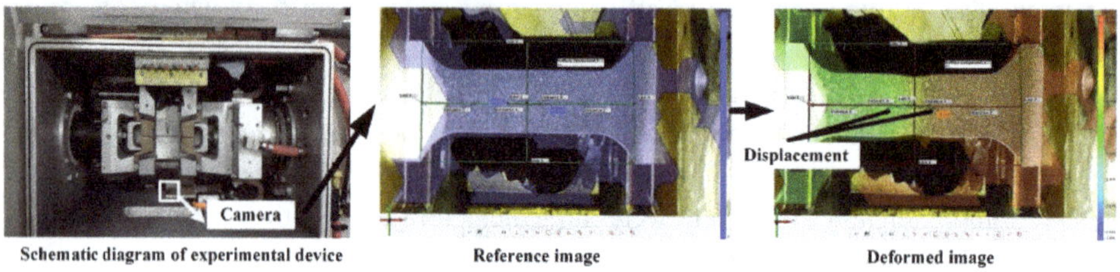

Fig. 3. Schematic diagram of experimental devices and digital image process and analyze.

3. Results and Discussion

3.1. *Flow behaviors of 7075-T6 aluminum alloy at elevated temperatures*

The precise strain was obtained by DIC method. And the stress was simultaneously recorded by Gleeble simulator. The true stress-strain curves are shown in Figure 4. It indicated that the elevated forming temperature had significant influence on decreased ultimate tensile stress and increased fracture strain. When the deformation temperature elevated from 373 K to 523 K, the ultimate tensile stress decreased from 594 MPa to 250 MPa at a strain rate of 0.1 s^{-1}, and decreased from 590 MPa to 216 MPa at a strain rate of 0.01 s^{-1}. The flow stress at higher strain rate is larger than which at lower strain rate at the same deformation temperature. The elevated temperature also increased fracture stain from about 0.10 to 0.17. Work hardening behavior dominated at lower deformation temperature and higher strain rate. Further increasing temperature to 523 K causes significantly decreased stress and extended uniform straining, suggesting thermally related softening is more severe at 523 K. However, the anisotropy of 7075 aluminum alloy was not considered.

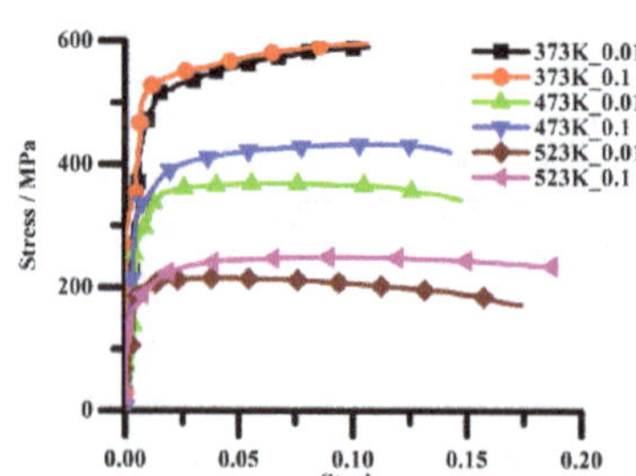

Fig. 4. The true stress-strain curves at different strain rates and temperatures.

3.2. *The Johnson-Cook constitutive model*

The constitutive equation is widely used to characterize the flow behavior of aluminum alloys. Liu et al. achieve excellent predicting performance of 6xxx series aluminum by using Arrhenius and Cowper–Symonds model [7]. The Johnson-Cook model multiplicatively decomposes the flow behavior into three functions representing the effect of strain hardening, strain rate and temperature. Flow stress is described by equations below:

$$\sigma = (\mathrm{A} + \mathrm{B}\,\varepsilon_e^{pn})\,(1 + \mathrm{C}\ln\dot{\varepsilon}^*)\,(1 - (\mathrm{T}^*)^{\mathrm{m}}) \tag{1}$$

$$T^* = \frac{T - 298}{T_{melt} - 298} \tag{2}$$

Where σ is effective stress, ε is the effective plastic strain, ε^* is normalized effective plastic strain rate ($\varepsilon^* = \varepsilon / \varepsilon_o$, ε is strain rate and $\varepsilon_o = 0.01\ s^{-1}$ is reference strain rate), n is work hardening exponent, Tmelt is melting temperature, A, B, C, m are the material constants. According to the true stress-strain curves at different temperatures and strain rates, the coefficients were calculated by non-linear least squares with Levenberg-Marquardt method. A=470 MPa, B=210 MPa and other values are shown in table 2. Comparison between the experimental and predicted flow curves are shown in figure 5, the Johnson-Cook model is consistent with experimental result. And the R-square value of fitting result is 0.9841.

Table 2. The coefficients of the Johnson-cook constitutive model for 7075-T6 aluminum alloy.

Temperature (K)	Strain rate (s^{-1})	n	C	m
373	0.01	0.3103	0.0403	0.7561
	0.1	0.3317	0.1202	0.7329
473	0.01	0.3000	0.0539	0.6987
	0.1	0.3086	0.0802	0.7015
523	0.01	0.2401	0.0374	0.7443
	0.1	0.2624	0.0412	0.8130

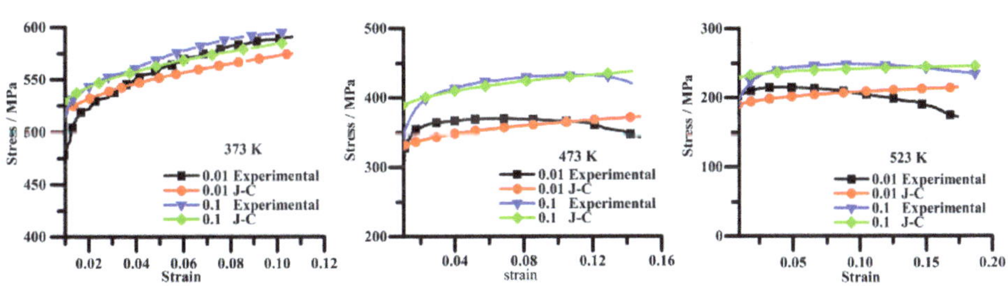

Fig. 5. The comparison between the experimental and predicted flow curves on different conditions.

4. Simulating and Experimental Analysis on Warm Stamping for 7075-T6 Aluminum Alloy

Warm stamping experiment was conducted by using heating furnace, the servo mechanical press with U-shaped tool (Figure 6). The AA7075-T6 sheets with a size of 280×150×1.5 mm were used in the experiments. The procedures of warm stamping process are shown in Figure 7. The furnace was heated to 373, 473 and 523K, respectively. Then sheet was placed in furnace for 10 min holding. And then sheet was rapidly transferred to the press to stamp and dwell for 10 s. Finally, the stamped part was taken out. Based on Johnson-Cook constitutive model and experimental conditions above, the FE model was developed to simulate the warm stamping process in LS-dyna software.

Fig. 6. The servo mechanical press with U-shaped tool.

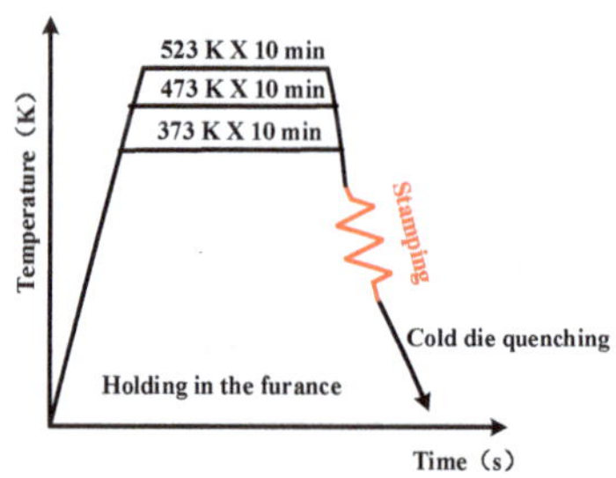

Fig. 7. The procedures of warm stamping.

The comparison of simulation and experiment on springback angle at various forming temperatures is shown in Figure 8 and Table 3. Both tool angles at flange corner and die corner areas are 120°. Springback angle reduced with forming temperature elevated. Simulating analysis also indicated this trend. The springback angles reduced to 3.30° at flange corner area and to 0.20° at die corner area at the forming temperature of 473 K. (The angles were 10.70° and 3.15° respectively at room temperature. And the angles were 1.40° and 0.05° respectively at the forming temperature 523 K).

The hardness of stamped parts was also measured by Vickers hardness tester at 3 kgf loading and 15 s dwelling. The results are shown in Figure 9. It indicated that final hardness decreased with elevated temperature. And the hardness decreased sharply with further elevated temperature above 473 K. The hardness of stamped parts was 180.6 HV at 473 K and 155.1 HV at 523 K.

Table 3. The comparison of simulation and experiment on springback angles of stamped parts.

	Flange corner springback angle/°		Die corner springback angle/°	
	Simulating	Experimental	Simulating	Experimental
Room temperature	--	10.70	--	3.15
373 K	5.925	5.30	1.228	2.25
473 K	4.612	3.30	1.125	0.20
523 K	0.703	1.40	0.005	0.05

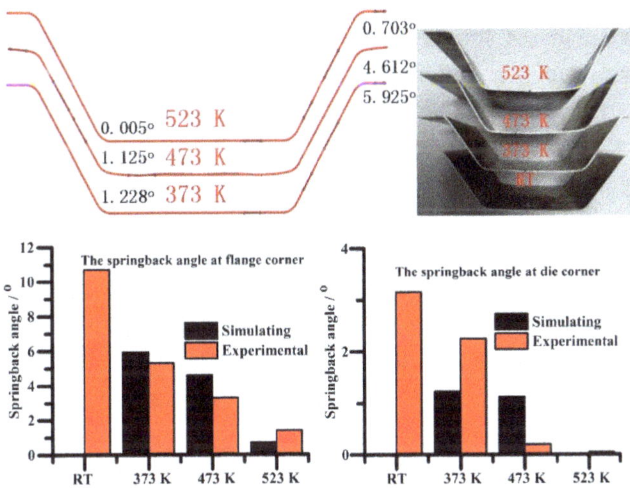

Fig. 8. The simulation and experiment comparison of stamped parts.

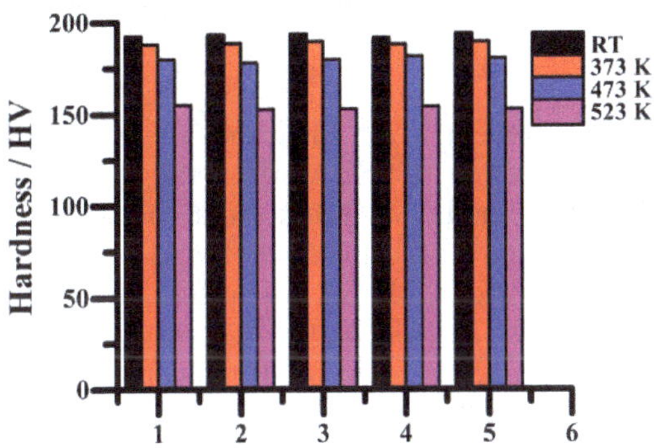

Fig. 9. The hardness of stamped parts.

5. Conclusion

(1) The flow curves of 7075-T6 aluminum alloy can be obtained by using Gleeble simulator with a set of DIC devices. It reduced the error of strain measurement brought in by mechanical rigidity.

(2) The Johnson-cook constitutive model was developed to predict the flow behaviors of 7075-T6 aluminum alloy at elevated temperatures. The model is consistent with experimental result and the R-square value of fitting result is 0.9841.

(3) The flow behavior influenced by deformation temperature and strain rates. When the deformation temperature elevated from 373 K to 523 K, the ultimate tensile stress decreased from 594 MPa to 250 MPa at a strain rate of 0.1 s^{-1}, and decreased from 590 MPa to 216 MPa at a strain rate of 0.01 s^{-1}. The elevated temperature also increased fracture stain from about 0.10 to 0.17.

(4) The optimal parameters for warm stamping were found by simulating and experimental analyses. It indicated that the process of 10 min heating and holding at 473 K, then stamping and in die quenching. The springback and hardness of stamped part were 4.16° and 180.6 HV, respectively. Further increased stamping temperature led to a sharp reduction of final hardness.

Acknowledgments

This research work was financially supported by the National Natural Science Foundation of China (grant no. U1564203 and U1760205) and National Major Science Technology Project of China (grant no. 2018zx04023001).

References

1. M. T. Ma, H. L. Yi, H. Z. Lu and X. R. Wan, On the lightweighting of automobile, *China Eng. Sci.* **11**, 20 (2009).
2. W. Huo, L. Hou, Y. Zhang and J. Zhang, Warm formability and post-forming microstructure / property of high-strength aa 7075-T6 al alloy, *Mat. Sci. Eng. A* **675**, 44 (2016).
3. W. Hui, Y. B. Luo, P. Friedman, M. H. Chen and G. Lin, Warm forming behavior of high strength aluminum alloy aa7075, *T. Nonferr. Metal. Soc.* **22**, 1 (2012).
4. T. A. Ivanoff, E. M. Taleff and L. G. Hector, Warm forming of aa7075-t6 with direct electrical resistance heating, *Tms Light Met.* **2015**, 223 (2015).
5. J. R. Yates, M. Zanganeh and Y. H. Tai, Quantifying crack tip displacement fields with dic, *Eng. Fract. Mech.* **77**, 2063 (2010).
6. N. A. Hoult, W. A. Take, C. Lee and M. Dutton, Experimental accuracy of two dimensional strain measurements using Digital Image Correlation, *Eng. Struct.* **46**, 718 (2013).
7. Y. Liu, Z. J. Zhu, Z. J. Wang, B. Zhu, Y. L. Wang and Y. S. Zhang, Flow and friction behaviors of 6061 aluminum alloy at elevated temperatures and hot stamping of a B-pillar, *Int. J. Adv. Manuf. Tech.* **3**, 1 (2018).

Study on Microstructure Evolution and Mechanical Properties of Aluminum Alloy 7075 in Stamping Process

Y. F. Jiang[1], H. Ding[1,a], H. J. Pan[1], M. H. Cai[1, b,†], Y. Liu[2] and Y. S. Zhang[2]
[1]*School of Materials Science and Engineering, Northeastern University, Shenyang, 110819, China*
[2]*State Key Laboratory of Material Processing and Die & Mould Technology, Huazhong University of Science and Technology, Wuhan, 430074, China*
E-mail: [a]dinghuaneu@163.com,[b,†]cmhing@126.com

In this paper, three forming processes of 7075 aluminum alloy sheet were discussed, including cold stamping, traditional hot stamping and novel hot stamping-quenching integrated process. Artificial aged T6 was applied during forming. Cold stamping was carried out at room temperature, while, hot stamping was completed at 430 °C. Microstructure were observed by optical microscope (OM) and scanning electron microscope (SEM), the mechanical properties were measured by Vickers hardness and tensile tests. The forming quality was investigated in the forming process. The relationship among schedules, microstructure and mechanical properties of aluminum alloy were also discussed. The results showed that hot stamping U-shaped parts have better forming quality. The precipitated η′phase and the second phase which is in rich of Fe-Mn were distributed in grain inhomogeneously in each of three forming process. The tensile strength of novel hot stamping-quenching integrated process is slightly lower than cold stamping and traditional stamping. While, the yield strength of hot stamping is higher than that of cold stamping. Furthermore, the elongation of hot stamping can reach above 13%.

Keywords: Hot stamping; Aluminum alloy; Mechanical properties.

1. Introduction

The traditional hot forming process is to perform thermal deformation at high temperature, followed by heat treatment. The disadvantage of this thermoforming process is the poor accuracy and time consuming. It may also lead to softening and the undesirable coarse phases. In order to solve the above problems, Wang proposed a novel hot forming called hot stamping-quenching integrated process with cold dies (HFQ) at 2005 for the first time. The aluminum alloy is heated to the solution heat treatment temperature for a certain period of time and quickly transferred to a cold die for hot forming, then held for 10-15s to obtain the desired shape. Finally the work piece removed from the dies to air cooling to the room temperature. In the end, ageing was carried out to get better mechanical properties. Hot forming and quenching are finished in one operation in the integrated process while the traditional hot forming is separate. This method can effectively solve the shortcomings of aluminum alloy materials, such as poor room temperature plasticity [6].

AA7075 is a precipitation-hardened Al-Zn-Mg-Cu aluminum alloy [7]. The main semi-finished products are plates, bars, profiles and forgings. Due to their high strength characteristics, they have been widely used in important structural parts of aircraft structures, such as aircraft skins, spars, bulkheads, landing gear, and hydraulic system components. At the same time, it also applied to important parts such as A-pillars, B-pillars, side beams, bumpers, roof longitudinal beams, door beams, and sidebars of automobiles. In the previous work, few people compared the formability of different forming process. In this study, cold stamping, traditional hot stamping and novel hot stamping process were investigated systematically.

2. Experimental

The as-received material was T6 temper AA7075 sheet with 1.5 mm thickness in rolled condition. The nominal chemical compositions of this alloy are listed in Table 1.The study included three forming processes. The desired shape is U shape. (i) Cold stamping: AA7075 was stamped at room temperature (25 °C) and then held for 15 s without any heat treatment followed. (ii) Traditional hot stamping: AA7075 was stamped at 430 °C, then held for 15s within cold-dies to reduce the temperature rapidly to the ambient temperature. Finally it was conducted by solution treatment at 480 °C for 60 min and artificial ageing at 120 °C for 24 h. (iii) Novel hot stamping process (HFQ): AA7075 sheet was quickly transferred to the hot stamping machine held for 15 s within cold dies after solution heat treatment at 480 °C for 60min, then cooled to room temperature. Finally aging at 120 °C for 24 h followed was carried out to obtain better precipitates.

The samples of microstructure and tensile experiment were performed at the central position of U-shaped piece and the location was marked in Fig. 1. Microstructure was conducted by Olympus DSX-500 optical microscopy (OM), scanning electron microscope (SEM) and transmission electron microscope JEM2100 (TEM). Transmission samples are polished with a solution of 70% methanol and 30% nitric acid at -30 °C and 12 V. Tensile test was performed at 25 °C with strain rate of 1.5 mm/min.

Table 1. Nominal chemical compositions (wt. %) of the studied AA7075 aluminum alloys.

Alloy	Si	Fe	Cu	Mn	Mg	Cr	Zn	Ti	Other each	Other total
AA7075	0.38	0.47	1.8	0.26	2.2	0.20	5.5	0.20	≤0.05	≤0.05

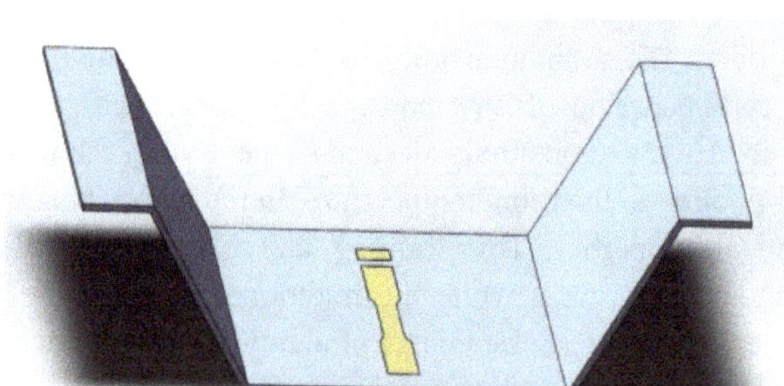

Fig. 1. The sample position of U-shaped pieces.

3. Results and Discussion

As shown in Fig. 2, from the forming appearance point of view, the rebound angle of cold stamping is significantly greater than the traditional hot stamping and novel hot stamping. Springback is a non-uniform release of the internal residual stress of the work piece during the forming process. This is due to the poor formability of the AA7075 at room temperature. After the hot stamping parts are quenched in cold dies, the thermal stress has little effect on the shape and dimension, and the dimensional precision of the forming parts is fine. Because the stamping shape is relatively simple, the three groups of stampings are not cracked in terms of surface quality. The hot stamping process has great advantages in the formability of aluminum alloy.

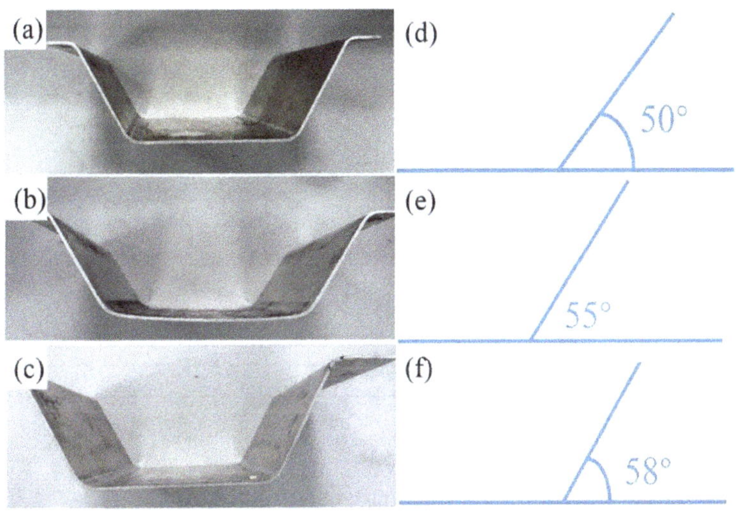

Fig. 2. Macrographs and the measured actual angles of U-shaped stamped parts: (a) (d) cold stamping; (b) (e) traditional hot stamping; (c) (f) novel hot stamping.

The microstructures of AA7075 in three forming processes are shown in Fig. 3. AA7075 is difficult to completely dissolve [8], the grains are significantly elongated which is due to the aluminum alloy plate being rolled during processing. The insoluble second phase is dispersedly distributed in grain boundaries and intragranular regions of the AA7075. The width of grain in three processes were counted, and it was found that the proportion of grain width of 15-30 μm in cold stamping was larger, while that of hot stamping was 0-15 μm. In order to analyze the morphology and composition of the second phase particles more specifically, the microstructure of the AA7075 was characterized by SEM and EDS. AA7075 mainly has many massive and lamellar second phases in cold stamping process (Fig. 3e, f), while the traditional hot stamping and novel hot stamping are mainly punctiform (Fig. 3d). The reason is caused by diffusion and dissolution in hot stamping process. From the analysis of the Table 2, it can be seen that the second phase in forming process is mainly rich in Fe and Mn, and it is significantly different from the matrix composition in the backscattered electron image.

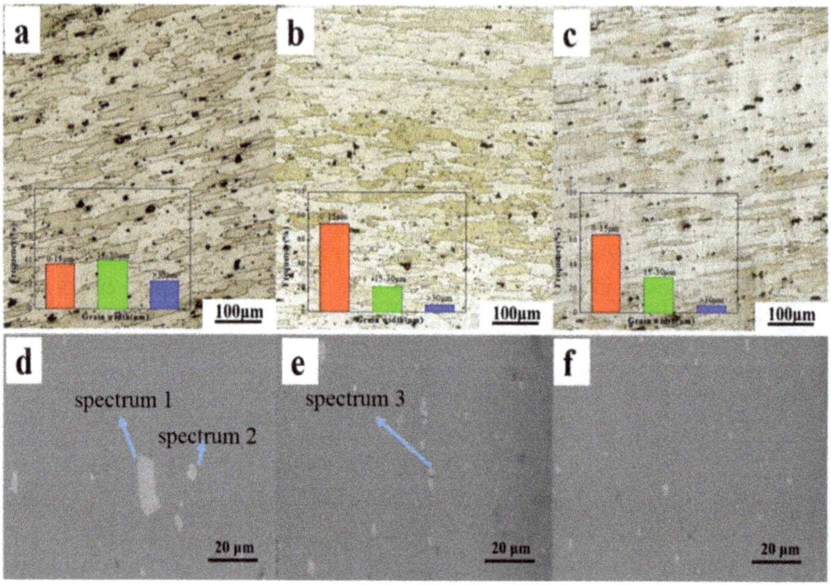

Fig. 3. The evolution of microstructure by OM and SEM: (a) (d) cold stamping; (b) (e) traditional hot stamping; (c) (f) novel hot stamping.

Table 2. The eds anlysis of matrix and second phase.

	Zn	Mg	Cu	Mn	Fe	Cr	Al
Matrix	3.803	3.059	1.558				Bal
Spectrum1	1.960		2.918	8.635	14.170	9.306	Bal
Spectrum2	4.028		5.922	2.032	19.150		Bal
Spectrum3	2.669			3.488	25.002	1.049	Bal

Precipitation strengthening is the main strengthening method of 7075 alloy, the shape, number, and size of the precipitated strengthening phase all affect the mechanical properties of the aluminum alloy [4]. As shown in Fig. 4, High-density precipitates diffusely distributed throughout the matrix which are different under three forming process. After calibration of diffraction spots, these precipitate phases are mainly η (η′)-MgZn2, the η′-phase was described as a slightly strained version of the η-phase in current research which cannot be distinguished [9]. The strengthening phase precipitated during the aging process. It seems that the η (η′) phase in cold stamping is a bit coarser than hot stamping, but not obvious. It may be caused by inadequate solution and aging, which can not affect mechanical properties much.

Fig. 4. TEM micrographs of the strengthening precipitates: (a) cold stamping; (b) traditional hot stamping; (c) novel hot stamping.

The results of the tensile tests are shown in Table 3. The tensile strength of novel hot stamping process can reach 597 MPa, which is a bit lower than cold stamping and traditional stamping. The yield strength of traditional hot stamping and of novel hot stamping is similar, sight higher than cold stamping. The elongation of traditional hot stamping and novel hot stamping is obviously higher than that of cold stamping, which can be caused by strengthening phase η (η'), fine and great amount of precipitates is beneficial to strengthen the materials [10]. The width of grain may have some effect on mechanical properties of aluminum alloy, but it does not dominate. The elongation of hot stamping can reach 14%.

Table 3. The mechanical properties of three forming processes after stamping.

	Tensile strength (MPa)	Yield strength (MPa)	Elongation rate (%)
Cold stamping	612	523	9.4
Traditional hot stamping	613	541	14.8
Novel hot stamping	597	532	13.4

4. Conclusion

The forming accuracy of the novel hot stamping process is higher than traditional hot stamping process and cold stamping process. The η (η') phases were similar among three forming process. The ultimate tensile strength and yield strength of traditional stamping process is a bit higher than the other two forming process, which can reach 613 MPa and 541 MPa. The plasticity after cold stamping is poor and rebound is serious. The elongation of cold stamping is only 9.4%. The elongation of traditional hot stamping and novel hot stamping is similar which can reach approximately 14%. The novel hot stamping process is finished in one operation, more efficient.

References

1. S. Mahabunphachai and M. Koç, Investigations on forming of aluminum 5052 and 6061 sheet alloys at warm temperatures, *Mater. Des.* **5**, 2422 (2010).
2. X. B. Fan, Z. B. He, W. X. Zhou and S. J. Yuan, Formability and strengthening mechanism of solution treated Al–Mg–Si alloy sheet under hot stamping conditions. *J. Mater. Process. Tech.* 179 (2016).

3. H. Wang, Y. B. Luo, Peter .F, M. H Chen, L. Gao, D. W. Ma, M. Wang and Z. H. Hu, Warm forming behavior of high strength aluminum alloy AA7075, *T. Nonferr. Metal. Soc.* **1**, 1 (2012).
4. X. Fan, Z. He, S. Yuan and P. Lin, Investigation on strengthening of 6A02 aluminum alloy sheet in hot forming-quenching integrated process with warm forming-dies, *Mater. Sci. Eng. A*. **18**, 221 (2013).
5. Y. Liang, T. H. Gao, M. H. Dai and P Hu, Investigation of interfacial heat transfer mechanism for 7075-T6 aluminum alloy in HFQ hot forming process, *Appl. Therm. Eng*. 266 (2017).
6. D. Lorenz and K. Roll, Modeling and analysis of integrated hot forming and quenching processes, *Adv. Mater. Res*. 787 (2005).
7. S. Polaka, P. Kaczyńskia, Z. Gronostajskia, K. Jaskiewicza, J. Krawczyka, M. Skwarskia, M. Zwierzchowskia and W. Chorzępab, Warm forming of 7075 aluminum alloys, *Procedia Eng*. 2399 (2017).
8. Q. Chen, X. S. Xia, B. G. Yuan, D. Y. Shu and Z. D. Zhao, Effects of solid solution treatment on microstructure and mechanical properties of 7A09 aluminum alloy, *Heat Treat. Met*. **1**, 38 (2014).
9. J. Q. Su, T. W. Nelson, R. Mishra and M. Mahoney, Microstructural investigation of friction stir welded 7050-T651 aluminum, *Acta Mater*. **3**, 713 (2003).
10. Y. Chen, H. Ding, S. Malopheyev, R. Kaibyshev, Z. H. Cai and W. J. Yang. Influence of multi-pass friction stir processing on microstructure and mechanical properties of 7B04-O Al alloy. *J. Mater. Process. Tech.* **5**, 1157(2012).

Research on Hot Stamping Forming Limit Diagram of 6061 Aluminum Alloy Sheet

W. Huang, Y. L. Wang† and Y. S. Zhang
State Key Laboratory of Materials Processing and Die & Mould Technology, Huazhong University of Science and Technology, Wuhan, 430074, P.R. China
†E-mail: wangyilin@hust.edu.cn

Forming limit diagram is an important criterion for judging the success of plastic forming of sheet metal. In order to study the formability of 6061 aluminum alloy sheet at warm state, the forming limit diagrams (FLD) of the 6061 aluminum alloy sheet at warm state (room temperature to 300 °C) was tested on the hot sheet formability test machine. In this study, a new cruciform biaxial tensile testing technology was employed to evaluate the thermal limit strains of 6061 aluminum alloy. The FLD calculating model of 6061 aluminum alloy sheet was established based on the experimental data. It can be applied to the finite element simulation analysis of hot stamping parts, thus guiding the optimization design of stamping parts and the optimization of process parameters.

Keywords: 6061 aluminum alloy; Forming limit diagram; Hot stamping.

1. Introduction

In recent years, as the automotive industry develops toward lighter weight, the proportion of aluminum alloy stampings in auto parts has been increasing. Aluminum alloys have the advantages of higher specific strength and lower density, but they have poor formability at room temperature and are difficult to form body structural member with complex shapes. One way to solve this problem is to use the warm forming of the aluminum alloy. As the temperature increases, the plasticity of the aluminum alloy increases, and the formability increases, so that complex shape parts can be formed.

The forming ability of the metal sheet can be represented by a forming limit diagram. The forming limit diagram is the connection of the forming limit points when the sheet material is unstable under different strain paths. It is a very important performance indicator and process parameter in sheet metal forming, reflecting the maximum deformability of the sheet before the plastic instability in the deformation process. It has important guiding significance for the optimal design of the die and the optimization of process parameters for the stamping process of automotive parts.

The international standard [1] uses the hemispherical convex bulging method to obtain the forming limit diagram. However, this method is only suitable for measuring the forming limit diagram at room temperature. It is necessary to redesign the experimental system for the forming limit diagram at warm state. At the same time, it has disadvantages that it is difficult to observe the occurrence of cracks and make accurate judgement on forming limits. Therefore, in this paper, the cross-stretching method combine digital correlation image (DIC) method are used to measure the forming limit diagram of aluminum alloy at warm state, which overcomes the experimental deviation caused by the friction and lubrication effects of the traditional method and can more accurately judge the moment when sheet reaches forming limit.

2. Cruciform Specimen and Equipment

2.1. *Cruciform specimens*

Many studies have investigated and discussed the appropriate shape of the biaxial tensile specimen, but there is still no uniform standard for it. Therefore, biaxial tensile specimens need to be redesigned. The design is based on principles that stress and strain in the central deformation area are uniform and stress concentration outside the center area should be avoided [2]. In this study, through the method of finite element simulation, a total of three biaxial tensile specimens as shown in Fig. 1 were designed.

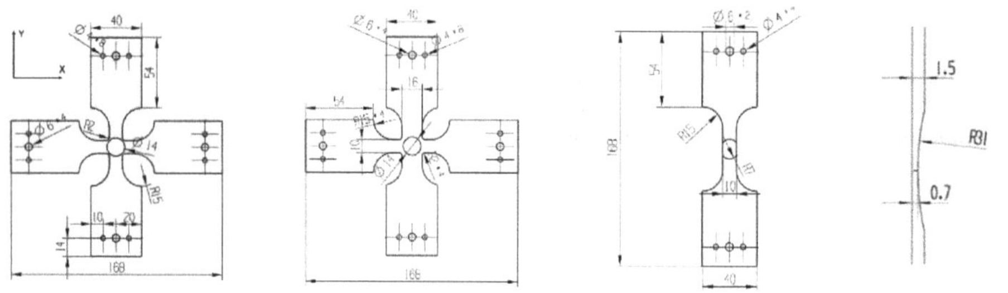

(a) (b) (c)

Fig. 1. Biaxial tensile specimen geometry.

Biaxial tensile specimen is made of 1.5mm thickness of 6061 aluminum plates, The chermical composition of the 6061 aluminum alloy is shown in Table 1.

Table 1. Main chemical compositions of 6061 aluminum alloy [wt. %].

Material	Si	Cu	Fe	Mg	Zn	Ni	Mn	Ti	Cr	Al
6061AL	0.676	1.495	0.423	1.104	0.114	0.0058	0.078	0.027	0.165	Bal

2.2. *Experimental equipment*

2.2.1. *Biaxial tensile testing machine*

In this experiment, a self-made biaxial tensile device was used and it was installed on the universal stretching machine as shown in Fig. 2. This device can be installed on a uniaxial stretching machine. The machine moves down to compress and transmit the force to specimen to achieve two-way simultaneous stretching.

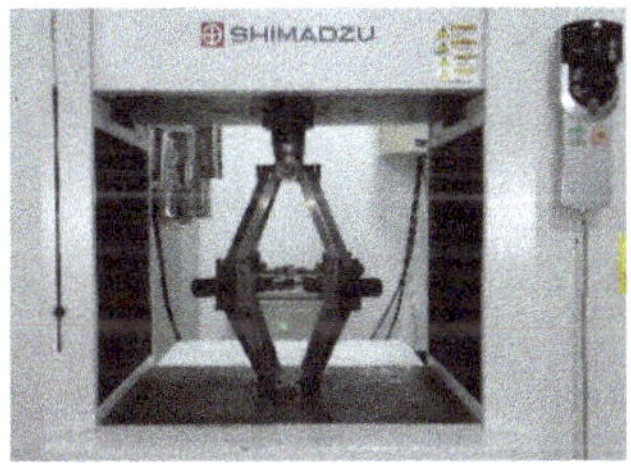

Fig. 2. Biaxial tensile testing machine.

2.2.2. *Heating device*

In this experiment, a high frequency induction heater, as shown in Figs. 3 and 4 was used to conduct electromagnetic induction heating on 6061 aluminum alloy samples. Its working principle is that the high current generated by the high-frequency machine flows to the induction coil, and a strong magnetic field changing instantaneously is generated in the coil. When the metal is put into the coil, a large eddy current will be generated, and due to the internal resistance of the metal, large Joule heat is generated to heat the metal. Advantages of this heating method are:

(1) High heating speed, short heating time and high efficiency.

(2) The machine is easy to operate and it controls the heating temperature and heating time conveniently.

2.2.3. *Experimental devices of DIC*

In this study, the strain around the fracture area was measured by digital-image correlation (DIC) measurement technology instead of by using traditional grids, which suffer from limited data and precision. Digital-image correlation (DIC) is a strain-measurement technology. The measurement of strain using the digital-image correlation method has the advantage of high precision, wide range, and 2D data integrity [4].

Experimental devices of DIC are shown in Fig. 4, mainly including a mobile phone, three-dimensional rotating tripod, light source and other components. The mobile phone is used to record the deformation information of the specimen; The three-dimensional rotating tripod is used to fix the mobile phone; The light source is used to reinforce the light.

Fig. 3. The high frequency induction heater.

Fig. 4. The coil of high frequency induction heater.

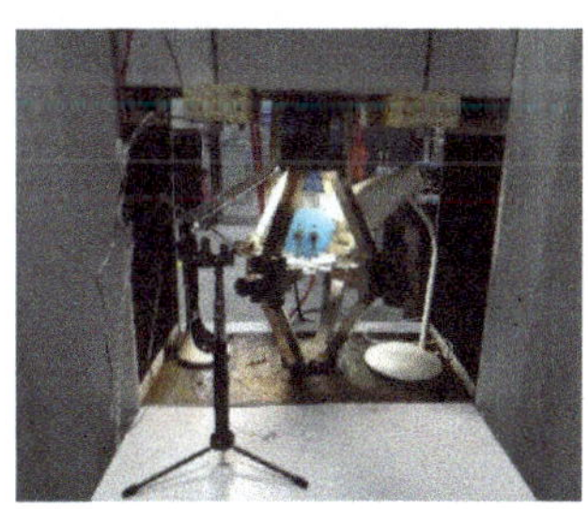

Fig. 5. Experimental devices of DIC.

3. Experimental Procedure

3.1. *Biaxial tests*

Specimens made of 6061 aluminum alloy are installed on a cross tensile device which is controlled by AG-100KN material high-temperature testing machine. The machine compressed the biaxial tensile device in the vertical direction, made the samples to stretch at the same speed on the X and Y directions, until the crack can be observed. The whole process of stretch was recorded by smartphone. The experiments were performed at temperature of 200 °C and at strain rate of 2 mm/min.

Six strain loading paths were used in this test:

(1) Stretching specimen (a) biaxially until it breaks.
(2) Stretching specimen (a) uniaxially in the X direction until it breaks.
(3) Stretching specimen (a) uniaxially in the Y direction until it breaks.
(4) Stretching specimen (b) biaxially until it breaks.
(5) Stretching specimen (b) uniaxially in the Y direction until it breaks.
(6) Stretching specimen (c) uniaxially until it breaks.

3.2. *Strain measurement*

In this study, the black paint and white paint were sprayed on the cross tensile specimens to make random speckle specimens. The test process was recorded in a series of images recorded via video camera. Those images were imported into DIC post-processing software - GOM correlate. It can process the images to get the strain value.

4. Results

Through the establishment of the DIC strain measurement system, the strain contour for various strain paths are obtained, as shown in Fig. 6. It can be seen from the strain contour that the position where the ultimate strain is the largest is at the intersection of biaxial tensile specimens and the middle of the uniaxial tensile specimens.

Select the maximum strain at the point on the specimen 0.5 seconds before breaking as its ultimate strain. Forming limits under different strain paths were secondary strain as the Y axis, the forming limit diagram of the 6061 aluminum alloy at 200 °C was obtained, as shown in Fig. 7.

Table 2. Forming limits under different strain paths.

Strain paths	1	2	3	4	5	6
ε_1	0.23	0.18	0.61	0.42	0.96	0.20
ε_2	0.12	0.05	-0.30	-0.24	-0.31	-0.04

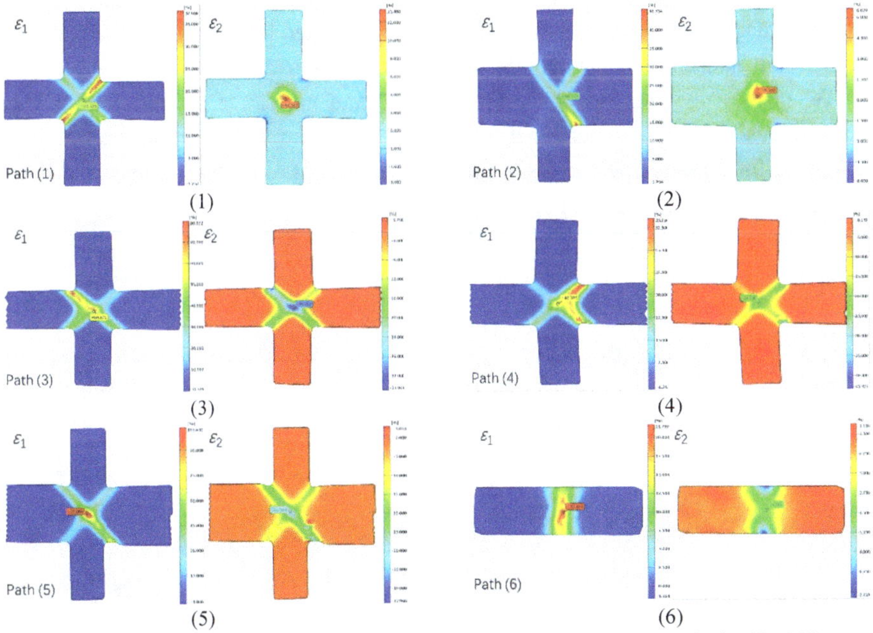

Fig. 6. Primary and secondary strain contour under different strain paths obtained by DIC.

In order to verify the reliability of the forming limit diagram obtained by biaxial tensile testing combined with DIC strain measurement system in this paper, the forming limit diagram of 6061 at warm state obtained by the convex bulging method [5] was compared on the same coordinate system for comparison. As shown in Fig. 8, it can be seen that the forming limit diagram obtained by the two methods do not differ greatly. Thus, the reliability of this method can be proved.

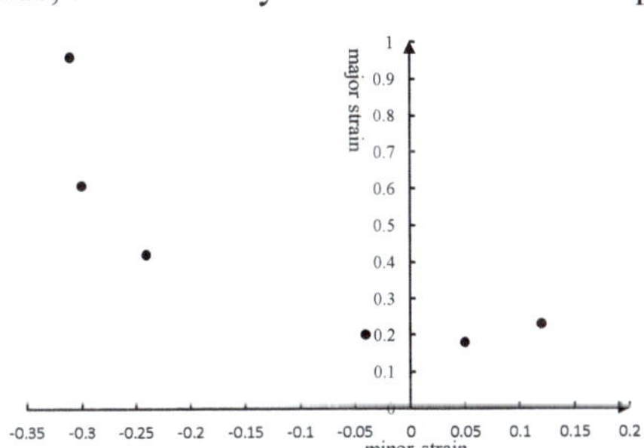

Fig. 7. The forming limit diagram of the 6061 aluminum alloy at 200°C.

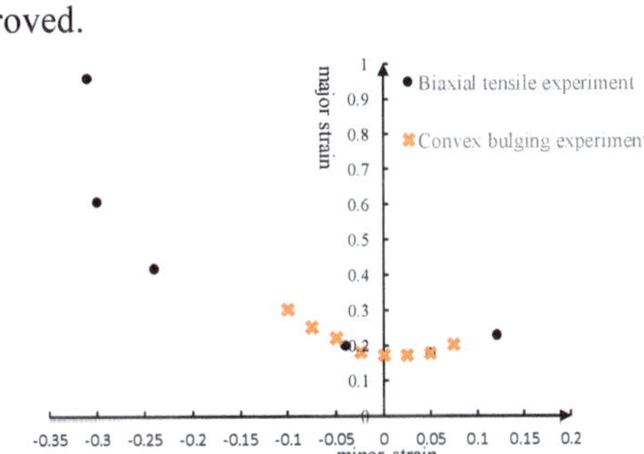

Fig. 8. Comparison of two methods obtaining the FLD.

5. Conclusion

Thermal biaxial tensile testing technology and a digital-image correlation measurement system were adopted to analyze the forming limit of 6061 aluminum alloy at high temperatures. The main conclusions drawn are as follows:

(1) The method used in this experiment can obtain the 6061 aluminum alloy's warm forming limit diagram which is similar to the results obtained by the conventional convex bulging method. Therefore, this method is relatively reliable.

(2) When the specimen was heated at 300 °C and more, the ultimate strain value measured by this method is generally low. The main reason for this is that the fluidity of spray painting for making speckles is not as good as that of aluminum alloys under high temperature. This is also the content needs to be improved in subsequent studies.

Acknowledgments

This research work was financially supported by the National Natural Science Foundation of China (grant no. U1564203 and U1760205) and National Major Science Technology Project of China (grant no. 2018zx04023001).

References

1. GB/T 15825.8-1995. Sheet metal formability and test methods-Forming limit diagram (FLD) test, CN-GB (1995).
2. R. Xiao and X. X. Xiao, Biaxial tensile testing of cruciform specimens, *Phys. Chem. inspection* **53**, 538 (2017).
3. M. H. Zhou and C. M. Liu, Research on semi-solid die casting processing of forming ZL112Y aluminum alloy kickstand of generator of the type of JH70 motorcycle, Master thesis, Chongqing University (2003).
4. R. Xiao and L. H. Liang, Biaxial tensile testing of cruciform slim superalloy at elevated temperatures, *Mater. Des.* **94**, 286 (2016).
5. J. D. Jin and Y. Chang, Research on warm forming limit of automotive lightweight aluminum alloy sheet, Master thesis, Dalian University of Technology (2014).

Springback Analysis of 7075 Aluminum Alloy V-shape Parts After Hot Stamping

X. M. Yang
School of Mechanical Engineering, University of Science and Technology Beijing, Beijing, 100083, China
E-mail: dryangxm@163.com

B. Y. Wang
Beijing Key Laboratory of Lightweight Metalforming, School of Mechanical Engineering, University of Science and Technology Beijing, Beijing, 100083, China
E-mail: bywang@ustb.edu.cn

In this paper, the springback of the aluminum alloy AA7075 under hot stamping conditions was characterized under V-bending conditions. And the influence of the initial blank temperature and the radius of the punch fillet on springback of V-shape parts was studied. The results show that the springback angle of V-shaped parts decreases rapidly with the increase of initial temperature of the sheet, the minimum springback angle at 400 °C is 0.15°, which is reduced by 96.61 % compared with that at room temperature. The springback angle increases with the radius of the punch. The results of which were used to verify finite elements simulations of the processes in order to analyze the reason of springback. And finite element simulation results and experimental results have a good agreement.

Keywords: Aluminum alloy; Springback; Hot stamping; Finite element simulation.

1. Introduction

As the pillar of the national economy, the automobile industry is closely related to many industries, which promotes the rapid development of related industries and induces the emergence of a large number of emerging industries and high-tech industries. However, the increase in the size of the auto industry has also led to the emergence of many environmental issues such as energy consumption and environmental pollution. Lightweight is the most direct and effective method to solve these problems [1].

Aluminum alloy has the advantages of low density, high specific strength, good corrosion resistance, high recycling rate, heat treatability, etc. It is an ideal lightweight material for automobiles [2]. However, it is difficult to form complex shapes and high precision produces at room temperature because of the low ductility of aluminum alloys, and there are often various rebound phenomena in the cold forming such as flanging back, bullet, cut edge rebound, bend rebound, etc [3]. For many years, hot forming methods have been studied and applied to improve the formability of aluminum alloy.

Although hot stamping process can significantly reduce springback, can not completely eliminate springback, many scholars have analyzed and studied the springback phenomenon in hot stamping process. Wang et al. used U-shape dies to study the springback behavior of AA5754 aluminum alloys in hot stamping and found that higher sheet temperatures and mold temperatures can reduce springback [4]. Keum and Han studied springback behavior of 5052 and 1050 aluminum alloys at different temperatures. It was found that the springback angle gradually decreased with increasing temperature. This phenomenon was especially noticeable after the temperature exceeded 150°C [5]. Li et al. simulated the springback behavior of AA6061 U-shape parts under hot stamping condition and found that the springback amount dropped by 25.7 % as the temperature of the blank increased from 465 °C to 565 °C [6]. Moon et al. studied the effect of mold temperature on springback. It was found that the combination of hot dies and cold punches can reduce springback by 20 % [7]. Furthermore, much research work has been conducted to study the springback using finite element analysis software [8-10].

AA7075 aluminum alloy has a more mature application in the 7xxx series aluminum alloy and has good comprehensive mechanical properties. Therefore, the AA7075 aluminum alloy was tested for springback on the basis of V-shape die in this paper. By analyzing the changes of temperature and punch radius, the springback law of 7075 aluminum alloy in hot stamping was studied. Then use the finite element simulation software for numerical simulation.

2. Experiment

The material used in the tests was a 7075-T6 aluminum alloy sheet with a thickness of 2 mm. The chemical composition is shown in Table 1.

Table 1. Chemical composition of 7075 aluminum alloy.

Element	Zn	Cu	Mg	Mn	Cr	Ti	Fe	Si	Al
wt.%	5.4	1.4	2.2	0.04	0.19	0.02	0.22	0.07	Rem

The springback phenomenon of AA7075 aluminum alloy after hot stamping was studied, using a cold V-shaped tool. 60 KN presses were used to perform the V-shape bending tests. In terms of the process of hot formability test, a specimen with a size of 180 mm×50 mm×2 mm is placed in a heating furnace and heated to a specified temperature for a certain period of time. The blank is then rapidly transferred to the blank holder, and the 60 KN presses worked to complete the stamping. The blank holder is supported by a nitrogen gas spring and exerts a fixed blank holder force on the blank. After the experiment, the angle of $\Delta\alpha$ and $\Delta\beta$ is analyzed and discussed.

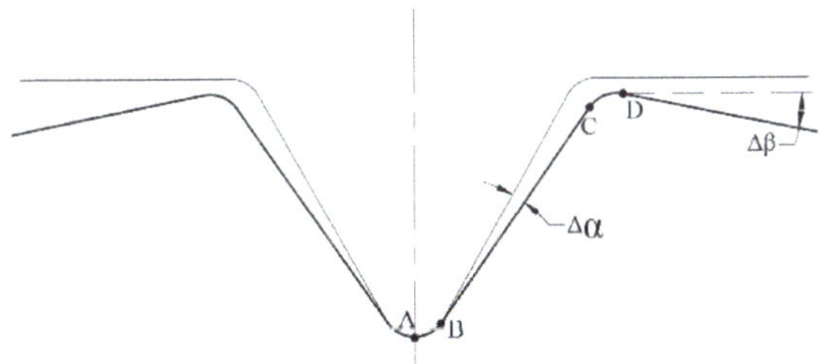

Fig. 1. Angles defining springback for V-shape bending.

3. Results and Analysis

3.1. *Effect of temperature on springback*

Figure 2(a) shows the influence of the sheet temperature on springback when the punch radius is 10 mm. Figure 2(b) shows the comparison between the experimental results of the bending angle Δα and the FE simulation results. It can be seen that the results obtained by using the finite element simulation and the V-shape test have a good agreement. It can be seen that, under the cold forming conditions, the V-shape part has obvious springback, the springback angle Δα of the forming zone is 4.43°, and the springback angle Δβ in the blank holder zone is 5.0335°. In the hot forming condition, the springback angle has significantly reduced. As the sheet temperature increases, the rebound angle rapidly decreases, the springback angle Δα is 0.15° at 400 °C, which is 4.28° reduced compared to room temperature, and the reduction rate is 96.61 %. This is due to the fact that after the temperature increases, the yield strength of the material decreases, the elastic recovery ability after forming is weakened, so that the springback angle of the V-shaped member is reduced.

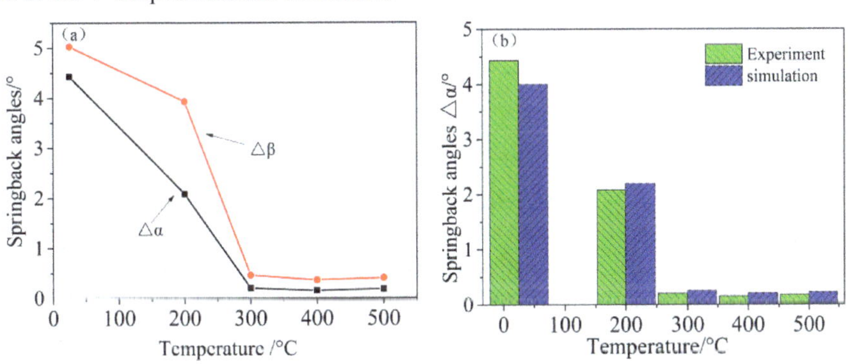

Fig. 2. Results of experiment and simulation ((a) Influence of temperature on the springback angles (b) Comparisons between the FE simulation and experiment results).

Figure 3 shows the distribution of shear stress along the length of the sheet. The region AB is the forming bending region. When the sheet material is deformed in this region, the tangential stress states of the upper and lower layers are compressive stress and tensile stress respectively; with the increase of temperature, the stress of the upper and lower layers has a more obvious decreasing trend. It can be seen that the stress changes from 424 MPa at 25 °C to 178.35 MPa at 400 °C in Figure 3. And the region CD is the blank holder region.

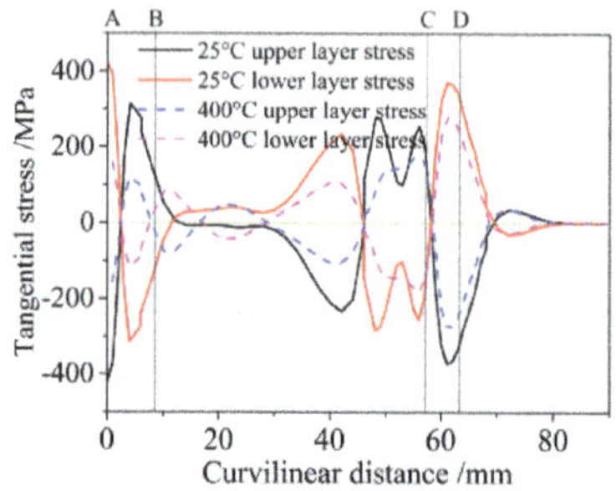

Fig. 3. Tangential stress distributions on the upper and lower layers for 25 °C and 400 °C.

3.2. *Effect of punch radius on springback*

Figure 4 shows the effect of punch radius on the springback angle at different temperatures. It can be seen that the springback angle increases with the radius of the punch at different temperatures. When the punch radius is 5 mm, the springback angle is the smallest, and even $\Delta\alpha$ is a negative value.

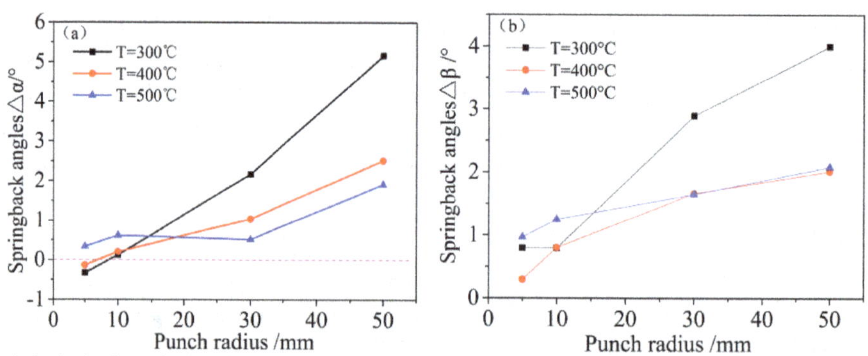

Fig. 4. Springback angle changing curves ((a) Springback angle changing on $\Delta\alpha$ (b) Springback angle changing on $\Delta\beta$).

Tangential stress distributions along the upper and lower layers of the V-shape part are shown in Figure 5 for 400 °C and a punch radius of 10 mm and 50 mm. As the punch radius increases, the plastic deformation area increases from AB region at R=10 mm to AB_1 region at R=50 mm, so that the number of stress changes in the deformation area increases, and the residual stress inside the sheet increases. The residual stress release after unloading increases the rebound angle.

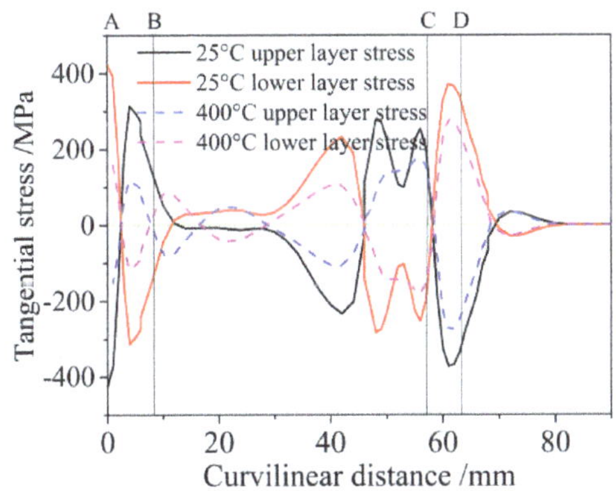

Fig. 5. Tangential stress distributions on the top and bottom layers for a punch radius of 10mm and 50mm.

4. Conclusion

In this paper, the influence of initial temperature and punch radius on springback angle of 7075 aluminum alloy V parts is studied by using V die.

(1) With the increase of the initial temperature of the sheet, the springback angle first decreases rapidly, and the minimum value is 0.15° when the sheet temperature is 400 °C.

(2) With the increase of punch radius, the springback angle increases.

(3) The simulation value is in good agreement with the experimental value. Through the finite element simulation, it is found that the springback angle of the V-shape parts has a great relationship with the tangential stress.

Acknowledgments

This work was supported by the Joint Funds of the National Natural Science Foundation of China (Grant No. U1564202). This work was also supported by the Beijing Laboratory of Modern Transportation Metal Materials and Processing Technology.

References

1. H. J. Kim, C. McMillan, G. A. Keoleian and S. J. Skerlos, Greenhouse gas emissions payback for lightweighted vehicles using aluminum and high-strength steel, *J. Ind. Ecol.* **14**, 929 (2010).
2. M. T. Ma, J H. You, H. Z. Lu and Z. W. Wang, Research process of aluminum alloy automotive sheet and application technology, *Eng. Sci.* **9**, 4 (2010).
3. E. Fakir, L. L. Wang, D. Balint, J. P. Dear, J. G. Lin and T. A. Dean, Numerical study of the solution heat treatment, forming, and indie quenching (HFQ) process on AA5754, *Int. J. Mach. Tool. Manuf.* **87**, 39 (2014).
4. A. Wang, K. Zhong, O. E. Fakir, J. Liu, C. Y. Sun, L. L. Wang, J. G. Lin and T. A. Dean, Springback analysis of AA5754 after hot stamping: experiments and FE modelling, *Int. Adv. Manuf. Technol.* **89**,1339 (2017).

5. Y. T. Keum and B. Y. Han, Springback of FCC sheet in warm forming, *J. Ceram. Process. Res.* **3,** 159 (2002).
6. W. K. Li, L. H. Zhan and J. Zhao, Effect of hot stamping process on forming quality of 6061 aluminum alloy U-shaped parts, *The Chinese Journal of Nonferrous Metals* **26**, 1159 (2016).
7. Y. H. Moon, S. S. Kang, J. R. Cho and T. G. Kim, Effect of tool temperature on the reduction of the springback of aluminum sheets, *J. Mater. Process. Technol.* **132**, 365 (2003).
8. H. Laurent, R. Grèze, M. C. Oliveira, L. F. Menezes, P. Y. Manach and J. L. Alves, Numerical study of springback using the split-ring test for an AA5754 aluminum alloy, *Finite Elem. Anal. Des.* **46**, 751 (2010).
9. F. Lan, J. Chen and J. Lin, A method of constructing smooth tool surfaces for FE prediction of springback in sheet metal forming. *J. Mater. Process. Technol.* **177**, 382 (2006).
10. Y. E. Ling, H. P. Lee and B. T. Cheok, Finite element analysis of springback in L-bending of sheet metal, *J. Mater. Process. Technol.* **168**, 296 (2005).

Study of Springback for High Strength Aluminium Alloys Under Hot Stamping

Z. H. Cai, P. Batthyány, S. Dhawan, Q. L. Zhang, Y. H. Sun, X. Luan and L. L. Wang[†]
Department of Mechanical Engineering, Imperial College London,
London, SW7 2AZ, United Kingdom
[†]E-mail: Liliang.wang@imperial.ac.uk

M. M. Gharbi
Schuler Pressen GmbH, Bahnhofstrasse 41,
Goeppingen, 73033, Germany
E-mail: Mohammad.Gharbi@schulergroup.com

Commercial aluminium alloys, AA6082 and AA7075, offer high stiffness to density as well as strength to density ratios, and are suitable for structural sheet-metal fabrication. The automotive industry tends to favour these aluminium alloys over steel for a reduction in the weight of components. However, the application of these alloys has been impeded by their poor formability at ambient temperatures. Therefore, a traditional hot stamping technique was employed to form AA6082 and AA7075, to study the springback of a U-Shape workpiece by conducting experiments at varying forming temperatures and forming speeds. It was found that springback was reduced at elevated temperatures compared to that at ambient temperatures, particularly a 16° (90%) reduction for AA7075 component. Increasing forming speed can reduce the level of springback as well, with an observed 15% reduction when comparing slow and fast forming.

Keywords: Hot stamping; AA6082; AA7075; Springback; Forming temperature; Forming speed.

1. Introduction

According to its low-carbon economy roadmap and closely related targets, the European Council proposes a 60% reduction of emission levels from 1990 by 2050 in the transport sector [1, 2]. These targets have increased the pressure on both the consumers and manufacturers in the automotive industry to achieve a substantial reduction in emissions. A feasible solution to increase the fuel efficiency is by a reduction of overall weight using high strength aluminium alloys. A study by Drucker Worldwide has shown that using 200 kg of aluminium (up to 150kg today) in a car could reduce CO_2 emissions by up to 16 grams per kilometer travelled and predicted an increase of aluminium for lightweight applications in the automotive industry by as much as 30% in the next 8 years [3].

The main challenge for high-strength aluminium alloys, is their low formability at room temperature and the subsequent springback. Springback is mainly due to the elastic recovery of sheet metal after forming, resulting in a no net shape. Studying springback for the purposes of prediction has been investigated in cold forming conditions and is well documented in handbooks [4, 5]. However, springback investigations at warm and hot forming conditions are relatively few in number [6, 7]. Common industry practice is to use predictions in simulating the forming process and over bending the part to achieve the final dimensional accuracy as close as possible to the desired target.

Hot stamping with Uni-form platform is a novel forming technology. By tailoring the heating and applying in-die-quenching, it aims to drastically shorten the production time and reduce energy consumption [8]. Hot stamping helps to heavily reduce and, in some cases, eliminate the amount of springback experienced by the formed part. This mitigates the necessity of springback reducing methods such as over-bending.

2. Experiment Details

The following explains the experimental setup used in studying the springback of AA6082 and AA7075 alloys using hot stamping.

2.1. *Material*

The raw material included two aluminium alloys, AA6082 and AA7075, with a thickness of 2 mm, the detailed chemical compositions are shown in Table 1. Material composition and processing factors while producing the raw material can have an effect on experimental outcomes. However, as the samples for this study have been cut from the same sheet, they can be assumed as constant. In addition, the sample dimensions are specified in Figure 1.

Table 1. Chemical composition (% mass) of AA6082 and AA7075.

	Si	Fe	Cu	Mn	Mg	Cr	Zn	Ti
AA6082	0.7-1.3	0.5	0.1	0.4-1	0.6-1.2	0.25	0.2	0.1
AA7075	0.4	0.5	1.2-2	0.3	2.1-2.9	0.18-0.28	5.1-6.1	0.2

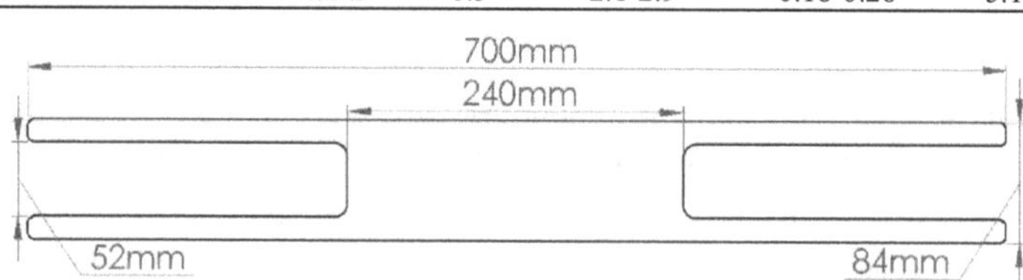

Fig. 1. Sample dimensions.

2.2. *Testing procedure*

In order to investigate parameters such as forming temperature and forming speed for a U-shaped component, the remaining forming parameters of blank holding force, heating rate and quenching rate are kept the same in this study. In addition, the process diagram in Figure 2 illustrates the steps necessary during hot stamping combined with a body-in-white (BIW) chassis production line.

For accurately measuring the springback, the angles were extracted by first scanning the edges of the formed samples, then using image processing software to define the outlines of the edges and finally using a drawing software to define vertical and horizontal lines along the edge as seen in Figure 3 to determine the size of angles α and β, and the associated springback.

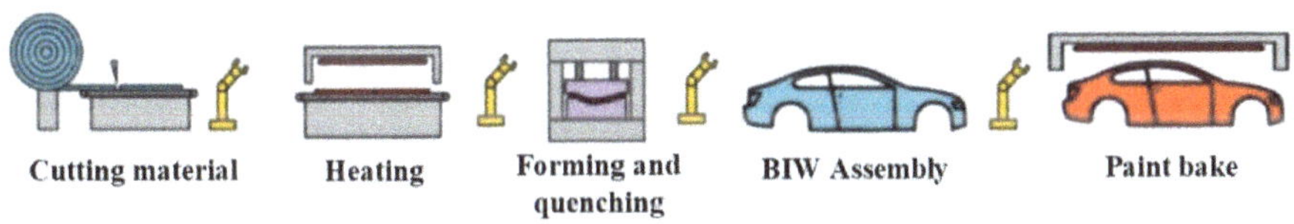

Fig. 2. Process diagram of the hot stamping forming process.

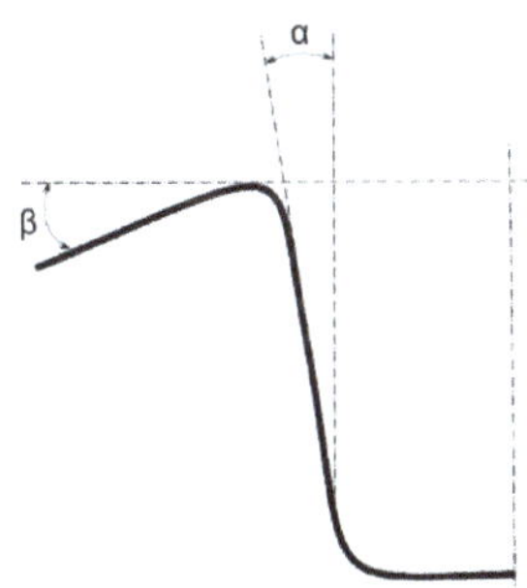

Fig. 3. Springback angles α and β.

Fig. 4. Uni-form system with hot stamping.

2.3. *Forming equipment*

The automated forming system, Uni-form, was developed within the author's group [8], which allowed accurate repeatable trials of the hot stamping forming cycle by utilising a 100 tonnes hydraulic press (supported by a set of gas spring cushion), a conveyer belt and a contact heating station as seen in Figure 4. Additionally, with an online multi-objective platform SMARTFORMING (*www.smartforming.com*), Uni-form allowed users to conduct simulations simultaneously and work remotely.

3. Results and Discussion

3.1. *Forming temperature effect*

Figure 5 shows the effect of forming temperature from 200 °C to 500 °C in increments of 50 °C on AA6082 and AA7075 springback angles. Springback was only observed at warm forming temperatures above 200 °C as the specimens formed in cold (room temperature) conditions lost their dimensional accuracy and fractured. This was mainly due to the amount of elastic deformation sustained during forming. For warm/hot forming conditions, the flow stress of a material decreases with increasing forming temperature. As decreasing flow stress results in decreased elastic deformation, a reduction in springback is observed.

A decreasing springback with increasing forming temperature is seen for both AA6082 and AA7075 U-shaped samples due to a decrease in flow stress at elevated temperature resulting in decreased plastic deformation and residual stresses. The main contrasting observation between the two alloys is that the maximum observed springback for AA7075 (14°), is an order of magnitude greater than the maximum springback observed for AA6082 (2°). This is due to inherently higher tensile strength of AA7075 compared to AA6082. Springback in AA6082 is very limited, ranging from 2° to 0° which can be contributed to the lower tensile strength.

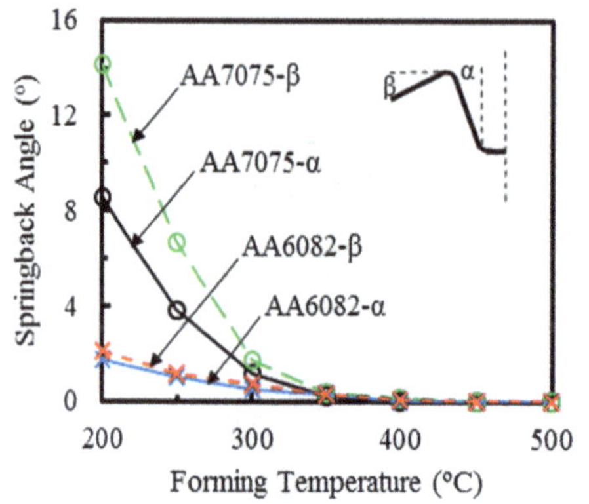

Fig. 5. Effect of forming temperature on springback for AA6082 and AA7075.

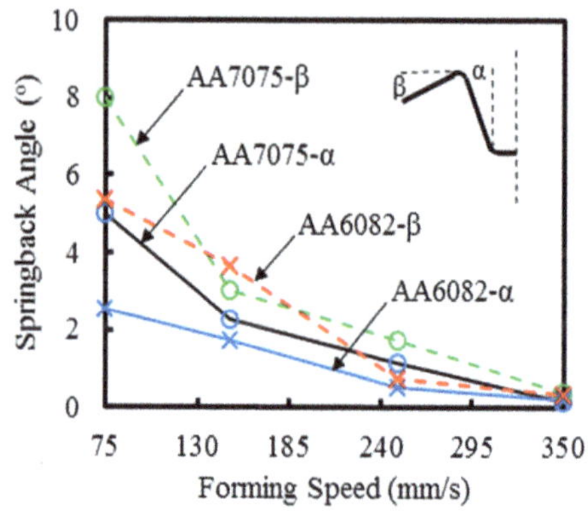

Fig. 6. Effect of forming speed on springback for AA6082 and AA7075.

3.2. *Forming speed effect*

Figure 6 shows the effect of forming speed (75, 150, 250 and 350 mm/s) on AA6082 and AA7075 springback angles. For AA6082, α and β were reduced below 0.5° at forming speeds greater than 250 mm/s. Springback in AA7075 shows a similar decreasing trend with increasing forming speed as observed in AA6082, it is almost negligible at forming speeds of 350 mm/s. Although AA7075 experiences greater springback at lower forming speeds, when comparing the absolute magnitude of springback reduction, AA7075 again shows an overall significantly greater reduction than AA6082.

AA6082 and AA7075 exhibit viscoelastic behavior at warm/hot forming conditions, which results in increased flow stress with increased strain rate and hence an increase in springback is expected due to the increased elastic deformation. This trend has been proven for forming components at low and medium forming speeds (0.5-10 mm/s) [7]. However, the high forming speeds investigated by Wang et. al [6] and the results of this investigation (Figure 6) exhibit the opposite trend. Decreased springback is an indication for decreased residual stresses and elastic deformation. Even though flow stress is increased, the higher forming speeds also result in greater strain rate hardening. This combined with using room temperature dies (colder than the forming temperature), results in fast cooling of the blank towards the end stages of forming, introducing an increased yielding tension in the blank. Increased tension applied to the blank results in the reduction in bending moment stresses (reduces the stress gradient) and therefore a subsequent observed reduction in springback.

4. Conclusion

In this study, a reduction of springback from 2° and 14° to 0.6° and 2° with increasing forming temperatures can be confirmed in AA6082 and AA7075 respectively. This is due to a decrease in flow stress and hence reduced plastic deformation. This study also shows that by using forming speed greater than 250 mm/s, springback can be reduced to below 0.73° in AA6082 and 1.72° in AA7075, which is due to increase in flow stress with increased strain rate and hence an increase in springback. Therefore, by understanding and tackling commonly encountered limitations, such as springback, of commercial aluminium alloys, their weight-saving potential can be expanded with more widespread industrial applications.

References

1. A Roadmap for moving to a competitive low carbon economy in 2050, in *European Commission*, (Belgium, Brussel, 2011).
2. Proposal for a Regulation of the European Parliament and of the Council setting emission performance standards for new passenger cars and for new light commercial vehicles as part of the Union's integrated approach to reduce CO_2 emissions from light-duty vehicles, in *European Commission*, (2017).
3. Ducker Worldwide, Aluminium content in cars, Ducker Worldwide - Study, (2016), https://www.ducker.com/news-insights/ducker-worldwide-study-aluminium-content -cars-public-summary
4. T. Altan and A. E. Tekkaya, *Sheet metal forming: processes and applications*, (ASM international, 2012).
5. J. Hu, Z. Marciniak and J. Duncan, *Mechanics of sheet metal forming*, 2nd edn. (Butterworth-Heinemann, 2002).
6. A. L. Wang, K. Zhong, O. El Fakir, J. Liu, C.Y. Sun, L.L. Wang, J. G. Lin and T. A. Dean, Springback analysis of AA5754 after hot stamping: experiments and FE modelling, *Int. J. Adv. Manuf. Technol.* **89**, 1339 (2017).
7. H. S. Kim and M. Koç, Numerical investigations on springback characteristics of aluminium sheet metal alloys in warm forming conditions, *J. Mater. Process. Tech.* **204**, 370 (2008).
8. X. Luan, Q. L. Zhang, O. E. Fakir, L. L. Wang, and M. M. Gharbi, Uni-form: a pilot production line for hot/warm sheet metal forming integrated in a cloud based smart forming platform, in *International Conference on Advanced High Strength Steel and Press Hardening, World Scientific*, 492 (2017).

An Investigation on the Dynamic Property of Aluminum Alloy Sheets

J. P. Zhang[1,2,a,†], G. Fang[1,2,b] and M. T. Ma[1,2,c]

[1]*China Automotive Engineering Research Institute Co., Ltd., Automotive Lightweight Engineering Research Center, Chongqing, 401122, China*

[2]*Chongqing CAERI special car Co., Ltd., Chongqing, 401122, China*

E-mail: [a,†]zhangjunping@caeri.com, [b]fanggang@caeri.com, [c]mamingtu@caeri.com

The dynamic property of 6016 aluminum alloy sheet is essential when applied as automobile components. To investigate the dynamic property of 6016 aluminum alloy sheet, the tensile tests are carried out under quasi-static loading and dynamic loading respectively. The strain rates involve 0.001 /s, 0.1 /s, 1 /s, 10 /s, 100 /s and 500 /s. On the basis of the strain-stress curves, the strain rate sensitivity of the aluminum alloy sheet is discussed. Furthermore, different constitutive equations such as Jonson-Cook and C-S model are applied to fit the curves and the fitting accuracy is discussed as well. Finally, the fracture surfaces of specimens under different tensile speed are investigated.

Keywords: 6016 aluminum alloy sheet; Strain rate sensitivity; Constitutive equation; Fracture surface.

1. Introduction

Nowadays, lightweight has become the most essential technology for automobile research and development for the sake of energy saving and emission reduction. Aluminum alloy sheet is one of the most potential lightweight materials widely applied to body-in-white due to its much lower density than that of the steel. Besides, the application of aluminum alloy sheets is beneficial for the pedestrian protection [1, 2]. The dynamic mechanical property is different from the quasi-static mechanical property for most of the metallic materials. In the car collision simulation, the description of dynamic mechanical behavior of materials largely affects the accuracy of results [3-6]. Therefore, this work mainly investigates the dynamic mechanical property of 6016 aluminum alloy sheet which is used as the automobile hood. The tensile tests are carried out under quasi-static loading and dynamic loading respectively. The strain rates involve 0.001 /s, 0.1 /s, 1 /s, 10 /s, 100 /s and 500 /s. On the basis of the testing curves, the strain rate sensitivity of the aluminum alloy sheet is discussed. Meanwhile, the flow behavior of the alloy is described with different constitutive equations and the accuracy is compared. Finally, the SEM micrographs of the specimens under different tensile speeds are investigated to discuss the effects of the strain rates on the fracture models.

2. Experiments

2.1. *Materials*

Commercial 6016 aluminum alloy sheet with the thickness of 1.0mm is selected in the present work, which is mainly applied to the engine hood. The alloy is a heat treatable material which is beneficial for dent resistance due to the second strengthening of mechanical property in the baking process. In the raw material production, after quenching it is pre-aged in order to maintain stable mechanical property which is essential for forming and baking hardening effect. The chemical composition is shown in Table 1.

Table 1. Chemical composition of 6016 aluminum alloy sheet (wt.%).

Si	Fe	Cu	Mn	Mg	Cr	Zn	Ti	Al
1.01	0.15	0.009	0.07	0.51	0.013	0.015	0.015	Bal.

In this work, 6016 sheet has been experienced 2 % pre-strain tension and heat treated at 185 °C for 20 min in order to make the mechanical property of the material be consistent with that of its actual state. The pre-strain tensile process was introduced to characterize the forming process, and the heat treatment is applied to simulate the hood baking process. After the processes, the specimens are employed to quasi-static and various strain rates tensile test.

2.2. *Dynamic tensile test*

Dynamic tensile test is carried out through hydraulic servo-controlled tensile machine HTM5020. Its highest tensile speed is 20 m/s. The geometry and size of the sample in dynamic tensile test are shown in Fig. 1.

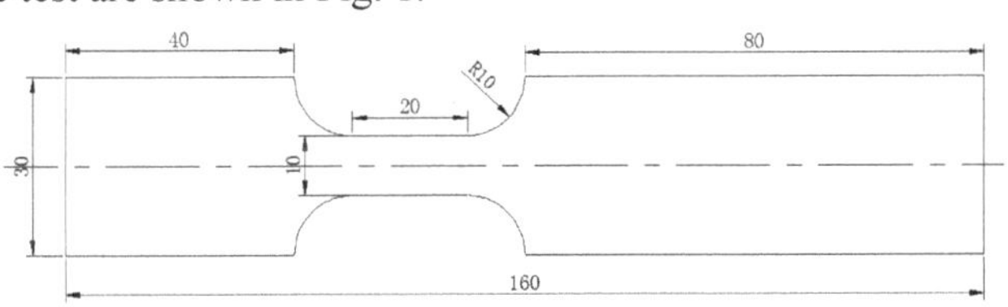

Fig. 1. The geometry of the sample in dynamic tensile test (unit: mm).

In this paper, the samples with the strain rate of 0.001/s are tested in CMT5305 tensile tester. The experiments with the other four strain rates, 1/s, 10/s, 100/s and 500/s are carried out in the HTM5020. Before experiments, the gauges of the samples are sprayed with speckles by alternately using white and black oil paint. Sequences of images of the sample deformation are recorded by ultra-high camera and they are transformed to strain distribution by digital image correlated software. The stress is obtained via the transformation of the force which is directly recorded from HTM520 force transducer. To study the strain rate sensitivity and the dynamic flow behavior of the alloy sheet, the true plastic strain-true stress curves are applied.

2.3. *Fracture investigation*

After dynamic tensile test, the fractures of the samples tested at different strain rates are respectively investigated by SEM. The effects of strain rate on the fracture modes are studied.

3. Results and Discussion

3.1. *Strain rate effect*

For numerical simulation and constitutive relationship fitting, the input data mainly involve true stress and the corresponding true plastic strain. The engineering data are directly obtained from the experiment. Then there are transformed to true data and their elastic deformations are eliminated. Finally, the true stress-strain curves of 6016 aluminum alloy sheet at different strain rates are acquired and shown in Fig. 2.

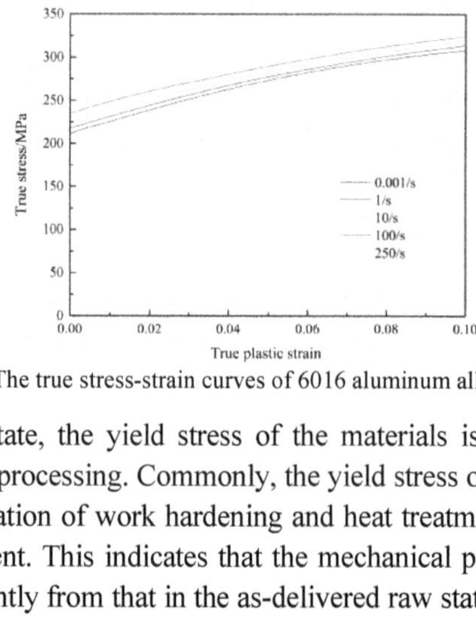

Fig. 2. The true stress-strain curves of 6016 aluminum alloy sheet.

In the quasi-static state, the yield stress of the materials is 211 MPa, which largely exceeds the value before processing. Commonly, the yield stress of 6016 is 120~140MPa. In the situation, the combination of work hardening and heat treatment contributes roughly 80 MPa strength improvement. This indicates that the mechanical property of the materials in actual state varies apparently from that in the as-delivered raw state for the components such as hood. As the strain rates increase, the flow stresses get improved as well. The material exhibits strain rate sensitivity. To investigate its strain rate sensitivity, the yield stresses and the flow stresses at the strain of 5 % and 10 % at different strain rates are compared. The variation of the stresses as the change of strain rates are shown in Fig. 3. The result shows that the material gets strengthened as the strain rates or the deformation increases. When the strain rate is lower than 10 /s, the strength improvement of the materials is small. As the strain rate increases over 10 /s, the strain rate effect becomes obvious.

During plastic deformation, the relationship between the stress and strain rate can be described by the power exponent equation show as Eq. 1:

$$\sigma = K\varepsilon^n \dot{\varepsilon}^m \tag{1}$$

In the equation, K is a constant coefficient. n denotes work hardening index. m represents the strain rate sensitivity. Under certain strain, the strain rate sensitivity can be calculated by Eq. 2:

$$m = d(\ln \sigma) / d(\ln \dot{\varepsilon}) \tag{2}$$

According to Eq. 2, the strain rate sensitivity index is calculated and plotted in Fig. 4. The alloy sheet displays the highest sensitivity at yield point. The index value was 0.0103. As the deformation develops, the value decreases. When the strain reaches 6%, the index value nearly keeps still. In the whole deformation process, the strain rate sensitivity of 6016 aluminum alloy sheet is among 0.006 to 0.01.

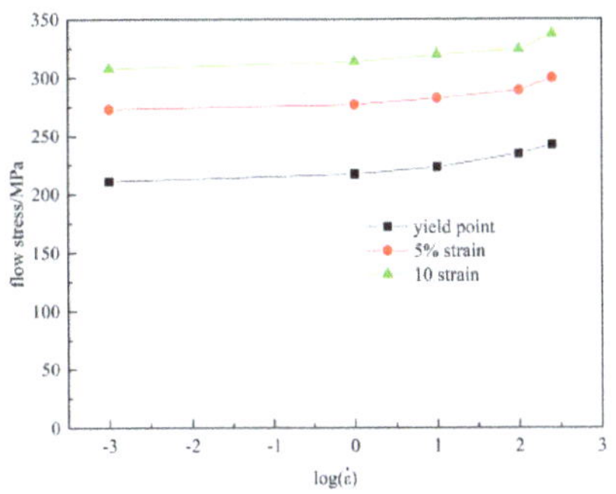

Fig. 3. Flow stresses at different strains and strain rates.

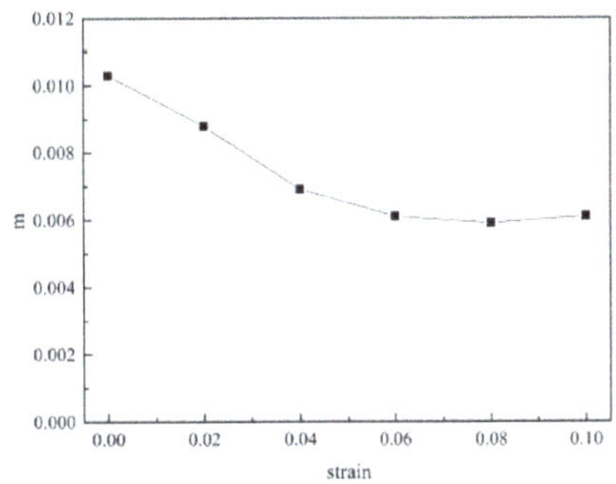

Fig. 4. Strain rate sensitivity at different strains.

3.2. *Constitutive equation fitting*

There are several constitutive models to describe the flow behavior of metallic materials, which are respectively Johnson-Cook, Cowper-Symonds and modified Johnson-Cook equations. Not considering the temperature effects, the equations are shown as Eq. 3 ~ 5:

$$\sigma = K\varepsilon^n[1+(\frac{\dot{\varepsilon}}{C})^{\frac{1}{P}}] \tag{3}$$

$$\sigma = (A+B\varepsilon^n)[1+C\ln(\frac{\dot{\varepsilon}}{\dot{\varepsilon}_0})] \tag{4}$$

$$\sigma = B\varepsilon^n[1+C\varepsilon^{n'}\ln(\frac{\dot{\varepsilon}}{\dot{\varepsilon}_0})] \tag{5}$$

For each equation, σ is stress and ε is strain, while n is work hardening index and $\dot{\varepsilon}$ refers to strain rate. In Cowper-Symonds equation as shown in Eq. 3, K is a constant coefficient. C and P is parameter defining the strain rate effect. For Johnson-Cook equation of Eq. 4, A, B and n are respectively parameters which describe the stress-strain shape of the referring curve with the strain rate of $\dot{\varepsilon}_0$. C is the key parameter which reflects the strain rate sensitivity of the materials. On the basis of Johnson-Cook equation, Xu makes some modification to better fit the flow curves of the material under different strain rates [6]. The equation is listed as Eq. 5. Compared to J-C equation, modified equation mainly adds a parameter after C to adjust the shape of the curve. For each fitting, the error is calculated according to Eq. 6.

$$Error = \frac{\sum_{i=1}^{s}\sum_{j=1}^{r}\left|\sigma_{ij}^{e} - \sigma_{ij}^{f}\right| / \sigma_{ij}^{e}}{r \times s} \tag{6}$$

In Eq. 6, r is the data point for each curve and s denotes the number of the strain rate. σ_{ij}^{e} and σ_{ij}^{f} respectively represent the tested stress and the fitting stress. The fitting results are shown in Table. 2. It illustrates that all the three constitutive equations precisely describe the flow behavior of 6016 alloy sheet. The errors are all lower than 5 %. Especially, Jonson Cook equation does the best fitting to the experimental data, with the error 2.54 %.

Table 2. The fitting results of different constitutive equations.

Equation	K	n	C	P	A	B	n'	error
Cowper-Symonds	374.8	0.1	194625	0.288				4.01%
Johnson-Cook		0.89	0.008		210.9	811		2.54%
Modified Johnson-Cook	374.8	0.1	0.0035				-0.12	3.44%

3.3. *Fracture investigation*

The fracture micrographs of the 6016 aluminum alloy samples tensioned at different strain rates are illustrated in Fig. 5. The fracture surfaces are full of dimples which indicate the ductile fracture is in charge for all the samples tested from quasi-static state to dynamic state. In the micrographs, bigger and deeper voids are discovered which are supposed to be the remains of the second-phase particles [4, 5]. During deformation the stress concentration is easily yielded around the second-phase particles. The crack initiation and propagation lead to the shearing and peeling of the particles and thus cause the voids formation.

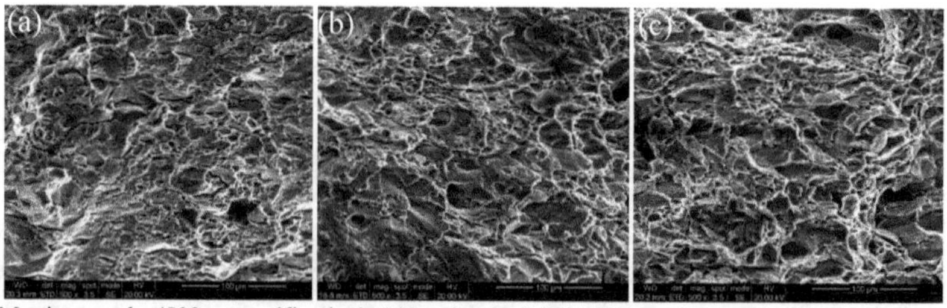

Fig. 5. SEM micrographs (500x magnification): (a) strain rate 0.001/s, (b) strain rate 1/s, (c) strain rate 250/s.

At the strain rate of 0.001/s, fewer dimples are investigated compared with the fracture surfaces with higher strain rates. Besides, some small cracks are also observed. As testing strain rates increase, the density of the dimples also enlarges. The results suggest that the ductility of the samples get better at higher strain rates. It is reported that higher strain rates lead to the increased macroscopic roughness and higher scatter on the location of voids nucleation sites on the fracture surface. However, in the present work, such phenomena are not inspected. The macroscopic roughness and the scatter of voids sites do not show obvious strain rate sensitivity.

4. Conclusion

This work discusses the effects of the strain rates on the flow behavior and fracture of the commercial 6016 aluminum alloy sheet on the basis of dynamic tensile testing and SEM micrograph investigation. The results indicate 6016 alloy exhibits strain rate sensitivity, higher strain rates tension leading to higher deformation resistance. Especially when the strain rates exceed 10 /s, the strain rates get improved. During plastic deformation, yielding point shows the largest sensitivity. Further deformation causes the strain rate sensitivity index decrease. The three constitutive equations accurately characterize the dynamic flow behavior of the materials and J-C model performs the best. Strain rates affect the fracture surfaces of the materials. The increase of the strain rates is beneficial for the ductility of 6016 aluminum alloy sheet.

Acknowledgments

This work was financially supported by The National Key Research and Development Program of China "Manufacturing technology of aluminum alloy panel for passenger car" (2016YFB0300805).

References

1. M. T. Ma, X. Y. Bi, J. H. You and H. Z. Lu, Research Progress of Property and Its Application of Aluminum Alloy Auto Sheet, *Mater. Mech. Eng.* **34,** 1 (2010).
2. W. S. Miller, L. Zhuang, J. Bottema, A. J. Wittebrood, P. D. Smet, A. Haszler and A. Vieregge, Recent development in aluminum alloys for the automotive industry, *Mater. Sci. Eng. A* **280**, 37 (2000).
3. A. S. Khan and H. Liu, Variable strain rate sensitivity in an aluminum alloy: Response and constitutive modeling, *Int. J. Plasticity* **36**, 1 (2012).
4. D. J. Zhu, B. Mobasher, Rajan S. D. Rajan and P. Peralta, Characterization of Dynamic Tensile Testing Using Aluminum Alloy 6061-T6 at Intermediate Strain Rates, *J. Eng. Mech.* **137**, 669 (2011).
5. R. Smerd, S. Winkler, C. Salisbury, M. Worswick, D. Lloyd and M. Finn, High strain rate tensile testing of automotive aluminum alloy sheet, *Int. J. Impact Eng.* **32**, 541 (2005).
6. K. Xu, C. Wong, B. Yan and H. Zhu, A High Strain Rate Constitutive Model for High Strength Steels, in *SAE World Congress*, (Detroit, Michigan, the USA, 2003).

Research on Influence of Precompression Process on Fatigue Properties of DP and TWIP Steel

Y. C. Ling[1,2,†], Y. Zhao[1,2] and T. Q. Fan[1,2]

[1]*China Automotive Engineering Research Institute Co., Ltd., Chongqing, 401122, China*
[2]*Chongqing Automotive Lightweight Engineering Research Center, Chongqing, 401122, China*
[†]*E-mail: lingyangchao@caeri.com.cn*

The effects of 5 % and 10% different pre compression shrinkage processes on the fatigue properties of DP590 and TWIP880 two steel products were studied in this paper. The effects of two different pre compression shrinkage processes on the fatigue properties of the two steels were investigated. The tensile test of material and cyclic loading fatigue test are carried out on the experimental research object of standard tensile specimen, and the basic parameters of the stress-strain curve and stress fatigue life curve of two materials under different pre compression shrinkage of the raw materials and 5 % and 10 % kinds of materials are obtained. The fatigue properties of two kinds of steels under the pre compression process are obtained through experimental study. Taking a wheel as the research object, two kinds of steel raw materials, 5 % pre compression and 10 % pre compression of DP590 and TWIP880 were applied to spoke and rim respectively. The static strength of the wheel under radial loading conditions was solved. The fatigue life of wheels is analyzed by stress fatigue analysis, and the influence of different compression amounts on fatigue life of wheels is studied.

Keywords: Pre compression process; Wheel steel; Fatigue performance; Cycle times.

1. Introduction

Wheels are important parts of automobiles, which play an important role in the safety, stability and compliance of automobiles during driving. The wheel is subjected to random loads from the outside world, so during the development of the wheel, the wheel has become the most concerned performance problem without fatigue failure during the service life. In the early stage of product design, it is of practical value to predict the fatigue life of the wheel structure and guide the structural design.

2. Material Base Performance Test

This paper mainly studies the properties of two steel grades of DP steel and TWIP steel. DP (dual-phase) steel is a high strength steel for low carbon steel or low alloy steel through critical heat treatment or controlled rolling. It is widely used in automobile side beams, side members, beams, struts, chassis reinforcement, oil tank supports and body structures, reinforced parts and collision avoidance parts. As a twin induced plastic steel (Twinning Induced Plasticity), TWIP steel is a hot spot of research in advanced and high strength steel at present, and it is also widely used in automobile structural parts.

2.1. *Tensile test*

Tensile test is a test method for determining the performance characteristics of materials under axial tensile load. The elastic limit, elastic modulus, yield strength, tensile strength, stress-strain curve and other tensile properties of the material can be determined by the data obtained by the tensile test.

DP590 in raw materials (DP590r), precompression 5% (DP590rpc 5%), pre compression 10% (DP590rpc 10%), TWIP880 in raw materials (TWIP880r), pre compression 5% (TWIP880rpc 5%), pre compression 10% (TWIP880rpc 10%), and two kinds of materials were tested in three process paths, and the materials tensile properties were obtained. Tensile properties include tensile modulus, Poisson's ratio and yield strength. The material tensile test machine used in this study is SANS CMT5305. As shown in Figure 1, the basic size of the tensile specimen is shown in Figure 2.

Fig. 1. Fatigue testing equipment.

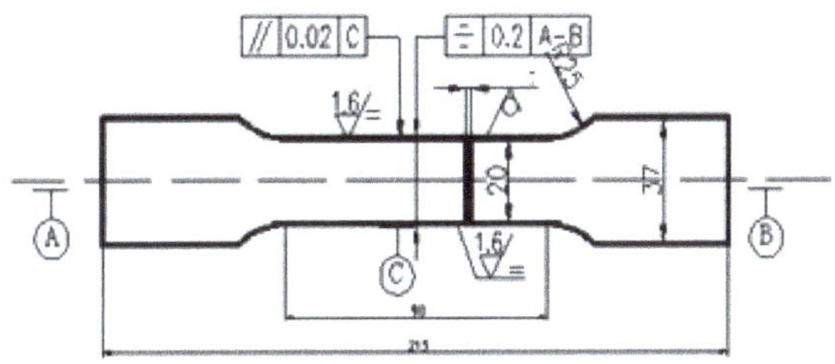

Fig. 2. Material simple sample size.

2.2. *Tensile and pressure fatigue performance test*

The tensile and pressure fatigue performance test is aimed at the high cycle fatigue performance test of two materials under the above three process conditions. The fatigue performance test parameters mainly include the stress ratio equal to -1, the test vibration frequency of 25-30 Hz, and the test loading waveform as the sine wave. The instrument tested is SDM 200 kN hydraulic servo fatigue test system. The fatigue test standard adopted is GB/T 3075-2008 "axial force control method for fatigue test of metallic materials". The test instrument and the simple sample before the experiment are shown in Figure 3.

According to the fatigue test, the S-N curve and the fatigue limit of the material can be obtained. The fatigue life curves of the two materials under three different process methods are shown in Figures 4.

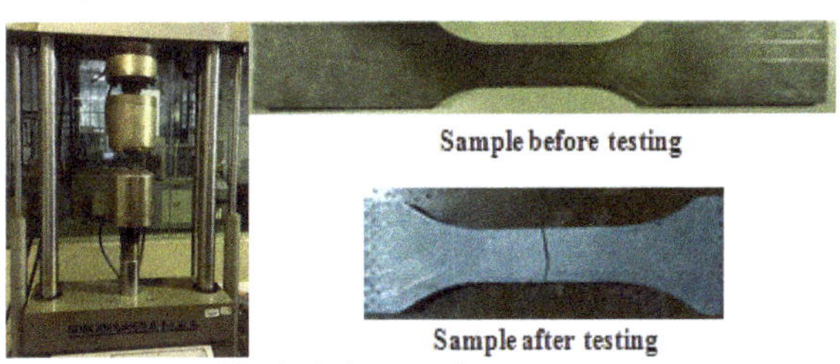

Fig. 3. Fatigue tension test.

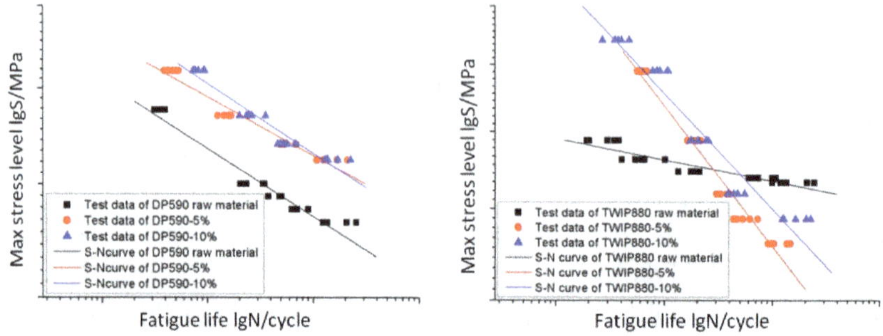

Fig. 4. SN curves of DP590 and TWIP880 under different pre compression strain rates.

According to the SN test curve, we can get the following conclusion: at the same stress level, the cycle times of the raw materials are the least, the 5% pre strain compression conditions are next, and the cycle times of the 10% and the strain compression are the greatest. TWIP880 has different life performance laws at different stress levels.

3. Analysis of Material Base Performance

This analysis mainly uses ANSYS and ncode design life Software Co simulation analysis, the overall flow of fatigue analysis is shown in Figure 5.

Through the experiment and analysis, the two methods can get the distribution of cycle times under different stress levels, as is shown in Figure 6.

The results of the test and analysis show that:

(1) The process method has a great influence on the fatigue properties of materials, and the material properties of different process paths are different.

(2) The stress fatigue life curve obtained by simulation is basically the same as that obtained from the test. It shows that it is feasible to use quasi static method to analyze strain fatigue life. This method can be used in engineering analysis.

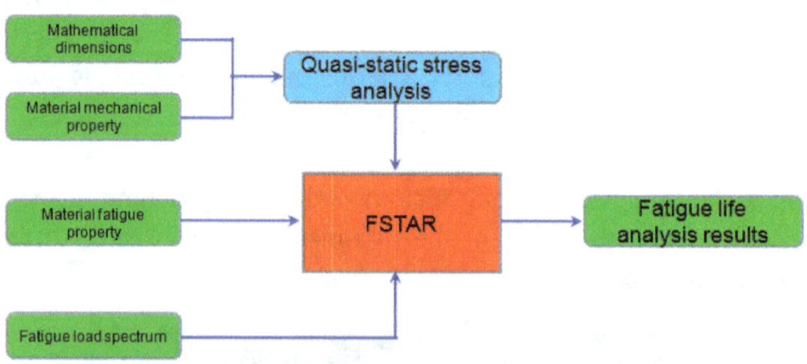

Fig. 5. Fatigue simulation process.

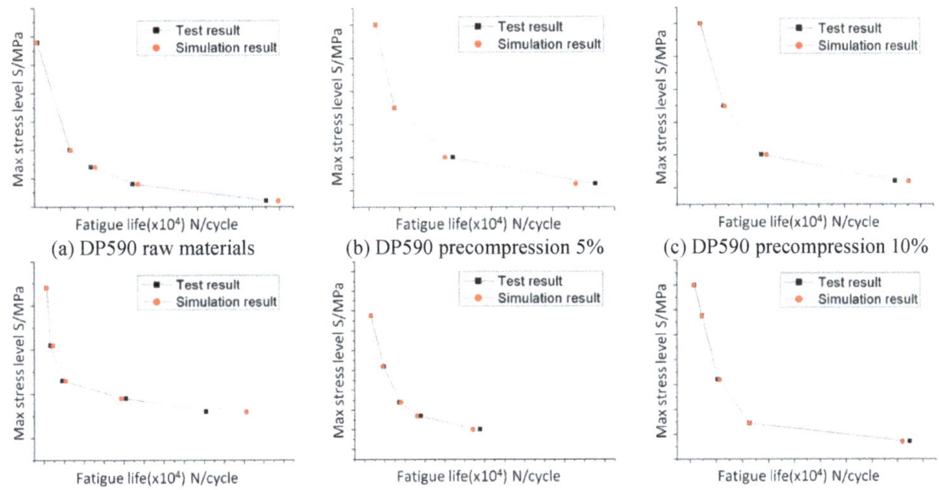

(a) DP590 raw materials (b) DP590 precompression 5% (c) DP590 precompression 10%

(d) TWIP880 raw materials (e) TWIP880 pre compression 5% f) TWIP880 pre compression 10%

Fig. 6. Comparison of stress life curves obtained in two ways.

4. Analysis of Wheel Life

The stress analysis and fatigue life prediction of a vehicle wheel are carried out and the working condition is analyzed. The materials of wheel spoke and rim are TWIP880r, TWIP880rpc 5%, TWIP880rpc 10% and DP590r, DP590rpc 5% and DP590RPC 10%.

The first step is to calculate the stress distribution of the wheel at the quasi static moment, and to simulate the pressure in different locations of the outer of wheel rim of the steel wheel to simulate the pressure of the wheels at different times. The loading diagram is shown in Figure 7.

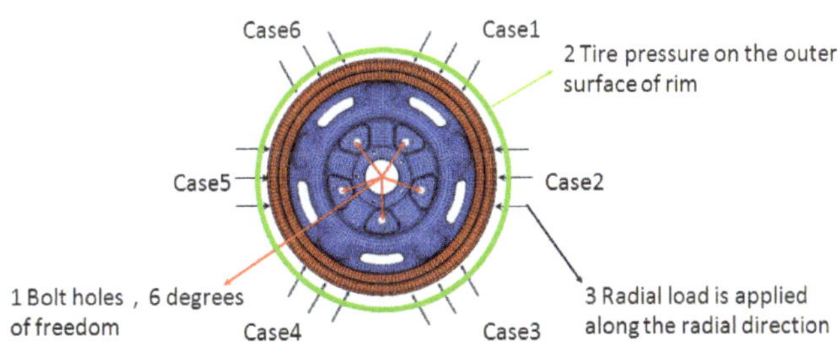

Fig. 7. Schematic diagram of wheel radial load analysis.

The stress distributions of two materials under the 6 loading sub steps of raw materials, pre compression 5 % and pre compression 10 % were calculated respectively.

After calculating the stress field distribution of 6 load parts under the three process methods, the fatigue load course of the wheel is formulated, and the fatigue life distribution of the wheels under the three process paths is calculated by the stress analysis module in the fatigue software ncode design life. The fatigue life distribution is shown in Figure 8. The fatigue life analysis shows that the number of radial fatigue cycles of the raw material is 1 million 187 thousand times, and the fatigue life of the pre compression

5% process is 1 million 72 thousand times, and the fatigue life of the pre compression 10% process is 1 million 314 thousand times, which meets the requirements of the design life.

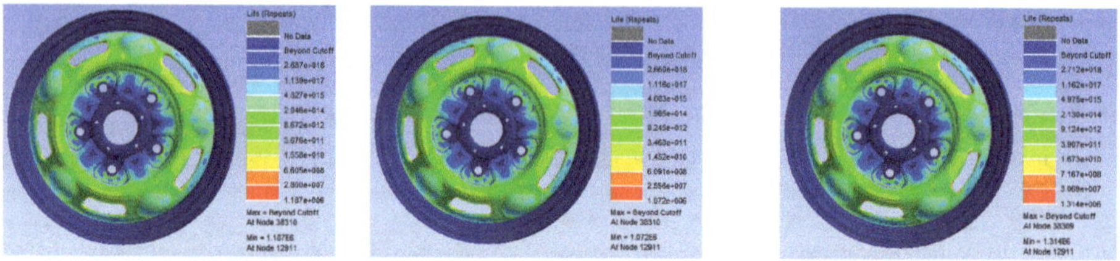

(a) DP590r+TWIP880r (b) DP590rpc5%+TWIP880rpc5% (c) DP590rpc10%+TWIP880rpc5%

Fig. 8. Fatigue life distribution under different process paths.

5. Conclusion

Through simple test, the tensile properties and stress fatigue life curves of two materials were obtained under three different process methods. The results show that the fatigue performance of the DP590 steel increases with the increase of pre pressure shrinkage, and the fatigue performance of TWIP880 increases with the increase of pre pressure shrinkage under some life span, but the SN curve is intersecting, and its regularity is not obvious.

The fatigue life of the wheel is analyzed by the tensile and stress fatigue life curves obtained from the fatigue test of simple sample materials as the basic material parameters. The analysis results show that the fatigue life of the wheels can meet the design requirements under the three different process paths. At the same time, it is concluded that the material pre compression process has great influence on fatigue performance of wheel parts.

Acknowledgments

This work is supported by project named research and application of automotive lightweight design and forming key technologies. Project num.: cstc2015zdcy-ztzx60010.

References

1. M. Ma and B. Wu, *Dual steel-physical and mechanical metallurgy*, 2nd edition. (Metallurgical industry Press, Beijing, 2009).
2. F. Hou, Research on US side impact regulations based on computer simulation of the car door impact bars, 47 (2005).

The Effect of Pre-compression Deformation on Tension-compression Fatigue Properties of Hot Rolled DP590

M. T. Ma[1,†], D. G. Kim[2], G. Y. Wang[1] and G. Liu[1]

[1]*China Automotive Engineering Research Institute Co., Ltd., Chongqing, 401122, China*
[2]*POSCO Global R&D Center, Incheon, 21985, Korea*
[†]*E-mail: mingtuma@126.com*

The tension-compression fatigue properties of dual-phase steel after pre-compression deformation of 0 %, 5 % and 10 % were investigated. The fatigue property under load ratio of r = -1 was measured. The fatigue limit was calculated by using of national standard and SAE standard. The microstructure and fatigue fracture were observed. The results show that the fatigue property under high stress and the fatigue limit of 50 % survival rate are obviously improved after the 5 % and 10 % pre-compression. But effect of 5 % pre-compression on the fatigue property is better than that of 10 % pre-compression. This result is related to the stress state produced by the pre-compression strain and the effect it has on the mechanical properties of the material. The typical fatigue fractures in three states are analyzed and compared, and the fracture morphology and fatigue property have a good correspondence.

Keywords: Dual phase steel; Pre-compression deformation; Tension- compression fatigue property; Fracture morphology.

1. Introduction

Dual phase steel which is mainly composed of martensite and ferrite has been widely used in the automotive industry due to its excellent formability [1-2]. Many dual phase steels are used in automotive structural parts which beared fatigue loads in using. Therefore, the influence of pre- compression deformation on the fatigue property of dual phase steel has special significance in application. A series of work [3-4] studied the fatigue property of dual phase steels. We have also studied the effect of process paths on dual-phase steel DP780, but the effect of pre-compression deformation on fatigue property of dual phase steel has not been reported. The purpose of this paper is to study the effect of pre-compression deformation on the fatigue of hot rolling dual-phase steel DP590. The fracture morphology is observed and analyzed.

2. Test Materials and Methods

The test steel is hot rolled dual phase steel DP590 with a thickness of 5.3mm.The chemical composition as follows (wt%): 0.078C, 0.77Si, 1.2Mn, 0.4Cr, 0.1Al, 0.002S, 0.006P. The sample is cut into a tensile fatigue test specimen with a gauge length of 12×5.3×12 mm after pre-compressed 0 %, 5 % and 10 % in a special fixture. It is shown in Fig. 1. Fatigue specimens are sampled parallel to the rolling direction. According to GB/T 3075-2008 Metallic materials-Fatigue testing-Axial-force-controlled method, the fatigue test was carried out in the SDM 200 kN .Test parameters: Stress Ratio R= -1; Frequency: 25~30 Hz, Load wave: sine wave; Specified cycle number for the fatigue limit: 1×10^7 cycles. The metallographic structure was observed on the Leica-DMI3000 type, and the fatigue fracture was observed by SEM (FE quanta 20 type).

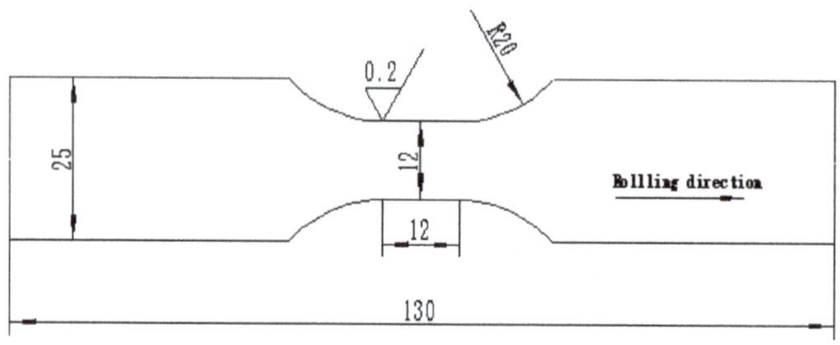

Fig. 1. The fatigue sample.

3. Test Results and Analysis

3.1. *Effect of pre-compression on the tensile-compressive flow curves*

The tensile-compression flow curve after 5 %, 10 % pre-compression is shown in Fig. 2.

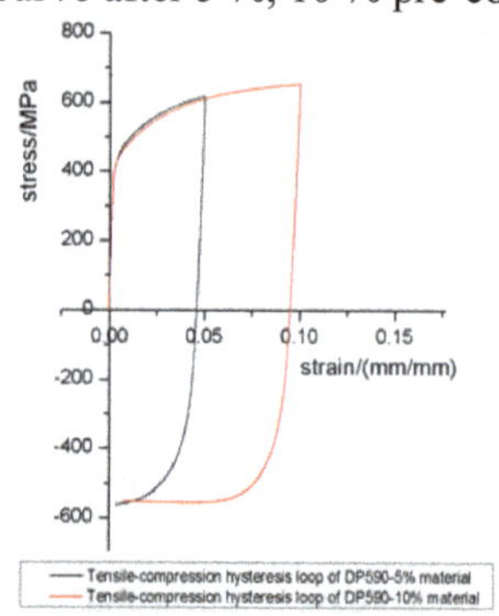

Fig. 2. Rheological properties of tensile and compression of different pre-compressed DP590.

The related mechanical properties of materials after each strain history are compared with raw materials. Its result is listed in Table.1. As seen from the figure, after tension 5 % and then compressing about 4 %, the compression yield strength of material is 310 MPa. After tension 10 % + compression 5 %, the compression yield strength of material is 290MPa. Two states and the associated strain history have an obvious Bauschinger effect. In this type of dual-phase steel, the Bauschinger effect is due to the large difference in strength between the martensite and ferrite phases, which produced the plastic strain incompatibility between the martensite and ferrite phases during deformation, resulting in interphase stress, which affects the flow stress during subsequent tension and compression [5]. Pre-compression deformation of material can also make the ferrite to genarate significant work hardening, all of which have a significant effect on fatigue strength.

Table 1. Mechanical properties of DP590 with three states.

Code	R0.2 (MPa)	Rm (MPa)	A, %
DP590-0%	418.8	636.3	25.4（A80）
DP590-5%	398.0	639.1	31.4（A50）
DP590-10%	400.0	662.9	18.6（A50）

Note: Table notes.
DP590-0 % is tension of hot rolling;
DP590-5 % is tension after 5 % pre-compression deformation;
DP590-10 % is tension after 10% pre-compression deformation.

3.2. *Effect of pre-compression on fatigue performance*

The tensile-compression fatigue test results of raw materials, pre-compression 5 %, pre-compression 10 % under different stress amplitudes are shown in Table 2. The load ratio is -1.

Table 2. Fatigue results of DP590 in different states.

DP590		Pre-compression 5%		Pre-compression 5%	
Max tensile stress amp.i/MPa	Fatigue life N/cycle	Max tensile stress amp.i/MPa	Fatigue life N/cycle	Max tensile stress amp.i/MPa	Fatigue life N/cycle
338	40079	360	47411	360	82453
338	32308	360	43157	360	93139
338	35546	360	53478	360	94983
338	36181	360	50671	360	76810
338	31540	360	39240	360	73614
300	210957	335	125793	335	201471
300	339751	335	143923	335	362184
300	235354	335	159553	335	234764
300	353399	335	245819	335	245819
300	204239	335	168724	335	264900
294	353824	360	47411	320	526124
294	383600	320	694102	320	468280
294	507545	320	529716	320	498738
294	377362	320	491001	320	571971
294	489266	320	541142	320	686100
288	710558	320	493530	312	2252005
288	683338	312	1087029	312	1703971
288	631856	312	2103036	312	1358462
288	712789	312	1260585	312	1297297
288	906281	312	1258431	312	1376017
282	1241626	312	1683040	281	>107
282	1314098	285	>107	281	>107
282	2057421	285	>107	—	—
282	1357100	—	—	—	—
282	2540019	—	—	—	—
276	>107	—	—	—	—
276	>107	—	—	—	—

The relationship between the stress amplitude and the cycle life logarithm (S-N curve) processed by the national standard is shown in Fig. 3. It shows that pre-compression significantly increases the cycle life of the material, especially in high stress areas, the effect is very obvious. The P-S-N curve processed by the SAE-China J3202-2013 rapid fatigue test method is shown in Fig. 4. It shows the fatigue limits at different survival rates (50 %, 95 %, 99 %). It can be seen that the pre-compression deformation can significantly increase the fatigue limit, and the fatigue limit of fluctuations of DP steel (50 % and 99 % survival rates) in three states are roughly the same. The fatigue limits (50 % survival rates) with different processing methods were compared with test values which are listed in Tab. 3. 5 %, 10 % pre-compression deformation significantly increases the fatigue limit, but the good effect of 10 % pre-compression deformation on fatigue limit is slightly lower than that of 5 % pre-compression deformation.

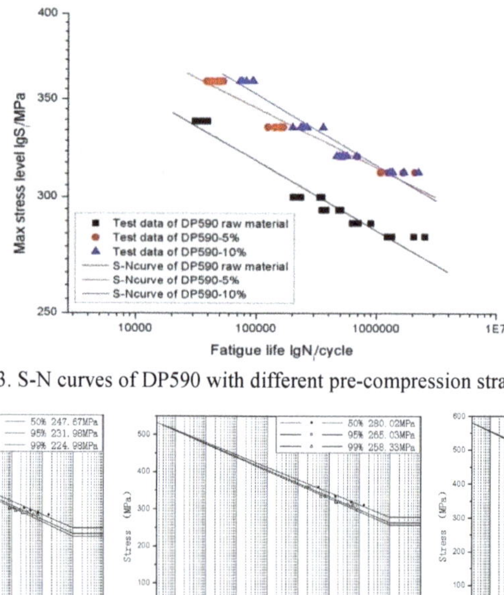

Fig. 3. S-N curves of DP590 with different pre-compression strain values.

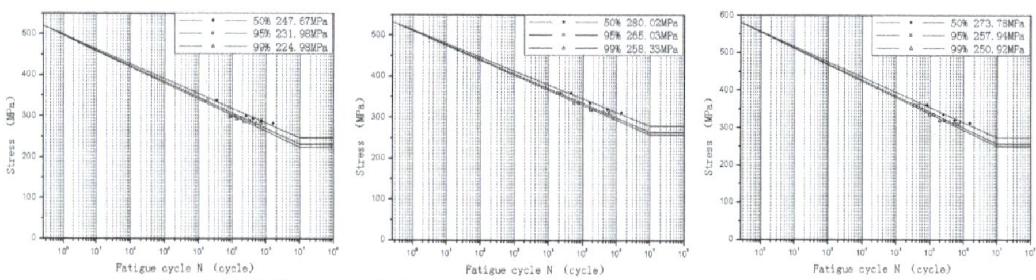

Fig. 4. P- S-N curves of DP590 with different pre-compression strain values.

These results show that pre-compression deformation can make DP steel have obvious strengthening effect, but excessive pre-compression deformation will introduce higher residual stress in DP steel, which may contribute to fatigue crack generation or crack propagation. Therefore, compared with the 5 % pre-compression deformation, the fatigue limit of 10 % pre-compression deformation is slightly reduced.

Any fatigue fracture process involves initiation of crack, crack propagation, and final fracture. These three different regions are clearly visible seen in the fatigue fracture. The fracture morphology of the raw material at a stress amplitude of ±300 MPa is shown in Fig. 5.

Table 3. Fatigue limits of DP590 in different states.

Material status	Fatigue strength of specimens reached 1×10^7 cycles, MPa	Fatigue strength obtained with S-N curve equation, MPa	Fatigue limit obtained with SAE-China J32022013, MPa
DP590raw material	276	255.31	247.67
DP590 with pre-compression strain 5%	285	286.14	280.02
DP590 with pre-compression strain 10%	281	282.09	273.78

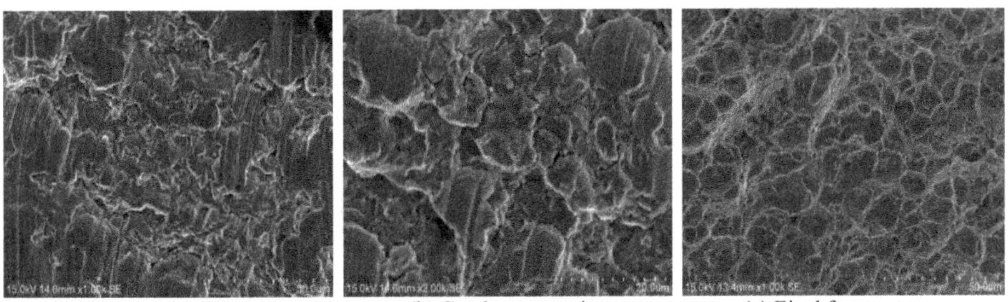

(a) Crack initiation zone (b) Crack propagation zone (c) Final fracture zone

Fig. 5. Fracture morphology of raw materials at a stress amplitude of ±300 MPa.

The zone of crack initiation has dimples, tears and traces of friction. The propagation zone has secondary cracks and fishbone patterns. The final fracture zone has a large number of shallow dimples. The fatigue fracture morphology of DP steel pre-compressed with 5 % stress amplitude ±360 MPa is shown in Fig. 6.

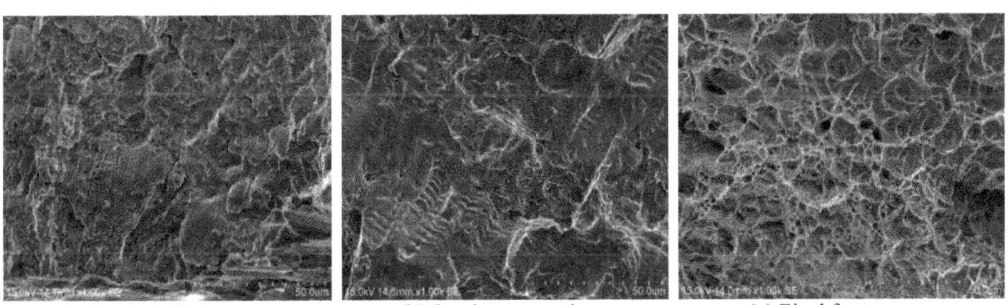

(a) Crack initiation zone (b) Crack propagation zone (c) Final fracture zone

Fig. 6. Fatigue fracture morphology of 5 % pre-compressed DP steel stress amplitude ±360 MP.

The shape of the three zones is obviously better than the raw material, and there is obvious fishbone or tire pattern in the propagation area, which is more obvious and deeper than the pattern in Fig. 5. It shows that during the crack propagation energy absorption is larger than that in raw material at ±300 MPa stress amplitude. The final fracture zone has a deep dimple whose the stress amplitude of the fracture is relatively high, which is consistent with the high fatigue life of the pre-compression in the high stress zone. The fatigue fracture morphology after 10 % pre-strain has similar three-zone fracture characteristics, but the ductility characteristics under high stress are lower than that of 5 % pre-compression.

4. Conclusion

Pre-compression strain of 5 % and 10 % can significantly improve the fatigue life in high-stress zone and fatigue limit of dual-phase steel, but the 5 % pre-compression effect is most significant. Compared with 5 % pre-compression, the fatigue limit of 10 % pre-compression is slightly lower than that of 5 % pre-compression.

The effect of pre-compression on fatigue life is related to the improvement of mechanical properties after pre-compression, including reasonable work hardening and corresponding Bauschinger effect, but when the pre-compression strain reaches 10 %, it may lead to excessive work hardening. The interphase residual stress introduced by pre-compression is too high, which affects the ductility of the material or reduces the crack initiation and propagation resistance, affecting the fatigue life. These effects of pre-compression strain are consistent with the fracture morphology. The ductility and energy absorption in fatigue fracture morphology of the 5 % pre-compression sample is more better than that without pre- compression strain. For example under the stress amplitude ±360 MPa, the fatigue crack propagation has obvious fish bone. The fatigue fringe of the crack propagation zone is not obvious fish bone at 10 % pre-compression strain, and is the fatigue fracture platform. In the final fracture zone, it is not a dimple, and is a tearing plus platform, that is, the brittleness feature is increased.

Acknowledgments

Work supported by grant U1564203 of the National Science Foundation.

References

1. M. T. Ma, *Advanced automotive steel*, (Chemical Industry Press, 2008).
2. M. T. Ma and B. R. Wu, *Dual phase steel-physical and mechanical metallurgy*, 2nd edn. (Metallurgical industry press, 2009).
3. B. Yan, Behavior of Advanced High Strength Steels For Automotive Application, in *44th MWSP Conference Proceedings*, (Orlando, America, 2002).
4. POSCO, *2018 automotive steel data book*, p63-71, (Product Application Center, Posco technical research lab, 2018).
5. M. T. Ma, Z. P. Duan and Y. Tomota, *Bauschinger effect in metal alloys and its application in industrial production*, (Metallurgical industry press, 1994).

Dual Phase Steel Development at TANGSTEEL and Its Application

M. S. Xia[†], C. Y. Liu, L. X. Liu, G. L. Li, G. Zhao, B. Q. Ai and X. H. Wang
HESTEEL TANGSTEEL R&D Center, Tangshan, Hebei, 063016, China
[†]*E-mail: xmswaterloo@163.com*

This paper investigates the development of dual phase steel in terms of chemical additives used; their impact on the resulting microstructure and tensile property; as well as their fatigue property. For study of fatigue damage; S-N tests were undertaken according to GB/T 3075-2008 for DP600 with thickness of 1.0mm; Dynamic strain-stress curves for DP590 were obtained by fitting the data with Swift-Hockett-Sherby constitutive equation; the forming limit curve (FLC) traditionally used to evaluate formability is not sufficient to predict potential edge crack, hence, the edge stretchability is studied hole expansion test that can emulate the stress/strain conditions formed during hole flanging under hole expansion test. The result of the Dynamic strain-stress curves obtained has shown that DP590 could achieve a maximum strain rate as high as 1000 /s, and with this increase of the strain rate, the strength also increases accordingly. This opens up dual phase (DP) steel for use as the first generation automotive steels apart from martensitic steel.

Keyword: Dual phase steel; Product development; Dynamic strain-stress curves.

1. Introduction

To increase the passenger safety and decrease the gas consumption, more and more high strength steels are applied in the car body manufacturing. Dual phase (DP) steel is one of the popular candidates for automotive makers because of such following characteristics as low yield strength ratio, high initial work hardening rate and good uniform elongation etc. Its stable bake hardening and non-aging property also makes it to be an ideal substitute for bake hardening (BH) steels. In recent years, automotive industry has gained great momentum and the annual car output reaches as high as 27 million in 2017 which push the robust demand for automotive steel. Many steel companies home and abroad are devoted to the research and development of automotive steels.

HBIS (Hebei Iron and Steel Group) TANGSTEEL started hot commissioning automotive steel production lines at Jan 2015 which includes one PLTCM, one CAL and three CGLs. The development of this three years has enabled TANGSTEEL to be an important automotive steel maker and supplier. The newly-built lines are devoted to high strength steel production. After three years' development, TANGSTEEL has become a qualified steel supplier for many car makers. Its product mix includes 1st generation steels and some of the 3rd generation steel like quenching and partitioning steel. In this paper, dual phase steel will be detailed in terms of production and application.

2. Line Characteristics and Chemistry

The product positioning of new plant is mainly on high strength steels. Its equipment represents the most advanced level of current metallurgy technology. For example, the cooling rate for the rapid cooling section of CAL can reach as high as 100 °C/s (the strip thickness is 1.0 mm) which makes the lean chemistry design for dual phase steel at CAL possible. For lower steel grades with tensile strength of 780 MPa or below, low carbon content (≤ 0.10 % weight percentage) and the addition of Mn, Si is preferred. The addition of Mn is to improve the hardenability of residual austenite and is beneficial to keep martensite at room temperature. Si can play the roles like expanding the intercritical temperature range, promoting the ferrite recrystallization and suppressing the formation of cementite during the austempering period [1]. In order to improve strip stretchability, bendability and flangeability, the elements like Al, Nb and Mo will be added individually or in combination during steelmaking. It has been studied that Mo is effective to get the beneficial annealed texture and homogenize the microstructure [2]. As for steel grades with tensile strength of 980 MPa or above, low carbon content is always matched with the addition of Mo, Cr and B [3]. On some applications, medium carbon content is taken to meet both strength and manufacturing requirement. The chemistry for investigated dual phase steels is shown in Tab.1.

Table 1. Chemistry design of investigated dual phase steels.

Steel grade	Chemistry
DP450	C(≤0.10%), Mn, Si
DP590	C(≤0.10%), Mn, Si
DP780	C(≤0.10%), Mn, Si
DP980	C(≤0.10%), Mn, Mo, Cr
DP1180	C(0.15-0.20%), Mn, Si, Cr, Nb, Ti

3. Microstructure and Tensile Property

The optical microstructure of DP450-DP1180 is shown in Fig. 1. The samples were sanded, polished and etched with 2 % nital. It can be found in Fig. 1 that the increase of steel strength is concurrent with the increase of martensite content and the decrease of ferrite content. In the matrix of DP450, DP590and DP780, the matrix is predominantly composed of ferrite and martensite. The martensite content varies from 15 % to 30 %. During production, the parameters at annealing process can be adjusted to attain some bainite to lessen the harmful effect caused by the marked hardness difference between ferrite and martensite. SEM microstructure observation of DP780 is shown in Fig. 2. It can be found that the matrix consists of ferrite, martensite and certain amount of lath bainite. DP980 is completely composed of ferrite and martensite, the content of which is 55 % and 45 % individually, as shown in Fig. 2 (b). Such microstructure composition results from the addition of Mo, B and Cr. For example, Mo addition can suppress the recrystallization of deformed austensite and grain growth during hot rolling process and as a result reduce the grain size of the final product. It is beneficial for the microstructure homogenization to improve the bendability and flangeability. As for DP1180, ferrite volume fraction is about 10% and the rest phase is martensite. Tensile strain-stress curves for different grades are shown in Fig. 3. It can be found for the investigated steels that the strength is propotioanl to martensite content. All curves show continuous yielding behavior.

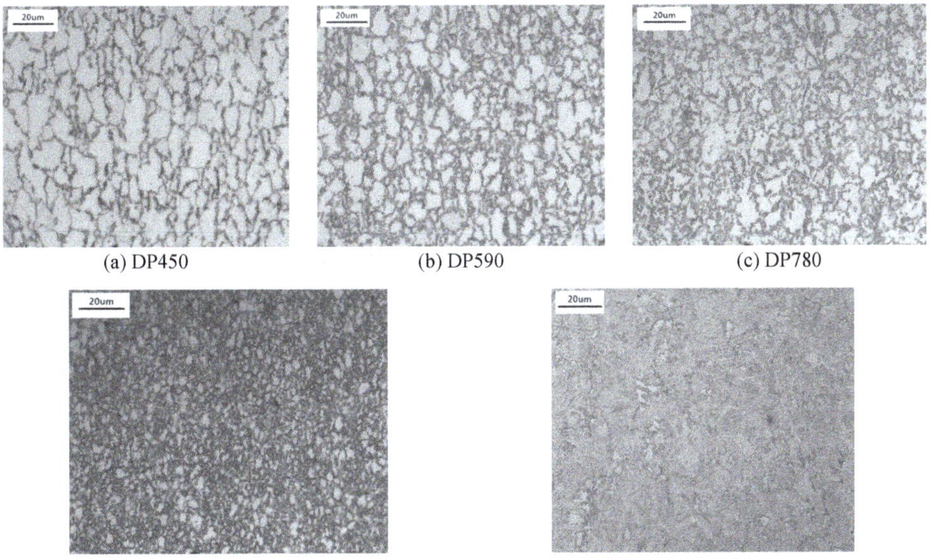

(a) DP450 (b) DP590 (c) DP780

(d) DP980 (e) DP1180

Fig. 1. Microstructure of dual phase steels DP450-DP1180 (Optical microscope).

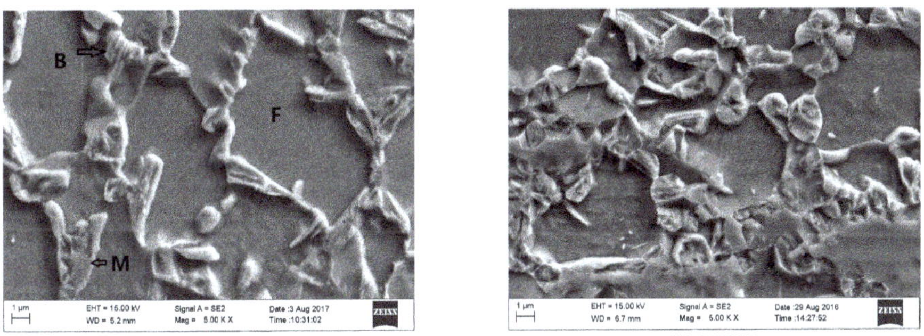

(a) DP780 (b) DP980

Fig. 2. Scanning electronic microstructure (F: ferrite, M: martensite, B: bainite).

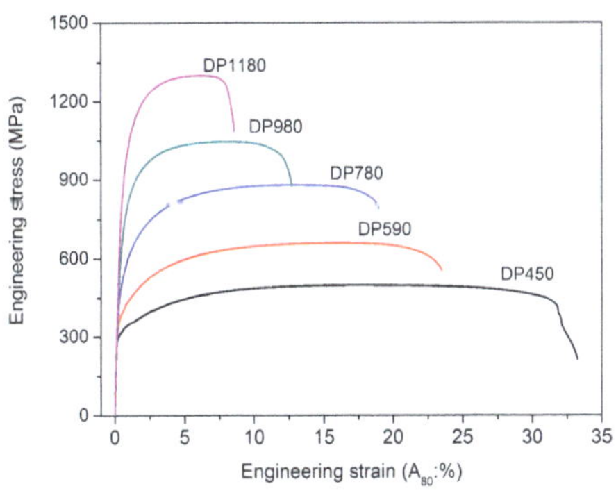

Fig. 3. Engineering stress-strain curves of dual phase steels (orientation 90°).

4. Data for Customers' Application

TANGSTEEL has built database necessary for the car body design and components' property evaluation. These data include static and dynamic strain-stress curves, fatigue property and welding process window etc.

4.1. *Fatigue property*

Fatigue damage is so critical for automotive structural parts which are usually underwent alternative stress load. Because of its typical microstructure composition, dual phase steel exhibits different failure characteristics from conventional steels like HSLA. The microcrack usually initiates from the interface between ferrite and martensite and propagates along the interface. S-N results tested according to GB/T 3075-2008 for DP600 with thickness of 1.0mm is shown in Fig. 4. The fatigue strength is 217 MPa which is about 62 % of its yield strength (350 MPa). Generally speaking, fatigue life with the cycles less than 2x106 shows a linear relation to stress amplitude.

$$\log N = a + b\sigma \tag{1}$$

where, *N*: fatigue life, *a* and *b*: linear fitting coefficient, *σ*: stress amplitude (MPa).

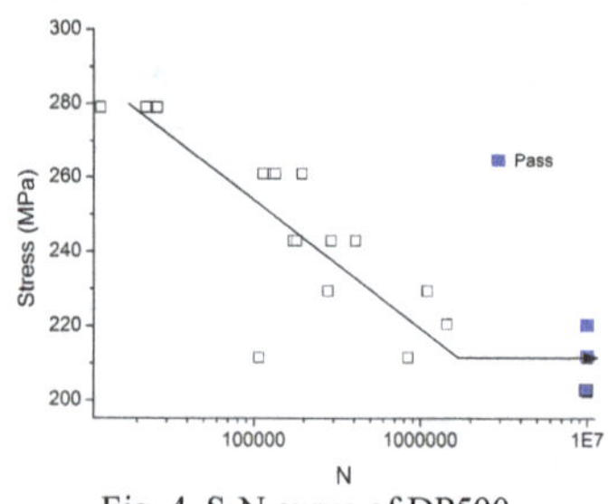

Fig. 4. S-N curve of DP590.

4.2. *Dynamic strain-stress curves*

Dynamic strain-stress curves for DP590 are shown in Fig. 5 in which the maximum strain rate is as high as 1000 /s. It can be found that with the increase of the strain rate, the strength also shows an obvious increase. The curves were obtained by fitting the data with Swift-Hockett-Sherby constitutive equation. The technique for recording the data can be referred to the standard ISO26203-2-2011or GB/T 30069.2-2016.

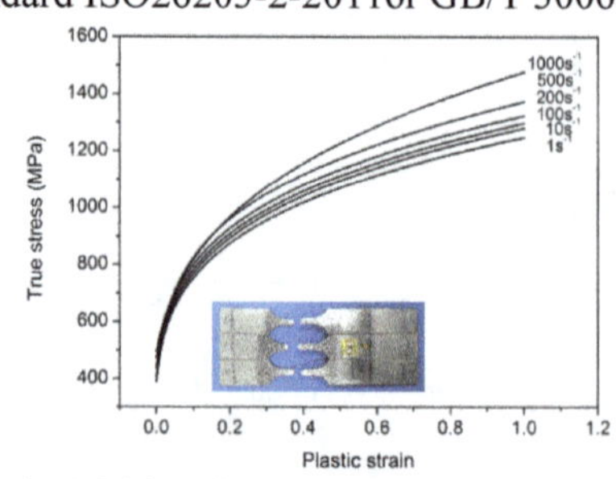

Fig. 5. High strain rate tensile curves of DP590.

4.3. *Hole expansion ratio*

Dual phase steels will be underwent flanging to some degrees in its applications which is different from low carbon Al-killed steels and high strength low alloy steels. The cracks always initiate from edges because of flanging and lead to the parts rejection. The parameters governing the dual phase steel flangeability can be phase constitution and its morphology, the hardness difference between ferrite and martensite and edge quality after cutting etc [4]. However, the forming limit curve (FLC) traditionally used to evaluate formability is not sufficient to predict potential edge crack because FLC represents the limit of stretching but not bending. Therefore, the edge stretchability is studied using hole expansion test that can be emulate the stress/strain conditions formed during hole flanging. The hole flange stretchability can be evaluated by the following formulae.

$$\lambda=(d-d_0)/d_0 \tag{2}$$

Where d and d0 (mm) are final and prehole diameters respectively.

The hole expansion ratio for different steel grades is shown in Fig.6. The relationship between hole expansion ratio and steel strength is similar to the results from Sadagopan [5]. The higher the strength, the lower the hole expansion ratio. The higher strength mainly results from the higher martensite volume fraction which provides more sources for cracking during deformation.

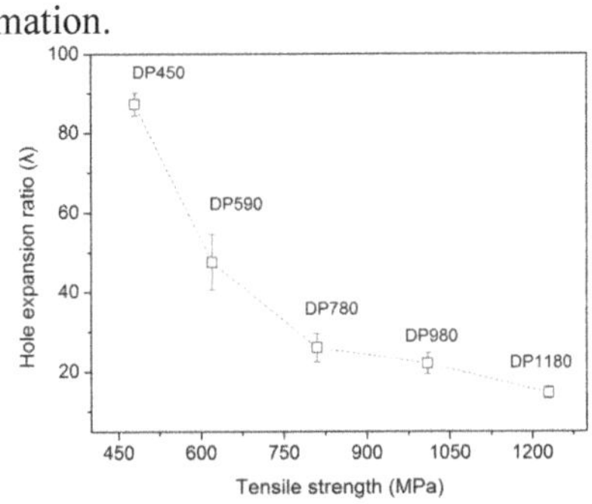

Fig. 6. The relationship between hole expansion ratio and tensile strength of dual phase steels.

5. Typical Applications

TANGSTEEL has made great progress in the development of automotive steels. Dual phase steel is one of them and has been widely applied in many structural parts as shown in Fig. 7-9. And forward product development can be carried out to meet the customer specific requirement. The front bumper assembly is an example. Based on the process which includes slitting, hole punching, roll forming, seam welding, bending and cutting, DP980 with low carbon content and microalloying elements was developed which shows good bendability and weldability.

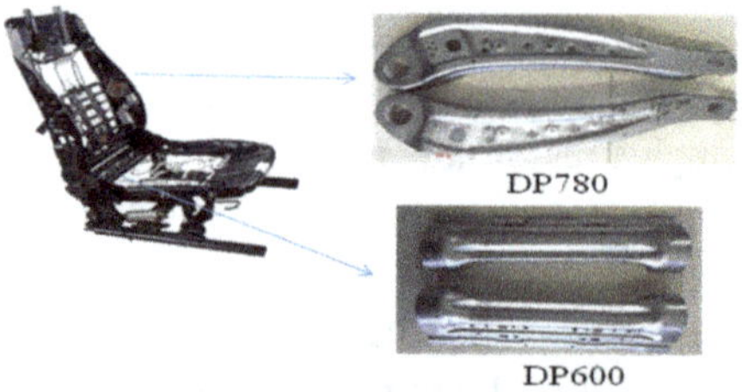

Fig. 7. Car seat assembly (DP600 and DP780).

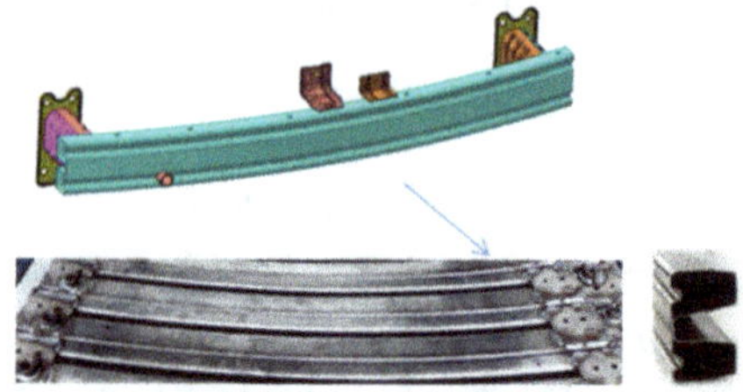

Fig. 8. Front bumper assembly (DP980).

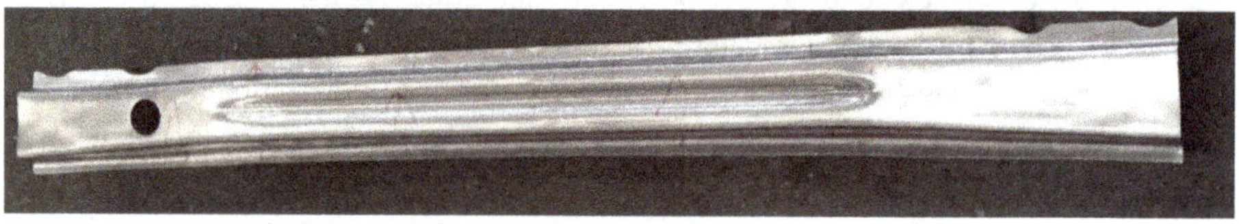

Fig. 9. Rear reinforcement beam of with cold forming (1180DP).

6. Conclusion

Currently, TANGSTEEL is capable of producing the first generation automotive steels apart from martensitic steel. Dual phase steels have been widely applied by many OEMs. Database for these steels is well constructed according to the customer demand.

References

1. J. Drumond, O. Girina, J. F. daSilva Filho, N. Fonstein and C. A. Silvade Oliveira, Effect of silicon content on the microstructure and mechanical properties of dual-phase steels, *Metallography, Microstructure, and Analysis* **1**, 217 (2012).
2. J. Y. Kang, H. C. Lee and S. H. Han, Effect of Al and Mo on the textures and microstructures of dual phase steels, *Mater. Sci. Eng. A* **530**, 183 (2011).
3. X. P. Shen and R. Priestner, Effect of boron on the microstructure and tensile properties of dual phase steel, *Metall. Trans. A* **21,** 2547 (1990).
4. Y Zhang, Y Liu and X Wu, Fracture behavior of a dual phase sheet steel in hole expansion, in *ASME 2007 International Mechanical Engineering Congress and Exposition* (San Diego, USA, 2007).
5. S. Sadagopan and D. Urban, Formability characterization of a new generation of high strength steels (AISI/DOE road program, www.autosteel.org), (2003). https://www.osti.gov/servlets/purl/1001171

Effect of Overaging Temperature on Microstructure and Mechanical Properties of Cold-rolled DP980 Steels

Y. Y. Xiao[a,†], H. Zhan[b], L. Cui[c] and J. C. Jin[d]
Technology Center of Ma'anshan Iron and Steel Company, Ma'anshan, China
E-mail: [a,†]xiaoyangyang1002@126.com, [b]40588205@qq.com,
[c]cuilei1219@163.com, [d]jinjiachun1988@163.com

The ULVAC-CCT-AY-II annealing simulator was used to simulate the production process of 980MPa grade dual phase (DP) steel. The characteristics of microstructure in experimental steel under different overaging temperature were observed by using scanning electron microscopy and transmission electron microscopy, and mechanical properties were measured by uniaxial tensile tests. With the increasing of the overaging temperature, the martensite content changes little, and the microstructure of the martensite is gradually changed from lath to island. After the overaging temperature increased to 340 °C, the martensite begins to decompose and some carbides precipitate. With overaging temperature increase, the tensile strength of the experimental steel decreases monotonically and the yield strength decreases first and then increases. The experimental steel has optimum mechanical properties when the overaging temperature is 290 °C, and it's PSE can reach 13.8 GPa•%.

Keywords: Dual-phase steel; Continuous annealing; Overaging Temperature; Microstructure and properties.

1. Introduction

The increasing strict demand for automobiles of lower weight, emission and higher safety has been pushing the application of lightweight and high strength automobile steel [1, 2]. Dual-phase steel, based on phase transformation strengthening, featured by its low yield ratio, high initial work hardening rate, good combination of strength and ductility, etc., has evolved into one of the advanced high strength steel for auto sheet forming [3]. Among the advanced high strength steels for automobiles, dual-phase steel is one of the most studied and widely used high-strength steels. Currently, only a few suppliers in China have the stable supply capacity of cold-rolled DP980 steel. With the development of automotive lightweight, the application of cold-rolled dual-phase steel with higher strength will become wider. But at present, domestic DP980 has problems such as high cost, difficulties in hot and cold rolling, low pass rate, etc [4]. Steel producers have to kept making efforts to improve the cost effectiveness, weldability and process feasibility of DP980 due to those widely recognized problems in DP980 production [5, 6] .Cold-rolled dual-phase steels are generally produced by continuous annealing which mainly includes dual-phase hearting, two-stage cooling and isothermal overaging treatment. The role of the overaging process is to reduce the hardness of martensite and improve the comprehensive mechanical properties by martensite tempering in duplex steel [7].

In this work, the effect of overaging temperature on microstructure and mechanical properties of cold-rolled DP980 was studied by using continuous annealing simulator and raw material for the experiment was cold-rolled DP980 by industrially produced.

2. Experimental Method

2.1. *Experimental steel and its chemical composition*

The test material in this paper is industrially produced cold rolled DP980 with a thickness of 1.4 mm. The chemical composition is shown in Table 1. The Microstructure of test material after cold-rolling is shown in Figure 1.

Table 1. Chemical Composition of the experimental steel (wt.%).

C	Si	Mn	Al	Cr	Nb	Ti	Fe
0.07~0.13	0.3~0.5	2.0~2.8	0.03~0.06	0.2~0.5	0.02~0.05	0.02~0.04	Balance

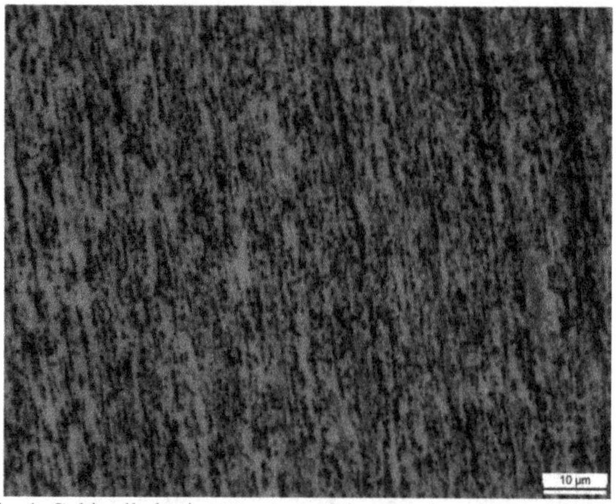

Fig. 1. Cold-rolled microstructure of the experimental steel (500×).

2.2. *Annealing simulation process*

Rectangular samples of 250×450 mm were taken from the cold-rolled sheet for continuous annealing simulation aiming at studying the effect of overaging temperature on the microstructure and mechanical properties of cold-rolled DP980 steels. The simulation was done on the continuous annealing simulator ULAVC-CCT-AY-II. Firstly, heating the samples to the Annealing temperature of 790 °C and kept for 120 s, then slowly cooling the samples to Start Cooling temperature of 690 °C followed by a fast cooling to Flash Cooling temperature. The Flash Cooling temperature and Overage temperature was set to 260 °C, 290 °C, 310 °C and 340 °C. The specific experiment numbers and parameters were shown in Table 2.

Table 2. Continuous annealing process parameters and number of the experimental steel.

No.	Strip speed (m/min)	Annealing temperature T1(°C)	Stating cooling temperature T2(°C)	Flash cooling temperature T3(°C)	Overaging temperature T4 (°C)
1	120	800	690	260	260
2	120	800	690	290	290
3	120	800	690	310	310
4	120	800	690	340	340

2.3. *Microstructure and mechanical property analysis*

Metallographic specimens taken from the annealed samples were grinded, polished, etched by 4 % nital and then observed under metallographic microscope AxioVert.A1 from German company Zeiss and FE-SEM FEI QUANTA 600. Specimens with the typical microstructure were further grinded and processed by twin-jet electro-polishing device before observation under TEM TECNAI G220 for finer microstructure analysis. Meanwhile, standard A80mm tensile specimens were taken from the annealed samples along normal of the rolling direction according to GB T/228-2002. Tensile testing machine Zwick/roell 050 was used to analyze the specimens' mechanical properties.

3. Experimental Result and Discussion

3.1. *Effect of the overaging temperature on the microstructure of DP980 steels*

The microstructures taken by SEM of the experimental steel at different overaging temperature were shown in Figure 2. It can be seen that there is obvious unevenness in the microstructure of the experimental steels. The convex part is the harder phase of martensite and the concave part is the soft phase of ferrite. The martensite morphology gradually changes from lath to island with the increasing of overaging temperature, and the martensite island grain boundary changes from clear and smooth to a finer and densely distributed polygonal structure. After low overaging temperature treatment at 260 °C, the lath martensite morphology is relatively obvious (Figure 2(a)). The edge of the martensite island in the microstructure gradually becomes blurred as the overaging temperature increases. The martensite island of the experimental steel began to decompose and some of the particulate carbides precipitated when the overaging temperature was increased to 340 °C (Figure 2(d)).

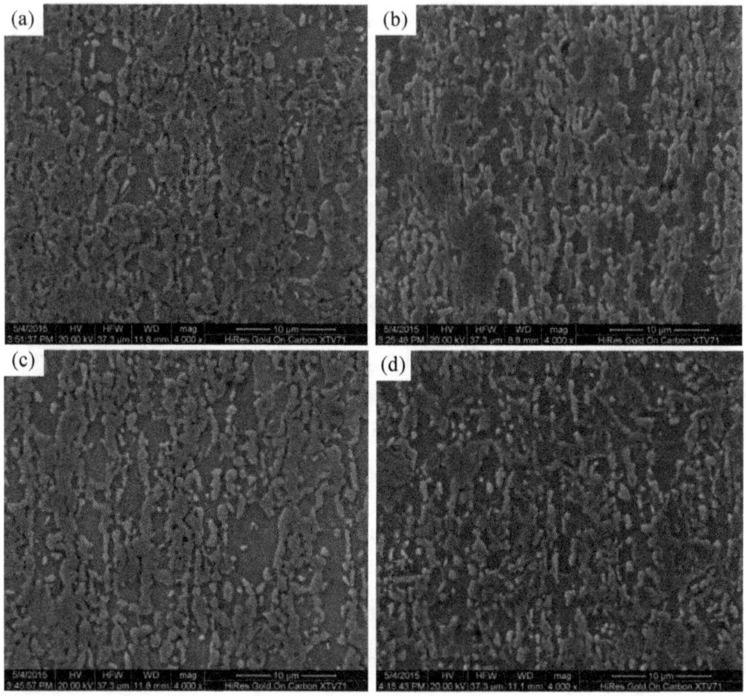

Fig. 2. SEM micrographs of the experimental steel under different overageing temperature: (a) 260 °C, (b) 290 °C, (c) 310 °C, (d) 340 °C.

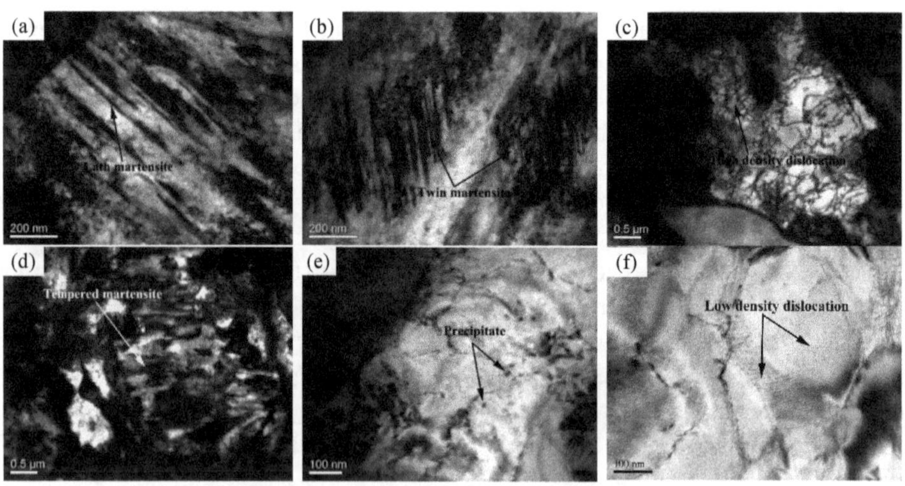

Fig. 3. TEM micrographs of experimental steel overaging at 260 °C and 340 °C.

The transmission electron micrograph of experimental steel under the condition of overaging temperature at 260 °C was shown in Figure 3(a)-(c), and overaging of 340 °C was shown in Figure 3(d)-(f). As shown in Figure 3 (a)-(b), the structure of martensite lath is relatively clear, and there are also twin martensite characteristics in the microstructure when the overaging temperature is low. And there is almost no carbide precipitation at low overaging temperature, so it shows Carbon-riched twin martensite has a good tempering stability. Comparing Figure 3(c) with Figure 3(f), we can find that the dislocation density in the ferrite matrix is greatly reduced with the overaging temperature increasing. As shown in Figure 3(d) and Figure 3(e), the martensite exhibits a tempered characteristic and part of carbides precipitated when the overaging temperature is 340 °C.

3.2. *Effect of the overaging temperature on mechanical properties of DP980 steels*

The yield strength, tensile strength, uniform elongation (AU) and elongation at break (AT) of the experimental steel at different overaging temperatures were shown in Table 3. The engineering stress-strain curves of the experimental steel under different overaging temperatures were shown in Figure 4. From the stress-strain curves, it can be seen that the stress-strain curves of the continuously annealed samples at different overaging temperature exhibit continuous yield behavior without yield extension.

It can be seen from Figure 5 and Table 3, with the increasing of overaging temperature, the tensile strength tends to decrease monotonously, and the highest tensile strength reaches 1122.4 MPa. The yield strength decreases first and then increases, but the overall is improved. The uniform elongation is basically about 10 %. The general trend of the yield ratio is also increased with the increasing of the aging temperature. When the overaging temperature is 290 °C, the experimental steel has a best comprehensive mechanical properties, and its PSE (AT×σb) reaches 13.82 GPa•%.

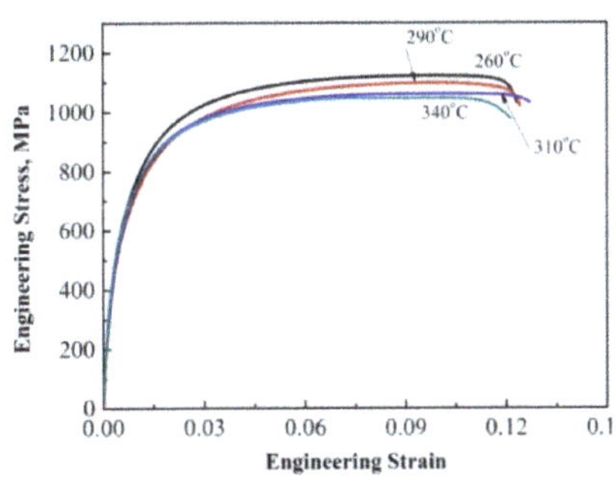

Fig. 4. The engineering stress-strain curves of experimental steel under different overaging temperature.

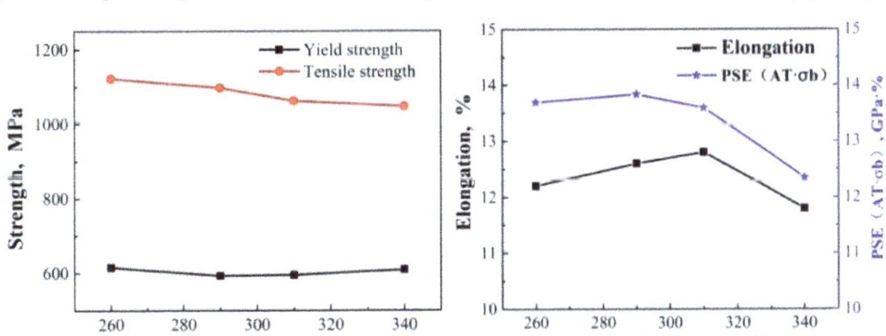

Fig. 5. Mechanical properties of experimental steel under different overaging temperature.

Table 3. Mechanical properties value of experimental steel with variable overaging temperature.

Overaging temperature	σ_s(MPa)	σ_b(MPa)	AT(%)	AU(%)	AT×σ_b(GPa%)	AU×σ_b(GPa%)	σ_s/σ_b
260 °C	615.5	1122.4	12.2	9.6	13693.3	10775.0	0.55
290 °C	592.7	1097.3	12.6	10.2	13826.0	11192.5	0.54
310 °C	595.3	1061.4	12.8	10.6	13585.9	11247.7	0.56
340 °C	608.9	1046.9	11.8	9.4	12353.4	9840.9	0.58

3.3. *Discussion*

The mechanical property of dual-phase steel depends on its microstructure, and the microstructure depends on the chemical composition and production process. The microstructure of experimental steel changes as the overaging temperature changed. The mechanism of dual-phase steel's low yield strength is that there is a large amount of supersaturated carbon inside the martensite which will generate a large number of movable dislocations in ferrite by cause severe lattice distortion and interaction with ferrite. When the overaging temperature increases, the martensite became soft, and a large number of dislocations at the interface between martensite and ferrite cancel each other out. The fine carbide precipitated inside the ferrite pinned the dislocation and destroyed the original stress state and increases the difficulty of dislocation motion caused an increase in yield strength. With the overaging temperature increases, the precipitation and purification of ferrite, and the martensite may partially soften, in addition that carbide particles precipitate and grow at the grain boundaries, the strength of the martensite matrix decreases. These can all lead to a decrease of tensile strength.

At low overaging temperature (~260 °C), the temper softening phenomenon of the experimental steel is not obvious. The morphology of martensite is clear and the carbon enrichment is high. So the experimental steel has a high yield strength and tensile strength when the overaging temperature is low. The strength of the ferrite matrix decreases with the increasing of the overaging temperature, but when the overaging temperature rises to 310 °C, fine carbides are precipitated in the microstructure which increases the difficulty of dislocation motion, and then the yield strength begin to rise. When the overaging temperature reached 340 °C, the regular lath morphology of martensite is gradually destroyed, and due to the outward diffusion of carbon atoms and the gradual elimination of internal strength, martensite softens. So the tensile strength is gradually reduced. When the overaging temperature is 290 °C, the experimental steel has a best comprehensive mechanical property with a higher yield and tensile strength, higher yield ratio and elongation.

4. Conclusion

(1) With overaging temperature increase, the tensile strength of the experimental steel decreases monotonically and the yield strength decreases first and then increases

and the highest tensile strength reached 1122.4 MPa. The experimental steel has an optimum mechanical properties when the overaging temperature is 290 °C, and it's PSE can reach 13.8 GPa•%.

(2) At low overaging temperature (~260 °C), the morphology of martensite is lath-like. With the increasing of the overaging temperature, the martensite content in the experimental steel was almost constant, the microstructure of the martensite was gradually changed from plate to island. The martensite of the experimental steel began to decompose and carbide began to precipitation when the overaging temperature reached 340 °C.

(3) The overaging temperature has a significantly effect on the morphology of martensite and the precipitation of carbides, which ultimately affects the mechanical properties of the experimental steel. With the increasing of overaging temperature, the dislocation density in the ferrite matrix is greatly reduced. When the overaging temperature exceeds 310 °C, the martensite exhibits tempering characteristics, the tensile strength gradually decreases.

References

1. K. Q. Wang, R. D. Liu and X. W. Xu, The reseach on microstructure and properties of 980 MPa grade cold-rolled dual-phase steel, *Technologies in Ansteel* **3**, 21 (2012).
2. S. Q. Ying, Y. H. Zhang and G. X. Cao, Application of lightweight automobile and high strength steel, *Automobile Technology & Material* **10**, 11 (2012).
3. M. T. Ma and B. R. Wu, *Dual-phase Steel: Physical and Mechanical Metallurgy*, 2nd ed. (Metallurgical Industry Press, 2009).
4. H. Zhan, Y. Zou and K. Zhou, Investigation on continuous annealing process and strength-toughness behavior of high strength cold-rolled DP980 steel, *Steel Rolling* **34**, 69 (2017).
5. X. D. Zhu, The effect of continuous annealing on mechanical properties on cold-rolled martensitic steel sheets, *Iron and Steel* **47**, 84 (2012).
6. S. Kim and S. Lee, Effects of martensite morphology and volume fraction on quasi-static and dynamic deformation behavior of dual-phase steels, *Metall. Mater. Trans. A* **31**, 1753 (2000).
7. X. Zhu, L. Wang, S. K. Ji and C. Wang, Effect of overaging temperature on microstructure and mechanical properties of Si-Mn Dual-phase steels, *T. Mater. and Treat.* **24**, 50 (2003).

The Influence of Pre-compression Deformation on Tension Property of DP590

M. T. Ma[1,†], D. G. Kim[2], G. Y. Wang[1] and G. Liu[1]
[1]*China Automotive Engineering Research Institute Co., Ltd., Chongqing, 401122, China*
[2]*POSCO Global R&D Center, Incheon, 21985, Korea*
[†]*E-mail: mingtuma@126.com*

The effects of pre-compression deformation 5 %, 10 % on tension property of DP590 are investigated. The pre-compression deformation 5 %, 10 % were carried out on specific fixture in order to prevent buckling deformation. The samples were machined to 5 time standard tensile samples after pre-compression deformation. Tensile flow curves were measured. The metallographic structures are observed by OM. The results show that the tension yield strength after 5 % pre-compressed strain decreases, the n value increases when the strain is 2 %, but the n value decreases when the strain is 2-6 %, and the total elongation increases slightly. When the pre-strain was 10 %, the n value had a similar variation, but the yield strength increased slightly and the total elongation decreased significantly. The effects of pre-compression strain on mechanical properties of dual phase steel are discussed considering microstructure, fracture morphology, Bausinger effect and working hardening of dual phase steel.

Keywords: Dual Phase steel; Pre-compression deformation; Bauschinger effect; Flow characteristic.

1. Introduction

The development of automotive industry leads to the problem of energy consumption, emission and safety. The lightweight is the direct and effective way to energy-saving and emission-reduction [1-2]. The lightweight must ensure safety, which leads to development and application of advanced high-strength steel and ultrahigh-strength steel [3-4]. The dual phase steel belonging to the first generation advanced steel has been widely used for automotive lightweight, and a great deal of research has been done on dual-phase steel [5-6]. A lot of high-strength steel has been applied to automotive stamping parts, which is subjected to tensile, compressive or compound stress state during the forming and using. Therefore, it is significant to study the effect of pre-deformation on property of material and the function of the parts and predicting the failure mode and position of the parts. However, there are few reports in this respect. The purpose of this paper is to study the effect of pre-compression deformation on the quasi static tension properties of DP590 and the cause of influence produced has been discussed.

2. Steels and Methods Used in the Testing

The testing steel is hot rolled dual phase steel DP590, with a thickness of 5.3 mm. The composition analyzed by CAP6300 is listed in Table 1. The microstructure of test materials with three directions was observed by microscope Leica-DMI3000 type. The pre-deformation sample is cut from parallel to the rolling direction, which size is 38×5.3×160 mm (Fig. 1).

Table 1. The composition of hot rolled dual phase steel DP590.

Material	C	S	Si	Mn	Cr	P	Ti	Al
DP590	0.078	0.0021	0.77	1.20	0.408	0.0066	0.0031	0.10

The total pre-compression shrinkage volume is 5% and 10%, and the pre-compression is the stroke control volume. The tension samples were machined as 12.5×5.4×60 mm after 5 % pre-compression. The tension samples were machined as 12.5×5.6×60 mm after 10 % pre-compression (Fig. 2). The pre-compression was carried out in a special fixture (Fig. 3). The tension test was carried out on MTS810. The mechanical properties and flow curves of various strain histories were measured, and tensile fracture was observed by SEM (FE quanta 20 type).

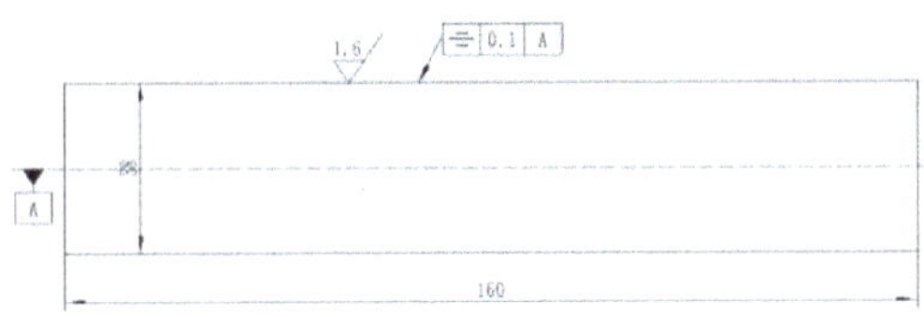

Fig. 1. The size of pre- compression sample (Thickness: 5.3 mm).

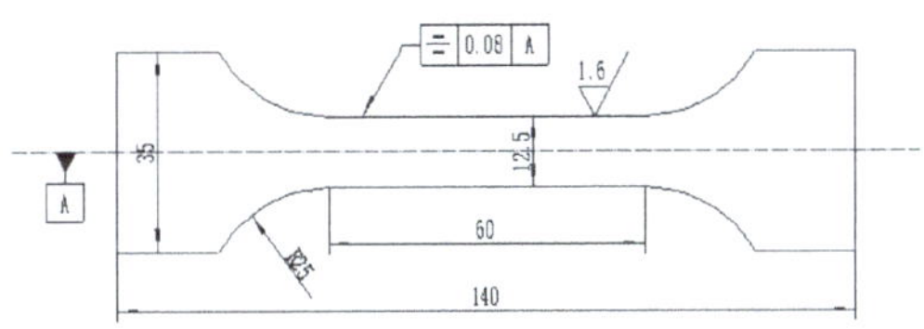

Fig. 2. The tension sample after pre-compression (pre-compression 5 %, d=5.4 mm; pre-compression 10 %, d=5.6mm).

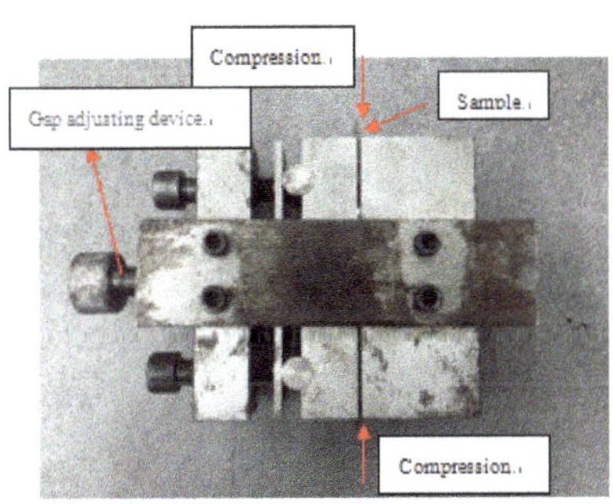

Fig. 3. The special fixture of compression.

3. Test Result and Analysis

3.1. *Mechanical properties after pre-compression*

The mechanical property of DP590 was listed in Tab. 2. The result shows that the mechanical properties of the three directions are basically the same, only a slight discrepancy in the r value. It indicates that the mechanical properties of the steel are isotropic. The microstructure of the three directions was observed. The section schematic diagram and the microstructure were shown in Fig. 4. It shows that the microstructures of three sections were composed of the ferrite matrix and martensite, and the distribution of the two phases was similar, and thus the mechanical properties of the three directions were similar. Therefore, the samples for researching the effect of pre-compression on the properties of dual phase steel are cut paralleling to rolling direction.

Table 2. Mechanical property of DP590.

Code	E, MPa	R0.2, MPa	Rm, MPa	n1	n2	r	A80, %
DP590-0	193803	418.8	36.3	0.16	0.17	0.63	5.4
DP590-45	195741	419.5	27.7	0.15	0.18	0.89	5.0
DP590-90	199696	425.5	34.5	0.13	0.17	0.79	4.5

Note:

E is Young's modulus, n1 is calculated between 0.2% and 2% strain, n2 is calculated between 2% and 6% of strain , r is Plastic strain ratio.

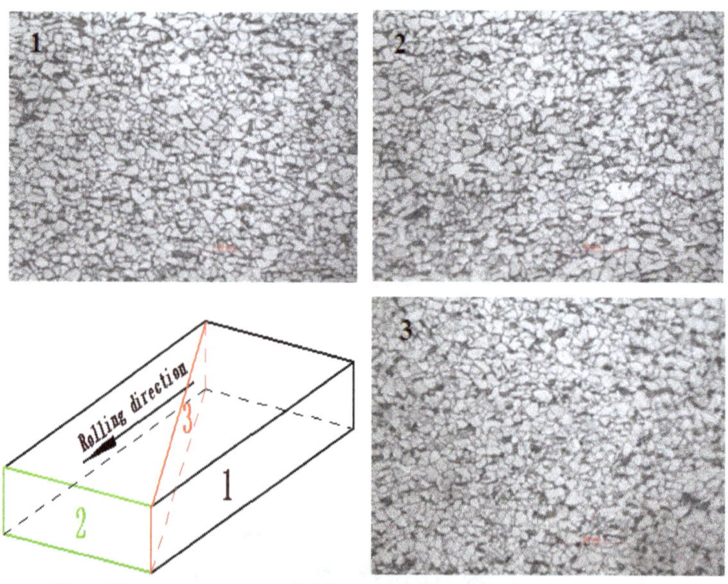

Fig. 4. The microstructures of different directions and observation sections.

3.2. *Mechanical properties after pre-compression*

The mechanical properties of DP590 after pre-compression strain 5 % and 10 % are shown in Table 3. The result shows that after pre-compression strain 5 % during the tension, the modulus of elasticity remains the same, yield strength decreases slightly, tensile strength variation is a little, n1 obviously rises, n_2 decreases, and the elongation has some increasing. The decrease of yield strength and the rise of initial working hardening rate n_1 are related to the Bauschinger effect. The Bauschinger effect refers to some dependence of the flow behavior (flow stress and working hardening rate) of metal alloy on the strain history and stress state, i.e. in the plastic deformation, the flow stress (e.g. compression) of the reverse deformation is often less than the flow stress (e.g. tensile) of the initial direction strain. But after the stress reversal, the work hardening rate has a short rising [7]. This description is in accordance with the experimental results in Table 3. The increase of elongation may be related to the shortening of the gauge length and the data of the same material A80 should be lower than that of A50. 5% the pre-compression strain has little effect on the tensile strength of subsequent tension. The tensile strength of the dual phase steel can be calculated by the mixture law [5].Compared with the original sample, the thickness of the sample with 5% pre-compression increased by 0.1mm. Depending to 60 gauge lengths, the strain of sample only has 2%. At this amount of deformation, the martensite morphology is basically maintained, but the apparent volume fraction of martensite slightly increases. As a result, the tensile strength only has a little variation. The mechanical properties of the 10% pre-compression sample shows that the modulus of elasticity is the same as the raw material, the yield strength and tensile strength increase obviously, the n1 rise, the n2 decline, the total elongation drop, which is related to the apparent work hardening in the material after 10% pre-compression, and this kind of work hardening can be inherited, leading to the subsequent reverse deformation, the Bauschinger effect of the is covered by large work hardening. Work hardening can cause the yield strength and tensile strength of the material to rise, and the ductility of the material to decrease.

Table 3. Mechanical property of DP590 after pre- compression 5 % and 10 %.

Code	E,MPa	$R_{0.2}$,MPa	R_m, MPa	n_1	n_2	A_{50},%
DP590-5 %	195728	398.0	639.1	0.27	0.10	31.4
DP590-10 %	198830	438.6	662.9	0.26	0.08	18.6

The flow curves of raw materials, pre-compression 5% and pre compression 10% are shown in Figure 5. The diagram shows that after the compression pre-strain, the shape of the flow curve changed obviously, the uniform elongation rate decreased obviously, and the working hardening rate after yield increases obviously.

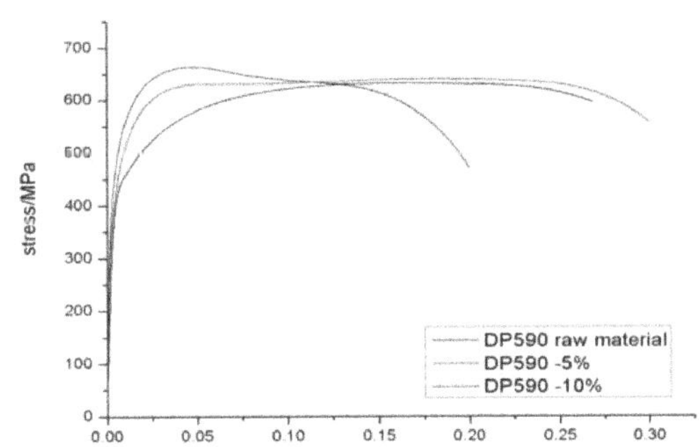

Fig. 5. The flow curve of DP590 with raw material, pre- compression 5 %, 10 %.

The metallographic structures of samples with different pre-compression are shown in Figure 6. No more difference can be found. In order to further research the effect of pre strain on the tensile process, tensile fracture has been observed. Generally, the tensile fracture is divided into three zones: the fiber area, the radiation area and the shear lip from inside to outside. The three zones of the fracture of the double phase steel after stretching is shown in Figure 7.

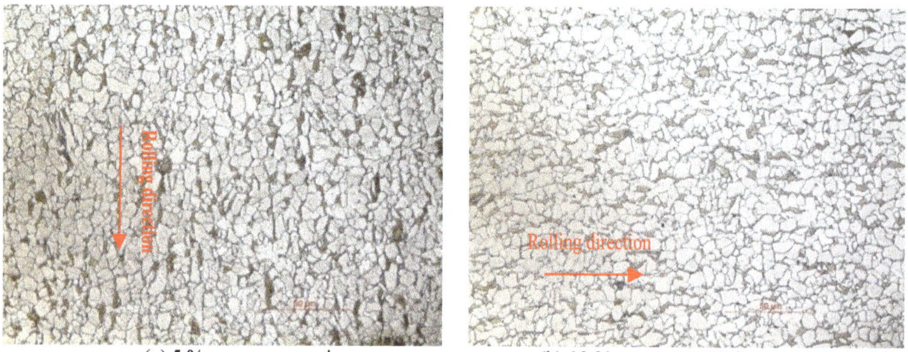

(a) 5 % pre- compression (b) 10 % pre- compression

Fig. 6. The microstructures of rolling direction after pre- compression.

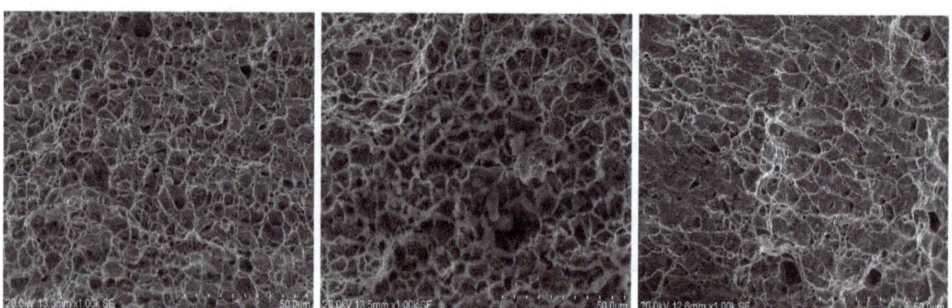

(a) Fibrous region (b) Radial region (c) Shear lip

Fig. 7. Tensile fracture morphology of three areas of DP590.

The fiber and radiation areas are composed of dimples. The fiber zone is the zone where the material begins to crack and gradually expand at the inclusion under the tensile stress of three axes. The radiation zone is a rapidly expanding zone after core cracking during stretching. Generally, the fracture morphology of the fiber zone is more obvious than that of the extended zone, such as shallow dimple and flat fracture. The shear zone is the area with large deformation before fracture, which is made up of stretched-out dimples. The deformation before the fracture is large and consists of dimples formed by tensile deformation. The comparison of fiber area and radiation area between three states of dual phase steel tensile fracture is shown in Figure 8.

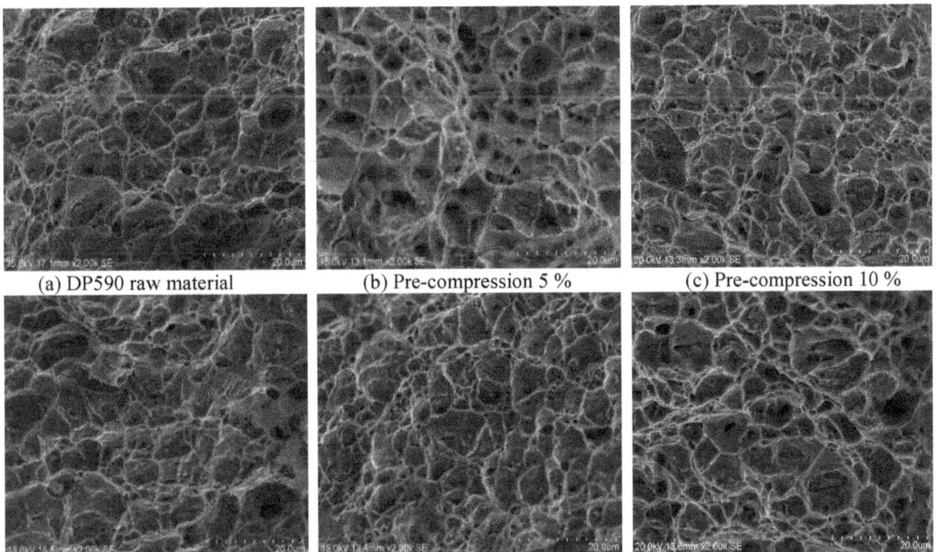

(a) DP590 raw material (b) Pre-compression 5 % (c) Pre-compression 10 %

(d) DP590 raw material (e) Pre-compression 5 % (f) Pre-compression 10%

Fig. 8. Tensile fracture morphology of DP590 with strain history: (a-c) Fibrous region, (d-f) Radial region.

The fracture morphologies of simples with different pre-compression are shown in Figure 8. Figure 8 shows that the fibrous region of the fracture of simples with different pre-compression is composed of dimples with inclusion at the bottom. The 5 % pre-compressed dimples are deeper, indicating more deformation in the fiber area under this pre-compressed strain. The dimples of simples with 10% pre-compression are shallow and thin, indicating lower elongation. The radiation area is the rapidly expanding area of the crack. The fluctuation of the dimples in both pre-deformation states is smaller and more flat than that in undeformed ones. The morphology of simple with 10% pre-compression has larger dimples, fewer tearing edges, indicating that the ductility of this part is lower. The fracture morphology and mechanical property test results are consistent. The pre-compression strain of 5 % endows the material with reasonable reinforcement. And in the fiber area and radiation area of the fracture, the fracture morphology shows good ductile. The ductility of the material with 10 % pre-compression decreased obviously. The fracture morphology shows poor ductile, which is related to large work hardening, obvious enhancement and work hardening heredity.

4. Conclusion

The mechanical properties of DP590 of hot-rolled dual phase steel are basically isotropic. The microstructures in the three directions are approximately the same. The yield strength of material after 5 % pre-compression decreases when it is tension, n value increases during the strain 0.2 %-2 %, n value decreases during the strain 2 %-6 %. When the pre-compression was 10 %, the n value had a similar change, but the yield strength increased slightly and the total elongation decreased significantly. The change of mechanical properties is closely related to the characteristics of martensite and ferrite composition of dual phase steel, and the changes of fracture morphological in fiber area and radiation area during stretching after pre-compression strain. Considering the comprehensive effect of Bauschinger effect and work hardening, the effects of different pre-compressive strains on mechanical properties are illustrated.

Acknowledgments

Work supported by grant U1564203 of the National Science Foundation.

References

1. M. T. Ma, H. L. Yi and H. Z. Lu, On the automobile lightweight, *Eng. Sci.* **11**, 02 (2009).
2. M. T. Ma and Y. S. Zhang, *Research progress in hot stamping of ultra high strength steel in Trace research on advanced automobile manufacturing technology*, ed. By C-SAE, Beijing: 15-75, (Beijing Institute of Technology Press, 2016).
3. M. F. Shi, Advanced High Strength Steel, Properties, Performances and Application, in *China-America Automotive Materials Seminar*, (Detroit, America, 2003).
4. M. T. Ma and M. F. Shi, Advanced high strength steel and application in automotive industry, *Iron and Steel* **70**, 68 (2004).
5. M. T. Ma and B. R. Wu, *Dual phase steel-physical and mechanical metallurgy*, 2nd edn. (Metallurgical industry press, 2009).
6. M. T. Ma, D. G. Kim, L. F. Song, Y. Zhao and G. Fang, Comparison research for spring backs of dual-phase steels with three different strength grades, in *Advanced High Strength Steel and Press Hardening in proceedings of 3 international conference on Advanced High Strength Steel and Press Hardening (ICHSU)*, (Xi'an, China, 2016).
7. M. T. Ma, Z. P. Duan and Y. Tomota, *Bauschinger effect in metal alloys and its application in industrial production*, (Metallurgical Industry Press, 1994).

Study on Low Temperature Brittleness and Hydrogen Embrittlement of Automotive Dual Phase Steel Sheets

G. Fang[1,2,†], L. F. Song[1], Y. Feng[1], J. P. Zhang[1] and M. T. Ma[1]

[1]*State Key Laboratory of Vehicle NVH and Safety Technology, Chongqing, 401122, China*
[2]*China Automotive Engineering Research Institute Co., Ltd., Chongqing, 401122, China*
[†]*E-mail: fanggang@caeri.com.cn*

Dual phase steels are widely applied in automotives due to its light-weight and safty potential. However, strength improvement brings in the embrittlement crisis. In this paper, low temperature brittleness and hydrogen induced delayed embrittlement of DP780 and DP980 were studied. The traditional secondary working brittleness test was replaced by layered lamination impact test to evaluate the low temperature brittleness. Four stress states and an acid environment were designed to analyze their hydrogen induced delayed embrittlement. The results indicate that the brittle to the ductile transition temperature of the two dual phase steels is between -40 °C to -50 °C and DP780 exhibits better ductility. Both the two materials show good anti-hydrogen induced delayed embrittlement properties.

Keywords: Dual phase steels; Low temperature brittleness; Hydrogen induced delayed embrittlement.

1. Introduction

The rapid development of the automobile industry has greatly improved people's travel mode and quality of life, bringing about the development of the whole society. However, with the increase of vehicle production and quantity, problems such as energy crisis, environmental pollution and automobile accidents are becoming more and more prominent [1]. In recent years, a series of grade evaluation and regulations involving fuel consumption, emission and automobile safety have been introduced, and the automobile industry has been driven rapidly into the era of light-weight, intelligence and security.

The use of the advanced high strength steel instead of traditional mild steel is an important direction for the development of vehicle lightweight. As one kind of advanced high strength steels, dual phase steel is mainly composed of martensite and ferrite. It has the characteristics of high strength and strong working hardening ability, which has been widely used in the key parts and assembly of automobile parts, car body, suspension and steering system [2, 3] and so on. China has a vast territory and a complex climate and environment. Whether the toughness of the material will decrease obviously when the car is in service under extreme cold or acid conditions, and whether the phenomenon of hydrogen induced fracture in the acid condition and complex driving condition needs further study [4].

The impact energy absorption of the low temperature brittle material decreases with the temperature decreasing. When the temperature is lower than a certain temperature, the impact absorption work decreases obviously, and the material changes from the ductile state to the brittle state. The delayed fracture is a phenomenon of brittle failure after a certain period of time under the action of static stress. It is the result of the interaction of material, environment and stress, and is a form of hydrogen induced material deterioration. With the increase of strength grade of high strength steels for automobile, the low temperature brittleness and delayed fracture sensitivity of high strength grade steel have become an important factor restricting the popularization and application.

2. Test Materials and Test Methods

2.1. *Test materials*

The testing materials are DP780/1.6 mm and DP980/1.4 mm which are widely applied to automobile body's key safety parts. Their chemical compositions and mechanical properties are shown in Table 1 and Table 2 respectively.

Table 1. Chemical compositions of DP780 and DP980 (wt %).

	C	Mn	Si	Cr	Mo	V	Ni	Cu
DP780	0.070	2.4	1.05	0.1	0.03	0.01	0.1	0.1
DP980	0.115	2.6	0.35	0.05	0.01	0.01	0.1	0.1

Table 2. Mechanical properties of DP780 and DP980.

	$R_{p0.2}$/MPa	R_m/MPa	r	n	A/%
DP780	489	791	0.16	0.9275	22
DP980	745	1023	0.09	0.7875	12

2.2. *Test methods*

2.2.1. *Low temperature brittleness test*

Steel sheets Secondary work embrittlement test (SWE) is a test method to evaluate the ability of automobile steel plates to withstand certain impact load at different low temperatures after stamping. But for some high strength steels and super high strength steels, the edge of sample cup is easy to crack because of the influence of its formability. As shown in Fig. 1, the round billet was stamped into the cup-shape specimens. However, lots of samples cracked before the drawing position had been reached. It seriously prohibited the following cold impact process. Thus this test method is not suitable for embrittlement evaluation of the DP steels.

In the present work, lamination impact method of the specimens at different low temperatures was applied to investigate the ductile and brittle transition temperature (DBTT for short) of DP steels. The transition temperature of impact energy A_{KV} of material is defined as the ductile brittle transition temperature.

Fig. 1. Round billet and cup.

Fig. 2. Cracked cup.

The low temperature impact test of materials is evaluated by Charpy impact test with reference to ISO148-1:2009 [5] and GB/T229-2007 [6]. According to the standard requirements, the size of the impact sample (length × height × thickness) is 50 mm × 10 mm × 10 mm, and the smallest sample thickness is 2.5mm. However, in the body-in-white, the thickness specifications of the DP steels are generally less than 2.0 mm. On the basis of ensuring the length of 55 mm and the height of 10 mm, the thickness of the sample can be as close to 10 mm as possible by lamination. During the test, a special fixture is used to make the center line position of the thickness of the specimen keep in the center position of the pendulum blade, ensuring the accuracy and consistency of the test results. V shape notched specimen was machined perpendicular to the rolling direction of the sheets as shown in Fig. 3 and 7 plates were laminated and closely fixed by two bolts. The laminated samples with the thickness 9.8 mm were cool treated by dry ice + alcohol and impacted within 5 s. The test temperature range is -60 °C to room temperature. Before formal test, the effect of the lamination number on the impact energy A_{KV} was studied and the result was depicted in Fig. 4.

The test results show that the impact work of the same material has a linear relationship with the number of different laminations, so the linear conversion of different thickness can be carried out, and the plates with different thickness can be compared.

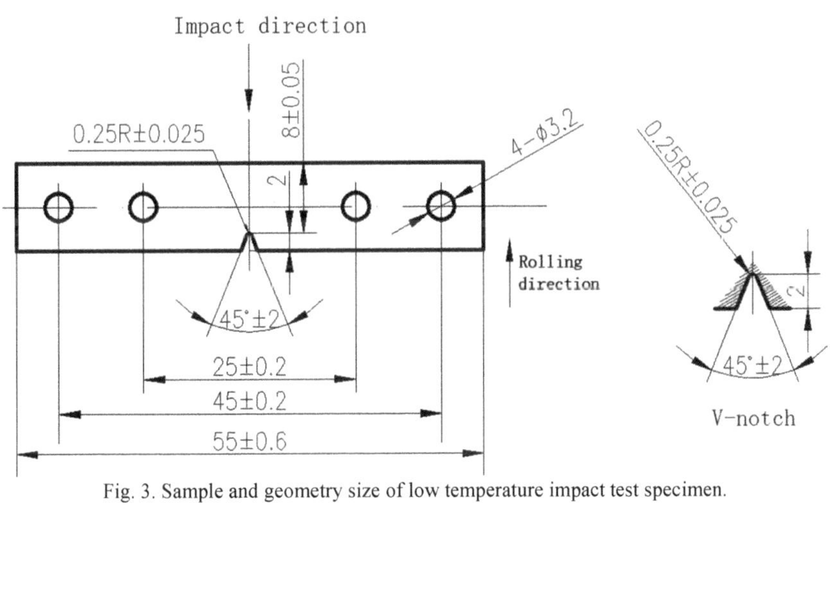

Fig. 3. Sample and geometry size of low temperature impact test specimen.

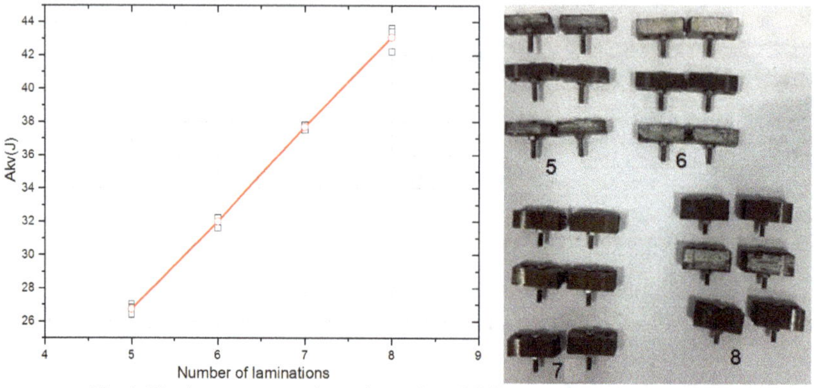

Fig. 4. The impact test results and samples of different thickness specimens.

2.2.2. *Hydrogen induced delayed fracture test*

The hydrogen induced delayed fracture is a phenomenon of sudden brittle failure after a certain period of time under static stress. This phenomenon is a kind of environmental embrittlement that occurs in the interaction of the environment and stress. The delayed fracture of high strength steel is usually related to hydrogen in the environment. Therefore, it is called hydrogen induced delayed fracture.

This experiment uses continuous bending + acid soaking test method to test and evaluate the hydrogen embrittlement performance of the tested steels. The specific test conditions are: 1) clamp the specimens as shown in Fig. 5 with the fixture as depicted in Fig. 6. Different stress state was achieved by controlling the span. Four stress groups were set and each group included three specimens. 2) Place the specimens with various stress state into the acid solution with 1N standard HCL solution and keep for 300h. 3) Take out the specimens and investigate the crack.

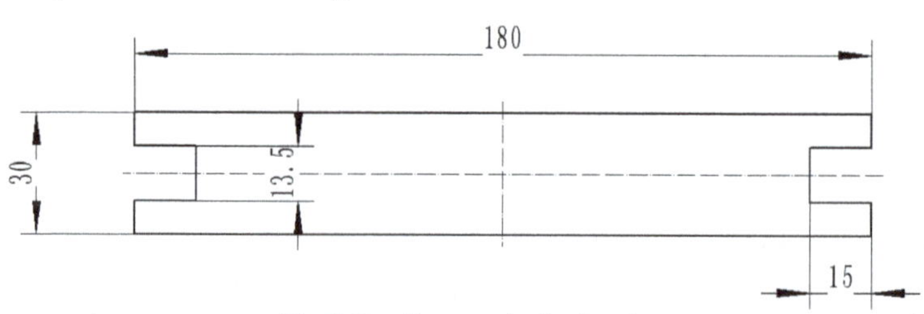

Fig. 5. Bending sample size (mm).

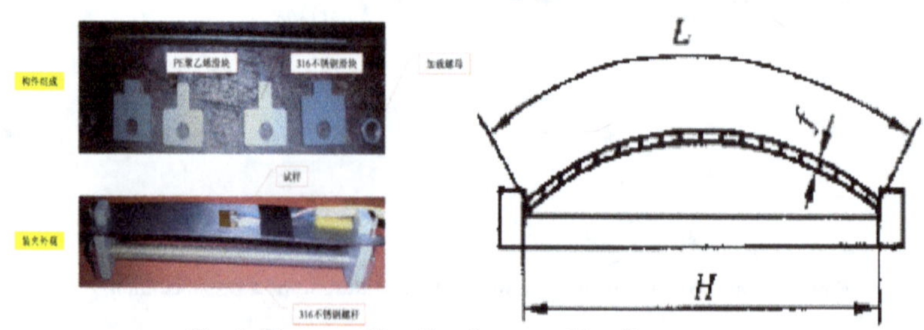

Fig. 6. Diagram of bending fixture and bending process.

3. Test Results and Analysis

3.1. *Low temperature impact test of material*

In order to evaluate the toughness and brittleness of metal sheet DP780/1.6 mm and DP980/1.4 mm at low temperature, a series of temperature impact tests were carried out from 20 to -60 to observe the impact absorption energy and the change of fracture morphology with the temperature. The relationship between the impact energy and the temperature is shown in Fig. 7. It can be seen that with the decrease of temperature the impact energy of the material decreases, and the obvious turning point appears between -40 °C and -50 °C, indicating that the steel plate appears obvious transition from toughness to brittleness at the temperature range.

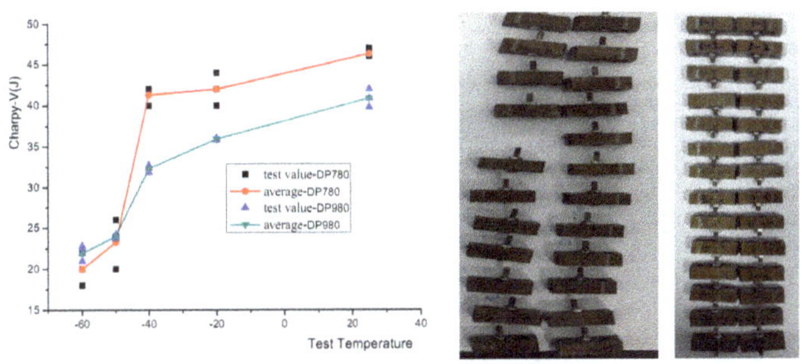

Fig. 7. DP780/1.6 mm and DP980/1.4 mm low temperature brittleness test results and test samples.

The ductile brittle transition temperature t_T of materials is often used as an important design criterion and criterion for brittle fracture prevention in practical engineering. The test [7~9] shows that the regression analysis of the relationship between the impact power A_{KV} and the temperature changes with the Eq. (1) Boltzmann function, with a small residual and good correlation, can well describe the relationship between A_{KV} and the temperature.

$$A_{KV} = \frac{A_1 - A_2}{1 + \exp\left(\dfrac{t - t_T}{t_R}\right)} + A_2 \tag{1}$$

In Eq. (1), A_1 is the lower platform energy. A_2 is the upper platform energy. t_T is the ductile and brittle transition temperature. The higher the t_T is, the more easily the ductile brittle transition happens. t_R is the range of the ductile brittle transition temperature zone of the material. The fitting result of A_{KV} is shown in Table 3 that the upper and lower platform energies of DP980 are smaller than DP780, showing a poor low temperature impact toughness.

Table 3. Fitting parameters for A_{KV}.

Materials	A_1/J	A_2/J	t_T/°C	t_R/°C
DP780	21.67	43.22	-42.73	6.05
DP980	22.97	36.39	-43.24	5.38

Macroscopic and microscopic analysis of fracture can determine the location of crack source, crack propagation direction, fracture properties and material properties. The fracture surfaces of DP780 with the impact temperature of 20 °C, -20 °C, -40 °C, -50 °C, and -60 °C respectively shown in Fig. 8 from the left to the right. The macro analysis of the fracture is to have a preliminary understanding of the fracture. It provides a basis for judging the failure mode of the specimen from the upper map. At the temperature of -50 °C and -60 °C there is almost no shear lip and fiber area on the DP780 fracture surface. The area of brittle fracture area is very large, and there is no macroscopic plastic deformation. The fracture surface is relatively flat and shows more crystalline form. It can be concluded that the fracture properties of DP780 steel at these two temperatures are brittle fracture. With temperature decreasing, the shear lip and fiber area gradually decrease and disappear, and the fiber area is gradually replaced by the radiation zone.

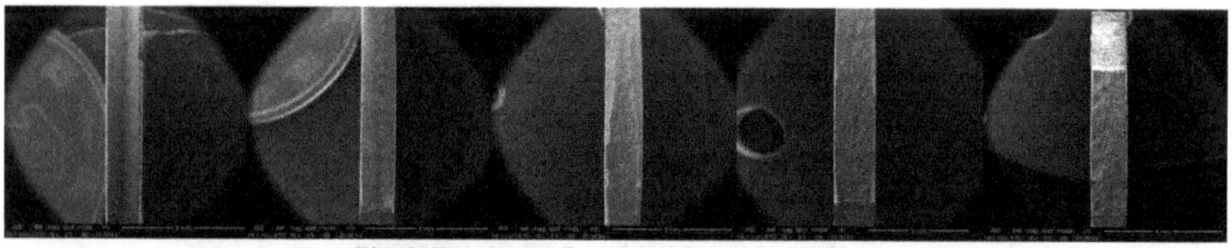

Fig. 8. Fracture at five temperature for DP780.

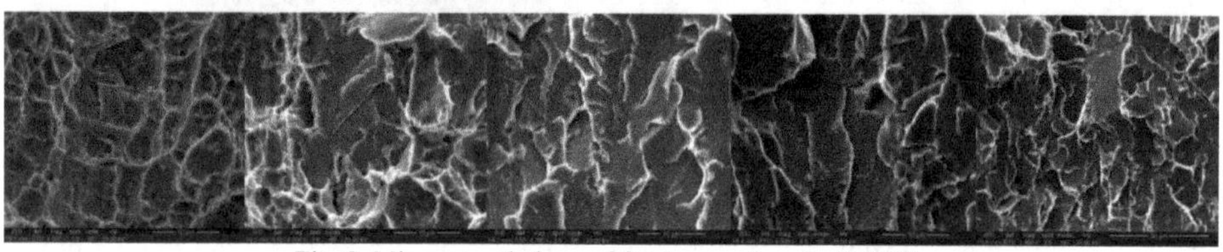

Fig. 9. Micro-morphology of fracture surface for DP780.

As can be seen from Fig. 9 with the magnification of 4000x, with the decrease of temperature, the dimples and tearing edges of the fracture gradually decrease, the fracture morphology at 20 °C, 0 °C and -20 °C is microporous, and the fracture is covered with dimples of different sizes, but with the decrease of temperature, the dimples become smaller and shallower. In general, ductile deformation is in domain. The obvious cleavage steps and tongue like patterns can be observed at -40 °C, -50 °C and -60 °C. The cross section is crystallized and there are many small planes with strong reflective light. It shows that the plastic deformation is very small, and there are not too many signs of macro plastic deformation. In this situation, brittle fracture is in charge.

3.2. *Hydrogen induced delayed fracture test*

This test uses a special bending fixture and controls the bending strain by adjusting the span of the sample, as shown in Fig. 10. During the bending process, the DIC device was used to detect the strain at the minimum bending radius of the sample. Four stress groups were set and each group included three specimens. The specimens with various stress state were placed into the acid solution with 1N standard HCL solution an kept for 300h before investigation.

According to the results of Table 4, two kinds of materials exhibit good hydrogen resistance delayed fracture under the stress level and hydrogen environment.

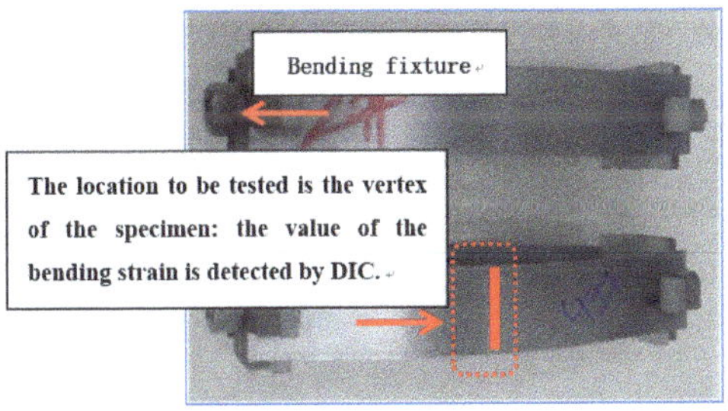

Fig. 10. Bending loading process.

Table 4. Hydrogen induced delayed fracture test results of DP780 and DP980.

Material	Stress/MPa	Test conditions	Test result
DP780/ 1.6 mm	320	Soak in a 300 hour acid environment	Pass
	360	Soak in a 300 hour acid environment	Pass
	400	Soak in a 300 hour acid environment	Pass
	440	Soak in a 300 hour acid environment	Pass
DP780/ 1.6 mm	400	Soak in a 300 hour acid environment	Pass
	450	Soak in a 300 hour acid environment	Pass
	500	Soak in a 300 hour acid environment	Pass
	550	Soak in a 300 hour acid environment	Pass

4. Conclusion

(1) For high strength steel and ultra-high strength steel, the toughness and brittle transition temperature test of material cannot be tested by secondary working brittleness test because of its poor formability. The ductile and brittle transition temperature of the material can be effectively obtained by the method of laminated impact test.

(2) By using a reasonable lamination connection, the impact power of the same material with different layers of the same material is linear, so the linear conversion of different thickness can be made to make the plate with different thickness comparable.

(3) The ductile to brittle transition of DP780 and DP980 occurs at -40 °C to -50 °C It is found that the upper and lower platform energies of DP980 is smaller than that of DP780 when Boltzmann function is used to analyze the relationship between the impact work A_{KV} and the temperature change. The ductility at low temperature of DP980 is worse than that of DP780.

(4) DP780 and DP980 exhibit good anti-hydrogen induced delayed fracture properties.

Acknowledgments

The authors are grateful for the financial supports from National key research and development plan (2017YFB0304405).

References

1. J. G. Li, J. D. Zhang, H. L. Huang and S. Y. Zhang, Research status and development trend of high strength steel for automotive use, *Mater. Rev.* **26**, 397 (2012).
2. M. T. Ma and B. R. Wu, *Dual phase steel-physical and me-chanical metallurgy*, 2nd edn. (Metallurgical Industry Press, 2009).
3. G. Pocock, High strength steel use in Australia, Japan and the US, *Struct. Eng.* **84**, 27 (2006).
4. F. X. Diao, Y. Zhang, L. Ni and R. S. Yang, Study on low temperature brittleness of 1000 MPa grade dual phase ultra high strength steel for automobile use, *J. Plasticity Eng.* **24**, 161 (2017).
5. EN ISO 148-1:2009, Metallic materials-Charpy pendulum impact test-Part 1: Test method.
6. GB/T 229-2007, Metallic materials-Charpy pendulum impact test method.
7. Y. D. Zhou, S. Zhou, C. Ding, J. W. Zhang, H. T. Li and J. Fang, Research on low temperature toughness and brittleness of QP980, *J. Plasticity Eng.* **22**, 103 (2015).
8. American Society for Testing and Materials, Standard Test Methods for Wetting and Drying Compacted Soil-Cement Mixtures, West Conshohocken: ASTM, USA, ASTM D559-03.
9. Y. M. Wu, Y. Q. Wang and Y. J. Shi, Effects of low temperature on properties of structural steel, *J. Univ. Sci. Techn. Beijing* **11**, 442 (2004).

Researches on Secondary Work Embrittlement Properties of High Strength IF Steel

L. B. Pan†, S. Z. Zhou, F. Y. Sun, X. D. Liu and Z. J. Deng
Wuhan Iron and Steel Company, Baosteel
No 28, Yejin Avenue, Qingshan District, Wuhan, Hubei, 430080, China
†E-mail: panlibo@baosteel.com; pan97181@126.com

To study the secondary work embrittlement of High Strength IF steel, the chemical compositions, ductile-brittle transition temperature (DBTT) and microstructures of typical steel samples with different chemical compositions were analyzed. The results showed that there were large amounts of fine and irregular-shaped FePTiNb precipitates distributing around grain boundary area for steel samples with high DBTT. Usually Nb and Ti were supposed to precipitate combining with elements C, N and S. The emergence of FePTiNb precipitates was caused by excess addition of Nb and Ti. The optimized calculation was carried out according to the precipitating theory and sequence of Nb and Ti in High Strength IF steel, and then the optimized High Strength IF steels with reduced Nb and Ti addition were produced in production line. There was almost no FePTiNb precipitate in the optimized products, and the DBTT also dropped dramatically. It could be concluded that besides high content of P, the content of Nb and Ti in High Strength IF Steel also played important roles in secondary work embrittlement.

Keywords: High Strength IF Steel; Secondary Work Embrittlement; DBTT; Optimization.

1. Introduction

Interstitial free steel is one of the important products for automotive industry. According to safety requirement and light-weighted strategy, high strength IF steel becomes an alternative to produce some auto parts with solution strengthening elements addition such as P, Mn and Si based on IF steel. However, because of the absence of solute interstitial elements and P addition, it is more sensitive to secondary work embrittlement (SWE). Ductile-to-brittle transition temperature (DBTT) test is a common method to evaluate SWE. Due to car safety, many car producers made requirements for the application of high steel IF steel in regard with DBTT. Usually, DBTT should be no higher than -40 °C according to those requirements. For the past decades, some researchers tried to find out the main mechanism how chemical composition and process parameter affect DBTT, and how to improve SWE performance. James [1] found that P content is far higher in grain boundaries area than other area in high strength steel which made it easy to subject to crack. It was showed [2-4] that certain amount of B addition could be beneficial to SWE. Study in literatures [5-6] indicated that coiling temperature decreasing could result in DBTT drop. Fusato [7] developed a new type of high strength IF steel with good SWE performance through grains refining. Debao Wang [8] tested the DBTT of high strength IF steel and analyzed the effect factors on DBTT. From most of the viewpoints, the

DBTT should be easily affected by P content and B addition for diminishing the effect of P. However, in some cases, even though the steel has more B and less P, DBTT is yet not acceptable, which means there will be some other factors having effect on DBTT instead of P and B. This study will focus on different chemical composition especially the content of Nb and Ti in high strength IF steel and analyze the correlation between DBTT and these elements.

2. Materials and Experiments

Two high steel IF steels A1 and B1 from the continuous annealing production line with different strength grades were investigated in this study to analyze chemistry effect on SWE. The chemical composition was shown in Table 1. Both materials had ultra-low carbon content, and Si, Mn and P were applied to provide high strength through solution.

Table 1. Materials chemical composition.

Steel No.	Thickness (mm)	Chemical Composition (wt. %)								
		C	Si	Mn	P	Nb	Ti	S	N	B
A1	1.5	0.0014	0.079	0.38	0.038	0.031	0.04	0.0074	0.0023	0.0001
B1	0.8	0.002	0.077	0.81	0.069	0.037	0.047	0.0061	0.0022	0.0011

To evaluate the secondary work embrittelment of these two steels, firstly steel samples were deep drawn to six-inch cup, and immersed in liquid bath with certain constant temperature. Then DBTT tests were conducted in a drop-testing machine by using testing criterion in accordance to the specifications in standard GB/T 24173-2009 (Testing Method of Secondary Work Embrittlement Property for Steel Sheet) procedure. The DBTT results for the two steel samples A1and B1 were -20 °C and -30 °C, respectively, which indicated that two steel were very sensitive to SWE with DBTT above -40 °C. Therefore, these two steels failed to meet the standard of car producer which required DBTT of high strength IF steel no higher than -40 °C. Microstructural analysis of the samples was conducted on Transmission Electron Microscopy (TEM) as shown in Fig. 1 and Fig. 2, respectively.

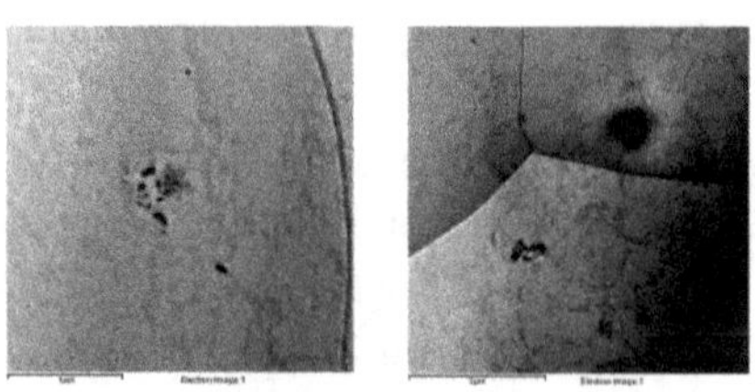

Fig. 1. TEM photos of steel A1.

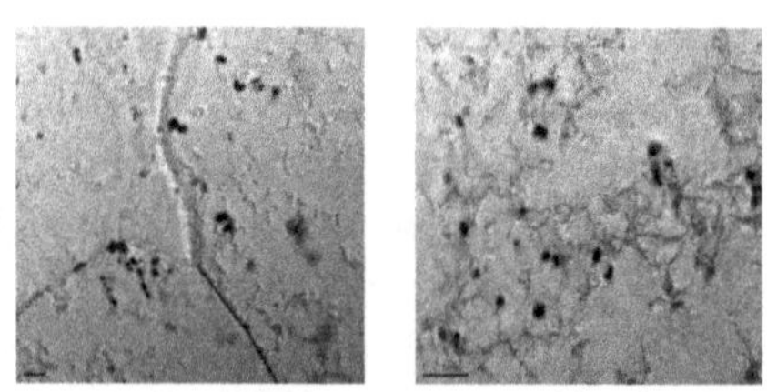

Fig. 2. TEM photos of steel B1.

3. Analysis and Discussion

According to DBTT results above, both steels were sensitive to SWE. Traditionally, P was regarded as a detrimental element to SWE sensitivity, while B addition was more likely to counteract the effect of P. However, comparing to Steel A1 with less contained

P and less B addition, Steel B1 seemed to have no big difference in SWE sensitivity even though there were more B addition supposed to reduce the negative effect of P, which means in this case P and B contents were not the most important factors to affect SWE sensitivity. In microstructure analysis, more attentions had been paid to the grain boundaries area to observe if there was much segregation or precipitation. From Figure 1 and Fig. 2, it could be seen that there was much precipitates around the grain boundary area for Both Steel A and B. Through engery-dispersive X-ray spectroscopy, these precipitates can be identified, and the result for steel A1 was shown in Fig. 3.

Similar result could be found for steel B1, it can be known that a large amount of fined precipitates distributing around grain boundary area were recognized as FePTiNb precipitates. These precipitates were in irregular shape, which became the form of P segregation, and made the materials sensitive to SWE. However, according to the design principle of high strength IF steel, P was supposed to be soluted in the High strength IF steel to effectively improve strength without affecting plasticity. It was known that in IF steel, Ti and Nb were added to stabilized C and N to obtain the absence of solute interstitial element. The occurrence of FePTiNb in this case indicated that probably there was too much Ti and Nb added in which part of them was enough to stabilize C, N. It was known that Nb and Ti were easily combined with different element such as C, N and S to form carbide or nitride precipitates. And these elements had different cohesive sequence according to their combination activity. Normally, In IF steel, if Nb and Ti were added, Nb will take the priority to be combined with C, and Ti will be combined with N and S. The aimed Nb and Ti contents could be calculated as as in Eq. (1) and Eq. (2).

$$Nb = 7.75 \times \%C \tag{1}$$

$$Ti = 3.42 \times \%N + 1.5 \times \%S \tag{2}$$

Based on this, in steel A1, the contents of C, N and S were 0.0014%, 0.0023% and 0.0074%, respectively; therefore, the ideal content of Nb should be about 0.011%, and the aimed content of Ti should be 0.019%. In the same way, it could be obtained that the ideal contents of Nb and Ti in steel B1 should be about 0.0155% and 0.0167%. Compared to the actual contents, the ideal Nb and Ti contents for both steels were far more less, which meant the high strength IF steels investigated in this study contained two much extra Ti and Nb according to the chemical composition provided in Table 1. And it resulted in formation of FePTiNb precipitates beside carbide and nitride (carbonitride) precipitates and dramatically increased the SWE sensitivity.

4. Material Optimization

According to analysis above, these steels could be optimized by reducing the contents of Ti and Nb while keeping similar content of the rest elements. The optimized steels for A1 and B1, named A2 and B2, respectively, were tried out in the same production line with similar process technology and parameters. The chemical composition of the optimized steels was shown in Table 2. During practical manufacturing in production line, the content for every element should be designed within a certain range; therefore, the contents of Ti and Nb should be a bit more than ideal content through the calculating method above considering the fluctuation. From Table 2, it can be seen that the contents of Ti and Nb in both A2 and B2 had been reduced. Moreover, the decreasing of Ti and Nb contents also made the steel products more costly.

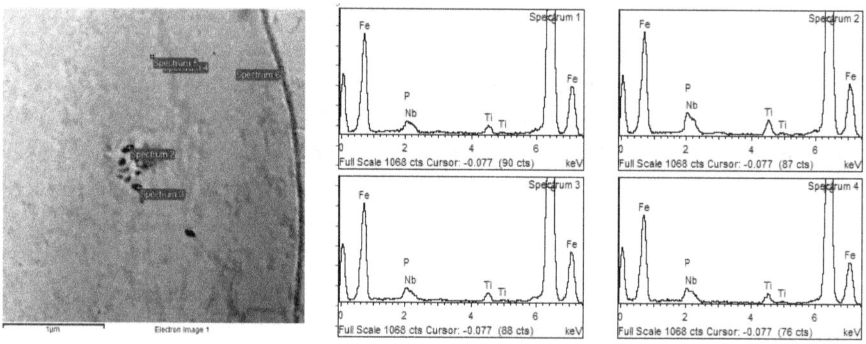

Fig. 3. Energy-dispersive X-spectroscopy result for steel A1.

Table 2. Chemical composition of optimized materials.

Steel No.	Thickness (mm)	Chemical Composition (wt. %)								
		C	Si	Mn	P	Nb	Ti	S	N	B
A2	1.5	0.001	0.074	0.364	0.033	0.017	0.022	0.007	0.0023	0.0002
B2	0.8	0.0022	0.081	0.63	0.066	0.021	0.024	0.0036	0.0016	0.001

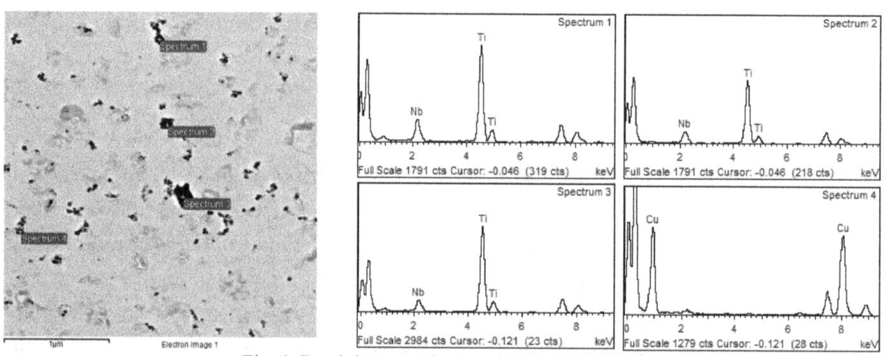

Fig. 4. Precipitates analysis result for steel A2.

DBTT test also was conducted in the same condition and procedure and the result indicated that there was an obvious drop on DBTT. The optimized steel A2 had DBTT at -45 °C while previous steel A1 at -25 °C, meanwhile, the optimized steel B2 also achieved the DBTT of -60 °C in comparison with previous DBTT at -30 °C. Microstructure analysis was performed on TEM in the same way, and it indicated that there was only a small amount of FePTiNb precipitates found in both optimized steels since there were still a bit excess Ti and Nb existing. Fig. 4 showed the precipitate analysis in steel A2 while Steel B2 showed the same case, which confirmed the previous conjecture about excess Ti and Nb content affecting DBTT. Most precipitates were carbides and nitrides. From above analysis, it can be concluded that DBTT is not only affected by P and B addition as described in previous researches, but also is strongly influenced by excess Ti and Nb content.

5. Conclusion

The secondary work embrittlement is one of the key properties for high strength IF steel. By analyzing microstructures and DBTT of high strength steels with different Nb and Ti contents, it can be concluded as follows.

(1) Excess Nb and Ti addition in high strength steel could cause the formation of FePTiNb precipitates, which will increase the SWT sensitivity and result in a high DBTT.

(2) Other than P and B element, the addition of Nb and Ti is also critical for high strength IF steel in terms of secondary work embrittlement properties.

References

1. J. R. Fekete, D. C. Strugala and Z. Yao Advanced sheet steels for automotive application, *JOM.* **1**, 17 (1992).
2. S. H. Li, Progress in research of High strength IF steel, *Shanghai metal.* **5**, 66 (2007).
3. U. Heidtmann, *International symposium for LC and ULC sheet steels.* **1**, 39 (Aachen, Germany, 1998).
4. N. Takahashi, Metallurgy of continuous-annealed sheet steel, *TMS-AIME.* 133 (1982).
5. N. T. Kyokai, *International Forum for Physical Metallurgy of IF Steels,* (Iron and Steel Institute of Japan, 1994).
6. K. G. Chin, *Developments in the Annealed sheet steel* (TMS, 1992).
7. Fusato KITANO, et al. New type of IF-high strength steel with superior secondary work embrittlement. *ISIJ Int.* **11**, 1042 (2001).
8. D. B. Wang, et al. Assessment and study on the secondary work embrittlement of IF steel. *Anhui Yeji.* **1**, 7 (2014).

Modeling & Simulation

Effect of Load Histories on the Quenching and Partitioning Process

Z. J. Wang[1], H. C. Geng[1], S. Q. Li[2], Y. S. Zhang[1], B. Zhu[1], Y. L. Wang[1,†], H. Q. Li[3] and T. Wu[3]

[1]*State Key Laboratory of Materials Processing and Die and Mould Technology, Huazhong University of Science & Technology, Wuhan, 430074, China*

[2]*School of Mechanical Science and Engineering, Huazhong University of Science & Technology, Wuhan, 430074, China*

[3]*Vision Hot Stamping Technology Co. Ltd., Dongguan, 523460, China*

[†]*E-mail: wangyilin@hust.edu.cn*

The hot stamped high strength steel parts are now widely used in the automobile industry, but the fully martensitic parts with low ductility cannot match the demand of crashworthiness. The quenching and partitioning (Q&P) process can improve the ductility of parts without great strength losing. Usually, the Q&P process are carried out under the static state and the load histories such as stress path and temperature history are not considered. In this paper, the effect of load histories on the Q&P process are investigated. The stress station during the stamping process can affect the martensite start transformation temperature Ms. An analytical equation linking the Ms and the principal stressed was used to predict the influence of the complex stress state on Ms. The changing of Ms then affect the volume of retained austenite and the carbon partitioning during the Q&P process. The effect of temperature loading history on the interface migration, carbon diffusion was investigated. The interface first moved from austenite plate to the martensite plate and then moved to the opposite direction. The heating rate during the partitioning process can greatly affect the microstructure and mechanical properties of the final part. The results of this paper can be referred to the industry application of quenching and partitioning process.

Keywords: Quenching and partitioning; Load histories; Interface migration; Retained austenite.

1. Introduction

In recent years, the demand for safety and energy saving in the automotive industry has been continuously increasing [1]. The hot stamping process of Advanced High Strength Steels (AHSSs) can effectively reduce the weight of parts with only a small loss in strength [2]. Moreover, the hot stamping of traditional boron steels leads to fully martensitic parts with very low post forming plasticity, which evidently reduces the overall mechanical properties and the application of hot stamping parts [3].

The quenching and partitioning (Q&P) process is currently being investigated to replace the direct quench in traditional hot stamping with the aim of improving the post-forming ductility of high strength structural parts. There are two Q&P approaches: one-step and two-step [4, 5]. This paper will focus on the two-step Q&P process because hot stamping can be more productive with this process. This Q&P process consists of quenching of the part to a temperature between Ms and Mf, the part is then heated to a partitioning temperature and held for a number of seconds when carbon diffuses from supersaturated martensite to austenite [6]. After partitioning, the part is quenched to room temperature, where a significant portion of austenite is stabilized [7]. The final martensite-austenite duplex microstructure exhibits high strength (about 1500MPa) and much greater elongation (14.8%) than typical hot stamped boron steel (6.6%) [8].

In this paper, hot stamping-quenching-partitioning processed U-shaped parts were obtained, and the full-process simulation was conducted. The effect of stress on Ms was implanted into the quenching simulation to obtain a more accurate volume fraction of retained austenite after the first die quenching. The effect of heating strategies on microstructure and mechanical properties was investigated in this paper. The micro-scale interface migration, carbon diffusion and retained austenite volume fraction simulations were coupled into the macro-scale partitioning simulation using a user-defined subroutine.

2. Method

2.1. *Material*

The chemical composition of the studied material was given in Table. 1. Fe makes up the balance, and the composition is provided by the material supplier. The blank was first fully austenitized at 930 °C for 5 minutes and then quickly transferred to the heated die to be stamped and quenched for 20 s (die was heated to the set temperature using heating rods). After the quenching process, the blank was transferred to a hot air furnace at 425 °C for partitioning and then finally quenched to room temperature. As depicted in Fig. 1, 7 elements were taken as representative elements to analyze and explore the carbon diffusion, interface migration and retained austenite volume change during the whole simulation.

Table 1. Chemical composition of the studied material.

C	Si	Mn	P	S	Cr	Nb
0.28	1.60	1.10	0.012	0.005	0.99	0.029

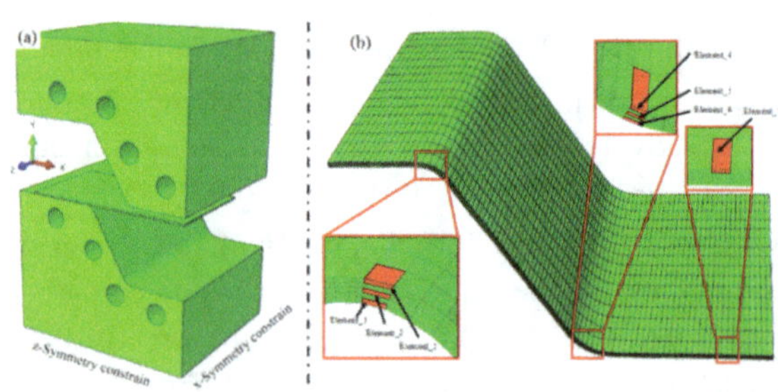

Fig. 1. a) 1/4 simulation model and b) location of elements with history output.

2.2. *Simulation method*

In this paper, the simulation includes four steps. The full-process simulation procedure is shown in Fig. 2. At the quenching simulation stage, the effect of stress on the Ms was evaluated using the following equation:

$$\Delta M_s = a\sigma_1 - b\sigma_3 \text{ , } \sigma_1 \geq \sigma_3 \tag{1}$$

where a and b are material constants and σ_1 and σ_3 are the principal stresses. The equation was implanted into the simulation using a user-defined subroutine 'quench.for'. The initial austenite phase volume and austenite domain were calculated in this stage. Finally, the simulative results of the quenching stage were transferred to the partitioning simulation. In this paper, the specimens were partitioned in a hot air furnace, and the temperature can be calculated using the following equation:

$$T(t) = T_{air} + (T_{init} - T_{air})\exp\left(-\frac{hA}{\rho c V}t\right) \tag{2}$$

where T_{air} is the temperature of hot air in the furnace, T_{init} is the initial temperature of the specimen, h is the effective heat transfer coefficient, t is the time, A is the contact area with the hot air, V is the volume of the specimen, and ρ and c are the density and specific heat, respectively.

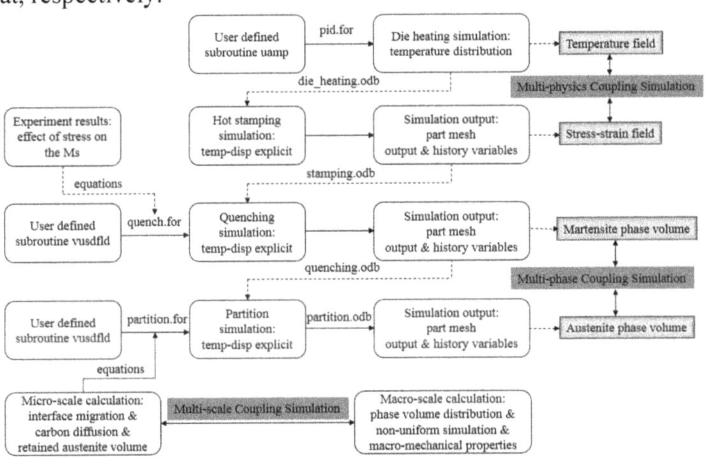

Fig. 2. Visual presentation of full-process simulation procedure.

3. Results and Discussion

3.1. *Effect of stress on Ms changing and carbon partitioning*

The stress emerging from hot stamping can affect Ms, which will influence the initial volume fraction of retained austenite before the partitioning process. In this section, the effects of stress on simulation results were investigated. A comparison was made between the results with and without considering the stress effects. Fig. 3 shows the effects of stress on austenite domain, interface migration distance and retained austenite volume fraction of Element_1. The austenite domain without considering stress effects was larger than that in consideration of stress effects during the partitioning process. The austenite domain affected the carbon diffusion as well as the interface migration. As a result, a larger interface migration distance was gained without considering stress effects, as shown in Fig. 3 (a). It took a longer partition time to reach the maximum value for the one without taking stress effect into account, as shown in Fig. 3 (b). However, the final volume fraction of retained austenite turned out to be the same under both conditions.

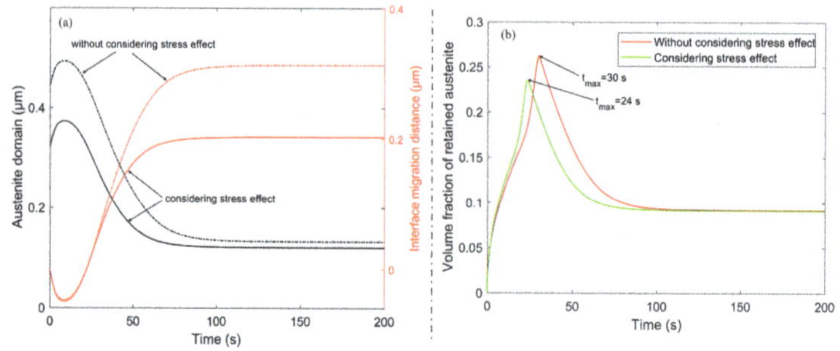

Fig. 3. Effects of stress on a) austenite domain, interface migration distance and b) retained austenite volume fraction of Element_1.

3.2. *Effect of heating strategies on partitioning process*

Many partitioning experiments were conducted under ideal conditions, such as in a salt bath, and the partition temperature was considered to be quasi-statically. However, it is very difficult to apply the salt bath method in industrial applications. Therefore, the method of partitioning in hot air furnace was chosen in this paper. It took time to increase the temperature of the specimen from quenching temperature to partition temperature, and this process was defined as dynamic partitioning. The effective heat transfer coefficient h was calculated by fitting the test data into Eq. (2). The experimental temperature was obtained by using a thermocouple, which was welded to the specimen at the Element_7 location.

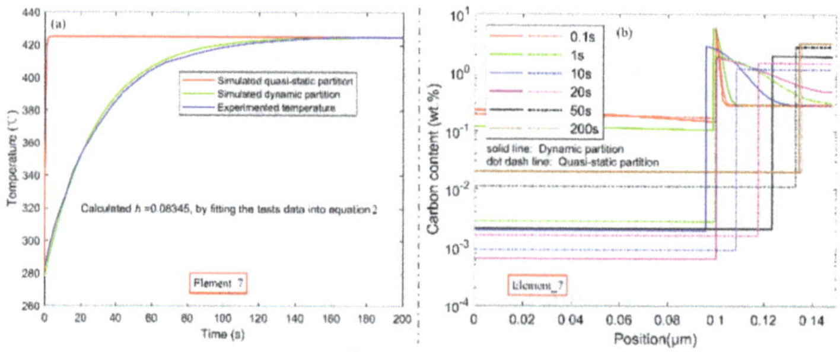

Fig. 4. a) Heating histories of dynamic and quasi-static partitioning; b) contrast between predicted evaluations of carbon distribution under dynamic and quasi-static partitioning conditions

The simulative heating histories of dynamic and quasi-static partitioning are shown in Fig. 4(a). The temperature increased to the set temperature quickly under the quasi-static partitioning process, which meant higher Gibbs free energy in the beginning. The carbon distribution in both austenite and martensite turned homogeneous in less time and the interface migration velocity was higher in the beginning under the quasi-static partitioning condition. However, the interface migration distances turned out almost the same after partitioning for enough time under both conditions, as shown in Fig. 4(b).

4. Summary

In this paper, the effect of load histories on the Q&P process are investigated, and the following conclusion is obtained:

(1) The stress station during the stamping process can affect the martensite start transformation temperature Ms. An analytical equation linking the Ms and the principal stressed was used to predicting the influence of the complex stress state on Ms. Stress emerging from stamping increased Ms, which resulted in initial austenite domain decreasing and non-uniformity throughout the thickness.

(2) The effect of temperature loading history on the interface migration, carbon diffusion was investigated. The carbon distribution in both austenite and martensite turned homogeneous in less time, and the interface migration velocity was higher in the beginning under the quasi-static partitioning condition.

Acknowledgments

This research work was financially supported by the National Natural Science Foundation of China (grant no. U1564203 and U1760205) and National Major Science Technology Project of China (grant no. 2018zx04023001).

References

1. R. M. Wu, X. J. Jin, C. L. Wang and L. Wang, Effect of intercritical annealing on microstructural evolution and properties of quenched & partitioned (Q&P) steels, *J. Mater. Eng. Perform.* **25**, 1603 (2016).
2. A. Turetta, S. Bruschi, and A. Ghiotti, Investigation of 22MnB5 formability in hot stamping operations, *J. Mater. Process. Tech.* **177**, 396 (2006).
3. H. Karbasian and A. E. Tekkaya, A review on hot stamping, *J. Mater. Process. Tech.* **210**, 2103 (2010).
4. P. X. Liu, B. Zhu, Y. L. Wang and Y. S. Zhang, Coupled model for carbon partitioning from martensite into austenite during the quenching process in Fe-C steels, *Metall. Mater. Trans. A* **64**, 4325 (2016).
5. J. G. Speer, E. D. Moor, K. O. Findley, D. K. Matlock, B. C. DeCooman and D.V. Edmonds, Analysis of microstructure evolution in quenching and partitioning automotive sheet steel, *Metall. Mater. Trans. A* **42A**, 3591 (2011).
6. C. Wang, H. Ding, M. H. Cai and B. Rolfe, Multi-phase microstructure design of a novel high strength TRIP steel through experimental methodology, *Mater. Sci. Eng. A* **610**, 436 (2014).
7. Y. Sakuma, O. Matsumura and H. Takechi, Mechanical properties and retained austenite in intercritically heat-treated bainite-transformed steel and their variation with Si and Mn additions, *Metall. Mater. Trans. A* **22A**, 489 (1991).
8. H. P. Liu, X. W. Lu, X. J. Jin, D. Han and J. Shi, Enhanced mechanical properties of a hot stamped advanced high-strength steel treated by quenching and partitioning process, *Scripta Mater.* **64**, 749 (2011).

Theoretical Study of the Carbon Content of Retained Austenite After Quenching-partitioning

C. N. Jing[†], Z. H. Xing, Y. M. Tu, X. Y. Ding and F. Guo

School of Materials Science and Engineering,
Shandong Jianzhu University, Jinan, 250101, China
[†]E-mail: jcn@sdjzu.edu.cn

The carbon content of retained austenite has a significant effect on mechanical properties of quenching-partitioning steel. The partitioning fitting parameter was proposed and its variation with partitioning time was established. The carbon content was calculated by analysing the partitioning process and the behaviour of elemental carbon therein. A theoretical model relating key parameters governing the quenching-partitioning process and carbon content was established. Results show that the absolute error between calculated and experimental results can be less than 1 %, and the partitioning fitting parameter has a significant effect on the calculated carbon content. Ultimately, 60 % of the carbon in retained austenite comes from the partitioning processing, and the changes of it exert a decisive influence on the stability of retained austenite.

Keywords: Quenching and partitioning; Retained austenite; Carbon content; Theoretical model.

1. Introduction

At present, with the development of light alloys and composite materials, and their increasing application in automobile components [1-3], it has become a significant challenger to the market share of automotive steel. To reduce the effects on the environment, while maximising vehicle safety at the same time has become a difficult issue facing the automobile industry [4-7]. Quenching and partitioning (Q&P) steel is a type of high strength-plastic steel with a TRIP effect, which has a martensite and austenite microstructure [8-10]. Its production process is proposed by Speer et al. [11] Q&P steel, which has broad application prospects in the automotive industry, can offer the double guarantee of automotive steel thinning and automobile safety [12, 13] based on the characteristics of its microstructure.

Under constrained carbon equilibrium (CCE) conditions, most of the carbon partitions from martensite into austenite during distribution processing [14, 15]. While carbon has a great effect on the stability of retained austenite. Therefore, it is important to control key parameters of the Q&P process to control the carbon content and to ensure the desired ultimate mechanical properties of Q&P steel considering the effect of each added element on the properties [16, 17]. Meanwhile, there is little literature on the mathematical model governing the relationship between the carbon content of retained austenite at room temperature and the partitioning time of a Q&P steel. The relationship between partitioning time and partitioning fitting parameter is proposed based on the analysis of Q&P processing in this research. A theoretical model relating the critical parameters of Q&P processing and the carbon content in retained austenite at room temperature is established. It is hoped that the key parameters of Q&P processing can be obtained from this model, and that it will provide a theoretical reference for actual production.

2. Experimental Description

Fig. 1 shows typical Q&P heat-treatment processing as used on the experimental materials. A theoretical model for the carbon content, after partitioning, in retained austenite was established by using the data [18-22] (Tables 1 and 2). Subsequently, the calculated carbon contents, during Q&P processing, were compared to those obtained experimentally to verify, or otherwise, the efficacy of the model. The determination of characteristic temperature M_S is based on experimental results available in the literatures or another empirical Eq. 1 [23]:

$$Ms(°C) = 539 - 423w(C) - 30.4w(Mn) - 7.5w(Si) + 30w(Al) \tag{1}$$

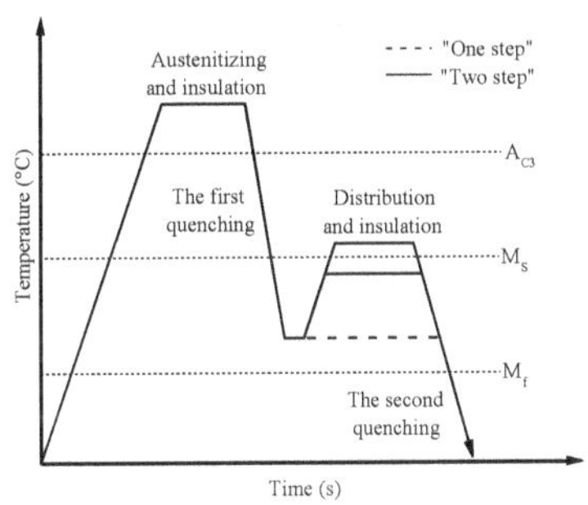

Fig. 1. Schematic representation of the quenching and partitioning heat-treatment process.

Table 1. Chemical composition of the experimental steel (wt %).

NO.	C	Mn	Si	Al	Cr	Ref.
1	0.19	1.59	1.63	0.036	-	[15]
2	0.21	4.00	1.60	-	1.0	[16]
3	0.25	3.00	1.50	-	-	[17]
4	0.30	3.50	1.60	-	-	[18]
5	0.20	2.50	1.50	-	1.01	[19]

Table 2. Main parameters and characteristic temperature.

NO.	Quenching/Partitioning temperature (°C)	Partitioning time (s)	MS (°C)	Ref.
1	200/400	10	372	[15]
2	260/400	100	372	[15]
3	210/450	300	273	[16]
4	243/400	100	325	[17]
5	180/400	5	293	[18]
6	220/400	50	293	[18]
7	275/450	500	338	[19]
8	275/350	1000	338	[19]

3. The Changes in the Carbon Content of the Retained Austenite

In Q&P processing, the microstructure is almost all austenitic when the experimental materials are heated to A_{C3} or beyond. Carbon atoms in austenite can only realise short-range partitioning and are retained in unit cells as the rapid quenching and cooling processes are carried out. Some of the austenite is transformed into martensite when the temperature is below M_S and carbon atoms are ultimately retained in the martensite lattice. The first quenching martensite inherited the original austenite composition in Q&P processing and it has the same mass fraction of carbon as the original austenite grains. So, the carbon contents are equal to the average carbon content of the experimental material, namely:

$$C_{fM} = C_A = C_0 \tag{2}$$

where, C_{fM} is the carbon content of the first quenching martensite, C_A is the carbon content of the austenite, and C_0 is the average carbon content of the experimental materials.

In the primary and secondary quenching processing, carbon content of martensite and austenite does not change substantially: only the variation of the carbon content during partitioning was considered in the calculation of the carbon content of the retained austenite. Therefore, the calculation of the carbon content of retained austenite at room temperature can be calculated by use of Eq. 3:

$$C = C_A + \Delta C_P \tag{3}$$

where, C_A is approximately equal to the average carbon content of the material, and ΔC_P is the change in carbon content during distribution processing.

For simplicity, the diffusion of carbon in austenite grains, and the influence of other elements such as manganese, silicon, on its diffusion behaviour are neglected in the calculation of ΔC_P during partitioning because the diffusion behaviour of each element is quite complicated [24]. A method of calculation of ΔC_P is proposed (Eq. 4) on the basis of other published work [25] and a parameter ε is added in the formula for these reasons.

$$\Delta C_P = \frac{2 \cdot N_C \cdot m_C}{\rho \cdot d \cdot A_{\gamma\alpha'}} \times \varepsilon \tag{4}$$

where, N_C is the number of carbon atoms diffusing from martensite to austenite during partitioning, m_C is the mass of a carbon atom (1.993 × 10-26 kg), ρ is the mass density of steel (7850 kg m^{-3}), d is the width of the austenite phase. The austenite grows in different sizes due to the different composition and parameters, and d needs to be determined by experiment. $A_{\gamma\alpha'}$ is the area of the martensite-austenite interface, ε is the partitioning fitting parameter, which is related to the distribution time.

The relationship between partitioning time and partitioning fitting parameter, as shown in Fig. 2, was obtained through the analysis of the data provided elsewhere [26- 28] and from Fig. 2, Eq. 5 can be obtained. It can be seen from Fig. 2 that the fitting parameter varies with the partitioning time obviously, which may be a representation of a major term in the distribution process, and it needs further study.

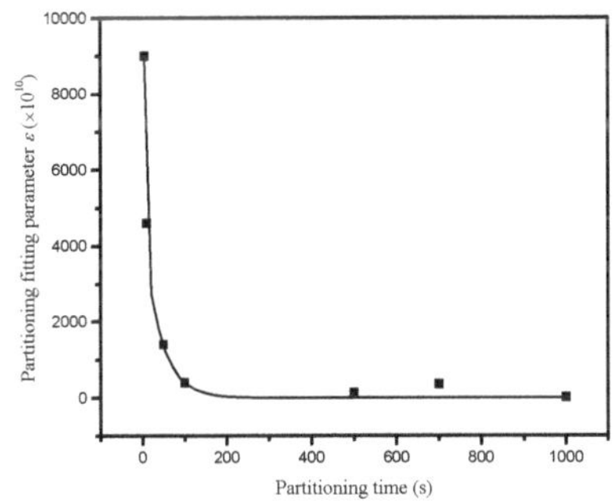

Fig. 2. The relationship between partitioning fitting parameter and partitioning time.

$$\varepsilon = 5.7091\times 10^{10} \cdot t^{-0.9773} \tag{5}$$

Where, t is the diffusion time of carbon.

The literature [25] points out that austenite is sandwiched between martensite in the partitioning microstructure. Single martensite and austenite grains are treated as a rectangular parallelepiped when calculating $A_{\gamma\alpha'}$: the width of the austenite is generally [29] less than, or equal to 500 nm. Assuming that at the martensite-austenite interface, every m atoms of iron contain one carbon atom, and from the interface distance x, every n iron atoms contain one carbon atom during NC treatment. So, if the diffusion coefficient of carbon is D, the diffusion flux J can be obtained by Fick's law when the partitioning temperature is T, as shown in Eq. 6 and 7:

$$J = -D \cdot \frac{\mathrm{d}\left(1/(na^3) - 1/(ma^3) \right)}{\mathrm{d}x} \tag{6}$$

$$D = (0.07 + 0.06\% \times \frac{C_0 + C_1}{2})\exp(-\frac{Q}{\mathrm{R}T}) \tag{7}$$

where, a is the lattice constant of face-centred cubic austenite, and Q is the diffusion activation energy of carbon in austenite. Thus, the value of N_C is determined when the partitioning time is t:

$$N_C = J \times t \times a^2 \tag{8}$$

4. Calculated Results and Analysis

The diffusion coefficients of carbon in austenite, change in carbon content (ΔC_P), and the carbon content at room temperature were calculated (Table 3).

Table 3. Calculated diffusion coefficients, changes in carbon content (ΔC_P), and the carbon content at room temperature.

NO.	Diffusion coefficient（$m^2 \cdot s^{-1}$）	Change of carbon content (%)	Carbon content (%)	C_2 (%)
1	9.55×10-16	1.62	1.81	~1.10
2	9.53×10-16	1.10	1.29	~1.30
3	5.38×10-15	0.34	0.55	~0.60
4	9.54×10-16	1.10	1.40	~1.30
5	9.53×10-16	0.64	0.94	~0.73
6	9.53×10-16	0.71	1.01	~0.89
7	5.38×10-15	0.47	0.67	~0.68
8	1.28×10-16	0.74	0.94	~0.82

Where, the C_2 is the mean value of the austenitic carbon content at room temperature in the corresponding literature.

The carbon content of the retained austenite at room temperature mostly comes from ΔC_P, i.e., the partitioning processing. After calculating the values in Table 3, in the partitioning process, about 60 % of the carbon content in retained austenite is produced. It can be seen that carbon atoms can quickly move from the martensite to the martensite-austenite interface [30] due to dislocations in the martensite providing channels through which carbon can diffuse in the partitioning process. Carbon atoms diffused to the austenite and were then homogenized under the driving force of chemical potential gradient. All of these show that the diffusion is a process of increasing carbon content in austenite. Therefore, the carbon element partitioning in Q&P processing exerts a decisive influence on the room temperature stability of austenite.

In terms of chemical composition, test 4 has a greater carbon and manganese content than test 2; however, manganese atoms only have the ability to diffuse at a high temperature, or over a long insulation treatment time, since its activation energy of diffusion is much larger than that of carbon [31] as a substitution-type atom. So the influence of carbon on the partitioning effect is much larger than that of manganese at the same partitioning temperature (400 °C) and time (100 s). In addition, due to the limiting effect of grains near the interface on the carbon partitioning behaviour [32] and the different silicon contents between the two groups, the results show that the value of ΔC_P remains the same, although the carbon and manganese contents in test 4 are greater than those in test 2.

Fig. 3 shows the degree of fit between the calculated, and experimental, results. The absolute error μ between the experimental, and calculated, results, shown in Eq. 9, is analysed.

$$\mu = \left|(C - C_2)/C_2\right| \times 100\% \tag{9}$$

Although the model is not considered the composition and quenching time on the impact of carbon content, it can be seen from Fig. 3 that the trend in the calculated results is the same as that in the experimental results, which shows that the carbon content varies with partitioning time and Eq. 4 can better reflect the carbon-partitioning behaviour during Q&P processing. The differences between the calculated, and experimental, results can be about 0.01 wt %, and the absolute error μ can be less than 1 % during the fitting of the experimental results [18-22]; however, the calculated results are quite different from the experimental results for short partitioning times, and in particular, times of less than 50 s, which is reflected better in the absolute error curve. The main reason is the partitioning fitting parameter ε changes (Fig. 2) for short partitioning times, and the error is larger when selecting this value in subsequent calculations. The absolute error μ decreased and the absolute error curve gradually flattened as the partitioning fitting parameter ε became more accurate as the partitioning time increased. It can be seen that the trend in the absolute error between experimental, and calculated, results is similar to that of the partitioning fitting parameter combined with Fig. 2 and 3. Therefore, this also proved that the partitioning fitting parameter has a significant influence on the calculated the carbon content of retained austenite at room temperature, which is an indispensable parameter in the establishment of the model.

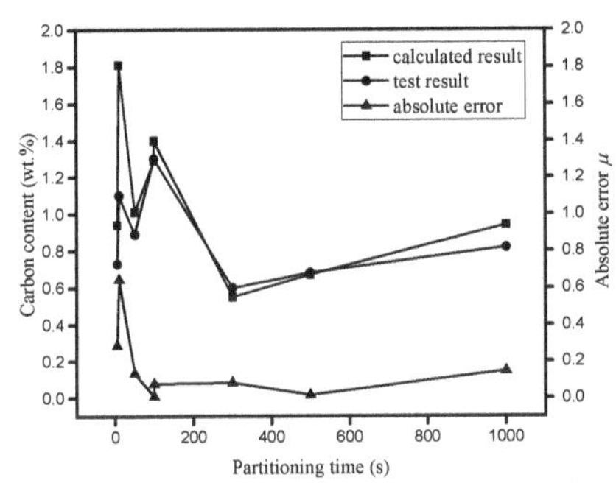

Fig. 3. Comparison of experimental, and calculated, results.

5. Conclusion

The partitioning fitting parameter is proposed and its variation with partitioning time is established. The effect of partitioning fitting parameter on the carbon content of retained austenite at room temperature is obvious. Ultimately, 60 % of the carbon content in retained austenite comes from the partitioning processes and the changes in carbon content have a decisive influence on the stability of austenite.

A mathematical model relating the carbon content of retained austenite at room temperature, and key partitioning parameters, was established. The model can fit the experimental results better and its absolute error was less than 1 %. The partitioning temperature, and duration of the quenching and partitioning process, can be pre-selected on the basis of the carbon content of retained austenite at room temperature, which provides a theoretical reference for the selection of parameters in actual production.

References

1. J. C. Liu, J. Hu, X. Y. Nie, H. X. Li, Q. Du, J. S. Zhang and L. Z. Zhuang, The interface bonding mechanism and related mechanical properties of Mg/Al compound materials fabricated by insert molding, *Mat. Sci. Eng. A* **635**, 70 (2015).
2. S. J. Guo, Y. Xu, Y. Han, J. Y. Liu, G. X. Xue and H. Nagaumi, Near net shape casting process for producing high strength 6xxx aluminum alloy automobile suspension parts, *T. Nonferr. Metal. Soc.* **24**, 2393 (2014).
3. C. Haase, C. Zehnder, T. Ingendahl, A. Bikar, F. Tang and B. Hallstedt, On the deformation behavior of κ-carbide-free and κ-carbide-containing high-Mn light-weight steel, *Acta Mater.* **122**, 332 (2017).
4. D. Q. Zou, S. H. Li and J. He, Temperature and strain rate dependent deformation induced martensitic transformation and flow behavior of quenching and partitioning steels, *Mat. Sci. Eng. A* **680**, 54 (2017).
5. E. Salamci, S. Candan and F. Kabakci, Effect of microstructure on corrosion behavior of dual-phase steels, *Kovove Mater.* **55**, 133 (2017).
6. Y. Akinay and F. Hayat, The influence of the heat treatment on mechanical and microstructure properties of Fe - Mn - C high-manganese steel, *Kovove Mater.* **54**, 91 (2016).
7. M. Šebek, P. Horňak, S. Longauer, P. Zimovčak and P. Zahumensky, Influence of annealing process on microstructure and mechanical properties of C-Mn dual phase steel, *Kovove Mater.* **53**, 341 (2015).
8. H. H. Bok, S. N. Kim, D. W. Suh, F. Barlat and M. G. Lee, Non-isothermal kinetics model to predict accurate phase transformation and hardness of 22MnB5 boron steel, *Mat. Sci. Eng. A* **626**, 67 (2015).
9. L. Cho, E. J. Seo and B. C. D. Cooman, Near-A c3, austenitized ultra-fine-grained quenching and partitioning (Q&P) steel, *Scripta Mater.* **123**, 69 (2016).
10. D. T. Pierce, D .R. Coughlin, D. L. Williamson, J. Kähkönen, A. J. Clarke and K. D. Clarke, Quantitative investigation into the influence of temperature on carbide and austenite evolution during partitioning of a quenched and partitioned steel, *Scripta Mater.* **121**, 5 (2016).
11. R. M. Wu, L. Wang and X. J. Jin, Thermal Stability of Austenite and Properties of Quenching & Partitioning (Q&P) Treated AHSS, *Physics Procedia* **50**, 8 (2013).
12. Y. J. Li, X. L. Li, G. Yuan, J. Kang, D. Chen and G. D. Wang, Microstructure and partitioning behavior characteristics in low carbon steels treated by hot-rolling direct quenching and dynamical partitioning processes, *Mater. Charact.* **121**, 157 (2016).
13. W. S. Li, H. Y. Gao, H. Nakashima, S. Hata and W. H. Tian, In-situ study of the deformation-induced rotation and transformation of retained austenite in a low-carbon steel treated by the quenching and partitioning process, *Mat. Sci. Eng. A* **649**, 417 (2016).
14. C. Y. Wang, J. Shi, W. Q. Cao and H. Dong, Characterization of microstructure obtained by quenching and partitioning process in low alloy martensitic steel, *Mat. Sci. Eng. A* **527**, 3442 (2010).
15. Y. Toji, G. Miyamoto and D. Raabe, Carbon partitioning during quenching and partitioning heat treatment accompanied by carbide precipitation, *Acta Mater.* **86**, 137 (2015).
16. E. J. Song, H. K. D. H. Bhadeshia and D. W. Suh, Interaction of aluminium with hydrogen in twinning-induced plasticity steel, *Scripta Mater.* **87**, 9 (2014).

17. I. Lonardelli, M. Bortolotti, W. V. Beek, L. Girardini, M. Zadra and H. K. D. H. Bhadeshina, Powder metallurgical nanostructured medium carbon bainitic steel: Kinetics, structure, and in situ, thermal stability studies, *Mat. Sci. Eng. A* **555**, 139 (2012).
18. A. J. Clarke, J. G. Speer, M. K. Miller, RE. Hackenberg, D. V. Edmonds and D. K. Matlock, Carbon partitioning to austenite from martensite or bainite during the quench and partition (Q&P) process: A critical assessment, *Acta Mater*. **56**, 16 (2008).
19. E. J. Seo, L. Cho and B. C. D. Cooman, Kinetics of the partitioning of carbon and substitutional alloying elements during quenching and partitioning (Q&P) processing of medium Mn steel, *Acta Mater*. **107**, 354 (2016).
20. M. G. Mecozzi, J. Eiken, M. J, Santofimia and J. Sietsma, Phase field modelling of microstructural evolution during the quenching and partitioning treatment in low-alloy steels, Comp. *Mater. Sci.* **112**, 245 (2016).
21. F. Hajyakbary, J. Sietsma, G. Miyamoto, T. Furuhara and M. J. Santofimia, Interaction of carbon partitioning, carbide precipitation and bainite formation during the Q&P process in a low C steel, *Acta Mater*. **104**, 72 (2016).
22. M. J. Santofimia, L. Zhao, R. Petrov, C. Kwakernaak, W. G. Sloof and J. Sletsma, Microstructural development during the quenching and partitioning process in a newly designed low-carbon steel, *Acta Mater*. **59**, 6059 (2011).
23. N. Maheswari, S. G. Chowdhury, K. C. H. Kumar and S. Sankaran, Influence of alloying elements on the microstructure evolution and mechanical properties in quenched and partitioned steels, *Mat. Sci. Eng. A* **600**, 12 (2014).
24. A. Arlazarov, O. Bouaziz, J. P. Masse and F. Kegel, Characterization and modeling of mechanical behavior of quenching and partitioning steels, *Mat. Sci. Eng. A* **620**, 293 (2015).
25. J. Min, L. G. H. Jr, L. Zhang, J. Lin, J. E. Carsley and L. Sun, Elevated-temperature mechanical stability and transformation behavior of retained austenite in a quenching and partitioning steel, *Mat. Sci. Eng. A* **673**, 423 (2016).
26. J. Sun, H. Yu, S. Wang and Y. Fan, Study of microstructural evolution, microstructure-mechanical properties correlation and collaborative deformation-transformation behavior of quenching and partitioning (Q&P) steel, *Mat. Sci. Eng. A* **596**, 89 (2014).
27. M. J. Santofimia, T. Nguyen-Minh, L. Zhao, R. Petrov, I. Sabirov and J. Sietsma, New low carbon Q&P steels containing film-like intercritical ferrite, *Mat. Sci. Eng. A* **527**, 6429 (2010).
28. X. Tan, Y. Xu, X. Yang and D. Wu, Microstructure–properties relationship in a one-step quenched and partitioned steel, *Mat. Sci. Eng. A* **589**, 101 (2014).
29. M. J. Santofimia, J. G. Speer, A. J. Clarke, L. Zhao and J. Sietsma, Influence of interface mobility on the evolution of austenite-martensite grain assemblies during annealing, *Acta Mater*. **57**, 4548 (2009).
30. J. Emo, P. Maugis and A. Perlade, Austenite growth and stability in medium Mn, medium Al Fe-C-Mn-Al steels, *Comp. Mater. Sci.* **125**, 206 (2016).
31. H. T. Jiang, B. T. Zhuang, X. G. Duan, Y. X. Wu and Z. X. Cai, *Int. J. Min. Met. Mater.* **20**, 1050 (2013).
32. S. Kim, J. Lee, F. Barlat and M. G. Lee, Transformation kinetics and density models of quenching and partitioning (Q&P) steels. *Acta Mater*. **109**, 394 (2016).

Damage Behavior of Boron Steels with Various Hardness Using the GTN and the Johnson-Cook Model

Q. F. Wang, F. X. Wu and B. T. Tang†
School of Mechanical & Automotive Engineering, Qilu University of Technology (Shandong Academy of Sciences), Jinan, 250353, China
†E-mail: 13626409179@163.com

C. C. Li
Institute of Engineering Mechanics, Shandong Jianzhu University, Jinan, 250353, China

In order to obtain high-strength steels with tailored properties, a set of heatable die was designed. Five kinds of hardness materials were obtained by adjusting the die temperature. Four kinds of tensile specimens were designed and five kinds of flow stress models were selected. It was found that the extended Swift model can satisfy five kinds of materials at the same time. The damage and fracture of the materials were studied by using the GTN model and the Johnson-Cook model. The material parameters were optimized by using the modeFRONTIER software. The force-displacement curves of the experiment were compared with those of the simulation, which shows considerable good fitting results.

Keywords: High strength steel; Tailored properties; Flow stress model; Damage parameters.

1. Introduction

In order to meet the requirements of energy conservation and emission reduction, lightweight components have become the mainstream trend in the development of the automotive industry. There are many ways to reduce weight, such as the development of new materials, structural optimization of components, and traditional riveting instead of glued joints. However, the safety of a car must be considered while it is lightweight. Maikranz-Valentin et al. [1], for example, considered an automotive B-pillar. It is believed that the upper part of the B-pillar should have a high energy absorption to transform the impact energy, while the lower part should ensure a high impact resistance, that is, a gradient distribution of performance. This will both ensure safety and reduce weight. Scientists have done a lot of work to obtain gradient performance materials. Maikranz-Valentin [1] used local heating to rapidly cool a specific area into martensite after austenitizing, while the other part did not austenitize in order to obtain large plasticity. Mori et al. [2] also heated the sheet to austenitization state by electrical resistance by means of local heating, but it was expensive. Munera DD [3] welded the materials USIBOR1500P and DUCTIBOR500P together to meet mass reduction requirement. George [4] designed a B-pillar mold for laboratory use. The die was divided into two parts. One part was heated by heaters and the other part was cooled by water channels. The gap between the two parts was left as a transition area, and the mechanical properties of the two parts were studied. Zimmermann et al. [5] performed a partial treatment of the hot stamped parts to make the area softer.

In this paper, a set of die that can be heated is designed. Martensite, upper bainite, and lower bainite are obtained by changing the temperature of the die. Ferrite/pearlite is obtained by cooling in the open furnace. Metallographic analysis and hardness measurement were identified to five individual materials. The specimens under four different stress states were designed for tensile test and flow stress models suitable for this process were found. GTN model and Johnson-Cook model were used to study the ductile fracture process. The damage parameters of different hardness materials under different stress states were obtained.

2. Experiment

2.1. *Material description*

The B1500HS produced by Shanghai Baosteel used in the experiment had a thickness of 1.6mm. The chemical composition is shown in Table 1:

Table 1. Chemical composition of B1500HS cold - rolled steel sheet(wt%).

Alloy element	C	Si	Mn	Cr	Mo	B
Mass fraction(%)	0.23	0.25	1.35	0.19	0.04	0.0032

2.2. *Acquirement of different hardness materials*

In order to obtain materials with different hardness, the original material was placed in a chamber electric furnace SX2-4-10G at 930 °C, fully austenitized after 3 minutes maintenance, and then rapidly placed in a die with different heating temperatures. Ten test points were selected for each sample and the hardness values were averaged. The averaged hardness results in Table 2 show that with the increase of die temperature, the material hardness decreases. For the convenience of analysis, the following conditions of die temperatures 500 °C, 450 °C, 400 °C, and 25 °C, are denoted as T500, T450, T400 and T25, respectively. The condition of cooling in the open furnace is called O25.

Table 2. Measured Vicker's hardness HV0.1.

Die temperature /°C	HV0.1										Average value
25(T25)	515	525	519	515	521	530	536	531	541	538	527
300	472	479	446	474	475	441	447	456	465	470	463
350	471	456	446	450	450	469	439	461	450	461	455
400(T400)	402	393	398	393	401	410	390	415	412	402	401
450(T450)	329	326	330	326	324	311	313	320	325	318	322
500(T500)	282	281	284	283	276	286	281	280	275	271	280

2.3. *Tensile test*

In order to study the mechanical properties of the various materials under different stress conditions, five kinds of heat treated materials are designed as uniaxial tensile specimen, cut-out specimen, specimen with central hole and pure shear specimen. For each type of tensile test of each material, five test results were averaged to reduce the error. The tensile prestress is 90N and the strain rate is $0.002s^{-1}$. Figure 1 shows the tensile test specimen.

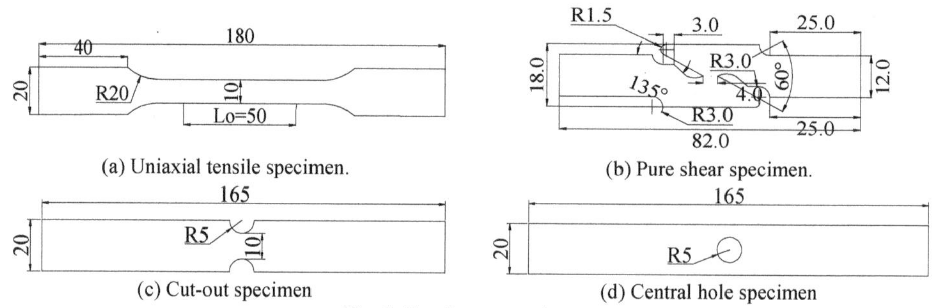

(a) Uniaxial tensile specimen. (b) Pure shear specimen.
(c) Cut-out specimen (d) Central hole specimen

Fig. 1. Tensile test specimen.

3. Flow Stress Model

The flow stress curve can be obtained from a standard uniaxial tensile test. Because of the necking phenomenon, the obtained true strain of tensile test is less than 0.2. In order to obtain the flow stress curve under large strain, an extrapolation model is needed for estimation. In order to select an accurate material flow model, five constitutive equations were adopted: Voce+Voce, Voce++, modified Voce, extended Swift, and Ghosh. Table 3 shows the mathematical expressions of the five models.

Table 3. Expressions of flow stress model.

Flow model	Mathematical expressions
Ghosh	$\sigma = A(\varepsilon_p + B)^n - C$
Extend Swift	$\sigma = K(\varepsilon_p + A)^n + \frac{B \cdot \varepsilon_p + C}{\varepsilon_p + D}$
Modified Voce	$\sigma = A + B(1 - \exp(-C \cdot \varepsilon_p)) + D \cdot \varepsilon_p$
Voce++	$\sigma = \sigma_s + A\left[1 - \exp(-m \cdot \varepsilon_p)\right] + B\sqrt{\varepsilon_p} + C \cdot \varepsilon_p$
Voce+Voce	$\sigma = \sigma_s + A_1\left[1 - \exp(-m_2 \cdot \varepsilon_p)\right] + A_2\left[1 - \exp(-m_2 \cdot \varepsilon_p)\right]$

4. Ductile Damage Models

4.1. *GTN model*

Yield function of GTN model [6]:

$$f = \left(\frac{\sigma_e}{\sigma_M}\right)^2 + 2q_1 f^* \cosh\left(\frac{3q_2\sigma_m}{\sigma_M}\right) - \left(1 + q_3 f^{*2}\right) \tag{1}$$

$$f^* = \begin{cases} f & f \le f_c \\ f_c + \frac{1/q_1 - f_c}{f_f - f_c}(f - f_c) & f_c < f < f_f \\ 1/q_1 & f > f_f \end{cases} \tag{2}$$

σ_e, σ_m are the Mises equivalent stress of the material and the flow stress of the matrix material; σ_m is the hydrostatic pressure. The constant q_1, q_2 and q_3 is the description of the mutual influence between voids. f_c represents the critical volume void fraction, f^* is the effective porosity and f_f is the change in the volume fraction of the volume fraction of pores when the material breaks, mainly from the expansion of existing pores and the initiation of new pores. The hole volume fraction changes as follows:

$$df = (1-f)d\varepsilon^p : I + \lambda d\overline{\varepsilon_m}^p \tag{3}$$

λ is a function of the equivalent plastic strain $\overline{\varepsilon_m}^p$, which is generally believed to follow a normal distribution:

$$\lambda = \frac{f_n}{S_n\sqrt{2\Pi}}\exp\left[-\frac{1}{2}\left(\frac{\overline{\varepsilon_m}^p - \varepsilon_n}{S_n}\right)^2\right] \tag{4}$$

f_n is the nucleation volume fraction, S_n is the standard deviation of the distribution, ε_n is the average equivalent plastic strain.

4.2. *Johnson-Cook model*

The Johnson-Cook model is divided into two parts, one is the plastic constitutive model and the other is the damage fracture model. Eq. 5 is the plastic constitutive model 7, and Eq. 6 is the damage fracture model.

$$\sigma = (A + B\varepsilon_p^{\ n})[1 + C\ln(\dot{\varepsilon}_p / \dot{\varepsilon}_o)][1 - (\frac{T - T_{room}}{T_{melt} - T_{room}})^m] \tag{5}$$

A is the yield stress under quasi-static experiment, ε_n is the equivalent plastic strain, B, C, n and m are model parameters, T is the test temperature, T_{melt} is the material's melting point, and T_{room} is the room temperature during testing.

$$\varepsilon_f = [D_1 + D_2\exp(D_3\sigma^*)][1 + D_4\ln(\varepsilon_p / \varepsilon_o)](1 + D_5T) \tag{6}$$

D_1, D_2, D_3, D_4 and D_5 are the material constant and σ^* is the stress triaxiality. The Johnson-Cook damage fracture model not only considers the stress state but also considers factors such as strain rate and temperature, which makes the model more applicable.

4.3. *Determination of model parameters*

The GTN model was used to simulate tensile process of specimen with cut-out and central hole, respectively. Initially, Tvergaard's classical values of q_1 = 1.5 and q_2 = 1, $q_3 = q_1^2$ =2.25 were used 8. Parameter S_n is generally simplified to be 19. The inverse finite element method coupled with optimization algorithm is used to obtain ε_n, f_n, f_c and f_f. The simulated force-displacement data of different specimen is compared with that of experiment.

The Johnson-Cook model was used to simulate pure shear cases. Since quasi-static stretching was used at room temperature, the parameters involving strain rate and temperature were all default values. The Johnson-Cook plastic constitutive model consists of three parts. When considering only the work hardening, the model can be simplified as:

$$\sigma = A + B\varepsilon_p^{\ n} \tag{7}$$

A, B and n can be fitted with tested plastic strain and stress curves. Similarly, D_1, D_2 and D_3 can be obtained by inverse finite element method.

Figure 2 shows the comparisons of force-displacement curves between the experiments and simulations. It can be seen from the figure that GTN damage model can well predict the material failure process of tensile predominated stress state, no matter what kind of material is used. The Johnson-Cook damage model is more suitable to predict the force-displacement evolution in case of pure shear stress state.

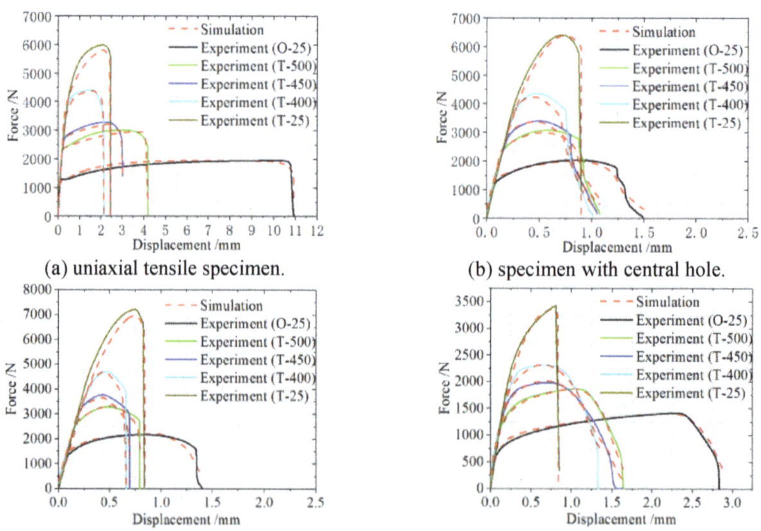

(a) uniaxial tensile specimen. (b) specimen with central hole.

(c) cut-out specimen. (d) pure shear specimen.

Fig. 2. Comparisons of force-displacement curves between experiment and simulation.

5. Conclusion

(1) Specimen with different hardness values are obtained verified by metallographic observation and hardness values.

(2) Tensile specimens under four different stress states were designed to study the fracture properties of the material. Five different flow stress models were used to simulate the standard tensile test data, indicating that the extended Swift model is more suitable for the plasticity flow of five materials.
(3) The tensile process of the standard tensile, central hole tensile, and cut-out tensile tests were simulated using the GTN model, while the Johnson-Cook model was used to simulate the pure shear tensile test. The comparison of force-displacement curves between simulation and experiment shows quite well good fitness between each other.
(4) The fitting formula of the plasticity and damage parameters with die temperature is obtained. This can provide a fast way to find the proper plasticity and damage model to hot stamped components with tailor distributed hardness.

References

1. M. Maikranz-Valentin, U. Weidig, U. Schoof, H. H. Becker and K. Steinhoff, Components with optimised properties due to advanced thermo-mechanical process strategies in hot sheet metal forming, *Steel Res. Int.* **79**, 92 (2008).
2. K. Mori, T. Maeno and K. Mongkolkaji, Tailored die quenching of steel parts having strength distribution using bypass resistance heating in hot stamping, *J. Mater. Process. Tech.* **213**, 508 (2013).
3. D. D. Munera, F. Pinard and L. Lacassin, Very and Ultra High Strength Steels Based Tailored Welded Blanks: A Step Further Towards Crashworthiness Improvement, *J. Mater. Manuf.* **115**, 796 (2010).
4. R. George, A. Bardelcik and M. J. Worswick, Hot forming of boron steels using heated and cooled tooling for tailored properties, *J. Mater. Process. Tech.* **212**, 2386 (2012).
5. F. Zimmermann, J. Spoerer, P. Maier-Komor, G. Rank and W. Volk, Partielles Flamanlassen von pressgehärteten Stählen und die Auswirkungen auf Oberfläche und Verzug, in *Merklein, M. (Ed.), Proceedings of the 8th Erlanger Workshop Warmblechumformung*, (Erlangen, Germany, 2013).
6. V. Tvergaard and A. Needleman, Analysis of the cup-cone fracture in a round tensile bar, *Acta Metall.* **32**,157 (1984).
7. G. R. Johnson and W. H. Cook, A constitutive model and data for metals subjected to large strains, high strain rates and high temperatures, in *7th International Symposium on Ballistics, Am Dref Prep Org (ADPA)*, (Hague, Netherlands, 1983).
8. V. Tvergaard, On localization in ductile materials containing spherical voids, *Int. J. Fracture* **18,** 237 (1982).
9. C. K. Oh, Y. J. Kim, J. H. Baek, Y. P. Kim and W. Kim, A phenomenological model of ductile fracture for API X65 steel, *Int. J. Mech. Sci.* **49**, 1399 (2007).

FE Simulation of Aluminium Galling Induced Forming Tool Failures Under Multi-cycle Loading Conditions: A Feasibility Study

Y. Hu[1,a,†], Y. Zheng[1], X. Yang[1], D. J. Politis[1], L. Wang[1], M. M. Gharbi[2,b], and J. Cui[3,c]
[1]Department of Mechanical Engineering, Imperial College London, London, SW7 2AZ, UK
[2]Schuler Pressen GmbH, Bahnhofstrasse 41, 73033 Göppingen, Germany
[3]CALT-Imperial Advanced Aerospace Structure Manufacturing Technology Laboratory, Imperial College London, London, SW7 2AZ, UK
E-mail: [a,†]y.hu14@imperial.ac.uk, [b]mohanmad.gharbi@schulergroup.com, [c]j.cui@imperial.ac.uk

Galling is a severe adhesive wear phenomenon, which can cause damage of workpiece and tool surfaces, misalignment of tools and thus shorten the tool-life. In the present study, the friction and galling behaviors of a tool steel are studied by using pin-on-disc tests to simulate the adhesive wear process at the workpiece-tool interface for sheet metal forming process. The evolution of the coefficient of friction follows a pattern of increase to saturation. At the running-in state, the coefficient of friction gradually increases, followed by a relatively stable coefficient of friction at the steady state after a certain number of forming cycles. This phenomenon is caused by the interactive response between friction, wear and aluminium transfer from an adhesive wear tribo-system, which is often neglected by metal forming researchers, and constant friction coefficient values are normally used in the finite element (FE) simulations. Meanwhile, most of the current FE simulations consider single-cycle loading processes, where the dramatic tool morphology changes and the interactive effects of friction are neglected. Therefore, a novel interactive friction model is developed to characterize the evolutions of coefficient of friction and galling on the tool surface, to enable the tool-life to be predicted. The friction model is then implemented into the FE simulation of a sheet metal forming process for feasibility study.

Keywords: Galling; Sheet metal forming; Interactive friction model; Multi-cycle loading.

1. Introduction

Unlike many other transient tribological phenomena that occur in a short period, galling process has a long running in state, which may go through hundreds of forming cycles before reaching the steady state [1]. This long duration makes the tribological variables, such as galling area, wear rate and coefficient of friction, variate between each forming cycle [2]. In this work, a multi-cycle stamping modeling is conducted to address these variations. The evolution of galling area and coefficient of friction between workpiece and the die material as well as the effects of key operation parameters have been studied. This quantitative model not only deepens the understanding of the evolutionary tribological conditions at the tool-workpiece interface but also enables a tool-life prediction in terms of critical galling area and coefficient of friction.

2. Interactive Friction Model

During the dry sliding between an aluminium alloy and cast iron, the aluminium transfer process is found to be dependent on sliding distance and contact load. Furthermore, it is revealed that this transfer process has induced the transition in the frictional mechanisms and thus affects the evolution of the coefficient of friction. These dependencies and interactions can be drawn as follows. i) Sliding distance dependency: the aluminium transfer experiences a transition from a running in state, in which the transfer amount gradually increases until to a dynamic equilibrium, namely, the steady state. Meanwhile, the coefficient of friction starts off at a low initial value and increases to a constant when the tribo-system reaches the steady state. ii) Contact load dependency: a high contact load will shorten the running in state and increase the saturated transfer amount as well as wear rate at the steady state.

Based on these analyses, an interactive friction model is developed to enable the prediction of evolutions of the coefficient of friction and the galling area. The origins of the coefficient of friction can be classified as i) aluminium-cast iron contact and ii) aluminium-aluminium contact (Eq. 1). To represent the variation of galling area in the formation progression of transfer layer, a normalised galling area, $\bar{f}(t)$ is introduced, which is defined as the galling area at a certain sliding time ($f(t)$) by the saturated galling area at the steady state ($f(t_{sat})$) (Eq. 2). The coefficient of friction, as a function of the transfer progress, can be predicted, where the coefficient of friction generated at the clean surface is gradually replaced by the saturated value with the increase of normalised galling area.

$$\mu(t) = (1 - \bar{f}(t))\mu_{Al-CI} + \bar{f}(t)\mu_{Al-Al} \tag{1}$$

$$\bar{f}(t) = f(t) / f(t_{sat}) \tag{2}$$

At the equilibrium state, although wear particles are continually generated, the morphology of the interface is relatively stable [3, 4]. An important conclusion is drawn from these observations where the wear particle generation and ejection mechanisms play a critical role in balancing the wear particles and lumps (third body) at the contact. To describe this dynamic balance, the concept of third body density, ρ, is introduce, where it is a measurement of the number of third body elements (N) in a unit contact area (A) (Eq. 3). The variation of $f(t)$ is the main consequence of the accumulation of third body at the interface, which can be expressed by (Eq. 4), where b and λ_1 are model parameters.

$$\rho = N / A \tag{3}$$

$$f(t) = b(\rho_i(t))^{\lambda_1} \tag{4}$$

In the running in state, there is a competition between the rate of generation ($\dot{\rho}_s$) and ejection ($\dot{\rho}_e$) at the contact. This competition results in the accumulation rate of entrapped third body at the interface, $\dot{\rho}_i$ (Eq. 5). The calculation of $\rho_i(t)$ is shown by (Eq. 6). The expression of $\dot{\rho}_s(t)$, $\dot{\rho}_e(t)$ and $\rho_i(t_{sat})$ cab be written by (Eq. 7), (Eq. 8) and (Eq. 9), where m, n and B are model parameters. In simulation, the blank temperature varies in a narrow temperature range between 400-500 °C [5], where the effect of temperature on galling is not sensitive. Therefore, the temperature parameter, C_T, can be treated as a constant in this study.

$$\dot{\rho}_l = \dot{\rho}_s - \dot{\rho}_e \tag{5}$$

$$\dot{\bar{\rho}}_l(t) = \dot{\rho}_l(t) / \rho_i(t_{sat}) = (1/\rho_i(t_{sat}))(\dot{\rho}_s(t) - \dot{\rho}_e(t)) \tag{6}$$

$$\dot{\rho}_s(t) = c_T BPv(t_{sat})\left[n\dot{\bar{\rho}}_l(t) + 1\right] \tag{7}$$

$$\dot{\rho}_e(t) = c_T BPv(t_{sat})\bar{\rho}_l(t) \tag{8}$$

$$\rho_i(t_{sat}) = B / mP^{1-\lambda_2} \tag{9}$$

Table 1. Model parameters.

Parameters	m	n	B	C_T	b	$\mu_{Al\text{-}Al}$	$\mu_{Al\text{-}Cl}$	λ_1	λ_2
Value	0.1	15	0.09	10.5	0.43	0.20	0.55	1.6	0.7

3. Numerical Simulation of Hot Stamping

Finite element (FE) simulations are conducted using the commercial software PAM-STAMP. As shown in Fig. 1, the simulation model comprises of a punch, die, AA6082 tailor welded blanks (TWBs) and blank holder pairs. The tools are modeled using rigid elements and a stepped surface is made on the punch facing towards the TWB. A friction coefficient of 0.2 is chosen to account for surface interaction between the sliding sheet and the die assembly [5].

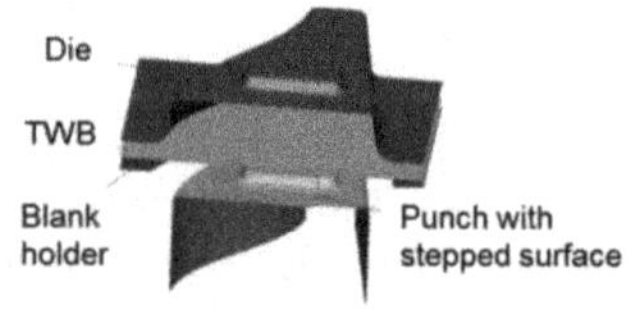

Fig. 1. PAM-STAMP simulation model.

The data obtained from PAM-STAMP is imported into a MATLAB programme, in which the multi-cycle forming calculation is carried out by running the friction model cyclically. Galling area and coefficient of friction are set as outputs. After the calculation, the result from the MATLAB programme is imported back to PAM-STAMP and thus the distribution of results relating to each parameter can be visually analyzed.

4. Results and Discussion

4.1. *Distribution characteristics of tribological conditions*

The tribological conditions of sheet metal forming have the following characteristics : i) uneven in spatial distribution and ii) uneven in historical distribution, which indicates that these influential parameters will vary with location on the tool/work-piece surface as well as forming phases/number of cycles. In this simulation, different die areas present clear variation of tribological conditions. As shown in Fig. 2, a medium contact pressure (0-3 MPa) and a high sliding distance (17-60 mm) are applied on the blank holder areas. On the contrary, the stepped feature on the bottom of die generates a high contact pressure (10-122 MPa) but short sliding distance (<10 mm) due to die closure. The entry and exit of radius areas experience a medium contact pressure (3-6 MPa) as well as a high sliding distance (25-52 mm).

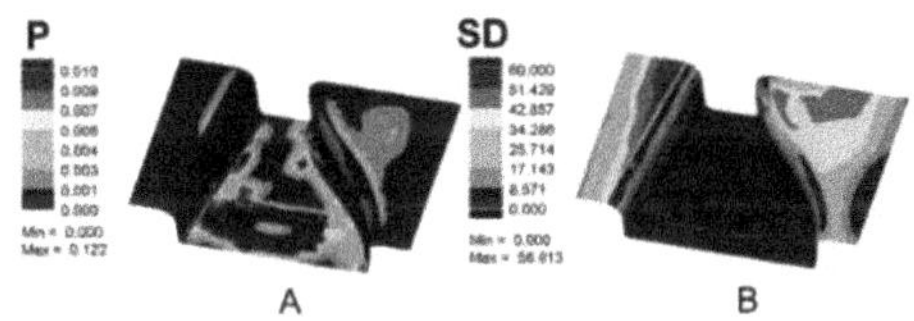

Fig. 2. Distribution characteristics of pressure, *P* (A) and sliding distance, *SD* (B) on the die surface at the end of one forming cycle.

4.2. *Effect of forming cycles*

At the same location with same contact pressure, the galling area increases with the increase of forming cycle. Based on the general observations described by Holm-Archard equation, there is a linear relationship between the wear volume and the amount of entrapped wear particles, which is the source of aluminium transfer layer. Thus, the increase of entrapped wear particles results in an increase of the galling area.

The different tribological conditions also result in a variation of accumulation of galling area. It can be seen in Fig. 3 that the contact pressure and sliding distance have a combining effect on material transfer. Due to the combining of high contact pressure and long sliding distance, initial galling is generated on the entry of radius area at 5 cycles. Then the blank holder area starts to generate galling layer due to high sliding distance (10 cycle). Although the bottom area is subject to short sliding distance, the stepped feature generates high contact pressure, which strengthens the galling accumulation. At the end (300 cycles), both the blank holder areas and the stepped feature area accumulate a severe galling area that close to 100%.

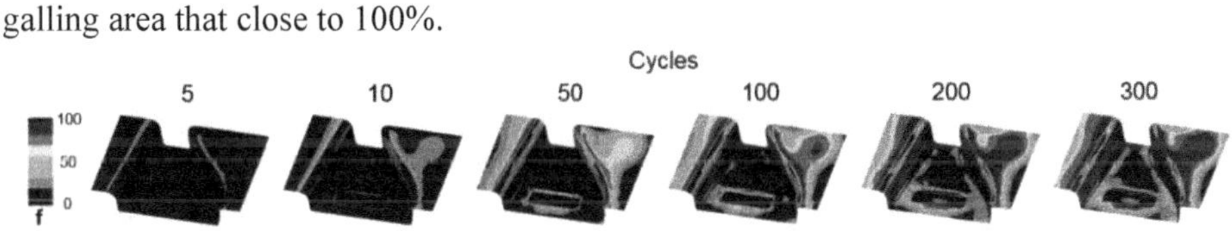

Fig. 3. Distribution of galling area under different forming cycles.

4.3. *Effect of contact pressure*

Fig. 4 shows the evolutions of galling area and coefficient of friction of two points on the die with increasing forming cycles. As shown in Fig. 4A, point A and B are two elements, which locate at entry and exit of the radius area respectively with large difference of contact pressures (A: 0.83 MPa; B: 3.93 MPa) but similar sliding distances (A: 36.06 mm;

B: 36.08 mm). Under the high contact pressure at the entry, point B has a larger generation rate of galling area with forming cycles increasing (Fig. 4B). At this point, the contact reaches the steady state of galling area earlier than Point A (B: no. of cycle=136, A: no. of cycle=250) and has a high saturation galling area (B: 96%; A: 92%). The galling area increase will simultaneously increase the friction at the contact and change the distribution of COF. Coefficients of friction of point A and B increase in a similar pattern as galling area. At the beginning clean contact, the coefficients of friction start at a constant value of 0.2. When they reach the steady state, a pressure-independent coefficient of friction of 0.55 is obtained.

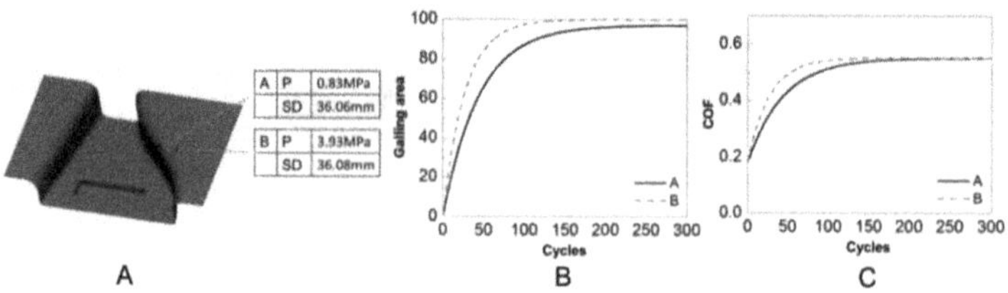

Fig. 4. Effect of contact pressure on galling area and coefficient of friction. (A) the location of point A and B; (B) galling area evolution of point A and B; (C) coefficient of friction evolution of point A and B.

5. Conclusion

In the present research, an interactive friction model is implemented into a stamping simulation and enables a tool-life prediction. The dependency of pressure and sliding distance (forming cycles) can be drawn as follows: i) the aluminium galling area gradually increases with increasing forming cycles until a steady state and ii) high contact pressure will shorten the running in state and increase the saturated transfer amount.

References

1. D. D. Olsson, N. Bay, and J. L. Andreasen, Analysis of pick-up development in punching. *CIRP Ann-Manuf. Techn.* **1**, 185 (2002).
2. J. Heinrichs and S. Jacobson, The influence from shape and size of tool surface defects on the occurrence of galling in cold forming of aluminium, *Wear.* **9**, 2517 (2011).
3. M. B. D. Rooij, G. V. D. Linde, and D. J. Schipper, Modelling material transfer on a single asperity scale, *Wear.* **1**, 198 (2013).
4. W. Österle, A. Dmitriev, and H. Kloß, Possible impacts of third body nanostructure on friction performance during dry sliding determined by computer simulation based on the method of movable cellular automata, *Tribol. Int.* 128 (2012).
5. J. Liu, A. Wang, and Y. Zheng, Hot stamping of AA6082 tailor welded blanks for automotive applications. *Procedia Eng.* 729 (2017).

Modelling of Materials Properties for Metal Forming Simulation

Z. Guo[1,†], J. P. Schillé[1] and P. Lasne[2]
[1]*Sente Software Ltd, Surrey Technology Centre, 40 Occam Road, Guildford, GU2 7YG, UK*
[2]*Transvalor S.A., Parc de Haute Technologie, 694, av. du Dr. Maurice Donat, 06255, Mougins cedex, France*
[†]*E-mail: z.guo@sentesoftware.co.uk*

This paper reviews the recent development of material models via which many of the material properties required in metal forming simulation can be calculated. These models are based on sound physical principles and can calculate a wide range of physical, thermophysical, and mechanical properties including flow stress curves, as well as phase transformation and/or precipitation kinetics. The calculated material data can be easily transferred to CAE tools for the simulation of metal processing. A case study is presented here on the simulation of hammer forging process of an aluminum alloy. The simulated results demonstrate good agreement with experimental observations. Simulations of casting, welding and heat treatments in other industrial alloys are also discussed.

Keywords: Materials modelling; Metal forming; CAE simulation; JMatPro®; Forge®.

1. Introduction

Computer-aided engineering (CAE) simulation tools are increasingly being used in metals industry to speed up processing and manufacturing, where material data have been an integral part ever since such simulations were made possible. The traditional way of obtaining such data is through experimentation, which is expensive and time-consuming. It is therefore of no surprise that lack of material data has been a common problem in CAE simulation. Moreover, such simulation practice has left materials design outside the optimization loop of product design and manufacturing, which reduces the potential design space and may result in suboptimal end products.

The ultimate solution to problem above is to merge the materials design loop and the processing optimization loop into one complete design space. The fundamental technical challenge here is the development of material models to calculate the properties that are essential for processing design and simulation, which has been the focus of our research over the past two decades [1-3]. The models developed are based on thermodynamics, phase transformation and/or precipitation kinetics, deformation and strengthening mechanisms, and microstructure-property relationships. The properties that can now be calculated include temperature dependent physical and thermophysical properties, as well as temperature and strain rate dependent strength and flow stress curves. These material models have been implemented into a computer software JMatPro® and the calculated material data can be easily transferred to CAE tools for simulations of casting [4-8], welding [9-11], forming [12-15] and heat treatments [16, 17] in various metallic systems e.g. Fe- [9-10,12-17], Ni- [5,7], Al- [4,6], Ti- [11] and Mg- [8] based alloys.

This paper first briefly reviews the development of material models based on which the material data required in processing simulation can be calculated. A case study is then presented on the simulation of hammer forging of an aluminum alloy. The current status of linking materials modelling with CAE processing simulation in other industrial alloys is also discussed.

2. Materials Data for Metal Forming Simulation

The forming operation investigated here is a hammer forging process of an aluminum alloy 7175 with nominal composition (wt. %) Al-2.5Mg-5.6Zn-1.6Cu-0.23Cr-0.2Fe-0.1Mn-0.15Si-0.1Ti. The material models employed by JMatPro® to calculated the relevant properties are introduced first, followed by example calculations based on this alloy chemistry.

2.1. *Physical and thermophysical properties*

JMatPro's ability to calculate physical and thermophysical properties has been well documented [2, 3], following procedures described in Fig. 1. Fig. 1(a) shows the calculated phase amount as a function of temperature, whereas the composition of each phase is calculated at the same time, e.g. that of the AL phase in this alloy, Fig. 1(b). For each individual phase, properties such as molar volume, thermal conductivity, Young's modulus and Poisson's ratio are calculated using simple pair-wise mixture models.

$$P = \sum_i x_i P_i^o + \sum_i \sum_{j>i} x_i x_j \sum_v \Omega_{ij}^v (x_i - x_j)^v \tag{1}$$

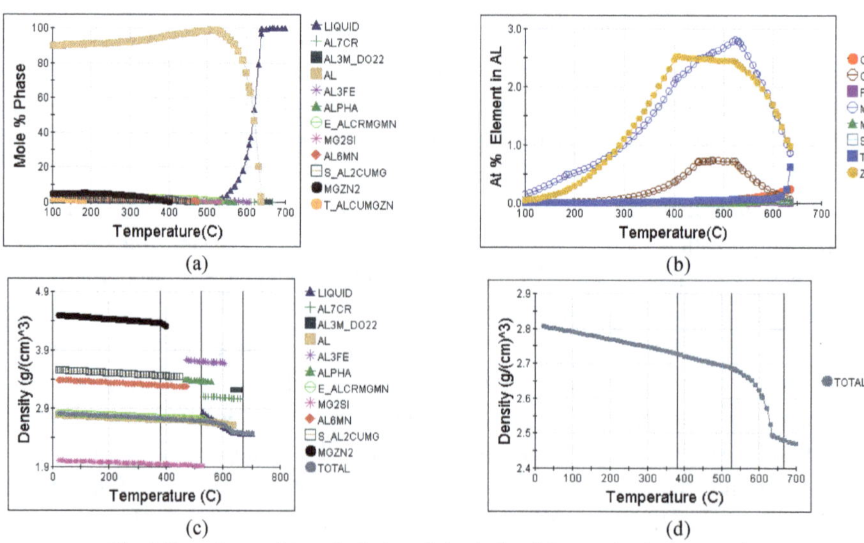

(a) (b) (c) (d)

Fig. 1. Procedures of the calculation of physical and thermophysical properties.

Where, P is the property of the phase, P_i^o is the property of the phase in the pure element, Ω_{ij}^v is a binary interaction parameter dependent on the value of v, x_i and x_j are the mole fractions of elements i and j in the phase. Both P_i^o and Ω_{ij}^v are temperature dependent. Once the property of the individual phase is defined, the property of the final alloy can be calculated using mixture models that can account for the effect of microstructure on the final property [18-20]. The information in Figs. 1(a) and 1(b) on phase amount and composition is from thermodynamic equilibrium calculation, which can also be obtained via non-equilibrium calculation or modelling of phase transformation kinetics [21].

2.2. *Mechanical properties – strength and flow stress curve*

The calculation of strength and flow stress curve is based on strengthening and deformation mechanisms [22-26]. The model looks at material flow from the viewpoint of the deformation mechanisms in operation. Two competing mechanisms are considered in this model, either dominated by dislocation glide (DDG, responsible for strain/work hardening) or dominated by dislocation climb (DDC, responsible for creep or flow softening). Strength vs temperature plot and stress-strain curve both describe the material flow, so they share a similar switch of deformation mechanisms: as temperature switch T_t in Fig. 2(a) and strain switch ε_t in Fig. 2(b), respectively. The model determines whether deformation should be either in DDG mode or DDC mode automatically, depending on the temperature, strain and strain rate.

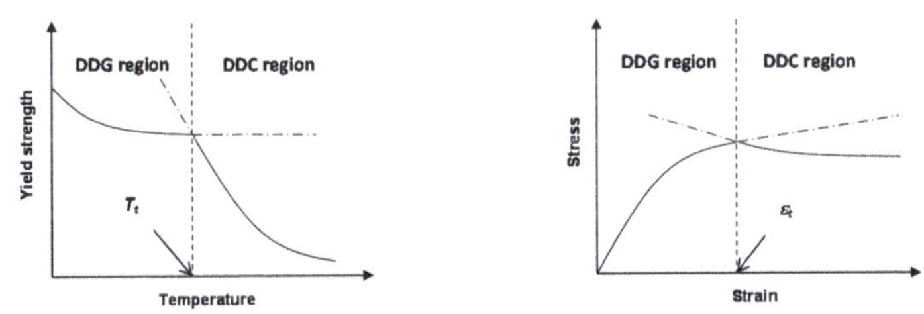

(a) Yield strength vs temperature. (b) Stress-strain curve.

Fig. 2. Change of deformation mechanism from dominated by dislocation glide (DDG) to dominate by dislocation climb (DDC) in (a) high temperature strength curve and (b) stress-strain curve.

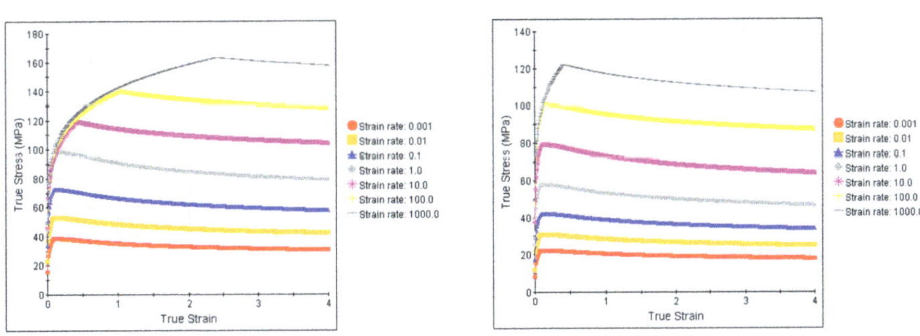

(a) (b)

Fig. 3. Procedures of the calculation of physical and thermophysical properties.

The flow stress curves of alloy 7175 at two temperatures and a range of strain rates are shown in Fig. 3, where the change of ε_t as a function of temperature and strain rate is clearly demonstrated. Extensive validation has been carried out for various series of aluminum alloys and good agreement between calculations and experimental results has been achieved over a wide range of temperature and strain rates [25].

3. Simulation of Hammer Forging for Aluminum Alloy

Using material data generated from JMatPro® for aluminum alloy 7175, it is possible to simulate its hammer forging process using commercial CAE package Forge®.

3.1. *Description of the experimental procedures*

The hammer forging process is described in Fig. 4. The cylindrical billet is heated in a furnace at 380 °C. After transfer to the hammer, the part is formed on the MONTBARD LG 1000 drop hammer. The operation contains 7 blows of 22 kJ, with the falling mass equal to 1250 kg. The falling and rebound velocities, the forces on dies, and the variation of the dimensions of the part are followed during all the process.

Fig. 4. Experimental hammer forging process containing 7 blows.

3.2. *Model for hammer press*

The simulation of hammer forging process is carried out to predict the die filling and the evolution of the dimensions of the part during the blows. The accuracy of the simulation results depends on the models used to describe the material behavior as well as values for the boundary conditions (friction and thermal exchange). To represent hammer kinematics, one considers the mechanical energy (kinetic, potential) and the elastic dissipated energy due to the press stiffness. Fig. 5 present the simple model used to represent the hammer where two different velocities were considered: the velocity of the falling mass of the hammer (V_m) and the velocity the forging die (V). The rigidity of the press is modeled by an equivalent spring.

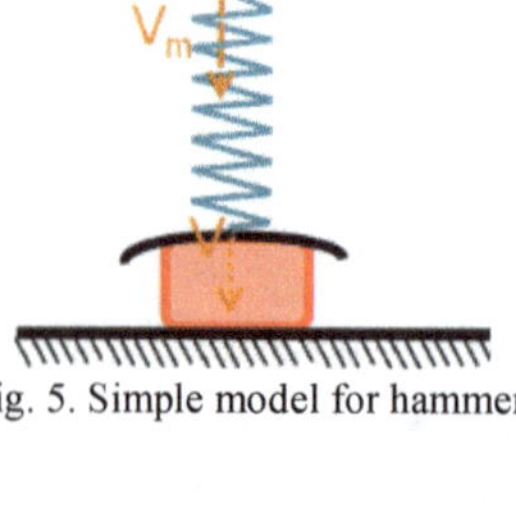

Fig. 5. Simple model for hammer.

3.3. *Thermo-mechanical coupling*

Mathematical models for both temperature field calculation and mechanical equilibrium are solved in the software. The mathematical treatment of heat transfer follows: [27].

$$\rho c dT / dt - div(k \cdot grad(T)) - r\sigma : \dot{\varepsilon} = 0 \tag{2}$$

Where ρ, c, and k are density, specific heat, and thermal conductivity, and r is the fraction of plastic work transformed into heat. The mechanical processing generates heat by plastic work (the $r\sigma : \dot{\varepsilon}$ term), which affects the temperature field, and consequently modifies the material parameters due to their temperature dependence.

The aluminum rheological behaviour is considered as elasto-visco plastic. The calculation of total strain generated during processing can be decomposed into various individual strain contributions as follows: [28].

$$\dot{\varepsilon} = \dot{\varepsilon}^{e} + \dot{\varepsilon}^{p} + \dot{\varepsilon}^{\theta} \tag{3}$$

Where ε^{e}, ε^{p} and ε^{θ} are the elastic, plastic, and thermal strain contributions, respectively. The elastic deformation follows the classical linear equations with Young modulus and Poisson ratio. The visco-plastic behaviour is described by:

$$\sigma_{eq} = \sigma_{o}\left(\bar{\varepsilon}, \dot{\bar{\varepsilon}}, T\right) \tag{4}$$

Where σ_{eq} is the equivalent stress, invariant of the stress tensor and σ_{o} is the yield stress of the alloy considered as a function of accumulated strain $\bar{\varepsilon}$, instantaneous strain rate $\dot{\bar{\varepsilon}}$ and temperature T. The dependency of flow stress curves with strain, strain rate and temperature are calculated from JMatPro®.

3.4. *Simulation vs experimental results*

One of the simulation results is the global evolution of the shape of the billet after each blow. The control and prediction of the evolution of the blows is important because it allows enhanced use of the hammer press. As presented in Fig. 6, the good correlation between numerical prediction and the experimental measurements confirms that the models used are correct. The material behavior has a real influence on this prediction and the material data generated by JMatPro® models reflects the material behavior with good accuracy.

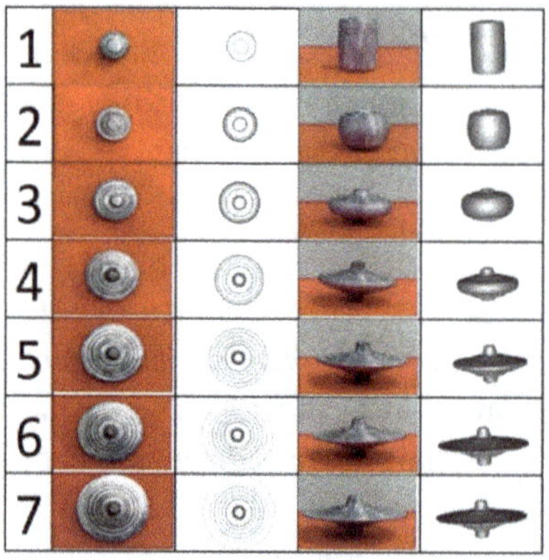

Fig. 6. Comparison of the billet experimental shape with simulation results.

4. Discussion and Conclusion

The application of material data calculated by JMatPro® in processing simulation is not something new. The first simulations of this kind are found in casting simulations, where the calculated property data has been fed into packages such as MAGMASOFT [4,6], PROCAST [7], and ABAQUS [8] to perform the simulation of heat transfer, fluid flow and/or stress distribution in various metallic systems.

With the recent model development on the calculation of flow stress curves, such applications have been extended to the simulation of welding, forming and heat treatments [9-17]. A good example [14] is the hot stamping simulation of a part made of steel 22MnB5. The simulation is carried out with coupled analysis of heat transfer, deformation mechanics and phase transformations using Deform-HT®. The simulated springback shows good agreement with experimental observations. Another case study carried out recently is on the hammer forging process of a 30CrMo4 steel using Forge® [15]. The predicted evolution of flash thickness at the end of each hammer blow again shows good correlation with experimental measurements. The success in such cases studies serves as a testimony of the reliable material data generated by JMatPro®.

In summary, the development of material models has reached a stage where many of the material data essential for processing simulation can be reliably calculated. The ability of JMatPro® in providing such data has been gradually recognised by the CAE community, and the calculated material data have been used for the simulation of casting, welding, forming and heat treatment processes. This clearly demonstrated that it is now possible to incorporate alloy design into the optimization loop of product design and manufacturing: we are moving one-step closer towards true virtual design.

References

1. Sente Software Ltd., (UK, 2012), http://www.sentesoftware.co.uk/downloads/ articles-and-papers aspx.

2. N. Saunders, Z. Guo, X. Li, A. P. Miodownik and J. P. Schillé, Using JMatPro to model materials properties and behaviour, *J. Met.* **12**, 60 (2003).
3. Z. Guo, N. Saunders, A. P. Miodownik and J. P. Schillé, Material properties for process simulation, *Mater. Sci. Eng.* **1**, 7 (2009).
4. Z. Guo, N. Saunders, A. P. Miodownik and J. P. Schillé, Modelling of materials properties and behaviour critical to casting simulation, *Mater. Sci. Eng.* **6**, 465 (2005).
5. K. D. Carlson, S. Ou and C. Bechermann, Feeding of high-nickel alloy castings, *Metall. Mater. Trans. B* **6**, 843 (2005).
6. Z. Guo, N. Saunders, E. Hepp and J. P. Schillé, Modelling of material properties - A viable solution to the lack of material data in casting simulation, in *the 5th Decennial International Conference on Solidification Process,* (Sheffield, U.K. 2007).
7. X. Zhang, X. Cao, H. L. Luo, D. Feng and S. P. Li, Simulation of Ni3Al-based alloy and investment casting process of its thin wall parts, *Israel J. Chemistry* **47**, 363 (2008).
8. H. Hao, D. M. Maijer, M. A. Wells, S. L. Cockcroft, D. Sediako and S. Hibbins, Development and validation of a thermal model of the direct chill casting of AZ31 magnesium billets, *Metall. Mater. Trans. A* **35**, 3843 (2004).
9. A. Krasovskyy, S. Sönnichsen and D. Bachmann, On the residual stresses in multi-pass welds: coupling of welding simulation and fatigue analysis, *Procedia Eng.* **10**, 506 (2011).
10. A. A. Bhatti and Z. Barsoum, Development of efficient three-dimensional welding simulation approach for residual stress estimation in different welded joints, *J. Strain Anal. Eng.* **47**, 539 (2012).
11. R. Turner, R. M. Ward, R. March and R. C. Reed, The magnitude and origin of residual stress in Ti-6Al-4V linear friction welds: an investigation by validated numerical modeling, *Metall. Mater. Trans.* **1**, 186 (2012).
12. M. S. Chae, G. D. Lee, Y. S. Suh, K. H. Lee and Y. S. Kim, Mechanical and forming characteristics of high-strength boron-alloyed steel with hot forming, Transactions of Materials Processing, *T. Mater. Process.* **18**, 236 (2009).
13. Seo, S. J. Yoon, C. H. Suh and H. Y. Kim, Numerical Modeling of Hot Press Forming Process of Boron Steel Tube, in *Proceedings of the 10th International Conference on Numerical Methods in Industrial Forming Processes*, (Pohang, Korea, 2010).
14. Z. Guo, G. Kang and N. Saunders, Modelling of materials properties used for simulation of hot stamping, in *the 13th Int. Conference on Metal Forming,* (Toyohashi, Japan, 2010).
15. Z. Guo, P. Lasne, N. Saunders and J. P. Schillé, Introduction of materials modelling into metal forming simulation, in *the 17th International Conference on Metal Forming*, (Toyohashi, Japan, 2018).

16. J. Fu, Y. T. Li and H. P. Qi, Microstructure simulation and experimental research of as-cast 42CrMo steel during quenching process, *Adv. Mater. Res.* 19 (2011).
17. A. D. da Silva, T. A. Pedrosa, J. L. Gonzalez-Mendez, X. Jiang, P. R. Cetlin and T. Altan, Distortion in quenching an AISI 4140 C-ring - Predictions and experiments, *Mater. Des.* **42**, 55 (2012).
18. Z. Fan, P. Tsakiropoulos and A. P. Miodownik, A generalized law of mixtures, *J. Mat. Sci.* **29**, 141 (1994).
19. Z. Fan, A microstructural approach to the effective transport properties of multiphase composites, *Phil. Mag. A.* **73**, 1663 (1996).
20. A. P. Miodownik, N. Saunders and J. P. Schillé, unpublished research.
21. Z. Guo, N. Saunders and J. P. Schillé, Modelling phase transformations and material properties critical to simulation of heat treatment distortion in steels, *Netsu Shori.* **2**, 753 (2009).
22. X. Li, A. P. Miodownik and N. Saunders, Modelling of materials properties in duplex stainless steels, *Mater. Sci. Tech.* **18**, 861 (2002).
23. Z. Guo, N. Saunders, A. P. Miodownik and J. P. Schillé, Quantification of high temperature strength of nickel-based superalloys, *Mater. Sci. Forum.* **546-549**, 1319 (2007).
24. Z. Guo, N. Saunders, J. P. Schillé and A. P. Miodownik, Modelling high temperature flow stress curves of titanium alloys, in *MRS Int. Mater. Res. Conference*, (Chongqing, China, 2008).
25. Z. Guo, N. Saunders, A. P. Miodownik and J. P. Schillé, Hot deformation of aluminum alloys, *Sente Software Ltd.* (2016).
26. Z. Guo, N. Saunders and J. P. Schillé, Age hardening of aluminum alloys, *Sente Software Ltd.* (2018).
27. J. L. Chenot, M. Bernacki, L. Fourment and R. Ducloux, Advanced numerical methods for FE simulation of metal forming processes, in *AIP Conference Proceedings*, (New York, American, 2010).
28. P. Lasne, M. Barbelet, O. Jaouen and F. Costes, Simulation of the heat treatment of an automative cast part, in *ASME 2010 Int. Manufacturing Sci. and Eng. Conference*, (Pennsylvania, USA, 2010).

The Thermodynamics Calculation of Nb-V Bearing Press Hardened Steel During Hot Stamping Process

Q. J. Zhao[1,2,†], Y. Zhao[1,2], M. T. Ma[2], G. Fang[1,2] and T. Q. Fan[2]

[1]*State Key Laboratory of Vehicle NVH and Safety Technology, Chongqing, 401122, China*
[2]*Chinese Automotive Engineering Research Institute Co., Ltd., Chongqing, 401122, China*
[†]*E-mail: zhaoqingjiang@caeri.com.cn*

In order to improve the toughness and hydrogen induced fracture resistance of press hardened steel (PHS), novel Nb-V bearing compositions are developed on the basis of 22MnB5. The commercial software JmatPro is employed to study the phase transformations of novel PHS, complying with the heat treatment conditions during hot stamping process. The austenite decomposition, martensite transformation, precipitation of (Ti, Nb, V) (C, N), and continuous cooling transition (CCT) curves are studied. The results indicate that, B is the most important element for PHS hardenability, while Nb also affects this slightly. The final precipitation of three samples is (Ti, Cr) C, (Nb, Ti, Cr) C and (Nb, V, Ti) C, respectively. When only Nb is added, NbC has stronger precipitated capability than TiC. When both Nb and V are added, TiN is preferred to be precipitated, and the interactions of Ti, Nb and V are complicated.

Keywords: Press hardened steel; Nb-V bearing; Martensite transformation; Precipitation; CCT.

1. Introduction

Press hardening steel (PHS) has been widely applied on automobile, which trends to meet more and more serious safety requirement. Volvo has designed a PHS frame for XC90, in which the percent of PHS accounts for 38 %. The Body-in-white of automobile is the first and most important barrier to protect driver as well as passenger. The application of PHS on automotive part aims to promise driver or passenger's safety as much as possible. Thus minimum deformation (intrusive displacement) under the impact force is the advantage of PHS due to its ultra-high strength. This means that deformation and force are both essential factors during impact performance evaluation. Therefore, toughness (force multiplies displacement) is the actual factor characterising impact fracture resistance of automobile, which is directly related to the safety requirement rather than ultimate strength. As a result, the improvement of impact resistance of PHS will focus on how to increase its toughness.

The aim of this research is to improve the toughness of PHS, including increase of yield stress or ultimate stress with little loss of ductility and improvement of hydrogen induced delayed fracture resistance. In this paper, Nb and V are employed to give expected solution via precipitation. On the basis of different compositions and assigned heat treatment, martensite transformation, precipitation and continuous cooling transition (CCT) are calculated and analysed via JmatPro software.

2. Design Concept

In order to improve the toughness of PHS as well as its corresponding toughness automotive part. One hand, a better match between strength and elongation should be found. The other hand, hydrogen induced delayed fracture should be avoided, which has taken placed in many types of martensite steel.

Addition of microalloy has been proved as an effective method in matensite steel, such as niobium (Nb) and vanadium (V). The most remarkable effect of Nb in steel is the Nb(C,N) particles hinder grain boundary and prohibit austenite coarsen and recrystallization, thus grains are refined and subject to a rising yield stress according to Hall-Petch equation. The increase of yield stress contributes to the improvement of toughness because it expands the area enclosed by force-displacement curve. Another explanation is that total area of grain boundary is enlarged thus crack propagation is weakened.

V is an effective element used in strengthen mechanism of martensite steel (spring or gear steel). The precipitated V(C, N) with very small size (~10nm [1]) disperses at dislocations and improve the ultimate strength of steel. However, the emphasis of adding V is to utilise its brilliant hydrogen trapping capacity depending on its much smaller size and larger number density [2].

On the basis of above analysis, various compositions of PHS are designed and listed in Table 1. The effects of Nb and Nb+V will be investigated via thermodynamic calculation in comparison with reference composition S1.

Table 1. Chemical compositions of PHS (wt %)

	C	Si	Mn	P	S	Al	Cr	B	Ti	N	Mo	Nb	V
S1	0.24	0.24	1.32	0.01	0.002	0.03	0.13	0.002	0.03	0.003	0.002	0	0
S2	0.24	0.24	1.32	0.01	0.002	0.03	0.13	0.002	0.03	0.003	0.002	0.06	0
S3	0.24	0.24	1.32	0.01	0.002	0.03	0.13	0.002	0.03	0.003	0.002	0.04	0.02

3. Processing Scheme

JmatPro is an available thermodynamics software to calculate both complete equilibrium and typical or industrial non-equilibrium processes. It is used to calculate phase transformation during austenization and quenching processes. The heat treatment parameters are from industrial hot stamping process, which is as follows:

(1) Stage 1: Reheat the steel sheet (ferrite + pearlite) in electrical furnace at 930 °C for 280 s;

(2) Stage 2: Transfer the hot sheet into modulus using 8-10 s, and the temperature of sheet is about 780 °C before entering modulus;

(3) Stage 3: The sheet is held in modulus for 8 s and the temperature is 200 °C when it is collected.

The thermodynamics calculations will be carried out in accord with this scheme, in which the phase transformation is driven by Gibbs free energy, regardless of mechanical deformation in this process.

4. Result and Discussion

4.1. *Martensite transformation*

Figure 1 shows the phase diagram of S1 indicating the decomposition of austenite during the entire process. It can be seen in Fig.1 (a), the starting temperature of austenite is about 320 °C, and the volume fraction of martensite experiences a rapid increase, accounting for 90 % at 200 °C and above 95 % at 150 °C, respectively. Finally, the austenite completes decomposition at 100 °C, accompanied by very little ferrite, bainite and austenite. The phase diagrams of S2 and S3 are almost the same with that of S1. This means Nb and V have limited effect on martensite transformation. Boron (B) is reported to have close relationship with the hardenability of PHS. Then B is deleted from S1 and the related phase diagram is shown in Fig.1 (b). As seen in Fig.1 (b), Ms is raised to 380 °C, but martensite transformation rate is strongly lowered, less than 90 %. The bainite and ferrite is emerged during the process. It can be demonstrated that, B is of extreme importance for martensite transformation.

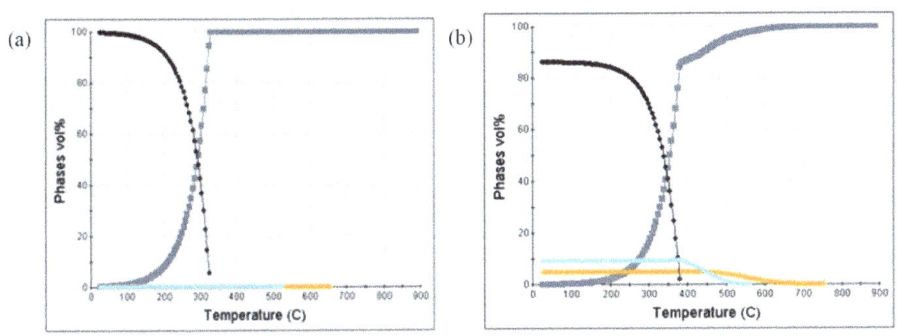

Fig. 1. Phase diagram of S1 indicating the decomposition of austenite during the entire process: (a) with B; (b) without B.

4.2. *Precipitation of secondary phase*

The secondary phase M(C, N) can be regarded to precipitate from austenite according to Fig. 2. As shown in Fig. 2 (a), carbon(C), niobium (N) and titanium (Ti) simultaneously precipitate at Stage 2 and probably form Ti (C, N). After entering Stage 3, the precipitation of C is accelerated while the percentage of N in M(C, N) rapidly declines, which contributes to the much less content of N. This means the preferred precipitated period of C and N is Stage 3 and Stage 2, respectively. This also can be explained that C in M(C, N) is released from the decomposition of austenite. Below 400 °C, chromium (Cr) starts to precipitate, which is nearly coherent with martensite transformation. In Fig. 2 (b), Nb prefers to be conjunct to C and rapidly precipitates in Stage 2, and NbC has stronger precipitated capability than TiC, which has been demonstrated by Hong's experiment [3]. At Stage 3, the precipitation of TiC is promoted because of rapid quenching, and the group precipitations of (Ti,Nb)(C,N) have been reported in many references [4- 6]. Han et al [7] found the effective grain size of TiC is less than 10 nm when a medium manganese was quenched from 900 °C. The precipitation of N is prohibited all the cooling process. In Fig. 2(c), after adding 0.02% V, TiN is preferred to form rather than other particles in Stage 2, while NbC, VC and TiC gradually are generated in Stage 3. TiN has a lower solubility between 700 °C- 900 °C [8, 9], and surely precipitates first, but why it does not precipitate in Sample 2 prior to NbC, or whether V retards the precipitation of NbC, there still are not clear explanations. The final precipitation of three samples is (Ti, Cr) C, (Nb, Ti, Cr) C and (Nb, V, Ti) C, respectively. (Nb, Ti, Cr) C is also found under TEM and EDS as shown in Fig. 2 (d). According to the EDS result, the atomic mass percent of Nb, Ti and Cr in this cubic precipitation is 7.38 wt %, 7.78 wt % and 1.86 wt %, respectively.

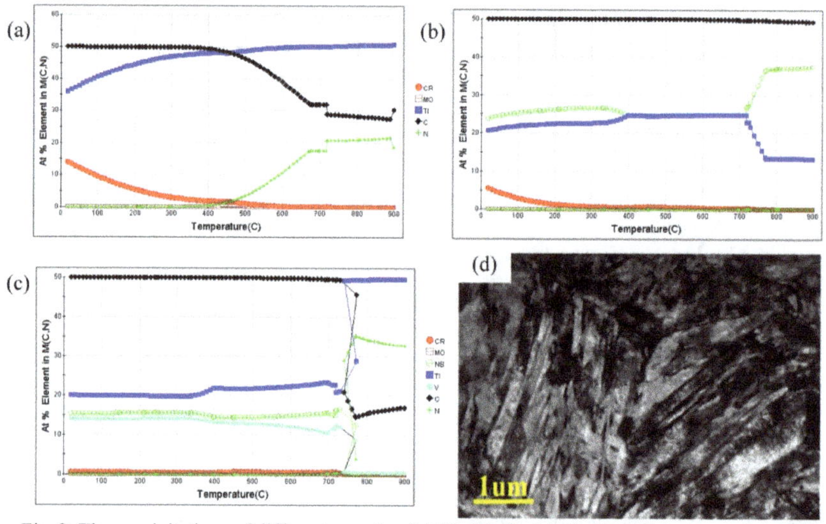

Fig. 2. The precipitations of different samples: (a) S1; (b) S2; (c) S3; (d) TEM observation of S3.

4.3. *CCT curves*

Figure 3 illustrates the continuous cooling transition (CCT) curves of three samples. In Fig. 3(a), the critical cooling rate for 99.9 % martensite is above 100 K/s, which is much higher than experimental results [10, 11]. This is mainly because that the higher content of martensite regarded as the end of phase transformation in this calculation. The Ms and Mf is 390 °C and 280 °C, respectively. As shown in Fig. 3(b) and (c), the addition of Nb and V improves the hardenability of 22MnB5, and the critical cooling rate of matensite is about 100 K/s. However, the effect of V on the hardenability can be neglected according to Fig. 3(c). In industrial practice, the deformation has significant influence on the phase transformations due to the mechanical stability of austenite, for example, the areas of ferrite, pearlite and bainite are magnified thus hardenability is decreased, and the critical cooling rate rises from 20 K/s to more than 60 K/s [12]. The driving force of phase transformation during hot stamping should include energy from both temperature and deformation, therefore, the calculation of CCT curve here can be partly referenced for industrial processing parameters. The end quenching experiments are also carried out to valid the prediction of thermodynamics calculation, which is illustrated in Fig. 4. The addition of Nb and V greatly improves the hardenability of steel, in which the hardenability layer is increased from 3.7 mm to 11 mm. The sample of steel is cylinder, it can be concluded that the novel steel will lower down the threshold of cooling capability of industrial module of PHS and save the actual cost. Moreover, the local strength of component undergone bigger strain and poorer contact with module wall will be surely increased due to the better hardenability of novel steel.

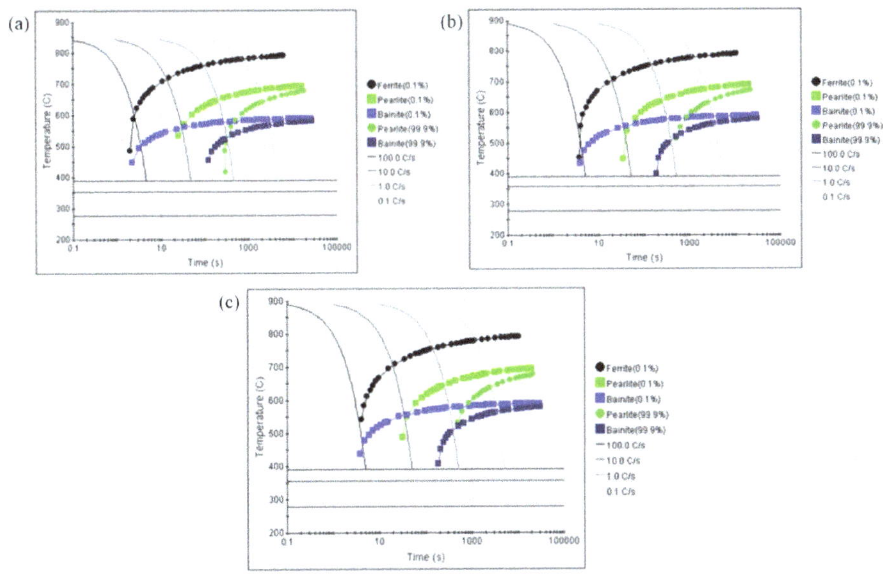

Fig. 3. The predicted CCT curves of three samples: (a) S1; (b) S2; (c) S3.

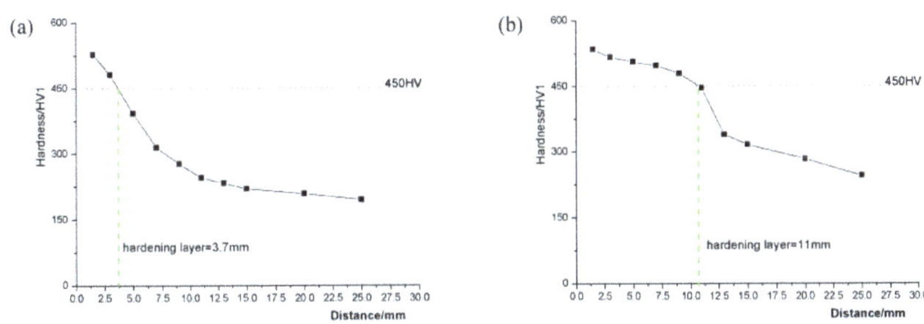

Fig. 4. The hardenability testing in end quenching experiment: (a) S1; (b) S2.

5. Conclusion

The thermodynamics of Nb and Nb-V bearing PHS as well as 22MnB5 are calculated via JmatPro software during both austenization and quenching processes. Martensite transformation, precipitation and CCT curves are obtained to initially analyze the composition design of PHS. Some conclusions are as follows:

(1) Toughness of PHS is directly related to the safety performance of automotive part. Nb is used to refine austenite grain size and improve the toughness of PHS. The

addition of V is mainly applied to form effective hydrogen traps because of its very small precipitated particles.

(2) The austenite can be completely transformed into martensite at 100 °C. B has significant effect on the hardenability of PHS rather than Nb and V.

(3) The final precipitation of three samples is (Ti, Cr) C, (Nb, Ti, Cr) C and (Nb, V, Ti) C, respectively. When only Nb is added, Nb prefers to be conjunct to C and rapidly precipitates in Stage 2, and NbC has stronger precipitated capability than TiC. When both Nb and V are added, TiN is preferred to be precipitated, and the prohibition of NbC precipitation caused by V is not concluded. The cubic precipitation of (Nb, Ti, Cr) C can also found under TEM and identified using EDS.

(4) The critical cooling rate for 99.9% martensite is above 100 K/s, this is mainly because that the higher content of martensite regarded as the end of phase transformation. Both simulation and experiment indicate the addition of Nb and V greatly improves the hardenability of steel, and the experimental results show much better effect.

Acknowledgments

The authors are grateful for the financial supports from National key research and development plan (2017YFB0304405).

References

1. Y. Li, D. N. Crowther, P. S. Mitchell, P. S. Mitchell and T. N. Baker, The evolution of microstructure during thin slab direct rolling processing in vanadium microalloyed steels, *ISIJ Int.* **42**, 636 (2002).
2. S. Yamasaki and H. K. D. H. Bhadeshia, Modelling and characterisation of Mo2C precipitation and cementite dissolution during tempering of Fe-C-Mo martensitic steel, *Mater. Sci. Tech.* **19**, 723 (2003).
3. S. G. Hong, H. J. Jun, K. B. Kang and C. G. Park, Evolution of precipitates in the Nb-Ti-V microalloyed HSLA steels during reheatin, *Scripta Mater.* **48**, 1201 (2003).
4. S. G. Hong, K. B. Kang and C. G. Park. Strain-induced precipitation of NbC in Nb and Nb–Ti microalloyed HSLA steels, *Scripta Mater.* **48**, 163(2002).
5. R. D. K. Misra, H. Nathani, J. E. Hartmann and F. Siciliano, Microstructural evolution in a new 770MPa hot rolled Nb–Ti microalloyed steel, *Mater. Sci. Eng. A* **394**, 339 (2005).
6. M. Charleux, W. J. Poole, M. Militzer and A. Deschamps, Precipitation behavior and its effect on strengthening of an HSLA-Nb/Ti steel, *Metall. Mate. Trans. A* **32**, 1635 (2001).

7. Y. Han, J. Shi, L. Xu, W. Q. Cao and H. Dong, Effects of Ti addition and reheating quenching on grain refinement and mechanical properties in low carbon medium manganese martensitic steel. *Mater. Des.* **34**, 427 (2012).
8. A. J. DeArdo, New Challenges in the thermomechanical processing of HSLA Steels, *Mater. Sci. Forum* **49**, 426 (2003).
9. B. Aronsson, in *Proc. Symp. On 'Steel strengthening mechanisms'. Zurich, Climax Molybdenum Company* 77 (1969).
10. A. Bardelcik, C. P. Salisbury, S. Winkler, M. A. Wells and M. J. Worswick, Effect of cooling rate on the high strain rate properties of boron steel, *Inter. J. Impact Eng.* **37**, 694 (2010).
11. M. Nikravesh, M. Naderi and G. H. Akbari, Influence of hot plastic deformation and cooling rate on martensite and bainite start temperatures in 22MnB5 steel, *Mater. Sci. Eng. A* **540**, 24 (2012).
12. M. Nikravesh, M. Naderi, G. H. Akbari and W. Bleck, Phase transformations in a simulated hot stamping process of the boron bearing steel, *Mater. Des.* **84**, 18 (2015).

Effects of Nb-bearing Steel on the Crashworthiness of Hot-stamping B-pillar

Y. Lin[1], J. P. Lin[1,a,†], H. Z. Lu[2], Y. J. Li[1] and J. Y. Min[1,b]
[1]*School of Mechanical Engineering, Tongji University, Shanghai, 201804, China*
[2]*CITIC METAL Co., Ltd., Beijing, 10004, China*
E-mail: [a,†]*Jplin58@tongji.edu.cn,* [b]*Junying.Min@tongji.edu.cn*

Element Nb has been proved with advantages of grain size refinement and mechanical property improvement. This paper focused on the effect of element Nb on the crashworthiness of hot stamping B-pillars from 22MnB5 steels. Both Nb-bearing and Nb-free 22MnB5 were studied. It was found that the impact toughness of Nb-bearing 22MnB5 was increased by 31.07 % compared with Nb-free 22MnB5. This difference was reflected by the equivalent plastic strain to failure in the FEM simulation of this study. Simulations of side impact of 22MnB5 B-pillars with hot stamping outer were carried out and it was found that the Nb-bearing 22MnB5 could delay the fracture during impact. Energy absorption of impact regions was increased by 19.26 % and the maximum impact velocity was decreased by 8.43 %. The crashworthiness of B-pillar can be effectively improved by applying Nb-bearing 22MnB5.

Keywords: Hot stamping; Nb-bearing material; B-pillar; Side impact.

1. Introduction

Element Nb has been proved with advantages of grain size refinement and mechanical property improvement. However, the studies in the past focused more on the benefits of Nb on material property itself rather than on the crashworthiness of automotive parts. Since the safety performance of automotive parts were directly concerned in automotive industry, it is significant to understand the connection between material property and automotive part performance. Regarding this issue, this paper studied the effects of Nb-bearing steel on the crashworthiness of automotive part by taking hot stamping B-pillar as an example.

2. Comparisons of Mechanical Properties of Nb-bearing and Nb-free 22MnB5

Nb-bearing and Nb-free 22MnB5 provided by MaSteel were selected as representative materials to study the effects of Nb element on crashworthiness. Nb contents of 0.04 % was presented in Nb-bearing steel. The thickness of both sheets was 1.4 mm. The heat treatment was processed with heating temperature of 950 °C and dwell time of 5 min. The sheet was cooled by die quenching to simulate practical hot stamping process. The mechanical properties were examined by quasi-static tensile test, high speed tensile test and Charpy impact test. Each test was repeated three times for repeatability verification.

MTS tensile test machine and servo-hydraulic high speed tensile testing machine were used for quasi-static tensile test and high speed tensile test respectively, and the strain rates were 10^{-4}/s and 100 /s, respectively. Digital Image Correlation system was used to measure stresses and strains in experiments. Two tensile tests shared with the same specimen shape, which is shown in Fig. 1. The specimen of Charpy test was machined according to GB/T 229-2007 standard [1] and tested at room temperature. Since the thickness of sheet was smaller than the minimum thickness required in the standard, five specimens were fixed by using bolts according to study in Ref. [2]. Specimens for Charpy test are shown in Fig. 2.

The comparative results of yield stresses, tensile stresses, elongations at fracture and impact toughnesses are list in Table 1. It can be concluded that the mechanical properties of Nb-bearing and Nb-free 22MnB5 in both quasi-static tensile and high speed tensile conditions had little difference, while the impact toughness and impact energy of Nb-bearing 22MnB5 had 31.07% increase compared with that of the Nb-free ones. The capability of energy absorption during impact was increased by applying Nb-bearing steel.

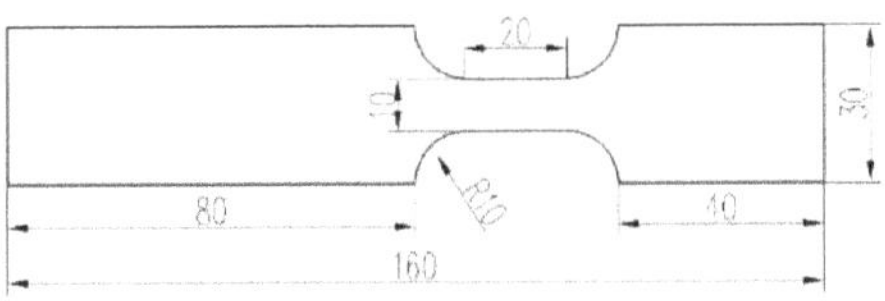

Fig. 1. Specimen for tensile test.

Fig. 2. Specimen for Charpy test.

Table 1. Comparisons of mechanical properties of Nb-bearing and Nb-free 22MnB5.

		Nb-bearing 22MnB5	Nb-free 22MnB5	Effects
Quasi-static tensile test (Strain rate 10^{-4}/s)	Yield stress (MPa)	1091	1087	3.68 %
	Tensile Stress (MPa)	1619	1634	0.92 %
	elongation at fracture	0.1076	0.1071	0.47 %
High strain rate tensile test (Strain rate 100/s)	Yield stress (MPa)	1280	1258	1.74 %
	Tensile Stress (MPa)	1604	1628	1.47 %
	elongation at fracture	0.1595	0.1615	1.24 %
Charpy impact test	Impact energy (J)	36.7	28	31.07 %
	Impact toughness (J/cm^2)	0.6554	0.5000	31.07 %

3. Reiteration of Difference in Charpy Impact Test by FEM Simulation

The FE model of Charpy impact test was established by HyperMesh and performed by using explicit solver LS-DYNA version 971, which is shown in Fig. 3. The two kinds of material properties were defined by MAT 24. The values of density, Young's modulus and Poisson's ratio were taken as 7850 kg/m3, 206 GPa and 0.3, respectively. The stress-strain curves in different strain rates were measured by high speed tensile test. Since toughness failure cannot be directed reflected by fracture in uniaxial tensile test [3], the equivalent plastic strain to failure was selected as the parameter to reflect toughness of material in researches [4-5]. The punch was defined as rigid shell with the initial speed of 5.24 m/s and the initial impact energy was 300 J, which was the same as those of experiment. The contact of punch and specimen was automatic surface to surface and friction coefficient was set to 0.2 [6].

Curves of internal energy of specimen versus time are shown in Fig. 4. The impact energies were 36.1 J and 28.8 J for Nb-bearing and Nb-free 22MnB5, respectively, which matched well with experimental results. The corresponding equivalent plastic strains to achieve these impact energies were 0.12 and 0.103, respectively.

Fig. 3. FE model of Charpy impact test.

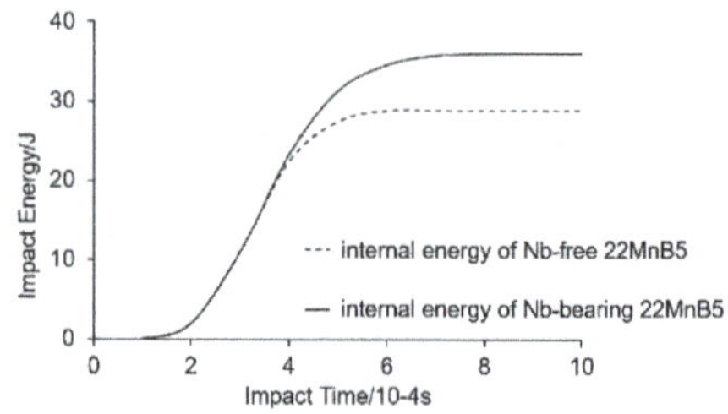

Fig. 4. Curves of internal energy versus time.

4. Effects of Nb-bearing Steel on the Crashworthiness of Hot-stamping B-pillar

The B-pillar structure in this study is shown in Fig. 5, which included hot-stamping outer and cold forming reinforcements and inner. Materials of these parts were 22MnB5, DP590 and DP780, respectively. For Nb-free B-pillar, all material were Nb-free steels and for Nb-bearing B-pillar, the material of outer was replaced to Nb-bearing 22MnB5 and the rest remained unchanged. The side impact FE model of B-pillar is shown in Fig. 6, the material definition was the same as that of Charpy test. The impact speed of punch was 54 km/h, and fracture of outer was observed during impact process.

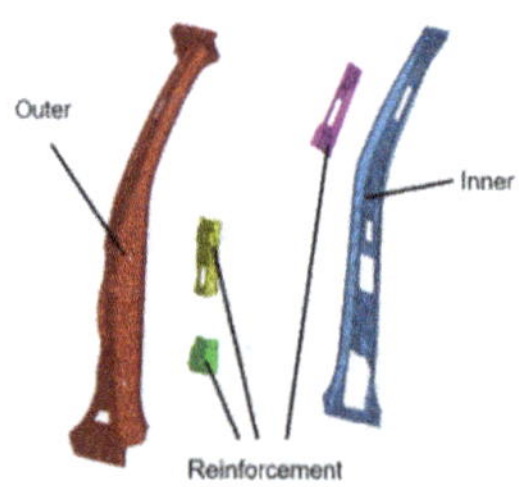

Fig. 5. Structure of B-pillar.

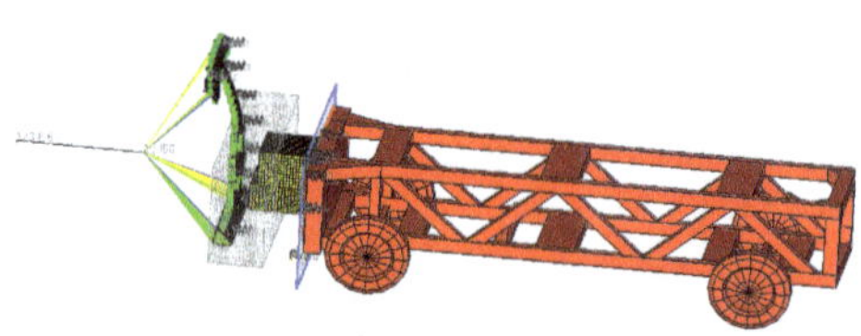

Fig. 6. FE model of side impact of B-pillar.

The plastic deformation and fracture of B-pillar was observed at the outer lower as shown in Fig. 7 (a). The comparisons of local energy absorption at deformation region and intrusions at the critical point of fracture are shown in Fig. 7(b) and (c), respectively. It can be seen from Fig. 8 that energy absorption of Nb-bearing B-pillar is 19.42 % more than that of the Nb-free one at deformation region, which is benefit from the higher impact toughness of the material. Comparisons of intrusion of four measuring points are shown in Fig. 9, where Nb-bearing B-pillar had an obvious increase of intrusion than the Nb-free one. The maximum improvement was 3.34 mm, which was 6.63 % more than that of the Nb-free one. It meant that applying Nb-bearing 22MnB5 contributed to the delay of the fracture of outer in high speed impact and produced a more than 3 mm margin for outer to stand impact and absorb more energy that would transfer to car body and passengers.

Comparison of maximum impact velocities of inner are shown in Fig. 10, where velocity of Nb-bearing B-pillar 8.43 % decrease at the peak value compared with Nb-free B-pillar. It was a good result that less kinetic energy transferred to inner of B-pillar and most of it was dissipated during the plastic deformation of outer. Energy absorptions of B-pillar assemblies are shown in Fig. 11. Nb-bearing B-pillar had 5.7 % improvement on energy absorption, which indicated a better safety performance. It can be concluded that applying Nb-bearing material can effectively slow down the impact, decrease the damage bought to passengers and improve the crashworthiness of hot-stamping B-pillar.

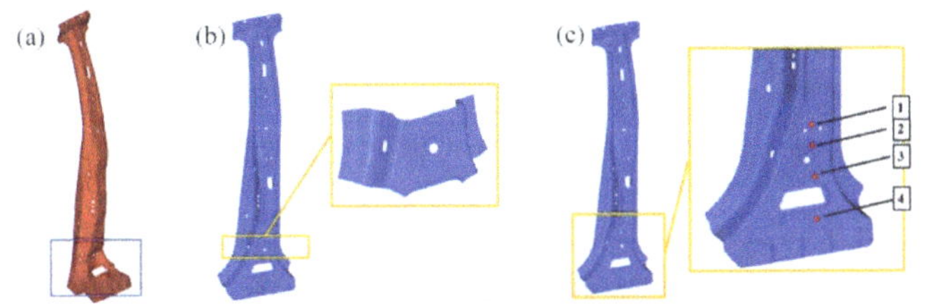

Fig. 7. (a) Location of deformation region of B-pillar and corresponding, (b) measuring region of local energy absorption, (c) four measuring points of intrusion.

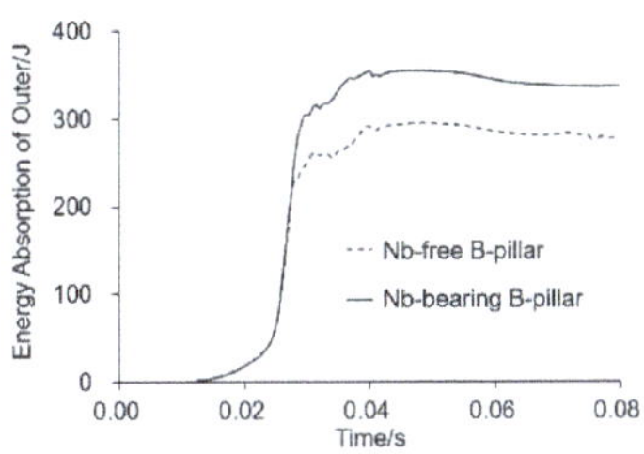

Fig. 8. Energy absorptions of outers at deformation region.

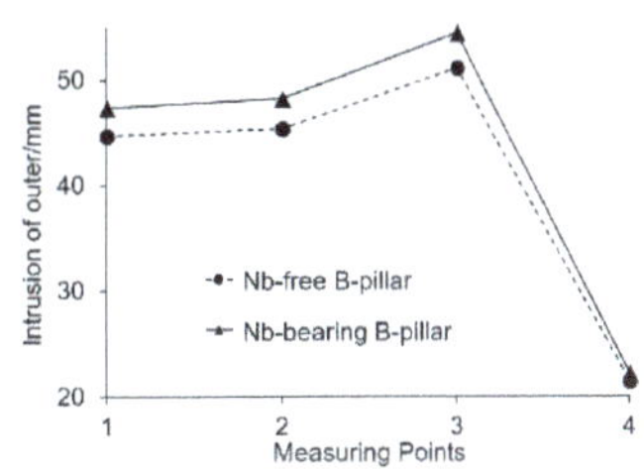

Fig. 9. Intrusions of outers at deformation region.

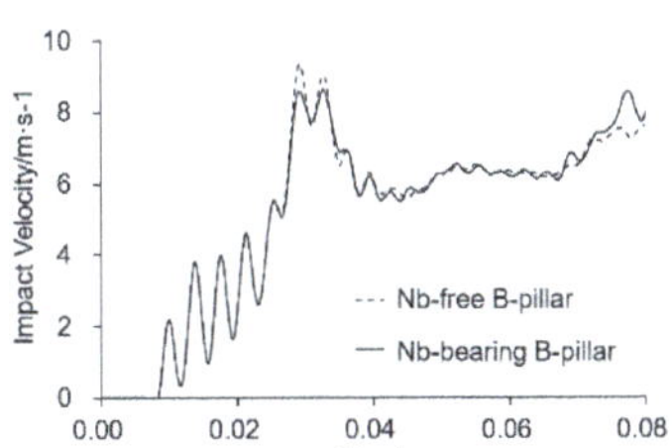

Fig. 10. Maximum impact velocities of inners.

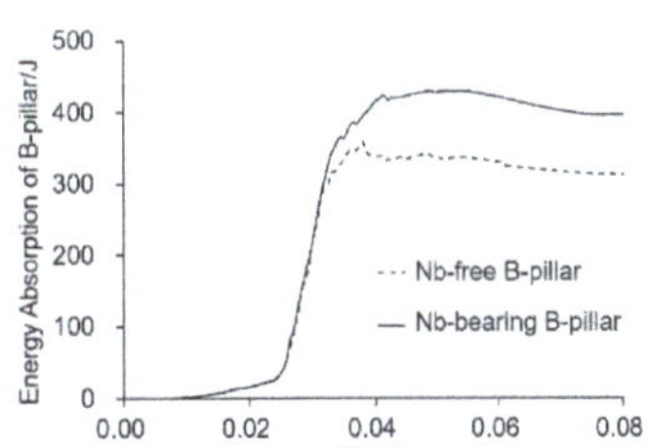

Fig. 11. Energy absorptions of B-pillar assemblies.

Moreover, there was only outer applied with Nb-bearing material and not every region of outer was involved in deformation during impact, the proportion of energy absorption improvement of B-pillar assembly was less compared with that of local regions. Increasing the amount of Nb-bearing material involved in deformation should effectively increase energy absorption during side impact and improve the crashworthiness of hot stamping B-pillar. Therefore, how to increase the contact region and the material involved during deformation should be considered in B-pillar structure design in future study.

5. Conclusion

Based on the experiment and FEM simulation, the difference of Nb-bearing and Nb-free 22MnB5 and the effects of material property improvement brought by element Nb on the crashworthiness of B-pillar were studied. The conclusions drawn are as follows:

(1) Quasi-static and high speed tensile tests and Charpy impact test were processed to identify the differences in mechanical properties of Nb-bearing and Nb free 22MnB5. No differences in tensile test were observed and the impact toughness of Nb-bearing steel increased 31.07 % compared with Nb-free material.

(2) The difference of impact toughness was reflected by the equivalent plastic strain to failure in FEM simulation. The equivalent plastic strain to failure of Nb-bearing and Nb-free 22MnB5 were 0.12 and 0.103, respectively.

(3) Simulation of side impact of a B-pillar with hot-stamping outer was processed and found Nb-bearing 22MnB5 could delay the fracture of part during impact process. Energy absorption at deformation regions increased by 19.26 % and the maximum impact velocity decreased by 8.43 %. The crashworthiness of B-pillar could be improved effectively by applying Nb-bearing material.

Acknowledgments

The support from CITIC METAL Co.Ltd. and Shanghai Key Laboratory for R&D and Application of Metallic Functional materials (Tongji University) is gratefully acknowledged.

References

1. GB/T 229-2007, Standardization Administration of the People's Republic of China. (2007).
2. S. Hlebová and L. Pešek, Toughness of ultra high strength steel sheets, *Mater. Sci. Forum*. **57**, 782 (2014).
3. J. K. Huang and X. H. Don, Numerical simulation and experimental verification of ductile fracture criteria of metals, *Mater. Sci. Technol.* **450**, 4 (2010).
4. Z. W. Li and W. J. Ju, Numerical simulation of relationship of impact absorbed power and failure strain of anchored bolt with high strength, *Coal. Min. Technol.* **50**, 1 (2013).
5. X. D. Xu, H. L. Li and T. Zhang, Parameters for the material failure model based on Charpy impact test, *Explor. Shock.* **56**, 36 (2016).
6. P. Oscar and R. L. Eduardo, Impact performance of advanced high strength steel thin-walled columns, in *Proc. the world congress on engineering (WCE)*, (London, U.K., 2008).

Prediction Method for Dynamic Bending Toughness of 22MnB5 Hot Stamped Steel

Y. Feng†, X. M. Wan, Y. Zhao, L. F. Song, M. T. Ma and J. P. Zhang
China Automotive Engineering Research Institute Co., Ltd., Chongqing, 401122, China
†E-mail: fengyi@caeri.com.cn

Based on the theoretical analysis of basic material mechanics, a prediction model for bending energy absorption has been established. In order to investigate the energy absorption characteristics of 22MnB5 hot stamped steel under plane bending strain state, on the basis of German VDA 238-100 standard, the limit tip cold bending performance test of 22MnB5 hot stamped steel samples with different grain sizes was tested. The results show that the prediction value of the sample's bending energy absorption increased with the decrease of grain size which is consistent with the general regularity. The difference between test results and prediction results of bending energy absorption value is also small which shows that the model is practical and accurate.

Keywords: Hot stamped steel; Bending energy absorption; Prediction method.

1. Introduction

The hot stamping technology combines the high temperature stamping and the quenching strengthening which can obtain the 1500MPa strength grade parts, and perfectly solve the contradiction between the super high strength and the good formability [1-3]. As a safety component, hot stamped parts often bear the angular bending stress along the surface of the vehicle during the collision process. The hot stamped part should have enough strength and excellent toughness to meet the requirements of the collision and energy absorption under the stress state of the plane from the safety considerations [4]. In recent years, the automobile manufacturing industry has been working to establish the relationship model between the limit cold bending performance of automobile steel and the collision performance of the part matrix. Based on the basic material mechanics theory, a dynamic bending energy absorption prediction model with the combination of the limit tip cold bending performance index and the strain rate index is proposed and applied to the 22MnB5.

2. Model Establishment

First, the engineering stress, engineering strain, true stress, true strain and other parameters of the material under the applied tensile load are obtained from the basic material mechanics theory.

$$\sigma_e = \frac{P}{A_0} \tag{1}$$

$$\varepsilon_e = \frac{L - L_0}{L_0} \tag{2}$$

$$\varepsilon_t = \ln(1+\varepsilon_e) \tag{3}$$

$$\sigma_t = \sigma_e(1+\varepsilon_e) \tag{4}$$

In the Eq. (1) ~ Eq. (4), P is the external loading force, L is the length of the standard distance, A_0 is the cross section area of the sample, L_0 is the original length of the standard distance; σ_e is the engineering stress; ε_e is the engineering strain; σ_t is the true stress; ε_t is the true strain. The relationship between tensile energy absorption value and related stress and strain parameters can be obtained which is shown as Eq. (5) ~ Eq. (6).

$$J = \int PdL = \int \sigma_e \cdot A_0 d[L_0(\varepsilon_e+1)] = A_0 L_0 \int \sigma_e d\varepsilon_e \tag{5}$$

$$J = A_0 L_0 \int \frac{\sigma_t}{\exp(\varepsilon_t)} d[\exp(\varepsilon_t) - 1] = A_0 L_0 \int \sigma_t d\varepsilon_t \tag{6}$$

Based on above analysis, a dynamic deformation energy absorption calculation model is created which is shown as Eq. (7). $\sigma(\varepsilon, \dot{\varepsilon})$ is the plastic constitutive model for the dynamic deformation of the material. Further, the dynamic equation of steel is described by Cowper-Symonds equation commonly used in collision which is shown as Eq. (8).

$$J = A_0 L_0 \int \sigma(\varepsilon, \dot{\varepsilon}) d\varepsilon \tag{7}$$

$$\sigma(\varepsilon, \dot{\varepsilon}) = K\varepsilon^n (1+\frac{\dot{\varepsilon}}{C})^p \tag{8}$$

The formula of dynamic deformation and energy absorption of material can be obtained by substituting Eq. (8) into Eq. (9). It can be seen that the dynamic deformation energy absorption is actually the product of quasi static deformation energy absorption and strain rate sensitive factors.

$$J = A_0 L_0 \int K\varepsilon^n (1+\frac{\dot{\varepsilon}}{C})^p d\varepsilon = A_0 L_0 (1+\frac{\dot{\varepsilon}}{C})^p \int K\varepsilon^n d\varepsilon \tag{9}$$

The functional relationship between K, N, elongation rate (δ) and the limit tip cold bending angle of the material is further analyzed, and the Eq. (9) can be converted into Eq. (10).

$$J = A_0 L_0 (1+\frac{\dot{\varepsilon}}{C})^p \int_0^{\delta(\theta)} K(\theta)\varepsilon^{n(\theta)} d\varepsilon = A_0 L_0 (1+\frac{\dot{\varepsilon}}{C})^p \frac{K(\theta)}{n(\theta)+1} \left|\varepsilon^{n(\theta)+1}\right|_0^{\delta(\theta)} \tag{10}$$

3. Test Verification

The above model was tested and verified. The material grade is 22MnB5, and its chemical composition is shown in Table 1.

Table 1. Chemical composition [wt %].

Brand	C	Si	Mn	P	S	Al	Cr	B	Ti
22MnB5	0.22	0.26	1.40	0.0103	0.0037	0.039	0.17	0.002	0.020

All the samples were austenitized at 880~970 °C and the austenitizing time is 5 minutes. Then all the samples were quenched form austenitic structure to martensitic structure with different grain sizes which is due to the austenitizing temperature. As shown in Figure 1, the tensile and bending tests of the quenched samples were made to obtain the relationship between the tensile mechanical properties (K, N, δ), the limit tip cold bend angle (θ), grain size and bending energy absorption. The results of the tests are shown in Table 2.

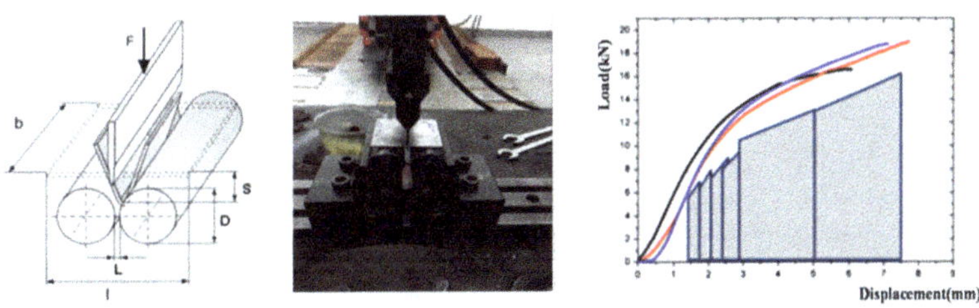

Fig. 1. Test process and bending curve.

Table 2. Tensile mechanical properties under different austenitizing temperatures.

No.	$R_{p0.2}$ (MPa)	R_m (MPa)	δ (%)	Grain size (mm)	Bending angle (°)	Bending toughness (J)
1#	1085	1620	8.50	11.5	58.3	185
2#	1045	1560	5.00	15.5	52.5	100
3#	1025	1550	6.50	24.0	49.4	121
4#	977	1440	6.25	34.0	48.6	96
5#	970	1475	5.75	45.0	36.5	85

According to Table 2, the model of the corresponding function relationship between K, N, elongation rate (δ) and the maximum cold bending angle (θ) of the 22MnB5 is obtained which is shown as Eq. (11) ~ Eq. (13).

$$K = -2.1 \cdot \theta^2 + 187.6 \cdot \theta - 898.2 \tag{11}$$

$$n = -0002 \cdot \theta^2 + 0.015 \cdot \theta - 0.09 \tag{12}$$

$$\delta = 0.012 \cdot \theta^2 - 0.996 \cdot \theta + 26.8 \tag{13}$$

The Eq. (11) ~ Eq. (13) is substituted into (10), and then the C and P values (C=6114, P=1.12) of Eq. (10) are obtained according to the measured bending energy absorbing values in Table 2. The theoretical prediction model of bending energy absorption corresponding to 22MnB5 under different strain rate conditions can be established, and the prediction results are shown in Figure 2. It can be seen that under certain grain size conditions, with the increase of strain rate, the bending energy absorption value of the 22MnB5 will also increase. In addition, as the grain size decreases, the bending energy absorption value of the material increases, which is consistent with the general law between toughness and matrix structure [5]. As shown in Figure 3, through the comparison of the impact test data and the predicted data, it can be seen that the model established in this paper can be used to predict the dynamic bending itoughness of 22MnB5 accurately.

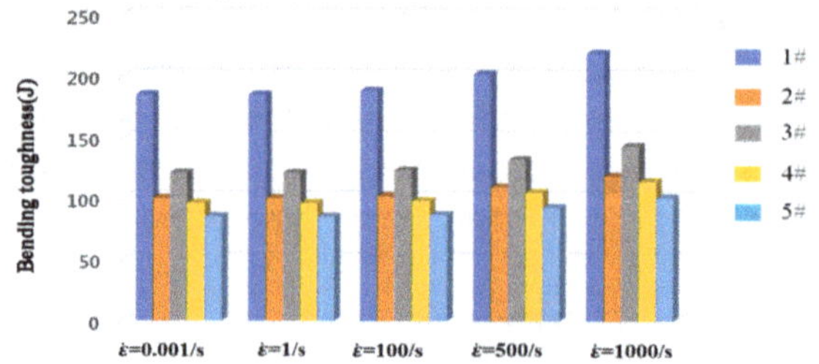

Fig. 2. Prediction of bending toughness of 22MnB5 under different strain rates.

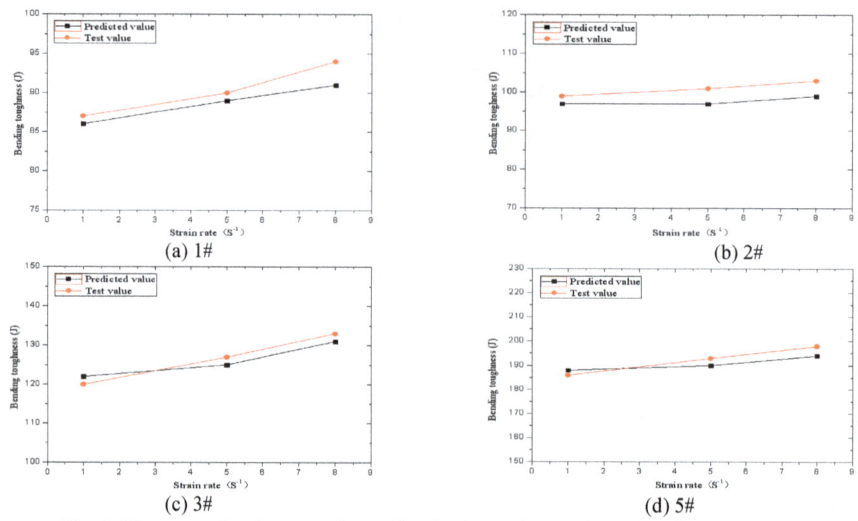

Fig. 3. The comparison between the predicted value and test value of bending toughness.

4. Conclusion

Based on the theoretical analysis of basic material mechanics, a prediction model for bending energy absorption has been established. In order to investigate the energy absorption characteristics of 22MnB5 hot stamped steel under plane bending strain state, on the basis of German VDA 238-100 standard, the limit tip cold bending performance test of 22MnB5 hot stamped steel samples with different grain sizes was tested. The results show that the prediction value of the sample's bending energy absorption increased with the decrease of grain size which is consistent with the general law. The difference between test results and prediction results of bending energy absorption value is also small which shows that the model is practical and accurate.

Acknowledgments

This work is supported by the capital and technical support provided by the national key R & D project "Electric vehicle lightweight common technology platform" (2016YFB0101607).

References

1. H. Karbasian and A. E. Tekava, A review on hot stamped, *J. Mater. Process. Tech.* **210**, 2103 (2010).
2. P. Rozenak and I. M. Robertson, HVEM studies of the effects of hydrogen on the deformation and fracture of AISI type 316 austenitic stainless steel, *Acta Metall. Mater*. **38**, 31 (1990).
3. J. Bian, H. Z. Lu and W. J. Wang, Impact of alloying design on the crash relevant material properties of press hardening steel based on Mn-B concept, in *ICHSU*, (Changsha, China, 2015).
4. M. T. Ma, H. L. Yi and H. Z. Lu, On the lightweight of automobile, *Eng. Sci.* **11**, 20 (2009).
5. W. S. Choi and B. C. DeCooman, Characterization of the Bendability of Press-hardened 22MnB5 Steel, *Steel Res. Int.* **5**, 824 (2014).

Investigation for a Hot-stamped B-pillar of a SUV Based on Dynaform

X. Li[a,†], Z. Z. Chen[b] and H. Liu[c]
School of Materials Science and Engineering, University of Shanghai for Science and Technology, Shanghai, 200093, China
E-mail: [a,†]397679183@qq.com, [b]zzhchen@usst.edu.cn, [c]709335553@qq.com

In response to the increasingly strict automobile exhaust emission standards, automakers have been increasingly adopting hot-stamped components to improve fuel efficiency and crash behavior. In this paper, the hot stamping for a B-pillar reinforcement of a SUV in B1500HS steel was studied by numerical simulation and experiment. The formability analysis for the B-pillar is emphasized firstly to determine the appropriate hot stamping process. Then, a finite element model was established and simulated by Dynaform. According to numerical simulation analysis, a set of optimal parameters is determined. Additionally, stress-strain states of the deformation zone and FLD of hot stamping were studied. Finally, the accuracy of the results and validity of the simulation method were verified by experiment. The results of micro-structure observation show that the micro-structure of B-pillar after hot stamping is lath martensite, and the Vickers hardness of B-pillar is more than 466 HV. This study verifies the feasibility of hot stamping process based on numerical simulation by experiment and provides important guiding significance for hot stamping production.

Keywords: Hot stamping; B-pillar reinforcement; Formability; Numerical simulation.

1. Introduction

Nowadays, high strength steels are wildly used in car body to realize the requirements for low weight and high security. The hot stamping can significantly improve the forming properties of UHSS and mechanical properties of components. In hot stamping, forming and quenching are performed in one step in which phase transformation and plastic deformation happen simultaneously and affect each other [1, 2].

In this paper, B-pillar, one of the typical automobile safety structural components, was selected as the study object. The thermal-mechanical-metallurgical coupling analysis model is set up by Dynaform to analyze the distribution of stress-strain and the change of temperature field in the hot stamping of B-pillar. Comparing with hot stamping experiment, that improves the actual production of hot stamping B-pillar.

2. Finite Element Numerical Simulation

2.1. *Finite element model*

The B-pillar reinforcement selected in this paper is shape-complicated and large-sized. There are a large number of bulges and grooves in the B-pillar, which affect the accuracy of the forming. Traditional cold forming process is difficult to produce the B-pillar. According to the structural analysis of B-pillar, the direct hot stamping process is adopted.

After calculating the blank size of the B-pillar by Dynaform and adding certain trimming allowance, a three-dimensional slice model of B-pillar Metal Sheet Blank is established. As shown in Fig. 1, the red line is the trimming line. The integral algorithm is selected to ensure the calculation precision. The initial thickness is 1.2 mm, surface area is 0.43 m^2. The final hot stamping finite element (FE) model of B-pillar is composed of die, punch, blank holder and blank, as shown in Fig. 2.

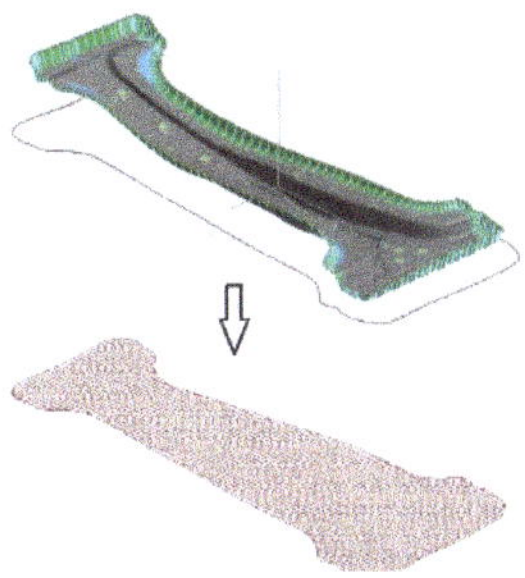

Fig. 1. Parting line and blank model.

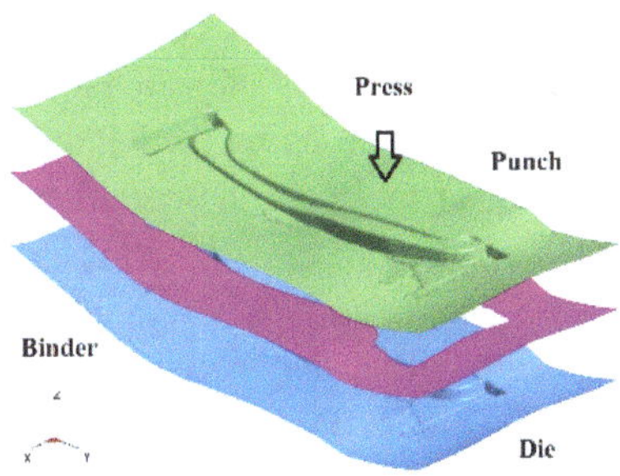

Fig. 2. B-pillar model for hot stamping simulation.

The material of hot stamping B-pillar is B1500HS steel and the chemical composition of B1500HS steel is shown in Table 1. Fig. 3 shows the true stress–strain curves of B1500HS at different temperatures at a strain rate of 10 s^{-1}. With the increase of temperature, the formability improves while the yield stress of B1500HS decreases.

Table 1. Chemical compositions of the boron steels B1500HS (wt %).

C	Si	Mn	Cr	M0	B	S	P
0.23	0.25	1.25	0.19	0.04	0.003	0.006	0.015

The *MAT_244 structure material model is used to characterize the thermodynamic behavior of B1500HS. The temperature dependent Young's modules and shear modules of the material are shown in Fig. 4. This material model is developed for both shell and solid models. It is mainly suited for hot stamping processes where phase transformations are crucial. It has five phases and it is assumed that the blank is fully austenitized before cooling.

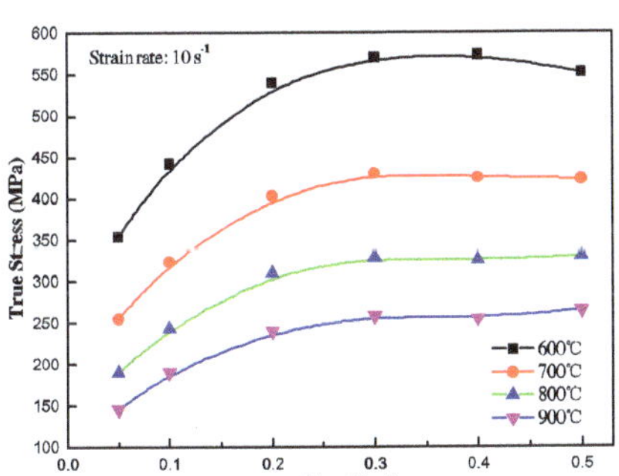

Fig. 3. True stress and strain curves at strain rate of 10s^{-1}.

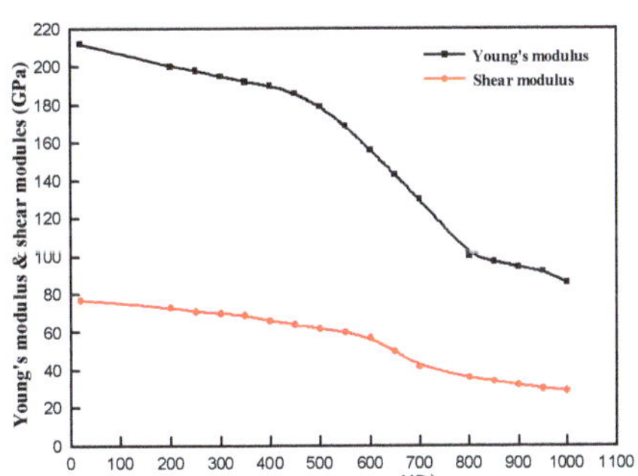

Fig. 4. Young's modulus and shear modules.

The hot blank will generate radiation and convection with environment hot stamping process. In this paper, 6*MAT_THERMAL_ ISOTROPIC_TD_LC, a hot material model is used to define the thermo-physical properties of B1500HS. Thermal conductivity is

32W/m-K, friction coefficient is 0.35 and it is affected by the contact condition. The temperature of blank is set to 930 °C. The stamping speed is 300 mm/s, and the quenching time should be 15 s to achieve full martensite transformation.

2.2. *Results and discussion*

The differences of stress-strain field caused by different parameters will result in various properties [3]. The variation of stress and strain in the section A-A (Fig. 5) can be seen in Fig. 6. The top corner A is subjected to the pressure stress, the tensile stress and the radial compressive stress caused by the bending, resulting in a certain degree of thinning. With the increasing of the blank holder force, the bottom corner B is subjected to the more tensile stress and compressive stress in the direction of thickness, which leads to the smaller deformation and the fracture easily. Therefore, the flange and the bottom of the B-pillar need to fill in the process of hot stamping to improve die lubrication conditions and reduce the production of wrapping defects.

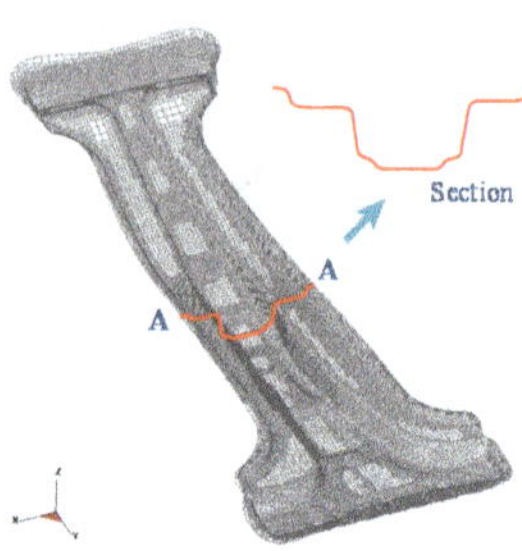

Fig. 5. Cross section A-A.

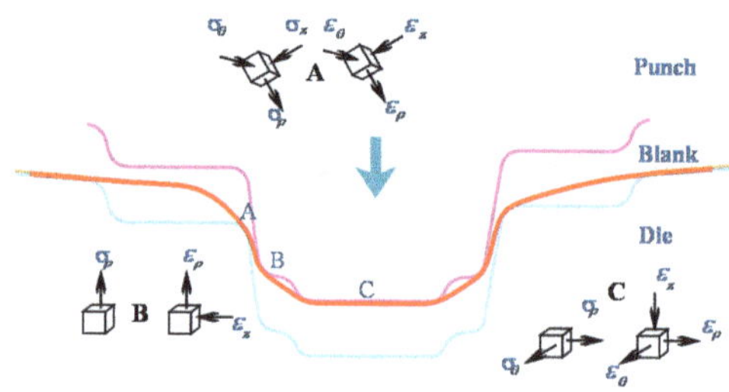

Fig. 6. Equivalent stress and strain distribution of B-pillar in section.

The temperature affects mechanical properties and the microstructure of the hot stamped parts significantly [4]. Fig. 7 is the temperature distribution of hot stamped B-pillar in different stage. The heat transmission begins by the contact between the blank and die. The maximum temperature of blank is above 590 °C. Then, the temperature of blank decreased from 930 °C to 560 °C (Fig. 7b). The outline of the B-pillar begins to be clear. At the same time, there are several low temperature regions, which are distributed in the middle side wall and the edge of the B-pillar. After quenching, the temperature distribution of B-pillar is inhomogeneity, and the temperature of the B-pillar is below 150 °C. This is because the B-pillar has a relatively fast heat transfer due to the close contact with the die. It is believed that martensite transformation is complete.

The forming limit diagram (FLD) analysis results of hot stamped B-pillar is shown in Fig. 8. The dangerous locations of crack and wrinkle can be judged according to the FLD. The overall working area of the B-pillar is well formed, without cracks and excessive thinning [5]. The material at the top area of the B-pillar is easily piled up to produce wrinkles. The wrinkle in B-pillar will be completely cut off in subsequent trimming process. Therefore the overall formability of the B-pillar is satisfied.

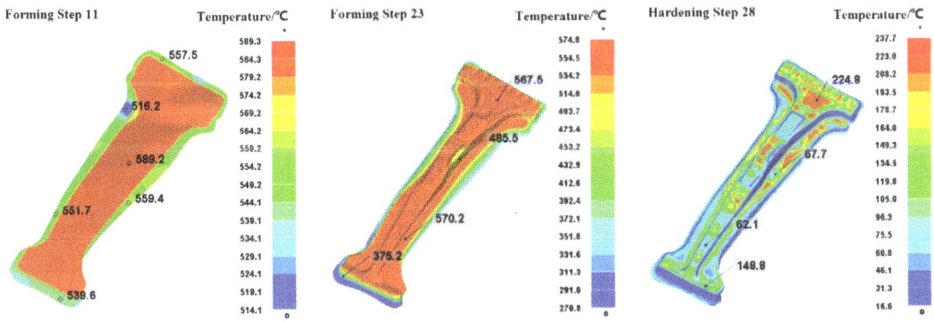

(a) before deformation (b) forming stage (c) quenching stage

Fig. 7. Temperature distribution in different stage.

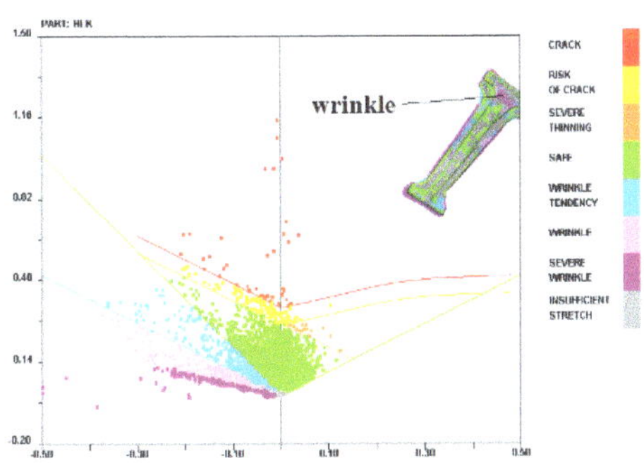

Fig. 8. FLD of hot stamped B-pillar.

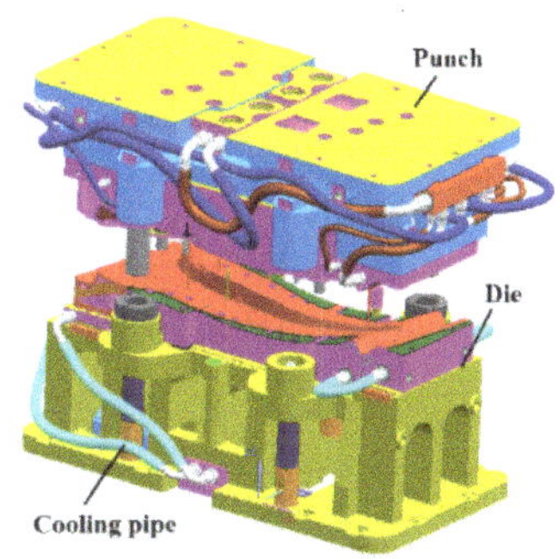

Fig. 9. Experimental apparatus for hot stamping.

3. Experimental Validation

Experiments on hot stamping are performed to evaluate the accuracy of simulation results. The blank was put into furnace to be heated until it austenitize completely. Then, the hot blank was transferred quickly to the die to form and quench. The various parameters of field trials were consistent with the simulation parameters. The cooling pipe is arranged inside the hot stamping die (Fig. 9). Fig. 11 demonstrates the microstructures of two investigated locations (Fig. 10) after quenching, lath martensite can be obtained.

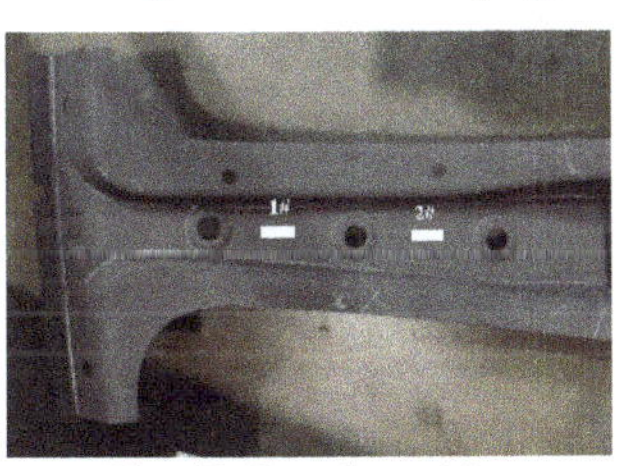

Fig. 10. Sample positions.

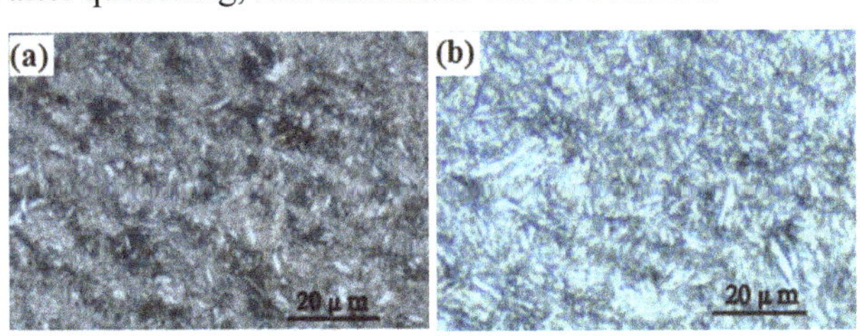

Fig. 11. Optical micrographs of B-pillar at the different location.

The cross section of the B-pillar selected includes the characteristics of the bottom, flange and side wall of the B-pillar. The thickness of the section A-A in Fig. 12 is measured and compared with the thickness distribution nephogram. As shown in Fig. 13, the thickness of B-pillar is between 1.1~1.2 mm and distributed uniformly, which is coherent with the conclusion of the above stress-strain analysis. The maximum thinning rate is lower than the standard 25 %, which is in the range of forming safety [6].

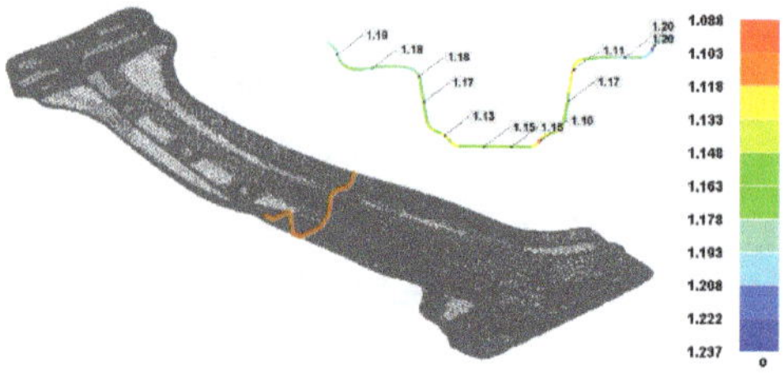

Fig. 12. Locations of thickness measurement on B-Pillar.

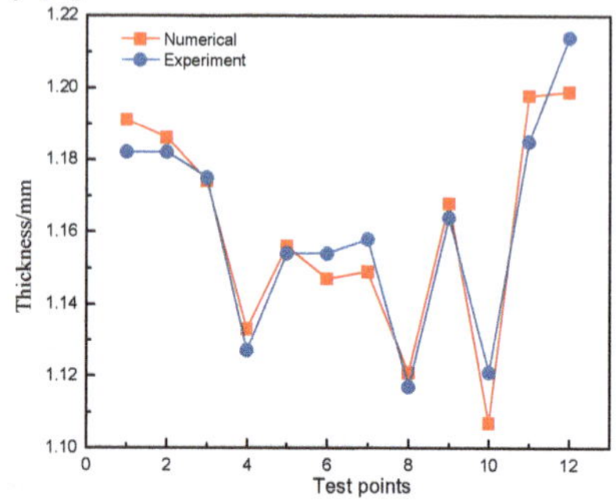

Fig. 13. Experimental and simulated thickness.

The Vickers hardness distribution of 7 typical positions of B-pillar was measured by MH-3 digital micro-hardness tester. The hardness values at different positions of the hot stamping B-pillar are shown in Table 2. It is obvious that the average hardness of B-pillar is over 466 HV. The result of experiment agrees well with the numerical simulation results, which meet the requirements of the hot stamping B-pillar.

Table 2. Vickers hardness for hot-stamped B-pillar.

Testing points	1	2	3	4	5	6	7
Hardness-experiment/HV	471	462	456	458	451	467	474
Hardness-simulated/HV	474	467	452	466	457	470	481
difference	3	5	4	8	6	3	7

4. Conclusion

This paper is based on B-pillar reinforcement and a FE model established by Dynaform focused on the hot stamping process. The best process parameters of B-pillar are determined. The stress-strain distribution of B-pillar and thinning of sheet during hot stamping process are analyzed. The simulation results show that the rounded corners and straight wall areas of B-pillar tend to be thinning due to the larger tensile stress and compressive stress. The FLD shows that the overall deformation of the hot stamped B-pillar is sufficient. The hot stamping experiment was carried out and the microstructure distribution and hardness were obtained. The results showed that the metallographic structure of the B-pillar is lath martensite and the thickness distribution is basically uniform, whereas the Vickers hardness is above 466 HV. It can be concluded that the results of the test were in agreement with the simulation results.

References

1. Y. Shen, Y. Song, L. Hua and J. Lu, Influence of plastic deformation on martensitic transformation during hot stamping of complex structure auto parts, *J. Mater. Eng. Perfor.* **4**, 1 (2017).

2. W. Wang, L. Zhang, M. Guo, L. Hung and X. Wei, Non-isothermal deformation behavior and FE simulation of ultrahigh strength BR1500HS steel in hot stamping process, *Int. J. Adv. Manuf. Tech.* **9-12**, 1 (2016).
3. H. Karbasian, A. E. Tekkaya, A review on hot stamping, *J. Mater. Process. Tech.* **210**, 2103 (2010).
4. B. He, L. Ying, X. Li and P. Hu, Optimal design of longitudinal conformal cooling channels in hot stamping tools, *Appl. Therm. Eng.* **106**, 1176 (2016).
5. M. Ganapathy, N. Li, J. Lin and D. Bhattacharjee, Investigation of a new hot stamping process with improved formability and productivity, in *Int. Conference on the Tech. of Plasticity*, (Cambridge, UK, 2017).
6. H. Hou, H. Li and L. He. Effect of technological parameters on microstructure and accuracy of B1500HS steel parts in the hot blanking, *Int. J. Adv. Manuf. Tech.* **1-2**, 1 (2017).

The Forming Process Analysis, Experimental Verification and Quality Optimization of the Variable Section Frame Longeron with QSTE650 High Strength Steel

Y. Q. Xu†, L. C. Cui, L. W. Zhao, M. Tang and D. X. Liu
Anhui Jianghuai Automobile Group Corp., Ltd, Hefei 230601, China
†E-mail: zlw.jszx@jac.com.cn

With the emission upgrade and the release of new regulations such as GB1589, it is urgent to develop the automobile lightweight technology to meet the demand of energy conservation and emission reduction in the automobile industry. As the frame structure of the frame, the longeron material upgrade is very important. With the upgrade of the longeron material to QSTE650, the high strength steel rebound, the precision quality of the plastic difference and the cracking risk become the research hotspot. Based on the CAE simulation platform, this paper analyzes the opening, twisting and warping elastic modulus of the variable section longeron with QSTE650, and forms the compensation scheme based on physical verification and multiple iterative optimization. Through experimental analysis and organization observation, the solutions of longeron neck, cracking and wrinkling were obtained, and the design planning of longeron was established to serve the automobile industry.

Keywords: Finite element simulation; Elastic modulus; Type surface compensation; Resistance to tear hole spacing L; Cracking defects.

1. Introduction

With the emission upgrade and the release of new regulations such as GB1589, it is imperative to carry out the light weight of the frame under the background of the national overrule. As the frame structure of the frame, the longeron material upgrade is very important. How to ensure the precision and quality of the longeron is the key of the frame technology [1-3].

At present, B510Lmaterial is still widely used in frame longeron, and the yield strength is close to 355 Mpa. Considering that the frame needs to adopt double longeron, it has become a bottleneck of light weight of the frame. Due to the high strength of QSTE650 materials, the thickness of the longeron is greatly reduced under the condition of the same frame stiffness, which is the main measure of the lightweight scheme. With the upgrade of the longeron material to QSTE650, the high strength steel rebound, the precision quality of the plastic difference and the cracking risk become the research hotspot [4-7]. Based on the CAE simulation platform, this paper analyzes the opening, twisting and warping elastic modulus of the variable section longeron with QSTE650, and forms the compensation scheme based on physical verification and multiple iterative optimization. Through experimental analysis and organization observation, the solutions of longeron neck, cracking and wrinkling were obtained, and the design planning of longeron was established to serve the automobile industry.

2. CAE Simulation and Process Analysis of the Longeron Forming

Traditional evaluation of the deep drawing process is based on experience and analogy design method. After the die design is finished, the specified plate is used to punch the parts in the mold. According to the various quality problems of the parts, such as rupture, wrinkling, lack of rigidity and excessive rebound, the mould and stamping process parameters are modified repeatedly until the qualified parts have been rushed out. Based on CAE simulation platform technology, this problem can be solved [3, 4].

The longeron material adopts QSTE650, the yield strength of the material is higher, the processing hardening phenomenon is serious, the rebound quantity is large, the precision control is difficult, the chemical composition and mechanical properties of material are shown in Table 1. The longeron considers the engine layout, and the longeron abdominal surface X= (-600~-400) mm and X= (600~800) mm have 25 mm steps respectively. The longeron is designed to meet the common platform of multiple models, and the width of longeron wing surface are designed respectively 110mm, 180mm and 140mm in the transition of X= (-500~100) mm and X= (2900~3300) mm. In the modeling of CAE, the size of the material is 5966 mm×350 mm×5 mm, the friction coefficient is 0.15, and the part is rigid 50. The simulation results are as follows.

Table 1. Mechanical properties of QSTE650.

Y.S./MPa	T.S./MPa	Elongation %	Bending tests
≥650	700~880	≥14	d=2a

(a) Using convex mold and avoid empty plate steps, compensation opening rebound, stripper plate bench height X=1mm, crush width is Y=26mm, the other steps do chamfering to alleviate the indentation, the CAE analysis results as shown in Fig. 1.The opening is about 1mm, and the accuracy standard is (0-2)mm.

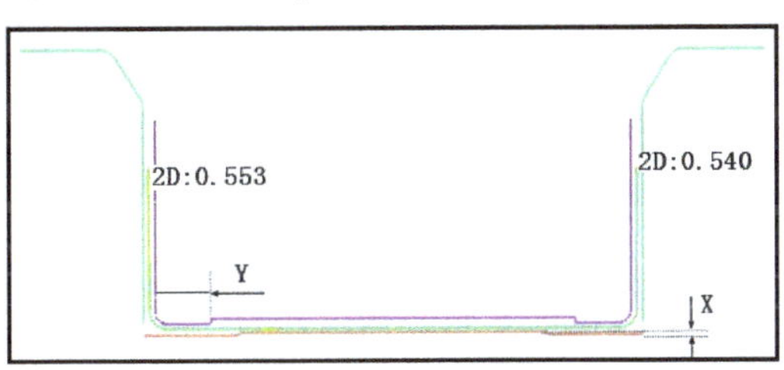

Fig. 1. The opening analysis.

(b) Part of the material is squeezed and part of the material is pulled in the forming of the longeron. According to the equivalent volume theory of the stamping deformation, the material is thickened and the material is thinned, and the elastic deformation will be restored after forming, leading to the warping. Fig. 2 is not compensation model surface forming CAE simulation results.

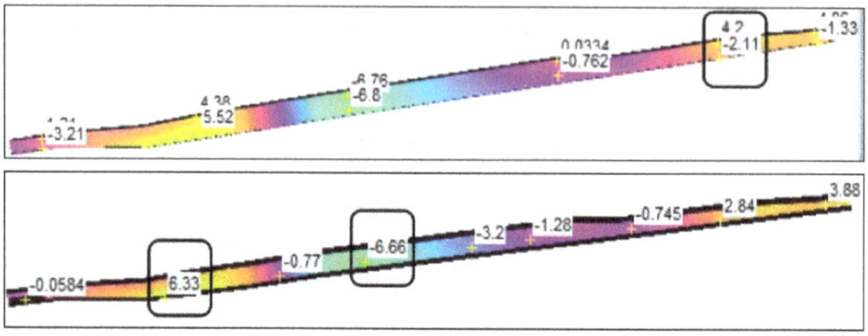

Fig. 2. The twisting and warping analysis.

According to QSTE650 mechanical properties and tensile test results, elongation decreased and plastic toughness is poor. The inconsistency between the inner and outer layers causes the segregation to be serious. The above cause more prone to delamination crack defect in sheet forming.

From the structural analysis of the longeron, the process risk is as follows: (1) the crack of the connecting hole and end in the corner of parts; (2) the neck and crack in the the change section of parts; (3) the thickening area is wrinkled and pulled.

3. Physical Verification of Longeron

Based on existing mold tooling similar to QSTE650 longeron modelling, the section longeron is verified, as shown in Fig. 3. It obtains the opening, twisting and warping elastic modulus, judges the neck and crack in the the change section of parts and accounts forming tonnage.

Fig. 3. Section sheet forming verification.

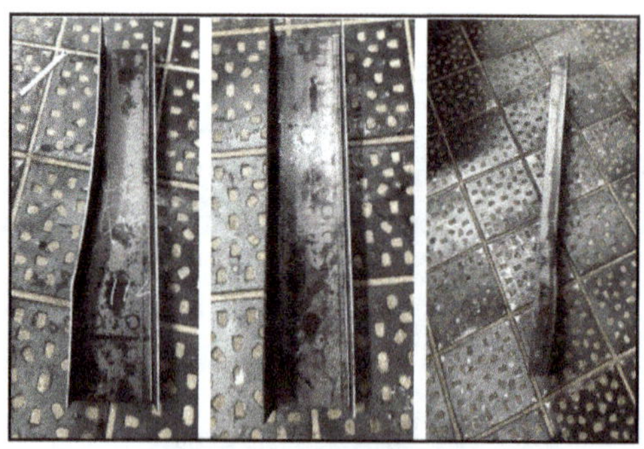

Fig. 4. Sheet forming results.

(a) The forming force of the longeron is calculated from the section sheet metal forming, and the pressure of the main cylinder is 4200T;

(b) Observe the forming state of the sheet material at the variable section, and there is no necking and cracking in the change section of parts, and the process is satisfied, as shown in Fig. 4;

(c) Stripper plate bench height is X=0.5 mm, and crush surface width is Y=20 mm. The opening of cross section is 3 mm with measuring the opening of part, as shown in

Fig. 5, consistent with CAE analysis. The opening controllable under increasing the amount of compensation and reducing the cylinder pressure of equipment.

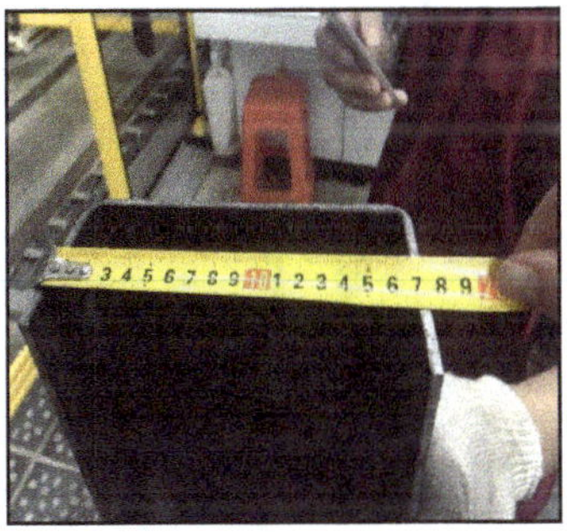

Fig. 5. The opening rebound 3mm.

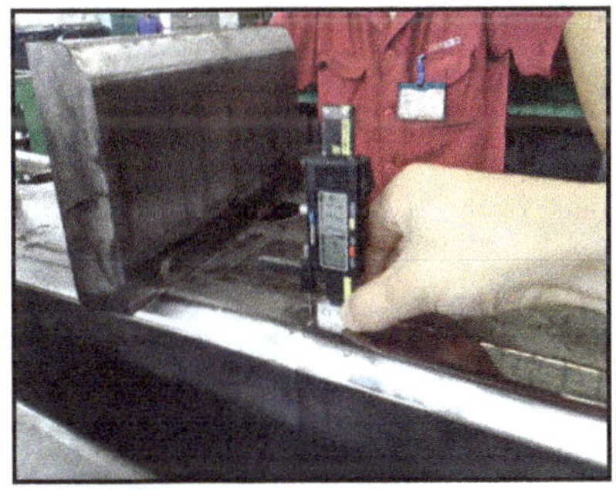

Fig. 6. The assembly surface difference≤0.1mm.

Based on the existing mold tooling conditions, the panel board of hardness and smoothness is improved through the set piece of PVD surface treatment. Because of the deformation is small, the panel board can be directly assembled production. The result of verification is good by 30000 part production, and form R Angle position without mark, as shown in Fig. 6.

4. QSTE650 Experimental Verification

For identifying the crack process risk of the connecting hole and end in the corner of parts with QSTE650 longeron, it establish longeron process design planning, to guide product design. This chapter verified QSTE650 standard sample process based on the standard of U/V bending experiment.

The experimental verification process is as follows: sample making → punching → bending → quality detection. Consider 180 longeron structure design and U bending test device, sample size is designed to 300×88×5 mm, as shown in Fig. 7. Based on mechanical properties and punching device φ10 mm.

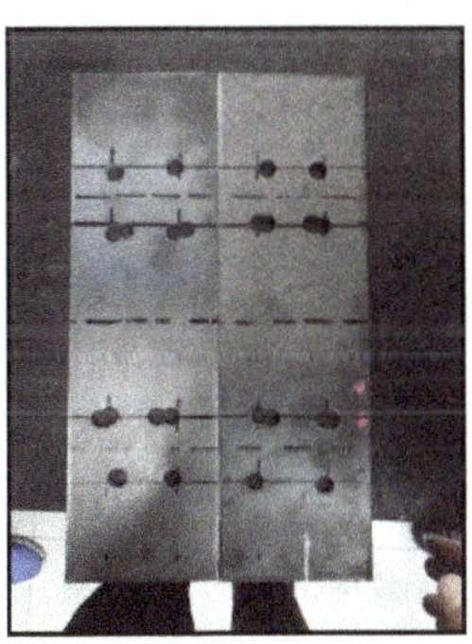

Fig. 7. The sample making.

Fig. 8. The bending device.

The bending experiment of R=2mm and U type bending experiment R=1.0t/1.5t were verified as shown in Fig. 8.

(a) The appearance analysis of the bending test sample from R=2mm was analyzed, and there was no cracking problem at the corner.

(b) From R=1.0t/1.5t U bending test sample end analysis, the shear section bright band appeared cracking in the inner part of R, while the bright band had no crack at the outside of R, as shown in Fig. 9.

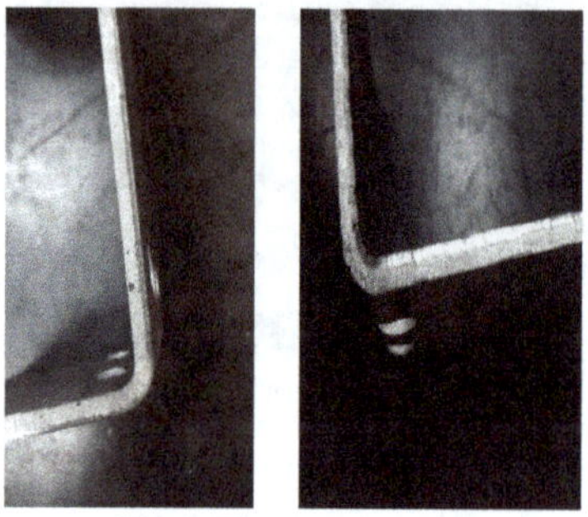

Fig. 9. The corner of sample.

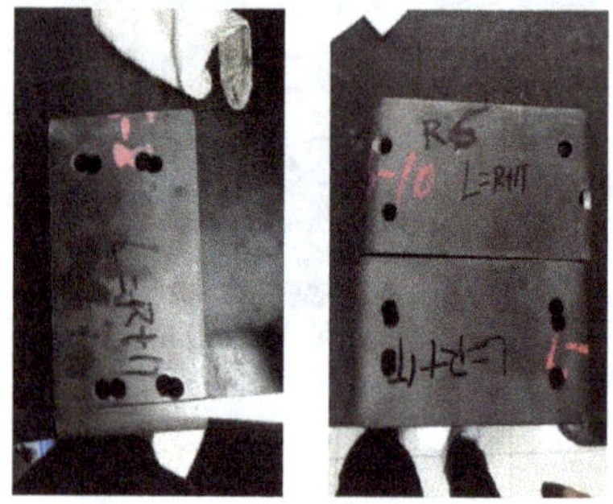

Fig. 10. The bending results.

Through U type L=R+1t/2t bending experiment verification, sample punching euphotic zone on the outside of R by L=R+1t/2t no cracking problem occurred, as shown in Fig. 10. Sample punching euphotic zone on the inside of R by L=R+2t have occasional cracking phenomena, and connected to the corner office.

Therefore, according to the design of L=R+2t, and ensuring that the punch light belt is on the outside of R, the process is satisfied. Considering the longitudinal beam process planning, the blanking process of the QSTE650 longeron blanking is used for the blanking (1 pay molds, the right and left parts) → the punching + two end shear (2 die, the left and right pieces are separated).

5. QSTE650 Experimental Verification

Cut along the hole diameter, in low power and high power drill hole sampling observation, it can be seen punching blanking euphotic zone and tear tape. Euphotic zone of height is larger (level accounted for about 2/5 or more), torn belt surface uneven, tearing the pits, tearing the serrated fracture (Fig. 11).

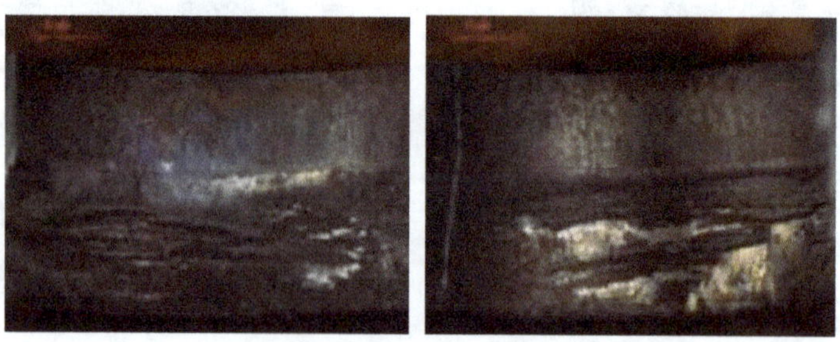

Fig. 11. The shape of hole wall.

Was observed in the vertical wall of hole on the cross section of crack, crack usually in tear with serrated fracture to development (Fig. 12 and Fig. 13), near the crack or end have not found the inclusion of crack, crack near homogeneity segregation has not been found. A lot of tiny carbonized titanium precipitates can be observed on the split surface (Fig. 14 and Fig. 15).

Fig. 12. The punching crack.

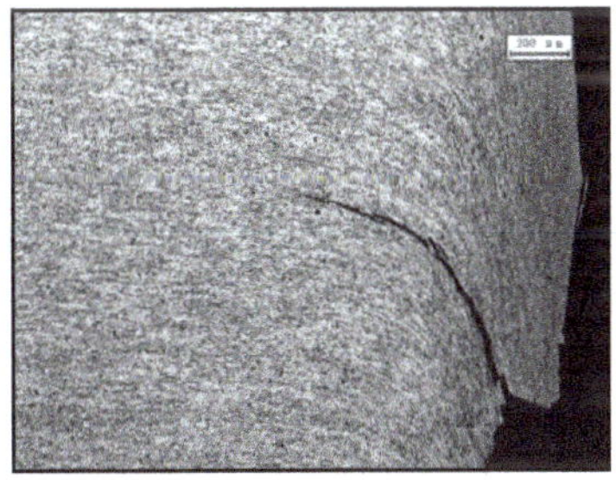

Fig. 13. The crack tissue.

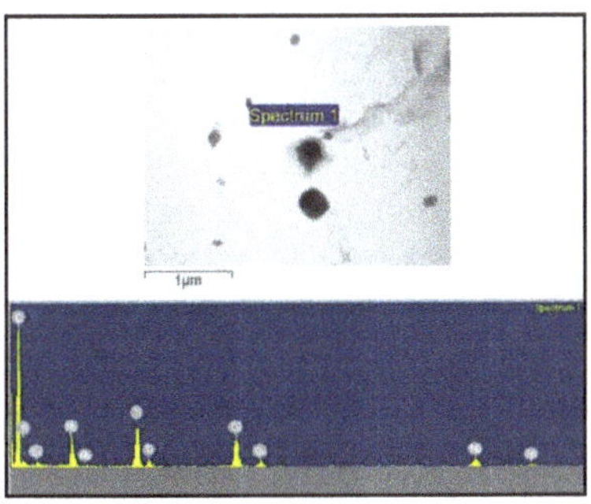

Fig. 14. Precipitate phase of the split surface.

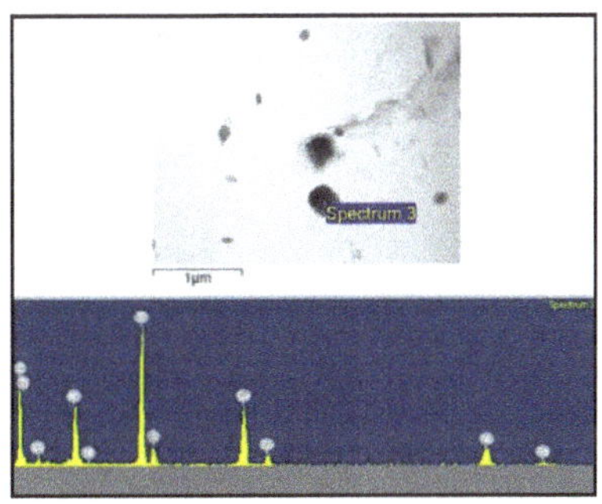

Fig. 15. Analysis of precipitation phase energy spectrum.

The crack in the hole wall of the QSTE650 longeron is in the cutting zone, which is also developed at the sawtooth fracture. Under the action of load before metal fracture, the cracks formed by cavity nucleation, growth and polymerization are the main forming mechanism and root cause of fracture. Most of these holes are nucleated in the second phase of particles, which lie in the separation of particles or particles, and then the holes will grow under the plastic strain of the surrounding metals.

6. Conclusion

Based on the CAE simulation platform, this paper analyzes the opening, twisting and warping elastic modulus of the variable section longeron with QSTE650, and forms the compensation scheme based on physical verification and multiple iterative optimization. Through experimental analysis and organization observation, the solutions of longeron neck, cracking and wrinkling were obtained, and the design planning of longeron was established to serve the automobile industry.

References

1. J. Gao, X. B. Wei and X. M. Wei, The car left and right rear wheel cover design formability analysis and process the outer plate. *Mod. Manuf. Eng.* **8**, 64 (2013).

2. F. Y. Xu, *Die design and manufacture application handbook.* (Beijing: China Machine Press, 1991).
3. Y. Q. Liu, Quantitative for springback of unloading and trimming in sheet metal stamping forming, *Chin. J. Mech. Eng.* **16**, 116 (2013).
4. Y. L. Zhang and Y. Li, Numerical simulation and analysis of solid-liquid two-phase flow in centrifugal pump, *Chin. J. Mech. Eng.* **1**, 53 (2013).
5. X. F. Wang and Y. C. Zhang, *Cold stamping die design and manufacture.* (Beijing: Beihang University Press, 2012).
6. W. L. Chen, Sheet forming CAE analysis course. (Beijing: China Machine Press, 2005).
7. R. Yao, W. B. Zhu and P. Hung, Accuracy analysis of stewart platform based on interval analysis method, *Chin. J. Mech. Eng.* **3**, 29 (2013).

Numerical Simulation of Hot Stamped 7075 Aluminum Alloy

B. Q. An[†], Y. Z. Geng, Y. B. Li, Z. Y. Han, Y. P. Sun and X. L. Yu
LingYun Industrial Corporation Ltd., Zhuozhou, Hebei, China
[†]E-mail: anbaoqin@lygf.com

Hot stamping of aluminum alloy is a new technology used in manufacturing aluminum alloy sheet. Hot stamping experiment of U-shape 7075 aluminum alloy component was conducted in this work. A thermal-mechanical coupled finite element analysis model was also established. The effects of binder force, initial temperature and forming speed on the thickness of a component were studied. Results showed that the component had a tendency to wrinkle as the binder force was lower than 10KN and there was no wrinkle with the binder force higher than 20KN. The minimum thickness of the component reduced dramatically with increasing binder force. As the initial forming temperature increased gradually, hardening exponent and the degree of uniform deformation decreased, resulting in local thinning and final decrease in thickness. The increase of forming speed caused the minimum thickness of the component to decrease and the temperature of the component after forming to increase. The thickness results of the finite element analysis were in good agreement with the experimental results. The validity of the established finite element model was verified and it could be used to guide the manufacture of aluminum alloy hot stamping.

Keywords: Aluminum alloy; Hot stamping; Thermal-mechanical coupling; Finite element analysis.

1. Introduction

There is an increasing demand for aluminum alloy vehicle body components in recent years due to the improved requirements for vehicle weight reduction. However, the low formability and springback of aluminum alloy sheet at room temperature seriously restrict the application of aluminum alloy to vehicle body components, especially the middle strength or high strength grade of aluminum alloy components which cannot even be processed at room temperature. To develop a new type of aluminum alloy with high plasticity or advanced forming technology is an effective way to obtain higher formability in the processing of aluminum alloy sheet [1-3]. Hot stamping of aluminum alloy has no special demand for the current aluminum alloy materials and has the advantage due to its quick production. Its process includes that the aluminum alloy sheet is treated with solid solution first, and then transferred to cool die for forming and quenching simultaneously to obtain supersaturated solid solution. Finally, the aging treatment precipitates the strengthening phase to obtain high strength [4-6]. Hot stamping of aluminum alloy combines hot forming and heat treatment and achieves the goal of "shape" and "performance" in the same set of die. This not only improves the strength of the component but also ensures its dimensional accuracy. The production process is also reduced.

Hot stamping of aluminum alloy which is still studied at the initial stage is a complicated thermal-mechanical-microstructural coupled forming process. The determination of its parameters is rather difficult due to the lack of experience. With the rapid development of computer technology, CAE has been widely used by stamping component enterprises. The numerical simulation of the stamping process by the finite element can predict the possible forming defects of the product, thus reducing die test times to shorten the production cycle. Meng et al. [7] have used thermal-mechanical coupled finite element method to study the effects of binder force, forming temperature and stamping speed on the hot stamping formation of 6111 aluminum alloy. Ma et al. [8, 9] have studied the hot stamping process of 6082 aluminum alloy by finite element analysis. They used the response surface method to analyze the influence of initial forming temperature and blank holder force on thinning rate and springback. The finite element simulation method can be used to predict the stress, strain and thickness distribution during the forming process, guiding actual stamping test and manufacture. It is necessary to use the finite element software to analyze the hot stamping process of aluminum alloy so as to understand more about it and improve the forming quality of the components.

2. Establishment of Finite Element Model

The material used for this study is 7075 aluminum alloy sheet with a thickness of 2mm. The chemical composition of this material is shown in Table 1. The initial forming temperature was 460 °C. Uniaxial tension tests were conducted on 7075 aluminum alloy at different temperatures and different strain rates using Gleeble 1500. The results are shown in Fig. 1.

Table 1. Chemical composition of aluminum alloy 7075 (wt%).

Si	Fe	Cu	Mn	Mg	Cr	Zn	Ti	Al
0.09	0.13	1.4	0.05	2.6	0.19	5.7	0.03	Bal

Fig. 2 displays the finite element model of hot stamped U-shape component. The model consists of punch, die, binder and blank. The blank is elastoplastic deformable body and the punch, the die and the blank holder are rigid bodies. The thermal properties of the blank and the tool are presented in Table 2.

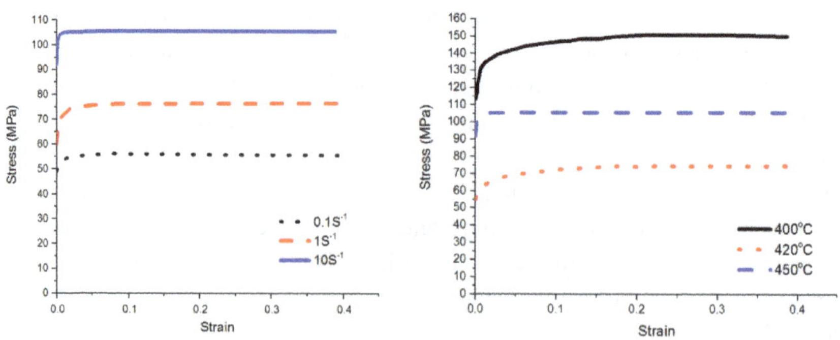

Fig. 1. Rheological curves at different temperatures and strain rates: (a) 450 °C, (b) 10 s^{-1}.

Table 2. Thermal properties of tool and blank.

Material	Temperature (°C)	Density (kg/m^3)	Thermal conductivity (W/mK)	Specific heat (J/kgK)
H13	25	7.78×10^3	36.5	440
7075	450	2.71×10^3	121.0	857

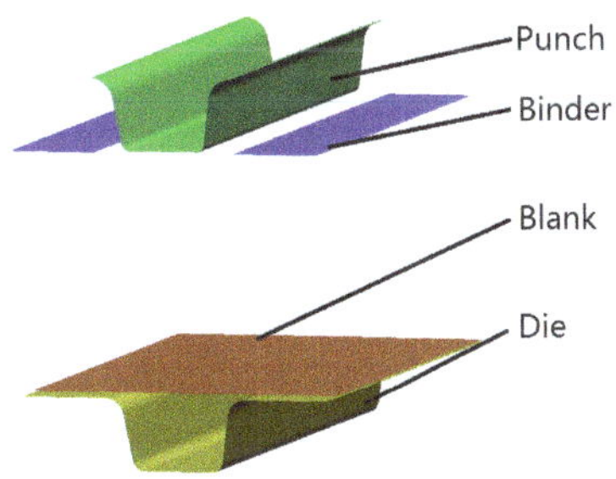

Fig. 2. Finite element model of hot stamped U-shape component.

3. Influence of Parameters

3.1. *Binder force*

Fig. 3 shows the effect of binder force on the maximum and the minimum thickness of a component after forming with forming temperature of 460 °C, friction coefficient of 0.15 and forming speed of 200 mm/s. It can be seen that the maximum thickness of the component is larger than the 2 mm initial thickness of the blank with the binder force less than 10 kN and there is a trend of wrinkling. When the binder force is greater than 20 kN, the maximum thickness is less than 2 mm and the components are not wrinkled after forming. The minimum thickness of the component decreases dramatically with the increase of binder force. When the binder force is less than 10 kN, the minimum thickness is over 1.98 mm and the change is not obvious. The reason is that the smaller binder force has less restriction on the inflow of the blank to the die, which makes the material enter the die easily in the forming process. When the binder force is 100 kN, the minimum thickness is 1.632 mm and the thinning rate is close to 20 %.

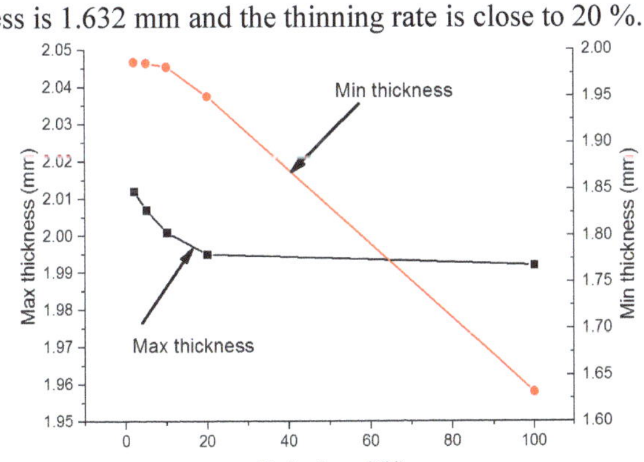

Fig. 3. Influence of binder force on the thickness.

3.2. *Initial forming temperature*

To study the influence of initial forming temperature on the thickness of the component, the initial forming temperatures were set at 460 °C, 420 °C and 350 °C respectively with the binder force of 100 kN, the friction coefficient of 0.15 and the forming speed of 200 mm/s. The influence of initial forming temperature on the thickness of the component can be seen in Fig. 4. The maximum thickness of the component, over 1.99 mm, is less affected by the initial forming temperature and there is no wrinkling trend. With the increase of initial forming temperature, the decrease of minimum thickness is due to the reduction in hardening exponent n. The hardening exponent reflects the ability of uniform deformation after the deformation of the material. As the temperature rises, the n value decreases and the degree of uniform deformation drops. With larger blank holder force, the material is partially thinned, resulting in the final reduction in thickness. It can be seen that although the material exhibits good ductility while forming at a high temperature close to solution temperature, the lower n value is not conducive to the uniform deformation and it is easy to produce local thinning or even fracture. The ductility of the material in the medium temperature forming is lower than that in the high temperature forming, whereas the larger n value can ensure more uniform deformation. Therefore, it is necessary to optimize the initial forming temperature according to the complexity and performance requirements of the component during hot stamping of aluminum alloy.

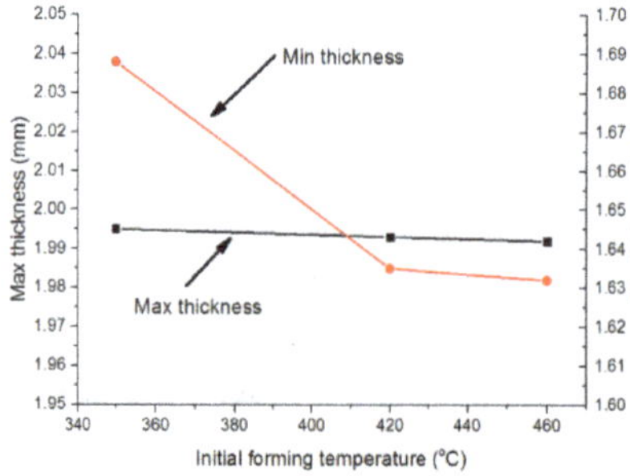

Fig. 4. Influence of initial forming temperature on the thickness.

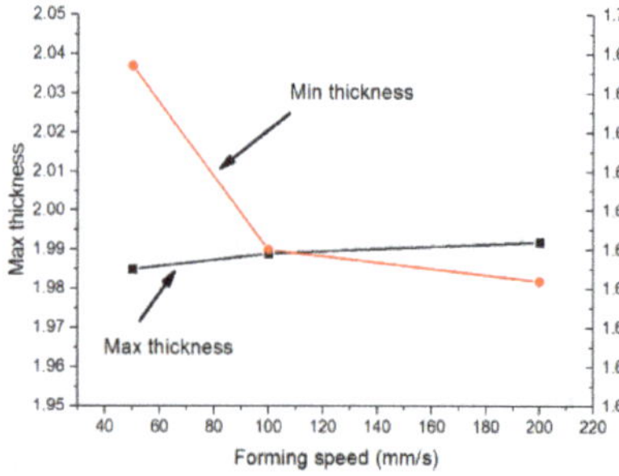

Fig. 5. Influence of forming speed on the thickness.

3.3. *Forming speed*

Fig. 5 displays the effect of forming speed on the thickness of the component after forming with temperature of 460 °C, friction coefficient of 0.15 and blank holder force of 100 kN. It is revealed that with the increase of the forming speed, there is small change in the maximum thickness of the component. This is because the larger blank holder force can guarantee no wrinkling trend after formation. The increase of the forming speed will lead to the decrease of the minimum thickness of the component. The reason is that the forming time is short and the temperature of the material drops less, while high temperature and low plasticity will lead to the local thinning under the constraint condition. The forming temperatures at different forming speeds are shown in Fig. 6. With the increase of forming speed, the temperature of the component increases after forming. The minimum temperature of the component after forming increases from 138 °C to 285 °C with the forming speed increased from 50 mm/s to 200 mm/s. Fig. 6 shows that the temperature of the flange region is the lowest because this region has the earliest contact with the binder. Due to the high thermal conductivity of aluminum alloy, the temperature during the forming process drops rapidly. For the forming of the complex components, high forming speed is suggested to use to meet the demands for the ductility of the material.

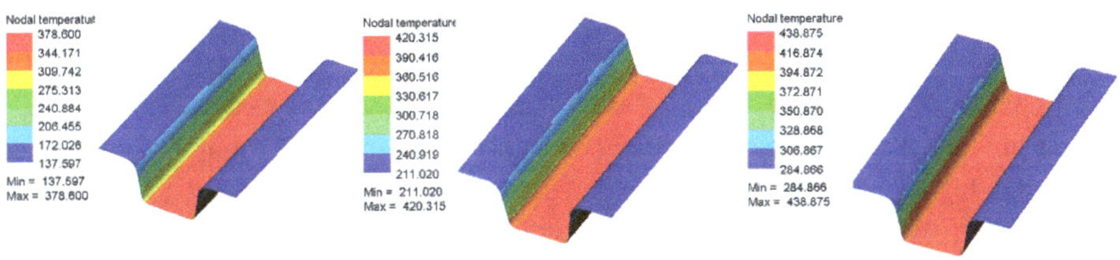

(a) 50mm/s (b) 100mm/s (c) 200mm/s

Fig. 6. Influence of forming speed on the temperature.

4. Forming Experiments

The devices used in the experiments include a press, a heating furnace and a forming die. The blank was heated to 475 °C and kept for 5 minutes and manually transferred to the forming die after solution treatment. The transfer time was 6-8 seconds and the initial forming temperature was about 460 °C. Fig. 7 shows the locations for measuring thickness of the forming component with binder force of 5 kN and forming speed of 50 mm/s. The thickness results of simulated and experimental components are given in Table 3. It can be seen that the thickness distribution of simulation basically matches the experimental results. The thickness of the component is basically unchanged at the bottom and the flange after formation. The thinning of the thickness occurs mainly in the corner of the punch and the die and the side wall, and the maximum thinning position is in the corner of the punch.

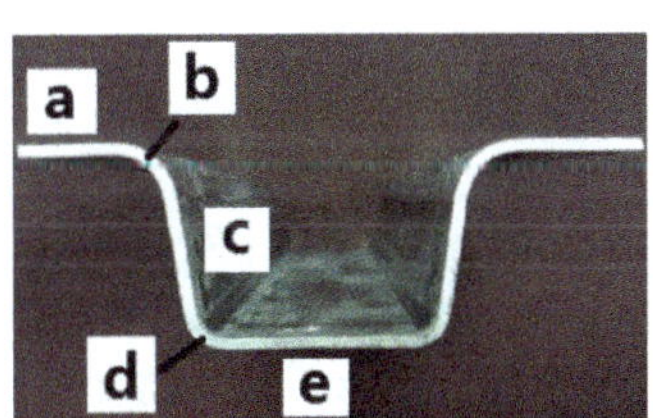

Fig. 7. Locations for measuring thickness of the component.

Table 3. Simulated and experimental values of thickness.

Locations	Thickness (mm)	
	Simulation	Experiment
a	1.995	2.00
b	1.994	1.98
c	1.991	1.98
d	1.988	1.97
e	1.997	2.00

5. Conclusion

Hot stamping model of U-shape 7075 aluminum alloy component was established in this paper. The effects of binder force, initial forming temperature and forming speed on the thickness of component were studied. Several key conclusions can be drawn:

(1) The maximum thickness of the component was larger than the 2 mm initial thickness of the blank with the binder force less than 10 kN and there was a trend of wrinkling. When the binder force was greater than 20 kN, the component was not wrinkled after forming. The minimum thickness of the component decreased dramatically with the increase of binder force. When the binder force was less than 10 kN, the change of the minimum thickness was not obvious. When the binder force was 100 kN, the minimum thickness was 1.632 mm and the thinning rate was close to 20 %.

(2) With the increase of initial forming temperature, the n value decreased and the degree of uniform deformation reduced. The material was partially thinned, resulting in the final reduction in thickness. For hot stamping of aluminum alloy, it was necessary to optimize the initial forming temperature according to the complexity and performance requirements of the component.

(3) The increase of the forming speed would lead to the decrease of the minimum thickness of the component and the increase of the component temperature after forming. The minimum temperature of the component after forming increased from 138 °C to 285 °C with the forming speed increased from 50 mm/s to 200 mm/s. For the forming of the complex components, high forming speed was suggested to use to meet the demands for the ductility of the material.

References

1. W. S. Miller, L. Zhuang and J. Bottema, Recent development in aluminum alloys for the automotive industry, *Mater. Sci. Eng. A* **280**, 37 (2000).
2. S. Toros, F. Ozturk and I. Kacar, Review of warm forming of aluminum-magnesium alloys, *J. Mater. Process. Tech.* **207**, 1 (2008).
3. S. Ji, D. Watson and Z. Fan, Development of a super ductile die cast Al-Mg-Si alloy, *Mater. Sci. Technol. A* **556**, 824 (2012).
4. M. S. Mohamed, A. D. Foster and J. Lin, Solution heat treatment in HFQ process, *Steel. Res. Int.* **79**, 160 (2008).

5. M. S. Mohamed, A. D. Foster and J. Lin, Investigation of deformation and failure features of AA6082: experimentation and modeling, *Int. J. Mach. Tool. Manu.* **53**, 27 (2012).
6. J. Zhou, B. Wang and J Lin, Optimization of an aluminum alloy anti-collision side beam hot stamping process using a multi-objective genetic algorithm, *Arch. Civ. Mech. Eng.* **13**, 401 (2013).
7. Q. Meng, B. Wang and L. Fu, The influence of process parameters during hot stamping of aa6111 aluminum alloy sheet, *Adv. Mater. Res.* **572**, 255 (2012).
8. W. Y. Ma, B. Y. Wang and J. Zhou, Multi-objective optimization of hot stamping of aluminum alloy blank, *J. Harbin Eng. Univ.* **36**, 1246 (2015).
9. W. Y. Ma, B. Y. Wang and J. Zhou, Analysis of hot stamping of aluminum alloy based on numerical simulation, *Aeronaut. Manuf. Tech.* **499**, 95 (2016).

Simulation on Direct Contact Heating of Aluminum Alloys

H. C. Geng, Z. J. Wang, K. Wang, Y. L. Wang† and Y. S. Zhang
State Key Laboratory of Materials Processing and Die and Mould Technology,
Huazhong University of Science & Technology,
Wuhan 430074, China
†E-mail: wangyilin@mail.hust.edu.cn

To improve the formability and productivity in cold stamping of aluminum alloys, warm forming and hot forming processes were developed, in which heating process plays an important part. This paper presents investigations on direct contact heating of aluminum sheets by the approach of modeling. The influence factors of working face temperature uniformity of the direct contact heating device were studied, as well as the continuous heating of aluminum sheets by direct contact heating process. Simulation results indicate that direct contact heating process can realize continuous quick heating of aluminum sheet, and the solution heat treatment of aluminum alloys can also be achieved. In the process of direct contact heating, the heating time is short, the temperature of the heated sheets is relatively uniform, and there is little difference of temperature distribution between different sheets in same heating condition. The direct contact heating process represents a feasible as well as time-saving alternative for the heating of aluminum alloys.

Keywords: Aluminum alloys; Hot forming; Contact heating; Temperature distribution.

1. Introduction

Nowadays, for the purpose of reducing fuel consumption and gas emission, transportation industries dedicate to reduce the weights of their products. Applying lightweight materials, such as aluminum and magnesium alloys, becomes an efficient approach to achieve weight target [1-3].

Aluminum alloys are popular candidates for automotive and aircraft industries [2]. But their application is restricted for their relatively low formability at room temperature and the springback after forming. To solve these defects, the warm forming and the hot forming techniques were proposed, among which the heating process plays an important role.

Traditional heating method by furnace has the disadvantages of long heating time and big energy consumption. Direct contact heating process is a sheet metal heating approach proposed recently, in which sheets are heated by contacting with hot metal plates, which are heated through magnetic induction or electrical resistance heating [4, 5]. In the literatures, direct contact heating process was only used to heat the steel sheets, there is no study on continuous heating of multiple sheets with direct contact heating, and the feasibility of industrial application has not been verified.

In this paper, a modeling for the direct contact heating of aluminum sheets was implemented in ABAQUS, and a user defined subroutine UAMP with PID control code was used in the heating simulation to model a real temperature control process. Simulation of continuous quick heating for multiple aluminum sheets was done, as well as the simulation of the heating and holding processes for solution heat treatment of aluminum alloys. Eventually, the heating characteristics of the technique was discussed, a sufficiently uniform temperature distribution of the sheets required by the forming process was obtained, and the feasibility of the application of direct contact heating on aluminum sheets in industrial production was also approved.

2. Model and Simulation Conditions

The simulation model used in this paper is shown in Fig. 1. The device is closed when heated. The dimension of the working face is 380×290 mm, and the parameters of the contact plates are set according to steel QRO-90, which has great chemical stability, higher rigidity and superior thermal conductivity at high temperature. When the device is heated, higher thermal conductivity results in more uniform working face temperature distribution. When the sheet is placed on the working face, 40 mm margin are left at each side of the working face. The total heating power was set to 16 kW, and the power was implemented to the model by heat flux over the inner surface of the heating holes in the contact plates. And there was also the insulation boards outside the contact plates to reduce the heat loss to surroundings. The interfacial heat transfer coefficient between the contact plate and aluminum sheet was set to 800 W/(m^2K) [6].

PID control code was used to control the surface heat flux, consequently, device temperature was controlled at the set value.

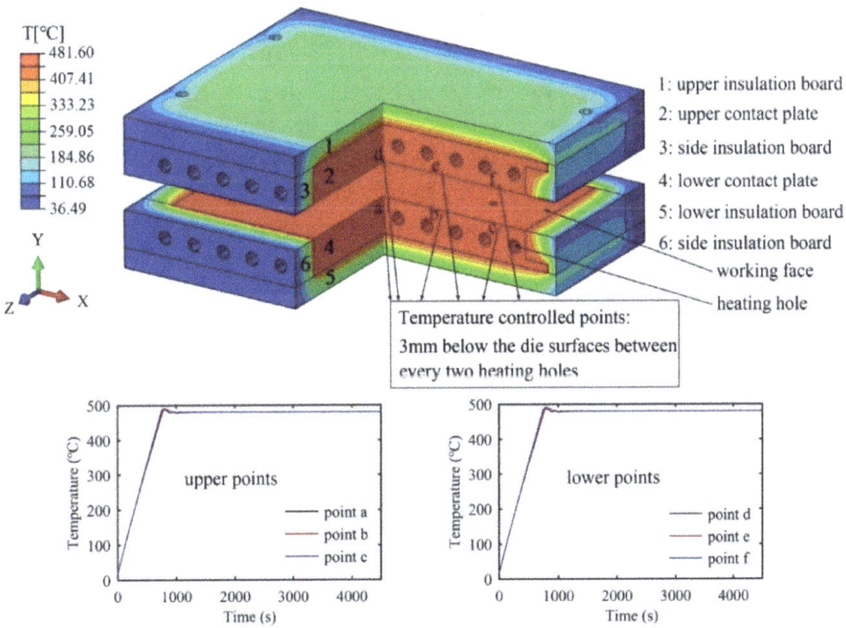

Fig. 1. 3/4 simulation model holding at 480 °C.

The modeling involved continuous quick heating process of multiple sheets at lower temperature, as well as the heating and holding processes at higher temperature for the solution heat treatment of aluminum alloys. The temperature in the modeling referred to the temperature in Fig. 2. Since narrow temperature range was required in heating process, the temperature uniformity was used as an important index to measure the effect of direct contact heating.

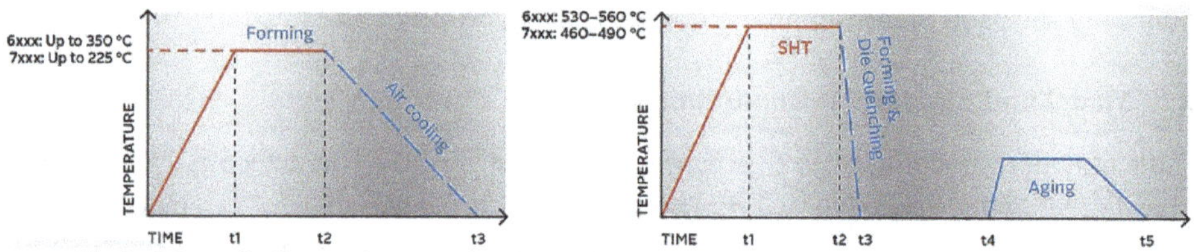

Fig. 2. Schematic diagram for warm forming (left) and hot forming (right) of aluminum alloys [7].

3. Results and Discussion

3.1. *Heating and holding processes of the device*

The temperature distribution uniformity of the working face has great influence on the temperature distribution of the heated sheets. To obtain a more uniform temperature distribution on the working face, numerous simulations with different structure parameters values were performed. It's found that the most uniform temperature distribution of the working face can be obtained when the diameter of heating holes is 16 mm, the distance between the heating holes is 23 mm, and the distance from working surface to the inner surface of heating holes is 14 mm. Eventually, ten heating holes are set in the upper and lower contact plate respectively, and temperature controlled points are set between every two heating holes.

The temperature distribution of the model after heating and holding processes is shown in Fig.1. When the device reaches target temperature, under the controlling of PID control code, the temperature fluctuates up and down at 480 °C and gradually stabilizes. The temperature in the working face is relatively uniform and the temperature difference is about 10 °C.

3.2. *Continuous quick heating of multiple sheets*

Fig. 3a shows the temperature distribution of lower contact plate after continuous quick heating of ten aluminum sheets with a thickness of 1.2 mm. The target temperature was set to 350 °C. And the temperature evolution of three points (locating at the surface center and the midpoint of the edges of sheets) in the working face is shown in Fig. 3b. It indicates that continuous quick heating leads to the increase of the surface temperature at the end of the heating holes, this is because heat consumption at the margin of the working face without contact with aluminum sheet is little. Fig. 4 shows the temperature distribution of the ninth and tenth sheet after heating. It's obvious that the difference of the temperature distribution between different sheets gets smaller after repeated heating processes. Eventually, the temperature difference of single sheet is less than 20 °C, which basically meets the heating requirement for warm forming.

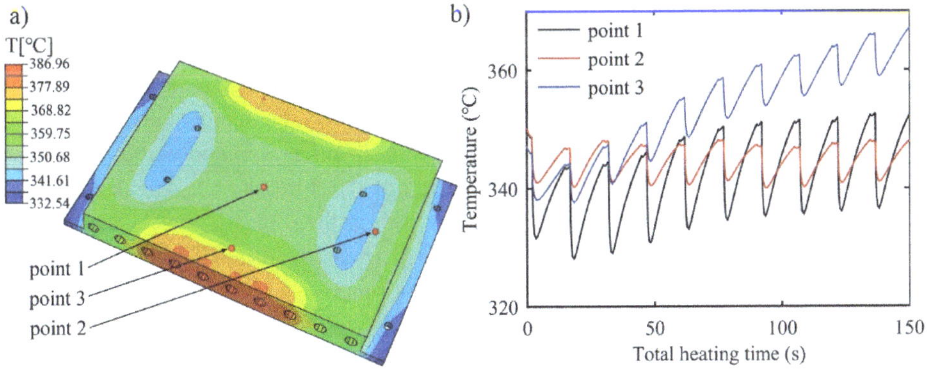

Fig. 3. (a) Temperature distribution of lower contact plate after continuous quick heating process of ten aluminum sheets, (b) Temperature evolution of three point in working face.

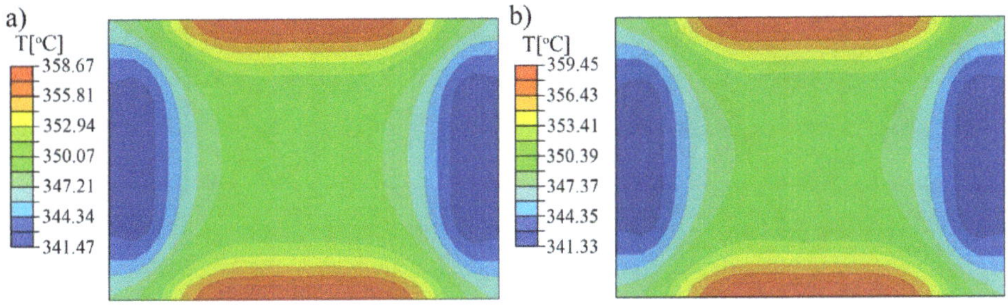

Fig. 4. Temperature distribution of sheets after continuous quick heating process. (a) The ninth sheet, (b) The tenth sheet.

Meanwhile, for the sheets with different thicknesses, the suitable heating time for single sheet in the continuous quick heating process is shown in Fig. 5 And extra 2 seconds is needed before the heating of each sheet for the opening and closing of the device. The heating time increases as the thickness of the sheets increases.

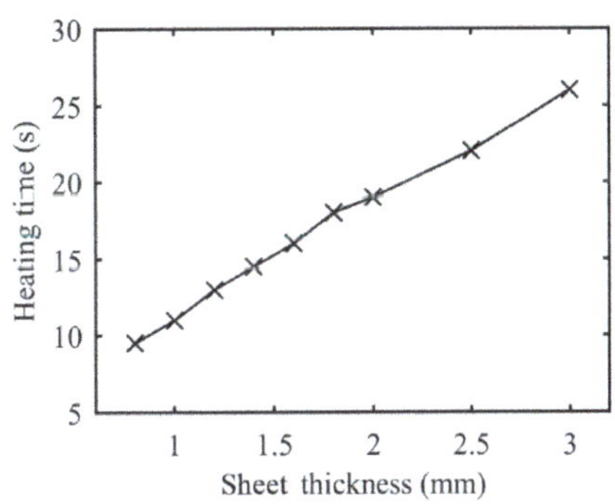

Fig. 5. The heating time for single sheet with different thicknesses.

3.3. *Heating and holding processes for solution heat treatment*

In hot forming of aluminum alloys, heating and holding processes for solution heat treatment is needed. Take AA7××× as an example, model the direct contact heating process of a sheet with thickness of 1.5 mm, the target temperature is 480 °C and the heating and holding time is 600 seconds with extra 2 seconds for opening and closing of the device before the heating process. After heating 3 sheets, the temperature distribution of lower contact plate and the temperature evolution of three points in working face are shown in Fig. 6. And the temperature distribution of the third sheet is shown in Fig. 7. The temperature difference in the single sheet is less than 20 °C, which ensures the solution heat treatment. And direct contact heating can also improve the heating velocity to the target temperature, and shorten the processing time.

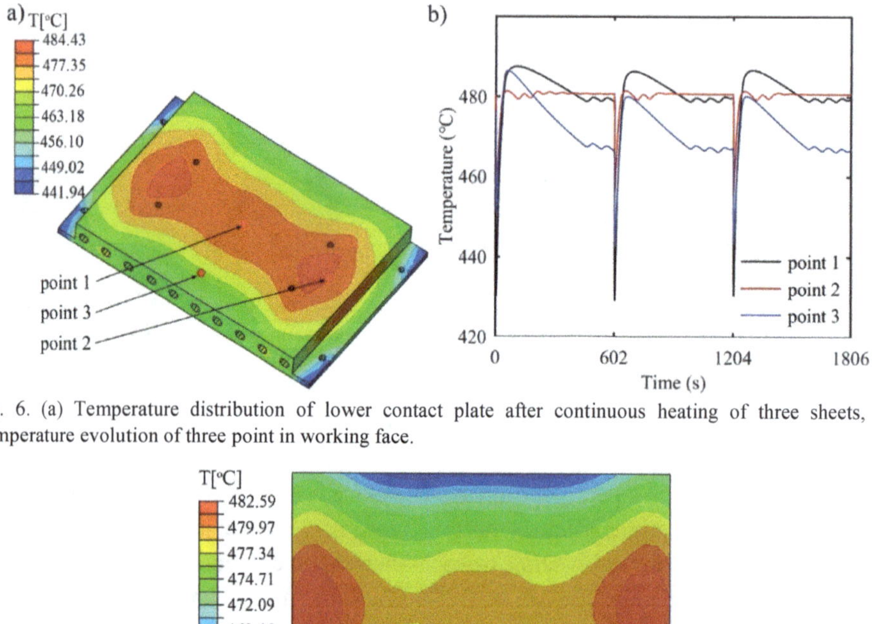

Fig. 6. (a) Temperature distribution of lower contact plate after continuous heating of three sheets, (b) Temperature evolution of three point in working face.

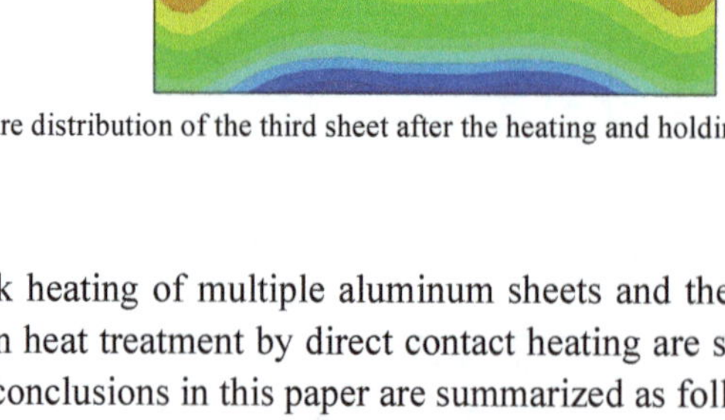

Fig. 7. Temperature distribution of the third sheet after the heating and holding of 600 seconds.

4. Conclusion

The continuous quick heating of multiple aluminum sheets and the heating and holding processes for solution heat treatment by direct contact heating are studied by a modeling approach. The main conclusions in this paper are summarized as follows:

(1) Direct contact heating process can be used for the heating of aluminum sheets. And the sheet temperature during the process is relatively uniform.

(2) The continuous quick heating process can be achieved by direct contact heating. For example, the 1.2 mm thick aluminum sheet can reach 350 °C within 13 seconds in continuous quick heating. And the continuous heating frequency depends significantly on the sheet thickness and the heating power of the contact heating device.
(3) The heating and holding processes by direct contact heating can realize the solution heat treatment of aluminum alloys. And the time required for the heating process is small, which can effectively save production time.

Acknowledgments

This research work was financially supported by the National Natural Science Foundation of China (grant no. U1564203 and U1760205) and National Major Science Technology Project of China (grant no. 2018zx04023001).

References

1. P. F. Bariani, S. Bruschi, A. Ghiotti and F. Michieletto, Hot stamping of AA5083 aluminium alloy sheets, *CIRP Ann. Manuf. Technol.* **62**, 251 (2013).
2. T. Maeno, K. Mori and R. Yachi, Hot stamping of high-strength aluminium alloy aircraft parts using quick heating, *CIRP Ann. Manuf. Technol.* **66**, 269 (2017).
3. X. B. Fan, Z. B. He, W. X. Zhou and S. J. Yuan, Formability and strengthening mechanism of solution treated Al-Mg-Si alloy sheet under hot stamping conditions, *J. Mater. Process. Tech.* **228**, 179 (2016).
4. N. Field, Austenitization of Ultra High Strength Steel by Direct Contact Heating for Hot Forming Die Quenching, Master thesis, University of Waterloo (2017).
5. J. N. Rasera, K. J. Daun, C. J. Shi and M. Souza, Direct contact heating for hot forming die quenching, *Appl. Therm. Eng.* **98**, 1165 (2016).
6. W. C. Xiao, B. Y. Wang, K. L. Zheng, J. Zhou and J. G. Lin, A study of interfacial heat transfer and its effect on quenching when hot stamping AA7075, *Arch. Civil Mech. Eng.* **18**, 723 (2018).
7. M. Vrolijk, D. Lorenz, H. Porzner and M. Holecek, Supporting lightweight design: virtual modeling of hot stamping with tailored properties and warm and hot formed aluminium, *Procedia Eng.* **183**, 336 (2017).

Analyzing Large Tooling Cooling Using CFD

B. F. Rolfe† and W. M. Gao
Institute for Frontier Materials, Deakin University
Geelong, VIC 3216, Australia
†E-mail: brolfe@deakin.edu.au

M. Pereira and Y. Wang
School of Engineering, Deakin University,
Geelong, VIC 3216, Australia

L. Pan and G. Rui
Wuhan Branch of Baosteel Central Research Institute, Baosteel,
Wuhan, 430080, China

Hot stamping is now the main forming process to produce the ultra-high strength components in an automotive body structure. Tooling in hot stamping has multiple cooling channels to maintain a lower temperature of the tooling. It is also important that the cooling channels are designed to ensure uniformity of the temperature distribution on the final part. This will stop part warpage. This paper will show the steps in analyzing a large tool, where the tool has been partitioned into sections. Moreover, suggestions will be made in how to balance the water flows in each section.

Keywords: Hot stamping; Tooling; CFD; Cooling channels.

1. Introduction

Recently, the requirements of vehicle weight reduction and improved crash performance have promoted the development of fabricating components using the hot stamping process. The cooling of the tooling is vital to achieve a fully quenched part. It is important that the tooling is designed with cooling channels that maintain a low and uniform tool temperature [1]. If this cannot be achieved, then the critical cooling rate may not be reached and a full martensitic part may not be possible. Furthermore, a non-uniform temperature distribution may lead to thermal warping of the final part.

This paper will investigate the water cooling of a large tool. The cooling channels in this tool are manufactured in several sections. This work will show the differences in water flow through the tools. The analysis considers both laminar and turbulent flows.

2. Water Flow Model Description and Functions

2.1. *Physical models*

The model for the water flow is composed of the flow channels formed by the main water supply pipes to water distribution pipes, channels in dies, the slots in the basal plate and the main water collection pipes (Figure 1). The functions of this model include:

(1) Simulating the water distribution through flow paths to evaluate the design of the water distribution pipes.

(2) Analyzing the water distribution through the channels of each die to assess the channel design.

(3) Calculating the pressure drops of water through the slots in the basal plates to exam the influence of them on the water flow.

(4) Providing the boundary conditions for future modelling of die cooling.

2.2. *Testing machine*

Two water flow models were created for the bottom tool (die) and top tool (punch), respectively, to identify the "uniformity" of water through channels. The top tool results will be shown. First, the water flow in top and bottom tools was modelled as a steady-state, lamina flow. The results were analyzed with the aim of determining the flow distribution through each channel and the Reynolds number (Re) (Eq. (1))) to calculate if the flow is turbulent. The calculated Re based on the lamina flow modelling result is between 30000 and 40000, indicating the water flow in channels is turbulent. Thus, the second model was solved as a turbulent flow.

$$\mathrm{Re} = \frac{\rho v d}{\mu} \tag{1}$$

The flow field was described locally by general conservation equations, including governing equations for mass, momentum, and realizable k-epsilon turbulence. The details of the computational model and boundary conditions are summarized in Table 1.

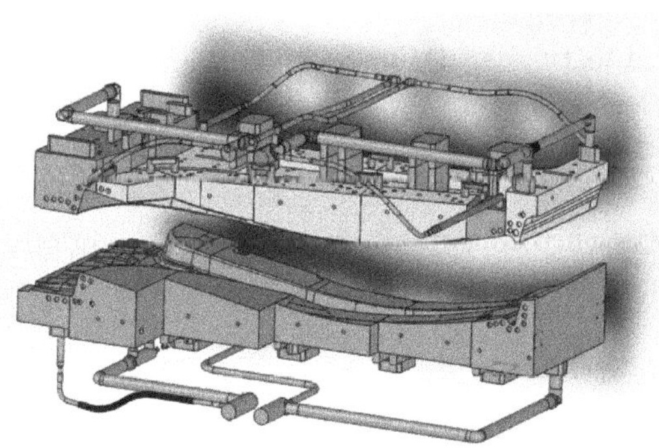

Fig. 1. Cooling system of hot-stamping and quenching for a B-pillar tool set.

Table 1. Computational model and boundary conditions.

Model	Steady, both lamina and k-ε turbulent flow Fluid: water Density: 998.2 kg/m3 Viscosity: 0.001003 kg/m-s
Water Inlet	Velocity inlet: 3.7 kg/s (220l/min) Or pressure: 300 kPa (3 bar) Turbulent kinetic energy 1.0 m2/s2 Turbulent dissipation rate 1.0 m2/s3
Outlet	Outflow
Die channel and pipe interior surfaces	No slip wall: Roughness height: 0 mm Roughness constant: 0.5 Channel diameter: 14mm

To evaluate the "uniformity" of water distribution, the pressure drop (ΔP_i) and mass rate (Q_i) through each channels are calculated. The average pressure drop through the die is calculated by Eq. (2). It should be noted that this is not the real water pressure drop through the die. The maximum pressure drop should be also determined. The deviation of each mass rate is calculated by Eq. (3), indicating the difference in mean water velocity (Q/A) between channels. For the die with the same size channels, the deviation is calculated by Eq. (4). The absolute values of the water rate deviation of each die are averaged to evaluate the water distribution uniformity of the die (Eq. (5)).

$$\overline{\Delta p} = \tfrac{1}{n}\sum \Delta P_i \tag{2}$$

$$\sigma_{u,i} = 100\% \frac{Q_i/A_i - \frac{1}{n}\sum Q_i/A_i}{\frac{1}{n}\sum Q_i/A_i} \tag{3}$$

$$\sigma_{u,i} = 100\% \frac{Q_i - \frac{1}{n}\sum Q_i}{\frac{1}{n}\sum Q_i} \tag{4}$$

$$\overline{\sigma_u} = \tfrac{1}{n}\sum \left|\sigma_{u,i}\right| \tag{5}$$

In the equations, ΔP_i, Q_i and A_i indicates the pressure drop and mass flow rate through channel i and the channel size, respectively.

It should be noted that the analysis of above parameters only represents the water distribution. In some cases, higher mass rate in some channels and lower rate through other channels may be benefit to approach a required quenching rate at die surface, as the die surface temperature and quenching rate depend on not only the energy removed from each channel, but also the heat transfer in dies. The main aim of this section is to provide the water mass rate through each die for future heat transfer modelling.

2.2.1. *Water flow in top tools*

The model for water flow in top dies (punch) is shown in Figure 2.

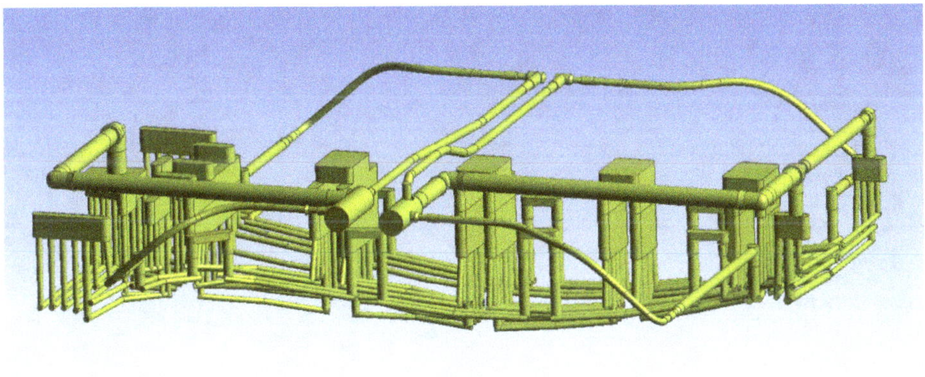

Fig. 2. Model of cooling system for top die (punch).

In the model the channels, the inlet and outlet of each channel were individually defined to enable outputting the pressure, velocity and mass rate through each channel (Figure 3). The top die is split into six sections (P1 - P6) and two flange sections (S1-S2).

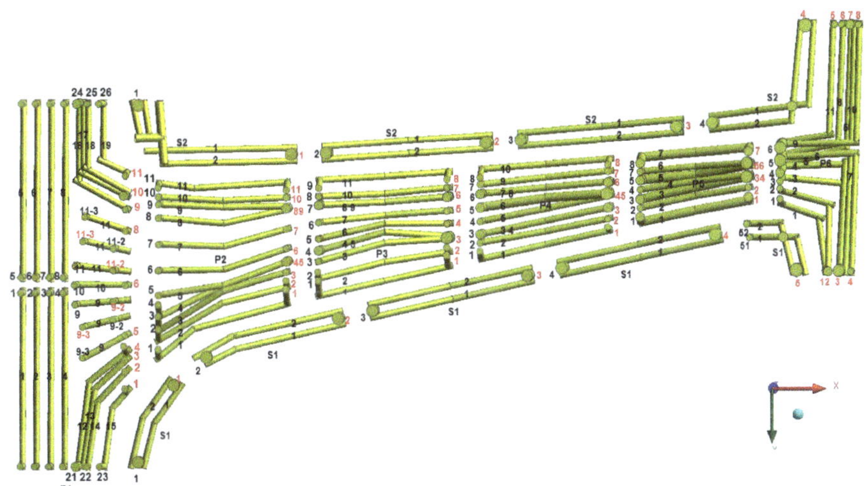

Fig. 3. Channels in top die (punch) sections (P1 - P6, S1 and S2) and inlets and outlets. Channels are labelled with the black numbers, inlets are labelled by the black numbers on the left-hand side of the channels, while outlets are in red numbers on the right-hand side of the channels.

3. Results of the Flow Model in the Top Tooling

3.1. *Velocity, mass rate and pressure drop*

The pressure, pressure drop, velocity and mass rate through each channel were calculated and analyzed with respect to the uniform water distribution between channels, see Figure 4. The results from lamina flow and turbulent flow simulations are summarized in Table 2 and Table 3, respectively.

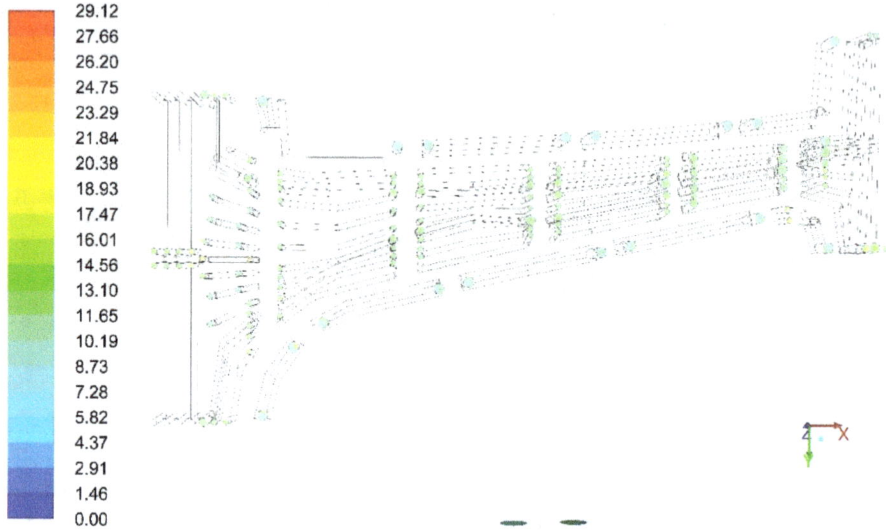

Fig. 4. Velocity of water through the inlets and outlets (for a 50 m/s inlet).

Table 2. Summary of results through top dies (lamina flow and mass rate at 3.7 kg/s).

Flow Path	Die Section	P1	P2	P3	P4	P5	P6	Total
Centre Q=2.61kg/s	ΔP (kPa)	19.24	15.59	27.62	18.38	15.79	18.66	115.28
	Mean Q (kg/s)	0.23	0.24	0.24	0.24	0.26 [b]	0.24	
	Min Q (kg/s)	0.37 Ch10	0.25 Ch11	0.27 Ch3	0.50 Ch6 [a]	0.26 Ch6	0.27 Ch6	
	Max Q (Kg/s)	0.18 Ch17	0.22 Ch4	0.23 Ch6	0.23 Ch5	0.25 Ch4	0.20 Ch8	
	Mean σ_U %	9.32	4.39	2.97	2.06	3.07	12.91	
	σ_U >5 % and Chs	-5.01% Ch1 -6.96% Ch5 -13.3% Ch11 -14.05% Ch13 -14.71% Ch9 -17.19% Ch17 10.69% Ch15 5.57% Ch19 59.54% Ch10	-6.95% Ch8 -7.71% Ch4 6.05% Ch1 6.36% CH11	11.96% Ch3	6.11% Ch6	6.77% Ch6	-9.46% Ch10 -12.48% Ch9 -13.63% Ch11 -13.75% Ch8 -21.70% Ch5 7.19% Ch4 9.14% Ch7 11.01% Ch1 13.64% Ch2 14.31% Ch3 15.73% Ch6	
Side S1	Flange S1	1	2	3	4	5		
Q=0.57kg/s	Mean Q (kg/s)	0.57	0.57	0.57	0.57	0.57		
	ΔP (kPa)	21.07	20.62	17.78	23.36	33.03		115.75
	σ_U>5 %							
Side S2	Flange S2	1	2	3	4			
Q=0.52kg/s	Mean Q (kg/s)	0.52	0.52	0.52	0.52			
	ΔP (kPa)	56.45	22.80	15.05	28.13			112.41
	σ_U>5 %							
Total: Q=3.70kg/s; ΔP=231.79kPa								

Table 3. Summary of results through top dies (turbulent flow and mass rate at 3.7 kg/s).

Flow Path	Die Section	P1	P2	P3	P4	P5	P6	Total
Centre Q=2.60kg/s	ΔP (kPa)	40.03	18.97	16.61	15.39	13.59	21.54	126.13
	Mean Q (kg/s)	0.23	0.24	0.24	0.24	0.26 [b]	0.24	
	Min Q (kg/s)	0.38 Ch10	0.25 Ch1	0.26 Ch3	0.50 Ch6 [a]	0.26 Ch6	0.27 Ch3	
	Max Q (Kg/s)	0.18 Ch17	0.22 Ch4	0.23 Ch4 5	0.23 Ch5	0.46 Ch7 [a]	0.18 Ch5	
	Mean σ_U %	9.68	4.16	2.63	1.70	3.43	13.34	
	σU >5 % and Chs	-5.05% Ch1 -6.19% Ch5 -10.26% Ch11 -11.35% Ch9 -21.91% Ch17 -22.54% Ch13 7.39% Ch15 7.52% Ch19 65.21% Ch10	-7.24% Ch4 -6.73% Ch8 5.25% Ch1	11.35% Ch3	5.30% Ch6	7.55% Ch6	-8.89% Ch10 -12.81% Ch8 -13.81% Ch9 -14.03% Ch11 -23.81% Ch5 6.35% Ch1 8.16% Ch4 10.39% Ch7 13.98% Ch2 15.67% Ch3 18.81% Ch6	
Side S1 Q=0.58kg/s	Flange S1	1	2	3	4	5		
	Mean Q (kg/s)	0.58	0.58	0.58	0.58	0.58		
	ΔP (kPa)	23.39	22.01	25.76	22.50	33.53		127.19
	σ_U>5 %							
Side S2 Q=0.52kg/s	Flange S2	1	2	3	4			
	Mean Q (kg/s)	0.52	0.52	0.52	0.52			
	ΔP (kPa)	42.16	18.34	19.78	34.71			115.00
	σ_U>5 %							
Total: Q=3.70kg/s; ΔP=230.79kPa								

4. Conclusion

The laminar and turbulent modelling produced approximately the same results. However, the analysis of the relationship of the results with the geometry of channels shows the turbulent modelling produced more rational results than the laminar.

The model showed that there was a large difference in the water distribution in the top Die Sections P1 and P6, while Die Sections P2, P4, and P5 and the side dies (S1-S2) gave relatively uniform water flow. The complex geometry of the head and the sill sections of the b-pillar led to large water flow differences.

4.1. *Die section P1*

In P1 (sill section), the average deviation in mass rate is up to 9.3%. There is less water flowing through channels 1, 5, 9, 11, 13 and 17, while more water flows through channels 10, 15 and 19. By reducing the diameters of channels 10, 15, 19 the water flow could be reduced to balance the flows in Die Section P1 (sill section). However, this would increase the complexity of the die build.

To increase the mass rate through channels 13 and 17 in Die Section P1 (sill section), a simple way would be to increase the diameters of the holes (with inlets numbered with 21 and 24, respectively) before the water flows into the parallel holes (12 and 13; 16 and 17).

4.2. *Die section P6*

In Die Section P6 (head of the b-pillar), the mean deviation is about 13%. The water flow through channels 5, 8, 9, 10 and 11 are low, and channel 5 is the lowest (-23.81%). The water flow through channels 1, 2, 3, 6 and 7 are high, and the highest is through channel 6 (+18.81 %).

4.3. *Die section P3*

In Die Section P3, there is 11.35 % more water flowing through channel 3, compared to the average mass rate.

The maximum deviation of water flow in other top dies is in ±8 %, which could be considered as normal.

5. Conclusion

This project analyzed the water flow through a large tooling set, in this case it was a b-pillar. Two models were created: laminar and turbulent to see how different were the results. The Reynolds number based on the laminar model indicated that the water flow was turbulent. The water flow distribution through the six die section and two flange sections were compared to evaluate the design of the pipes. The main differences were seen in Die Sections P1 and P6, where the complexity of the geometry gave a large difference in the water flow. Suggestions were made in how to balance the flow. The water flow provide a mass rate and pressure drop for a coupled fluid-thermo-mechanical model of the hot stamping process.

Acknowledgments

This work was supported by the R&D Center - WISCO.

Reference

1. W. Lim, H. Choi, S. Ahn and B. Kim, Cooling channel design of hot stamping tools for uniform high-strength components in hot stamping process, *Int. J. Adv. Manuf. Technol.* **70,** 1189 (2014).

Efficient Modelling of Cooling Channel Flow Conditions in Forming Tools for PHS

M. Düring
AutoForm Engineering B.V., Krimpen aan den IJssel, South Holland 2921LB, The Netherlands
E-mail: michael.duering@autoform.nl

C. Wang
AutoForm China, Melchers Industrial (Shanghai) Ltd., 12C, Sino Ocean Tower, 588 Yanan Road (E), 200001 Shanghai, China
E-mail: curlywang@sh.melchers.com.cn

In the hotforming community, a controversy exists regarding the adequate level of detail for modelling fluid flow conditions in cooling channels. Many consider thermally coupled forming simulations supported by CFD for the description of the cooling media flow behavior as a vital prerequisite for high accuracy simulation in numerical PHS-process assessments. This necessarily leads to substantially increasing efforts in modelling and computation time. Others consider many cooling channel layouts in forming tools generically as a standard geometry based design approach. Under such assumptions, flow conditions are sufficiently described in formulations coming from similitude models in fluid dynamics and their scaling. By means of a lower/upper border approximation, applying integral HTCs to the interface between the cooling media and the tool material in a realistic tool geometry, the sensitivity of the results variables — such as hardness — for variation of the flow behavior is determined. A basic formula based description is introduced for this use case, which is subsequently combined with multiple modelling options for cooling channels in FEA. It turns out that due to the complexity of the press hardening process, the details of cooling channel flow conditions do not carry a direct and proportional impact on part quality outcomes. This observation clearly indicates that engineers can derive descriptions of the cooling channel flow conditions from easy-to-use analysis tools that are fully integrated with the simulation of the forming process, rather than resort to extended, time consuming "offline" flow computations.

Keywords: Press hardening steel; Cooling channels; Heat transfer.

1. Introduction

In order to fulfill emerging requirements for the design of more energy efficient and safe vehicles, car body designs have undergone numerous assessments regarding potential for further simultaneous performance enhancement and weight reduction. Material-based lightweight design approaches have been manifold; one particular focus using steel has offered the application of the 22MnB5-grade, providing substantial benefits in terms of strength-to-weight ratio and impact resistance, as well as and energy absorption to weight ratio when combined with different strategies for tailoring of final part properties. Since this material is easy to integrate into existing production technologies, not least due to their good compatibility with thermal joining techniques, more efforts in production are justified. Nowadays, the press hardening technology has reached a remarkable level of maturity and acceptance in the market.

Nevertheless, there are still enough opportunities for improvement dealing with lean and efficient product and process design concepts. Extensive use of finite elements based virtualized solutions plays a crucial role in the development and validation of product and process. Therefore there needs to be a clear understanding of the importance, or lack thereof, of all the component models and features of these virtual engineering tools. Unnecessary functionality, overly complex functionality, and complicated methods of application of these tools run the risk of driving up engineering lead times and costs.

A very important extension that has recently become available is the ability to integrate cooling channels in the design of tools; through fully coupled thermo-mechanical simulations that model their effectiveness and influence, it is now possible to identify hot spots on tool surfaces - and therefore to countermeasure these. This extended capability also enables the engineer to further investigate the opportunities and limitations of the influence of flow conditions inside the cooling channels. This paper addresses the level of sensitivity accompanying varying flow conditions in a practically proven tool and PHS process set-up.

2. State of the Art Press Hardening Tool

Valuable results in experiments can be obtained best by applying a proven prototype tool set and appropriate process parameters as input for all kind of virtualization, modelling and simulation. The considered forming process is proven multiple times to be stable and reliable in terms of the quality of the outcome. The blank used is of typical dimensions - 700×230 mm^2, thickness of 1.6 mm. Since the virtual experiment deals dominantly with thermally induced effects, the mechanical material model employs a regular Hill-based yield surface and does not handle with temperature depended r-values even though possible in the code. In principle, temperature dependent r-values would provide accuracy without additional computation time. Apart from the material model, the code is linked to a hardness prediction routine which refers to a Dynamic TTT. Based on this input data - the CAD data and the process parameter provided - a generic reference model was created (Figure 1).

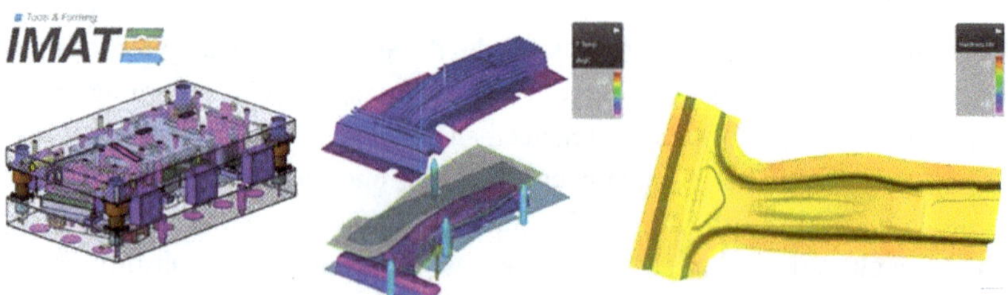

Fig. 1. (a) Tool geometry, (b) FE-model, (c) Reference part results.

The discretization for the input of the implicit solver was done automatically in form of a triangle mesh. The element section is an elastic-plastic-shell with 11 integration points - as typically used for final validation cycles. Thus, the initial maximum element size assigned the blank geometries mesh is 20 mm, refinement upon contact was enabled. For the rigid tools a maximum element size of 5 mm was determined. The thermal contact conditions between the rigid tools and the blank mesh are represented by an advanced gap and pressure dependent heat transfer formulation. The heat transfer conditions at the interface were ideally adapted - as typical for productive tool sets.

Characteristic measures can be derived from the given data and enable a better understanding regarding in particular the thermal conditions during the forming and quenching operation (Table 1). Chosen average parameters describing a typical forming cycle were considered for the model generation such as a handling time of the blank - which has been initially heated to 930 °C - of 5 seconds, a waiting time in the open tool set of 5 seconds and a quenching force of 1,000 kN for also 5 seconds before releasing the final part. In our model we were able to generate a perfect parameter set by using Systematic Process Improvement functionality which is not commented in detail.

Table 1. Characteristic measures of the reference tool geometry.

	Die	Punch	Binder
Volume [m^3]	0.048	0.014	0.048
Length - cooling channels [m]	13,450	8,120	-

It turns out that under the initial reference conditions - referring to the default heat transfer coefficient (HTC) of 10.0 mW/(mm^2 K) at the interface between cooling channel and the tools - a good quality part can be obtained through simulation. The hardness is nearly evenly distributed all over the surface area. No critical hotspot could be detected. Furthermore, no potential issues are formed in terms of splits and wrinkles. As typical for press hardening processes, thermal distortion is less critical. During modelling the contact area between the tool surfaces and the blank was defined as ideally adapted. This can be considered as an elegant option to represent perfectly spotted dies. Gap and pressure dependent formulation of the HTC implemented in the software leads to more realistic results as long as efforts in spotting activities are considered, tracked and documented. To a certain extent this is a regular use case that is also beneficial for further explanations.

3. Cooling Channel Flow Conditions

The impact of the position of the cooling channels embedded in the tool on final part quality has been the focus in recent publications [1, 2]. It has been shown that even though an intentionally bad distribution of cooling channels in the tool bodies can lead to a pronounced inhomogeneity of the temperature field at the surface area of the tools - hotspots - under regular process conditions, such inhomogeneity produces only limited influence on the quality of the final part. This is valid in particular for the hardness. It seems that the order of heat conduction in the tool and the sheet itself, combined with damping effects of the heat transfer at the contact interfaces form a stable thermal system.

While on the one hand there is the opinion that there is almost no influence of the flow conditions in the cooling channels, some consider flow condition control as a vital prerequisite to achieving high quality parts. The latter consider that extensive CFD studies towards accurate prediction of tool surface temperature distribution are crucial to predicting final part quality outcomes from forming simulation. This is naturally linked to remarkable additional efforts spent in terms of installing equipment, modelling and computation.

By means of an "upper bound – lower bound" approach, the influence of the flow condition for the given real geometry has been tested. In order to create a worst case scenario one can integrally increase the HTC at the interface between the cooling medium and the tool. This means that the cooling channels are not even segmentally designed with varying HTCs assigned to several segment groups for channel center lines describing arches or edges. This leads to less effort for such experiments and additionally pronounces the results. So, the idea was to generate a set of integral flow characteristics that finally results in reasonable HTCs as widely occurring in real tools.

For the expected range of parameters, good and bad flow conditions, it has been proven that there is a less pronounced influence on the punch temperatures after quenching detected. Nevertheless, due to the interface conditions between punch and blank there was a far less pronounced impact on the blank itself observed. Translating that into the final hardness of the part after cooling the effects as a result of the extreme variation of the flow condition of the cooling channels led to negligible effects in terms of hardness evolution.

4. Analytical HTC Calculations for Modelling

The results discussed indicate that an easy approach for the determination of the HTC is possible. Since the necessary geometric definition of cooling channels is directly linked to the heat transfer conditions between the channels surfaces and the tool material, it is useful to do some preliminary calculations regarding various flow conditions of the cooling media from the fluid mechanics point of view prior setting up highly sophisticated models in FEA-models. Although there are many descriptions and formulations of heat transfer from cooling fluids toward surrounding material in a tubular shape, in our further explanations we refer to most basic knowledge. This easily showcases the process of defining appropriate parameters for modelling and computation. This basic approach can be expanded by taking into account many other influential variables such as pressure for instance.

The preliminary calculation also distinguishes between different states of flow in the cooling channel, which directly affect the heat transfer coefficient between the cooling channel surface and the tool material: the laminar and the turbulent current state. Hereby a highly turbulent state is the most desirable one in terms of high heat transfer rates although it creates a need for more powerful pumping equipment. For highly turbulent flow, the constraint of a Reynolds number Re 15,700 under a given cooling channel diameter must be fulfilled and will be kept constant. Furthermore, the length of the cooling channels is measured individually for the punch and the die, the diameter of the cooling channels will be varied between values between 6 and 10 mm. One should recall that in the case of specific heat transfer rates — as typical for certain flow conditions and geometric dimension of the cooling channels — the necessary length/diameter ratio can be roughly calculated by balancing incoming energy amounts with the extracted heat amounts by means of water cooling to be designed.

The literature [3] indicates how the relationship between the diameter of the cooling channels and the corresponding flow rate can govern the Reynolds number as a measure for the cooling medium's flow state (1).

$$v = \frac{\mathrm{Re}\, v}{d} \tag{1}$$

The Nußelt number - as a measure for heat transfer conditions, is derived from the Péclet number (2).

$$Pe = Pr\,\mathrm{Re} \tag{2}$$

The Nußelt number has been calculated for the less desired laminar state of the current in a straight tube (3), and the Prandtl number Pr is equal to 7.

$$Nu = [3.65 + \frac{0.0668 Ped / l}{1 + 0.045 (Ped / l)^{2/3}}](\frac{\eta_t}{\eta_w})^{0.14} \tag{3}$$

For the recommended turbulent states, equation 4 applies (4).

$$Nu = 0.04 Pe \frac{\mathrm{Re}^{-1/4}}{1 + 1.5\,\mathrm{Re}^{-1/8}\,\mathrm{Pr}^{-1/6}(\mathrm{Pr} - 1)} \tag{4}$$

The relationship between the relevant heat transfer coefficient and the Nußelt number is formulated as given in equation (5).

$$Nu = \alpha l_0 / \lambda \tag{5}$$

Under given conditions, the corresponding heat transfer coefficients are easily calculated and can be modified as a characteristic parameter in models deploying cooling channels.

5. Validity Proof by FE-Experiments

Since there are plenty of formulas offered in more or less recent publications, there might be questions regarding the accuracy of such an approach. Employing the most basic formulations above, there are reasonable ways to prove their validity. One simple way is to keep the energy values retrieved from the water cooling constant whilst varying the remaining cooling channel current parameters. Therefore, in our scenario we keep the Reynold number of 15.700 constant. This implies that in reality the cooling devices are capable of handling such conditions in terms of the cooling channel's current. As presented in previous studies this leads under the assumption of a cooling channel diameter of 8 mm to an average flow rate of 1.75 m/s. As the formulations indicate the extracted energy values and the Reynold numbers stay constant as long as the requirement of a specific ratio between cooling channel diameter and heat transfer coefficient is fulfilled. Such a set of diameter and HTC-combinations are displayed in Table 2.

Table 2. Experimental parameter set-up for FE-analysis.

Run	1	2	3	4	5
Current condition			turbulent		
Reynolds number []			15,700		
Diameter cooling channels [mm]	10	9	8	7	6
Average flow rate [m/s]	1.40	1.56	1.75	2.00	2.34
HTC [W/m^2K]	8,000	8,890	10,000	11,430	13,335
HTC [mW/mm^2K]	8.000	8.890	10.000	11.430	13.335

In all cases the corresponding FE-simulation results are supposed to be the same. As long as the proposed formula approach is correct or there is only negligible sensitivity of the model with regards to this aspect, the equal length of the cooling channels and the diameter to HTC combinations should lead to the same energy balance. Thus, results obtained in representative spots of the simulation will vary in small tolerance ranges which can be expressed in form of values of the standard deviation. Even though the base data set is rather small, an exemplary evaluation of these results is given. Referring to studies which have been presented in the past, 3 major result variables where extracted in 6 different spots. The temperature at the surface of the punch geometry, the temperature in the blank/part - both after quenching - as well as the hardness in the part after cooling were of particular interest in the assessment. The measured spots in the geometry are displayed in Figure 2.

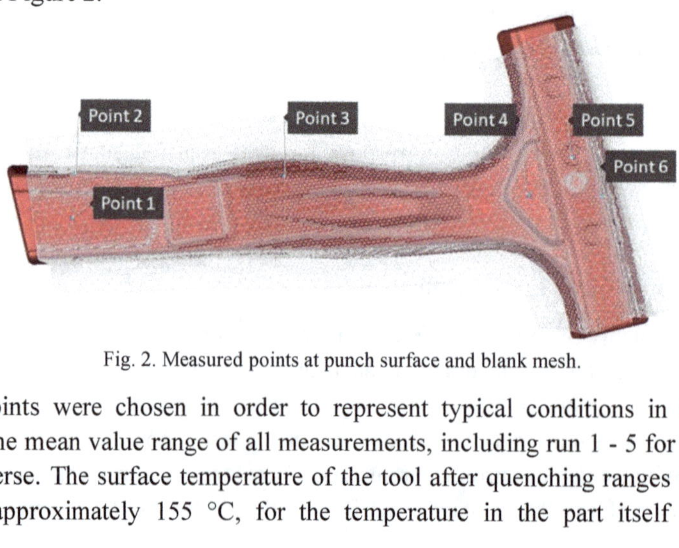

Fig. 2. Measured points at punch surface and blank mesh.

The points were chosen in order to represent typical conditions in the process. Therefore the mean value range of all measurements, including run 1 - 5 for every single spot, is diverse. The surface temperature of the tool after quenching ranges from around 67 °C to approximately 155 °C, for the temperature in the part itself a minimum temperature of 119 °C can be determined while the maximum reaches a value of 314 °C (Figure 3).

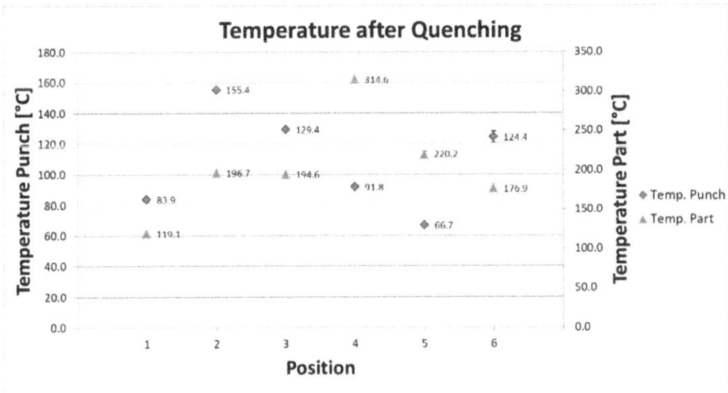

Fig. 3. Resulting temperatures after quenching.

The more interesting fact is that the standard deviation is almost negligible except for point 6. Whilst the standard deviation in points 1-5 does not exceed values of 3 °C a value of 6 °C can be calculated after value extraction in point 6. This applies to both, the temperature of the punch surface and the blank itself even though the values have to be related to the base values.

The interface between the punch surface and the blank can already be considered as a damping layer which affects the temperature evolution in the blank. Also the cooling curve of the blank is affected which influences the hardness measures in point 1 - 6 correspondingly. Additionally, values become more equalized since heat fluxes inside the sheet become by trend more dominant. The extracted hardness values in all points, measured in HV, range between 449 and 480 (Figure 4).

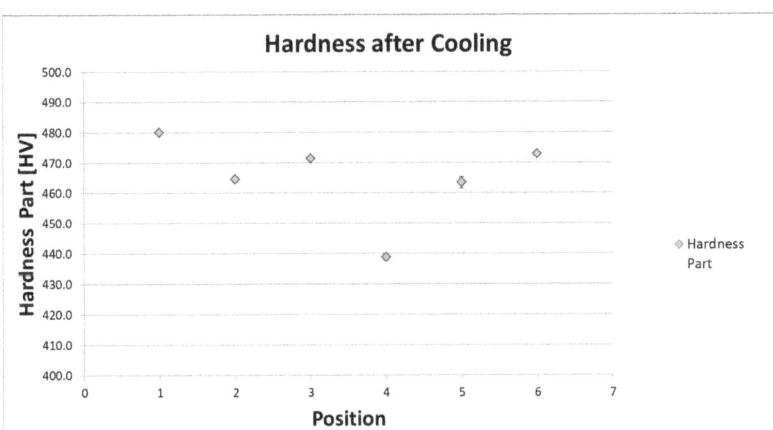

Fig. 4. Resulting hardness after cooling.

As one can expect, the values for the standard deviation are even more negligible. Finally, this proofs that there is a sufficient level of accordance since the energy values remained at the same level for the computations - as predicted in the formula based calculations.

6. Conclusion

As it has been already proven in previous publications, the assumption that formula based approaches for the description of heat transfer under particular current condition inside the cooling channels is valid. In "upper bound - lower bound" comparisons considering various HTCs derived from varying water current conditions it was already obvious that there are only minor effect of such process variations on the temperature distribution in the final sheet after quenching and the hardness in the part after cooling. Nevertheless, there are manifold options for formula based options offered to determine such relationships. Exemplarily, one of the basic ones has been selected to assess of such combination of "offline"-calculations with the FE-software for formability. The main objective was to determine the level of accuracy of such solutions. Therefore, energy neutral constellations of diameter to HTC-ratios were generated. Setting up such parameters pairs in a given FE-model and computing them sequentially must leads to equal results. The level is expressed by the standard deviation. It has been shown that for the given formula based approach the standard deviation of the temperature on the punch surface, the temperature and the hardness in the sheet for 6 selected spots considering 5 different diameter/HTC-ratios were negligible. This provides proof that even basic approaches are valid or the aspect of cooling channel flow condition does not require particular attention during the modelling efforts.

References

1. T. Brenne, M. Stippak and M. Düring, Parametric FEA-Study on the Impact of Cooling Channel Design on Final Part Quality, in *Proc. CHS² 2017*, (Atlanta, USA, 2017).
2. K. Kidambi, T. Brenne and M. Düring, Impact of Flow Conditions in Cooling Channels on Thermal Cycling, in *Proc. IDDRG 2018*, (Waterloo, Canada, 2018).
3. Gesetzliche, *Grundwissen des Ingenieurs* (Veb Fachbuchverlag, 1967).

Investigation on 22MnB5 Thin-wall Welded Rectangular-tube by Progressive Induction Quenching

C. Y. Wu, Y. Liu, B. Zhu and Y. S. Zhang†
State Key Laboratory of Materials Processing and Die & Mould Technology, Huazhong University of Science and Technology, Wuhan, 430074, China
†E-mail: zhangys@mail.hust.edu.cn

22MnB5 steel has been widely used for its good formability at high temperature and high strength after quenching. Then, a so-called progressive induction quenching process matched with 22MnB5 has been proposed to make thin-wall welded rectangular-tubes. However, about 20 ~ 45 mm bending deformation occurs in the direction of length. In this paper, in order to reduce the bending deformation of thin-wall welded rectangular-tube, the reason of deformation is analyzed firstly. Both the hardness of inner and outer surface and tensile strength of rectangular-tube were measured. The microstructure at weld zone was observed to investigate the effect of weld joint. The relationship of temperature distribution and deformation of rectangular-tube was investigated by means of ABAQUS simulation. From the analysis, it can be seen that although the weld microstructure is large in grain size, it has little influence on the deformation. The cooling rate and non-uniform cooling will cause bending deformation, but the main reason is non-uniform cooling.

Keywords: 22MnB5; Thin-wall welded rectangular-tube; Progressive induction quenching.

1. Introduction

Rectangular metal section tube is widely used in automobile, ship, building, subway, aviation, aerospace due to its characteristics of uniform force, stable structure, easy to install and so on [1]. The tensile strength and the yield strength of 22MnB5 in the annealed state are about 600 MPa and 400 MPa respectively, and the elongation is 25 %, but the tensile strength after austenitizing at high temperature and quenching can reach more than 1500 MPa [2]. The seamless rectangular section tube of high strength steel is widely used in the field of high strength requirement. Progressive induction quenching is shorter heating time and higher heating efficiency than traditional heating method [3]. Electromagnetic induction heating originates from electromagnetic induction phenomenon. It is a green and environment-friendly heating technology that inductive current is produced in conductor by alternating magnetic field, and the conductor is heated without pollution [3].The martensite microstructure of the tube was obtained by quenching after austenitizating with induction heating. Heating frequency, holding time and cooling temperature are all the key factors affecting the final quality of pipe.

This subject comes from the practical problems in production. The rectangular-section tube with weld joint of 6 meters long has resulted in bending deformation about 20 ~ 45 mm after progressive induction quenching. This paper mainly studies the effect of cooling rate on deformation.

2. Experiments

2.1. *Material*

The experimental material was taken from 22MnB5 thin-wall welded rectangular section tube with bending deformation in production. The chemical compositions of 22MnB5 are shown in Table 1. The tensile strength of 22MnB5 after austenitizing and then quenching can reach more than 1500 MPa.

Table 1. Chemical composition of 22MnB5 [wt %].

C	B	Mn	Si	P	S	Cr	Ni	Mo	Ti
0.23	0.019	0.95	0.17	0.01	0.002	0.24	1.50	0.21	0.023

2.2. *Condition of experiment*

The section size of the rectangular-tube with a length of 6 meters is 60 mm×40 mm, and the wall thickness is 1.2 mm. Steel tubes were heated by four induction heaters with a total length of 2 meters. The tube was water quenched directly after austenitization in medium frequency induction heaters. The moving speed of the tube was 2000 mm/min. The processing diagram is shown in Fig. 1. The last furnace could heat the sample to 850°C.

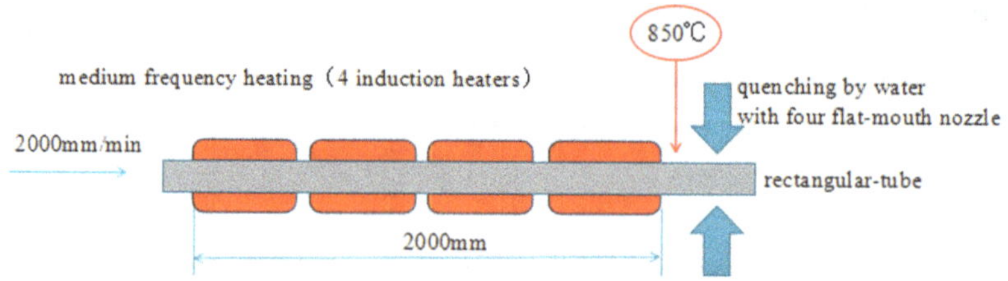

Fig. 1. Process diagram.

The hardness of tube after heat treatment was measured on Vickers hardness meter. Four surfaces of the rectangular-tube as show in Fig. 2(b) were selected to test the hardness. In order to detect the accuracy of hardness, a group of tensile tests measured by the high temperature rupture test machine were added. The tensile test specimen is show in Fig. 3(a). The metallographic structure of weld and non-weld was photographed by Leica microscope to analyze the effect of weld on the deformation of steel tube.

3. The Experimental Results

3.1. *Hardness and tensile strength*

Fig. 2(a) shows that the hardness of inner and outer surfaces of tubes at different positions is almost the same. The maximum and minimum hardness values are located at section 4, the maximum value is 566 HV1, the minimum value is 540 HV1, and the difference is 26 HV1.

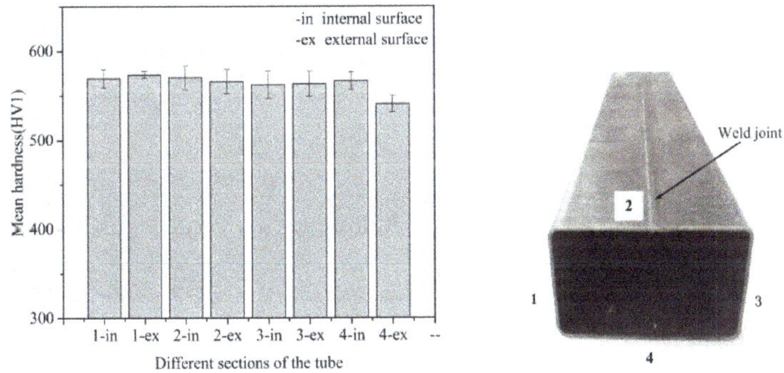

(a) hardness (b) different positions

Fig. 2. Hardness of internal and external surfaces of the tube (a) at different positions (b).

As Fig. 3(b) shows, the maximum tensile strength is 1830.5 MPa, and the minimum tensile strength is 1730.9 MPa, which is more than 1500 MPa. This indicates that the strength requirement of 22MnB5 was meet by medium frequency induction heating and quenching. The maximum elongation is 6.726 %, the minimum elongation is 4.360 %, and the average elongation is 5.276 %, which is smaller than theoretical value.

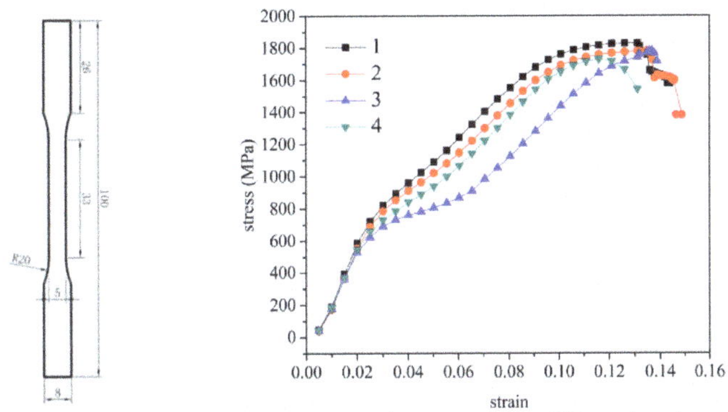

(a) specimen (b) stress-strain curves at four different locations

Fig. 3. Stress-strain curves at different locations obtained by tensile test with (a) specimen.

3.2. *Microstructure of weld joint*

Fig. 4(a) is a metallographic structure of the weld joint. The weld section should be divided into weld joint, fusion zone and heat affected zone strictly. Because of high heating speed and short residence time at high temperature, welding is not conducive to the homogenization of austenite. At the same time, the peak temperature of the heat affected zone near the fusion line is very high, which is close to the melting point of the weld metal, and it is easy to cause the grain overheating and serious growth [4]. Fig. 4(a) shows that the lath martensitic structure is larger than that in Fig. 4(b). But this kind of large structure is so small that it has little impact on overall performance. Due to the short time of induction heating, the inhomogeneous austenite is transformed into fine martensite after quenching, which causes high strength.

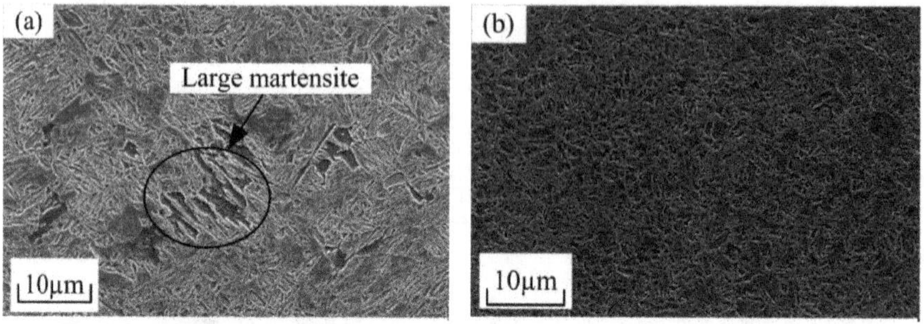

Fig. 4. Structure of the rectangular-tube at different places a) weld joint, b) normal place.

4. The Thermo-mechanical Coupling Finite Element Analysis

The effect of non-uniform cooling on deformation is mainly considered by using the thermo-mechanical coupling module of ABAQUS [5]. From the contents of part 2, we could see that the speed of 2000 mm/min passes through the induction heating furnace of 2000 mm, the last induction furnace was heated above 850 °C, and the time from heating to holding was 60 s. The temperature distribution applied in the simulation work is shown in Fig. 5.

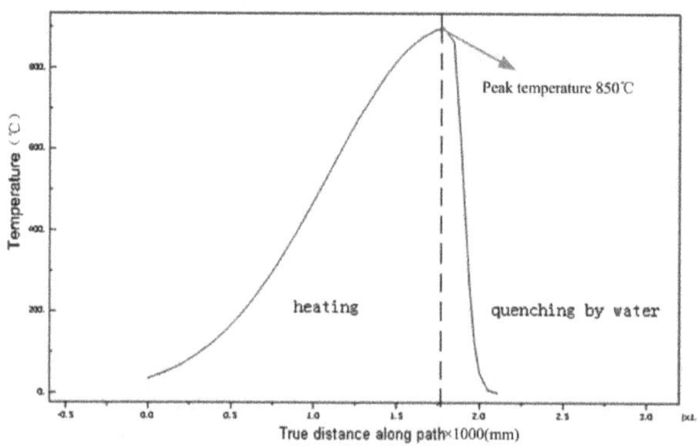

Fig. 5. 22MnB5 heating-cooling process curve.

Material parameters are shown in Table 2. The length of model is 3000 mm, the size of rectangular-section is 60 mm×40 mm and the wall thickness is 1.2 mm. The Young's modulus, Poisson's ratio and specific heat capacity vary with time.

Table 2. Material parameters.

Conductivity ($W \cdot m^{-1} \cdot °C^{-1}$)	Mass density ($tonne \cdot mm^{-3}$)	Thermal expansion coefficient (°C)	Specific heat ($J \cdot kg^{-1} \cdot K^{-1}$)
32	7.83×10^{-9}	1.45×10^{-5}	4.44×10^{8}

The boundary conditions were set to one end fixed to simulate the roll constraint. Deviation of steel tube after deformation could be observed. The temperature function is shown in Eq. 1.

$$T = a\exp\left(-\left((x-b)/c\right)^2\right) \tag{1}$$

Where T is temperature, x is the length of steel tube, a is the peak coefficient for temperature, b is displacement coefficient and c is heating or cooling rate coefficient. First, the influence of cooling rate on deformation was investigated. Different cooling rates can be obtained by changing the coefficient c. The tube deformation at different cooling rates is shown in Fig. 6. As the deformation under different cooling rates shows, cooling rates have little effect on bending deformation. After multiple sets of coefficient c calculation, when c=5~10, the cooling rate v=400 °C/s~500 °C/s, the bending deformation is the smallest.

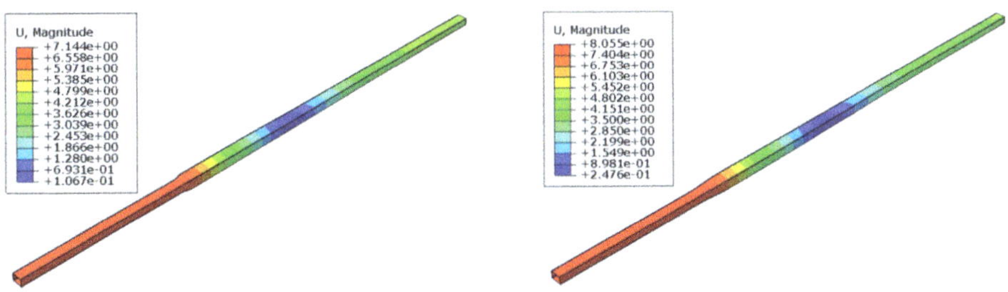

(a) c=10, v=450 °C/s (b) c=300, v=75 °C/s

Fig. 6. The bending deformation of tube at (a) c=50, v=450 °C/s and (b) c=300, v=75 °C/s.

Different temperatures were set on different surfaces of the tube in the cooling stage. The deformation caused by non-uniform is shown in Fig. 7(a) and by uniform cooling is shown in Fig. 7(b), respectively. As the figure shows, the bending deformation caused by the non-uniform cooling is very obvious and by the uniform cooling is small, respectively. Both of the set temperatures were the same during heating.

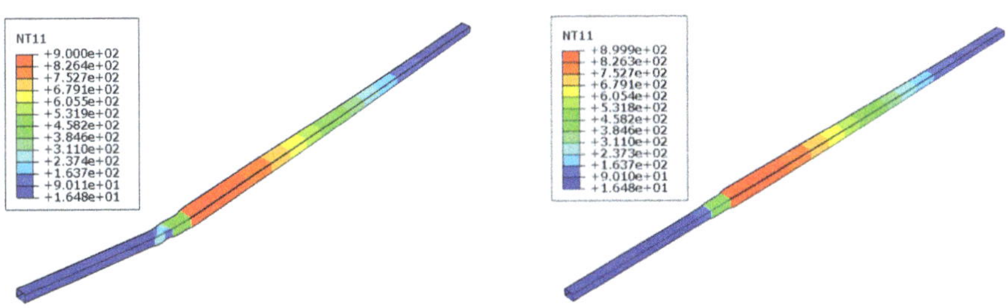

(a) non-uniform cooling (b) uniform cooling

Fig. 7. Deformation caused by (a) non-uniform cooling and (b) uniform cooling.

From the above two factors simulated, it is obvious that the uniform cooling has a great influence on the bending deformation.

5. Cooling Spray Pipe Designing

From the simulation results, obvious bending deformation of tube could be formed by non-uniform cooling. In order to solve the problem of non-uniform cooling, the spray pipe as shown in Fig. 8 was designed. The nozzle clearance was 10 mm, and the diameter of each nozzle was 2 mm. In order to prevent the water from spraying out into the heating zone, there was 45° between nozzle and the moving direction of steel tube. The nozzle was linearly distributed along the spray pipe to ensure the cooling uniformity.

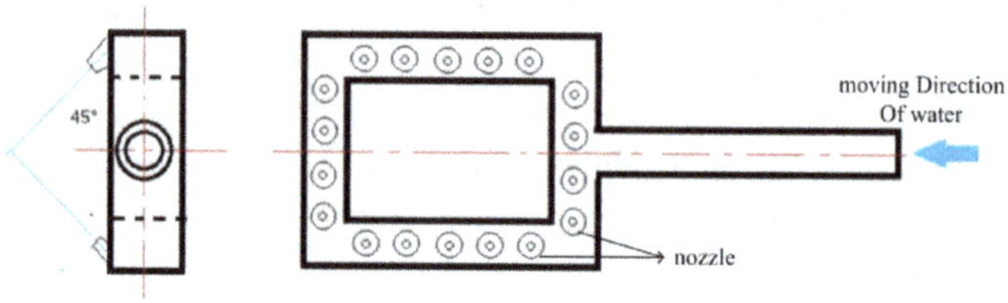

Fig. 8. The designed spray pipe (each nozzle is 10 mm apart; nozzle in the moving direction of steel tube is 45°).

The improved cooling water spray pipe was applied to the actual production, and the bending deformation of the rectangular-tube was improved. The bending displacement is reduced to about 4 mm, compared with the initial 20 ~ 45 mm.

6. Conclusion

(1) The progressive induction quenching process was proposed to produce thin-wall welded rectangular-tube. After process, the Vickers hardness of final part reached about 540 HV1, the tensile strength is above 1700 MPa, and fully martensitic microstructure can be obtained.

(2) The thermal-mechanical coupling simulation works have been done for the progressive induction quenching process. From the results of deformation distribution obtained from the simulation works, both cooling rate and cooling uniformity have influences on deformation, but the main influencing factor is the latter. With the decrease of cooling rate, the deformation decreases first and then increases. When the cooling rate is 400 °C/s ~ 500 °C/s, the bending deformation is the smallest.

(3) The bending deformation is reduced to 4 mm because of the designed spray pipe which solve the non-uniform cooling. It is necessary to cool the long tube uniformly to reduce its bending deformation.

Acknowledgements

This research work was financially supported by the National Natural Science Foundation of China (grant no.U1564203 and U1760205) and National Major Science Technology Project of China (grant no.2018zx04023001).

References

1. J. Wang, S. Afshan and N. Schillo, Material properties and compressive local buckling response of high strength steel square and rectangular hollow sections, *Eng. Struct.* **130**, 297 (2017)
2. H. Zhou, S. P. Zeng and S. G. Yuan, Development and application of induction heating quenching technology, *Heat Treatment Technology & Equipment* **3**, 9 (2008).
3. H. Liu and J. Rao, Temperature field Simulation of steel pipe in the induction quenching and tempering process, in *Proceedings of the 6th International. Conference on Physical and Numerical Simulation of Material Processing*, (Guangxi, China, 2012).
4. H. Zhao and H. Liu, Microstructure and properties of TIG welded 22MnB5 ultra high strength steel, *Transactions of the China Welding Institution* **2**, 67 (2014).
5. J. Chen, G. Zhang and Z. Zhao, Numerical simulation of continuous induction quenching process of 45 steel plain shafts, *Heat Treatment of Metals* (2016).

Identification of Hot Stamping Fully Martenstic Microstructure SEM Photograph with Support Vector Machine

F. K. Hu, Z. J. Zhu, K. Wang, B. Zhu† and Y. S. Zhang
State Key Laboratory of Materials Processing and Die and Mould Technology, Huazhong University of Science & Technology, Wuhan 430074, China
†E-mail: zhubin26@hust.edu.cn

Hot stamping has attracted significant attention to produce boron steel in automobile industry for its excellent shape accuracy and small springback. A qualified hot stamping component must have fully martensitic microstructure to ensure the strength reached 1500MPa. Then in the current paper, a methodology to identify the fully martensitic microstructure SEM photograph with Support Vector Machine is proposed. At first, different heat treatment experiments are implemented to obtain different microstructures, and the SEM photographs are collected. Then, the Gray-Level Co-occurrence Matrix of those SEM photographs are calculated, and the features of different SEM photographs are extracted to train the Support Vector Machine (SVM) classification. The experimental results show that GLCM based feature extraction achieves high accuracy, and an automated system has been developed to identify fully martensitic microstructure SEM photograph without human intervention.

Keywords: Martensite; SEM photograph; Support Vector Machine; Gray-Level Co-occurrence Matrix.

1. Introduction

With the rapid development of the automotive industry, lightweight car bodies become the focus of automobile research. The hot stamping process of high strength steel, as one of the important technologies to realize the lightweight of automobiles, has gotten increasingly applications. As it is known, to ensure that the strength reaches 1500 MPa, the microstructure of a qualified hot stamped component must be entirely martensite. Therefore, it has become a topic worthy of study that how to determine the obtained parts have fully martensitic microstructure.

Microstructural information is most often conveyed via images, which are obtained through various microscopy techniques [1], such as metallographic microscope, scanning electron microscope, etc. These microstructural images themselves are rich in fairly accurate information, however, they have been always evaluated by human experts, which leads inevitably to subjective. What's worse, in some companies, those who conduct microcosmic analysis of materials often do not have antique expertise and judge by experience alone, sometimes contributing to serious consequences. There are also other methods for determining the composition of a material (i.e. X-ray diffraction), but their cost is often too high and only the content of the ingredients can be obtained, while the distribution cannot be determined.

In order to check objectively and accurately whether the hot stamping parts are qualified, martensitic feature extraction and Support Vector Machine classification experiments were conducted and a methodology based on it to identify the fully martensitic microstructure SEM photograph was proposed.

2. Methodology and Experimental Procedure

It is well-known that the three-dimensional morphology of martensite usually is a plate or lath, but it is usually needle-shaped in metallographic observations (two-dimensional). Even so, the microstructure of martensite is superficially complex and has proven difficult to describe in any simple way [2]. Therefore, in order to be able to recognize the full martensitic microstructure more accurately, a machine learning method was introduced and a martensitic SEM micrographs recognition system based on texture features was developed (Fig. 1).

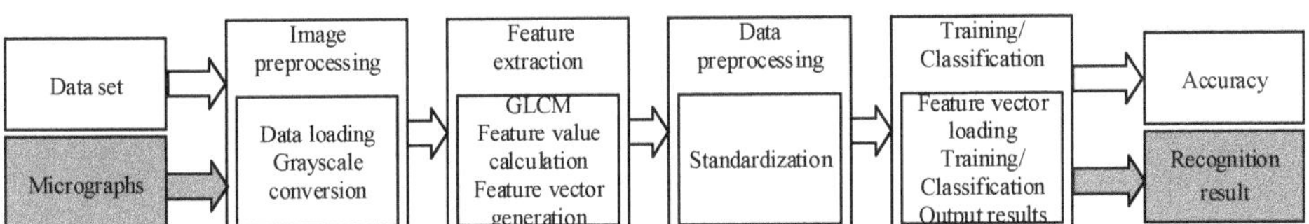

Fig. 1. Flow of martensitic SEM micrographs recognition system.

2.1. *Material and dataset*

The experimental material is boron steel, and the detailed data of its components is shown in Table 1.

The dataset consists of 38 SEM micrographs of different microstructures (a half is fully martensite microstructure and the rest is not), which are obtained through different heat treatment experiments. About half of them are fully martensitic microstructure (Fig. 2(a)), while others are not, but include ferrite, bainite and other microstructures (Fig. 2(b)), and their temperature curve during the forming process is shown in Fig. 3.

When making a data set, each SEM micrograph is divided into multiples with a size of 112*112 pixels (Fig. 4 (a)). The reason for this is as follows: Firstly, the entire SEM image will have more or less impurities, and after screening the undesirable parts can be screened. Secondly, if the image is divided too small, its texture features cannot be well represented. Finally, it expands the capacity of the data set and can better suppress the occurrence of overfitting. Preparing for the subsequent classification training, all the divided images will be labelled (Fig. 4 (b)), and the first position in the label is 1 for full martensite and 0 for non-full martensite.

Table 1. Chemical composition of the 22MnB5 substrate in wt %.

C	Si	Mn	P	S	Cr	Ni	Mo	Ti	B
0.23	0.17	0.95	0.01	0.002	0.24	1.50	0.21	0.023	0.019

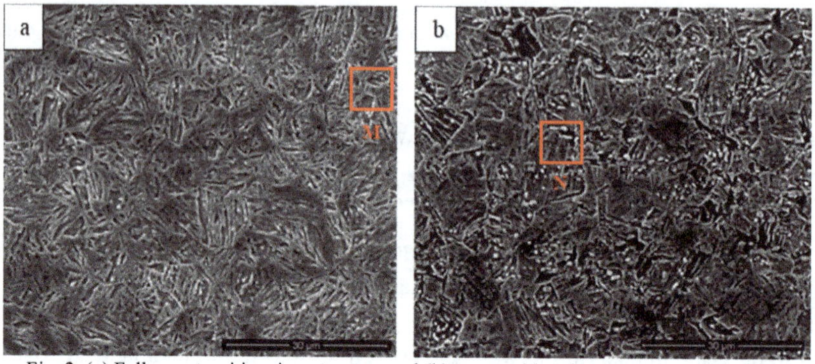

Fig. 2. (a) Fully martensitic microstructure and (b) martensitic microstructure with bainite.

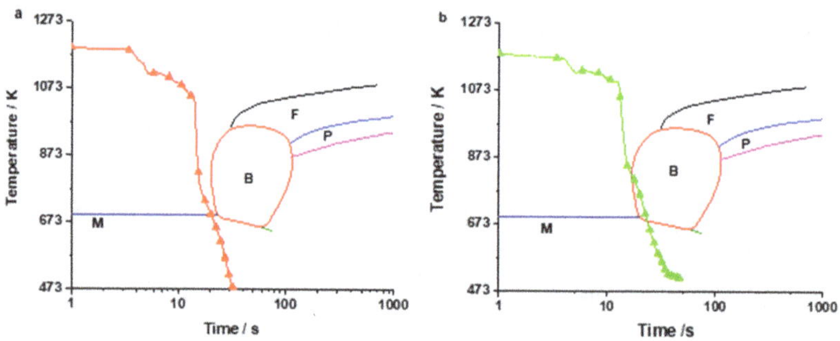

Fig. 3. The temperature curves (20MPa) of (a) fully martensitic microstructure and (b) martensite microstructure with bainite.

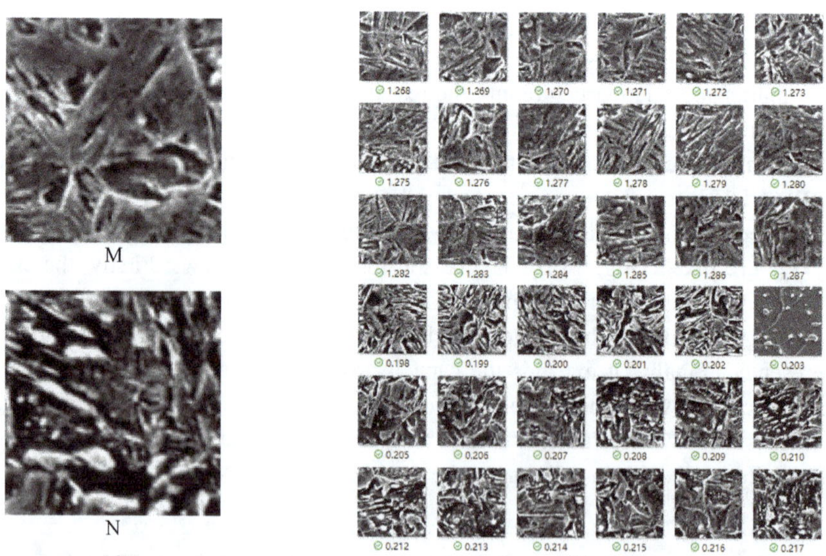

(a) (b)

Fig. 4. (a) Split SEM micrographs (M and N refer to Fig. 2) with a size of 112*112 pixels and (b) the divided images with labels.

2.2. *Feature extraction*

For the SEM micrographs of martensite, texture features are prominent, while color features are not obvious, so GLCM is considered to be a relatively reasonable way to describe characteristics. When calculating a GLCM, the relative frequencies of gray level pixel pairs separated by a distance d in a particular angle θ forms the displacement vector (d, θ) [3]. Usually GLCM of four common directions (0°, 45°, 90°, 135°) are calculated and combined as eigenvectors so that the SEM micrographs feature values have rotation invariance. When referring to the distance, d = 1 is chosen to extract the detailed textural information of the SEM micrographs.

Further, there are 14 parameters calculated from the GLCM up to describe the characteristics [4]. Only a few parameters that characterize the texture are noticed in this experiment. They are Contrast, Dissimilarity, Homogeneity, ASM and Energy. Then GLCM features are integrated into support vectors and support vector machine is trained to achieve full martensite identification. Table 2 and 3 show detailed information of a support vector based GLCM features of fully martensite (referring to Fig. 4 (a) M) and martensite with other microstructures (referring to Fig. 4 (a) N). It is evident from the tables that for full martensite, the characteristic values contrast and dissimilarity are lower than those of martensite with bainite impurities, but the homogeneity, ASM and energy are much higher, those all reflecting the difference between them.

There are other methods for extracting texture features, such as Local Binary Patterns (LBP) and Histogram of Gradient (HOG). In section 2.4 the classification effects based on them are also explored.

Table 2. Detailed information of a support vector based GLCM features of full martensite.

Eigenvalues	0°	45°	90°	135°
Contrast	1.20171640e+07	2.12493080e+07	1.47447080e+07	2.54621800e+07
Dissimilarity	3.89504000e+05	5.24032000e+05	4.32220000e+05	5.57948000e+05
Homogeneity	2.00940155e+03	1.42013358e+03	1.80786579e+03	1.43724495e+03
ASM	1.06586000e+05	8.40500000e+04	9.95140000e+04	8.44080000e+04
Energy	3.26475114e+02	2.89913780e+02	3.15458397e+02	2.90530549e+02

Table 3. Detailed information of a support vector based GLCM features of martensite with bainite.

Eigenvalues	0°	45°	90°	135°
Contrast	2.02521840e+07	3.36461620e+07	2.77738240e+07	5.25720300e+07
Dissimilarity	5.14980000e+05	6.75430000e+05	5.91860000e+05	8.24094000e+05
Homogeneity	1.63529775e+03	1.10973653e+03	1.48209320e+03	1.01167081e+03
ASM	7.62540000e+04	6.05320000e+04	6.60700000e+04	5.38460000e+04
Energy	2.76141268e+02	2.46032518e+02	2.57040853e+02	2.32047409e+02

2.3. *Grayscale conversion and standardization*

The flow of martensitic SEM micrographs recognition system includes two pre-processing processes: grayscale conversion and standardization. Since grayscale co-occurrence matrix input must be a grayscale image, the SEM micrographs must be grayscaled. And it is clear from the tables that the change in the value range of the eigenvalue is too large, which will cause the subsequent high-dimensional iteration process to fail to find the hyperplane, resulting in non-convergence. To address this problem, the support vector must be standardized.

2.4. *Training and classification*

Support Vector Machines (SVM) are based on hyperplane classifiers, and they are frequently applied to classify samples in different categories called classes [5]. Judging whether a certain microstructure is full martensite is a binary problem, which SVM can achieve well. In the process of classification using SVM, the choice of kernel function has a great influence on the result. Out of the different types of kernel functions, the linear was selected because it is suitable for the low dimensionality of the gray level co-occurrence matrix (5x4) in this experiment, which is in line with the general conclusion.

There are also other ways to classify these SEM images, such as Random forest (RF) and K nearest neighbors (k-NN). Each of them can be combined with different feature extraction methods (GLCM, LBP, HOG) to form different classification methods. In order to explore which method works better, statistics on the accuracy of training and testing of methods mentioned above were performed.

3. Results and Discussions

Fig. 5 shows the classification effect of different methods. The accuracy and running time of the training and testing are the main considerations. Fig. 5(a), (b) and (c) show the effects of SVM, RF, and k-NN classifier based on different feature extraction methods, which indicates that the accuracy of the GLCM method is the highest. Fig. 5(d) shows the running time of the training and testing process of each method. It can be learned that when LBP is chosen as feature extraction method, the running time is shortest, less than 5 s, while that of HOG is longest, more than 45 s. The results are not difficult to understand. After all, the feature vector based on the LBP has a smaller dimension and less data, so the required computation time is naturally shorter. The running time of GLCM is about 5 s, very close to that of LBP. Overall, the classification based on GLCM works better, which reveals that the gray-level co-occurrence matrix can well represent the martensite texture features.

Fig. 5(e) shows the effect of different classifiers based on GLCM. The accuracy of them are all quite high, and the SVM classifier has the best performance. The above experiments show that the GLCM-based SVM method has better classification effect than other combinations and can well identify the SEM photographs of the fully martensite microstructure.

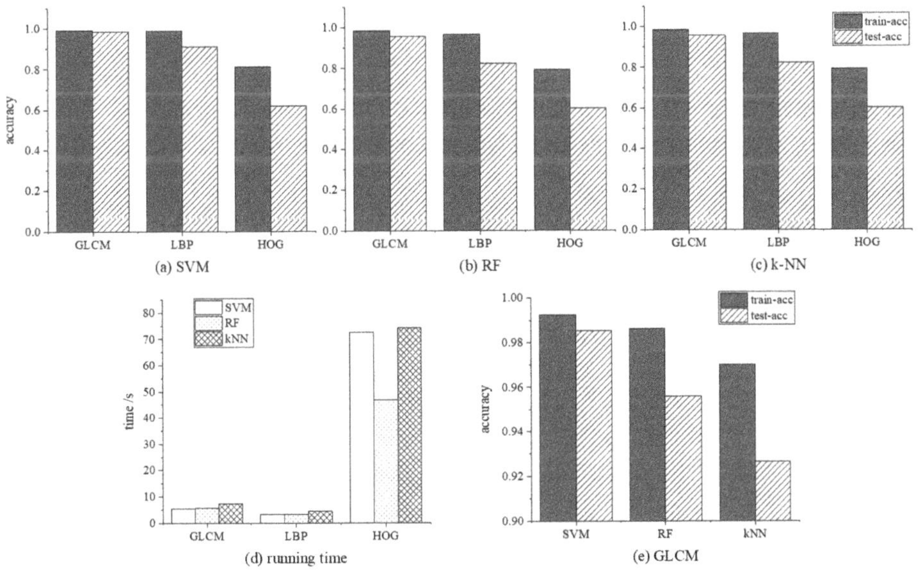

Fig. 5. The different effect of (a) SVM classifiers (b) RF classifiers (c) k-NN classifiers (d) running time of these classifiers and (e) the effect of different classifiers based on GLCM.

Through the above exploration, a method of martensitic recognition based on gray level co-occurrence matrix and support vector machine is proposed. Based on this, a system is developed for automatic identification of martensite. In fact, the application of machine learning methods in material organization will go far beyond this. The applications of these methods can not only identify certain simple microstructures, but also recognize multiphase microstructures, and even describe their tissue distribution and calculate the tissue content.

4. Conclusion

For the hot stamping process of high strength steel, it is vital to ensure that the microstructure of components get full martensite. In the present paper, in order to find a simple and effective method to identify the entire martensite, a series of experiments have been implemented and the conclusion could be givens as follows.

(1) An automated system has been developed to identify fully martensitic microstructure SEM photograph without human intervention. It takes about 10ms to achieve single image's recognition.

(2) Compared to local binary patterns and histogram of gradient, gray level co-occurrence matrix can better characterize the texture of martensite.

(3) Compared to RF and k-NN classifiers based on different feature extraction methods such as GLCM, HOG and LBP, SVM classifier based on GLCM can better identify the fully martensitic microstructure, and the accuracy of test has been up to 98%, which is much higher than that of the other classifications.

Acknowledgments

This research work was financially supported by the National Natural Science Foundation of China (grant no. U1564203 and U1760205) and National Major Science Technology Project of China (grant no. 2018zx04023001).

References

1. B. L. DeCost, T. Francis and E. A. Holm, Exploring the microstructure manifold: Image texture representations applied to ultrahigh carbon steel microstructures, *Acta Mater.* **133**, 30 (2017).
2. S. Morito, H. Tanaka., R. Konishi T. Furuhara and T. Maki, The morphology and crystallography of lath martensite in Fe-C alloys, *Acta Mater.* **51**, 1789 (2003).
3. K. Manivannan, P. Aggarwal, V. Devabhaktuni, A. Kumar, D. Nims and P. Bhattacharya, Particulate matter characterization by gray level co-occurrence matrix based support vector machines, *J. Hazard. Mater.* **223**, 94 (2012).
4. C. Malegori, L. Franzetti, R. Guidetti, E. Casiraghi and R. Rossi, GLCM, an image analysis technique for early detection of biofilm, *J. Food Eng.* **185**, 48 (2016).
5. J. Ortegon, R. Ledesma-Alonso, R. Barbosa, J. V. Castillo and A. C. Atoche, Material phase classification by means of Support Vector Machines, *Comp. Mater. Sci.* **148**, 336 (2018).

Process Design

Hot Stamping of Tailor Welded Blanks

C. H. Hung[1,a,†], C. H. Lee[1,b], C. K. Chiu Huang[2,c] and F. K. Chen[3,d]

[1]Department of Mechanical Engineering, National Chiao Tung University, Hsinchu, Taiwan
[2]Product Application Research Section, Green Energy & System Integration R&D Department China Steel Corporation, Kaohsiung, Taiwan
[3]Department of Mechanical Engineering, National Taiwan University, Taipei, Taiwan
E-mail: [a,†]chhung@mail.nctu.edu.tw, [b]loghot520@gmail.com, [c]185066@mail.csc.com.tw, [d]fkchen@ntu.edu.tw

With the continuous improvement of vehicle safety standards in various countries in recent years, current car body designs need to feature high strength and impact energy absorption characteristics to protect the human body from harm as a result of direct impact energy. Hence, in car structural components, two or more materials with differing mechanical characteristics are used. Concerning car structural components, parts made of high strength materials mitigate deformation, whereas those made of low strength materials absorb impact energy. This study examined the strength of various material combinations for the hot stamping of tailor welded blanks. Using finite element analysis software DEFORM, this study constructed tailor welded blank models. Specifically, U-shaped models were designed through hot stamping for tailor welded blanks. The strength of the high-strength regions exceeded 1500MPa, while that of the low-strength regions was maintained between 500 and 800 MPa by manipulating parameters such as material thickness and blank holder force. Experiments on the hot stamping of tailor welded blanks were conducted with the same conditions adopted in the analysis. The experiment results verified the accuracy and practicality of the analyzed method.

Keywords: Hot stamping; Finite element analysis; Tailor welded blanks.

1. Introduction

Hot stamping processes typically involve use of manganese-boron steels e.g. [1]. In hot stamping, steel sheets are heated to 930 °C in a heating furnace. The sheets are maintained at this temperature for 3 to 5 min for austenitization and then transferred to a press machine where they are formed, die quenched, and die opened after they are cooled to ≤ 200 °C. When the average cooling rate is higher than the critical quenching rate of 27°C/s, a metallurgical transformation from austenite to martensite occurs, and a tensile strength of over 1500 MPa is achieved.

Global automobile safety standards have been continually improving in recent years. In the improvement of automobile designs, not only lightweight materials, but also the safety of drivers and passengers must be considered. High-strength materials should be used in automobile designs, and greater emphasis should be placed on vehicles' crash-energy-absorption capacity, the purpose of which is to avoid direct transfer of crash energy to the human body. Designs of automobile structural components have begun incorporating tailored blanks, which feature a single automobile part with two mechanical functions. Studies e.g. [2], [3] have reported use of blank tempering, tool tempering, tool materials, tool surface areas, and welded blanks in the tailored hot stamping process. In side collisions, B pillars e.g. [4] are the main vehicle body parts that resist external deformation. They are conventionally designed to exhibit high strength to reinforce a vehicle's side strength. B pillars formed through the hot-stamping process and tailored blank designs feature high-strength upper parts for deformation resistance and high-toughness bottom parts for crash-energy absorption.

Choi et al. e.g. [5] used 22MnB5 steel to investigate the effect of using hot stamping on tailor-welded blanks (TWBs); their results indicated a heat-affected zone near the weld joint after laser welding of the blanks. High-strength martensite was formed in this heat-affected zone, which was therefore harder than other zones. However, no significant hardness difference was observed between the laser-welded blanks and the hot-stamped blanks. Tang et al. e.g. [6] adopted China Baowu steel sheets and small tailor-welded B pillars in an experiment to investigate the accuracy of LS-DYNA finite element analysis in TWB production processes. The analysis results for thinning rate in specific locations and the experimental results did not differ considerably, suggesting that finite element analysis can be used in actual TWB production processes.

This study investigated the material strength ratio in the hot stamping process of TWBs. Finite element software DEFORM was used to model a TWB. A U-shaped part was subjected to a hot stamping process so that the strength of the high-strength zone exceeded 1500 MPa and the low-strength zone was between 500 and 800 MPa. Strength control of low-strength zones was achieved by adjusting material thickness, blank-holder force, and other parameters. Using the same conditions as those of the simulation analysis, this study also conducted a series of experiments involving hot stamping of TWBs.

2. Experimental Process

2.1. *Materials and parameters*

In production of TWBs, this study used high-strength manganese-boron steels and low-strength alloy steels to represent materials with high-strength and low-strength zones, respectively. For the high-strength zones of tailor-welded materials, 15B22 steel (China Steel Corporation) was used to enable control of the quenching rate of the hot stamping process and thereby achieve complete martensitic transformation. Table 1 presents the material composition of the 15B22 steel. Materials with low-strength zones were primarily required to exhibit excellent weldability and consistent properties following the hot stamping process. Table 2 lists the composition of JSC440W steel (China Steel Corporation), which was used for the low-strength zones of tailor-welded materials.

Table 1. Material composition of the 15B22 steel (wt %).

C	Si	Mn	P	S	Ti	B	N	Al	Cr
0.224	0.243	1.087	0.016	0.002	0.044	0.0024	0.0039	0.041	0.187

Table 2. Material composition of the JSC440W steel (wt %).

Fe	Mn	Cr	Al	Nb	Ni	Si	C	P	S
98.927	0.76	0.02	0.034	0.006	0.01	0.08	0.09	0.066	0.007

2.2. *Model parameters setting*

The parameters of a U-shaped model were specified. Fig. 1 shows the geometric representation of the model. 15B22 was used for the high-strength tailor-welded steel sheet, whereas JSC440W was used for the low-strength one. The ambient temperature was 28 °C. The initial temperatures were 919 °C for 15B22 and 926 °C for JSC440W. The steel sheets and molds were solid elements, the coefficient of friction was 0.4, the holder force was 15–20 ton, the transposition time was 6–7 sec, the stamping time was 3sec, the holdup time was 10 sec, SKD61 was used as a die material, and the cooling-pipe parameter was 4770 W/m^{2}K. Fig. 2 presents the analysis model of the tailor-welded steel sheet U-shaped model.

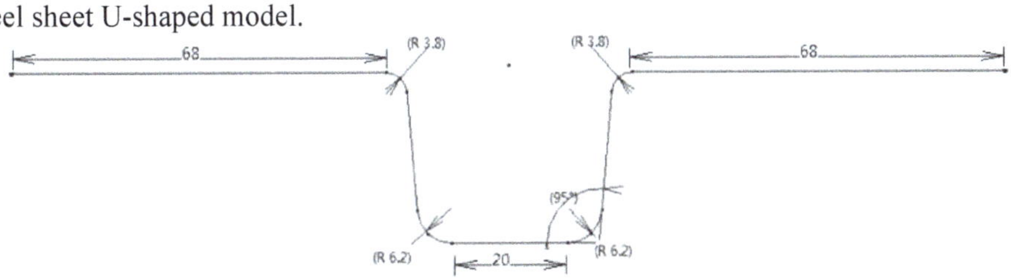

Fig. 1. Geometric representation of a U-shaped model.

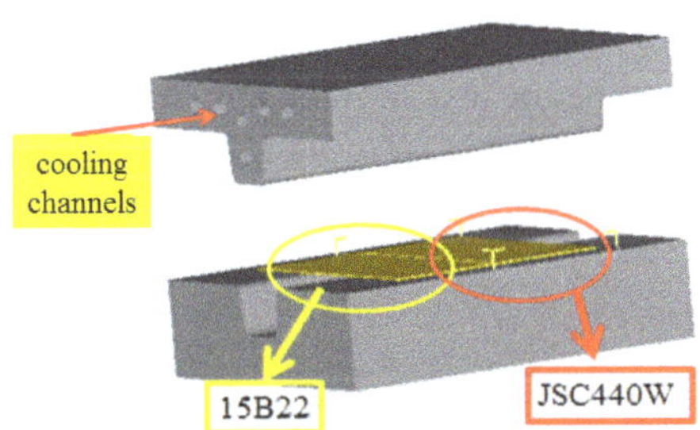

Fig. 2. Tailor-welded steel sheet U-shaped model.

2.3. *Die development*

The experimental die design in Fig. 3 illustrates the punch set, die set, upper die, lower die, guide pillars, locator block, stop block, and cooling channels. The dimensions of the SKD61 die were 180 mm×240 mm×90 mm. The upper die featured a locator block and cooling channels as represented in Fig. 3. For the lower die, two sides of alternating thicknesses were constructed, and a packing-block design was used to tightly lock the two lower die parts.

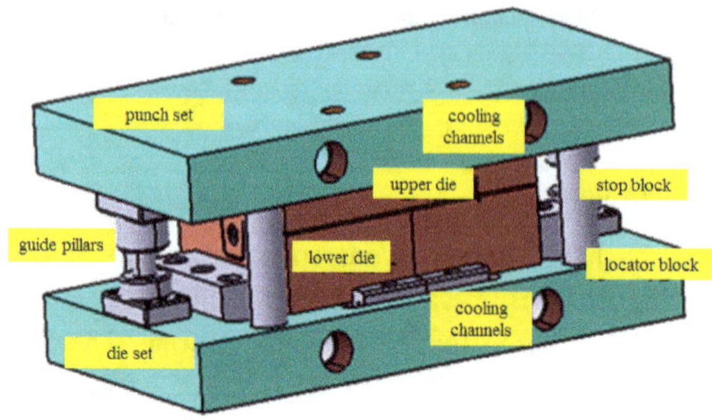

Fig. 3. Design of the U-shaped part.

2.4. *Experimental procedure and measurement*

High and low-strength hot-stamped tailor-welded steel sheets of the same thickness (both were 1.2 mm or were 1.6 mm) were joined using laser beam welding. The size of the laser-welded steel sheet was 140 mm × 140 mm. Thermocouple holes with a diameter of 1mm were drilled on the steel sheets to measure temperature changes during processing. The blank holder force and holdup time were altered to examine their effects on the hot-stamped tailor-welded steel sheets. After the steel sheets were sheared, their hardness was measured. Simulations were subsequently performed on the sheets, and the results were compared with the experimental results.

3. Results and Discussion

The tailor-welded steel sheets before and after the experiment and hardness measurement points, and cracks were identified in the thermocouple holes of the TWBs (Fig. 4). Table 3 lists the experimental parameters.

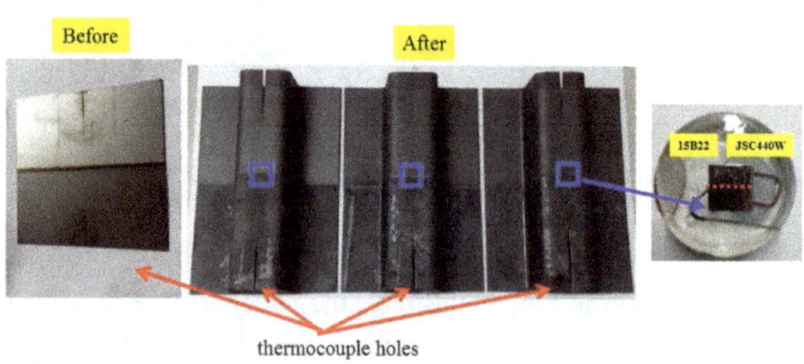

Fig. 4. The tailor-welded steel sheets before and after the experiment and hardness measurement points.

Table 3. Experimental parameter plan.

Types of materials	15B22-JSC440W	15B22-JSC440W	15B22-JSC440W
Blank-holder force (ton)	15	20	20
Hold time (s)		10	
Welded material thickness (mm)	1.2-1.2	1.2-1.2	1.6-1.6
Transfer time (s)	6	7	6

The welded 15B22 (high-strength zone) and JSC440W (low-strength zone) steels exhibited the same thicknesses. After the U-shaped part was quenched for each sample, samples of both types of materials had completed phase transformation. Three experiments were performed for each set of parameters. The results of the experiments were compared with the simulation results. Additionally, experimental and analysis results revealed that the material temperatures were below 200 °C at this point in the process. Overall, temperature simulation analysis trends were consistent with the experimental results. Temperature distributions of high-strength 15B22 steel and low-strength JSC440W steel are presented in Figs. 5 and 6, respectively. As both figures reveal, temperature differences were more noticeable in relation to sheet thickness than blank holder force.

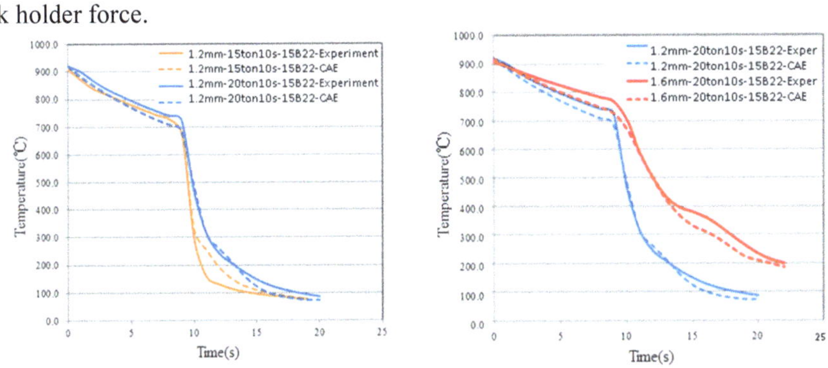

Fig. 5. Temperature simulation analysis and experimental results for high-strength 15B22 steel.

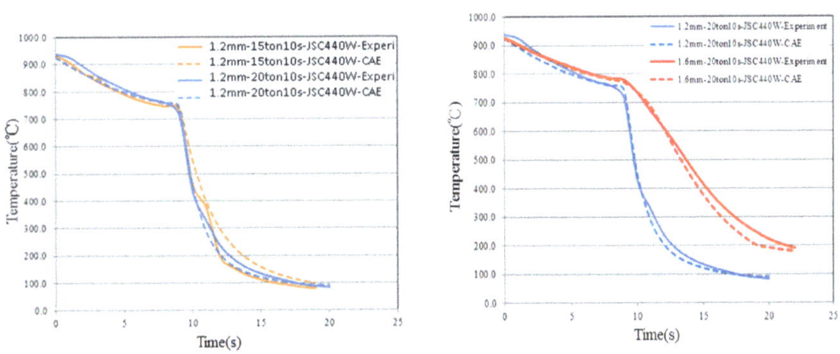

Fig. 6. Temperature simulation analysis and experimental results for low-strength JSC440W steel.

Fig. 7 shows the hardness measurement points on the thinner steel sheets (1.2 mm). When blank-holder force was increased, the hardness of high-strength 15B22 steel increased slightly from an average of 461 HV to 465 HV, and the hardness of low-strength JSC440W increased slightly from an average of 160 HV to 164 HV. When the steel sheets were thicker (1.6 mm), the hardness of high-strength 15B22 steel decreased slightly from an average of 465 HV to 463 HV, and the hardness of low-strength JSC440W decreased slightly from an average 164 HV to 156HV. Fig. 7 also illustrates the simulation analysis and experimental results. Its temperature differences compared to those of the thinner sheets were only reflected in the hardness of the low-strength sheet; the hardness of the high-strength sheet was almost not influenced by temperature.

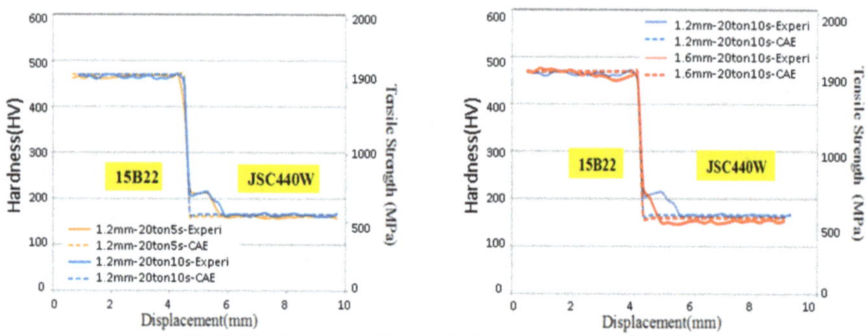

Fig. 7. Hardness simulation analysis and experimental results.

Using the equation (1) for converting material thickness and tensile strength proposed by Pavlina et al. e.g. [7], this study discovered that as the blank holder force increased, the average tensile strength of the 15B22, the high-strength sheet, increased from 1621 to 1636 MPa, whereas that of the JSC440W, the low-strength sheet, only slightly increased from 515 to 520 MPa. Moreover, as sheet thickness increased, the average tensile strength of the 15B22 and JSC440W declined slightly from 1636 to 1629 MPa and from 520 to 501 MPa, respectively. The value of the tensile strength converted from hardness is illustrated in the longitudinal axis in the right hand side of Fig.7.

$$TS = -99.8 + 3.734HV \text{ [7]} \tag{1}$$

4. Conclusion

In this study, parameters such as the hardness and blank holder force were altered to induce changes in the strength of a low-strength tailor-welded steel sheet. A series of hot stamping experiments was also performed on the tailor-welded sheet under the same conditions as the simulations. The simulation and experimental results indicated that all temperature distributions were less than 200 °C for the steel sheets after they were processed. Moreover, their temperature changes during the simulation were identical to those during the experiment. The temperature differences were more considerable in relation to sheet thickness than blank holder force. Comparisons of sheet thickness and blank holder force with hardness and with tensile strength both suggested that the temperature differences due to sheet thickness were only reflected in the hardness differences in the weak-strength steel sheet. The hardness of the high-strength sheet was almost not influenced by temperature. The simulation and experimental results were consistent, with the experimental results confirming the accuracy and usability of the simulation results. Whereas the tensile strength of the high-strength steel sheet increased to at least 1500 MPa, that of the low-strength regions was maintained between 500 MPa and 800 MPa.

Acknowledgments

The authors wish to thank the Ministry of Science and Technology of Taiwan for the grant under project #MOST106-2622-8-006-001, which makes this research possible.

References

1. K. Hyunok and A. Taylan, Hot stamping technology, *EWI Forming Center* (2014).
2. H. Karbasian and A. E. Tekkaya, A review on hot stamping, *J. M. Process. Tech.* **210**, 2103 (2010).
3. V. Uthaisangsuk, Hot stamping of ultra high strength steel: a key technology for lightweight automotive design, *Automotive Summit* (2014).
4. B. Macek, Optimization side crash performance using a hot-stamped B-pillar, *Great Designs in Steel*.
5. H. S. Choi, D. C. Ko and B. M. Kim, Effect of weld line orientation on formability of tailor welded blank of 22MnB5 at elevated temperature, *Adv. Mech. Eng.* **1**, 1 (2014).
6. B. Tang, Q. Wang, Z. Wei, X. Meng and Z. Yuan, FE simulation models for hot stamping an automobile component with tailor-welded high-strength Steels, *J. Mater. Eng. Perform.* **25**, 1709 (2016).
7. E. J. Pavlina and C. J. V. Tyne, Correlation of yield strength and tensile strength with hardness for steels, *ASM Int.* **17**, 888 (2008).

A Novel Tailored Hot Stamping Tooling Concept to Achieve Multiple Locally Tailored Zones

Y. Wang[†], M. Pereira and B. F. Rolfe
School of Engineering, Deakin University, Geelong, Australia
[†]E-mail: yanan.wang@deakin.edu.au

Y. Liu, Z. J. Zhu, J. Zhu, Z. J. Wang and Y. S. Zhang
State Key Laboratory of Materials Processing and Die & Mould Technology, Huazhong University of Science and Technology, Wuhan, 430074, China

In this paper, the details of a new purposely designed multi-block tailored hot stamping tool will be presented. The design of this tooling permits the precise control of temperature/cooling at 12 individual regions within the punch and die of tooling to produce straight channel-shaped parts (U-shape). This tooling was fabricated and tested, and this study presents the first results which assess the capability of this novel tooling concept. Five experiments of different combinations of heating and cooling blocks were conducted using high strength hot stamping steel 22MnB5 as the blank material. The corresponding blank temperature history, the temperature of each tool block, as well as hardness of flange, bottom and side wall of the U-shape part, were obtained and analyzed. The performance of the tool was assessed, and a few problems were identified for future improvement.

Keywords: Tailored hot stamping; Multi-block; Tailored zones.

1. Introduction

It is known that tailoring a hot stamping part, to achieve locally graded properties, can improve the crash behavior. Depending on the role of the structural part (carrying either bending or axial crash load), the best position for the local regions with lower strength and higher ductility can be different. The distribution of these local regions and their mechanical properties affects the crash behavior of the part in each loading case and therefore can be effectively designed to improve the crash performance.

Previously the authors numerically investigated the effect of the local distribution and mechanical properties of hardened and soft zones in tailored hot stamped parts on side impact crash behavior [1]. The analysis was based on thermal-mechanical-metallurgical finite element models of a hot stamping process and a representative side impact crash model. The results of that numerical study provided the first steps in order to improve the crash response of the tailored part by the optimized positioning of the hardened and soft zones. The next steps towards optimized locally tailored properties are to experimentally produce parts with such properties and aim to validate the numerical predictions. This is not a trivial task and requires specially designed hot stamping tooling.

2. Tooling Design and Tests

As can be seen from Fig. 1 that the entire tooling is divided into two halves (red and grey) which are separated by a 3 mm air gap. An array of channels that have been drilled in both punch and die on both halves can be used to accommodate either heating or cooling rods. In our tests, only the red half is used to test the tailoring performance by having either heating or cooling in various blocks in both punch and die, the grey half on the right remains water cooled at all time. The tooling test setup can be seen in Fig. 2. Five tests were conducted on the tooling in order to assess its performance, the summary of which can be seen in Table 1. The blank material is 22MnB5, and the furnace temperature is set as 900 °C The blank was transferred to the tooling manually before stamping was performed with 20s holding time. Please note "No heating or cooling" in Test No.3 and No.4 means neither heating nor cooling rods were used in the tooling, the channels were left empty. However water cooling should instead be used in these two cases as planned, which was tried initially and we have found the temperatures of neighboring blocks being heated only increased to about 300 °C and couldn't reach 400°C mark as planned. This problem was overcome by using insulation material on the punch and die to prevent heat from losing while the center blocks (P3 and D3) are being water cooled in Test No. 5. The temperature history of blank was recorded from being heated in the furnace in the beginning of the test to the end of stamping. After the U-shape part was formed, specimens (150 mm in length) were cut from the flange, sidewall and bottom of the part for hardness testing (HRC) as can be seen from Fig. 7 as an example.

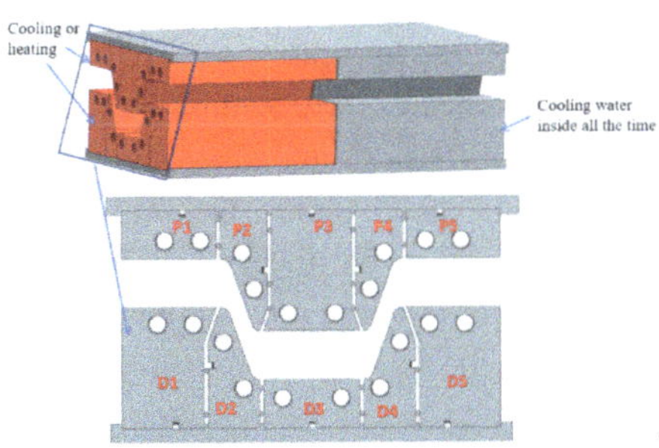

Fig. 1. Schematic illustration of the tooling with its cross section view.

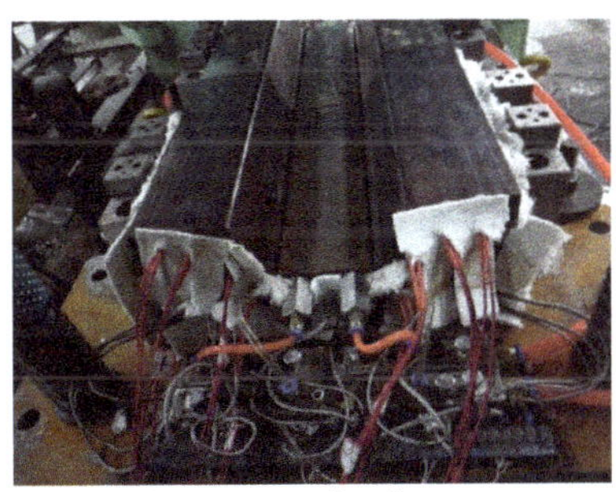
Fig. 2. Setup of tooling test.

Table 1. Summary of five tests conducted on the tooling.

Test. No.	P1 & D1	P2 & D2	P3 & D3	P4 & D4	P5 & D5	Temp. history
1	Cooling water	Cooling water	Cooling water	Cooling water	Cooling water	Yes
2	400 °C	400 °C	400 °C	400 °C	400 °C	Yes
3	400 °C	No heating or cooling	400 °C	No heating or cooling	400 °C	No
4	No heating or cooling	400°C	No heating or cooling	400 °C	No heating or cooling	No
5	400 °C	400 °C	Cooling water	400 °C	400 °C	Yes

3. Results

In Test No. 1, the entire tooling was water cooled, so when the hot blank at 900 °C got contact with cold tooling during stamping, its temperature quickly dropped to below 100°C as can be seen from Fig. 3, which caused quenching effect on the blank that hardened the material. The hardness of flange after hot stamping reaches about 40 HRC (Fig. 4), which is a significant increase compared with hardness around 10 HRC before stamping. The hardness curve for any part of the U-shape part should be similar to the flange since all blocks of the tooling were water cooled at the same condition. However as Fig. 4 shows the result curve for the bottom of the part is quite different from the flange, for which the only possible reason is the specimens were mixed up when they were cut from U-shape parts by external contractor.

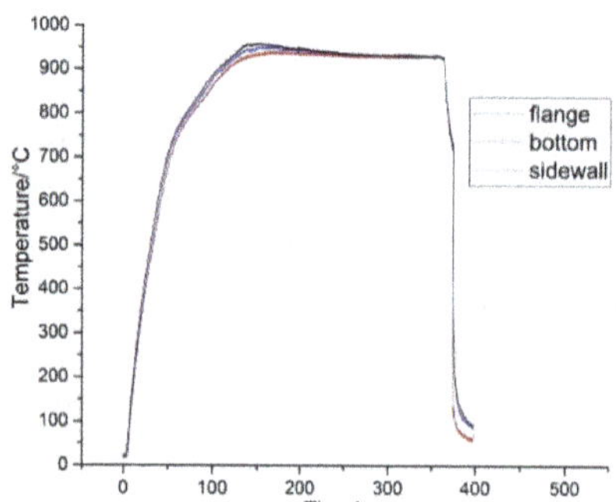

Fig. 3. Temperature history of blank for Test No. 1.

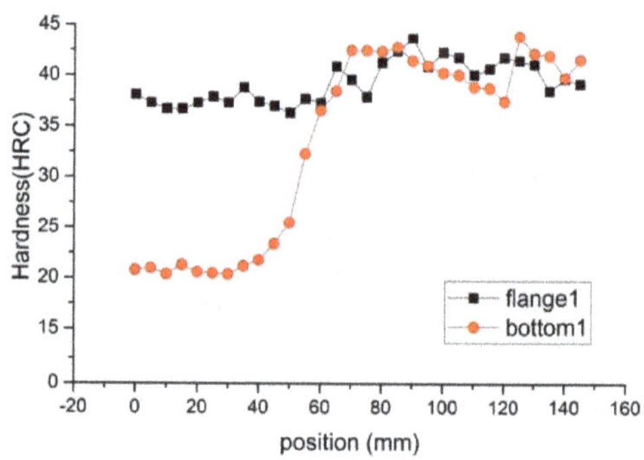

Fig. 4. Hardness testing results of blank in Test No. 1.

In Test No.2, the whole left half of the tooling was heated by heating rods to 400 °C as planned, the temperature for each block can be found in Table 2 which shows although the majority of blocks are around 400 °C, the blocks on either sides including P1, P5 and D1 demonstrate the most discrepancy, with 40 °C below in P5 being the most. This temperature distribution inhomogeneity will have an effect on the quenching and hardness of the part as a result. The temperature of blank almost instantly dropped to just below 500 °C and stayed there roughly during holding time before starting to decrease gradually in air (Fig. 5).

In theory, the right half of U-shape part should be harder than the left half as more martensite is produced due to cooling in comparison to the heating at 400 °C. The testing results in Fig. 6 generally reflect the theory, however the fairly big hardness variation in both left and right sides of the part is probably due to the uneven temperature distribution in the tooling which affects the quenching effect during stamping.

Table 2. Temperature of each block of tooling before stamping in Test No. 2.

Block No.	D1	D2	D3	D4	D5	P1	P2	P3	P4	P5
True Temp/°C	381.7	393.7	379.8	405.2	397.3	380.6	402.3	402.3	396.6	358.7

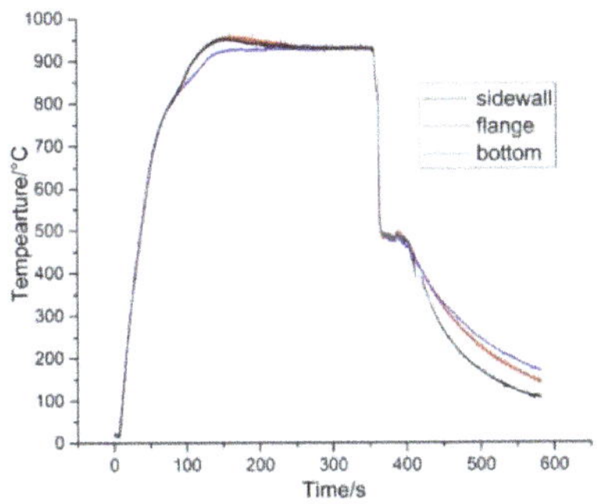

Fig. 5. Temperature history of blank in Test No. 2.

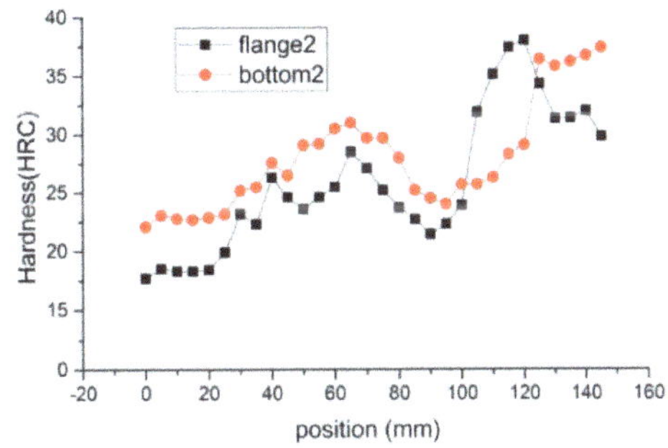

Fig. 6. Hardness testing resutls of blank in Test No. 2.

In test No. 3, although only the flange and bottom blocks were setup to be heated to 400 °C using heating rods in respective circular channels, as can be seen from the table on the left of Fig. 7, the neighboring blocks with no heating or water cooling applied such as P2 and D2 as well as P4 and D4 all had temperatures around 300 °C, which shows that the insulation effect of air gap between blocks is not satisfactory. The heating effect of heating rods wasn't very good either due to uneven temperature in heated blocks especially P5 which was 73 °C below designed 400 °C.

As right side of the tooling was water cooled all the time, the hardness of right side of U-shape part should be greater than the left side where tailored heating occurred. The general trend of curves on Fig. 8 reflects that, additionally the boundary between left and right part of tooling can also be seen at 75 mm mark. Again, the uneven temperature distribution in both left and right parts of the tooling caused variations of hardness results in the formed U-shape part.

The hardness testing specimens were cut from U-shape part as can be seen from Fig.7 right with 150 mm in length.

Block No.	True temperature
D1	356.4°C
D2	292.3°C
D3	355.9°C
D4	333.8°C
D5	374.3°C
P1	356.6°C
P2	297.4°C
P3	392.2°C
P4	291.3°C
P5	327.2°C

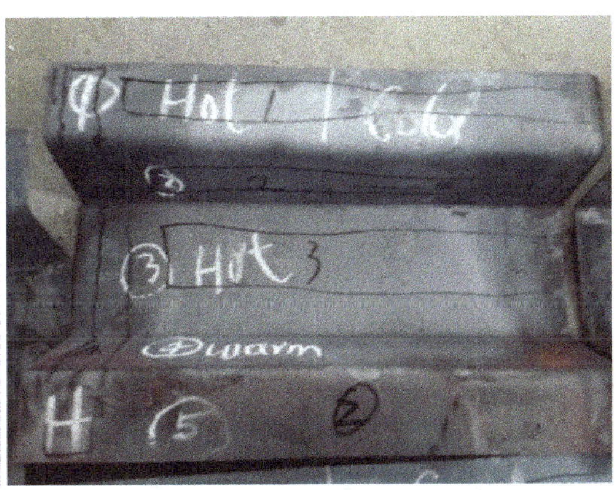

Fig. 7. Temperature of each block in tooling before stamping in Test No. 3 (left), and the U-shape part after stamping (right).

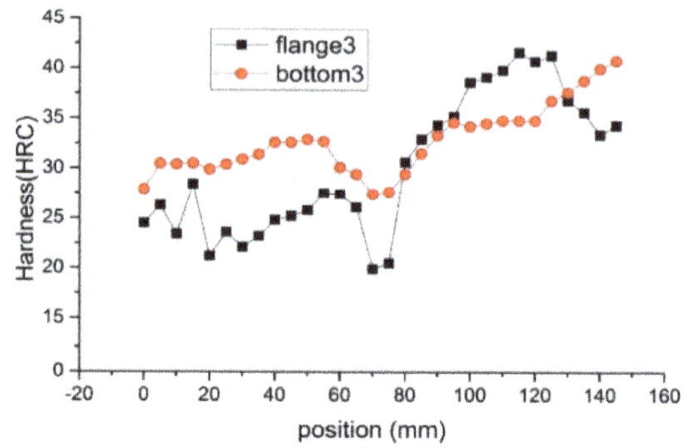

Fig. 8. Hardness testing results of blank in Test No. 3.

In Test No. 4, the sidewall blocks D2&P2 and D4&P4 were designed to be heated to 400 °C while no heating or water cooling was applied on the other blocks on the left of tooling. However due to the unsatisfactory insulation effect provided by air gaps between blocks, the neighbouring flange and bottom blocks were also heated to around 220 °C on average due to heat radiation, especially the block P3 reached 291.7 °C. The temperatures for each block can be seen from the table on Fig. 9 left.

Block No.	True temperature
D1	216.3°C
D2	341.4°C
D3	243.9°C
D4	400.1°C
D5	224.8°C
P1	222.7°C
P2	372.3°C
P3	291.7°C
P4	337.3°C
P5	203.8°C

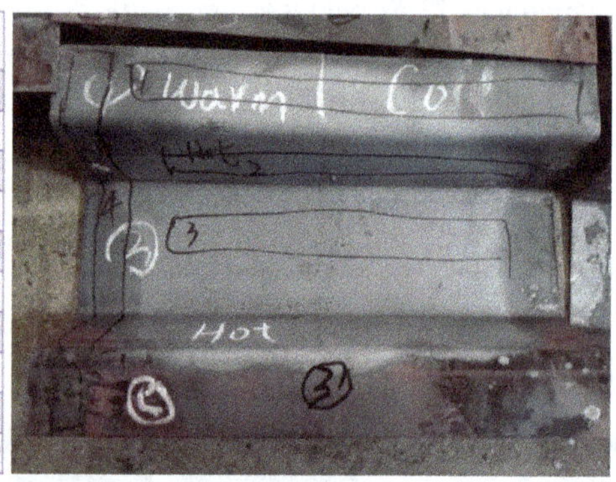

Fig. 9. Temperature of each block in tooling before stamping in Test No.4 (left) and the U-shape part after stamping (right).

The hardness results of flange and bottom of right half of U-shape part agreed very well, which are also greater than the left half, as can be seen from Fig. 10.

In Test No. 5, only the bottom blocks in punch and die were designed to be water cooled while the rest of the blocks from the left side of tooling were set up to be heated to 400 °C, the right side of tooling was water cooled all the time. As the table of Fig. 11 as well as the temperature history in Fig. 12 shows, the temperatures of water cooled blocks D3&P3 are 79 °C and 75 °C respectively, due to heat radiation from neighboring blocks. The heated blocks show uneven temperature distribution from 307.7 °C in P4 to 401.4 °C in D4.

The hardness of flange in Fig. 13 shows increase in value from left half to right half of the U-shape part, which is consistent with the theory where more martensite is produced due to lower temperature quenching. The bottom hardness results show a reduction in value from left half to right half, which is abnormal and not consistent with prediction.

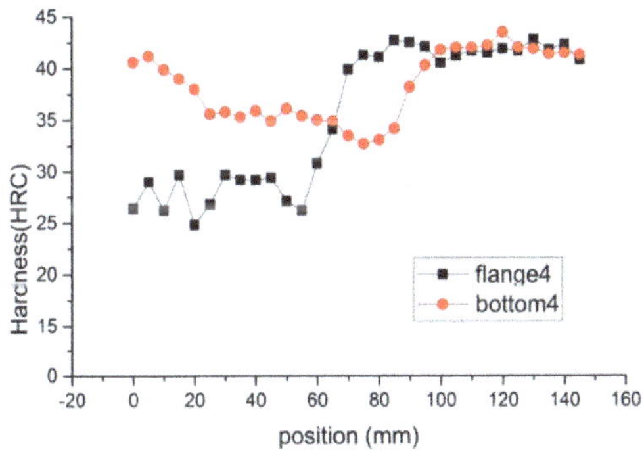

Fig. 10. Hardness testing results of blank in Test No. 4.

Block No.	True temperature
D1	385.7°C
D2	362.8°C
D3	78.9°C
D4	401.4°C
D5	391.5°C
P1	381.0°C
P2	357.2°C
P3	74.6°C
P4	307.7°C
P5	343.4°C

Fig. 11. Temperature of each block in tooling before stamping in test No.5 (left), and the U-shape part after stamping (right).

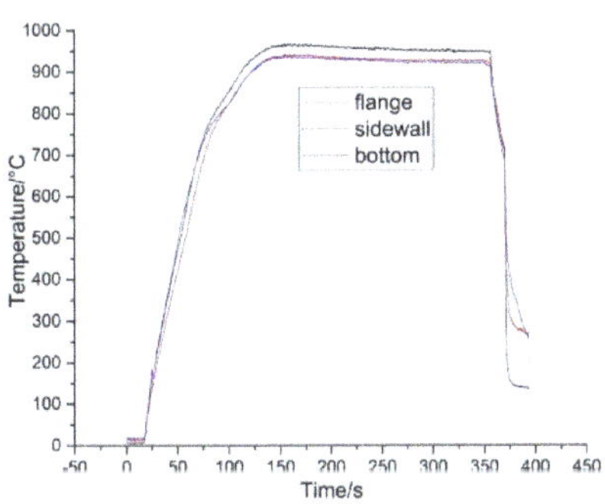

Fig. 12. Temperature history of blank in test No. 5.

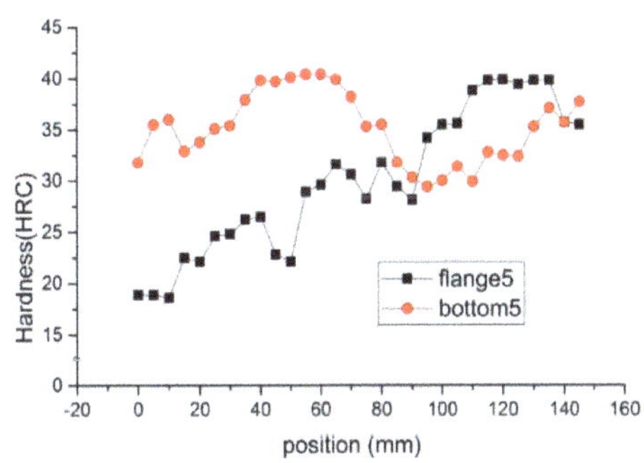

Fig. 13. Hardness testing results of blank in test No. 5.

4. Conclusion

There have been a few issues we have identified on the tooling that need to be fixed/improved in the future:

(1) The insulation effect of air gaps designed to isolate various blocks in the tooling was not satisfactory. This has caused the problems of blocks supposed to be cooled being

heated to substantial high temperatures, and blocks supposed to be heated couldn't reach the designed temperatures. The air gap's unsatisfactory insulation effect has compromised the tailoring effect of the tooling to a large extent. This issue can be addressed by adding insulation materials in the air gaps which was the case in Test No. 5 where asbestos were used although the temperature of water cooled block still rose to about 80 °C, which can be partially caused by inadequate cooling by cooling rods which is the second issue below.

(2) The cooling rods didn't work effectively in the tailored part of the tooling. We have observed that the water flow rate of cooling rods was very small in the tailored part of the tooling which is left side in this test. Though at the same time the water flow rate of the non-tailored part which is right side of the tooling was better. We believe this was because the heat from neighboring blocks as a result of poor insulation caused the interior of cooling rods to rust which restricted the water flow. Better insulation between blocks and better corrosion protection of cooling rods can be adopted to overcome this problem. We also found the plastic tubing used to connect cooling rods to water supply could be easily damaged by the heat dissipated from tooling. In this case heat resistant tubing and/or insulation at the end of cooling rods between can be applied to address such problem. The cooling effect can be further improved by using cooling machine which can supply cold water continuously rather than the same water was being recycled which became warm after some time in our tests.

(3) The uneven temperature distribution within blocks in both case of heating and cooling in the tooling has caused various degrees of quenching during stamping and inhomogeneous distribution of hardness in the formed U-shape part as a result.

(4) The temperature control system of heating rods didn't perform as expected. There is only one thermocouple in each block in the tooling, and one relay is used to control several heating rods in the control system, which together can cause inaccurate control of temperature. This can be improved by better design of the tooling such as introducing more thermocouples in each block as well as better temperature control system that uses one relay to control one heating rod.

Reference

1. A. Amir, X. Chen, M. Pereira, N. Xiao and B. Rolfe, The crash behaviour of hot stamped components – the effect of tailoring conditions, in *5th International Conference on Hot Sheet Metal Forming of High-Performance Steel (CHS2 2015)*, (Toronto, Canada, 2015).

A Novel Warm Stamping Process for Manufacturing Disc Springs

Y. H. Sun[1], K. H. Wang[1], D. J. Politis[1], Z. H. Cai[1], L. L. Wang[1,a,†], L. Wan[2,b] and Y. M. Wang[2]

[1]*Department of Mechanical Engineering, Imperial College London, London, SW7 2AZ, UK*

[2]*Capital Aerospace Machinery Company, Jingbeidong Road, Fengtai district, Beijing, 100076, China*

E-mail: [a,†]liliang.wang@imperial.ac.uk, www.imperial.ac.uk, [b]wanl002@163.com

Disc springs used in medical equipment, nuclear reactors and satellites have strict requirements for dimensional precision and quality. It is difficult to form disc springs accurately by traditional manufacturing processes as hardening and tempering heat treatments lead to considerable distortion which significantly affects the final geometrical precision. Adjustments of forming tools need to be completed iteratively in order to form the disc springs into the required size, which is time and energy intensive. In the present research, a warm stamping process for manufacturing disc springs was proposed, which aims to reduce cycle time and operational cost. The optimal processing window for the present forming technique is identified through thermo-mechanical analysis. Modified forming tools with cooling channels have been designed, fabricated and integrated into a dedicated pilot production line, *Uni-form*. Disc springs were successfully produced with required dimensional precision and strength by the present process.

Keywords: Disc spring; Spring steels; Warm stamping; Uni-form.

1. Introduction

Disc springs, sometimes named as Belleville springs or washers, are used in a variety of applications across a broad range of industries where there is a requirement for a high force within limited space. Disc springs are extensively used in precision instruments such as medical equipment, nuclear reactors and satellites which have strict requirements for their dimensional precision and performances compared to common disc springs [1]. Cold stamping followed with customizable heat treatment processes are traditional production methods for spring washers. However, a considerable deformation during the heat treatment is difficult to avoid. The deformation of disc springs affects the final dimensional accuracy, which significantly shortens the fatigue life [2, 3].

Warm stamping is a promising technology for manufacturing complex shaped components from high strength steels [4, 5]. It is well known that increasing the forming temperature offers significant advantages such as improved formability and minimized springback. The literature on warm stamping of disc springs is although warm stamping may offer a promising alternative solution to form disc spring in a cost-effective way.

In the present research, the primary goal is to manufacture disc springs from 60Si2Mn steel using the warm stamping process to attain the required strength and dimensional accuracy. The effect of forming temperature on the hardness and ductility of 60Si2Mn steel were studied through thermo-mechanical analysis. A novel forming tool with water-cooled channels was designed, manufactured and integrated into a dedicated production line, named 'Uni-form' with the proposed forming technique for disc springs verified experimentally. The geometric dimensions, surface roughness and load-deflection performance of the formed components were evaluated to confirm all the design requirements were successfully achieved.

2. Warm Stamping Process of Disc Springs

Fig. 1 schematically presents the processing routes of conventional manufacturing and warm stamping of disc springs. In the conventional method, as-delivered spring steel blank sheets, consisting of ferrite and perlite, are first cut from a coil and cold stamped in forming tools; the preformed parts are subsequently heated to 860 °C and soaked for 5 mins for austenitization, followed by oil quenching to achieve martensitic microstructure. As-quenched martensite normally has limited ductility and toughness, and therefore, the hardened disc springs must be tempered, which is the favored process to improve toughness and correct brittleness of the material dependent on tempering temperatures and periods.

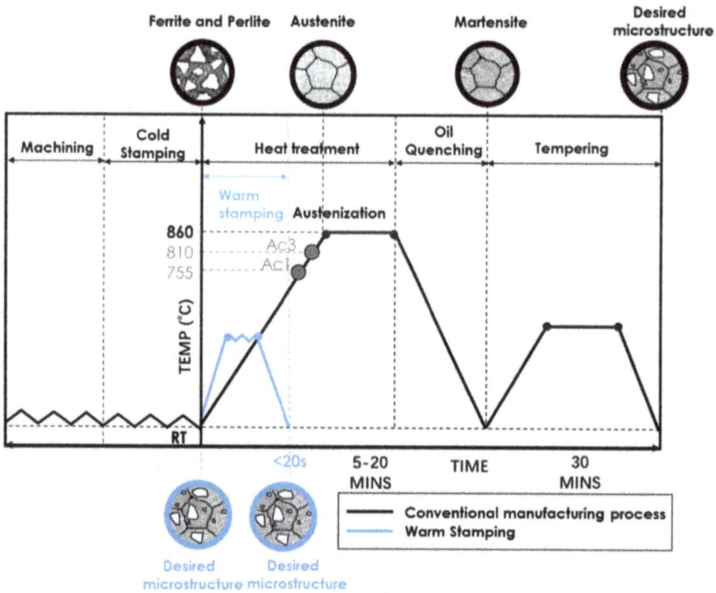

Fig. 1. The schematic diagram of the conventional manufacturing and warm stamping processes of disc spring.

During the warm stamping process, blanks are heated to an appropriate temperature, then formed and quenched simultaneously between cold dies. This process may avoid major negative microstructural changes thus enabling full strength to be retained for the formed components. The desired microstructure of a spring washer is tempered troostite, which has high toughness and reduced hardness for sustaining a load. Pre-heat treatments are necessary prior to warm stamping to achieve the desired microstructure.

3. Materials, Design Requirements and Experimental Procedures

The investigated material was commercial 60Si2Mn steel of chemical composition shown in Table 1. The as-received material was a 2.6 mm thickness steel sheet with an original microstructure consisting of ferrite and pearlite. A commercial disc spring was selected as an example to be formed by the proposed technique within required technical specifications as shown in the Fig. 2. According to the standards of 6769 DIN2093, 2.6mm thickness sheet material belongs to group 2 featuring a rectangular cross-section with square edges and permissible tolerances for spring geometry such as overall height, inside and outside diameter [6]. The hardness of formed disc springs must be within the range of 42-52 HRC to ensure it has excellent toughness and fatigue life. Good surface finish without any defect and decarburization layer is preferable.

Table 1. Chemical composition of 60Si2Mn steels (wt%).

Composition	C	Si	Mn	P	S	Cr
wt%	0.60	1.50	0.60	≤0.012	≤0.005	≤0.1

Uniaxial tensile tests at elevated temperature were conducted in a Gleeble 3800 system to characterize the thermal-mechanical properties of spring steels. The dog-bone shaped samples were heated to a target temperature by electrical resistance heating, immediately stretched to failure, and then air quenched to room temperature. Rockwell hardness measurements were conducted with microhardness tester before and after each test. Forming trials were subsequently conducted on a tool set with the separate cooling system to form disc springs via *Uniform* to verify the concept of warm stamping process [7]. Load-deflection performance and dimensional precision of the formed disc springs were evaluated according to the standard of FB/T 1972-2005.

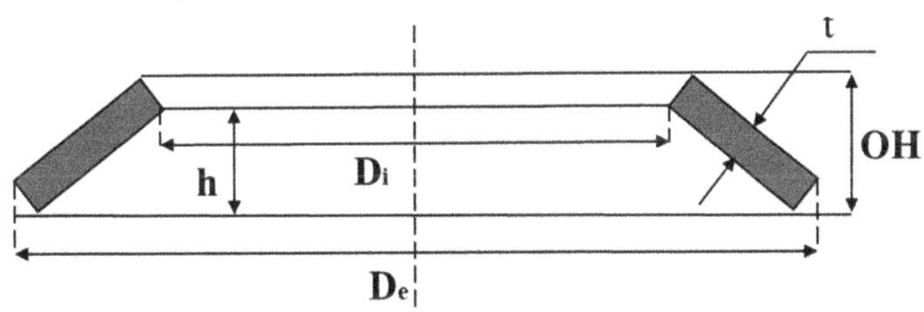

Fig. 2. Cross section detail of a disc spring with required geometric parameters.

4. Material Characterization

The effect of deformation temperature on flow stress of spring steel was investigated with true stress-strain curves plotted in Fig. 3. A clear reduction in flow stress with increasing temperature can be observed in addition to an improvement in failure strain. Moreover, compared to low ductility (less than 10 % elongation) at room temperature, excellent ductility, at over 1.15 tensile strain, can be achieved at the higher temperature 550 °C. The residual hardness of stretched specimens at elevated temperatures is also presented in Table 2. The hardness of the specimen revealed a decreasing trend with increased temperature. A remarkable reduction, of approximately 6.32 % in residual hardness at 550 °C was observed, which is consistent with the theory that higher temperatures accelerate the coarsening of carbides and thus some reduction in residual hardness occurs [8]. As such, the requirement of post-form strength (42-52HRC) was achieved at selected conditions, which indicates the proposed process would avoid a substantial change to the microstructure at the range from 450 °C to 550 °C.

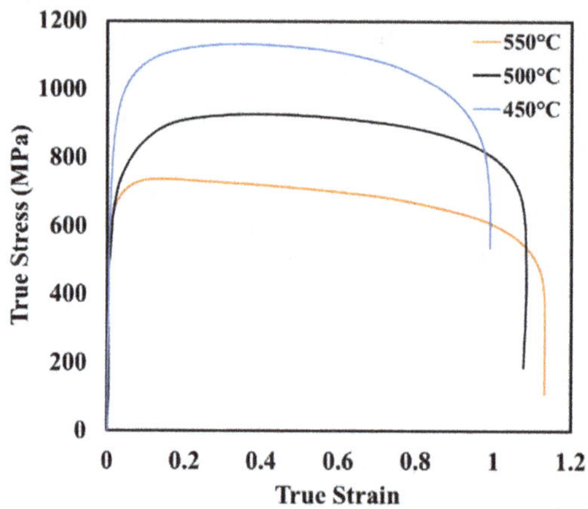

Fig. 3. Effect of the temperature on flow properties of spring steel at the strain rate 0.1/s.

Table 2. Hardness loss of spring steel after deformation.

Test condition	Hardness before deformation (HRC)	Hardness after deformation (HRC)	Hardness loss (%)
450 °C	48.7	48.0	1.43
500 °C	48.2	46.7	3.11
550 °C	49.0	45.9	6.32

5. Forming Tool Design and Forming Trials

In order to achieve a low cycle time whilst minimizing the cost of refining forming tools, a splitting forming die constructed from two separate parts, a cooling block and replaceable forming cap was designed and fabricated, as shown in Fig. 4(a). Rather than utilizing commonly used circular cooling tubes, the cooling block design makes use of square channels for the water flow, which can be milled efficiently from the top surface thus reducing manufacturing cost with the tools set assembly on the *Uniform*. In order to validate the proposed forming process, a disc spring was successfully formed, as shown in Fig. 4(c). Compared to the conventional manufacturing process, total cycle time was significantly reduced from longer than 30 minutes to less than 20s. The dimensions of the formed part at 500 °C was measured by coordinate measuring machine which indicated a good agreement with required geometric dimension. The deformation behavior of formed disc spring was evaluated as 11.2 kN load applied for five times and the results show the load capacity of the formed part also meet the requirements. The surface roughness of disc spring was also measured demonstrating an arithmetic average surface roughness, Ra, of 377.89 nm, which is significantly lower than required value (3.2 μm).

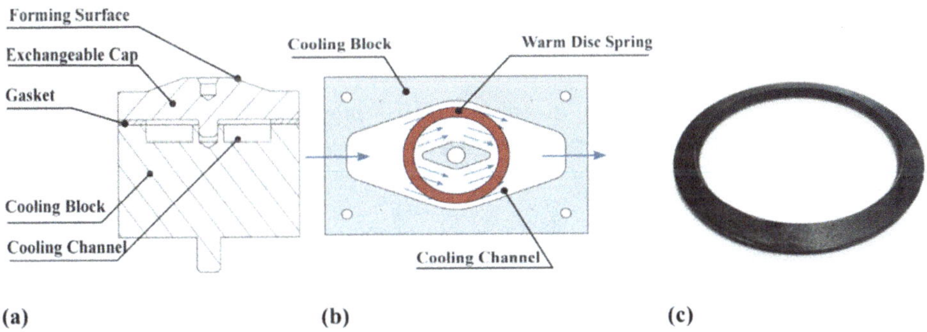

Fig. 4. (a) A section view of the forming tools; (b) A top view of the cooling channel; (c) Warm formed disc spring.

6. Conclusion

In the work, a novel processing route for manufacturing disc springs from 60Si2Mn steel was proposed. This forming technique enables a tremendous reduction of total cycle time from 30 minutes to less than 20 seconds and productivity improvement for a mass production setting. A separate forming tool set with a replaceable forming surface can significantly reduce the manufacturing cost for further tool refining. A disc spring was successfully formed using the proposed forming process with required dimensional precision, strength and surface roughness. Formed disc springs also indicated a good load and deflection capability for successful implementation.

Acknowledgments

Work partially supported by grant 2-4570.5 of the Swiss National Science Foundation.

References

1. A. S. V. Raja, S. Veeraghavanb, S. Badec and K. Srinivas, Performance comparison of belleville washer under compressive load using experimental and simulation methods, *Mater. Today: Proc.* **5**, 8486 (2018).
2. MW industries inc, Precision Disc Springs Custom and Catalog Products, https://www.mw-ind.com/wp-content/uploads/mps-disc-catalog.pdf.
3. W. Patangtalo, S. Aimmanee and S. Chutima, A unified analysis of isotropic and composite Belleville springs. *Thin-Walled Structures* **109**, 285 (2016).
4. K. Mori, S. Maki and Y. Tanaka, Warm and Hot Stamping of Ultra High Tensile Strength Steel Sheets Using Resistance Heating, *CIRP Annals* **54,** 209 (2005).

5. Y. Chang, C. Y. Wang, K. M. Zhao, H. Dong and J. W. Yan, An introduction to medium-Mn steel: Metallurgy, mechanical properties and warm stamping process, *Mater. Des.* **94**, 424 (2016).
6. DIN2093, Dimensionas and quality of conical disc springs. *SPIROL International Corporation*, (2016).
7. X. Luan, Q. L. Zhang, O. E. Fakir, L. L. Wang and M. M. Gharbi, A Pilot Production Line for Hot/Warm Sheet Metal Forming Integrated in a Cloud Based SMARTFORMING Platform, *Adv. High Strength Steel and Press Hardening*, 492 (2017).
8. T. Furuhara, K. Kobayashi and T. Maki, Control of Cementite Precipitation in Lath Martensite by Rapid Heating and Tempering, *ISIJ. Int.* **44**, 1937 (2004).

Research on the Deformation Temperature of Zinc Coated Press Hardened Steel in Hot Stamping Process

J. Zhu, K. Wang, Z. J. Zhu, B. Zhu and Y. S. Zhang†
State Key Laboratory of Materials Processing and Die and Mould Technology, Huazhong University of Science & Technology, Wuhan 430074, China
†E-mail: zhangys@mail.hust.edu.cn

For the benefit of corrosion resistance of zinc coated hot stamping parts, the zinc coated press hardened steel will be popular in the automotive industry. However, when the deformation temperature is too high, the Zn in the coating is liquid and the liquid Zn will result in the substrate cracking during the deformation stage, thus the appropriate deformation temperature is the key parameter in hot stamping process. In this paper, the tensile tests at different deformation temperature with Gleeble3500 were implemented, the stress-strain curves of Zn coated steel and uncoated steel were acquired, and the microstructure of the cracking is obtained. According to results, the appropriate deformation temperature range is selected between 690 °C~720 °C. Finally, the U-shaped parts were produced with the selected deformation temperature, and the results can meet the requirements of hot stamping parts.

Keywords: Zinc coated press hardened steel; Deformation temperature; Hot stamping process.

1. Introduction

In the past few years, the hot stamping parts with high strength have made great contribution to the automotive lightweight [1], however, some problems emerge in a long period of use. On the one hand, there are serious oxidation and decarbonization of the traditional high strength steel during the heating stage, which have terrible impact on surface quality. On the other hand, the traditional hot stamping parts with poor corrosion resistance are easy to be corroded in different environment. To solve the problems of the traditional high strength steel, the surface coating technology is considered [2], and the Zn coated high strength steel is a promising replacement [3].

As for the hot stamping process, there are two main differences between the traditional high strength steel and the Zn coated high strength steel. Firstly, the Zn coated steel has a lower austenite temperature. The austenite temperature of the traditional high strength steel is often higher than 920 °C, but it is lower than 900 °C of the Zn coated steel. Because the boiling point of Zn is 907 °C, and the Zn coating will boil and volatilize under a temperature over 900 °C. Secondly, compared with the traditional high strength steel, the Zn coated steel has a lower deformation temperature. When the deformation temperature is too high, the Zn is liquid in the coating and it will induce the substrate cracking during the deformation period, which is called the LMIE cracking. [4]. Therefore, it's significant to determine a suitable deformation temperature for the Zn coated high strength steel. This paper aims at the research on the deformation temperature of Zn coated high strength steel.

2. Experimental

2.1. *Experimental material*

The material used in the experiment is the Zn coated high strength steel with the thickness of 1.5mm. Its main chemical composition measured by the PDA-1000 is shown in Table 1, and the original cross section microphotograph is shown in Fig. 1, which shows the thickness of the original coating is about 10 μm.

Table 1. Chemical composition of Zn coated steel (wt%).

C	Mn	Si	Cr	Ti	Ni	Nb	B
0.20~0.25	1.02	0.24	0.16	0.023	0.029	0.012	0.001

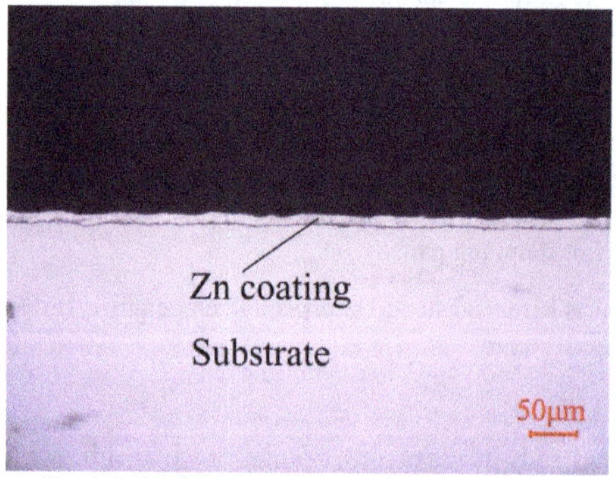

Fig. 1. Initial microstructure of Zn coated steel.

2.2. *Experimental method*

The experiment aims at seeking a suitable deformation temperature during the hot stamping process. To reach the requirements of accurate temperature control and large deformation, the experiment is implemented on the Gleeble-3500 thermal-mechanical simulator, and the specimen is shown in Fig. 2.

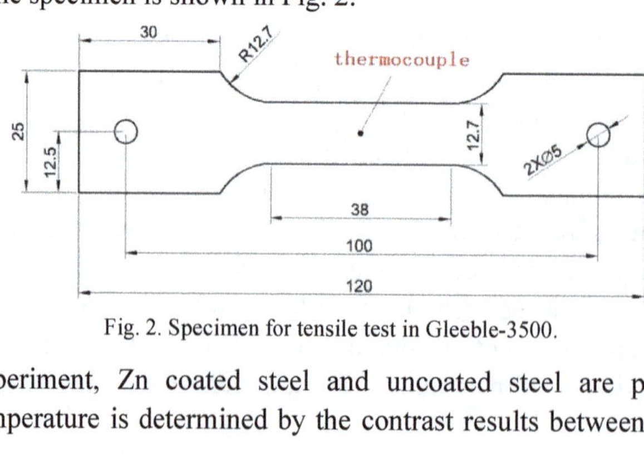

Fig. 2. Specimen for tensile test in Gleeble-3500.

In this experiment, Zn coated steel and uncoated steel are prepared, and the deformation temperature is determined by the contrast results between the two kinds of materials. By the way, the uncoated steel is obtained by grinding 15~20μm on both sides of the Zn coated steel. The experiment process is as follows: First, the specimen is heated to 880 °C and kept for 3minutes. Then, it is cooled to the deformation temperature with 30 °C/s cooling rate. Next, the specimen is stretched to fracture under the deformation temperature. Finally, it is cooled to the environment temperature rapidly. Four groups of deformation temperature are set as 690 °C, 720 °C, 750°C and 780 °C. The temperature process is shown in Fig. 3. When the deformation temperature is selected, the U-shaped parts are produced at different deformation temperature to verify the rationality of the selected deformation temperature.

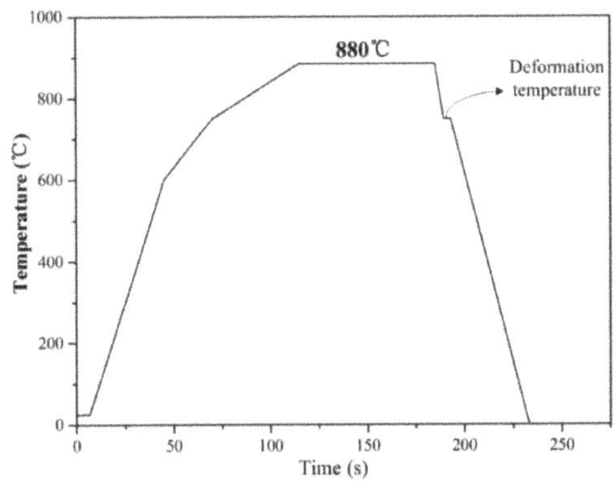

Fig. 3. Temperature process in the experiment.

3. Results and Discussion

3.1. *Comparison of the stress-strain curves*

Fig. 4 shows the comparison of stress-strain curves between the Zn coated steel and uncoated steel at different deformation temperature. On the one hand, when the deformation temperature is higher than 720 °C, the ultimate tensile strength (UTS) of Zn coated steel is lower than uncoated steel, and the cracking failure of Zn coated steel happened prior to the uncoated steel. Moreover, as the temperature increased, the cracking failure of Zn coated steel happened earlier. On the other hand, when the deformation temperature is lower than 720 °C, the cracking failure of uncoated steel happened prior to the Zn coated steel, as the temperature decreased, the difference of UTS between Zn coated steel and uncoated steel became smaller. As for the reason results in the difference, it is because the liquid Zn induces the substrate cracking, and the existence of liquid Zn depends on the deformation temperature [5]. When the deformation temperature is too high, a large amount of liquid Zn exists in the coating, as the tensile deformation continues, the liquid Zn crosses the coating/substrate interface and penetrates into the substrate along the austenite grain boundary [6]. Finally, the substrate cracking expands and the cracking failure happens in advance. As the deformation temperature decreases, the content of liquid Zn decreases, and the substrate cracking disappears until a critical temperature. From the results above, the critical temperature is between 690 °C~720 °C.

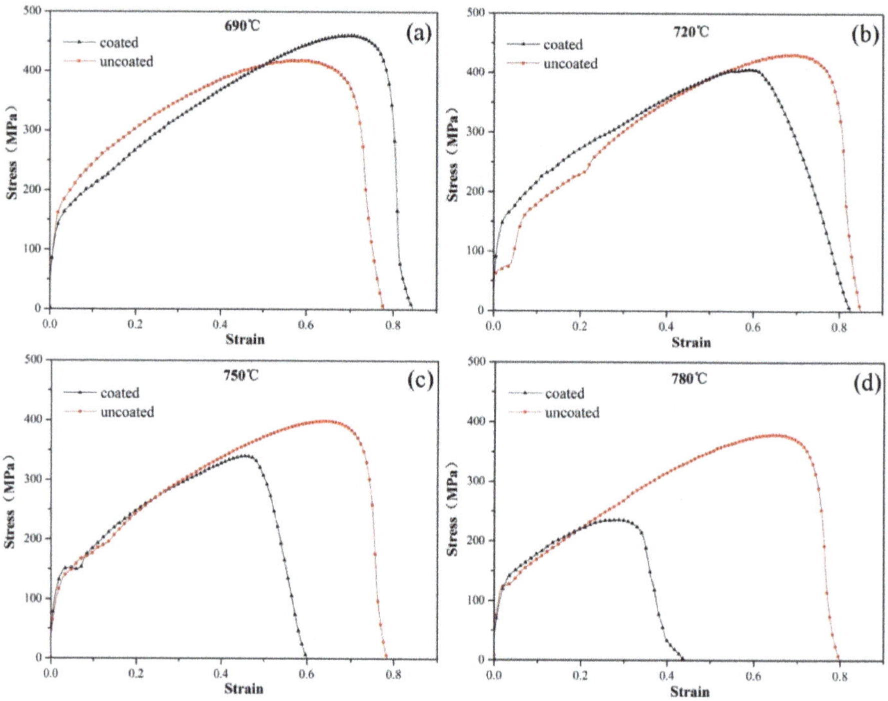

Fig. 4. Stress-strain curves of Zn coated steel and uncoated steel at different deformation temperature: (a) 690 °C; (b) 720 °C; (c) 750 °C; (d) 780 °C.

3.2. *Microstructure characterization*

Fig. 5 shows the comparison of microstructure between the Zn coated steel and uncoated steel at different deformation temperature. There is no cracking exists in the uncoated steel at any temperature and the boundary is very smooth. As for the Zn coated steel, the coating/substrate interface is no longer clear, but according to the color difference, the thickness of the coating is about 20μm. When the deformation temperature is too high such as 750 °C and 780 °C, the LMIE cracking emerges in the substrate, and the cracking becomes deeper as the temperature increases. When the deformation temperature is below 720 °C, no LMIE cracking exists in the substrate, but there is much tiny cracking in the coating. The results above confirms the results in Fig. 4, therefore, the critical deformation temperature is between 690 °C~720 °C.

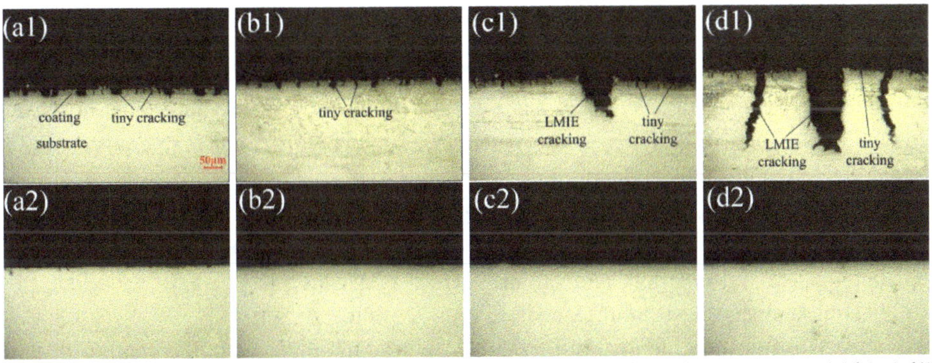

Fig. 5. Microstructure of Zn coated steel and uncoated steel at different deformation temperature: (a1,b1,c1,d1) coated; (a2,b2,c2,d2) uncoated; (a1&a2) 690 °C; (b1&b2) 720 °C; (c1&c2) 750 °C; (d1&d2) 780 °C.

3.3. *U-shaped parts experiment*

Fig. 6 shows the U-shaped parts produced at different deformation temperature. When the deformation temperature is 780 °C, the surface quality of the U-shaped part is very terrible, especially at the fillet, the coating spalling is serious. On the contrary, when the deformation temperature is 700 °C, the surface quality is good, and the coating is more complete than that at 780 °C. Fig. 7 shows the microstructure of the fillet at 780 °C and 700 °C. Only tiny cracking can be observed at 780 °C, and the coating is in good condition at 700 °C. Because the deformation of the U-shaped parts is not enough, no cracking appears in the substrate. Generally, the substrate hardness of the U-shaped part at 700 °C is 450HV, which reaches the requirements of hot stamping parts. Therefore, the deformation temperature between 690 °C~720 °C is reasonable.

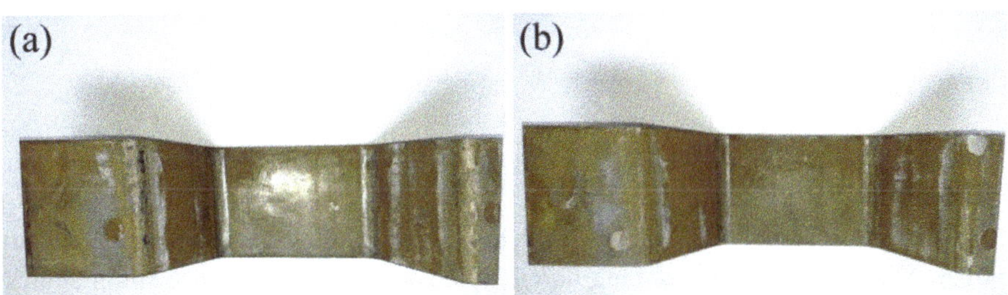

Fig. 6. U-shaped parts at different deformation temperature: (a) 780 °C; (b) 700 °C.

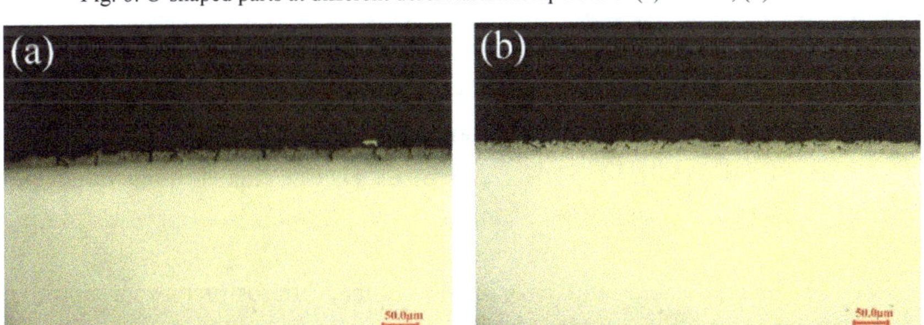

Fig. 7. Microstructure of U-shaped parts at different deformation temperature: (a) 780 °C; (b) 700 °C.

4. Summary

Because of the low melting point of Zn coating and the resulting problems, the hot stamping process of Zn coated steel has a great difference from the traditional hot stam ping process. This paper aims at the research on the deformation temperature of Zn coated high strength steel, and the following conclusion is obtained:

(1) Compared with the traditional high strength steel, a lower deformation temperature is required for the Zn coated steel. To eliminate the existence of liquid Zn, the appropriate deformation temperature is set between 690 °C~720 °C when the heating temperature is 880 °C, heating time is 3min.

(2) At each deformation temperature, there is much tiny cracking in the coating. When the temperature is below 720 °C, no cracking exists in the substrate. Only when the temperature is above 720 °C, the liquid Zn penetrates into the substrate and induces the substrate cracking.

(3) The U-shaped parts at different deformation temperature are obtained. When the deformation temperature is 780 °C, the surface quality is terrible with serious coating spalling, and tiny cracking exists in the coating. When the deformation temperature is 700 °C, the coating is complete and the substrate hardness is 450HV, which can meet the requirements of hot stamping parts.

Acknowledgments

This research work was financially supported by the National Natural Science Foundation of China (grant no. U1564203 and U1760205) and National Major Science Technology Project of China (grant no. 2018zx04023001).

References

1. H. Karbasian and A. E. Tekkaya, A review on hot stamping, *J. Mater. Process. Tech.* **210**, 2103 (2010).
2. K. Wang, Y. M. Jin, B. Zhu and Y. S. Zhang, Investigation on cracking characteristics of Al-Si coating on hot stamping boron steel parts based on surface strain analysis, *Surf. Coat. Tech.* **309**, 282 (2016).
3. A. Robert, G. Luckeneder and A. W. Hassel, Corrosion of press-hardened galvanized steel, *Corros. Sci.* **63**, 12 (2012).
4. P. J. L. Fernandes, R. E. Clegg and D. R. H. Jones, Failure by liquid metal induced embrittlement, *Eng. Fail. Anal.* **1**, 51 (1994).
5. C. W. Lee, D. W. Fan, I. R. SOHN and B. C. D. Cooman, Liquid-Metal-Induced Embrittlement of Zn-Coated Hot Stamping Steel, *Metall. Mater. Trans.* A **43**, 5122 (2012).
6. L. Cho, H. Kang, C. Lee and B. C. D. Cooman, Microstructure of liquid metal embrittlement cracks on Zn-coated 22MnB5 press-hardened steel, *Scripta Mater.* **90**, 25 (2014).

Effect of Tempering Treatment Process on Bendability of Press Hardening Steel

J. C. Jin[a,†], L. Cui[b], H. R. Gu[c] and Y. Y. Xiao[d]

Technology Center of Ma'anshan Iron and Steel Company, Ma'anshan, China

E-mail: [a,†]jinjiachun1988@163.com, [b]cuilei1219@163.com

[c]gohero1234@163.com, [d]xiaoyangyang1002@126.com

Usage of AHSS may reduce the weight of automobiles and improve the crash safety, consequently, AHSS have been used on automobile increasingly. As the typical representative of AHSS, boron alloyed steels have been the point of focus for the materials choice in hot stamping. However, the high strength and low toughness with full martensite microstructure of PHS induced the poor bending quality. In this paper, hot stamped steel 22MnB5 and Nb-V composite micro-alloyed PHS produced by Ma Steel was taken as the study object. The microstructure and mechanical properties and bendability were studied after hot stamping; subsequently experienced tempering process. The results indicate proper tempering treatment process can significantly improve bendability.

Keywords: 22MnB5; Nb/V alloyed; Hot stamping; Tempering.

1. Introduction

In the last few years, due to the increased demand for passenger safety, weight reduction, and lower gas emission, the application of AHSS has increased in industry. However, it was difficult to form successfully and the springback was hard to cancel by using AHSS [1]. Hot stamping is a proper way to overcome these difficulties. During the direct hot stamping, the tested steel is initially heated up to the austenite region, and then it is deformed at high temperature while it is cooled down simultaneously during the stamping operation [2, 3].

At present, Mn-B series hot forming steel is the most widely used, which has high strength, however, it also has a range of shortcomings such as insufficient cold bending performance. The cold bending property is mainly related to the structure state of the material, such as the grain size and the microstructure of the material, which have an obvious effect on that property. Therefore, the toughness of materials can be improved by refining grain and softening microstructure.

In this paper, the cold bending performance of 22MnB5 in hot forming steel at different heat treatment temperatures after hot forming is compared and the best subsequent heat treatment process is given.

2. Experimental Materials and Methods

2.1. *Experimental materials*

The composition of the experimental steel (the reference is A) is shown in Table 1 below, and the experimental materials were obtained by steelmaking, hot rolling, cold rolling and annealing.

Table 1. Chemical composition of the tested steel (wt. %) [4, 5].

Material	C	Si	Mn	P	S	Ti	B
A	0.20-0.25	≤0.5	1.0-1.5	≤0.02	≤0.01	≤0.05	0.002-0.005

2.2. *Experimental methods*

The quenching test of test steel mold is done in industrial production line, while hot forming process carried out is controlled according to table. After hot forming, after hot forming, the sample is processed into 6mm × 6mm (long × wide) sample (as shown in Fig. 1 below) and is put into the furnace for heat treatment test. The heat treatment process is shown in Table 3 below. After heat treatment, the surface of sheet with oxidized iron is removed by pickling.

Table 2. Hot forming process.

Hold temperature (°C)	Holding time of temperature (s)	Holding time of pressure (°C)
930	300	8

Table 3. Heat treatment process.

Heat treatment temperature (°C)	Heat treatment time (min)				
100	10	20	30	60	120
150	10	20	30	60	120
200	10	20	30	60	120
250	10	20	30	60	120
300	10	20	30	60	120

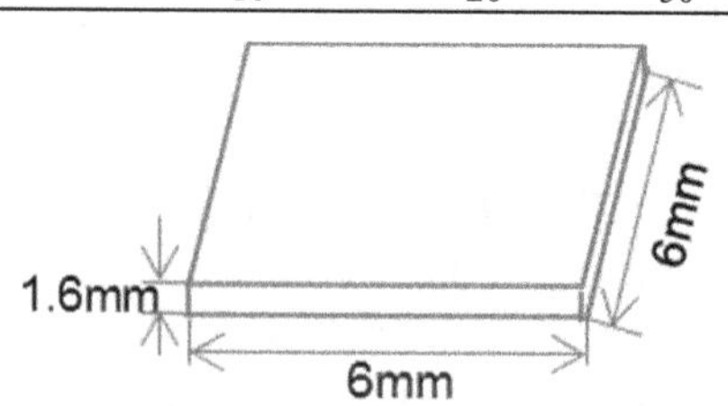

Fig. 1. The size of experimental material.

Three point bending experiments are carried out for different experimental steel materials obtained respectively, and three point bending experiments are carried out on the cold bending test machine, as shown in Fig. 2. The experimental standard is VDA238-100.

The 10 mm×10 mm thickness samples cut from the experimental steel after different heat treatments are grinded, mechanically polished and eroded by 4 % nitric acid alcohol. The microstructure is observed under optical microscope.

Fig. 2. Bending equipment.

3. Experimental Results and Discussion

3.1. *Effect of heat treatment technology on microstructure and micro-hardness of experimental steel*

After quenching of flat sheet dies, the microstructures of experimental sheets under the condition of different heat treatments (in this paper, the microstructures of test steels at 100 °C, 200 °C and 300 °C and kept the heat for 10min and 120 min respectively are given) are shown in Fig. 3, while micro-hardness changes are shown in Fig. 4.

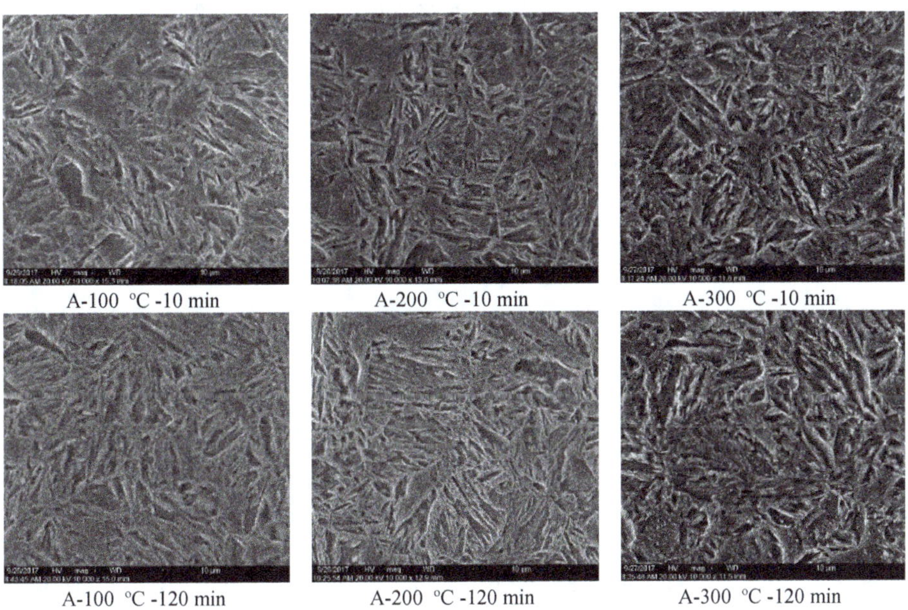

Fig. 3. Effect of heat treatment process on microstructure.

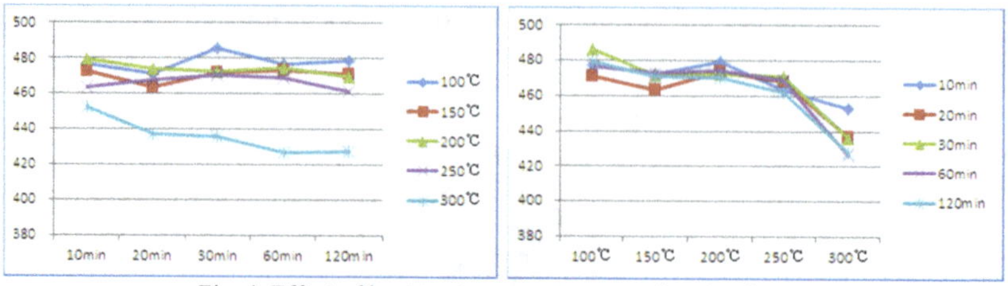

Fig. 4. Effect of heat treatment process on micro-hardness.

The above experimental results show that under the same heat treatment time, the higher the heat treatment temperature is, the greater the amount of carbide is, while under the same heat treatment temperature, the longer the heat treatment is, the deposition variation is not obvious. Therefore, in the range of 100 °C ~ 300 °C, the main factor influencing the degree of quenching tempering of martensite is heat treatment temperature, rather than the heat treatment time.

On the other hand, from the analysis of test steel micro-hardness can be found: (1) when heat treatment temperature is within 100 °C ~ 250 °C and the heating temperature is constant, with the increase of heat treatment time, micro-hardness has no obvious change; When the heat treatment temperature of 300 °C, with the increase of heat treatment time, the micro-hardness decreases. (2) Under the condition of different heat treatment time, with the increase of heating temperature, the micro-hardness decreases, and when the temperature is greater than 250 °C, the micro-hardness decreases obviously.

3.2. *Effect of heat treatment process on three-point bending performance*

According to test standard of VDA238-100, the three point bending performance measured under different heat treatment process is seen in Fig. 5, the abscissa 1~5 in Fig. 5 (a) represents 100 °C ~ 300 °C respectively, and the abscissa 1~5 in Fig. 5 (b) represents 10 min ~120 min respectively.

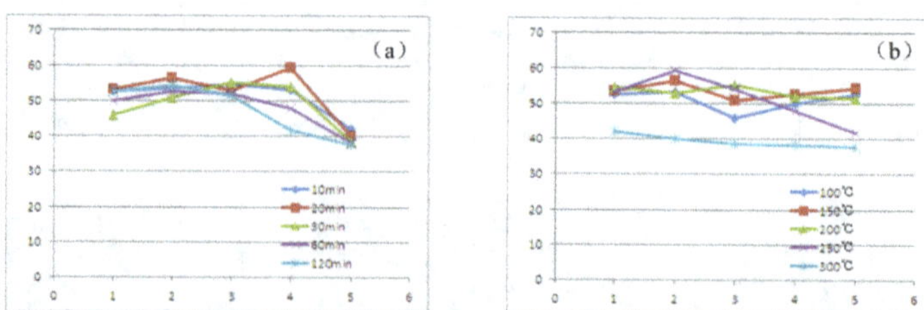

Fig. 5. Effect of heat treatment process on bendability.

From the Fig. 5 (a), under different heat treatment time, with the increase of heat treatment temperature, we can see that the three point bending angle on the whole shows the trend of first increasing and then decreasing. When the heat treatment temperature is higher than 250 °C, three point bending performance decreases obviously, and maximum bending Angle is 59.26 °. The corresponding heat treatment process is 250 °C to + 20min.

From the Fig. 5 (b), we can see that when the heat treatment time reaches 120 min and the heat treatment temperature is between 100 °C and 200 °C, the three point bending properties of test steel is more than 50 °, while after the heat treatment temperature is higher than 250 °C, three point bending Angle is only about 40 °, which means that the large amount of carbide precipitated is not conducive to the improvement of three-point bending performance of components; On the other hand, when the heat treatment is 300°C, three point bending performance is also obviously low, which further proves the reduction of three point bending performance is associated with plenty of carbide precipitation. From test results can also be concluded that, the heat treatment process after the hot forming can eliminate the internal stress of the martensite, thus improve the cold bending property, however, when the heat treatment temperature and time exceeds a certain value, the dominant factor is the precipitation of carbides, which make the three point bending performance reduced.

4. Conclusion

By studying the influence of heat treatment technology on the three-point bending performance of experimental steel, the following conclusions are obtained:

(1) After quenching, the carbide precipitation in hot forming steel is mainly associated with heat treatment temperature, and when the temperature is less than 250 °C, the influence is smaller, while when the temperature is higher than 250 °C, the carbide precipitation increases significantly.

(2) With the increase of heat treatment temperature, the micro-hardness shows a decreasing trend, and when the heat treatment temperature of 300 °C, the micro-hardness is less than 450 HV.

(3) Through the study on the influence of heat treatment technology on three-point bending performance, the best heat treatment to improve the performance of three point bending can be obtained, and the treatment scheme is 250 °C +20 min.

References

1. M. Naderi, M. Ketabchi, M. Abbasi, W. Bleck and J. Mater, Analysis of microstructure and mechanical properties of different high strength carbon steels after hot stamping, *J. Mater. Process. Tech.* **6**, 1117 (2011).
2. M. Eriksson and M. Oldenburg, Testing and evaluation of material data for analysis of forming and hardening of boron steel components, *Model. Simul. Mater. Sci. Eng*. **10**, 277 (2002).
3. M. Merklein and J. Lecher, Investigation of the thermo-mechanical properties of hot stamping steels, *J. Mater. Process. Tech.* **1-3**, 452 2006).
4. J. E. Morral and J. B. Cameron, Boron hardenability mechanisms, *Metal. Soc. A. I. M. E.* (1980).
5. M. Suehiro, K. Kusumi, T. Miyakoshi, J. Maki and M. Ohgami, Properties of aluminium-coated steels for hot forming, *Nippon Steel Technical Report* **88**, 16 (2003).

The Influence of Vanadium on the Mechanical Properties of Ultra-high Strength Boron Steel Treated by Quenching and Partitioning Process

W. Ding[1,2], Y. Gong[1,2,a,†], J. Y. Hong[4], X. J. Jin[1,2], X. F. Ding[3,b] and L. Wang[4]

[1]*Shanghai Key Laboratory of Material Laser Processing and Modification, School of Materials Science and Engineering, Shanghai Jiao Tong University, Shanghai 200240, China*

[2]*Institute of Advanced Steels and Materials, School of Materials Science and Engineering, Shanghai Jiao Tong University, Shanghai 200240, China*

[3]*School of Design, Shanghai Jiao Tong University, Shanghai 200240, China*

[4]*State Key Laboratory of Development and Application Technology of Automotive Steels, Baosteel Research Institute, Shanghai 201900, China*

E-mail: [a, †] yugong@sjtu.edu.cn, [b]ccding@sjtu.edu.cn

Recent development of hot press forming technique shows successful application of quenching and partitioning (Q&P) hot stamping process to improve the mechanical properties (tensile ductility and crash toughness) of hot stamped steel components. The present project investigated the Q&P hot stamping process of a vanadium bearing ultra-high strength boron steel with medium silicon content in order to reveal the effects of vanadium on the partitioning stage by comparing the corresponding mechanical properties of vanadium free steel. Typical Q&P parameters, such as partition time, were studied. Our results show that compared with the V free steel, the total elongation initially decreased faster during partitioning from 15 s to 35 s and then increased slowly from 35 s to 65 s for the V bearing steel, which resulted from the complicated interactions between several metallurgical factors, which will be discussed and verified experimentally in detail in the present project. Finally, our results show that the boron steels alloyed with vanadium are stronger than those without vanadium addition under all the studied Q&P processing.

Keywords: Vanadium; Q&P hot stamping; Retained austenite; Ductility.

1. Introduction

Hot stamping process is widely used in the production of ultra-high strength steel components in the automobile industry, e.g. A-pillar, B-pillar, bumper, roof rail, rocker rail and tunnel. The final microstructure by this process is full martensite, so the strength can reach very high level, which is usually up to 1500 MPa [1]. However, because of the full martensitic microstructure, the hot stamping parts possess very low ductility, which reduces the overall mechanical properties and limits the application of the hot stamping parts [2-5]. Recent researches reported that Q&P (quenching and partitioning) process can provide good ductility and at the same time doesn't deteriorate the strength of the steel with the aid of the TRIP (transformation induced plasticity) effect of retained austenite [6-10]. So there's a new trend to introduce Q&P process to hot stamping process. However, according to Bernd et al. [11], there was no retained austenite after Q&P treatment for the traditional 22MnB5 hot stamping steel. When the Si content was increased from the traditional 0.25 wt.% to 0.5-1.5 wt.%, the Q&P treatment can really introduce retained austenite into the final microstructure. Other work [10-12] related to

TRIP steels also highlighted the importance of Si, which indicated that Si can not only inhibit cementite formation, but also stabilize austenite. So in this study the silicon content was selected as 0.8 wt%. Moreover, in our newly designed boron steel, we added microalloying element vanadium, which can further improve the mechanical properties through precipitation hardening and grain refinement strengthening mechanisms [11-14].

In this work, the main aim is to study the effect of partition time on the final mechanical properties for the newly designed 0.8 wt% Si boron steel, and the influence of V addition on the overall mechanical properties is also evaluated. So this work can provide new insights into the design of the novel Q&P hot stamping process.

2. Materials and Experimental Procedure

The materials used in the present study are based on the traditional 22MnB5 hot stamping steel, and the nominal compositions are shown in Table 1. There are totally two types of newly designed steels, one is named reference steel, and the other one is vanadium alloyed steel. For the vanadium alloyed steel, vanadium was added to further improve the mechanical properties of the steel through precipitation hardening effect.

Table 1. Chemical compositions of the newly designed boron steels (wt.%).

ID	C	Si	Mn	Ti	P	S	B	V	Cr
Reference	0.22	0.8	1.5	0.022	0.0064	0.0014	0.0024	-	0.18
V alloyed	0.22	0.8	1.5	0.022	0.0064	0.0014	0.0024	0.05	0.18

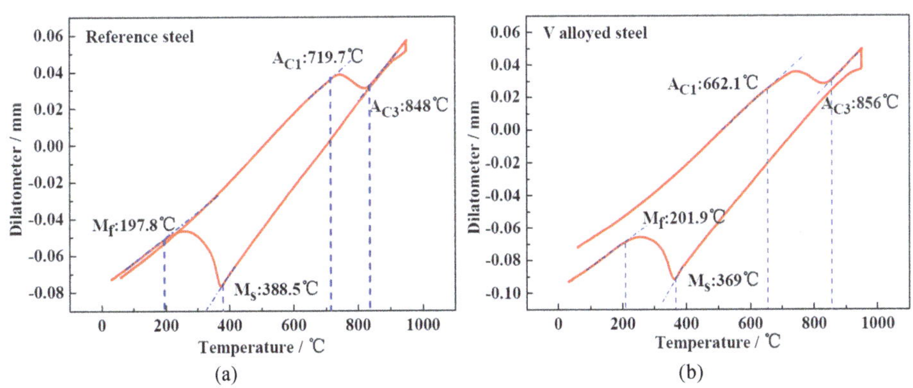

Fig. 1. Dilatation curves of the two investigated steels, (a) reference steel, and (b) V alloyed steel.

The materials were melted and rolled by Baosteel. The thickness of the as received cold rolled steel sheets is about 1.5 mm. For designing the Q&P heat treatment, it is necessary to first determine the relevant critical temperatures, such as M_s, M_f, and Ac_3 temperatures. In order to properly select the soaking temperature, quenching and partitioning temperatures, dilatometry tests were performed to get these critical temperatures. The dilatation curves of these two steels are shown in Fig. 1. From these curves, the M_s, M_f, and Ac_3 temperatures can be obtained and are listed in Table 2. According to these parameters, the Q&P heat treatment process is determined as shown in Fig. 2. Firstly, heat the blank to make it fully austenitized at 950 °C for 2 minutes, then quench it to a certain temperature (around 300 °C) where the microstructure consists of martensite and austenite and keep it at this temperature for different times (0 s, 15 s, 35 s, 65 s), thus the partition of carbon from martensite to austenite to stabilize the austenite can be realized, and finally the samples were quenched to room temperature.

Table 2. Critical temperatures of the two kinds of steels measured by dilatometry.

	Ac_1/ °C	Ac_3/ °C	M_s/ °C	M_f/ °C
Reference steel	719.7	848	388.5	197.8
V alloyed steel	662.1	856	369	201.9

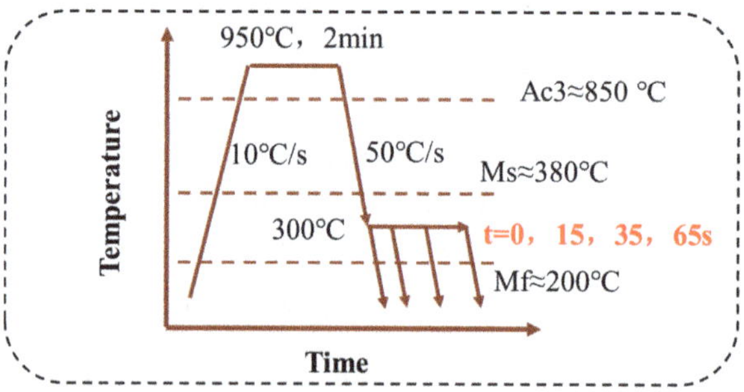

Fig. 2. Schematic illustration of the Q&P heat treatment process.

Tensile tests were conducted on a Zwick (BTC-T1-FR020 TN.ASO) universal testing machine with a constant strain rate of 4×10^{-4} s^{-1}. The size of tensile sample gage section is 16 mm×5 mm×1.5 mm and the loading direction is parallel to the rolling direction. The microstructure and austenite volume fraction were characterized and measured using scanning electron microscopy (SEM) and saturation magnetization (SM) method respectively. Transmission electron microscopy (TEM) with energy dispersive spectrometer (EDS) was used to study the size distributions of the precipitates. SEM samples were etched with 4% nital solution. Carbon replica extraction samples were used for TEM observations using JEOL JEM 2100 F field emission transmission electron microscopy operated at 200 kV.

3. Results and Discussion

3.1. *Microstructure evolution during quenching and partitioning heat treatment*

The final microstructures of these two boron steels after Q&P treatments are shown in Fig. 3. The prior austenite grain boundaries (PAGBs) can be seen clearly, which are highlighted using dark red lines. The martensitic laths within the prior austenite grains are arranged in specific orientations. There should be retained austenite in the microstructure, but it can't be distinguished from the martensitic matrix due to its very low volume fraction. No obvious difference can be observed in the SEM microstructures of the samples after different partition times either for the reference steel or for the V alloyed steel.

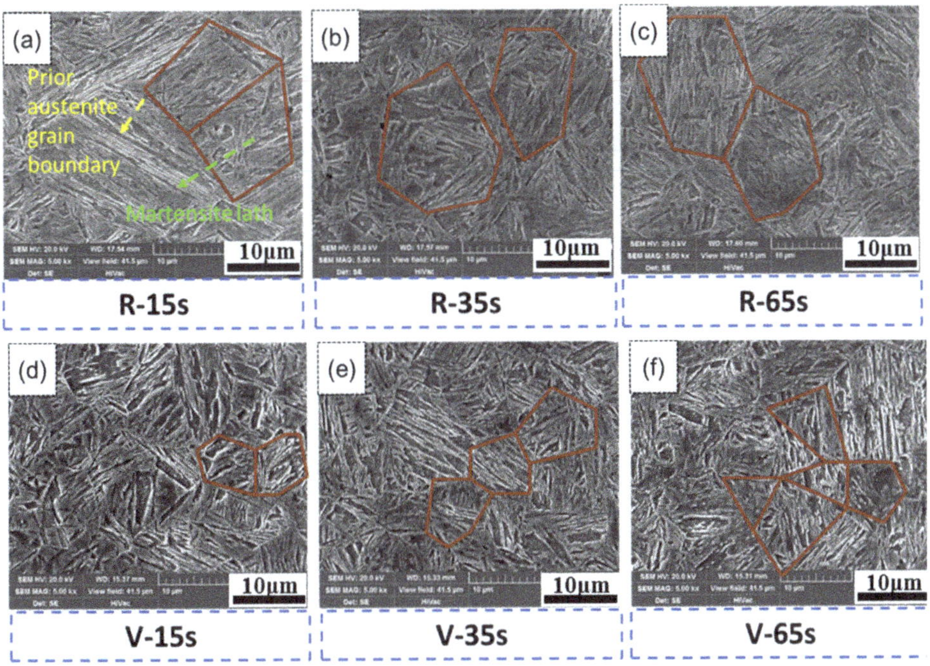

Fig. 3. SEM microstructures of the two steels after Q&P heat treatments. (a-c) Microstructures of the reference steel after partitioning at 300 °C for 15 s, 35 s and 65 s respectively. (d-f) Microstructures of the V alloyed steel after partitioning at 300 °C for 15 s, 35 s and 65 s respectively.

In order to evaluate the effect of Q&P treatment, it is essential to measure the volume fraction of retained austenite (RA). Here the saturation magnetization (SM) method was carried out to evaluate the volume fraction evolution of RA. In the magnetization measurement, the saturation magnetization can be obtained from the magnetization curve. The difference in saturation magnetizations of specimens with and without austenite is directly related to the volume fraction of non-ferromagnetic retained austenite. This is due to the fact that ferrite, as well as martensite and cementite, are ferromagnetic at temperatures below the Curie temperature (for pure ferrite $T_C = 770$ °C and for cementite $T_C = 210$ °C), while the austenite is paramagnetic [16]. The volume fraction of RA can be calculated by equation (1).

$$f_\gamma = 1 - \frac{M_s(c)}{M_s(f)} \tag{1}$$

where f_γ is the volume fraction of retained austenite, M_s is the saturation magnetization, $M_s(c)$ is the M_s value of the austenite-containing sample, and $M_s(f)$ is the M_s value of the austenite-free sample. In the present research, the volume fraction of RA in the reference steel is measured. $M_s(f)$ was obtained using the water quenched (partition time is 0 s) sample with the assumption that the water quenched sample has full martensite microstructure and contains no retained austenite. $M_s(c)$ values were measured using samples partitioned for 15 s, 35 s and 65 s respectively. So at each partition time we can get one f_γ value.

The magnetization measurements give a relation between measured magnetization (M) and applied magnetic field (H), as shown in Fig. 4 (a). It can be found that the water quenched sample has the maximum saturation magnetization, and with partition time increasing, the saturation magnetization decreases gradually. The larger the saturation magnetization, the smaller the RA volume fraction. One can also find that the saturation magnetizations of samples partitioned for 35 s and 65 s are nearly the same, which indicates that 35 s is enough for the carbon in martensite to completely partition to austenite. The RA volume fractions in reference steel samples partitioned for 15 s, 35 s and 65 s are calculated according to equation (1). The results are shown in Fig. 4 (b) and listed in Table 3. The maximum RA volume fraction (6.5 %) is obtained in the condition when the partition time was 65 s. When the partition time increases from 15 s to 35 s, the volume fraction of RA increases quickly while it increases slowly when the partition time increases from 35 s to 65 s.

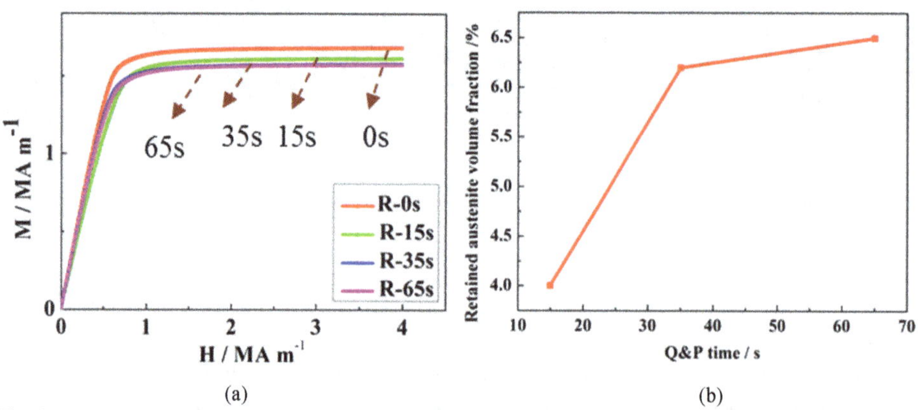

Fig. 4. Measurements of volume fractions of retained austenite in the reference steel, (a) magnetization curves of the reference steel samples partitioned at 300 °C for 0 s, 15 s, 35 s and 65 s respectively, and (b) calculated volume fractions of the retained austenite.

Table 3. Volume fractions of retained austenite measured by saturation magnetization (SM) method in the reference steel.

Partitioning time / s	15	35	65
Volume fraction of RA/%	4.0	6.2	6.5

Figure 5 shows the vanadium precipitates observed in the V alloyed steel partitioned for 15 s by TEM. The morphology of these precipitates can be seen clearly (as shown in Fig. 5(a)), which is spherical, ellipsoidal or cubic. The composition of the precipitate in the yellow dashed circle in Fig. 5(a) was evaluated by means of EDS analysis, which indicates that the precipitate is $[(V_x, Ti_y) (C_i, N_j)]$, where x, y, i and j are the atomic ratios of these four kinds of elements. However, the accurate values of x, y, i and j can't be determined because EDS is just a kind of semi-quantitative analysis method, moreover, for the carbon replica extraction sample, the weight percent of C can be affected by the carbon film itself.

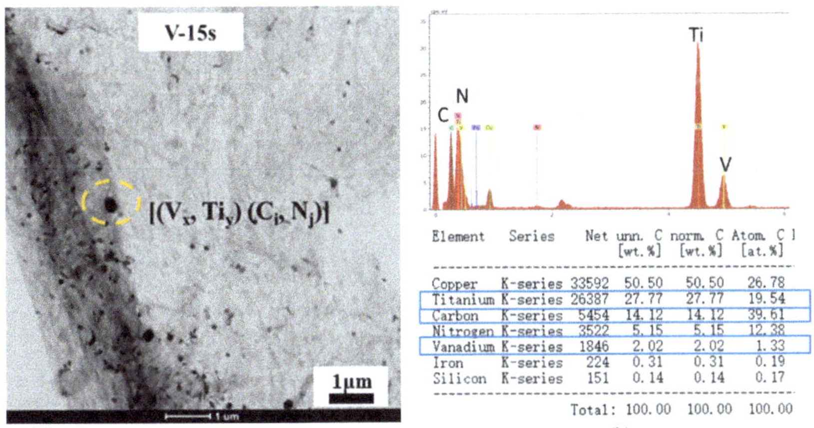

(a) (b)

Fig. 5. Morphology and composition of the precipitates in the V alloyed steel. (a) TEM image of V containing precipitates in the V alloyed steel partitioned at 300 °C for 15 s, and (b) composition of the particle in the yellow dashed circle measured by EDS.

3.2. *Mechanical properties*

The engineering stress–strain curves are shown in Fig. 6. The corresponding mechanical properties are listed in Table 4. One can find that compared with the water quenched samples (partition time is 0 s), samples after Q&P treatments have larger total elongations for both reference and V alloyed steels. The decrease of ultimate tensile strength is around 225-300 MPa after Q&P treatment for both steels, which can be seen clearly in Fig. 8. The data points in the blue dashed circles represent the experimental results of the water quenched samples.

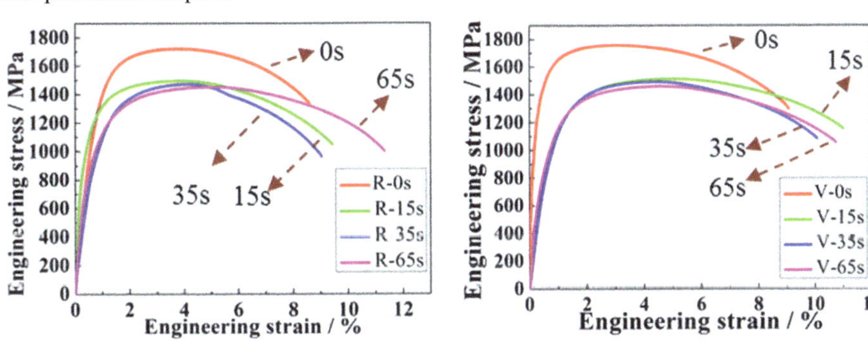

(a) (b)

Fig. 6. Engineering strain-stress curves of the investigated steels after different partitioning times, (a) reference steel, and (b) V alloyed steel.

Figure 7 shows the fracture morphology of the tensile test specimens. For both reference and V alloyed steels, many small dimples can be observed on the fracture surfaces of the samples treated by Q&P process with partition time 15s, while for water quenched samples, cleavage fracture surfaces can also be found. The transition from cleavage fracture to ductile fracture suggests that Q&P treatment can really improve the toughness of the newly designed boron steels.

Table 4. Mechanical properties of the investigated steels after partitioning for different times.

	Yield strength (MPa)	Ultimate tensile strength (MPa)	Total elongation (%)
R-0s	1078	1720	8.6
R-15s	1104	1495	9.4
R-35s	863	1471	9
R-65s	875	1451	11.3
V-0s	764	1756	9
V-15s	805	1513	11
V-35s	990	1491	10.1
V-65s	941	1458	10.7

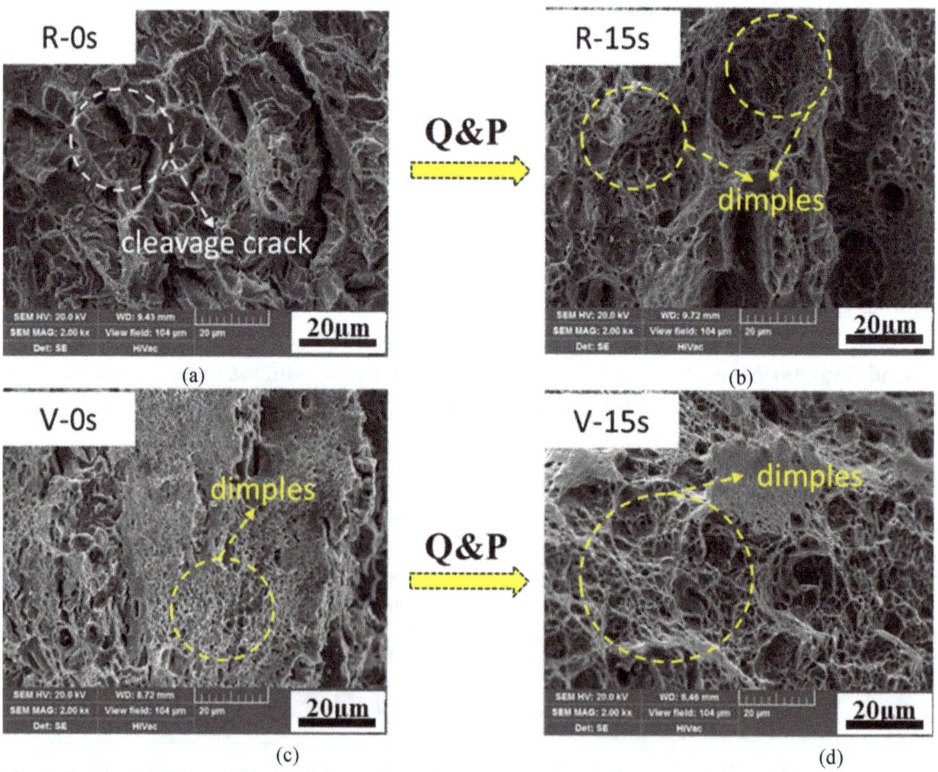

Fig. 7. Fracture surfaces of the tensile tested samples with different heat treatments, (a) reference steel water quenched, (b) reference steel with partition time 15 s, (c) V alloyed steel water quenched, and (d) V alloyed steel with partition time 15 s.

3.3. *The effect of partitioning time on mechanical properties*

According to Fig. 4, Fig. 6, Fig. 8 and Table 4, one can find that partition time has significant influence on the final mechanical properties through its effect on RA volume fraction. For samples treated by Q&P process, the volume fraction of RA increases from 4 % to 6.5 % when the partition time increases from 15s to 65 s. As a result, the tensile strength decreases gradually, however, the total elongation decreases firstly when partition time increases from 15 s to 35 s and then increases when the partition time increases from 35 s to 65 s for both of the two steels, which can be seen clearly in Fig. 8. For the reference steel, the maximum elongation (11.3 %) can be obtained when the partitioning time is 65 s, while the maximum elongation (11 %) has already been reached when partitioning time is 15 s for the V alloyed steel.

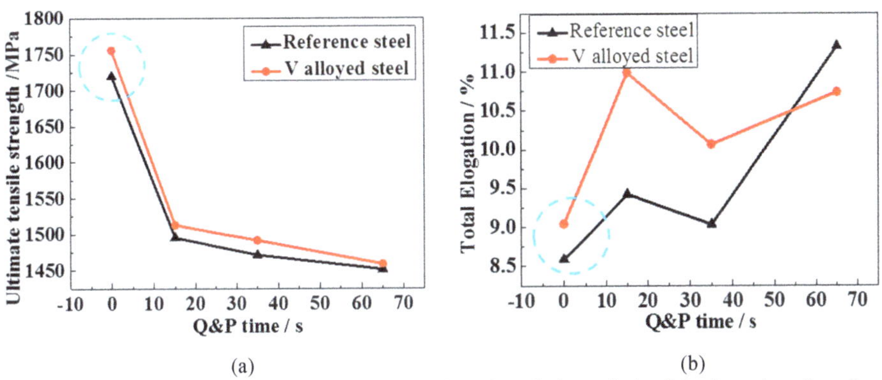

Fig. 8. Mechanical properties (ultimate tensile strength and total elongation) of the investigated steels with different partitioning times, (a) ultimate tensile strength, and (b) total elongation.

3.4. *The effect of V addition on mechanical properties*

From the above analysis in Fig. 5, we can find that vanadium carbides will precipitate in the V alloyed steel. According to references [12-15], nano-sized precipitates can not only inhibit the grain coarsening to improve the grain refinement strengthening effect but also play an important role in the precipitation strengthening. The size distribution and morphology of precipitates have great influence on the final strengthening effect. The variation of total elongation with partitioning time, as shown in Fig. 8, resulted from the complicated interactions between several metallurgical factors, e.g. the competition between vanadium and retained austenite for carbon, which is still under investigation for the present project and will be published in our another paper.

From Fig. 7 one can find that there are a few shallow small dimples on the fracture surface of the V alloyed steel after water quenching, while almost no dimple can be found in the reference steel after water quenching. For samples treated by Q&P process for 15 s, the dimples are smaller and deeper in the V alloyed steel than those in the reference steel, which suggests better toughness of the V alloyed steel. In Fig. 8 one can also find that the boron steels alloyed with vanadium are stronger than those without vanadium addition under all the studied Q&P conditions due to the two kinds of strengthening mechanisms mentioned above.

From the above discussion, one can make a conclusion that V addition can improve the toughness and strength of the newly designed boron steel.

4. Conclusion

From this work, the following conclusions can be drawn:

(1) Compared with the traditional hot stamping process, quenching and partitioning (Q&P) treatment can improve the ductility and toughness of the newly designed steels, at the same time, the strength is also high enough to meet the requirement of industrial application.

(2) With partitioning time increasing, the strength decreases gradually, however, the total elongation decreases firstly when the partitioning time changes from 15s to 35s and then increases from 35 s to 65 s for both of the two steels.

(3) Compared with the reference steel, there exist nano-sized (V, Ti) (C, N) precipitations in the V alloyed steel. The boron steels alloyed with vanadium are stronger than those without vanadium addition under all the studied Q&P conditions.

Acknowledgments

The authors are grateful to the financial support of the National Key Research and Development Program of China (No. 2017YFB0304401), National Natural Science Foundation of China (U1564203, Nos. 51571141 and 51201105). The authors also gratefully acknowledge the support provided by Shanghai Key Laboratory of Materials Laser Processing and Modification, Shanghai Jiao Tong University.

References

1. H. Karbasian and A. E. Tekkaya, A review on hot stamping, *J. Mater. Process. Tech.* **210**, 2103 (2010).
2. A. Turetta, S. Bruschi and A. Ghiotti, Investigation of 22MnB5 formability in hot stamping operations, *J. Mater. Process. Tech.* **177**, 396 (2006).
3. M. Merklein, J. Lechler and M. Geiger, Characterisation of the Flow Properties of the Quenchenable Ultra High Strength Steel 22MnB5, *Annals of the CIRP* **55**, 229 (2006).
4. K. Zhao, Y. Chang, P. Hu and Y. C. Wu, Influence of rapid cooling pretreatment on microstructure and mechanical property of hot stamped AHSS part, *J. Mater. Process. Tech.* **228**, 68 (2016).
5. P. F. Bariani, S. Bruschi, A. Ghiotti and A. Turetta, Testing formability in the hot stamping of HSS, *CIRP Ann.- manuf. Technol.* **57**, 265 (2008).
6. H. P. Liu, X. W. Lu, X. J. Jin, H. Dong and J. Shi, Enhanced mechanical properties of a hot stamped advanced high-strength steel treated by quenching and partitioning process, *Scripta Mater.* **64**, 749 (2011).

7. H. P. Liu, H. E. Sun, B. Liu, D. Z. Li, F. E. Sun and X. J. Jin, An ultrahigh strength steel with ultrafine-grained microstructure produced through intercritical deformation and partitioning process, *Mater. Des.* **83**, 760 (2015).
8. H. P. Liu, F. Sun, H. Sun, B. Liu, Y. Wang and X. J. Jin, Analysis of microstructure and mechanical properties of ultrafine grained low carbon steel, *Metall. mater.* **31**, 1099 (2016).
9. H. P. Liu, X. J. Jin, H. Dong and J. Shi, Martensitic microstructural transformations from the hot stamping, quenching and partitioning process, *Mater. Charact.* **62**, 223 (2011).
10. A. J. Clarke, J. G. Speer, D. K. Matlock, F. C. Rizzo, D. V. Edmonds and M. J. Santofimia, Influence of carbon partitioning kinetics on final austenite fraction during quenching and partitioning, *Scripta Mater.* **61**, 149 (2009).
11. B. M. Linke, G. Thomas, H. Ansger, S. Ilaria, A. Inigo and M. Arribas, Impact of Si on Microstructure and Mechanical Properties of 22MnB5 Hot Stamping Steel Treated by Quenching & Partitioning (Q&P), *Metall. Mater. Trans. A* **49**, 54 (2018).
12. F. Perrerd and C. Scott, Vanadium precipitation during intercritical annealing in cold rolled TRIP steels, *ISIJ Int.* **47**, 8 (2007).
13. Y. Q. Wang, S. J. Clark, V. Janik, R. K. Heenan, D. A. Venero, K. Yan, D. G. McCartney, S. Sridhar and P. D. Lee, Investigating nano-precipitation in a V-containing HSLA steel using small angle neutron scattering, *Acta Mater.* **145**, 84 (2018).
14. H. B. Wu, B. Ju, D. Tang, R. R. Hu, A. M. Guo, Q. Kang and D. Wang, Effect of Nb addition on the microstructure and mechanical properties of an 1800 MPa ultrahigh strength steel, *Mater. Sci. Eng. A* **622**, 61 (2015).
15. J. Takahashi, K. Kawakami, J. I. Hamada and K. Kimura, Direct observation of niobium segregation to dislocations in steel, *Acta Mater.* **107**, 415 (2016).
16. L. Zhao, N. H. V. Dijk, E. Bruck, J. Sietsma and S. V. der Zwaag, Magnetic and X-ray diffraction measurements for the determination of retained austenite in TRIP steels, *Mater. Sci. Eng. A* **313**, 145 (2001).

Developing New Steels for Hot Stamping via Partial Austenization

B. F. Rolfe[1,†], H. Beladi[1], H. Kong[2], E. J. Pavlina[3] and M. H. Cai[4]

[1]*Institute for Frontier Materials, Deakin University, Geelong, VIC 3216, Australia*

[2]*School of Materials Science and Engineering, UNSW, Sydney, NSW 2052, Australia*

[3]*AK Steel Research and Innovation Center, Middletown, Ohio, 45005, USA*

[4]*School of Materials Science and Engineering, Northeastern University, Shenyang, 110819, China*

[†]*E-mail: brolfe@deakin.edu.au*

Hot stamping is now the main forming process to produce the ultra-high strength components in an automotive body structure. Automotive manufacturers continue to find ways to reduce weight while increasing crashworthiness, and this has driven the industry to seek new hot stamping solutions. Currently, the hot stamping community uses 22MnB5 boron steel, and there are moves within the industry to progress to 20MnB8 boron steel. The behaviour of 22MnB5 boron steel is well known, and therefore the next phase of hot stamping research is to investigate other potential steel alloys. In previous research de Souza [2] discovered that cold stamping of TRIP steels is difficult. The problem arises from the variability in the hardening behaviour of TRIP steel, which creates an uncontrollable amount of springback away from nominal geometry. Hot stamping creates an opportunity to stamp these TRIP steels without the uncontrollable springback effects. This paper introduces the idea of using partial austenization to increase the range of TRIP steels available for hot stamping, by varying the volume fraction of ferrite in the final alloy.

Keywords: TRIP steel; Hot stamping; Partial austenization.

1. Introduction

Recently, the requirements of vehicle weight reduction and improved crash performance have promoted the development of fabricating components using the hot stamping process. The current hot stamping process uses a 22MnB5 boron steel alloy sheet to achieve a fully martensitic quenched part. Recently the hot stamping community has been trialing 20MnB8 boron steel, again to create a fully martensitic quenched part [1]. The next phase of hot stamping is to develop a range of new steel alloys to increase the capacity of the hot stamping lines. The new steel alloys will be high strength and have some ductility at room temperature, but not enough to enable deep drawn parts. One of the potential steel alloys that hot stamping now enables is TRIP steel. Our previous work shows that TRIP steel is a difficult steel to control the springback, based on its material variability [2]. However, hot stamping avoids these issues by forming at temperatures where there is no retained austenite transformation [3]. There has been a growing amount of work in Q&P steels for hot stamping [4]. We now see that the range of Q&P steels can be increased by developing alloys that only partially austenitize during the heating phase [5]. The partial austenization enables a greater fraction of ferrite to be retained in the microstructure facilitates new flexibility in the final hot stamped part properties. This paper outlines the development of a one-step quench and partition steel for hot stamping.

2. Developing the Alloy Design of the Q&P TRIP Steel

The alloy was designed for a one-step Q&P process, which consisted of partially austentizing the steel blanks at 910 °C for 5 min, followed by rapid cooling to the partitioning temperature between the martensitic transformation starting (M_s) and finishing (M_f) temperatures (260 °C and 320 °C) for a given time. After the partitioning stage, the part is quenched, which further produces fresh martensite from the remaining (unstable) austenite.

Based on the literature, the presence of pre-existing ferrite at 910 °C (i.e., partial austenitizing at 910 °C, which temperature is commonly used for the hot stamping austentization) would be critical for the TRIP steel to obtain enhanced ductility. Additionally, a certain amount of austenite retained after the one-step Q&P processing would also give rise to the excellent strength and ductility through the TRIP effect [6]. Thus, the design of the TRIP compositions was based on partial austenitization at 910 °C and obtaining more retained austenite at room temperature after the Q&P processing. There were some typical requirements for the design work of the TRIP steel undergoing the Q&P heat treatment [7]:

- Inhibition of ferrite and/or pearlite formation during the first cooling process from the austenitizing temperature to the desired quenching temperature;
- Retardation of bainite formation during the isothermal holding (i.e., partitioning) process;
- Minimization of the carbide precipitation during the partitioning process.

Based on the above requirements, the alloy was designed to ensure partial austenitization at 910 °C. Secondly, the M_s temperature of the austenite was designed such that it should be higher than the isothermal holding temperatures (between 260 °C and 320 °C). Carbon concentrations of 0.15, 0.20, 0.25 wt.% were chosen for the alloy design. This range of carbon concentration was chosen to reduce the influence of the carbide formation in the current low-carbon TRIP steel. The manganese content of 1.5 wt.% was used for the alloy design, to reduce the bainitic transformation start temperature, inhibit the formation of ferrite, pearlite, and bainite, and stabilize the residual austenite. In addition, the total concentration of silicon and aluminium of 1.5wt.% was selected to restrict carbide precipitation during the partitioning process. Therefore, the austenite formation finish temperature (A_3), ferrite fraction at 910 °C, chemical compositions in austenite at 910 °C, and the Ms temperature of austenite at 910 °C were calculated as a function of C concentration (i.e., 0.15, 0.20, 0.25 wt.%), as well as the concentration of silicon and aluminum in steels (i.e., Si: 0.2 ~ 1.5 wt. %; Al: 1.3 ~ 0 wt.%) (Figures 1 and 2). All of the values were calculated at equilibrium conditions using Thermo-Calc software (Thermo-Calc Software AB, Sweden) with the steel database TCFE3.

Firstly, the A_3 temperature was calculated as a function of Si, Al, and C contents (Figure 1(a)) to limit the selection window. The equilibrium A_3 temperature decreased with an increase in the Si content or a decrease in the Al content in steel. As the C content decreased from 0.25 to 0.15 wt%, the A_3 temperature increased for given alloy compositions. Moreover, the Si content should be less than ~0.7, ~0.85, ~1.0 wt% when the respective C content was 0.25, 0.20, 0.15 wt.%. This was to ensure that the steel would be partially austenitized at 910 °C (i.e., A_3 temperature > 910 °C). Under these restrictions the ferrite fraction at 910 °C increased continuously as the Si content and/or the C content decreased (Figure 1(b)). After this investigation, the Si content was set between 0.3 and 0.5 wt% thus the Al content was around 1.0 ~ 1.2 wt%. This combination could also minimize the cementite formation during the partitioning processing [8]. As a result, approximately 20% of ferrite would be retained under this narrowed composition range, during partial austenization.

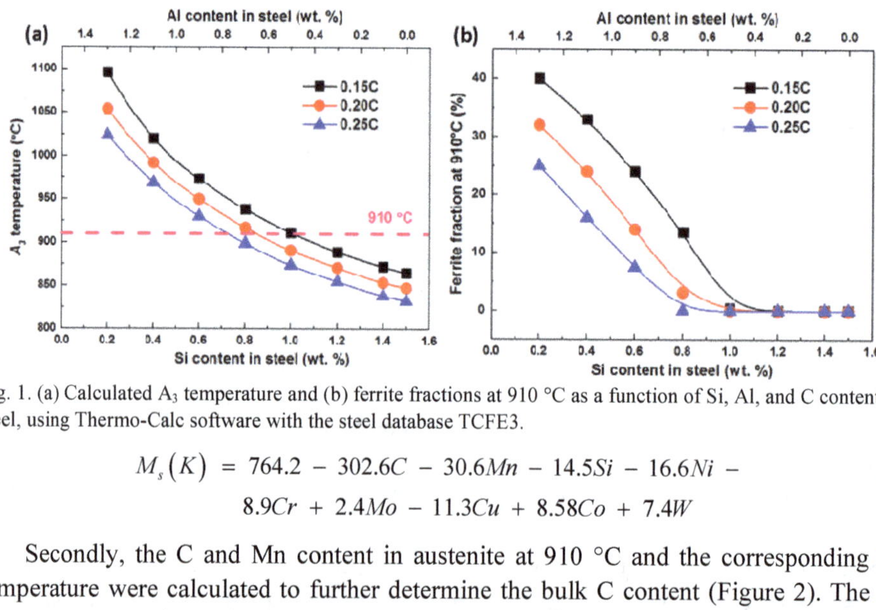

Fig. 1. (a) Calculated A_3 temperature and (b) ferrite fractions at 910 °C as a function of Si, Al, and C contents in steel, using Thermo-Calc software with the steel database TCFE3.

$$M_s(K) = 764.2 - 302.6C - 30.6Mn - 14.5Si - 16.6Ni - 8.9Cr + 2.4Mo - 11.3Cu + 8.58Co + 7.4W \quad (1)$$

Secondly, the C and Mn content in austenite at 910 °C and the corresponding M_s temperature were calculated to further determine the bulk C content (Figure 2). The M_s temperature was determined using the following empirical Eq. 1 proposed by Capdevila *et al.* [9]. Given the narrow range of Si content (0.3 ~ 0.5 wt.%), an increased C content and a decreased Mn content in the austenite at 910 °C could be obtained if the bulk C content of the steel alloy rose from 0.15 to 0.25 wt%. This would result in a reduced M_s temperature, suggesting relatively higher thermal stability of austenite. Therefore, the proposed bulk C content in the steel alloy was slightly decreased and set to be in the range of 0.2 ~ 0.25 wt%. This would not only enrich more carbon in austenite at 910 °C, but also provide sufficient carbon for partitioning.

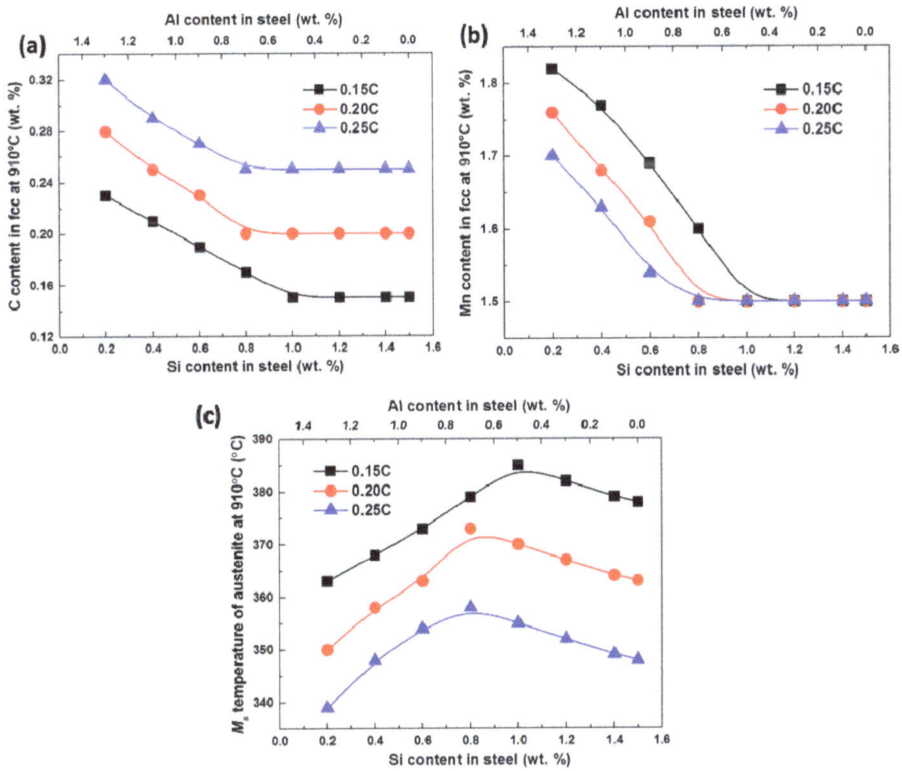

Fig. 2. (a) C and (b) Mn content in austenite at 910 °C, as well as the (c) corresponding M_s temperature, as a function of Si, Al, and C content in steel, using Thermo-Calc software with the steel database TCFE3.

Table 1. Chemical composition of the current TRIP steel, in wt%.

TRIP Steel	C	Mn	Si	Al
Final Thermo-Calc calculated composition	0.2-0.25	1.5	0.3-0.5	1.0-1.2
Actual melt composition	0.25	1.45	0.52	1.13

The final Thermo-Calc calculated chemical composition range is shown in Table 1. The final samples of the Q&P alloy were produced at Wuhan Iron Steel Corp. (WISCO) R&D Centre, China. Table 1 also lists the composition of the as-received hot-rolled slabs (i.e., melt composition) checked by the optical emission spectrometry (SPECTROMAXx from SPECTRO). The phase fractions at equilibrium condition as a function of temperature were calculated using Thermo-Calc software (Figure 3). The austenite formation start temperature (A_1) and finish temperature (A_3) given equilibrium were determined to be 731 °C and 995 °C, respectively. This TRIP alloy composition was predicted to be partially austenitized at 910 °C with a corresponding 22 % fraction of ferrite.

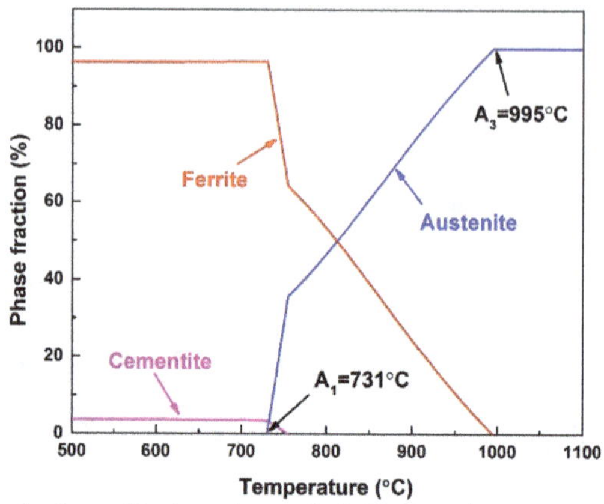

Fig. 3. Equilibrium phase fractions of ferrite and austenite in TRIP steel, calculated using Thermo-Calc software with the steel database TCFE3.

The carbon, manganese, silicon, and aluminium concentrations in austenite were also calculated as a function of temperature (Figure 4). As the intercritical temperature increased, the fraction of austenite increased, leading to a gradual decrease in the carbon and manganese concentrations in austenite. A slight increase in the aluminium concentration in the austenite from 0.55 to 1.13 wt% was observed, while the silicon concentration was nearly constant at 0.52 wt%. Thus, the chemical compositions in austenite at 910 °C were determined (Table 2) and the M_s temperature of the austenite was calculated to be 340 °C.

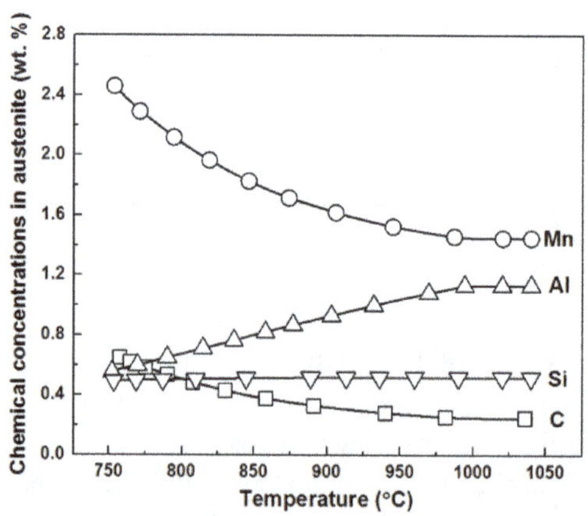

Fig. 4. Concentrations of C, Mn, Si, and Al in austenite as a function of temperature in TRIP steel through Thermo-Calc calculation.

Table 2. Chemical compositions in austenite at 910 °C calculated by Thermo-Calc, in wt%.

Austenite (910 °C)	C	Mn	Si	Al
	0.31	1.62	0.52	0.95

3. TRIP Alloy Properties

The TRIP alloy steel was reheated to different intercritical temperatures (840, 870, 910, and 960 °C) and held for 5 min. This was followed immediately by water-quenching. The as-received cold-rolled TRIP steel sheet was subjected to a one-step Q&P processing treatment using a fluidized bed furnace combined with the salt bath. The A_{c1}, A_{c3}, M_s, and M_f temperatures of the samples were 747 °C, 1004 °C, 385 °C, and 232 °C, respectively, determined using a DIL805 dilatometer (TA Instruments, Germany). The M_s temperature of the intercritical austenite at 910 °C was measured 351 °C. Phase transformation behaviour throughout the heat treatment was investigated using the dilatometer.

3.1. *Experimental validation of the thermodynamic calculation*

As the results of the Thermo-Calc software was for the equilibrium state, an experimental validation was carried out on the TRIP alloy samples by applying several intercritical annealing treatments (intercritical annealing temperatures of 840, 870, 910, and 960 °C). The as-received microstructure of the hot-rolled condition was mainly composed of polygonal ferrite (bright areas) and lamellar pearlite (dark areas) (Figure 5(a)).

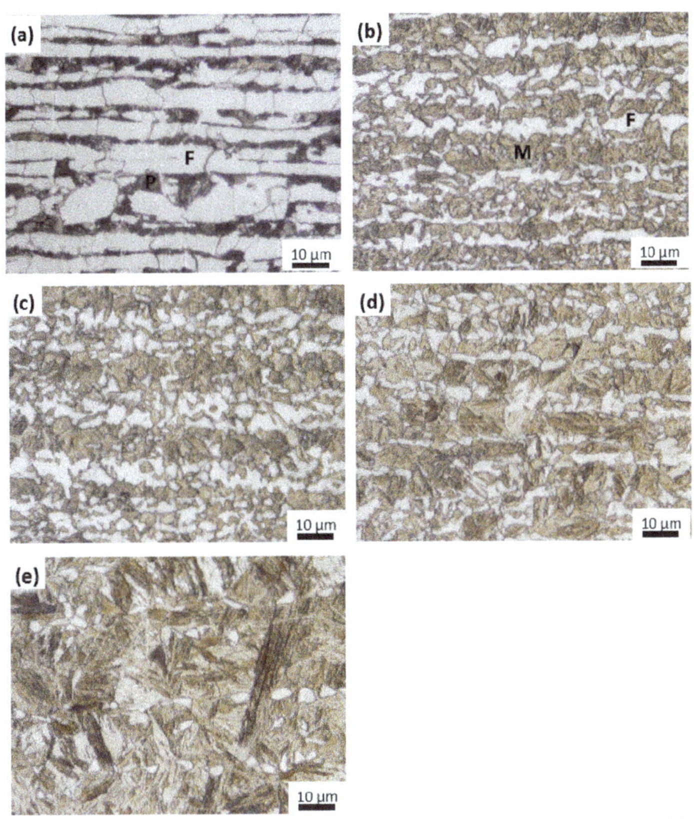

Fig. 5. OM micrographs of the (a) as-received hot-rolled microstructure and intercritically annealed microstructure at (b) 840, (c) 870, (d) 910, and (e) 960 °C followed by water-quenching. F, P, and M indicate ferrite, pearlite, and martensite, respectively.

After intercritical annealing, the microstructure consisted of lath martensite and polygonal ferrite (Figure 5(b)-(e)). The volume fraction of the ferrite was measured and compared with the theoretical values calculated through Thermo-calc (Figure 6). The slight discrepancy between experimental data and calculated results may be related to the presence of micro-segregation in the actual TRIP steel. It was shown that nearly 40 % of ferrite was formed at the temperature of 840 °C, similar to the theoretical value of 42 %. As the intercritical temperature rose from 840 °C to 910 °C, the experimental ferrite fraction decreased to ~26 %, which was slightly higher than the thermodynamically calculated fraction of ~22 %. Thus, it was proven that the steel indeed partially austenitized at 910 °C and the corresponding experimental fraction of ferrite was ~26 %. As a result, the experimental ferrite fraction was almost consistent with the theoretical values calculated from Thermo-calc.

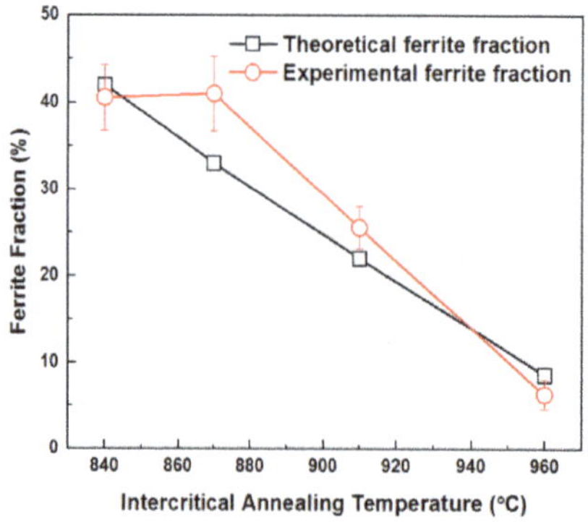

Fig. 6. Comparison between theoretical and experimental fractions of ferrite after intercritical annealing at temperatures of 840, 870, 910, and 960 °C, respectively.

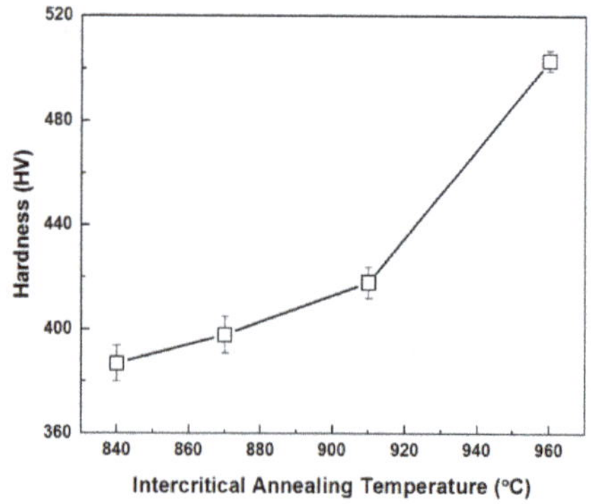

Fig. 7. Hardness as a function of intercritical annealing temperature.

Hardness measurement was performed in the intercritically annealed samples (Figure 7). The hardness of the present steel increased gradually from ~387 HV to ~418 HV as the intercritical annealing temperature increased from 840 °C to 910 °C. This was related to the slight increase in the martensite fraction in the duplex microstructure. The hardness value was increased remarkably to ~503 HV at the temperature of 960 °C, which was probably associated with the large amount of martensite (i.e., ~94 %) formed at this temperature. As a result, the measured hardness at each annealing condition was well corresponded to the microstructure characteristics.

4. Conclusion

The behaviour of 22MnB5 boron steel is well known, and therefore the next phase of hot stamping research is to investigate other potential steel alloys. Hot stamping creates an opportunity to stamp TRIP steels without uncontrollable springback effects. The partial austenization enables a greater fraction of ferrite to be retained in the microstructure facilitating a new opportunity for the properties of final hot stamped parts. This paper outlined the development of a one-step quench and partition steel for hot stamping.

A Q&P TRIP alloy was developed to ensure three conditions: 1) the inhibition of ferrite and/or pearlite formation during the first cooling process from the austenitizing temperature to the desired partitioning temperature; 2) the restriction of bainite formation during the isothermal holding (i.e., partitioning) process; and 3) minimization of the carbide precipitation during the partitioning process. The alloy was developed using Thermo-Calc, the final composition was cast and hot rolled. Samples from the cast composition indeed did achieve the desired properties, indicating that partial austenization may be an approach to create new steel alloys for hot stamping.

Acknowledgments

The authors would like to acknowledge the support of the Automotive Technology Cooperative Research Centre (Auto CRC) for Hui Kong's stipend and research support and to acknowledge the WISCO R&D Center for producing the Q&P TRIP alloy in their casting and hot rolling center.

References

1. T. Kurz, P. Larour, J. Lackner, T. Steck, and G. Jesner, Press-hardening of zinc coated steel – characterization of a new material for a new process, in *IDDRG2016 conference on "Challenges in Forming High-Strength sheets"*, (Linz, Austria, 2016).
2. T. de Souza and B. F. Rolfe, Understanding robustness of springback in high strength steels, *Int. J. Mech. Sci.* **68**, 236 (2013).
3. O. Bouaziz, H. Zurob and M. Huang, Driving force and logic of development of advanced high strength steels for automotive applications, *Steel Res. Int.* **84**, 937 (2013).
4. E. J. Seo, L. Cho and B. C. De Cooman, Application of quenching and partitioning processing to medium Mn steel, *Metall. Mater. Trans. A* **46**, 27 (2015).
5. X. Wang, L. Liu, R. D. Liu and M. X. Huang, Benefits of intercritical annealing in quenching and partitioning steel, *Metall. Mater. Trans. A* **49**, 1460 (2018).
6. B. C. De Cooman, Structure–properties relationship in TRIP steels containing carbide-free bainite, *Curr. Opin. Solid State Mater. Sci.* **8**, 285 (2004).
7. M. Santofimia, L. Zhao, R. Petrov, C. Kwakernaak, W. Sloof and J. Sietsma, Microstructural development during the quenching and partitioning process in a newly designed low-carbon steel, *Acta Mater.* **59**, 6059 (2011).
8. D. V. Edmonds, K. He, F. Rizzo, B. C. De Cooman, D. Matlock and J. Speer, Quenching and partitioning martensite-A novel steel heat treatment, *Mater. Sci. Eng. A* **438**, 25 (2006).
9. C. Capdevila, F. García Caballero and C. García de Andrés, Determination of M_s temperature in steels: A Bayesian Neural Network model, *ISIJ Int.* **42**, 894 (2002).

Research on Hot Blanking Process for Hot Forming of Press Hardened Steel

P. X. Liu† and H. C. Li
Research Institute of Iron and Steel,
Shandong Iron and Steel Group Rizhao Co., Ltd., Rizhao, China
†E-mail:qunxing985@126.com

Y. S. Zhang
State Key Lab. of Materials Processing and Die & mold Technology,
Huazhong University of Science and Technology, Wuhan, China
E-mail: zhangys@hust.edu.cn

To eliminate laser cutting for hot formed part, hot blanking process for hot forming of press hardened steel were developed. Hot blanking process is finished between forming and final quenching during the hot stamping. The effects of different blanking parameters on quality of blanked surface are studied in this paper. The results provide a reference for the actual hot blanking process and tool design.

Keywords: Hot blanking; Hot forming; Quality of blanked surface.

1. Introduction

Ultimate tensile strength of hot formed parts is up to 1500 MPa, and currently subsequent shearing process such as laser cutting [1], mechanical blanking is widely used. However, efficiency of laser cutting is low and the equipment investment of laser cutting is too high. Serious mold wear occurs during mechanical blanking.

Fracture morphology and fracture mechanism of hot formed parts in the mechanical blanking process were studied by Han X. et al. [2]. Hong-Seok Choi et al. proposed half-blanking in the hot forming quenching process, and subsequent mechanical blanking is performed to reduce blanking force and tools wear [3, 4]. Laumann T. et al. researched new process that controlling cooling speed with different mold materials was used to reducing strength of flange position, subsequent mechanical blanking was used [5]. However, in the mechanical blanking process, the strength of the steel still remains a high level. So the wear of the mold is still very serious and subsequent mechanical blanking is also used.

As shown in Fig. 1, hot blanking process for hot forming of press hardened steel is divided into three stages: hot forming, hot blanking and quenching. Hot blanking process is performed after hot forming, and is finished before finial quenching of the parts. In this process, flow stress of the steel with high temperature is lower compared to formed part with room temperature. And blanking force reduce, subsequent mechanical blanking is not needed.

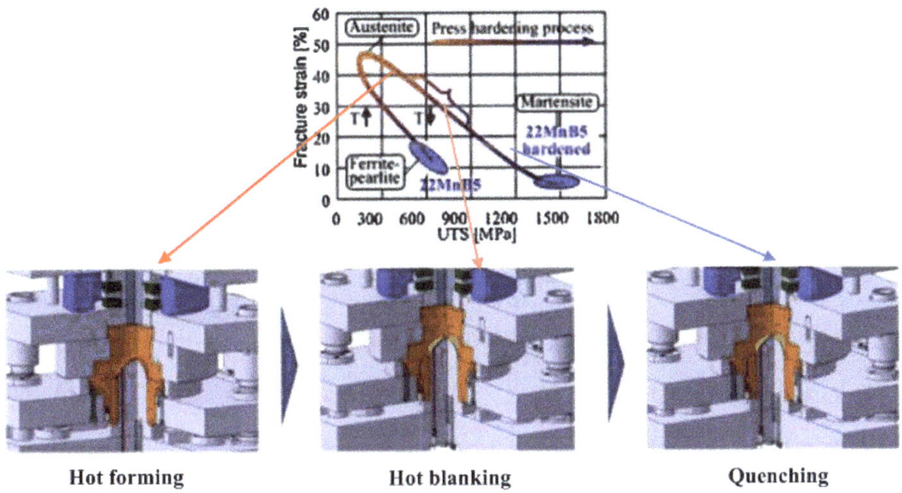

Fig. 1. Hot blanking process for hot forming of press hardened steel [6].

In this paper, a test tool was designed for simultaneous hot forming, hot blanking and quenching of press hardened steel, and different blanking parameters (blanking temperature and blanking gap) are used to research the influence on quality of blanked surface.

2. Experimental

The chemical composition of material used for this study with a nominal thickness of 1.5 mm is shown in Table 1. An electric furnace was used to heat the blanks to the austenizing temperature of 930 °C, with a total heating time of 5 min. And heated blank is transferred to tools for hot forming, before press holding with 10 second, the hot formed part is hot blanked with different blanking temperature and blanking cap. Before the press with tools close, the blanking temperature is recorded by FLIR320 infrared camera. As shown in Fig. 2, the temperature measurement of the hot punching process is completed by designing a reflecting device. Appearance and quality of blanked surface are analyzed by digital microscope (VHX-1000).

Table 1. Chemical composition in these experiments (wt%).

C	Si	Mn	Cr	Mo	P	S	Ti	B
0.23	0.17	0.95	0.24	0.21	0.010	0.002	0.023	0.0019

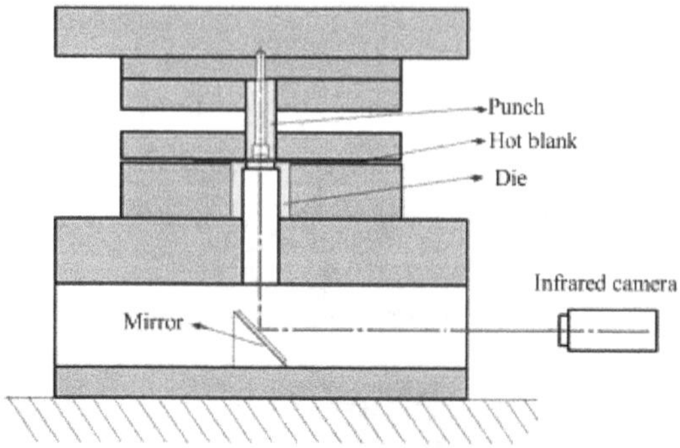

Fig. 2. Diagram of temperature recording during hot punching.

3. Results and Discussions

3.1. *Appearance of blanked surface*

Fig. 3 shows that the appearance of blanked surface with blanking temperature of 700 °C includes four regions: Roll-over, Burnish, Fracture and Burr. Different blanking parameters (blanking gap and blanking temperature) have different influence on the quality of blanked surface (Roll-over height, Burnish height, Fracture height and Burr height).

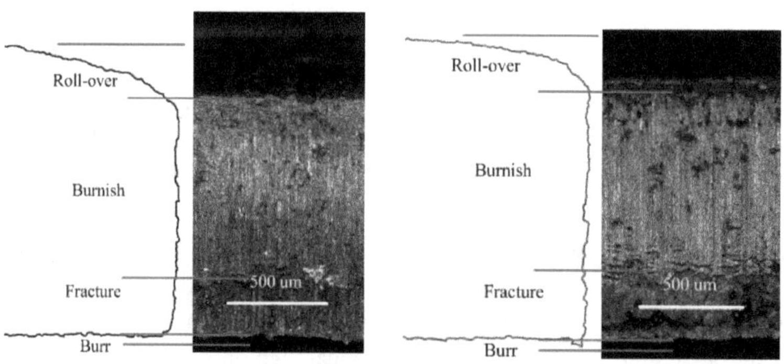

(a) trimming, 0.24 mm-10 mm/s (b) punching, 0.10 mm-10 mm/s

Fig. 3. Appearance of blanked surface (700 °C).

3.2. *Quality of blanked surface*

When the blanking temperature is constant, increasing the cutting gap results in a reduction of shear stress state in the blanking region (stress triaxiality increased). As shown in Fig. 4, the height of the roll-over, fracture, and burr zone increase with the increasing of blanking gap. However, burnish height increases as the gap decreases.

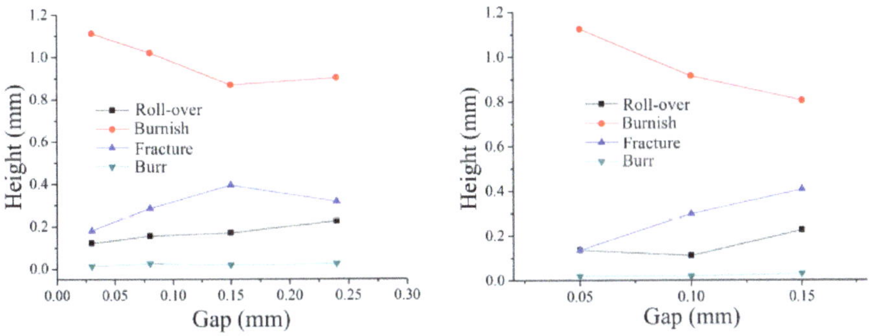

(a) trimming (b) punching

Fig. 4. Influence of blanking gap on quality of blanked surface (700 °C).

When the blanking gap is fixed, reducing the blanking temperature will increase the deformation resistance of the steel in the area to be blanked. Therefore, the height of the burnish zone becomes smaller as the temperature decreases, and the height of the fracture increases accordingly (shown in Fig. 5).

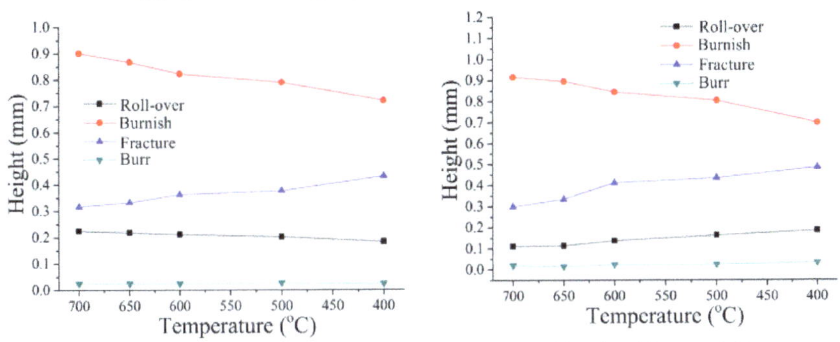

(a) trimming, blanking gap 0.24 mm (b) punching, blanking gap 0.10 mm

Fig. 5. Effect of blanking temperature on quality of blanked surface.

4. Conclusion

In this paper, hot blanking process for hot forming of press hardened steel is researched, and following conclusions can be drawn:

(1) Blanked surface appearance of hot blanking process includes four regions: Roll-over, Burnish, Fracture and Burr.

(2) The height of the roll-over, fracture, and burr zone increase with the increasing of blanking gap because of the reducing of blanked region's stress triaxiality.

(3) The height of the burnish zone becomes smaller as the temperature decreases because deformation resistance of the steel increase.

References

1. P. Liu, Y. Zhang, H. Liu and Z. Gui, Modeling for periodic striation and microstructure evolution in active gas melt laser cutting for phase hardened parts, *Int. J. Adv.Manuf. Technol.* **70**, 1421 (2014).
2. X. Han, K. Yang, S. Chen and J. Chen, Fracture Profile and Crack Propagation of Ultra-High Strength Hot-Stamped Boron Steel During Mechanical Trimming Process, *J. Mater. Eng. Perform.* **24**, 3845 (2015).
3. H. S. Choi, S. G. K, S. K. Lee, B. M. Kim and D. C. Ko, Improvement Technology of Tool Performance in Mechanical Trimming of Hot Stamped 22MnB5 Steel, *in 4rd International Conference on Hot Sheet Metal Forming of High-performance Steel*, (Lulea, Sweden, 2013).
4. H. S. Choi, B. M. Kim, D. H. Kim, D. C. Ko, Application of mechanical trimming to hot stamped 22MnB5 parts for energy saving, *International Journal of Precision Engineering and Manufacturing*, **15**, 1087 (2014).
5. T. Laumann, I. P. Anfruns, M. Grané and D. C. Padró, Hard cutting of tailored hardened 22MnB5, in *Tools and Technologies for the Processing of Ultra High Strength Steels*, (2010).
6. H. So, D. Faßmann, H. Hoffmann and R. Golle, An investigation of the blanking process of the quenchable boron alloyed steel 22MnB5 before and after hot stamping process, *J. Mater. Process. Tech.* **212**, 437 (2012).

Simulation and Experimental Study on Particularly Thick Patch Hot Forming Process

Y. B. Li†, H. Y. Feng, B. Q. An, S. Y. Chu, X. L. Yu and Y. P. Sun
LingYun Industrial Corporation Ltd., Zhuozhou, Hebei, China
†E-mail: liyanbo@lygf.com

The patch plate thermoforming technology is a technology in which two different shapes and sizes of thermoformed steel plates are welded and heated once to form a stamped one. It can not only improve the performance of the car, but also reduce the weight and cost, and is an advanced technology for realizing the lightweight of the car body. The special thick patch plate B column with a thickness of 3.9 mm in the patch plate area of a certain vehicle is being studied. Autoform and STAR CCM+ is for hot forming analysis and waterway cooling analysis. When the heating temperature is 950 °C, the heating time is ≤ 370 s, and the holding time is 15 s, the part's die temperature is < 235 °C, the thickness of the coating diffusion layer is < 13.5 μm, and the hardness of the parts is higher than 421 HV.

Keywords: Particularly thick; Patch; Process; CAE analysis.

1. Introduction

The new automobile collision regulation has been implemented in 2018 and the collision speed has brought into line with international standards, from 56 km/h to 64 km/h [1].The new evaluation system requires the safety evaluation star of the car to be displayed to the public. It increases requirements for vehicle safety. Thermoforming technology ushered in a new broad space. Hot stamping parts are from simple to complex development. TRB, soft zone, TWB, and Patch have emerged [2].

Patch thermoforming technology is to put two or more plates into a whole by spot welding, and then placed in a heating furnace to heat and hold for a certain time, together with a new type of thermoforming mold to achieve hot stamping forming and pressure quenching hot forming technology [3].The thermoforming of the patch is configured with different thicknesses in different areas to achieve different strengths, and the goal of optimizing the body performance and body weight reduction is achieved. Currently it is used in foreign and domestic joint venture high-end models [4].

Compared with the traditional thermoforming, the hot-forming technology of the patch plate is more difficult, especially the special thick-thickness patch plate with a thickness exceeding 2.5 mm has technical bottlenecks in the aspects of die temperature, part hardness, diffusion layer thickness control, etc. [5].

In this paper, a B pillar with a thickness of 3.9mm is taken as the object of study in a certain type of patch. The finite element model of the hot-formed B pillar of the patch is established in the Autoform and the forming process of the patch is accurately simulated.

Through the simulation results and the test results, the forming process of a 3.9 mm thick patch B-pillar was obtained.

2. B Pillar Patch Simulation Analysis

2.1. *Establishing a hot forming model*

The establishment of an extremely thick patch plate B-pillar model in CATIA is shown in Fig. 1. It contains B-pillar body 2.1 mm and patch 1.8 mm.The design of the patch is mainly based on the impact of the B-pillar assembly on the side impact performance of the vehicle body [6].

In this paper, Autoform software is used to set up the thermoforming model of the patch. The model of the sheet metal is shown in Fig. 2. There is a body and a patch board. Welding joints between the two plates are not shown.

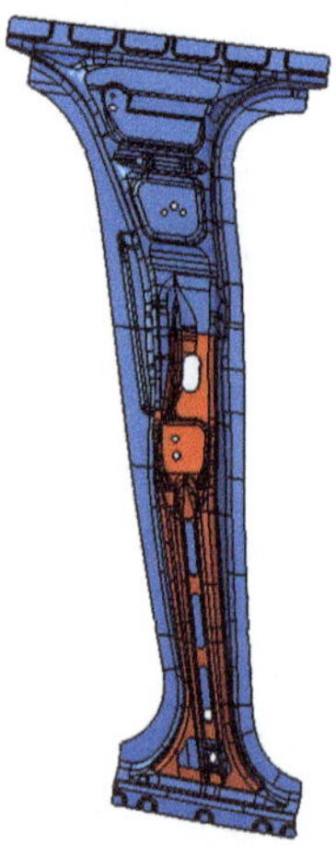

Fig. 1. B pillar model.

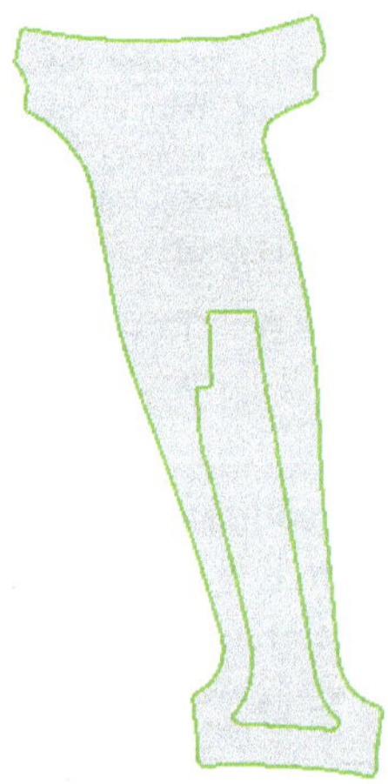

Fig. 2. B pillar blank model.

2.2. *Setting process parameters*

The thermoforming process of the patch plate B pillar is divided into heating, transfer, forming, and holding pressure quenching stages. In the heating stage, the sheet is heated to 950 ℃, and during the period of time before transfer and mold clamping, the sheet material is mainly convection and radiation heat exchange with the surrounding air, the heat transfer coefficient is 10 W/(m^2·K), and the coefficient of thermal friction between the sheet material and the mold is set to 0.6.

3. Process Research

3.1. *Effect of holding time on mechanical properties of parts*

The die temperature of conventional thermoforming products is 200 °C. In order to improve the mechanical properties of the products, the influence of dwell time on the die temperature is explored through Autoform software simulation analysis. Because the patch position of the patch B column contains the two layers of the body and the patch plate, the thickness is relatively large, which is not conducive to heat conduction. Therefore, the set test points are evenly distributed in the body and patch plate areas, as shown in Fig. 3.

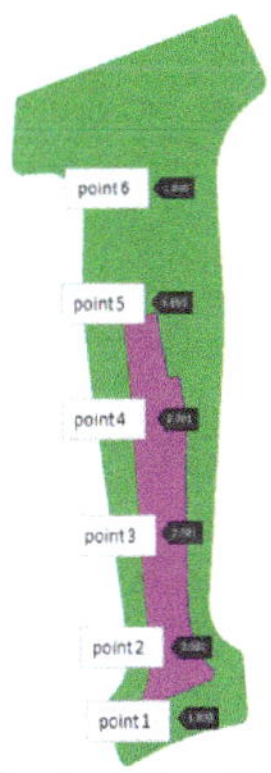

Fig. 3. Distribution of hardness test points.

Set the heating material to be sent to the mold for 10 s, the press down time is 1 s, the press reaches the bottom dead point in 4 s, that is, the first 15 s starts to hold pressure, the pressure holding time is set to 8 s, 10 s, 12 s, 15 s, as shown in Fig. 4-9.

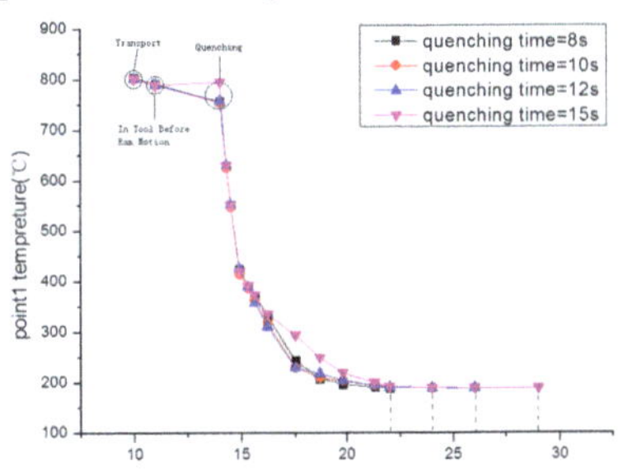

Fig. 4. Point 1 temp in different quenching time.

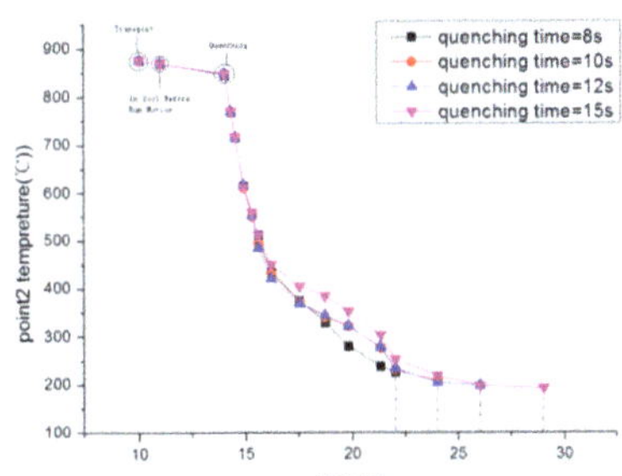

Fig. 5. Point 2 temp in different quenching time.

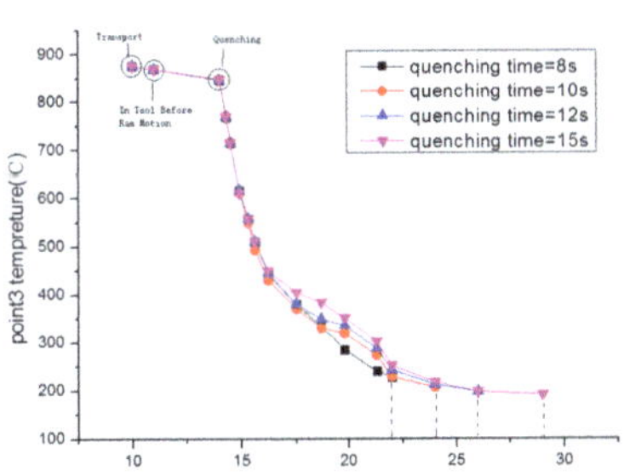

Fig. 6. Point 3 temp in different quenching time.

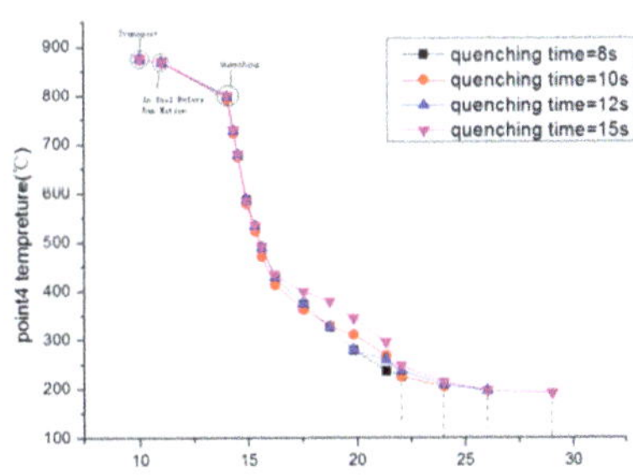

Fig. 7. Point 4 temp in different quenching time.

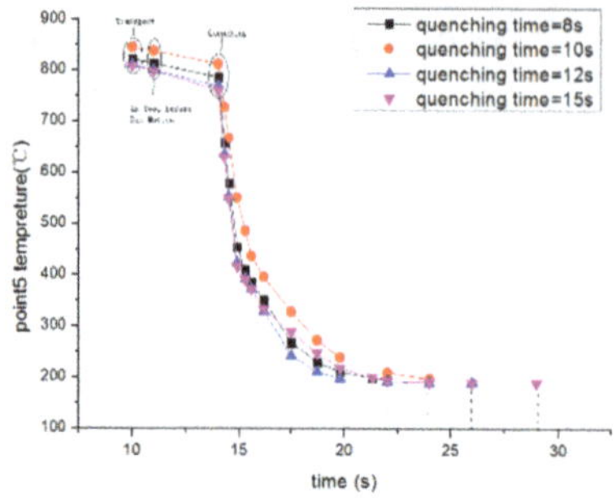

Fig. 8. Point 5 temp in different quenching time.

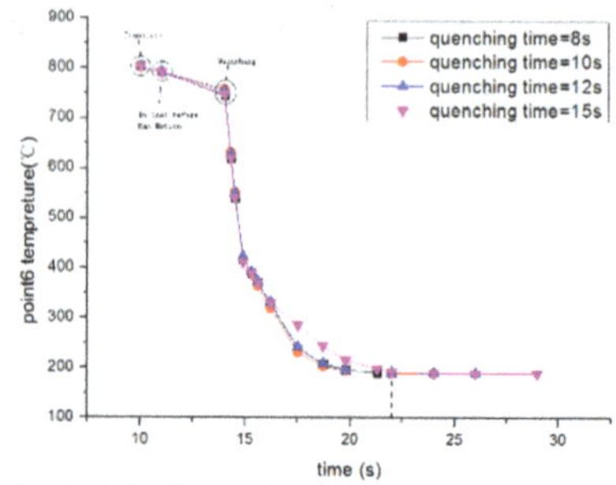

Fig. 9. Point 6 temp in different quenching time.

The result is through simulation analysis:

(1) When the pressure holding time is 8 s, the lowest temperature of the patch plate area is about 235 ℃;

(2) With the extension of holding pressure time, the lowest temperature of the re-pressure stage of the test point is obviously lower than 235 ℃, which can meet the requirements of the host plant.

Through Autoform software, the hardness distribution of the parts is compared and analyzed during the holding time of 10 s and 15 s. Other process parameters are unchanged, as shown in the Fig. 10-11.

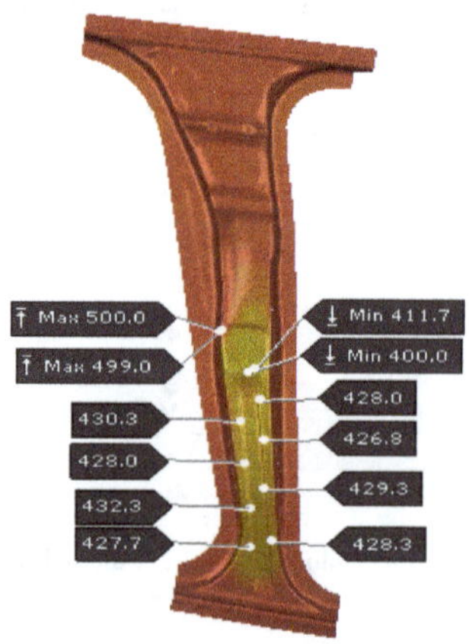

Fig. 10. 10 s patch hardness distribution.

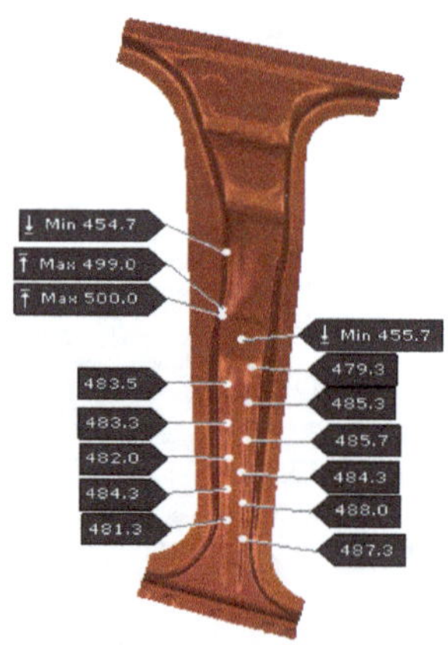

Fig. 11. 15 s patch hardness distribution.

Comparing the figure and the figure can be obtained: When the holding time is set to 15 s, the hardness of the patch plate is significantly higher than 421 HV, to meet the hardness requirements of the host plant parts.

3.2. *Effect of cooling methods on mechanical properties of parts*

Take the mold inserts in the area of the B pillar patch as an example. Design single-layer waterways and double-layer waterways according to the mold profile. The wall thickness of a single-layer water channel is approximately 10 mm, the wall thickness of a double-layer water channel is 10 mm, and the distance between two water channels is approximately 10 mm to 12 mm. STAR CCM+ software is used for cooling simulation analysis and the results are shown in Figs. 12 and 13. The solid insert material is Cr7V, thermal conductivity is 31 W/(m·K), water inlet speed is 2 m/s, inlet water temperature is 5 °C, and the heat transfer coefficient on the upper surface is 150 W/(m^2·K). The heat transfer coefficient of air contact is 10 W/(m^2·K).

Fig. 12. Single-layer water mold temperature field.

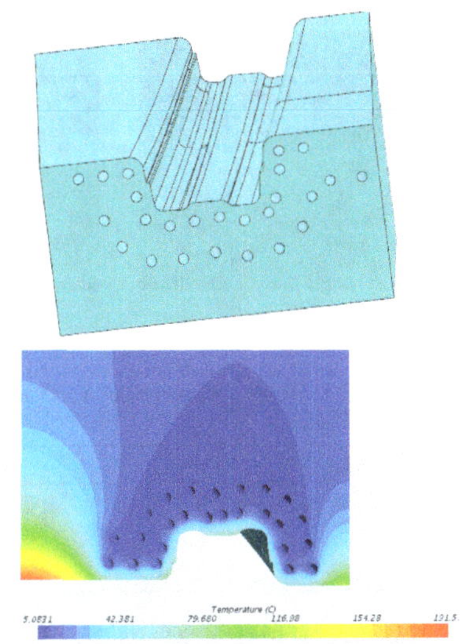

Fig. 13. Double-layer water mold temperature field.

3.3. *Experimental study on the effect of heating temperature on diffusion layer of coating*

The thickness of the coating diffusion layer will directly affect the welding performance. The thinner the diffusion layer, the better the welding performance. The thickness of the diffusion layer of specially thickened patch B pillar was studied with Arcelor USIBOR1500P AS60G/60G as the test material. The thickness of the body was 2.1 mm, the thickness of the patch plate was 1.8 mm, and the thickness of the coating was 30-50 μm, shown in Fig. 14.

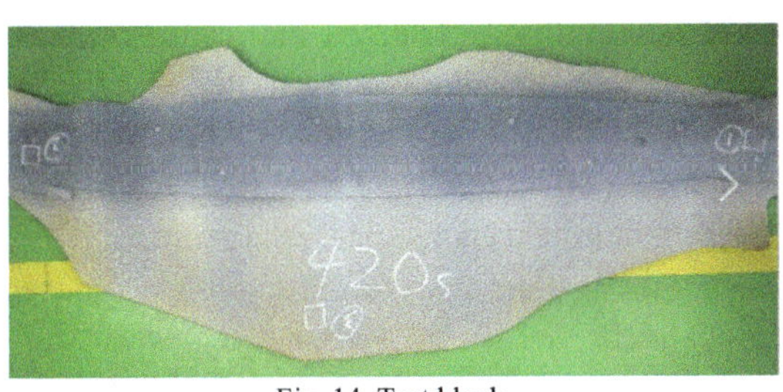

Fig. 14. Test blank.

The body plate and the patch plate are welded together. Five test points are selected on the plate. As shown in the figure, points 1 and 2 are respectively located at both ends of the patch plate, and point 3 is located on the body plate, points 4 and 5 located on the opposite side of point 1 and point 2, respectively. The heating temperature is 950 °C and the heating time is 320 s, 370 s and 420 s. The thickness of the diffusion layer is measured under a metallographic microscope as shown in Fig.15-17, and summary in Table 1. The meaning of the picture naming format 320 s-1 is: 320 s is the heating time and 1 is the test point.

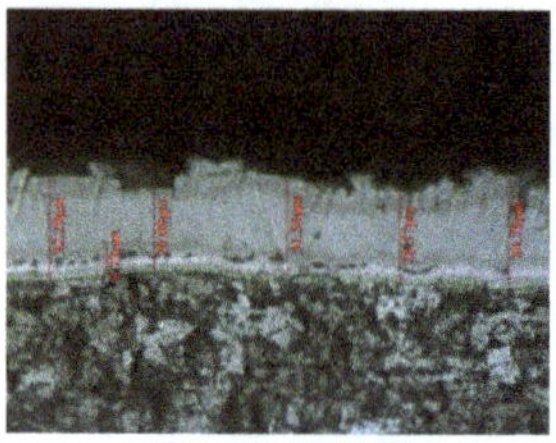

Fig. 15. 320s-1.

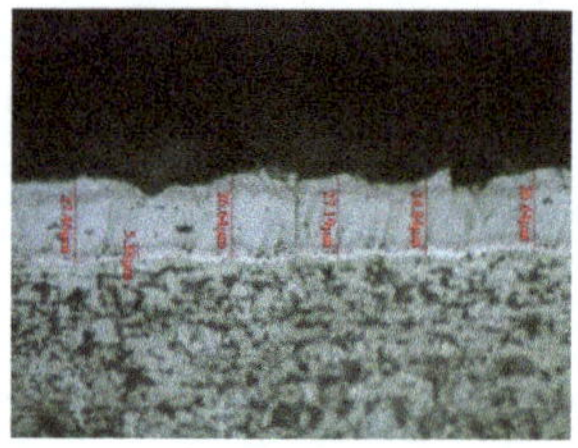

Fig. 16. 320s-2.

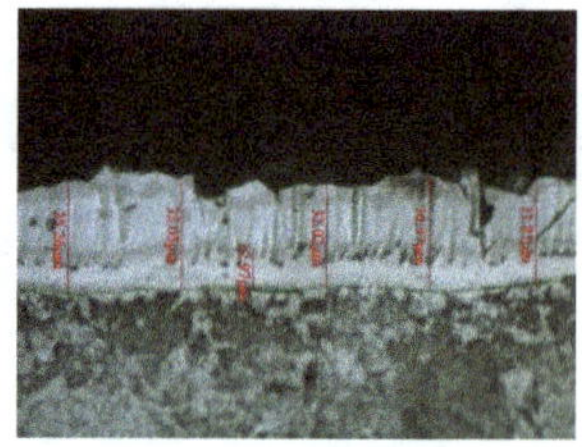

Fig. 17. 320s-3.

Table 1. Diffusion layer thickness (μm) for each test point at different heating temperatures.

Heating time	Point 1	Point 2	Point 3	Point 4	Point 5
320s	9.37	5.38	12.97	8.35	7.31
370s	8.63	9.45	13.41	8.63	8.07
420s	10.56	11.32	17.25	11.25	12.08

Through the above test results can be obtained:

(1) At the same heating temperature, the thickness of the diffusion layer increases with the heating time.

(2) At the same temperature and at the same heating time, the thickness of the diffusion layer at the location of the patch plate is lower than the thickness of the single layer plate.

(3) When the heating time is ≤ 370 s, the diffusion layer thickness can be controlled below 13.5 μm.

4. Conclusion

Using autoform software, thermo-compression analysis was performed on special thick patch plate B-pillars with a body thickness of 2.1 mm and a patch plate thickness of 1.8 mm. The temperature at the same test point of the part decreased with the prolongation of the holding time. When the pressing time is 8 s, the die temperature in the patch plate area can be reduced to 235 °C; when the dwell time is 15 s, the hardness of the part far exceeds 421 HV, which satisfies the technological requirements.

Using STAR CCM+ software to analyze the cooling of the patch plate area, the cooling effect of the double channel is obviously better than that of the single channel; the highest temperature point, the former is about 40 °C lower than the latter.

Through the test method to heat the coating and measure the thickness of the diffusion layer at the test point, when the heating temperature is 950 °C, the diffusion layer thickness increases with the heating time; when the heating time is ≤ 370 s the thickness of the diffusion layer can be controlled below 13.5 μm, which is conducive to the direct welding of parts.

References

1. China Automotive Technology and Research Center, *C-NCAP Management Rules*, (2012).
2. Q. H. Du and C. Huang, Application of steel plate hot stamping technology in bodywork, *Auto. Tech. Mater.* **3**, 23 (2014).
3. C. L. Wang, Y. J. Wu and W. L. Xu, Simulation and experiment of patchwork B-pillar hot forming, *Die and Mould Technology* **6**, 10 (2016).
4. J, Li, X. Q. Sun and Y. X. Chen, Domestic and foreign passenger car bodies Lightweight material application history and development forecast, *New Mater. Ind.* **12**, 31 (2015).
5. Y. L. Kang, *Theory and Technology of Processing and Forming For Advanced Automobile Steel Sheets*, (Beijing: Metallurgical Industry Press, 2009).
6. H. Karbasian and A. E. Tekkaya, A review on hot stamping, *J. Mater. Process. Tech.* **210**, 2103 (2010).

Study on Technology of Hot Stamping Based on Automotive Light-weight Engineering

M. Y. Zhu[†], L. C. Cui and J. L. Li
Technical Center, Anhui Jiang Huai Automobile Group Co., Ltd.,
Hefei, 230601, China
[†]*E-mail: gcy_jszx@jac.com.cn*

The hot stamping forming technology is studied by the technology of automotive lightweight. The formability of the hot stamping parts is studied in this paper. Besides, structure design, process design and tailored technology of hot stamping are analyzed and researched. The results show that the reasonable shape of structure design plays an important role in the maximum principal strain; which effectively overcome the drawback of crack and wrinkle that the absorbed wrinkle bead is designed in the region easy to wrinkle. Parts tailored technology of hot stamping makes the different regions of the same parts take on the different strength index. The development of automotive lightweight technology is promoted by the application of hot stamping technology effectively.

Keywords: High strength steel; Automotive lightweight; Hot stamping; Structure design; Tailored technology.

1. Introduction

Based on the idea of low carbon, green and energy saving and safe, automotive light-weight engineering becomes a mainstream of modern automobile design and manufacture. The technology of ultralight automotive can enhances fuel economy, energy saving and reduce emission. The car can run 1 km/L as the weight of car reduce 100 kg. The fuel consumption can reduce from 6 % to 8 % with the weight reduce 10 %, the result is the auto emission can reduce from 5 % to 6 %, which is shown in Fig. 1 [11]. Therefore, automobile lightening is an effective method for energy saving and emission reduction.

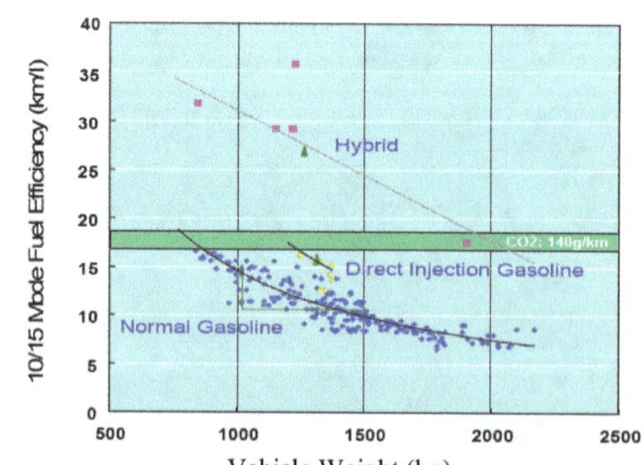

Fig. 1. Relationship between vehicle weight and fuel efficiency.

Recently, the automotive lightweight technology has been developed rapidly, and the hot stamping technology of high strength steel is one of the way for automobile lightweight. Hot stamping is the forming method of high strength steel parts to obtain uniform martensite, and the initial hardness of boron steel is 500~600 MPa that is heated to austenite state, transferred to the die quickly and to be high speed stamping forming, under the condition of certain pressure, the quenching cooling rate of mold parts in the main body is greater than 27 K/s, and pressure quenching for a period of time. The advantage of hot stamping technology is obatin ultra high strength of body parts, reduce body weight, improve safe and comfort of the body, improve the stamping formability and the precision of parts size. This paper study and analysis the structure design of hot stamping forming parts in order to realize the technology of automotive lightweight according to the structure characteristics of stamping parts. Fig. 2 is a typical case of hot stamping technology.

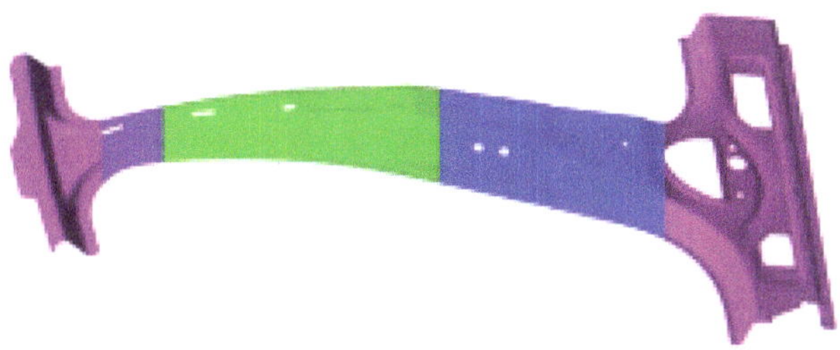

Fig. 2. Typical parts of Hot Stamping.

2. Hot Stamping Forming Performance

2.1. *Material model*

It is called the hardening of the material, the strength and hardness of the metal materia rised when the plastic deformation happened below the recrystallization temperature.

The hardening model of the material can be expressed as Eq. 1 [3]:

$$\sigma(T) = K(T)\varepsilon^{n(T)} \tag{1}$$

Where σ is true stress (MPa); n is hardenability value.

2.2. *Forming temperature*

The hot stamping of materials has its range of forming temperature. The relationship between optimum forming temperature and hardening is showed in Fig. 3.

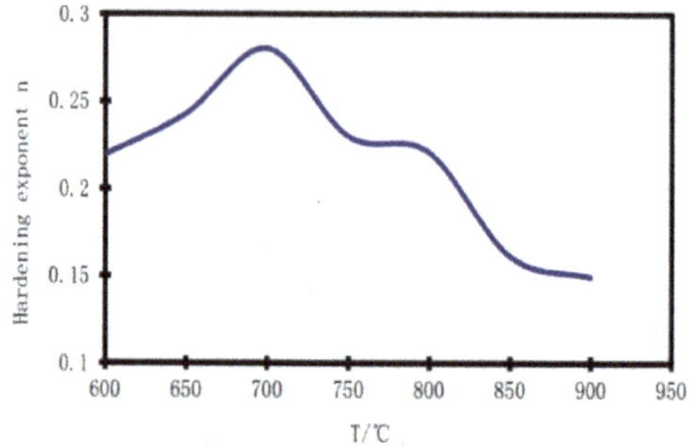

Fig. 3. Relationship between Optimum forming temperature and Hardening exponent.

2.3. *Forming boundary condition*

Hot stamping has its boundary conditions of forming temperature, and it is very important to the forming quality of the stamping parts on condition that the temperature boundary conditions is well controlled. The forming temperature boundary conditions of hot stamping parts is depicted in Fig. 4.

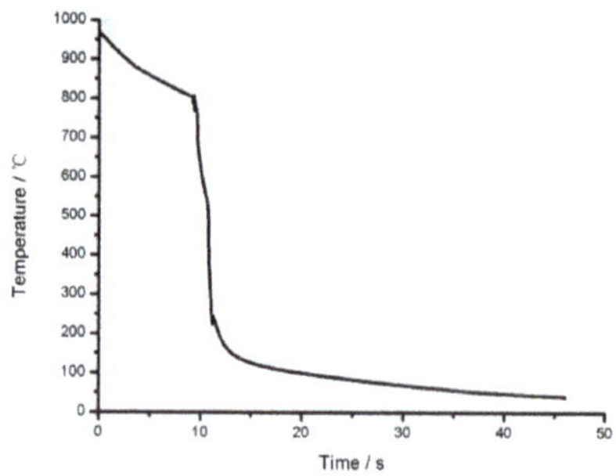

Fig. 4. Boundary condition of Hot stamping.

2.4. *Stress-strain curve*

The forming stress and strain curves of hot stamping parts will change correspondingly with different temperatures, which is shown in Fig. 5. It can be seen that stress-strain has close relation with temperature.

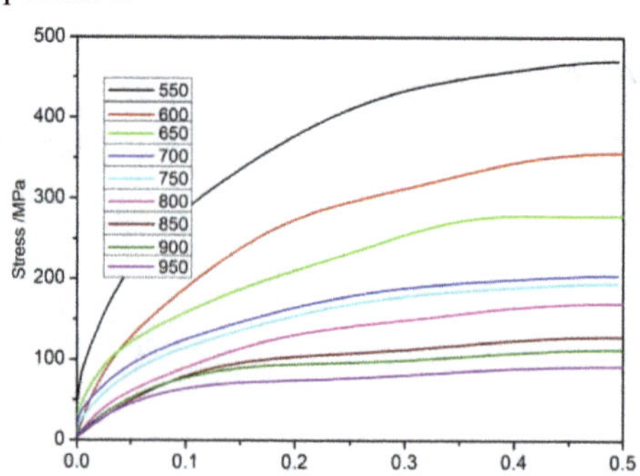

Fig. 5. The stress-strain curve of temperature dependent.

3. Advantage of Hot Stamping Forming Technology

Compared with cold stamping, hot stamping technology has the following advantages:

(1) The formability of the material is better, and it can produce the parts with complex shape.

(2) After hot forming, the strength of the parts is high and the stiffness is good.

(3) The surface hardness and wear resistance of the parts is improved effectively.

(4) After forming and quenching at high temperature, the size stability of parts is good and the springback is small.

(5) The safety performance of the automobile is improved;

(6) To make the body lighten to a greater degree.

4. High Strength Steel Hot Stamping Forming Parts Design

4.1. *Structure design*

The parts design of high strength steel hot stamping forming has its unique characteristics compared to the traditional cold stamping. Firstly, considering the shape and structure requirements, under the same collision conditions, the stiffness, strength and hardness of the parts fillet and sidewall are told the difference. Therefore, the shape and structure design of hot stamping parts have an important impact on the hot stamping forming quality. Under the same collision conditions, the stress and the strain of hot stamping parts with different structure shapes will change differently, as shown in Fig. 6 and 7.

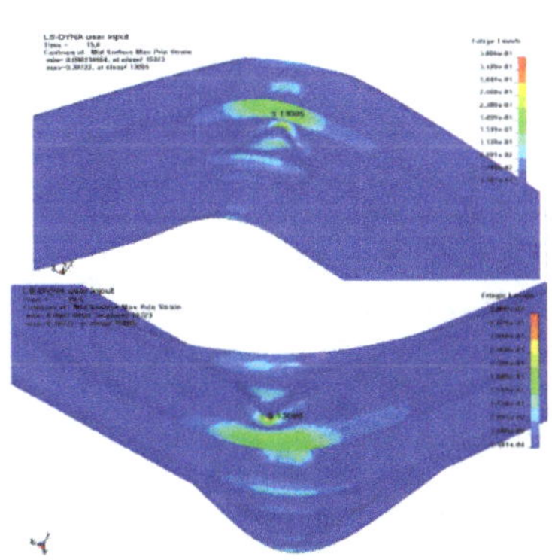

Fig. 6. The strain of hot stamping parts whose round radius is 5 mm.

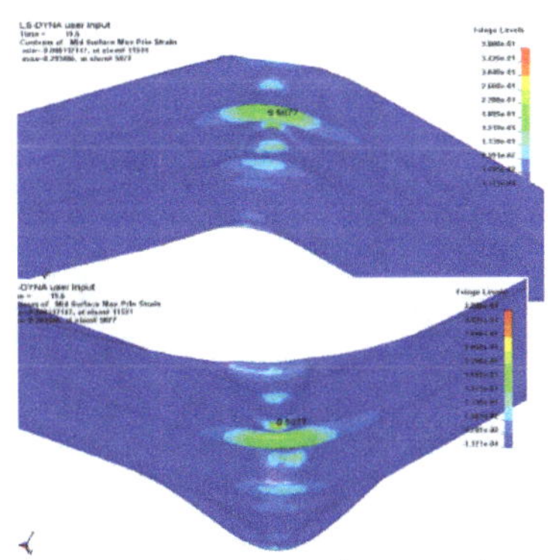

Fig. 7. The strain of hot stamping parts whose round radius is 8 mm.

Under the same collision conditions, that is, thc collision speed is 50 km/h is illustrated in Fig. 6. The fillet radius of the hot stamping part is designed to be 5mm, and the maximum principal strain is 0.387, as shown in Fig. 7. The maximum principal strain is 0.293 when the fillet radius of the hot stamping part is designed to be 8 mm.

The spacing size of the structural support block of hot stamping parts also has an effect on the forming stress and strain, as shown in Fig. 8 and 9. Under the same collision condition, the collision speed is 50 km/h, the spacing of supporting blocks are 600 mm and

800 mm, respectively. The maximum principal strain are 0.387 and 0.345 when the support block spacing are 600 mm and 800 mm.

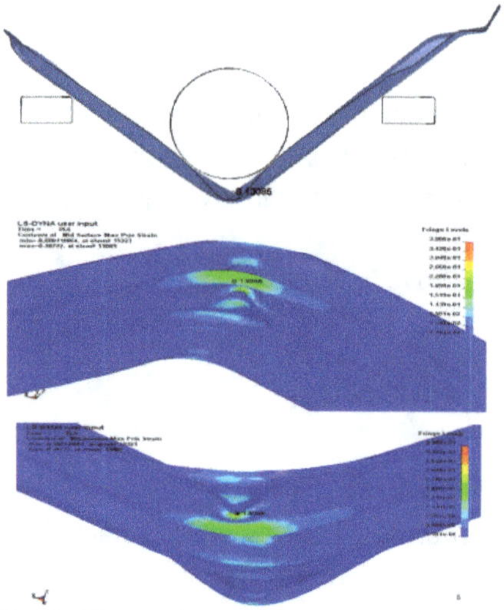

Fig. 8. The maximum main strain of support spacing is 600 mm.

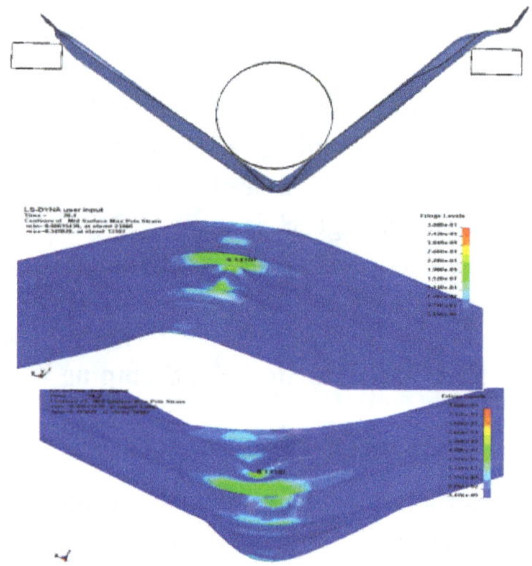

Fig. 9. The maximum main strain of support spacing is 800 mm.

Different structural design of hot stamping parts, such as round radius, support spacing, flange surface and so on can affect the maximum major strain. Hence, the reasonable structural shape design of hot stamping parts has an important impact on its hot forming quality. Table 1 shows the influence of its structural shape on the maximum principal strain.

Table 1. Table of hot stamping parts structure design.

Original condition		**Change content**	
Pre change project	**Max major strain**	**Post change project**	**Max major strain**
Support spacing 600 mm		Support spacing 800 mm	0.345
Round radius R=5 mm		Round radius R=8 mm	0.293
Flanged surface without extension	0.387	Flanged surface extension of 10 mm	0.362
Unchanged Round radius and Flanged surface		Round radius R=8 mm+ Flanged surface extension of 10 mm	0.287

4.2. *Process design*

The design and development of hot stamping parts is different from the traditional cold stamping, because of its special forming process, should fully consider the forming conditions of material sheet in the high temperature state, according to the forming limit and tribological characteristics of the material at high temperature, it is recommended to the design specifications of hot stamping parts, take B-pillar parts of the body as an example, fully consider the material flow in the hot stamping process, avoid forming cracks and wrinkles, prone to wrinkling defects in the illustrated arrow area, so the wrinkle absorbed bead is designed, as shown in Fig. 10.

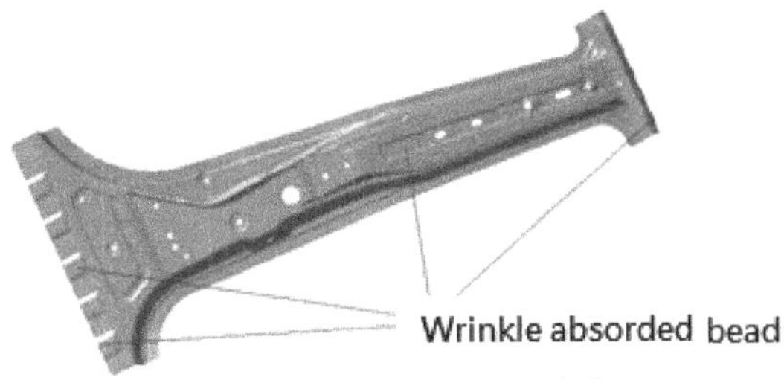

Fig. 10. B-pillar parts process design.

4.3. *Hot stamping forming technology*

(1) Improving the productivity in the hot stamping process, and reducing the pressure maintaining time of the workpiece on the premise of meeting the requirements of the process;
(2) Improving the die design and hot stamping forming process of one die and multiple parts;
(3) Carrying out the technology of hot-cutting edge of parts, reduce the laser machining quantity of forming parts as much as possible, and reduce the process cost;
(4) Typical parts such as flexible distribution technology of door B-pillar strength, including steel design, local heating process, mold cooling system, the application of laser welding plate;
(5) Researching on protective atmosphere of heating furnace, develop protective atmosphere with good anti-oxidation effect and cheaper price;
(6) A process technology for reducing oxidation and temperature drop of a workpiece from the furnace to the die;
(7) Studying on heating methods of parts, improving productivity, stable quality, reduceing energy consumption, including research on direct current heating, induction heating, box furnace and roller furnace heating and the combination of a variety of heating methods.

4.4. *Parts tailored technology*

By means of non-isothermal heating or partial heating of the mould, the temperature difference between the die and the body sheets can be reduced, thereby reducing the coefficient of thermal conductivity between the die and the body sheet. At present, it is more commonly used to carry out local heating and cooling on the die, so that the hot stamping part can produce full martensite structure with high strength and mixed structure with low strength, so that different strength indexes of different parts of the same hot stamping part can be obtained; taking the B-pillar as an example, the upper part of the B-pillar is required to have higher strength, certain rigidity is ensured, deformation of the cockpit is prevented, and the lower part is expected to have relatively low strength, so that the impact energy absorption effect can be achieved. As shown in Fig. 11 [11].

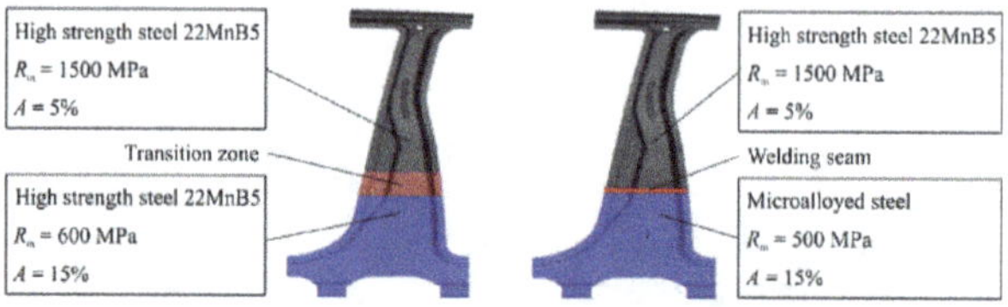

Fig. 11. B-pillar with tailored properties.

5. Application of Hot Stamping Forming Technology

Hot stamping technology in the application of the body, as shown in Fig. 12. A large number of hot forming parts are used in automotive body B-pillar, A-pillar , doorsill beam and so on. At present, the proportion of hot stamping parts used by domestic self-owned brands of automobiles is increasing continuously, and the number of hot forming parts of some models reached to 7-10 parts. With the collision, emission reduction, laws and regulations continue to be strict, which will increases the use of hot forming parts of self-owned brand cars.

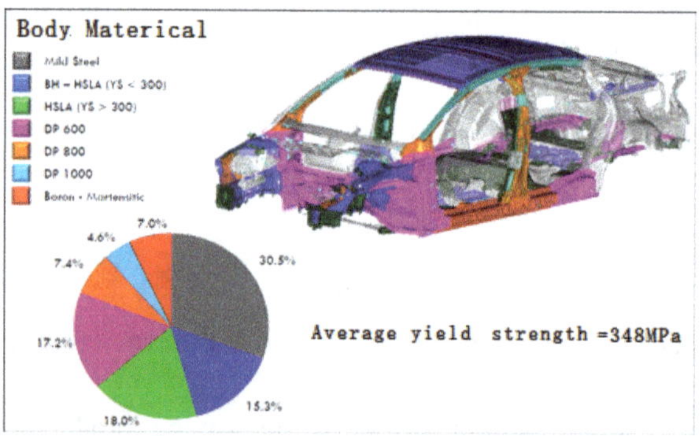

Fig. 12. Application of Hot Stamping Technology.

6. Summary

(1) The results of hot stamping technology show that the reasonable structure shape of hot stamping parts has an important influence on maximum major strain and stamping forming quality;

(2) Material flow in the process of hot stamping should fully be considered in the process design of hot stamping forming parts. Wrinkle absorption bead design in areas incline to wrinkles can avoid cracking and wrinkling defects;

(3) The application of tailored technology hot stamping parts make the different position of same parts have different strength index;
(4) The application of hot stamping technology promotes the rapid development of automobile lightweight technology effectively.

References

1. Hauger, Tailor Rolled Blanks for Automotive Light-weight Engineering, in *Automotive Circle International Conference*, (Frankfurd, Germany, 2008).
2. J. C. Benedyk, Light metals in automotive applications, *Light Metal Age* **10**, 34 (2000).
3. Z. Q. Lin, *The simulation of Body panel stamping*, (Machinery Industry Press, 2004).
4. R. George, A. Bardelcik and M. J. Worswick, Hot forming of boron steels using heated and cooled tooling for tailored properties, *J. Mater. Process. Tech.* **212,** 2386 (2012).
5. H. Hoffmann, H. So and H. Steinbeiss, Design of Hot Stamping Tools with Cooling System, *Cirp Ann. –Manuf. Techn.* **56**, 269 (2007).
6. B. Abdulhay, B. Bourouga and C. Dessain, Experimental and theoretical study of thermal aspects of the hot stamping process, *Appl. Therm. Eng.* **31**, 674 (2011).
7. K. Mori and D. Ito, Prevention of oxidation in hot stamping of quenchable steel sheet by oxidation preventive oil, *Cirp Ann. –Manuf. Techn.* **58**, 267 (2009).
8. K. Mori, S. Maki and Y. Tanaka, Warm and Hot Stamping of Ultra High Tensile Strength Steel Sheets Using Resistance Heating, *Cirp Ann. –Manuf. Techn.* **54**, 209 (2005).
9. H. Karbasian and A. E. Tekkaya, A review on hot stamping, *J. Mater. Process. Tech.* **210**, 2103 (2010).
10. K. Mori, S. Saito and S. Maki, Warm and hot punching of ultra high strength steel sheet, *Cirp Ann. -Manuf. Techn.* **57**, 321 (2008).
11. M. T. Ma, *The research progress of hot stamping technology*, (China Auto Engineering Institute, 2014).

Hot Stamping Process Research on Sill Beam with 22MnB5 Steel

Y. S. Wei[1,a,†], C. Feng[1] and M. T. Ma[2,b]
[1]*Great Wall Motor Co., Ltd., Technology Center, Baoding, Hebei province, 071000, China*
[2]*Automotive Light-weight Engineering Technology Center, Chinese Automotive Engineering Research Institute Co, Ltd., Chongqing, 401122, China*
E-mail: [a,†]*Weiyuansheng11@sina.com,* [b]*mingtuma@126.com*

In order to save energy and reduce emissions, increase fuel economy, as well as meet increasingly stringent safety regulations, the proportion to apply hot stamping parts increases for the car body year by year. The performance and production stability of hot stamping ultrahigh strength steels are particularly important for the development of hot stamping processes. This article describes in detail the design of hot stamping process parameters and research methods for process failure problems, and summarizes the issues and solution ideas that affect the hot stamping process crack failure problems, as well as the influence factors of various process parameters and failure modes. And reasonable suggestions to ensure the stability of hot stamping production performance are put forward.

Keywords: Hot stamping process; Stamping crack; Cooling uniformity.

1. Introduction

In 2017, the production and sales volume of automobiles in China reached 29.915 million and 28.887 million respectively. The vehicle ownership reached 200 million.

According to statistics, the annual growth rate of China's automobile sales during the 12 years from 2005 to 2017 was 14.4 percent. With the increase in the number of vehicle ownership, the external dependence of crude oil in China reached 65.4 percent in 2016. There into what auto gasoline consume are the above 70 percent for gasoline consumption of whole the country. In addition, with the establishment of China's international manufacturing power status, the development of the transportation industry, and the irrationality of the original industrial structure and energy-based fuel structure, it has caused tremendous pressure on the environment. Else, the pressure on the environment has been contributed by the increase of car ownership also.

Based on the dual pressure for environment and easing the external dependence of crude oil, the China Automotive Engineering Society released the "Energy-saving and New Energy Vehicle Technology Roadmap" to propose vehicle emission standards: By 2020, the average fuel consumption of new passenger vehicles will be 5 L/100 km, from the current 6.5 L/100 km decreased by 23 %. One of the ways to reduce the car's fuel consumption is to achieve a lightweight car. For conventional fuel vehicles, 75 % of the fuel consumption is related to the car's curb weight. The curb weight will increase the car's rolling resistance, climbing resistance, and acceleration resistance to increase fuel consumption. If the vehicle weight is reduced by 100 Kg, the fuel consumption per

100km will be reduced by 0.3-0.6 L, and the endurance mileage for electric vehicles will increase by 6-11 % [1, 2].

In order to meet the increasing requirements for vehicle safety and energy regulations, the way to improve vehicle safety and lightweight performance is to increase the proportion of super high-strength steel in white-body. Since the 1500 MPa martensitic steel to use by hot stamping has been born for 20 years, the proportion of applications on the car body has been increasing year by year.

2. Design of Hot Stamping Process

The hot stamping steel commonly used in automobiles is mostly 22MnB5, and is divided into uncoated bare boards and aluminum silicon coating boards and galvanized boards according to different purposes and requirements.

In order to overcome the problems of more spring back deformation, cracking, and large die wear during the stamping of ultra-high strength steel, hot stamping technology emerged from the 1980 s. The brief description of the hot stamping process is heating 22MnB5 or other hot stamping steel sheet and keeping temperature in the 900-950 °C range for austenitization and achieving the fully austenitic microstructure of high ductility, followed by hot stamping and cooled down below the martensitic transformation line in the die to rely on the rapid thermal conduction of the die, finally to form a martensitic microstructure for more than 98 % content, thus achieving the components of ultra-strong strength. The main process design is as follows:

(1) Set a reasonable heating temperature;
(2) Set a suitable quench cooling rate;
(3) Design die according to the formability analysis of parts;
(4) Calculate thermal conductivity and design heat transfer system of the die based on cooling rate requirements;
(5) Design post-processing operations.

2.1. *Austenitizing temperature and transfer time of steel sheet*

According to the CCT curve of 22MnB5 steel, the austenitizing heating temperature is set at 900-950 °C. Hot stamping temperature of 22MnB5 steel sheet is at about 800 °C. Fig. 1 shows the flow stress curves of 22MnB5 steel at different temperatures and strain rates. It can be seen from Fig. 1 that at 800 °C temperature, the yield strength and elongation of the steel sheet are equivalent to those of mild steel; it is advantageous for drawing and forming complex parts. According to the cooling rate curve of the steel sheet in the air, as shown in Fig. 2, it can be found that when the heating temperature is 950 °C, the allowable transfer time of the steel sheet is 4s [3]. According to different site conditions, generally require that the time to be transferred by steel sheet from the heating furnace to the die should be ≤ 5 s.

2.2. *Cooling speed of hot stamping components hardening*

From 22MnB5 continuous cooling transformation curve - Fig. 3, it can be found that if hot stamping parts obtained more than 90 % of the martensitic structure, its critical cooling rate should be greater than 27 °C/s [4, 5]. Therefore, in order to achieve a hot stamping part with a yield strength ≥ 1000 MPa and a tensile strength ≥ 1500 MPa, the cooling rate of the part in the die should be ≥ 30 °C/s.

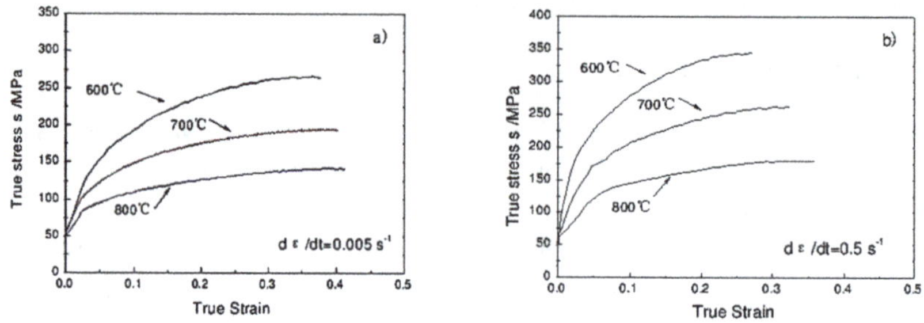

Fig. 1. Stress-strain curves of 22MnB5 at different temperatures and strain rates [1]: (a) $0.005s^{-1}$, (b) $0.5s^{-1}$.

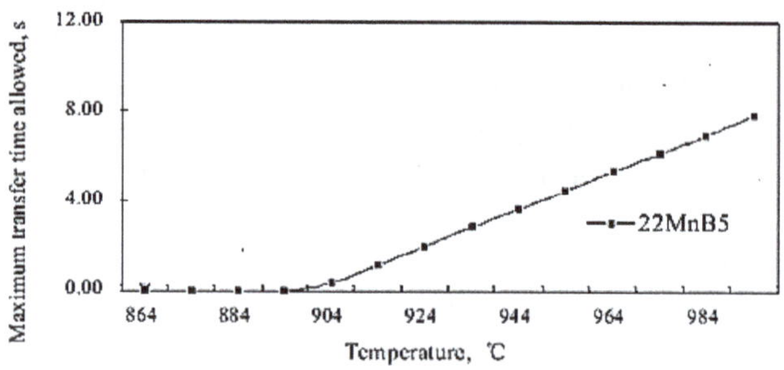

Fig. 2. Curve of the maximum allowable transfer time varying with heating temperature of the workpiece [1].

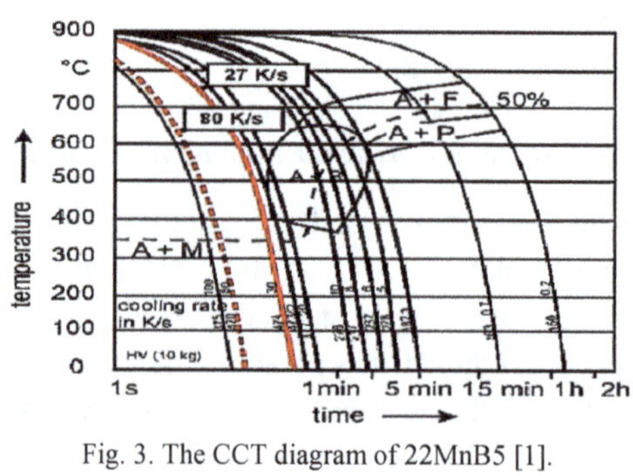

Fig. 3. The CCT diagram of 22MnB5 [1].

2.3. *Holding pressure time*

The forming pressure's holding time in the die is an important process parameter to guarantees the complete transformation of the martensitic microstructure and the mechanical properties of hot stamping parts. Figure 4 shows the metallographic structure of the hot stamping parts with different holding pressure time from 5 s to 60 s. Lath martensitic structure was obtained with little difference. Figure 5 is the relationship curve between the hot forming pressure holding time and the several mechanical properties of the stamping part, these properties include tensile strength, elongation, and strength plastic product. It can be seen that the tensile strength increases slowly after the pressure holding time ≥ 30 s, this indicate that as the holding time is increasing, the transformation of martensitic structure is more completely and the content of metastable structure is reduced, which ensures the stability of the performance.

(a) t=5s (b) t=30s (c) t=60s

Fig. 4. Metallographic structure of hot stamping forming parts with different holding time [1].

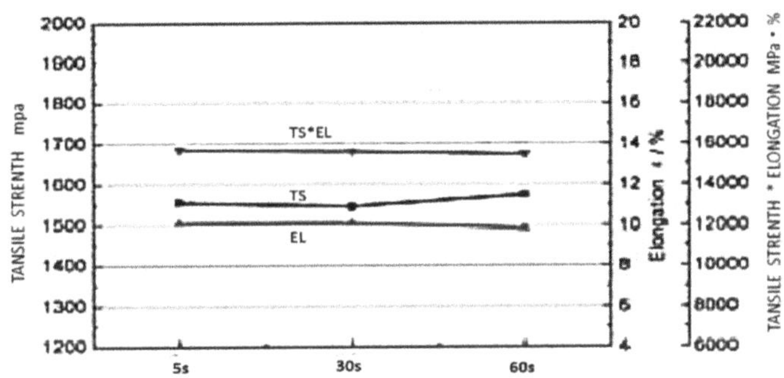

Fig. 5. Relationship of hot forming pressure holding time with tensile strength, elongation and strong plastic product.

3. Analysis of Root Reason about Failure Cases of Process and Verification of Results

The rear floor sill beam of some model has repeatedly appeared cracking during the hot stamping process, and the cracking appears on upper flanging of the concave part on beam and stretched position, as Fig. 6 had shown. The CAE analysis shows that the maximum amount of thinning at the cracking position is 11.57 %, which meets the requirement of thinning rate <15 %. In order to solve the problem of cracking of the stamping part in the forming process, the investigation of the cause is shown in Table 1. It is found that the poor cooling uniformity of the stamping part and that the radius of the arc of the punch in the failure part does not meet the requirements of the design as well as the grain coarsening of the austenite may be the cause of the failure of the hot forming fracture. Need further investigation and analysis.

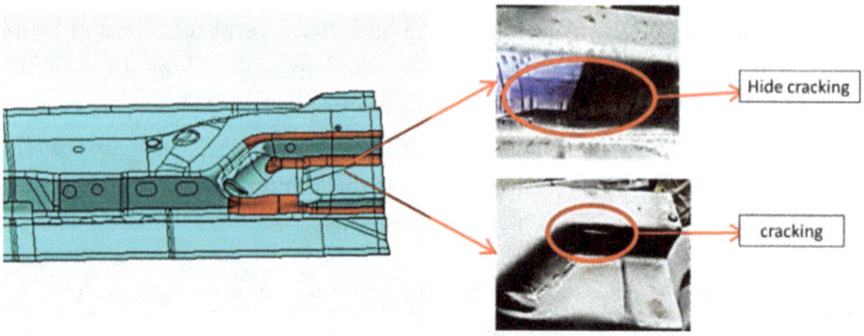

Fig. 6. Schematic diagram of hot forming cracks in the floor sill beam of a certain model.

Table 1. Failure investigation and results.

No	Checking the project	Result
1	The chemical composition of the steel sheet	Meets the requirement
2	Heating temperature, stamping temperature, transfer time	Meets the requirements
3	Surface roughness of cavity in die	Meet the requirements
4	The gap between punch and die	Meet the requirements
5	Stamping parts cooling uniformity	Cooling water path diameter and their distance from the surface is inconsistent other.
6	Failure point punch insert arc radius	The arc radius is 1mm smaller than the drawing requirement
7	Austenite coarsening at high temperature and mixed crystals	The heating time of the steel sheet exceeds the requirement of 160 S and bringing austenite coarsening.

3.1. *Effect of cooling uniformity in the forming process*

After investigation, the diameter of the cooling water path and the interval from the surface of punch's press block are inconsistent, and the water channel is not uniformly distributed (see Fig. 8).

The total of 5 waterways in which have 3 are 10 mm in diameter and 2 are 20 mm in diameter, and the interval of insert block' surface from the waterway is between 12.4 mm and 19.29 mm (see Fig. 7), the interval is not uniform. due to the inconsistent flow rate and flow volume of cooling water in the waterway of die, as well as the difference of vertical heat conduction distance between water path and the punch surface is existing, the surface temperature on the hot stamping part is uneven, so there is a difference in the forming performance and surface hardness of the steel sheet, and the stress concentration is formed in the cross section of which have the performance difference, fracture failure occurs when the stress level is greater than the tensile strength of the steel sheet. To verify the correctness of the analysis's opinion, the water channel in the punch' press block is re-designed, increasing the number of water channels to 6 be evenly distributed, total the water channel diameter be designed into 10 mm (as shown in Fig. 9), and ensured the distance between outer surface of the cooling water channel and the inserts block in punch is same. After replacing the new design, the cracking failure of the parts was significantly improved.

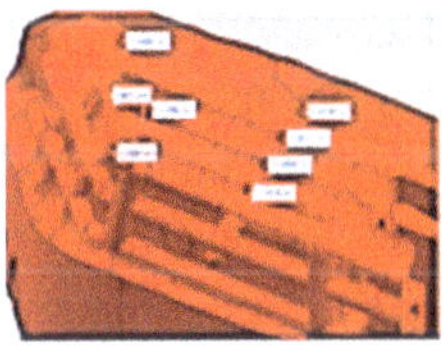

Fig. 7. Measuring distance of die's surface from waterway.

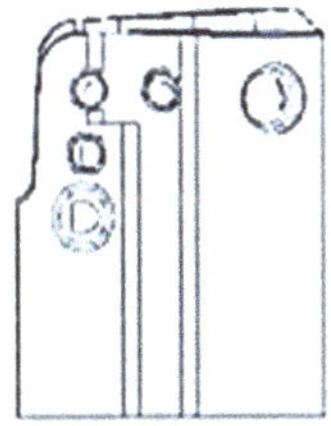

Fig. 8. Original design of cool water channel.

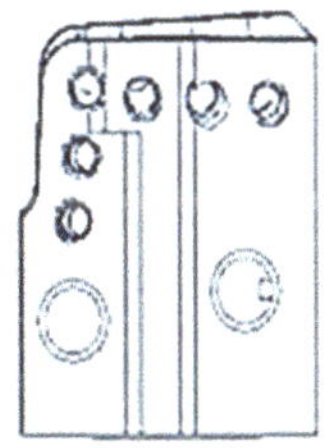

Fig. 9. Water channel.

3.2. *The effect on radius of the arc at the creaking position*

After inspection, the actual R-radius on the inserts block is 12.5 mm, which is less than the requirement of 13.5 mm for processing drawings. The small R corner radius forms a damping to the fluidity of the sheet in the die, and creates a stress concentration, as well as crack when the stress is above tensile strength of the steel sheet. The R corner radii of the four press block in punch were measured one by one, which were: 12.5 mm, 14 mm, 13.5 mm, and 14 mm. see Fig. 10-11. Finally, the radius of the R corner was uniformly adjusted to 14 mm, and no cracking was occurred by the verification of three batches.

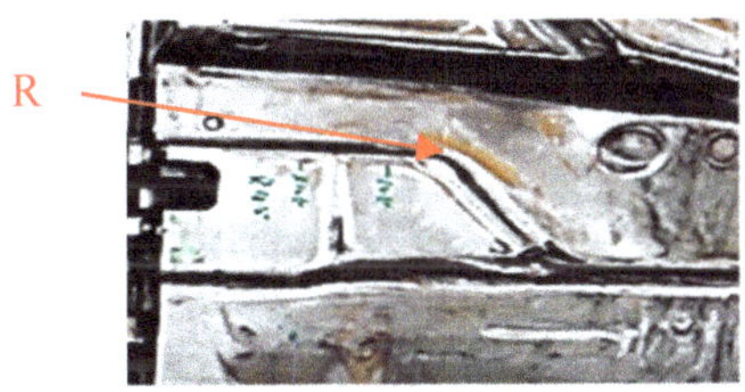

Fig. 10. R corner radii were 12.5mm In right press block in punch.

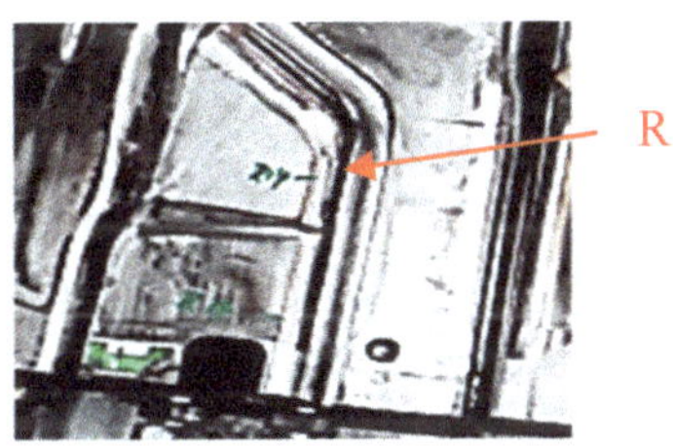

Fig. 11. R corner radii were 14mm In left press block in punch.

3.3. *Effect of austenitic grain coarsening*

When the steel sheet is heated by a long time in 900-950 °C temperature, it will causes the austenitic structure to be thick, and result in the coarse and uneven martensitic structure of the stamped part. It increases the brittleness of hot stamping part and uneven strain of the steel sheet also causes cracking.

By measuring metallographic structure, grain coarsening and mixed crystals are presence at cracking area in failure part. By strengthening the process control and strict process discipline, the problem of martensitic coarsening get hold of controlling.

4. Conclusion

In response to the increase in car ownership, increasing environmental pressure, and increasingly stringent safety regulations, automotive lightweight is an inevitable trend in the development of future automobiles. An important way to achieve vehicle safety and lightweight is to increase the proportion of super high strength steels in white-body. In this large situation, the application of hot stamping steel in the car body has been increasing and popularizing year by year. In order to ensure the stability of the production quality, full attention must be paid to the design of the production process, the design of the stamping mold, and the refinement of the processing. Strengthening the reliability and stability of the process parameters such as heating temperature, transfer time, quenching cooling rate is the guarantee of the stability of hot stamping part's quality. In die's design and processing, the cooling uniformity of the die surface must be attended by sufficiently.

During stamping mold design and construction, the heat conducting distance and the cooling capacity for cooling water channel should be strictly controlled. The arc radius of stamping die should be increased as much as possible to ensure that the fluidity of steel sheet in the die can compensate for the lack of plasticity, which is also the key to ensure the stability of formability.

References

1. X. Zhu, Investigation on hot stamping process of 22MnB5 high strength born steel, Master thesis, Shenyang University of Technology (2014).
2. M. M. Lu, Z. W. Gu, H. Xu and X. Lu, Optimization of hot stamping process for ultra-high Strength anti-collision beam, *J. Jilin Univ. (Department of Engineering)* **47**, 1834 (2017).
3. Y. Zhi, X. P. Li, C. Y. Peng and T. M. Jiang, Experimental study on the quenching process of BR1500HS high strength steel, *Modern Manuf. Eng.* **1**, 102 (2015).
4. L. Li, W. Zhang, C. L. Zhang and X. Wang, Research on hot stamping process and properties of automotive B-pillar reinforcing plate, *Auto. Tech. Mater.* **10**, 14 (2016).
5. Y. S. Wang, Development of hot stamping process and research of related mold materials, *Mold Making* **17**, 1 (2017).

Experimental Investigation of Hot Stamping Used Al-Si Coated Blank Heating Process

H. Y. Xie[†], Z. J. Ma, Q. W. Mao and Y. P. Wei
Shanghai Superior Die Technology Co., Ltd., Science & Technology Commission of Shanghai Municipality, Pudong district, Shanghai, 201209, China
[†]E-mail: huanyang.xie@ssdt.com.cn

In the hot stamping process, the blank heating process is very crucial because it can decide whether the final hot stamped part is acceptable or not. In this paper, two different heating processes including heating at a constant temperature of 930 °C and step forward heating are conducted and their influences to the final formed parts are compared. The results show that the transition zone between coating and basic material under step forward heating method is thinner than deserved by the other heating method. As the total thickness including coating and basic material is given, thus the thicker the transition zone the thinner the basic material. That is to say the blank heated by the step forward method possesses even higher strength which improves the collision performance.

Keywords: Hot stamping; Heating process; Step forward heating; Transition zone.

1. Introduction

With the increasing awareness of environmental protection, energy saving and emission reduction in automobile industry have been increasingly emphasized [1, 2]. As well known, lightweight is the most important and effective way of the energy saving and emission reduction. It has been studied that if the weight reduction in a car is 100kg, the gasoline will be reduced by 0.3~0.5 L per 100 km. Meanwhile, the carbon emission can be reduced by 8~11 kg per km [3]. In recent years, a novel forming technology called hot stamping technology has been considered a very promising way to solve this problem. In this technology, the high strength steels are heated in a furnace to complete austenitizing temperature (usually 930 °C) and keep a certain seconds, then the heated blanks are transferred to the hot forming die with the cooling system quickly to form the designated shape. As the cooling system works simultaneously with the forming process, microstructure evolution inside the formed parts from austenite to martensite occurred. Due to high strength and high hardness of the martensite phase, the final formed parts with excellent collision and occupant protection performances are deserved. Owing to these advantages, hot forming technology has been widely applied to produce auto body reinforcement parts, such as A/B pillar, door side impact beam, front/rear bumper, etc. [4].

Hot forming process mainly consists of four steps: decoiling, blanking, heating, forming and quenching, see Fig. 1. Decoling and blanking processes belong to ordinary machining and they are not commonly studied by researchers. In addition, the forming and quenching processes have been extensively investigated in recent years [5-8]. However, the report about the blank heating process and its influence in the final formed auto parts is relatively rare. Heating is an indispensable and influential process during hot forming. It can directly decide whether the final hot stamped parts are acceptable or not. Hence, heating should be well discussed to guarantee the auto parts quality.

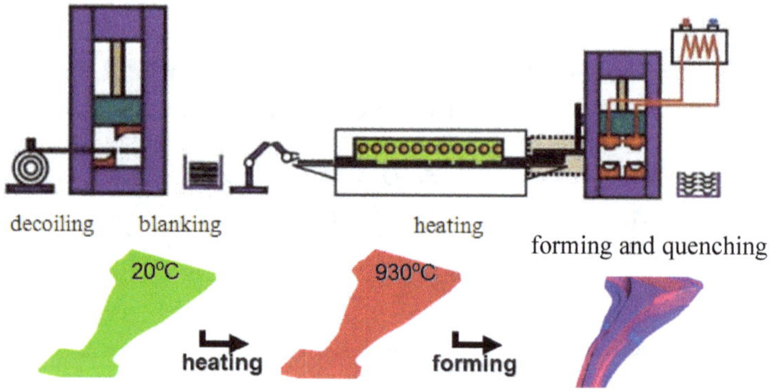

Fig. 1. Hot forming process.

Targeting to study the influence of the heating process to the final formed auto parts, this paper employs auto parts commonly used 22MnB blank as the research material. Two different heating processes including heating at a constant temperature of 930 °C and step forward heating are conducted and their influences to the final formed parts are also compared. Based on the results of the comparison, a relatively better heating method is proposed and recommended.

2. Experimental

1 mm thickness of 22MnB5 blank was used to perform the heating process. The heating equipment is Schwartz roller furnace. There are 19 heating zones (HZ) inside the furnace. Each HZ has a certain length and can be set one temperature, thus the whole furnace can be set 19 different temperatures theoretically. The HZ distribution and the blank moving direction are shown in Fig. 2.

Blank heating process can be described through the relationship between blank temperature and heating time (heating curve). One of the heating methods is heating at a constant temperature of 930 °C. In this heating process, all HZ are set 930 °C. Another heating method is step forward heating method. During this method, the temperature set was shown in Fig. 3.

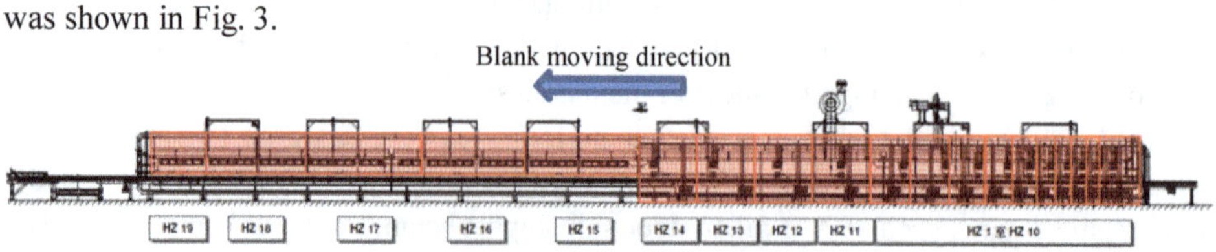

Fig. 2. Schwartz roller heating furnace.

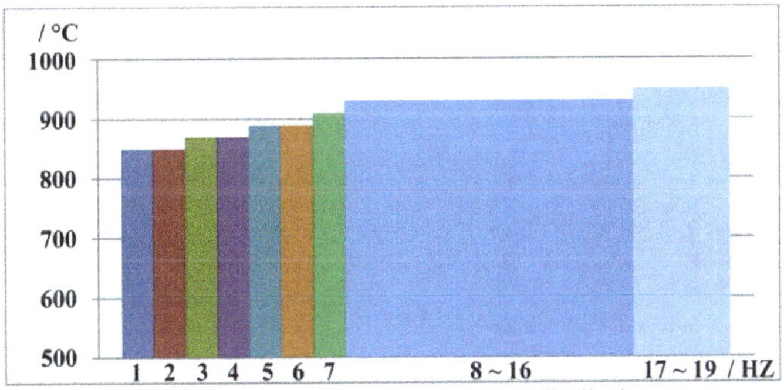

Fig. 3. Step forward heating temperature distribution.

3. Results and Analysis

Heating curves under two heating methods are shown in Fig. 4.

The heated blanks were hot stamped and the results of final parts are compared via metallographic, hardness, tensile performance, transition zone thickness on the basic blank analysis. Furthermore, heating used electricity consumption was easily compared.

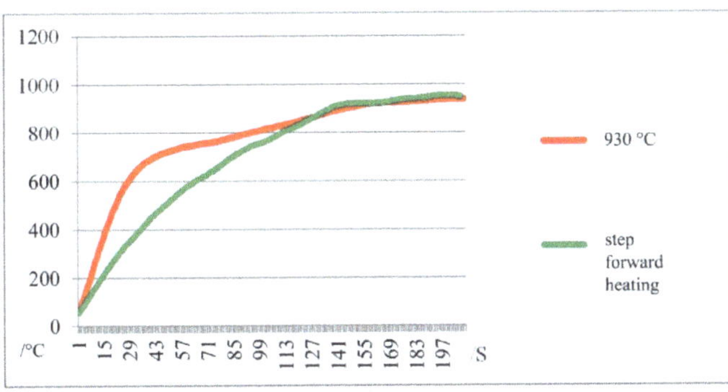

Fig. 4. Heating curve of 1mm thickness 22MnB5 blank.

3.1. *Metallographic analysis*

The samples were cut from the same position of two formed parts under 930 °C and step forward heating respectively. Fig. 5 shows microstructure from thickness direction under two different heating methods. It can be seen from Fig. 5 that microstructure in both samples is Martensite phase. Martensite phase uniformly distributes inside the formed parts. Coating thickness is 30 μm in both samples.

(a) (b)

Fig. 5. Microstructure: (a) 930 °C heating; (b) step forward heating.

3.2. *Hardness*

Four samples were selected to perform the Vickers hardness test. The indentation was conducted in the thickness surface. The results were shown in Fig. 6.

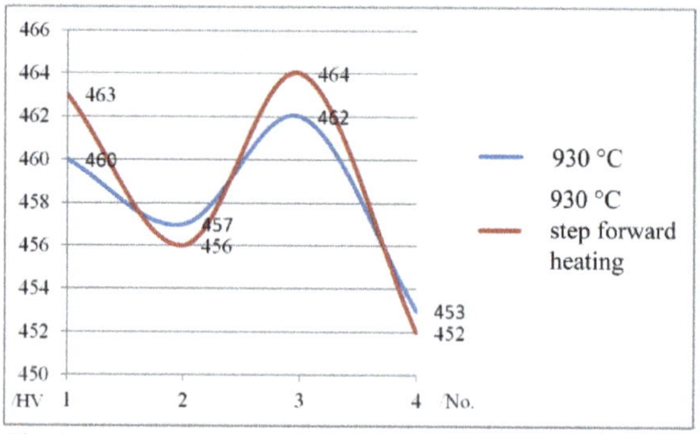

Fig. 6. The hardness of samples under two heating circumstances.

It is shown from Fig. 6 that hardness is very close under two heating circumstances. This is because these two heating methods seldom change the dislocation density inside the formed parts.

3.3. *Tensile performance*

Tensile samples were cut into the shape (see Fig. 7) from the formed parts. Tensile tests were carried out on the universal tensile testing machine.

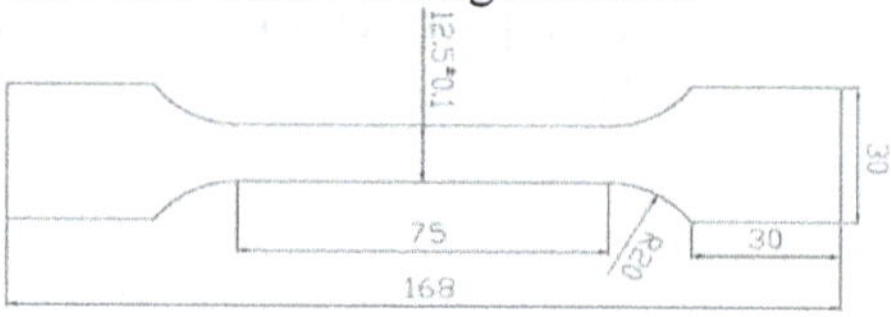

Fig. 7. Tensile sample.

Tensile strength, yield strength and fracture elongation rate were concerned. The results were listed in Table 1.

Table 1. Tensile results.

	tensile strength/MPa	yield strength/MPa	fracture elongation rate/%
930 °C	1539	1112	5
step forward heating	1554	1143	6

It can be seen from Table 1 that all the three indicators are higher when heating under step forward method. It is inferred that under 930 °C heating circumstance, the blank absorbs more heat, it may result in more aluminum atoms penetrate into the basic blank which leads to lower tensile performance.

3.4. *Transition zone thickness*

Due to different heating process, transition zone thickness on the basic blank was different. Fig. 8 shows the transition zone thickness as a results of 930 °C (Fig. 8(a)) and step forward heating (Fig. 8(b)).

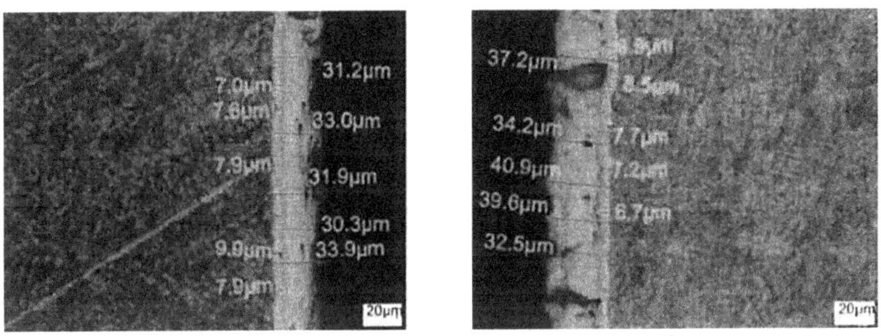

(a) (b)

Fig. 8. Transition zone thickness: (a) 8.1 μm; (b) 7.4 μm.

It is observed from Fig. 8 that the average transition zone thickness is 8.1 μm and 7.4μm respectively. Due to the diffusion of the transition zone and the aluminum atoms penetrate into the basic blank, the actual basic blank becomes thinner compared with original blank before heating. As the total thickness including coating and basic material is given, thus the thicker of the transition zone the thinner of the basic material. Thus, the thickness of the actual basic blank when heated under step forward heating method is thicker than under 930 °C heating method. That is to say the blank heated by the step forward method possesses even higher strength which improves the collision performance.

3.5. *Electricity consumption*

As can be seen in Fig. 3 that in HZ1~7, temperature in furnace is lower than 930 °C, only three HZ （17~19） the temperature is 950 °C. The electricity consumption is obviously lower than 930 °C heating circumstance.

4. Conclusion

This paper focuses on the influences of different heating process on the final formed parts, four conclusions are drawn as follows:

(1) Microstructure and hardness are almost the same under 930 °C and step forward heating method;
(2) Compared with 930 °C heating, the formed parts under step forward heating method possess higher mechanical property;
(3) Compared with 930 °C heating, the transition zone under step forward heating method is thinner which improves the collision performance;
(4) Compared with 930 °C heating, the step forward heating method consumes lower electricity power.

References

1. M. T. Ma and H. L. Yi, Application of high strength steel to manufacturing auto, *Heat Treatment* **26**, 9 (2011).
2. M. T. Ma, H. L. Yi, H. Z. Lu and X. M. Wan, On the lightweighting of automobile, *Engineering Sciences* **11**, 20 (2009).
3. F. B. Yao, Study on Microstructures and Properties of Resistance Spot-Welded Joints of Galvanized Steel Sheets with Trademark of CR590T/340Y DP, PhD thesis, Jilin University (2011).
4. Y. H. Guo, M. T. Ma, Y. S. Zhang, L. F. Song G. Fang and J. W. Jin, Numerical simulation and experiment of hot stamping for front bumper of automobile, *Forging & Stamping Technology* **38**, 46 (2013).
5. Z. W. Gu, C. Jiang, Z. D. Shan and H. Xu, Hot forming process for stamping parts of ultra high strength steels, *Automobile Technology & Material* **4**, 15 (2009).
6. N. Ma, Reaearch on boron steel for hot forming and its application, *J. Mech. Eng.* **46**, 68 (2010).
7. W. L. Xu, J. Ai, A. H. Luo and S. R. Guan, Introduction of sheet metal hot-forming, *J. Plasticity Eng.* **16**, 39 (2009).
8. A. Turetta, S. Bruschi and A. Ghiotti, Investigation of 22MnB5 formability in hot stamping operations, *J. Mater. Process.* Tech. **177**, 396 (2006).

Current Research and Challenges in Innovative Technology of Joining Dissimilar Materials for Electric Vehicles

H. Li† and X. S. Liu
College of Materials Science & Engineering, Beijing University of Technology, Chaoyang District, Beijing, 100124, China
†E-mail: hongli@bjut.edu.cn

Y. S. Zhang
Huazhong University of Technology, Wuhan, 430074, China
E-mail: zhangys@mail.hust.edu.cn

M. T. Ma
China Automotive Engineering Research Institute, Chongqing, 401122, China
E-mail: mingtuma@126.com

G. Y. Li
Central Iron & Steel Research Institute, Beijing, 100081, China
E-mail: cisrilgy@163.com

J. Senkara
Faculty of Production Engineering, Warsaw University of Technology, Warsaw, 00-661, Poland
E-mail: jacek.senkara@gmail.com

The emerging trend towards lightweight, high performance and emissions reduction is leading to the increasing use of multi-material hybrid structures in electric vehicles. Advanced high-strength steel, lightweight alloy, such as aluminum, titanium and magnesium, as well as plastics and carbon fiber reinforced polymers (CFRP) etc. are being used extensively in the design of complex structures, which is a challenging task using joining technologies due to the different physical and chemical properties of dissimilar materials. The paper reviews the advanced and alternative joining methods for CFRP to aluminum/steel, aluminum to steel and battery pack manufacturing, ranging from conventional adhesive bonding, weld-bonding and self-piercing riveting to new welding technologies, such as friction stir spot welding, ultrasonic spot welding, laser braze-welding and cold metal transfer welding. The influencing factors, process optimization, joining mechanism joints properties and numerical simulation are summarized and discussed. These methods have a very high potential in manufacturing of hybrid structures, and further research is required to completely exploit these methods in practical applications.

Keywords: Electric vehicle; AHSS; Aluminum; CFRP; Battery pack; Dissimilar material joining.

1. Introduction

Electric vehicles (EVs) have the advantages of reducing emissions and energy consumption as a function of renewable energy, which produce evident economic, environmental and social benefits. Due to the strict decarbonization policies associated with the demand for high price-performance ratio, EVs are proliferating globally at a rapid pace and it will become more affordable as the cost of batteries declines [1]. As shown in Figure 1, it is estimated there will be 140 million EVs on the road by 2035 (not including hybrids) or 8 % of the total fleet of 1.8 billion cars. This compares to a little over a 1.2 million EVs today and points to an increase of 11,500% in twenty years. At the same time, EV demand in China is to rise faster and further than previous projections. In 2017, over 579,000 EVs including battery electric (BEV) and plug-in hybrid (PHEV) were sold. It is targeting 3 million new energy vehicles sales in *Made in China 2025* released by the Ministry of Industry and Information Technology, which accounts for more than 80% in the domestic market [2].

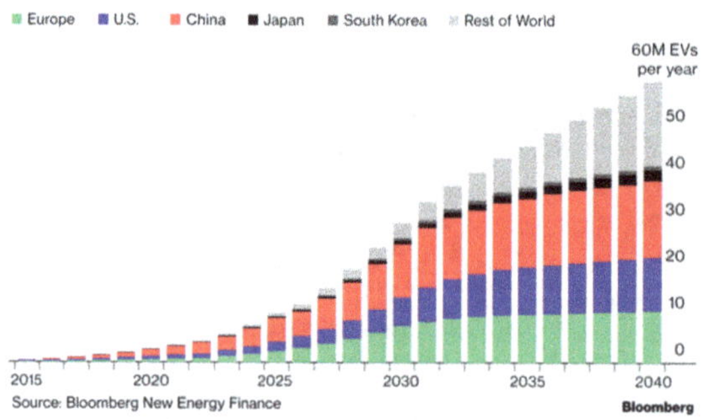

Fig. 1. Global Electric-car revolution set to take off.

The growth of the EVs market will have urgent requirement and positive consequences for lightweight materials. It is a common knowledge that a ten-percent reduction in overall vehicle weight can result in a 6 to 8 percent fuel economy benefit. The payoff is particularly beneficial in EVs including hybrid electric, plug-in hybrids, and battery electric vehicles. The first two categories inherently add weight and complexity in their dual propulsion. Pure electric vehicles are challenged by battery technology, cost, and range limitations, which is difficult to make a breakthrough in a short period.

Replacing traditional steel components with lightweight materials can directly reduce the weight of a vehicle's body and chassis by up to 50 percent, and therefore reduce a vehicle's fuel consumption. The leading lightweight material candidate for vehicle body include advanced high-strength steel (AHSS), aluminium, magnesium alloys, carbon fibers, plastic and polymer composites. Advancement in manufacturing technologies such of hot stamping, resin transfer molding, 3D printing, etc. are making it possible to produce lightweight parts in high volume. Actually innovative combinations of these materials have been used in new generations of automobiles. Figure 2 shows that the Audi A8 L features a unibody of varied materials, including steels, aluminium, polymers and magnesium etc., with the weight has been decreased by about 15 % and the torsion stiffness has been increased by 40 % [3].

The use of these advanced materials creates the need to develop reliable and cost-effective solutions for joining hybrid material parts and assemblies. However, each material has its unique attributes, while many are subject to degraded performance when exposed to high temperature joining processes. In order to ensure weight reduction without compromising any safety and performance of EV, the selection of a suitable joining method between typical dissimilar materials such as aluminium alloy steel and aluminium alloy/carbon fiber reinforced polymers (CFRP) is therefore important in structural application. Additionally, as a key component of electric vehicles, the battery pack for use in EVs consist of a large number of individual battery cells thus there is an increasing requirement for joining dissimilar materials.

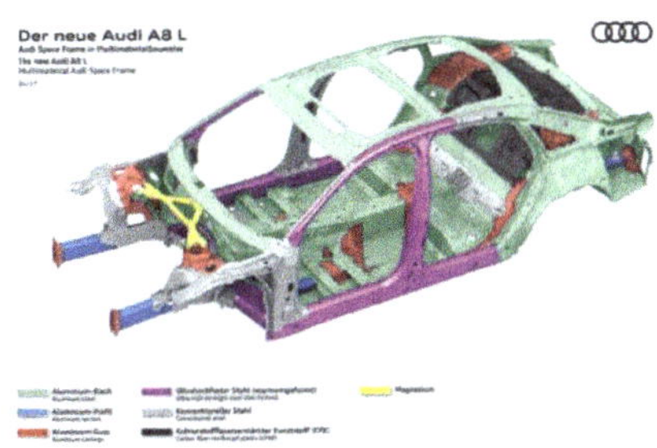

Fig. 2. Multiple materials used in the Audi A8 L. Tech Briefs, March 2018.

Much research work in the field of dissimilar lightweight materials joining is available in literatures with numerous results and applications. This paper reviews some of the popular methods for multiple material joining and identifies challenges and opportunities. Promising technologies for dissimilar materials joining, such as aluminium alloy/CFRP and aluminium alloy/high strength steel as well as joining of aluminium alloy battery housing cover for hermetic sealing are highlighted.

2. Multi-material Joining Technologies

2.1. *Joining of CFRP to aluminum or AHSS*

2.1.1. *Adhesive bonding*

Adhesive bonding is the process of binding two components using a suitable binder (*i.e.* an adhesive). Joining of CFRPs with aluminium alloys or high strength steel via adhesive bonding is the most conventional method as it could not only seal the joints but also prevent crevice and galvanic corrosion between dissimilar materials. It could offer lightweight structures with less stress concentration without bolt holes. Currently, the research is mainly focused on the adhesives type selection, joints design, surface pre-treatment, joining process optimization, joining mechanism, joints strength, failure mechanism and fatigue life evaluation based on FEM simulation method etc.

Most of research show that the change of temperature affects the mechanical properties of the adhesive epoxy, and thermal stresses are induced in the joints due to the big difference in the coefficients of thermal expansion between CFRP and Aluminium or steel. Zhang *et al.* [4] noted that in Al/CFRP single-lap adhesive joints, residual stress concentrates on adherents, and residual stress of glue coat is smaller than that in adherents, thus leading to the occurrence of joint failure in the glued layer. GH Abed [5] found that adhesives showed significant reduction in the strength and the stiffness coupled with nonlinear behavior as the temperature reaches the adhesive glass transition temperature (Tg). Based on experimental and numerical simulation results, it is possible to avoid CFRP/steel plate debonding at extreme temperature by using thinner and longer CFRP plates with high modulus. A. Pramanik *et al.* [6] analyzed the joining mechanism of CFRP/aluminum alloy bonding and deduced that adsorption theory based on diffusion and mechanical keying or interlocking contribute to the strength of adhesive joints.

Furthermore, hybrid joins by mechanical fastening or spot welding in combination with adhesive bonding can effectively improve the fatigue strength, torsional stiffness and crashworthiness of lightweight components and body structure [7]. Hwe Kweon *et al.* found that hybrid joining improves CFRP/7075-T62 aluminum alloy joint strength when the mechanical fastening is stronger than the bonding when the paste type adhesive is used [8]. On the contrary, the bolt joining contributes little to the hybrid joint strength when the strength of the bolted joint is lower than that of the bonded one when using film type adhesive. Kahlmeyer *et al.* [9] compared single-component (1C) adhesive with two-component (2C) crash-modified (CM) adhesives for joining CFRP/aluminum alloy. It was confirmed that the use of 2C adhesives with flexible curing ability shortens the cure time through induction heating and leads to a free process design with higher joint strength. There exist three different defect modes, as shown in Figure 3.

Fig. 3. Different defects after curing at 180 °C in the cathodic dip painting furnace (left, crack starts in the CFRP; middle, crack is located in the interphase between CFRP and adhesive; right, crack occurs within the adhesive).

Reitz *et al.* [10] confirmed that pretreatment using IR- and UV-laser prior to adhesive bonding leads to a significant increase of shear strength for adhesive bonded CFRP/AlMg3 single-lap joints. And IR-laser radiation leads to surface activation and reduces weakening of the fiber-matrix interface in near surface areas while UV-laser pretreatment leads to a thermal induced damaging of the fiber-matrix adhesion. Wu *et al.* [11] studied the effect of laser surface treatment on properties of CFRP/aluminum alloy. The results show that with the increase of pulse width, the amount of residual resin on CFRP surface decreases, while the surface free energy and surface roughness increase.

The bonding strength increased by 1.95 times compared with that of untreated samples. Hui *et al.* [12] found that ultrasonic vibration can enhance the strength and stability of adhesive CFRP/aluminum alloy bonded joint by 40% and 60%, respectively. A number of extensive research have been performed to calculate the overall joint strength, crack formation and residual stress in composite joints. Yang *et al.* [13] analyzed the influence of the lap length on the stress distribution and joint strength through experiments and finite element simulation. Duan [14] *et al.* investigated the influence of substrate thickness on the peeling stress distribution of L-shaped CFRP/Al joints through FEM modelling. However, adhesives with a lower curing temperature and similar thermal expansion coefficients with adherents should be selected for desired applications.

2.1.2. *Self-piercing Riveting*

Self-piercing Riveting (SPR) is a cold mechanical joining process used to join two or more sheets of materials by driving a rivet piercing through the top sheet or the top and middle sheets and subsequently lock into the bottom sheet under the guidance of a suitable die. Compared with other conventional mechanical joining methods, SPR has many advantages including no pre-drilled holes required, no fume, no spark and low noise, no surface treatment required, ability to join multi-layer materials and mixed materials and ability to produce joints with high static and fatigue strengths. Therefore, SPR is currently the main joining method for mixed-material lightweight automotive structures, especially for joining CFRP to aluminium or AHSS. It has been used widely by several automotive companies over the past 25 years. About 3600 self-piercing rivets are used in the Jaguar XJ. The process parameters, applications of SPR, the mechanical performance of SPR joints, failure mode and fatigue life predictions, as well as FEM modelling of SPR have attracted considerable interest and research [15].

Franco *et al.* [16] found that CFRP/2024-T6 SPR joints have an average tensile strength of 4480 N after heat treatment. Additionally, hybrid SPR-adhesive joints have a higher tensile strength than adhesive bonded joints equal to 17-32 %. Zhuang *et al.* [17] developed a SPR die for improving the tensile strength, shear strength and fatigue resistance of CFRP/Al joint. Landgrebe *et al.* [18] concluded that the damages e.g. delaminations, fiber or matrix fractures in the CFRP can be avoided completely. Mandel *et al.* [19] studied the long-term corrosion behavior of CFRP/6060-T6 SPR joint. After 17 cycles the pits coalesce and form a kind of trench alongside the overlap of the joint. Regarding to the fatigue behavior of CFRP/2024-T6 SPR joint, Franco *et al.* [20] verified the trial result with the FEM simulation results and it showed that the hybrid joint has good mechanical characteristics both under static and fatigue loads. The failure modes are shown in Figure 4, with debonding of composite panels occurred due to cracking of the top substrate and net section crack propagated across the blank. Rao *et al.* [21] proved that the tensile failure load and fatigue life of the lap-shear SPR joints in CFRP/AA6111 was highly influenced by the flushness of the rivet head against the top sheet of CFRP. Yang [22] confirmed that SPR/5754 joints with glass fiber reinforced composite failed in a bearing mode and exhibited improved joint strength, while the joints with carbon fiber reinforced composite exhibited slightly decreased joint strength but significantly reduced ductility owing to the brittleness of CFRP. Zhang [23] found that the failure mode of CFRP/5052 SPR joints are the fracture of the filament and the delamination of the upper sheet material, and the larger the thickness of the upper sheet, the larger the failure load and the energy absorption value. However, Lambiase *et al.* [24] pointed that SPR often cause thermal and mechanical damage of both the fiber and the matrix in the composite.

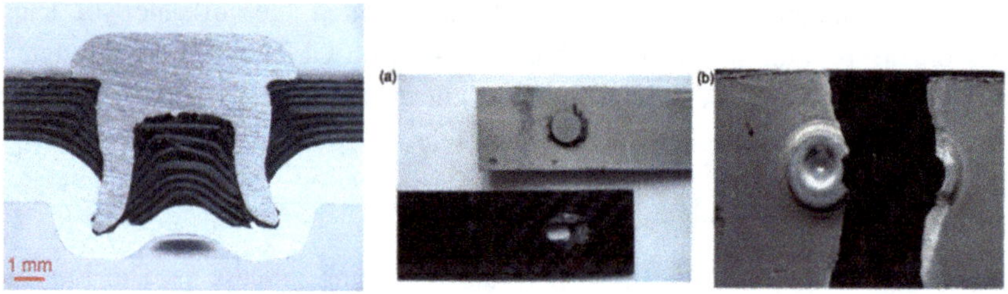

Fig. 4. The macrosection of joint created by semi-tubular self piercing rivet [25] and the fatigue failure modes: (a) debonding of composite panels and (b) net section cracking of the aluminium sheet [20].

In general, the important influential factors for SPR joint quality include materials characteristics (thickness and strength&ductility), rivet geometries and strength, die cavity depth and pip geometry, process parameters (surface pretreatment, setting force, with or without adhesive, loading direction and friction at interface etc.), resulting in the strength reliability. Based on current SPR developments, future research efforts are likely to focus on (1) SPR of new materials and mixed materials, including hot stamping steel, casting materials 7XXX Aluminium alloy and CFRP; (2) further understanding of the mechanical performance and fastening mechanism; (3) modelling of the SPR process and failure behavior; (4) developing non-destructive testing (NDT) for the defect visualization, such as computed tomographic CT measurement [26].

2.1.3. *Friction spot joining*

Friction Spot Joining (FSpJ) is an innovative friction-based joining process for metal-polymer or composite hybrid structures developed at Helmholtz-Zentrum Geesthacht, Germany [27]. The technique uses a three-piece non-consumable tool, comprising a pin, a sleeve and a clamping ring [28], see Figure 5. The pin and sleeve can move and rotate independently of each other while the clamping ring is the stationary part.

The feasibility of joining CFRP to Aluminium/Magnesium through FSpJ has been proved by a lot of works on process optimization including surface pretreatment, tool rotational speed, joining time, plunge depth and the joining force, temperature evolution on the surface and interface, as well as qualitative and quantitative analysis of joint microstructure, mechanical strength, fracture mode and fatigue properties. Goushegir *et al.* [29] measured the lap shear tensile strength of CFRP/AA2024 FSpJ joints, which is as high as 27 MPa. While it increased to 43 MPa when using Al clad specimens T3. The temperature distribution was measured during the joining process, and it was proved that the increase in the rotational speed of the FSpJ tool results in an increase of the peak temperature of 40-50 °C. The adhesion zone (AZ) is believed to be the weakest part of the joint area with the radial cracks. The plastically deformed zone (PDZ) is the strongest region and a thin transition zone (TZ) also exists.

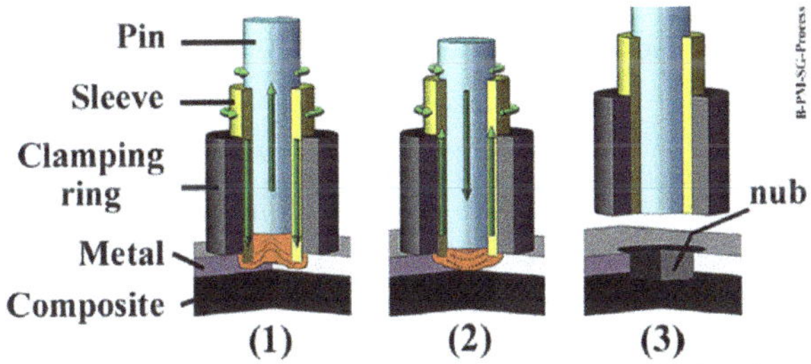

Fig. 5. Schematic description of the FSpJ process without the addition of film interlayer. (1) Sleeve plunging and plasticizing of the metal; (2) spot refilling and (3) joint consolidation [28].

Esteves *et al.* [30] proved that tool rotational speed was the main parameter which influencing the joint shear resistance. Higher heat input led to larger consolidated polymeric layer at the metal-composite interface, while higher joining force led to better wetting of the interface, increasing adhesive forces and joint mechanical performance. André *et al.* [31] successfully demonstrated the feasibility of FSpJ joining CFRP/2024-T3 with additional PPS film interlayer with the aim of improving the mechanical performance of the hybrid joints and exempting the required curing time of an adhesive. The result showed that the ultimate lap shear force of the joints were 20-55 % higher than the joints without interlayer due to the larger bonding area, better load distribution and improved micro-mechanical interlocking. Figure 6 showed PDZ existed in the central area of the joint where the molten film interlayer was squeezed out of. The joints showed a mixed adhesive-cohesive failure mode, however, cohesive failure is the dominative mode. Ogawa *et al.* [32] illustrated the surface treatments effectively attained the CFRP/AA5182 FSSW joints with improved static and fatigue strengths. Huang *et al.* [33] demonstrated the feasibility and the potentiality of acquiring a defect-free CFRP/2060-T8 FSpJ joint by using tapered thread pin with the triple facets through co-controlling shape and performance. Peng *et al.* [34] investigated the feasibility of FSpJ joining of ABS to Aluminium 6082 and confirmed that the joint underwent three fracture stages: crack initiation, stable growth and rapid expansion.

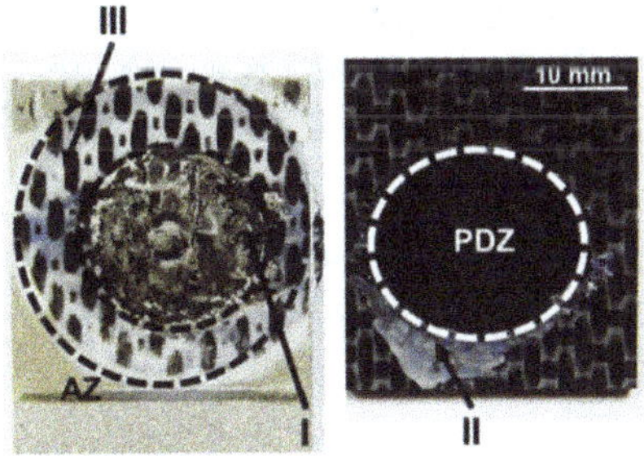

Fig. 6. Fracture surface of a spot joint indicating the bonding zones and the regions analyzed by SEM (L: 1900 rpm, 0.8 mm, 4 s, 0.2 MPa) [31].

With the aid of scanning electron microscopy, microanalysis and detailed microscopy are being applied for the determination of the polymer filling characteristics and the joints performance. Moreover, more detailed design of tool and experiment parameters combined with process computational modelling should be applied to better understanding the failure mechanism at the interface.

2.1.4. *Ultrasonic spot welding*

Ultrasonic spot welding is a solid-state joining technology that uses ultrasonic high-frequency vibration energy and standard air pressure to induce vibration on the lap surface of the weldment and thus achieve bond due to plastic deformation and softening of the material contact points [35, 36]. Ultrasonic spot welding has advantages including short welding time, high temperature upon local heating, low energy consumption, and is easy to automate.

Balle *et al.* [37] have demonstrated the good properties of similar materials e.g. Aluminium and copper alloy using ultrasonic spot welding. Moreover, the next-step is to join dissimilar materials such as non-metals CFRP, glass, ceramic to metals. Wagner *et al.* [38] characterized the bonding mechanism of ultrasonic spot welded CFRP/AA2024 and achieve a tensile shear strength up to 58 MPa using precipitation hardening Al alloy. Lionetto *et al.* [39] performed ultrasonic spot welding of CFRP through a novel low-melting thermoplastic coupling layer. Then further findings demonstrated that the CFRP/epoxy-PA6-AA5754 joints welded at 280 N presented an average tensile shear strength of 34.8 ± 3.9 MPa with an ultrasonic energy of 2300 J [40]. Welding pressure has the greatest influence on joint strength and is less dependent on ultrasonic energy. Figure 7 shows the application of ultrasonic spot welding process for joining of CFRP to Aluminium.

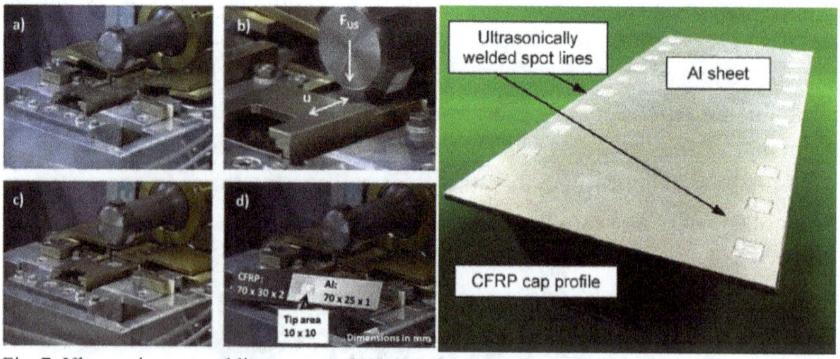

Fig. 7. Ultrasonic spot welding process [40] and ultrasonically welded AA5754/CF-PA66 [38].

Ultrasonic spot welding has great advantages in joining CFRP to metal since it does not cause damage to fiber structure and avoids the defects caused by fusion welding. Besides process optimization, further studies will need to be done to investigate the role of the thermoplastic interlayer in promoting the adhesion of the hybrid joint and the possible galvanic corrosion issues deriving from a direct contact between Al and CFRP at the welding interface.

2.2. *Joining of aluminum to AHSS*

In order to meet the requirements of lightweight and safety, the combination of aluminum alloy and high-strength steel are widely used in the hybrid auto body. However, the physical properties of aluminum alloy and high-strength steel are significantly different. Furthermore, the solid solubility of Fe in Al is extremely low, which brings great difficulty to the welding of dissimilar aluminum alloy/high-strength steel. Usually, shrinkage holes are likely to occur in welded joints. Defects such as cracks form hard and brittle intermetallic compounds in the weld, which greatly reduces the mechanical properties of the joint [41]. Currently, modified laser hybrid welding and friction stir spot welding (FSSW) can solve the above problems to some extent.

Regarding laser hybrid welding, Zhang *et al.* [42] adapted keyhole-mode laser braze welding of galvanized steel sheets and aluminum alloys for vehicles. The results showed that a penetrated weld was formed on the side of the aluminum alloy while a brazed joint was formed on one side of the galvanized steel. There exists a uniform interfacial reaction layer with a thickness of 1.5 to 13 μm. The tensile strength of the joint reached 162 MPa, and the fracture occurred in the heat-affected zone on the aluminum alloy side. Yi *et al.* [43] demonstrated that it is difficult to obtain a reliable joint if using direct laser braze-welding of high-strength boron steel and aluminum alloy. However, a tensile shear load of 3390N, which is close to the strength of the aluminum base material can be obtained after the blasting treatment of boron steel. Zhang *et al.* [44] analyzed the distribution of temperature field of laser braze welded B1500 HS high strength steel and 6016 aluminum alloy by finite element simulation. The results show that the surface temperature field is elliptical and unevenly distributed. The longitudinal section temperature field distribution is scooped. The molten pool depth becomes shallower while the temperature at the front of the laser spot is lower. In contrast, the molten pool depth becomes deeper while the temperature at the front of the laser spot is higher.

Subsequently, the research on friction stir spot welding of aluminum to steel is extensive, involving bonding mechanism, process optimization, joint strength measurement and IMC characterization, as well as fatigue and failure analysis etc. Process parameters are one of the key factors dominating the FSSW joint properties. Joaquin [45] experimented with a 0.65-1.5 mm stirring pin, and found that as the length of the pin decreases and the plunge depth increases while the maximum load increases. The fracture feature of the joints changes from interface fracture to mixed-mode fracture. Chen [46] studied the influence of the stirring speed and welding time on the joint strength. The results show that the welding time has a greater influence on the joint strength than the stirring speed. Piccini *et al.* [47] proved that the tapered stirring pin with a length of 0.3mm and a diameter of 9.6mm achieved the best effect for AA5052-LCS (low carbon steel) FSSW joint. Lyu *et al.* [48] demonstrate that the fracture load increases with the increasing of plunge force, and double-sided FSSW of AA5052/DC05 steel can sustain higher fracture load than single-sided FSSW regardless of the plunge force, which can be attributed to the formation of IMC layer. Two rotational directions result in similar fracture load of the welded part if the plunge force is less than 1.16 kN.

The schematic of FSSW system and the typical plug fracture, mixed fracture, and interfacial fracture are shown in Figure 8. Dong *et al.* [49] performed refilled FSSW for 200 ST aluminum alloy/ST06 Z galvanized steel sheets, the maximum tensile/shear fracture load reaches 4500 N and the maximum cross-tension fracture load is 359 N.

In particular, intermetallic compounds (IMC) identified to be Fe_xAl_y are observed to form in the joint when metallurgical joining of dissimilar aluminum to steel. However, IMC can be a double-edged sword. Thin IMC layer are proved to be beneficial for joint strength, which corresponds to the experimental results of Lan *et al.* [50], who demonstrate that IMC layer less than 1 μm formed in 6061-T6/TRIP 780 joint improved the welding joint intensity. But large amount of brittle IMC layer with a great thickness might aggravate crack initiation and significantly deteriorate joint. Lyu [48] found that when the thickness of the IMC is less than 7 μm, the strength of the double-sided FSSW joint increase. Ramachandran [51] identified the IMCs in AA5052 aluminum alloy/HSLA steel FSSW joint as FeAl, $FeAl_2$, Fe_xAl_3, which strengthened the stir zone (SZ) at the Al alloy side and alleviated the problem of softening in the stir zone of 5XXX series aluminum alloys.

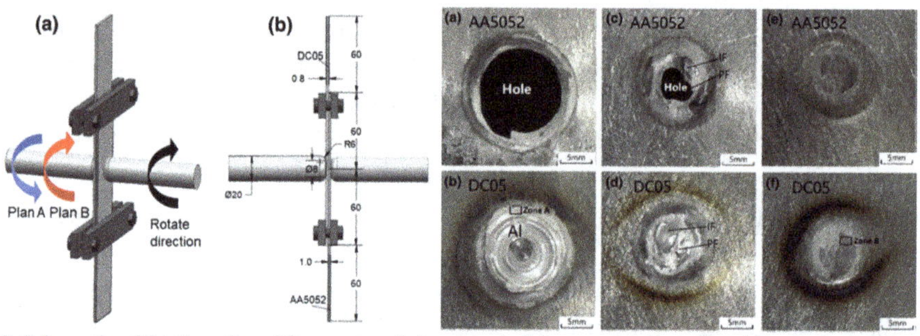

Fig. 8. Schematic of friction stir welding system *left* and typical fracture mode modes: (a) and (b) plug fracture; (c) and (d) mixed fracture; (e) and (f) interfacial fracture *right* [48].

Morever, the corrosion resistance of the Al/steel joint is also very important. Zhang [52] investigated the corrosion behavior of keyhole-free FSSW joints of 6082 aluminum alloy and DP600 galvanized steel in 3.5 % NaCl solution. The sequence of the corrosion resistance was WNZ (Weld nugget zone) > TMAZ (Thermo-mechanically affected zone) > BM (Base metal) > HAZ (Heat-affected zone). Ibrahim [53] demonstrated that the tensile shear strength of aluminum/galvanized steel weld was lower than that of the aluminum/ non-galvanized steel joint. However, the stress of the edge of the nugget is relaxed due to the resolidified Zn around the weld nugget, so that the aluminum/galvanized steel weld exhibits better fatigue performance.

At present, the research on the defects of the FSSW joint is still lacking, especially the unique “hooks” phenomenon corresponding to the places in the Al/steel interface, and the formation mechanism has not yet been accurately revealed. Limited by the automobile production line, FSSW has not really been applied for mass production, but it will definitely become one of the main joining methods for Al/steel hybrid structure in EVs.

Besides, cold metal transfer (CMT) joining of AA6061T6 to Boron steels with various coating are also investigated [54]. It was found that molten zinc from a galvanized coating enhanced the molten aluminum filler metal to the wet boron steel substrate. The FeZn coating from galvannealed boron steel was not fully molten, leading to poor joint quality. Furthermore, sound CMT weldbraze AA6061T6 to AlSi coated boron steel and AA6061T6 to bare boron steel could not be properly produced due to large formation of brittle $FeAl_3$ intermetallic during the welding process. Silvayeh et al. [55] investigated the influence of filler alloy composition and process parameters on the intermetallic layer thickness in single-sided CMT welding of 6014 T4 aluminum and galvanized dual-phase steel HCT 450 X + ZE 75/75steel blanks, see Figure 9. The results reveal that increasing the heat input and decreasing the cooling intensity tend to increase the IMC layer thickness. The silicon content of the filler alloy has the strongest influence on the IMC layer thickness, whereas the magnesium and scandium contents of the filler alloy influence the cracking tendency. In general, CMT process is especially suitable for welding thin aluminium alloy and high-strength steel sheet plates. The development of new welding wire mainly aims to find a suitable combination of alloy elements to avoid hot cracking of weld. The appropriate measures should also to taken to avoid electrochemical corrosion.

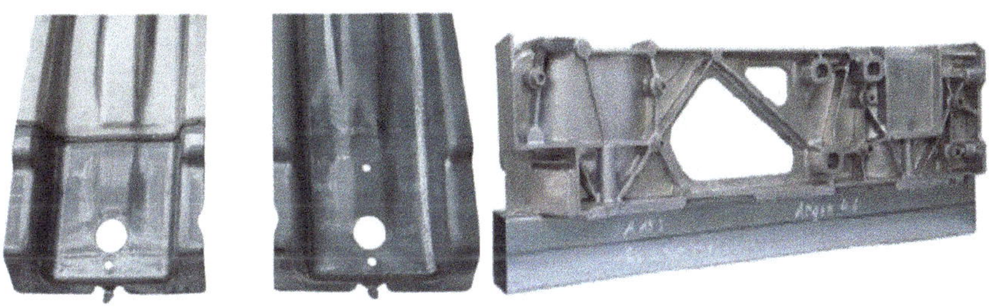

Fig. 9. Formed steel-aluminium blank *left* and steel-aluminium production-line with the CMT welding process from Fronius [56].

Not in the least, mechanical joining technologies especially the clinching and SPR are also appropriate for joining dissimilar aluminium to high-strength steel. Jin *et al.* [57] optimized the SPR process for aluminum alloy/high strength steel through DOE method. He *et al.* [58] summarized the finite element analysis of SPR technology. Considering that SPR technology is facing problems like weak interlocking and insufficient penetration problems due to the reduced formability of AHSS, Lou *et al.* [59] developed a new process named electroplastic (EP) SPR to reduce the deformation resistance of AHSS DP780 by applying a direct current during the riveting procedure. Compare with the traditional SPR joint, the tensile strength of the EP-SPR DP780/AA6061-T6 joint increased by 12.5-23.3 % when the Al plate is above. While when the DP780 is the upper plate, the tensile shear strength of the joint is decreased by 5.8 %, but the penetration risk of bottom Al plate can be reduced, and present a better energy absorption capability. In 2014, they [60] furtherly proved that the robustness and strength of the SPR joint can be improved by a hybrid joining method named rivet-welding (RW). The results showed that the electric current could not only improve the microstructure, but also increase the micro-hardness of the steel rivet and bottom DP590, and soften the AA6061-T6 around the rivet leg. Under long heating time, the IMC could be formed at the interface of trapped AA6061-T6 and bottom DP590. Huang *et al.* [61] studied the fatigue behavior of SPR joints of aluminum alloy AA6111-T4 and HSLA340 steel. It reveals that the mating sheet stiffness and load ratio have significant impact on the fatigue strength, and the dominant failure mode under tensile shear loading is eyebrow crack with an approximately semi-elliptical crack front. Chung and Kim [62] found that the fatigue endurance of a SPR joint with 5052 aluminum alloy and SPCC steel was governed by the strength of the lower sheet. Thus, it is desirable to use the stronger metal sheet as the lower sheet with regard to fatigue performance. Kang *et al.* [63] studied the corrosion performance of DP590/5052 aluminum alloy and DP590/7075 aluminum alloy SPR joints in alternate immersion test. Both groups have a mass loss before and after the corrosion test. As the corrosion time increases, the mass loss increases gradually, and the corrosion rate gradually decreases. The corrosion damage of the overlapping area is higher than that of other area, even the riveting area at the bottom of the plate corrodes into holes. Recently, ArceloMittal Gobel developed a new thermal integrated clinching process [64] for Next Gen. AHSS like GA 780, Usibor®1500 and Usibor®1500, aiming to avoid the coarse/brittle microstructure caused by extreme temperatures. A good combination of tension shear strength and cross-tension strength was achieved compared to cold clinching. The results from these innovative technologies show promising opportunity for joining of aluminium alloy to Next Gen. AHSS as an alternative solution when resistance spot welding fails.

2.3. *Joining of battery housing/cover*

To meet the requirement of lightweight, drive batteries and their peripheral units are posing new challenges, not only regard to their electrochemical capability but also in terms of the frame, housing, plug, cable and integration in the vehicle [65].

The heart of a battery pack is the pack of individual battery cells. In addition, the energy storage unit of an electric vehicle consists of components such as the charging system, cooling system, housing/cover, control system and high-voltage connecting system, as shown in Figure 8 (a).

The role of the housing is very important as it must not only have sufficient strength to improve the crashworthiness but also have good corrosion resistance and heat dissipation. Therefore, lightweight structural materials of high-strength are ideal. For prismatic Li-ion batteries, the housing is normally made of an aluminum ally. After welding, the housing is closed and sealed. Baumeister *et al.* [66] created a 20 kWh battery pack which exhibit a 10-15 % weight reduction by using sandwich materials mad of aluminium face sheets and a core of aluminium hybrid foam for the battery housing.

Currently the main technology for hermetic sealing of housing cover are TIG welding and laser welding. Zhao *et al.* [67] reported the TIG welding conditions regarding to the weldability of 5052 aluminium alloy employed in the housing cover of a particular model of BEV. However, for the production of EVs batteries, laser welding is becoming a manufacturing norm. Cheng [68] pointed out that as the electrical parts are already in the housing, only minimal heat can be tolerated and the welding seam must be gas tight, laser welding is the fastest and most reliable welding technologies. Figure 10 (b) shows actual welding in the two scenarios with a 3kW disk laser [68]. Bai [69] studied the effect of laser power, welding speed, focal distance, laser beam attitude and laser pulse energy on the morphology, penetration, fusion width, mechanical properties and microstructure of the 1060 aluminium alloy joint.

Furthermore, as carbon fiber composites show great advantages in lightweight vehicles, many automakers are turning to the design of composite battery pack housing and covers.

Choi *et al.* [70] developed a dual laminated structure for a EVs battery pack housing assembly with a closed cross-sectional area formed by composite molding to absorb impact energy. Liu *et al.* [71] adapted the idea of "replace steel with plastic" in EVs battery box and determine the product performance parameters by using Modal analysis. Mao *et al.* [72] summarized the application of composite in the EVs, and pointed out that economy composite materials with a lower price could be applied in the unload battery housing. Lu *et al.* [73] proposed a method for manufacturing battery pack composed of CFRP.

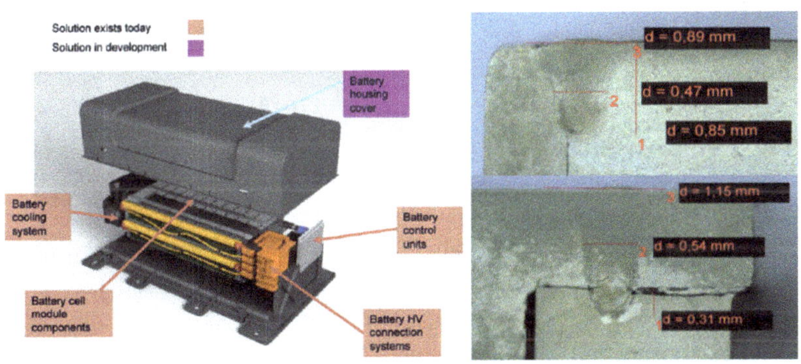

Fig. 10. (a) The typical units in a battery of an electric vehicle, *BASF* [65] (b) the two typical scenarios for hermetic sealing of the battery housing using laser welding, *butt joint (top) and overlap joint (bottom)* [68].

Regarding that the lightweight material of the battery pack relies more on aluminum alloy and carbon fiber composite materials, it can be deduced that the main issues in the joining the battery pack still relate to joint design, parameters optimization, and the reliable joint characteristics of CFRP to aluminum alloy.

3. Conclusion

The main goal of the paper is to review the latest literature relating to experimental and numerical simulation of joining technologies for lightweight materials used in electric vehicles and to provide a basis for further research and industrial application. The research and progress are described with particular reference to carbon fiber reinforced polymers (CFRP) to aluminium/steel, aluminum to steel and battery pack housing/cover hermetic sealing. Adhesive bonding, self-piercing riveting and new welding technologies, such as friction stir spot welding ultrasonic spot welding laser braze-welding and cold metal transfer (CMT) joining, etc. are introduced through summarizing the process monitoring, process optimization, joining mechanism, joints properties, fracture pattern, defects evaluation and fatigue behavior. Accurate and reliable prediction of fatigue life based on FEM simulation method is still a very difficult task. These methods have a very high potential in manufacturing of hybrid structures, but are still facing challenges such as long monitoring cycles, expensive equipment and robotic process automation. Further research is required to completely exploit these methods in practical applications.

Acknowledgments

Work partially supported by International Research Seed Funding (BJUT 2018-A12) and Youth Mentor Internationalization Capacity Development Program (BJUT 2017).

References

1. Y. Lang, Lightweight technology of body, *J. Aut. Appl. Technol.* **5**, 104 (2018).
2. Z. B. Yu, MIIT promote the concept of made in China 2025, *J. Commercial vehicles* **6**, 23 (2015).
3. D. Simanaitis, *Advanced automotive materials* (2018), https://simanaitissays.com/2018/03/23/advanced-automotive-materials.html.
4. K. Zhang, Z. Yang and Y. Li, A method for predicting the curing residual stress for CFRP/Al adhesive single-lap joints. *Int. J. Adhes. Adhes.* **46**, 7 (2013).
5. G. H. Abed, Effects of temperature on the adhesive bonding in steel beams reinforced with CFRP composites, PhD thesis, University of Southampton (2012).
6. A. Pramanik, A. K. Basak and Y. Dong, Joining of carbon fibre reinforced polymer (CFRP) composites and aluminium alloys-A review, *J. Comp. Part A* **101**, 1 (2017).
7. H. Li and Z. Li, Progress of high-performance composite joining technique for UlSAB material, *J. Aut. Technol.* **2**, 1 (2008).
8. J. H. Kweon, J. W. Jung and T. H. Kim, Failure of carbon composite-to-aluminum joints with combined mechanical fastening and adhesive bonding, *J. Compos. Struct.* **75**, 192 (2006).
9. M. Kahlmeyer and S. Böhm, Crash-modified two-component adhesives for joining CFRP and aluminium in automotive body in white, *J. Weld. World* **60**, 1 (2016).
10. V. Reitz, D. Meinhard and S. Ruck, A comparison of IR- and UV-laser pretreatment to increase the bonding strength of adhesively joined aluminum/CFRP components, *J. Composites Part A* **96**, 18 (2017).
11. Y. Wu, H. J. Kong and X. M. Ding, Effect of laser treatment on properties of CFRP/aluminum adhesive, *J. Fiber. Reinf. Plast./Compos.* **4**, 56 (2018).

12. H. Wang, H. Hao and K. Yan, Ultrasonic vibration-strengthened adhesive bonding of CFRP-to-aluminum joints, *J. Mater. Process. Technol.* **257**, 213 (2018).
13. X. L. Yang, The mechanical properties of adhesive bonded joints for lightweight automotives, Master thesis, Dalian University of Technology (2014).
14. Y. Duan, Peeling stress model and analysis of single-L carbon fiber reinforced plastic/polymer/Al composite component, *J. China J. Mech. Eng.* **48**, 44 (2012).
15. D. Li, A. Chrysanthou and I. Patel, Self-piercing riveting-a review, *Int. J. Adv. Manuf. Technol.* **92**, 1 (2017).
16. G. D. Franco, L. Fratini and A. Pasta, Analysis of the mechanical performance of hybrid (SPR/bonded) single-lap joints between CFRP panels and aluminum blanks, *J. Int. Adhes.* **41**, 24 (2013).
17. Y. M. Zhuang, Manufacturing method for carbon fiber composite plate and self-punching riveting die and method for carbon fiber composite plate and aluminum alloy plate, CN 105479771 A, 2016.
18. D. Landgrebe and R. Niegsch, Influence of process induced damages on joint strength when self-pierce riveting carbon fiber reinforced plastics with aluminum, *J. Key Eng. Mat.* **651**, 1493 (2015).
19. M. Mandel and L. Krüger, Long-term corrosion behaviour of EN AW-6060-T6 in an aluminium/carbon-fibre reinforced polymer self-piercing rivet joint, *J. Mater. Werks* **45**, 1123 (2015).
20. G. D. Franco, L. Fratini and A. Pasta, Fatigue behaviour of self-piercing riveting of aluminium blanks and carbon fiber composite panels, *J. P. I. Mech Eng. L-J Mat.* **226**, 230 (2012).
21. H. M. Rao, J. Kang and G.Huff, Impact of specimen configuration on fatigue properties of self-piercing riveted aluminum to carbon fiber reinforced polymer composite, *Int. J. Fatigue* **113**, 11 (2018).
22. J. Zhang and S. Yang, Self-piercing riveting of aluminum alloy and thermoplastic composites, *J. Compos. Mater.* **49**, 1493 (2015).
23. J. Zhang, X. C. He and W. Y. Ding, Mechanical properties of self-piercing riveted joint of carbon fiber composites and light alloy sheet, *J. Ordnance Mater. Sci. Eng.* **3**, 48 (2018).
24. F. Lambiase and D. C. Ko, Feasibility of mechanical clinching for joining aluminum AA6082-T6 and carbon fiber reinforced polymer sheets, *J. Mater. Des.* **107**, 341 (2016).
25. W. G. Drossel, R. Mauermann and R. Grützner, Numerical and experimental analysis of self-piercing riveting process with carbon fiber-reinforced plastic and aluminium sheets, *J. Key Eng. Mat.* **554**, 1045 (2013).
26. M. Jäckel, T. Grimm, R. Niegsch and W.G. Drossel, Overview of current challenges in self-pierce riveting of lightweight materials, in *Proc. MDPI*, (Brussels, Belgium 2018).
27. S. T. Amancio-Filho and J. dos Santos, European Patent 2329905B1, 2012.

28. S. T. Amancio-Filho, C. Bueno, J. F. dos Santos, N. Huber and J. E. Hage, On the feasibility of friction spot joining in magnesium/fiber-reinforced polymer composite hybrid structures, *Mater. Sci. Eng. A* **528**, 3841 (2011).
29. S. M. Goushegir, J. F. D. Santos and S. T. Amancio-Filho, Friction spot joining of aluminum AA2024/carbon-fiber reinforced poly (phenylene sulfide) composite single lap joints: microstructure and mechanical performance, *J. Mater. Des.* **54**, 196 (2014).
30. J. V. Esteves, S. M. Goushegir and J. F. D. Santos, Friction spot joining of aluminum AA6181-T4 and carbon fiber-reinforced poly (phenylene sulfide): effects of process parameters on the microstructure and mechanical strength, *J. Mater. Des.* **66**, 437 (2015).
31. N. M. André, S. M. Goushegir and J. F. D. Santos, Friction spot joining of aluminum alloy 2024-T3 and carbon-fiber-reinforced poly (phenylene sulfide) laminate with additional PPS film interlayer: microstructure, mechanical strength and failure mechanisms, *J. Comp. Part B* **94**, 197 (2016).
32. Y. Ogawa , Y. Xiong and H. Akebono, Fatigue properties of friction stir welds of treated Al to carbon fibre-reinforced plastic, *Sci. Technol. Weld. Joi.* **23**, 79 (2017).
33. Y. Huang, X. Meng and Y. Xie, Joining of carbon fiber reinforced thermoplastic and metal via friction stir welding with co-controlling shape and performance, *J. Comp. Part A* **112**, 328 (2018).
34. J. Peng, Study on friction stir welding process of ABS and aluminum alloy, Master thesis, Lanzhou University of Technology (2016).
35. C. Wen, Y. L. Li and C. Zhao, Ultrasonic spot welding of aluminum alloy and galvanized steel dissimilar metals sheets, *J. Trans. China Weld. I.* **36**, 39 (2015).
36. D. Zhao, Effect of welding parameters on tensile and fatigue properties of ultrasonic spot welded dissimilar joints of magnesium to titanium sheets, *J. China Jio. Mech. Eng.* **53**, 118 (2017).
37. F. Balle, G. Wagner and D. Eifler, Combined process of ultrasonic welding and precipitation hardening of aluminum alloy 2024/carbon fiber reinforced composites structures, in *ICAA12* (Yokohama, Japan, 2010).
38. G. Wagner, F. Balle and D. Eifler, Ultrasonic welding of aluminum alloys to fiber reinforced polymers, *J. Adv. Eng. Mater.* **15**, 792 (2013).
39. F. Lionetto, M. N. Morillas and S. Pappadà, Hybrid welding of carbon-fiber reinforced epoxy based composites, *J. Comp. Part A* **104**, 32 (2018).
40. F. Lionetto, C. Mele and P. Leo, Ultrasonic spot welding of carbon fiber reinforced epoxy composites to aluminum: mechanical and electrochemical characterization, *J. Comp. Part B* **144**, 134 (2018).
41. W. H. Zhang, Study on resistance spot welding of dissimilar materials of Aluminum alloy and high Strength Steel, PhD thesis, Jilin University (2011).
42. M. J. Zhang, G. Y. Chen and Y. Zhang, Research on microstructure and mechanical properties of laser keyhole welding-brazing of automotive galvanized steel to aluminum alloy, *J. Mater. Des.* **45**, 24 (2013).

43. Y. L. Yi, J. Chen and X. V. Amp, Study on laser welding-brazing method of aluminum alloy to Al-coated high strength 22MnB5 steel, *J. Work. Tech.* **15**, 49 (2016).
44. L. Y. Zhang, W. Dong and Y. F. Liu, Numerical simulation of thermal on laser welding and experimental study based on ANSYS, *J. Light Ind. Mach.* **35**, 45 (2017).
45. J. M. Piccini and H. G. Svoboda, Effect of pin length on Friction stir spot welding (FSSW) of dissimilar aluminum-steel joints, *J. Procedia Mater. Sci.* **9**, 504 (2015).
46. K. Chen and X. Liu, Effects of process parameters on friction stir spot welding of aluminum alloy to advanced high-strength steel, *J. Manuf. Sci.* **8**, 139(2017).
47. J. M. Piccini and H. G. Svoboda, Tool geometry optimization in friction stir spot welding of Al-steel joints, *J. Manuf. Process.* **26**, 142 (2017).
48. X. Lyu, M. Li and X. Li, Double-sided friction stir spot welding of steel and aluminum alloy sheets, *Int. J Adv. Manuf. Tech.* **96**, 2875 (2018).
49. H. Dong, S. Chen and Y. Song, Refilled friction stir spot welding of aluminum alloy to galvanized steel sheets, *J. Mater. Des.* **94**, 457 (2016).
50. S. Lan and X. Liu, Microstructural evolution during friction stir welding of dissimilar aluminum alloy to advanced high-strength steel, *Int. J. Adv. Manuf. Tech.* **82**, 2183 (2016).
51. K. K. Ramachandran, N. Murugan and S. S. Kumar, Friction stir welding of aluminum alloy AA5052 and HSLA steel, *J. Chem.* **6**, 4071 (2015).
52. Z. K. Zhang, Y. Yu and J. F. Zhang, Corrosion behavior of keyhole-free friction stir spot welded Joints of dissimilar 6082 aluminum alloy and DP600 galvanized steel in 3.5 % NaCl solution, *Metal. - Open Access Metall. J.* **7**, 338 (2017).
53. Ibrahim, Y. Uematsu and T. Kakiuchi, Fatigue behaviour of dissimilar Al alloy/galvanised steel friction stir spot welds fabricated by scroll grooved tool without probe, *Sci. Tech. Weid. Joi.* **8**, 670 (2015).
54. R. Cao, J. H. Sun, J. H. Chen and P. C. Wang, Weldability of CMT joining of AA6061T6 to boron steels with various coatings, *J. Weld. Joi.* **93**, 193 (2014).
55. Z. Silvayeh, R. Vallant and C. Sommitsch, Influence of filler alloy composition and process parameters on the intermetallic layer thickness in single-sided cold metal transfer welding of aluminum-steel blanks, *J. Metall. Mat. Trans.* **48**, 5376 (2017).
56. G. Trommer, *Cold joining-automotive industry supplier magna develops technology for joining steel and aluminum using the CMT welding process (Fronius International GmbH, 2017),http://weldingproductivity.com/article/cold-joining.html.*
57. X. Jin, Y. Li and M. Lou, Process optimization of self-piercing riveting aluminum to high strength steel using DOE method, *J. Auto. R. D.* **1**, 185 (2011).
58. X. He, F. Gu and A. Ball, Recent development in finite element analysis of self-piercing riveted joints, *Int. J. Adv. Manuf. Tech.* **58**, 643 (2012).
59. M. Lou, Y. B. Li and Y. T. Li, Behavior and quality evaluation of electroplastic self-piercing riveting of aluminum alloy and advanced high strength steel, *J. Manuf. Sci. Eng.* **135**, 011005 (2013).

60. M. Lou, Y. B. Li and Y. Wang, Influence of resistance heating on self-piercing riveted dissimilar joints of AA6061-T6 and galvanized DP590, *J. Mater. Pro. Tech.* **214**, 2119 (2014).
61. L. Huang, Y. Shi and H. Guo, Fatigue behavior and life prediction of self-piercing riveted joint, *Int. J. Fatig.* **88**, 96 (2016).
62. C. Chung and S. Kim, Fatigue strength of self-piercing riveted joints in lap-shear specimens of aluminium and steel sheets, *J. Fatig. F. Eng. M.* **39**, 1105 (2016).
63. Y. Kang, B. Y. Xing and X. C. He, Corrosion properties of self-piercing riveted joints with dissimilar metal of steel/aluminum in alternate immersion test, *J. Ord. Mater. Sci. Eng.* **41**, 1004 (2018).
64. A. Mark, Savoy and H. G. Armaki, Thermal integrated clinching process-innovative joining of Next Gen. of AHSS, *Great Designs in Steel,* (2018).
65. C. Dallner and V. Zeiher, Electric cars: Lightweight construction and more, *J. Gumni. Fast. Kunst.* **101**, 39 (2011).
66. J. Baumeister, J. Weise and E. Hirtz, Applications of aluminium hybrid foam sandwiches in battery housings for electric vehicles, *J. Proc. Mater. Sci.* **45**, 1099 (2015).
67. H. L. Zhao, J. P. Ch and X. T. Wang, Welding process analysis of aluminum alloy battery pack for electric passenger cars, *J. Auto. Tech.* **24**, 64 (2017).
68. P. Cheng, *For the production of electric car batteries, Laser welding is becoming a manufacturing norm*, (Trumpf Inc. 2017), http://weldingproductivity.com/ article/ building-batteries/.
69. J. Y. Bai, The research on laser welding process of battery bus bar for new energy vehicle, Master thesis, Changchun University of Science and Technology (2017).
70. C. H. Choi, Y. H. Hahm and H. M. Kang, Battery pack housing assembly for electric vehicle using plastic composite material, US 20120103714 A1, 2012.
71. S. L. Liu, S. L. Xia and J. Z. Zhao, Lightweight Design of Electric Vehicle Battery Box, *J. Auto. Appli. Tech.* **11**, 10 (2016).
72. Z. W. Mao, W. Li and Y. Q. Liu, Application and analysis of electric vehicle battery pack composite materials, *Chin. J. Power Source.* **40**, 977 (2016).
73. S. Y. Lu, Method for manufacturing carbon fiber composite electric vehicle battery pack, CN 106217903A, 2016 (in Chinese).

Research Progress on Plastic Forming of Aluminum Alloy Sheet

Y. L. Song[1,2,a,†], L. Hua[1,2,b], J. Lu[3,c], P. Geng[1,2,d] and D. G. Dai[1,2,e]

[1]*Hubei Key Laboratory of Advanced Technology for Automotive Components, Wuhan University of Technology, Wuhan 430070, China*

[2]*Hubei Collaborative Innovation Center for Automotive Components Technology, Wuhan 430070, China*

[3]*School of Materials Science and Engineering, Wuhan University of Technology, Wuhan 430070, China*

E-mail: [a,†]ylsong@whut.edu.cn, [b]hualin@whut.edu.cn, [c]lujuewhut@126.com [d]gpdy01@163.com, [e]daidingguo123@163.com

Automotive lightweight is a major national demand and also the development trend of automobiles. The usage of aluminum alloy sheet is one of the most effective ways to achieve lightweight automobiles. Compared with the steel sheets, the aluminum alloy ones have the advantages of low density, high specific strength, good energy absorption, high recycling efficiency, etc. However, due to their lower elongation and elastic modulus, the aluminum alloy sheets have relatively poor formability and large springback at room temperature. The following research work for 6000 and 7000 series aluminum alloy sheets have been carried out recently. For stamping at room temperature: The yield models and forming limit models of aluminum alloy sheet were established; the springback law under different deformation conditions were compared; according to the established material model and related deformation rules, the whole forming process of commercial vehicle cab roof with aluminum alloy sheet was achieved and the forming performance and forming precision were then realized. For stamping at high temperature: A dislocation-based unified viscoplastic constitutive equation was established based on the tensile test data of aluminum alloy; the servo stamping process based on jerk was put forward; the B-Pillar reinforcement was formed by servo-hot stamping process.

Keywords: Automobile lightweight; Aluminum alloy sheets; Cold stamping; Hot stamping.

1. Introduction

The application of lightweight materials is an important way for energy saving and emission reduction in automobile industry [1-2]. Aluminum alloy sheet has become one of the main lightweight materials for auto body panels and structural components because of the excellent properties including low density, high specific strength, good energy absorption and high recycling efficiency [3-4]. However, due to the low elongation and low elasticity modulus at room temperature, the aluminum alloy parts have many deficiencies in the forming process, such as difficult to form, easy to crack, large springback and low forming accuracy [5-6], which limits the wide application of aluminum alloy sheets in automobile industry. Therefore, it is necessary to investigate the forming and springback law at room and elevated temperature and realize the forming of aluminum alloy parts with high accuracy and high performance [7]. In this paper some research work on cold and hot stamping of aluminum alloy sheet in our group is introduced.

2. Cold Stamping of Aluminum Alloybody Panels

2.1. *Materials properties*

The material tested in this study was 6061 aluminum alloy sheet in T6 condition with a thickness of 1.5 mm. The chemical composition (wt%) of the alloy is given in Table 1.

Table 1. The chemical composition of the studied 6061-T6 aluminum alloy [8].

Composition	Si	Fe	Cu	Mn	Mg	Cr	Zn	Ti	Al
wt%	0.70	0.20	0.32	0.05	1.00	0.10	0.03	0.01	Bal.

Uniaxial tensile tests, anisotropy tests and forming limit tests were performed. Based on the results, the mechanical properties of 6061-T6 aluminum alloy were obtained. In order to investigate the influence of different yield models on AA6061-T6, the normalized yield surfaces of four yield models, i.e., Von Mises, Hill'48, Barlat89 and Yld2000 were calculated, as shown in Fig. 1. It can be found that the four yield loci differ distinctly in the equibiaxial area and Barlat89 model produces a yield locus with strongest anisotropy for AA6061-T6.

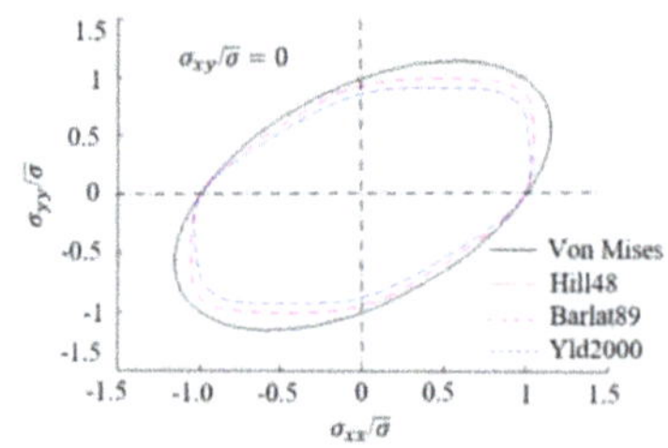

Fig. 1. Normalized yield surfaces for AA6061-T6 with different yield models [8].

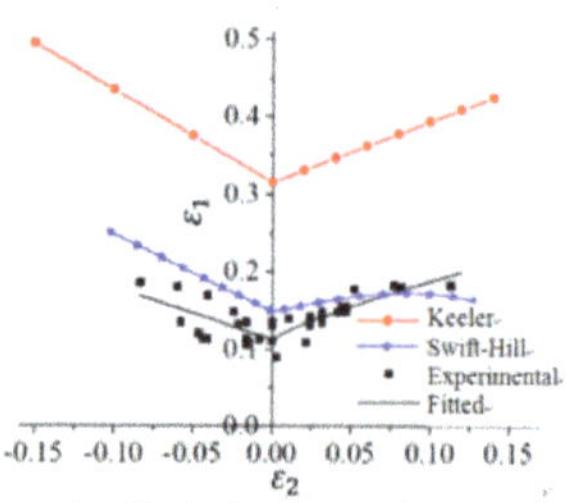

Fig. 2. Forming limit diagram of AA6061-T6 [8].

The forming limit diagram of AA6061-T6 using different forming limit models and fitted curve are shown in Fig. 2. The relative error of forming limit curve and the fitting curve of Swift-Hill and Keeler were 0.30%~38.29% and 103.22%~179.54%, respectively. The results would be more accurate inputting the experimental fitting curve in the simulation process.

2.2. *The springback law under different deformation conditions*

2.2.1. *The law of bulging springback*

Based on the Nakajima test, the bulging springback tests were performed. The models of specimens after the springback tests were created using a HandySCAN 700 3D laser scanner. The finite element simulation software is ETA/Dynaform 5.9. The reliability of the simulation was proved by comparing the shape of springback tests and the simulation results [8].

Fig. 3 shows the springback results of different width specimens. In path1, the springback distributes symmetrically about the line y=0, and increases gradually from the edge of sheet to the center. Besides, the springback increases with the increase of sheet width. In path2, the springback distributes symmetrically about the line x=0, but the biggest spingback is less than 0.12 mm which can be ignored.

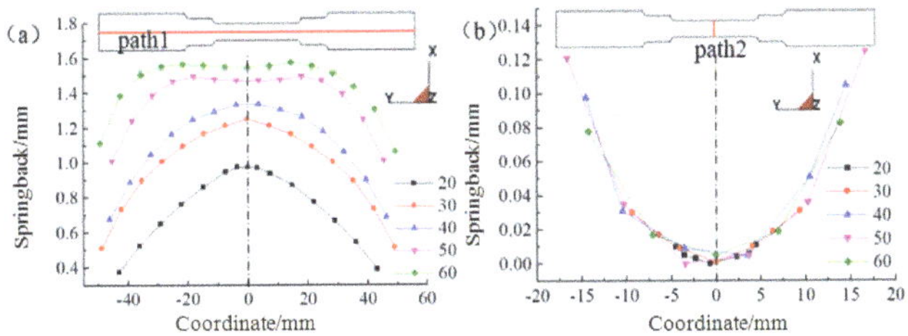

Fig. 3. Springback of different width specimens in: (a) path1; (b) path2 [9].

2.2.2. *The law of bending springback*

According to GB/T232-2010, the bending springback tests were conducted in INSTRON machine. The finite element model was created in DYNAFORM software.

Experiment and simulation results are summarized in Table 2. From Table 2, it can be seen that the springback angle increases with the increase of the bending angle. The difference between experiment and simulation results is little, and the biggest error are less than 10 %. Since there are many errors in the test, such as instrumental error, measurement error, finite element simulation error, etc. Therefore, the error is acceptable.

Table 2. Springback results of experiment and simulation [8].

Displacement of punch/f(mm)	Experiment/(°) (average value)			Simulation/(°)			Error /(%)
	Before/α_b	After/α_a	Springback angle/ $\Delta\alpha$	Before/α_b	After/α_a	Springback angle/$\Delta\alpha$	
18.232	85.5	72	13.5	82.360	67.639	14.721	9.04
24.406	105	90	15	104.014	88.746	15.268	1.79
30.503	122	103	19	121.263	104.07	17.193	9.51

2.3. ***The cold forming of large autobody covering parts***

2.3.1. *The cold forming of commercial vehicle cab roof*

Springback is a main defects affecting the forming precision in aluminum sheet metal forming.The method to solve the springback problem is full process simulation and springback compensation. The main process of this method is that adding the springback calculation after each process to simulate the stress relief of the part after unloading in actual production, and considering the next process to compensate the springback.

Compare the workpiece profile before the springback with the profile after the springback in the nth process, and obtain the springback amount of the nth process[8]:

$$s(x,y)=\sum p_{ij}x^{i}y^{j} \tag{1}$$

where s is the sprignback; x and y are the transverse and longitudinal coordinates of the panels, respectively; P_{ij} is the polynomial fitting coefficient (i, j= 0,1,2.....9).

Subtracting the springback amount of the (n+1) th process from the springback amount of the nth process, the influence of the (n+1) th process on the springback can be obtained, that is, $\Delta s = s_{n+1}-s_n$, which is the difference in the springback amount. For complex panels, the difference in springback amount is often not purely positive or negative, and there may be positive and negative in different areas of the entire profile. Therefore, the effect of the (n+1)th process on the springback amount of different profile areasin the nth processcan be analyzed bythe value of Δs [8].

Based on the precision transfer and prediction model, the springback amount of each process on the surface of the panel can be calculated. According to the calculated value, the mold profile can be compensated to realize the springback control. Fig. 4 shows the polynomial fitting springback amount of roof in different processes.

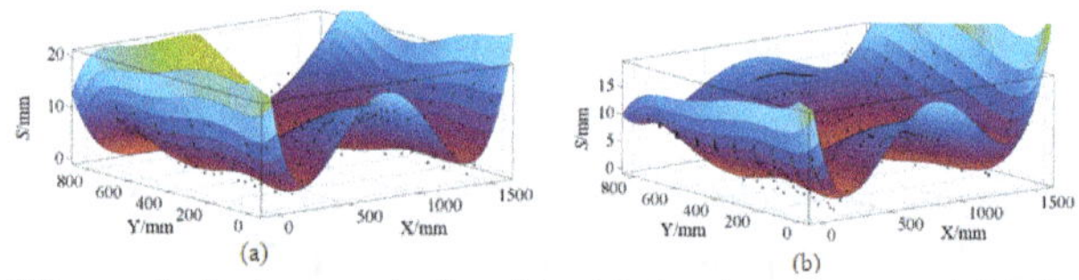

Fig. 4. Polynomial fitting springback amount of roof in: (a) drawing process and (b) flanging and punching process [8].

The springback in the last forming process of cab roof is illustrated in Fig. 5(a). The maximum springback amount is 19.943 mm. The springback amount after springback compensation is shown in Fig. 5(b). It can be found that the maximum springback amount has a significantly decrease and reduce to 0.876 mm.

Similar methods has been used for the stamping of the aluminum alloy outer and inner panels of the engine hood and cab door. Some experimental and simulated samples are shown as Fig. 6.

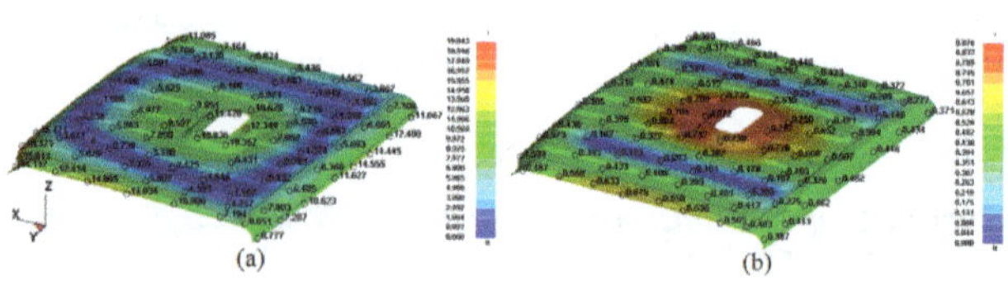

Fig. 5. Springback amount of cab roof (a) before and (b) after springback compensation [8].

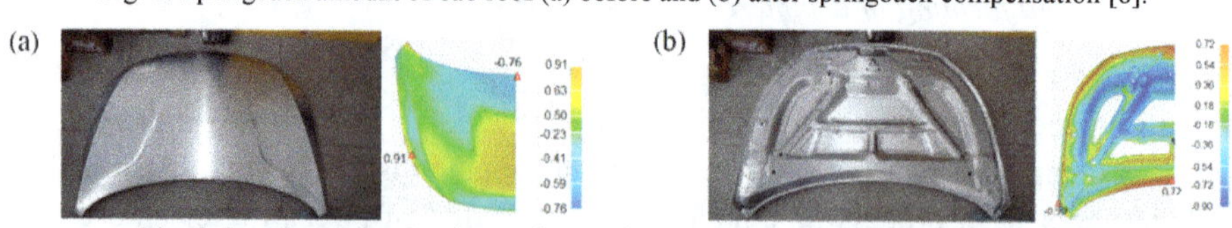

Fig. 6. Samples of the aluminum alloy engine hoods: (a) outer panel, (b) inner panel.

3. Hot Stamping of Aluminum Alloy Structure Parts

3.1. *A dislocation-based unified viscoplastic constitutive equation*

The material tested in this study was 7075 aluminum alloy sheet in T6 condition with a thickness of 2 mm. The chemical composition (wt%) of the alloy is given in Table 3. The hot uniaxial tensile tests were conducted at different temperatures of 300 °C - 450 °C and different strain rates of 0.01 s^{-1} - 1 s^{-1} based on the Gleeble 3500 thermal simulation machine.

The coupled dislocation densityviscoplastic constitutive equationsare shown in Eq. 2 [11-12]. This constitutive model considers the influence of dislocation density on the constitutive behavior of the material.

Table 3. Chemical composition of the studied 7075-T6 aluminum alloy [10].

Composition	Zn	Mg	Cu	Fe	Mn	Si	Cr	Ti	Al
wt%	5.28	2.66	1.52	0.35	0.092	0.07	0.22	0.029	Bal

$$\begin{cases} \dot{\varepsilon}_p = \left\langle \dfrac{\sigma - R - k}{K} \right\rangle^n \\ \dot{\bar{\rho}} = A(1-\bar{\rho})\left|\dot{\varepsilon}_p\right| - C\bar{\rho}^{\gamma_2} \\ \dot{R} = 0.5B\bar{\rho}^{-0.5}\dot{\bar{\rho}} \\ \dot{\sigma} = E\left(\dot{\varepsilon}_T - \dot{\varepsilon}_p\right) \end{cases} \tag{2}$$

$$\begin{cases} k = k_0 \exp\left(\dfrac{Q_k}{R_g T}\right); K = K_0 \exp\left(\dfrac{Q_K}{R_g T}\right); \\ B = B_0 \exp\left(\dfrac{Q_B}{R_g T}\right); C = C_0 \exp\left(\dfrac{-Q_C}{R_g T}\right); \\ E = E_0 \exp\left(\dfrac{Q_E}{R_g T}\right) \end{cases} \tag{3}$$

where, n, A and γ_2 are the constants; k, K, B, C and E are the temperature dependent material parameters which can be expressed Eq. 3 [11-12], in which, R_g is gas constant, T is temperature, Q is activation energy.

A genetic algorithm toolbox in software MATLAB was used to determine the values of material constants, and the results are listed in Table 4.

Table 4. The values of material constants for the constitutive model.

Material constant	Optimal value	Material constant	Optimal value	Material constant	Optimal value
K_0(MPa)	17.57733	Q_K(J/mol)	25874.49	n	4.147263
K_0(MPa)	0.101262	Q_k(J/mol)	37707.22	A	0.056522
B_0(MPa)	0.711031	Q_B(J/mol)	27153.85	γ_2	88456.75
C_0(MPa)	37942.36	Q_C(J/mol)	16478.41		
E_0(MPa)	3516.857	Q_E(J/mol)	0.001353		

Fig. 7 shows the comparison of the experimental and computed stress-strain data at different temperatures. It can be seen that the true stress-strain prediction curve of 7075-T6 high-strength aluminum alloy is close to the test curves. That is, the predicted true stress-strain curve is comparatively accurate.

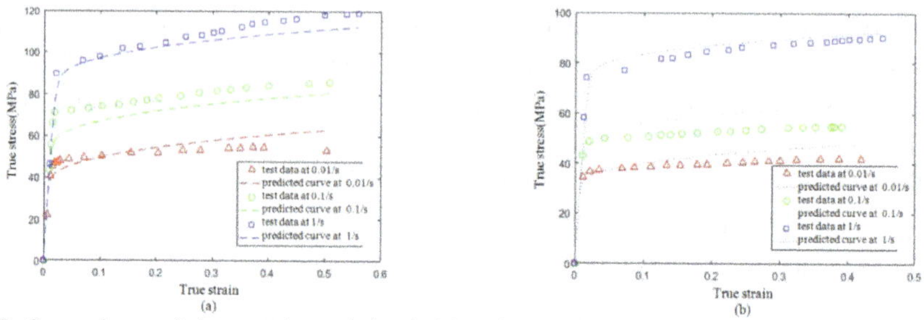

Fig. 7. Comparisons of the experimental (symbols) and computed (dotted curves) stress-strain data at different temperatures: (a) 400 °C; (b) 450 °C [10].

3.2. *Establishing processing maps*

Based on the dynamic materials model (DMM), the processing maps of 7075-T6 aluminum alloy sheet were established to optimize the hot processing parameters.The power dissipation efficiency η and instability parameter ξ can be expressed as [13]:

$$\eta = \frac{J}{J_{\max}} = \frac{2m}{m+1} \tag{4}$$

$$\xi(\dot{\varepsilon}) = \frac{\partial m}{\partial \ln \dot{\varepsilon}} \cdot \frac{1}{m(m+1)} + m < 0 \tag{5}$$

The processing map of 7075 aluminum alloy under strain values of 0.4, temperatures of 573-723 K and strain rates of 0.01-10 s^{-1} are shown in Fig. 8. It can be seen that there are two "safe" domains: 573-680 K/0.368-8.1 s^{-1}, and 695-723 K /0.05-1s^{-1}, which are suitable for hot stamping of aluminum alloy.

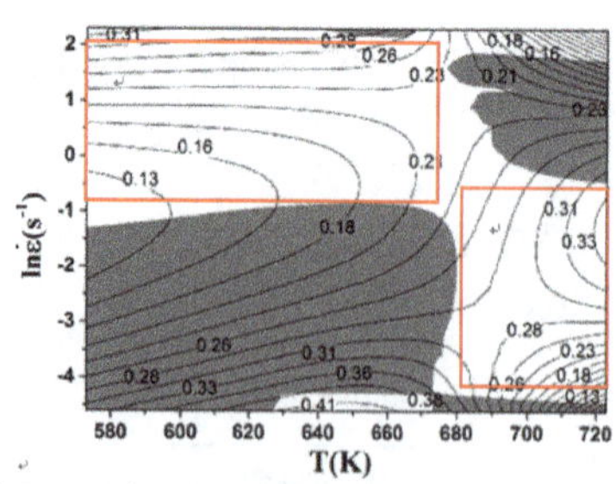

Fig. 8. Processing map of 7075 aluminum alloy under the strain of 0.4 [13].

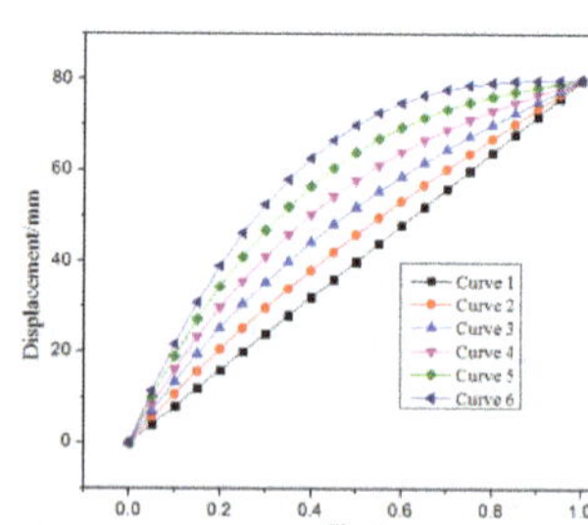

Fig. 9. Servo hot stamping process curves of B-pillar reinforcement [14].

3.3. *The jerk-based servo hot stamping process of B-pillar reinforcement*

The third derivative of displacement is jerk. The physical meaning of jerk is the change rate of acceleration. The acceleration may change smoothly when considering the effect of jerk, which not only has a beneficial effect on the forming of the panel, but also reduce the equipment impact. The jerk-based servo ram speed control method was proposed [14], the mathematical expression of the displacement amplitude is [14]:

$$x = At^3 + Bt^2 + Ct + D \tag{6}$$

The mathematical expression of jerk is [14]:

$$j = 6A \tag{7}$$

When x=0 and t=0, it indicates that the die has not yet started to move. x=1 and t=1 indicates that the die has been closed.

In the servo hot stamping simulation of B-pillar reinforcement of high-strength aluminum alloy, the mould closed height is 80 mm, and the stamping time is 1 s. According to the jerk-based servo ram speed control method, the servo curves of the high strength aluminum alloy were designed, as shown in Fig. 9.

3.4. *The hot stamping test of B-pillar reinforcement*

The hot stamping of B-pillars of AA7075-T6 under the servo process curves 2 and 3 are similar, Fig. 10a shows the hot stamping of B-pillar under curves 2. It can be seen that the B-pillar cracked at the side wall. Besides, the phenomenon of surface galling also appeared on the side wall. The hot stamping of B-pillar reinforcements of AA7075-T6 under the servo process curves 4, 5 and 6 are also similar, Fig. 10b shows the hot stamping fewer than 5. It can be found that the aluminum alloy hot stamping part under curve 5 did not crack, but the side wall also showed surface galling. No obvious scratches appeared in the rest of the area, the surface was smooth, and the appearance was good.

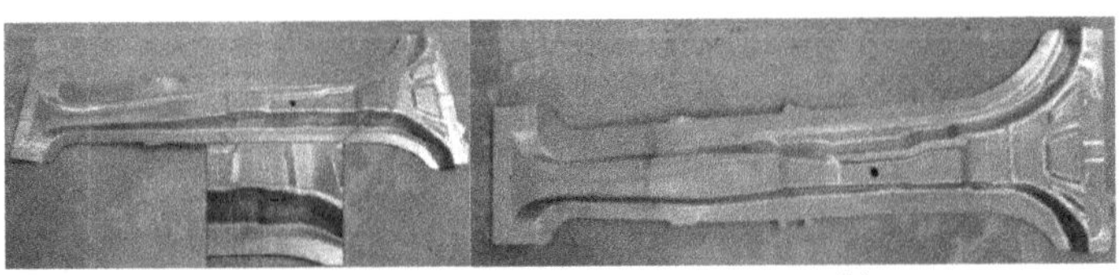

(a) (b)

Fig. 10. B-pillar reinforcement under different servo hot stamping process curves: (a) curve 2; (b) curve 5 [10].

The forming accuracy of B-pillar reinforcement was analyzed using HandySCAN 3D Laser Scanner. Fig. 11 shows the three-dimensional deviation of aluminum alloy B-pillar under curve 5.The maximum deviation is 2.19 mm, the minimum deviation is -2.14 mm, the average deviation is -0.34 mm to 0.43 mm and the standard deviation is 0.477 mm. It can be seen that the springback in four corners is relatively high. Springback is mainly affected by the stress and temperature. The residual stress is generated inside the part, while the dies were opened after the stamping. Therefore, the overall springback of the B-pillar reinforcement results from the residual stress, and the springback in lower stiffness area are relatively greater. When the aluminum alloy formed and quenched in the press, the cooling rates in different positions are different, causing the temperature distribution of the parts to be not uniform. The springback thus occurs in the effect of thermal expansion and contraction.

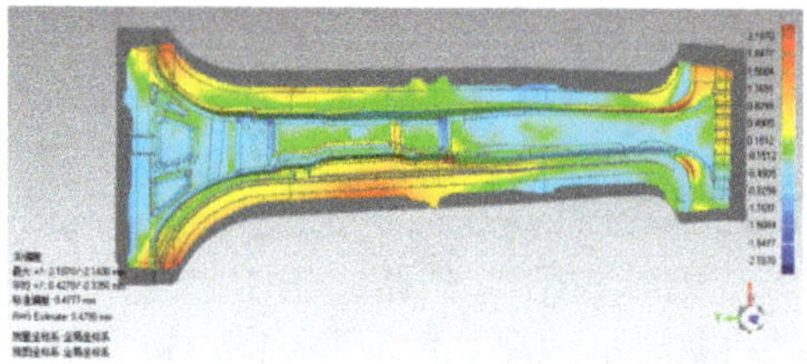

Fig. 11. Three-dimensional deviation of B-pillar of 7075-T6 aluminum alloy under servo process curve 5 [10].

4. Conclusion

The cold stamping of 6061-T6 aluminum alloy and the hot stamping of 7075-T6 aluminum alloy were investigated. The following conclusions can be obtained.

(1) For the cold stamping of 6061-T6 aluminum alloy, Barlat89 yield model is more suitable for the alloy. According to the full process simulation and springback compensation method, the springback of the commercial vehicle cab roof can be accurately predicted and the minimum springback can be obtained.

(2) For the hot stamping of 7075-T6 aluminum alloy, the dislocation-based unified viscoplastic constitutive equation was established. Jerk-based servos hot stamping process was proposed to form structural components. The parts have excellent surface performance and outstanding forming accuracy under the servo stamping process curve 5.

Acknowledgments

This work is supported by the National Natural Science Foundation of China (Grant no. 51675392), China Automobile Industry Innovation and Development Joint Fund (Grant no. U1564202), the 111 Project (Grant no. B17034), and the Innovative Research Team Development Program of Ministry of Education of China (Grant no. IRT17R83).

References

1. T. Dursun and C. Soutis, Recent developments in advanced aircraft aluminium alloys, *Mater. Des.* **56**, 862 (2014).
2. X. B. Fan, Z. B. He, S. J. Yuan and P. Lin, Investigation on strengthening of 6A02 aluminum alloy sheet in hot forming-quenching integrated process with warm forming-dies, *Mater. Sci. Eng.* **587**, 221 (2013).
3. M. Kumar and N. G. Ross, Influence of temper on the performance of a high-strength Al-Zn-Mg alloy sheet in the warm forming processing chain, *J. Mater. Process. Tech.* **231**, 189 (2016).
4. M. Kumar, N. Sotirov and C. M. Chimani, Investigations on warm forming of AW-7020-T6 alloy sheet, *J. Mater. Process. Tech.* **214**, 1769 (2014).

5. W. T. Huo, L. G. Hou, Y. S. Zhang and J. S. Zhang, Warm formability and post-forming microstructure/property of high-strength AA 7075-T6 Al alloy, *Mater. Sci. Eng.* **675**, 44 (2016).
6. K. Senthil, M. A. Iqbal, P. S. Chandel and N. K. Gupta, Study of the Constitutive Behavior of 7075-T651 Aluminum Alloy, *Int. J. Impact Eng.* **108**, 171 (2017).
7. M. R. Rokni, A. Zarei-Hanzaki, A. A. Roostaei and H. R. Abedi, An investigation intothe hot deformation characteristics of 7075 aluminum alloy, *Mater. Des.* **32**, 2339 (2011).
8. P. Geng, Simulation and precision control for the whole forming process of aluminum alloy automotive body panels, Master thesis, Wuhan University of Technology (2018).
9. P. Geng, Y. L. Song, P. Zhou, G. C. Liu, J. Lu and D. G. Dai, Research on bulging springback of AA6061-T6 based on Nakajima test, *J. Plasticity Eng.* **25**, 08 (2018).
10. D. G. Dai, Servo Hot Stamping Process and Property Regulationof High Strength Aluminum Alloy Automotive Parts, Master thesis, Wuhan University of Technology (2018).
11. J. Lin, Y. Liu, D. C. J. Farrugia and M. Zhou, Development of dislocation-based unified material model for simulating microstructure evolution in multipass hot rolling, *Philos. Mag.* **85**, 1967 (2005).
12. J. Lin and T. A. Dean, Modelling of microstructure evolution in hot forming using unified constitutive equations, *J. Mater. Process. Tech.* **167**, 354 (2005).
13. J. Lu, Y. L. Song, L. Hua, K. L. Zheng and D. G. Dai, Thermal deformation behavior and processing maps of 7075 aluminum alloy sheet based on isothermal uniaxial tensile tests, *J. Alloys. Comp.* (under review).
14. Y. L. Song, D. G. Dai, P. Geng and L. Hua, Formability of aluminum alloy thin-walled cylinder parts by servo hot stamping, *Procedia Eng.* **207**, 741 (2017).

Investigation on Fast Solution and Aging of 6061 Aluminum Alloy Sheet in Hot Stamping Processes and Its Negative Natural Aging Effect

Y. Liu, C. Y. Wu, L. Wang, K. Wang, Z. J. Wang, B. Zhu†, Y. L. Wang and Y. S. Zhang
State Key Laboratory of Materials Processing and Die & Mould Technology,
Huazhong University of Science and Technology, Wuhan, 430074, China
†E-mail: zhubin26@hust.edu.cn

Due to the demand for weight loss and energy saving, the need to manufacture automobile structural components from aluminum alloy sheets is apparent. Hot stamping technology has been developed to form complex parts of aluminum alloys. However, the longtime solution and aging of aluminum alloys is still a challenge. Aiming at this goal, the effects of solution temperature and time on the mechanical properties of 6061-T6 aluminum alloy were studied. The results showed that solution temperature had a significant effect on the solution process and solution heat treated at 560 °C for 10 min was recommended. Then the negative natural aging effect (keeping effect) of 6061 was also investigated and it indicates that artificial aging must be carried out within 3 hours after quenching. Finally, novel heating methods were employed in the solution and aging process to achieve an efficient hot stamping process. The results showed that parts with good mechanical properties can be obtained by solution at 560 °C for 5 min and aging or pre-aging at 192 °C for 5~10 min with the implementation of novel heating methods in hot stamping.

Keywords: 6061 aluminum alloy; Hot stamping; Fast solution; Aging; Negative natural aging effect.

1. Introduction

Lightweight of automobile has drawn more and more attention due to the demand for weight loss and energy saving. The mass of a car's body structure can be reduced by over 40% by using high performance sheet metal such as aluminum alloys [1]. Hot stamping process [2] has been developed for heat-treatable aluminum alloy forming, which can improve the formability of the alloy sheet, overcome the springback of the part and avoid thermal distortion of formed part during heat treatment. However, longtime solution heat treatment (SHT) and aging of aluminum alloys is still a challenge, which has limited the industrialization of aluminum alloys' hot stamping. Worse, some kinds of aluminum alloys have strict storage time limits before aging. Therefore, efforts should be devoted to reducing the SHT and aging time.

Li N. et al. [3] proposed the duplex ageing for AA6082: 15 min heating + 220 °C×5 min holding, and the results showed that full post strength guaranteed and more than 90% time reduction compared to standard aging. We haven't found other researches about fast solution and aging or pre-aging under hot stamping conditions. In order to reduce the solution and aging time, the effects of solution temperature and time on the mechanical properties of 6061-T6 aluminum alloy were investigated; then the negative natural aging effect (keeping effect) of 6061 was also investigated; finally, novel heating methods were used in the solution and aging process to achieve an efficient hot stamping process.

2. Materials

The 6061-T6 aluminum alloy was used in the current work and its main chemical compositions of the as-received alloy are given in Table 1. The original Vickers hardness was 112.3±4.4 HV1 and the ultimate tensile strength (UTS) was 337 MPa. Specimens were cut to dimensions of approximately 260 mm×170 mm×1.6 mm and tensile specimens as shown in Fig. 1.

Table 1. Main chemical compositions of the as-received6061-T6 aluminum alloy [wt%].

Chemical elements	Mg	Si	Mn	Cu	Fe	Cr	Ni	Zn	Ti	Al
Contents	1.1036	0.676	0.078	0.1495	0.4227	0.1647	0.0058	0.21	0.0273	Rem

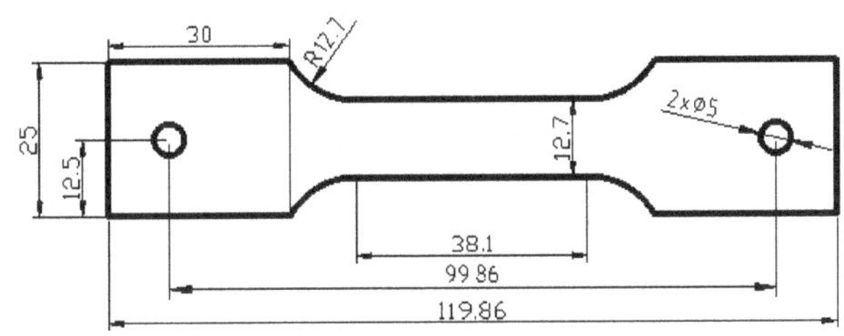

Fig. 1. Principle shape and dimensions of the hot tensile specimen (unit: mm).

3. Experiment

(1) Solution temperature and time. Tensile specimens were heated to 510~590 °C (interval: 10 °C) and soak for 5/10/20 min in a laboratory chamber furnace, and then they were water quenched. At last, the samples were aged at 200 °C for 3 hours in a furnace.

(2) Negative natural aging effect. Tensile specimens were heated to 560 °C and soak for 20 min in the same chamber furnace, and then they were water quenched. After storing at room temperature for 0/1/2/···12/14/16/···/24h, they were aged at 200 °C for 2 hours in a furnace.

(3) Fast solution. Four sheets with dimensions of 260 mm×170 mm×1.6 mm were sprayed with BN on both surfaces (B1~B4) and the other four sheets were bare (A1~A4). Samples A1 to A4 and B1 to B4 were heated to 560 °C and soak for 3/5/7.5/10 min in the same laboratory chamber furnace, respectively. And then they were quenched by cold flat dies. At last, the samples were aged at 200 °C for 2 hours in a furnace. Tensile samples as shown in Fig. 1 were cut from the sheets.

(4) Fast aging. Sheets with dimensions of 260 mm×170 mm×1.6 mm were solution heat treated at 560 °C for 15 min in the same chamber furnace and then hot formed and quenched in U-shape dies. After that, formed parts were pre-aged at 220 °C (actual temperature: 192 °C) for 3/5/7/10 min in a strong hot wind cyclic heating device as shown in Fig. 2. The hardness of the samples was tested after pre-aging and then

they were stored for five days at room temperature before a simulative paint bake process at 180 °C for 30 min in a furnace. Tensile samples as shown in Fig. 1 were cut from the parts.

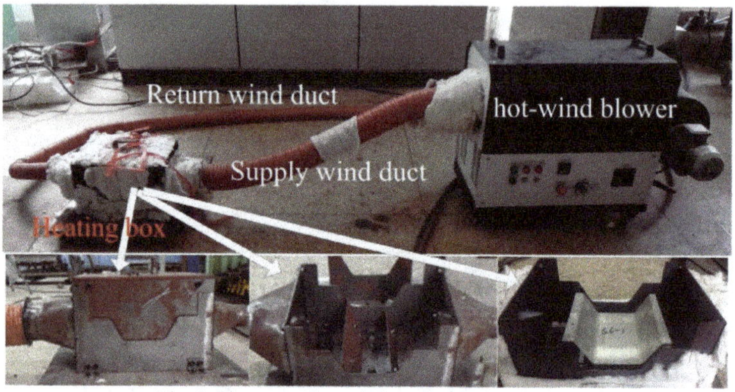

Fig. 2. The strong hot wind cyclic heating device.

Hardness and tensile strength of all the samples above were tested at last. Vickers hardness (HV) was tested with 1 Kg loading force and the stretching speed in tensile tests was 2mm/min. Please note that solution or aging time starts from closing the door of the furnace or device.

4. Results and Discussion

4.1. *Solution temperature and time*

Figure 3 shows the hardness and UTS of the samples after SHT (under different temperatures and time) and aging. When the SHT time was 5 min, the samples' hardness and strength increased with the increasing SHT temperature, only samples solution heat treated at temperatures higher than 570 °C could obtain strength higher than the as-received T6 condition. Given the hardness tests always have larger errors, the evaluating of the samples' mechanical properties was mainly based on the tensile tests. When the SHT time was 10 and 20 min, the samples' hardness and strength increased quickly with the increasing SHT temperature ranging from 510 °C to 550 °C; then the samples' hardness and strength increased very slowly or remained fairly steady with the increasing SHT temperature ranging from 550 °C to 590 °C. Only samples solution heat treated at temperatures higher than 550 °C could obtain higher strength than the as-received T6 condition. The samples' hardness and strength increased with the increasing SHT time when the SHT temperature ranged from 530 °C to 560 °C; when the SHT temperature was higher than 560 °C, the samples' hardness and strength remained fairly steady though the SHT time increased. The fairly steady hardness and strength means the completion of the solution process.

The precipitates (in any phase, e.g. β'' or β) are dissolved during the solution process and this process is controlled by diffusion. Solutes' diffusion rates increase with the increasing temperature, thus, it will take less time to complete the dissolution process. In addition, the solubility of the solutes is larger when the temperature is higher. Therefore, more solutes will dissolve into the matrix and higher strength will be obtained after aging. As a consequence, solution temperature had a significant effect on the solution process, and thus, affects the final mechanical properties of the alloy. So higher solution temperature should be used in SHT, but the temperature should be lower than the alloy's overburning temperature. According to Fig. 3, the solution temperature should be not lower than 560 °C. In D. Maisonnette et al.'s work, the solution temperature was 560 °C, and the overburning temperature of 6061 is 582 °C [4]. Therefore, the solution temperature should be 560~582 °C. Considering the temperature of the furnace may overshoot when its door closing after opening for a long time, 560 °C is recommended to be the solution temperature and 10 min should be a suitable solution time.

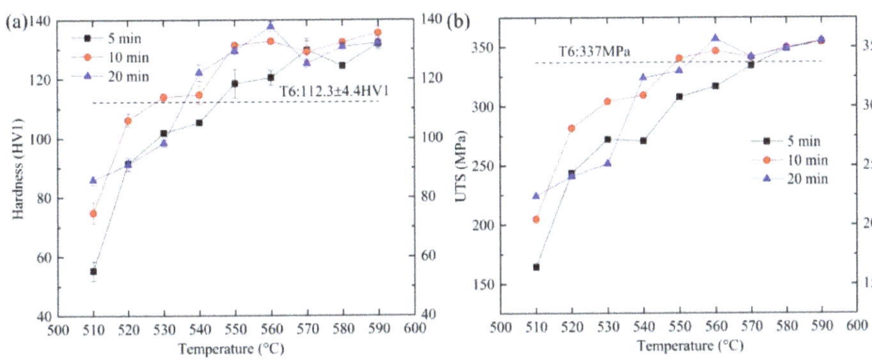

Fig. 3. (a) Hardness and (b) UTS of the samples after SHT (under different temperatures and time) and aging.

4.2. *Negative natural aging effect*

Figure 4 shows the experimental results of negative natural aging effect. The sample aged immediately after quenching had the largest hardness and tensile strength. Then the hardness and strength of the samples decreased with increasing storage time at the first 4 hours. The hardness and strength remained fairly steady when the storage time exceed 4 hour. The hardness and strength of samples which were stored for no more than 3 hours were higher than the as-received T6 condition whereas other samples had hardness and strength lower than the as-received T6 condition. According to literature [5], room temperature storage reduces the number of pre-β″ particles and the quenched-in vacancies, so the size of the pre-β″ particles increases and the final strengthening phase β″ is coarse. As a result, the hardness and the strength decrease. Consequently, formed parts made from 6061 should be artificial aged or pre-aged immediately after quenching; or the artificial aging process should be carried out with 3 hours after quenching at least.

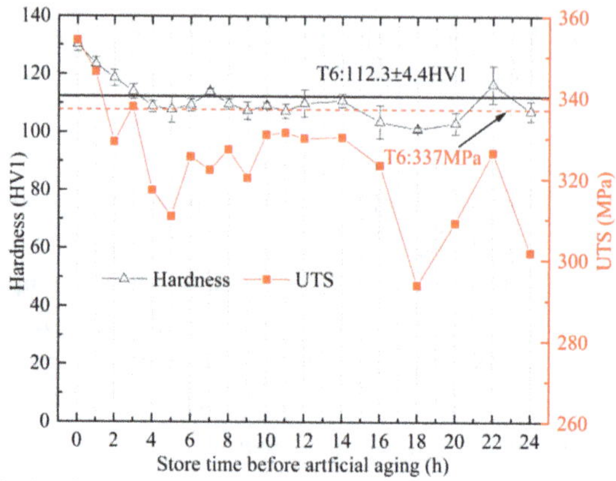

Fig. 4. Hardness and UTS of the samples which underwent artificial aging after being stored at room temperature for different time after quenching.

4.3. *Fast solution*

Figure 5 shows the experimental results of fast solution. The hardness and strength of A1~A3 increased quickly with the increasing SHT time while the hardness and strength of A4 increased a little than A3. A4 was considered to finish the solution process given the results in Section 4.1. It seemed that the sample with a solution time of 7.5 min (A3) nearly completed the solution, but it still needed about 2.5 min to complete the solution process. The strength of samples sprayed with BN (B1~B4) came to a fairly steady state after heating for 5 min and the hardness had almost the same tendency, indicating that the solution process had completed when heating for 5min. Besides, the hardness and strength of samples sprayed with BN (B1~B4) were lager than bare sheets samples (A1~A4) when heating for same time.

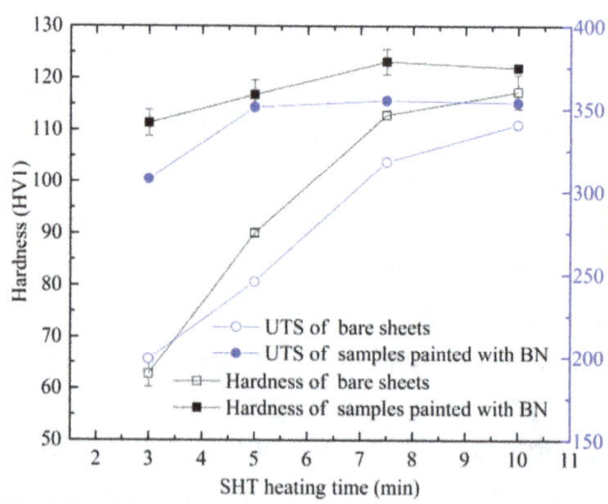

Fig. 5. Hardness and UTS of the samples (sprayed with/without BN) after being solution heat treated for different time and aging.

Aluminum alloy has a very low emissivity of about 0.09 [6] and thus, the heating rate is expected to be very low when heated in a radiation heating furnace. When sprayed with BN, the emissivity increased to about 0.75 at temperatures ranging from room temperature to 600 °C [7], so the heating rate increased significantly and the sheets took less time to reach the solution temperature. As demonstrated in Section 4.1, the temperature has a significant effect on the solution process. The temperature of sheets sprayed with BN was higher than the temperature of bare sheets when heating for same time, so the hardness and strength of B1~B4 was larger than A1~A4 under the same SHT time. When sheets sprayed with BN or other materials with high emissivity (e.g. graphite) during solution, the heating time is reduced and sheets are heated to the solution temperature faster, so the time of the whole solution process (including heating and dwelling) is reduced. By spraying BN on the sheets, the time of solution process decreased from 10 min to 5 min, reducing the solution time by 50%. Besides, BN has favorable lubrication and it can improve the formability of aluminum alloy sheet during hot stamping [1].

4.4. *Fast aging*

Figure 6 shows the experimental results of fast aging. The hardness of samples before baking increased with the increasing pre-aging temperature and the hardness after baking increased very slowly or remained fairly steady. The strength of the samples after baking increased with the pre-aging temperature, too. The strength of the sample pre-aged for 3 min was about 98.9% of the as-received T6 condition and the rest samples' strength was not less than the as-received T6 condition. A typical ageing process for 6061 involves heating and dwelling, so reducing the heating time is expected to reduce the aging time. Since aging is also a process controlled by diffusion, the temperature and time are the key factors that affect the aging process. The strong hot wind cyclic heating device can increase the heating rate and thus, reduce the aging time. The simulated paint baking process can be considered as part of the artificial aging process. As a result, aging or pre-aging at 192 °C for 5~10 min with the implementation of a strong hot wind cyclic heating device and undergo a paint baking process (or the second-step aging) can obtain higher hardness and strength than the as-received T6 condition.

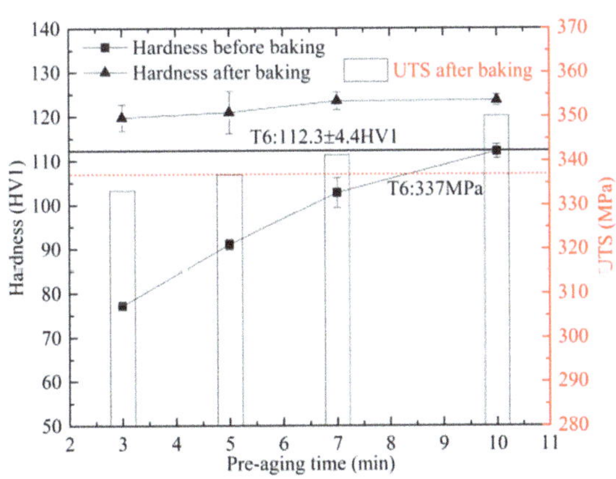

Fig. 6. Hardness and UTS of the samples pre-aged in the strong hot wind cyclic heating device for different solution time and then underwent a simulative paint bake process.

5. Conclusion

(1) Solution temperature had a significant effect on the solution process and the final mechanical properties of the alloy. Solution heat treated at 560 °C for 10 min were recommended.
(2) 6061 aluminum alloy has negative natural aging effect. Formed parts made from 6061 should be artificial aged or pre-aged immediately after quenching; or the artificial aging process should be carried out with 3 hours after quenching at least.
(3) By spraying BN on the sheets, the time of solution process decreased from 10 min to 5 min, reducing the solution time by 50%.
(4) Aging or pre-aging at 192 °C for 5~10 min with the implementation of a strong hot wind cyclic heating device and undergo a paint baking process (or the second-step aging) can obtain higher hardness and strength than the as-received T6 condition.

Acknowledgements

This research work was financially supported by the National Natural Science Foundation of China (grant no. U1564203 and U1760205) and National Major Science Technology Project of China (grant no. 2018zx04023001).

References

1. Y. Liu, Z. J. Zhu, Z, J. Wang, B. Zhu, Y. L. Wang and Y. S. Zhang, Flow and friction behaviors of 6061 aluminum alloy at elevated temperatures and hot stamping of a B-pillar, *Int. J. Adv. Manuf. Techn.* **96**, 4063 (2018).
2. R. P. Garrett, J. Lin and T. A. Dean, Solution heat treatment and cold die quenching in forming AA6xxx sheet components: feasibility study, *Adv. Mater. Res.* **673**, 6 (2005).
3. N. Li, J. Zheng, C. Zhang, K. Zheng, J. Lin and T. A. Dean, Investigation on Fast and Energy-Efficient Heat Treatments of AA6082 in HFQ processes for Automotive Applications, *MATEC Web of Conferences* **21**, 05015 (2015).
4. D. Maisonnette, M. Suery, D. Nelias, P. Chaudet and T. Epicier, Effects of heat treatments on the microstructure and mechanical properties of a 6061 aluminum alloy, *Mater. Sci. Eng. A* **528**, 2718 (2011)
5. C. D. Marioara, S. J. Andersen, J. Jansen and H. W. Zandbergen, The influence of temperature and storage time at RT on nucleation of the β″ phase in a 6082 Al-Mg-Si alloy, *Acta Mater.* **51**, 789 (2003).
6. W. L. Wolfe and G. J. Zissis, The infrared handbook, Technical report, Office of Naval Research, *Department of the Navy* (1985).
7. K. S. Jhajj, Heat Transfer Modeling of Roller Hearth and Muffle Furnace, Master thesis, University of Waterloo (2015).

Investigation on Hot Forming-quenching Integrated Process of High Strength AA7075 Aluminium Sheets

K. Huang, Y. L. Ren[†], Z. B. Xu, G. Xiao and Y. L. Zheng
Guangxi Alnan Aluminium Fabrication Co., Ltd.,
Guangxi Key Laboratory of Materials and Processes of Aluminium Alloys,
Nanning, 530031, China
[†]E-mail: renyuelu@alnan.com.cn

To solve the poor ductility of AA7075 aluminium sheets and serious spring-back in cold stamping at room temperature, an advanced hot forming-quenching integrated process for aluminium alloys, i.e., solution heat treatment - cold die Forming-Quenching (HFQ) is proposed and applied to a B-pillar outer panel in the automotive industry. In this paper, the HFQ of AA7075 is investigated experimentally. Firstly, the chemical composition is optimized in order to reduce the quench sensitivity. Secondly, the low melting point eutectic phase can be effectively eliminated by graded-homogenization. Thirdly, tensile tests, metallographic analysis and DSC are performed to optimize the suitable solution temperature. Fourthly, a series of hot tensile tests are conducted to investigate deformation characteristics at elevated temperatures. Fifthly, the CCT curve is simulated by JMatPro to determine the quenching cooling rate of the AA7075. Finally, the forming process of the AA7075 aluminium sheets is simulated by CAE software. The results show that the B-pillar component demonstrated limited spring-back with a high level of dimensional accuracy after the forming/quenching operation at appropriate solution temperature and an artificial aging process.

Keywords: Aluminium alloy AA7075; Hot stamping; Hot forming-quenching integrated process; Formability.

1. Introduction

With the rapid development of automobile industry, automobile lightweight has become the main direction of energy saving and emission reduction. Aluminium alloys have been widely used in automobile body parts because they are an important class of structural materials having excellent characteristics such as low density, high strength and excellent corrosion resistance. Recently, medium strength aluminium alloys, for example AA6016 and AA5182, are used in automobile stamping parts, but traditional cold stamping was hard to produce body parts with high strength because of the poor ductility and undesirable resilience, such as AA7075. To overcome the technical problems of the traditional cold stamping process, more and more research institutions and enterprises have paid attention to the hot forming-quenching integrated process (HFQ) [1, 2] and some of them have succeed in making high strength auto-parts by taking advantage of good ductility and forming ability at high temperature, which facilitates acquisition of complex shape.

Up to now, many research works have focused on HFQ due to the technological advantages of this process. Wang [4] studied the influence of temperature and punching speed on the AA2024 forming ability. Fakir [5] studied the forming properties of AA5754 subjected to HFQ process. Zhou [6] adopted the NSGA-II optimization algorithm with aim to obtain the key parameters of AA6111 during a thermal forming process. It is worth noting that the AA7075 is considered one of the most important materials that are used to produce auto B-pillar, however, the techniques are now still under the research and development stage in industry applications. B-pillar with AA7075 is not officially in mass production yet. In this paper, the forming ability of an AA7075-T6 with thickness of 2 mm with regard to availability of hot forming-quenching integrated process is explored, as shown in Fig. 1. Based on the customer's requirement of mechanical properties for B-column, we tried to set some parameters with aim to simultaneously increase the strength and maintain ductility of the AA7075-T6: tensile strength $R_m \geq 540$ MPa, yield strength $R_{p0.2} \geq 470$ MPa and elongation $\delta \geq 14\%$.

Fig. 1. Application of hot stamping component B- pillar.

2. Hot Forming-Quenching Integrated Process (HFQ)

2.1. *Composition design of materials*

According to customer requirements, the material 7075-T6 used in column B is required to be a thickness of 2 mm. In order to meet the key requirements of the material properties, the composition of the alloy is designed (see Table 1) and the principles of design are as follows:

- Using Zr instead of Cr and Mn to reduce quenching sensitivity of alloys;
- Increase Zn/Mg ratio with aim to increase the strength of the alloy;
- Strictly control Fe, Si content, and make sure the ratio Fe/Si ≥ 1.8, moreover, reduce content of impurity phase such as Al_7Cu_2Fe and Mg_2Si.

Table 1. Chemical composition of AA7075 aluminum alloy plate (mass fraction %).

Element	Si	Fe	Cu	Mn	Mg	Cr	Zn	Ti	Zr	Other		Al
										Uni.	Total	
ASTM B209M	≤0.40	≤0.50	1.2~2.0	≤0.3	2.1~2.9	0.18~0.28	5.1~6.1	≤0.20	—	≤0.03	≤0.10	allowance
AA7075 Sample	0.0313	0.0864	1.3858	0.0094	2.2631	0.086	5.5847	0.0515	0.0728	≤0.03	≤0.10	allowance

2.2. *Graded homogenization treatment*

Note the characteristics of semi-continuous casting, *i.e.*, because of the fast solidification process and cooling speed, it is easy to appear dendrite segregation and component segregation. Also, coarse non-equilibrium eutectics exist in grain boundary, which lead to negative influence on the thermal processing properties of the alloy. Based on the above considerations, we adopted a procedure of graded homogenization: (420±5)°C/(6±1)h + (455±5)°C/(10±1)h + (468±5)°C/(24±1)h.

As shown in Fig. 2, after graded homogenization, the layer-like amorphous and low melting-point eutectic microstructure on the grain boundary of the cast alloy is largely eliminated [7]. This desirable microstructure establishes the foundation for high strength, high toughness and good ductility of this alloy.

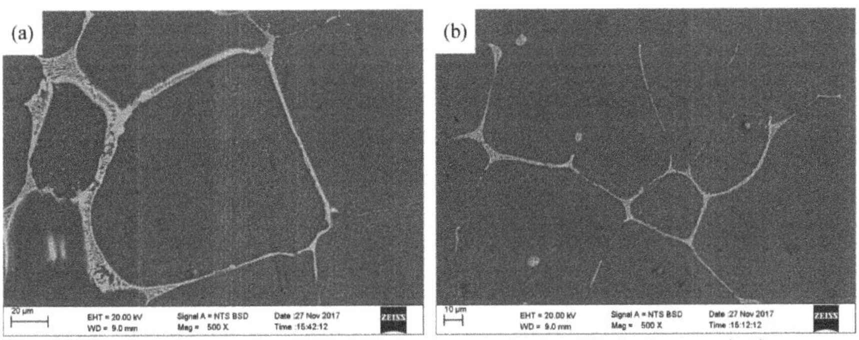

Fig. 2. SEM images of AA7075: (a) before homogenization; (b) after homogenization.

2.3. *Solution temperature*

As the key factor of solution quenching process, a proper solution temperature is set to ensure that the maximum numbers of second phases are dissolved into the matrix without causing any over-burn defects. In order to optimize the solution temperature, the mechanical properties, metallographic structure and thermal reaction were synthetically analyzed. According to the results of mechanical properties-solid solution-Aging, the solution temperature range was set to be 460~520 °C, and the aging system was 120 °C/24 h.

Figure 3 shows that both yield strength and tensile strength increase rapidly as the solution temperature increases. When the temperature is above 480 °C, the mechanical properties settle down stationary stage. With respect to the elongation, the value is 14 % when the temperature is higher than 490 °C. Especially at 500 °C, the elongation δ is 15 %. Fig. 4 shows a DSC curve of the AA7075 during a temperature range of 25 °C to 600 °C with a heating rate 2 °C/min. One can see that a curve inflection point appears during 533 °C and 541 °C, while no endothermic peak can be detected in temperature range of 490 °C~520 °C. According to the optical microstructure in Fig. 5, there is no evidence of over-burning in range of 490 °C~520 °C. After synthesizing each test result, the suitable solution temperature was selected to be 500 °C.

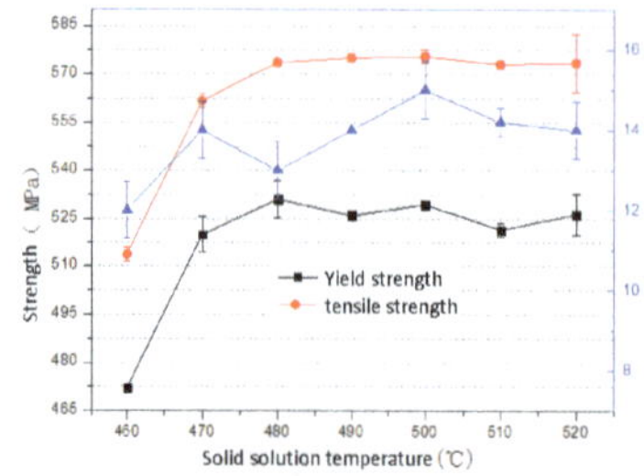

Fig. 3. Mechanical properties of AA7075 at different solution temperatures.

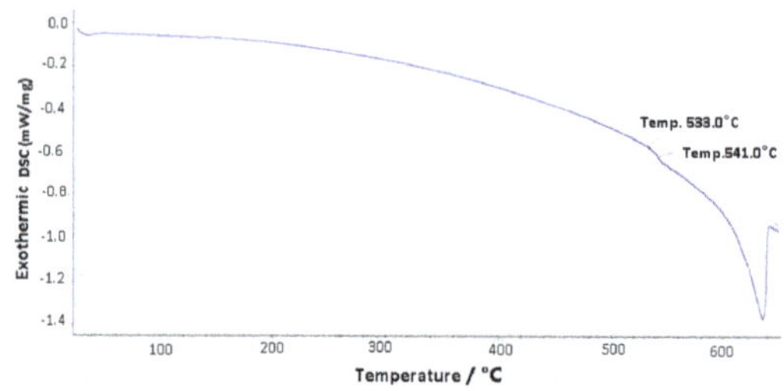

Fig. 4. DSC curve of AA7075.

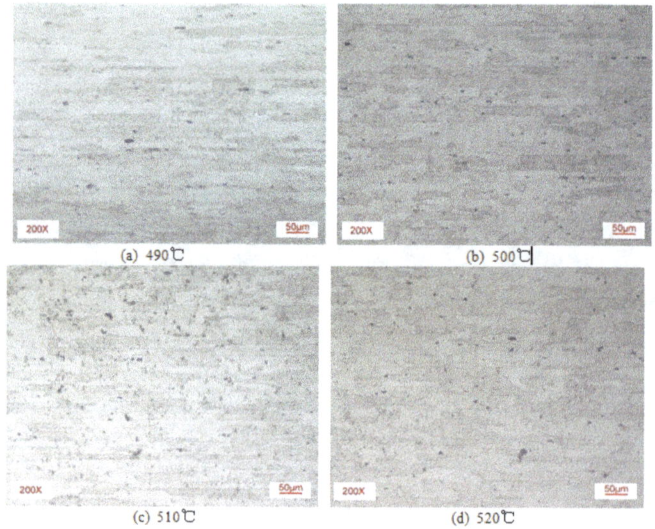

Fig. 5. Microstructure of AA7075 after solution treatment at different temperatures.

2.4. *Tensile test at high temperature*

In this paper, the influence of strain rate and deformation temperature on rheological stress during deformation was investigated by means of high temperature tensile test, trying to establish a foundation for FEM simulation of thermal stamping and optimization of forming process parameters for AA7075 aluminium alloy.

Figure 6a and 6b show that under given strain rates, the tensile strength decreases obviously with the increase of temperature, which indicates that the higher deformation temperature, the lower critical shear force of the aluminium alloy, note that the later facilitate the ability of stamping. On the other band, at a given strain rate, the elongation increased in initial stage and then declined. The elongation reaches a maximum value at 450 °C, which is the most favourable temperature for stamping.

Therefore, a temperature of 450 °C was selected for hot stamping deformation. In addition, a transfer time of 15s was set according to the result as shown in Fig. 7.

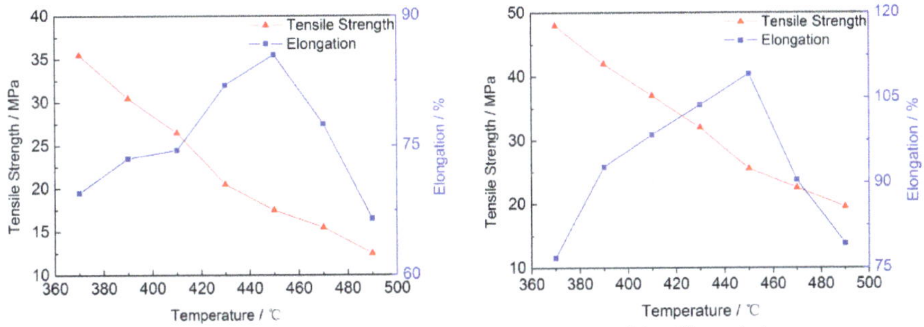

(a) ε=1 mm/min (b) ε=10 mm/min

Fig. 6. Mechanical properties of AA7075 as function of temperatures.

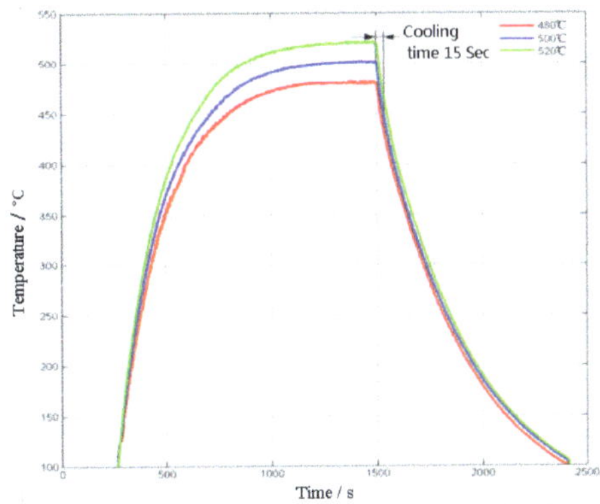

Fig. 7. Simulated cooling test of AA7075 in air.

2.5. *Cooling speed of quench*

Continuous cooling transition (CCT) curves of AA7075 aluminium alloy were calculated by JMatPro software simulation, as shown in Fig. 8. It can be seen that the critical quenching speed of precipitation η′ phase in AA7075 aluminium alloy is approximately 10 °C/s. Therefore, the quenching rate in hot stamping equipment should be controlled to 10 °C/s, which can not only ensure acceptable mechanical properties of the alloy but also deduce the residual stress due to lager cooling speed.

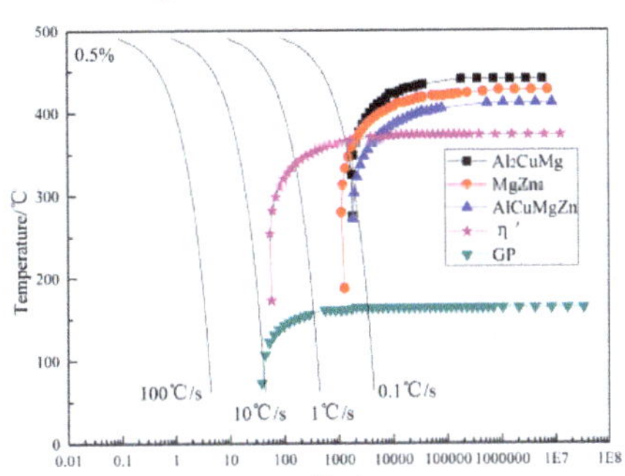

Fig. 8. CCT curves of AA7075 by JMatPro simulation.

3. Verification of Engineering

In this paper, the process parameters were validated by both CAE simulation and actual sample production. Fig. 9 shows the results of simulations by software Autoform, displaying distributions of temperature and thickness after forming. Also, sample testing was carried out on several parts of production, showing that the tensile strength is between 551 MPa and 553 MPa, yield strength is between 500 MPa and 502 MPA, elongation is large as 15 %. The mechanical properties fully meet requirements from customer.

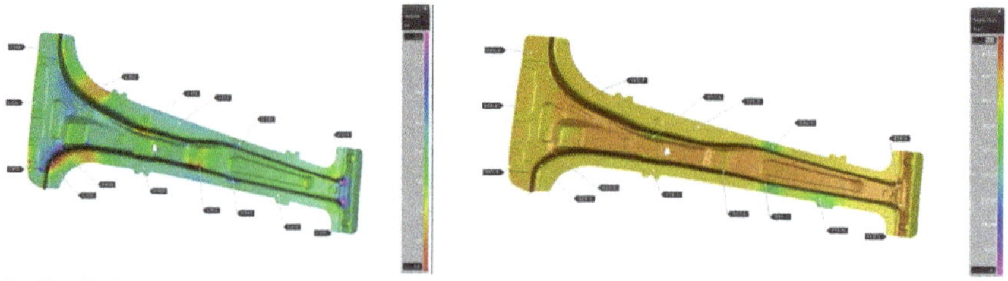

Fig. 9. Simulation result: (a) B-pillar thickness distribution cloud chart after hot forming (b) B-pillar temperature distribution after hot forming.

4. Conclusion

A series of conclusions can be drawn as follows:

(1) By designing alloy composition, a comprehensive performance of the alloy can be greatly improved by the graded homogenization treatment.

(2) An optimized solution temperature of 500°C was fixed according to results of mechanical properties, metallographic microstructure and DSC.

(3) The hot stamping temperature was set to 450 °C and the transfer time was less than 15s by high temperature tensile test and simulated cooling test.

(4) The quenching rate should be controlled within 10 °C/s by JmatPro Software simulation calculation.

(5) According to the simulation calculation of CAE software AutoForm and trial production, the process parameters can be directly applied to the actual mass production, which has great directive significance for automobile industry.

Acknowledgments

This work was supported by the Guangxi Key Scientific Research Programs (Project Number AA17202008), Nanning Scientific Research and Technological Development Program (Project Number GX20180128) and Nanning Special Fund for Talent Small Highland Programs (Project Number 2017001). The authors are grateful to Guangxi Alnan Aluminium Fabrication Co. Ltd., Guangxi Key Laboratory of Materials and Processes of Aluminium Alloys.

References

1. W. C. Xiao, B. Y. Wang and K. L. Zheng, An experimental and numerical investigation on the formability of AA7075 sheet in hot stamping condition, *Int. J. Adv. Manuf. Technol.* **92**, 3299 (2017).
2. J. Mendiguren, E. S. D. Argandoña and L. Galdos, Hot stamping of AA7075 aluminum sheets, in *IDDRG2016 conference on "Challenges in Forming High-Strength sheets"*, (Linz, Austria, 2016).
3. H. Wang, Y. B. Luo, P. Friedman, M. H. Chen and GAO Lin, Warm forming behavior of high strength aluminum alloy AA7075, *T. Nonferr. Metal. Soc.* **22**, 1 (2012).
4. L. Wang, M. Strangwood, D. Balint, J. Lin and T. A. Dean, Formability and failure mechanisms of AA2024 under hot forming conditions, *Mat. Sci. Eng. A* **528**, 2648 (2011).
5. O. E. Fakir, L. Wang, D. Balint, J. P. Dear, J. Lin and T. A. Dean, Numerical study of the solution heat treatment, forming, and in-die quenching (HFQ) process on AA5754, *Int. J. Mach. Tool Manu.* **87**, 39 (2014).
6. J. Zhou, B. Wang, J. Lin and L. Fu, Optimization of an aluminum alloy anti-collision side beam hot stamping process using a multiobjective genetic algorithm, *Arch. Civ. Mech. Eng.* **13**, 401 (2013).
7. X. M. Zhang, Y. H. Lu, S. D. Liu, W. J. Liu and H. P. Li, Effect of step-homogenization on microstructures and mechanical properties of 7055 aluminum alloy, *Chinese J. Nonferr. Metal.* **22**, 2154 (2012).
8. B. Han, Study on the plastic deformation behaviors of 7075z aluminum alloy at high temperarures, Master thesis, GuangDong University of Technology (2003).

Investigation on Quench Rate of 7075 Aluminum Alloy Under Hot Stamping Conditions

M. Y. Zhu, B. Zhu[†], Y. Liu and Y. S. Zhang
State Key Laboratory of Materials Processing and Die and Mould Technology, Huazhong University of Science & Technology, Wuhan 430074, China
[†]E-mail: zhubin26@.hust.edu.cn

Owing to the low density, high strength and stiffness, high-strength aluminum alloys have shown great potential in auto industries. Due to the poor plasticity at room temperature, hot stamping is seen to be a potential process to form complex shape products made from 7075 aluminum. Quenching rate is a very important parameter in hot stamping, which affects the mechanical properties of the formed parts. In the current paper, effects of die temperature which controls the quenching rate on the mechanical properties of 7075 aluminum alloy under hot stamping conditions were studied. Die temperatures of 25, 50, 100, 150 and 200°C were used in the investigation, and the corresponding quenching rates were 76 °C/s, 56 °C/s, 42 °C/s, 36 °C/s, 31 °C/s. Finally, the hardness and strength of the aluminum alloy sheet after hot stamping were tested. The results showed the best comprehensive mechanical properties of the 7075 aluminum alloy after artificial aging could be obtained when the die temperature was 100 °C.

Keywords: 7075 aluminum alloy; Hot stamping; Quenching rate; Mechanical properties.

1. Introduction

The ever-increasing demands of fuel economy and vehicle safety have put the transportation industry under great pressure. In transportation industry, significant weight reductions may be possible using precipitation-hardenable 7xxx-series aluminum alloy sheets with specific ultimate tensile strengths that are comparable to that of high strength steels but with a density only one-third that of steel [1]. The low formability of 7xxx series alloys at room temperature severely limits the types of structural components that can be formed, although there is some potential for warm forming at moderate temperatures such that the desired final temper is not significantly affected [2].

In order to improve the overall performance of the 7xxx series aluminum alloy in the manufacture of automotive parts, the hot stamping process is considered to be the ideal choice. During quenching with dies, the temperature of the dies rises slowly due to the heat transfer between the metals due to the higher temperature of the alloy after solution. Thus, the quench rate decreases as the temperature of the contact dies increases. The results of Liu et al. (2010) [3] show that aluminum alloy such as 7075 has a higher sensitivity to quenching rate. A lower quench rate leads to greater size of solute precipitation and the loss of excess vacancies [4]. The lower quench rate also leads to lower solute concentration when the material is quenched to room temperature [5], resulting in a loss of strength after subsequent ageing for the smaller volume fraction of hardening precipitates [6]. Deformation at elevated temperatures has been found to promote the formation of precipitates [7] that limit the age hardening potential of the as-quenched material.

The goal of the current work is to find suitable quenching rates under hot stamping conditions for the parts which can be used as automotive structural components considering the crashworthiness. So hot stamping experiments with different quenching rates which were controlled by the dies' temperature were carried out and the effects of different quenching rates on the mechanical properties of aluminum alloy plate were studied. It will provide a theoretical basis for practical engineering application.

2. Experimental

2.1. *Material*

The 7075 aluminum alloy was used in the study and its main chemical compositions of the as-received 7075-T6 aluminum alloys is given in Table 1.The thickness of the alloy sheets was 1.5 mm. The hardness of the as-received 7075-T6 aluminum alloys was 188 HV1 and its stress-strain curve at room temperature is shown in Fig. 1.

Table 1. Main chemical compositions of the as-received 7075-T6 aluminum alloy [wt. %].

Chemical elements	Mg	Zn	Mn	Cu	Fe	Cr	Si
Contents	3.161	5.895	0.1778	1.63	0.1309	0.2113	0.03

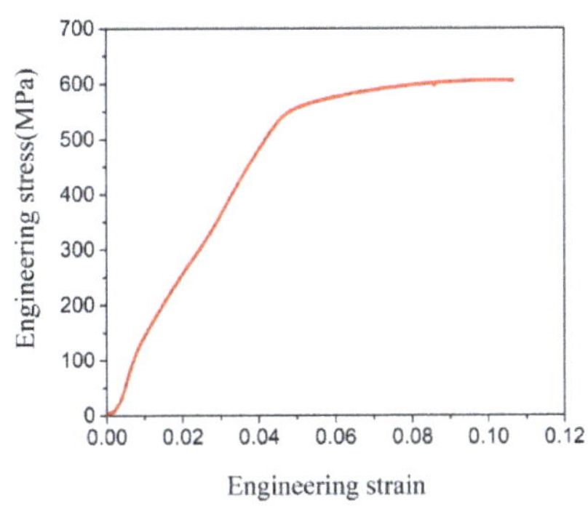

Fig. 1. Stress-strain curve of the as-received 7075 -T6.

2.2. *Quenching experiments*

Quenching experiments were performed in 2000 kN in a servo press and flat dies with heating control system as shown in Fig. 2. The servo press can adjust the holding pressure and holding time while the flat dies can be heated to different temperatures so different quenching rates were brought in.. Specimens were cut to dimensions of approximately 150 mm×150 mm. All specimens were heated to SHT temperature 480 °C and soak for 15 min, and then they were transferred to the dies with different temperatures to be quenched. Finally they were aged at 120 °C for 24 hours. The designed procedure is illustrated in Fig. 3. A K-type thermocouple was implanted into the specimens and the temperature history during quenching process was recorded. Vickers hardness (HV1) of specimens before and after aging was tested with 1 Kg loading force. Tensile samples as shown in Fig. 4 were cut from the aged specimens and tensile tests were also carried out on SHIMADZ AG.

Fig. 2. Flat dies with temperature control system.

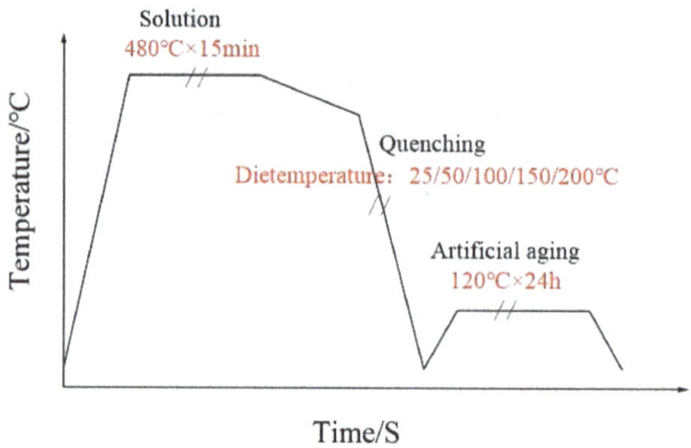

Fig. 3. Hot stamping process of 7075 aluminum alloy.

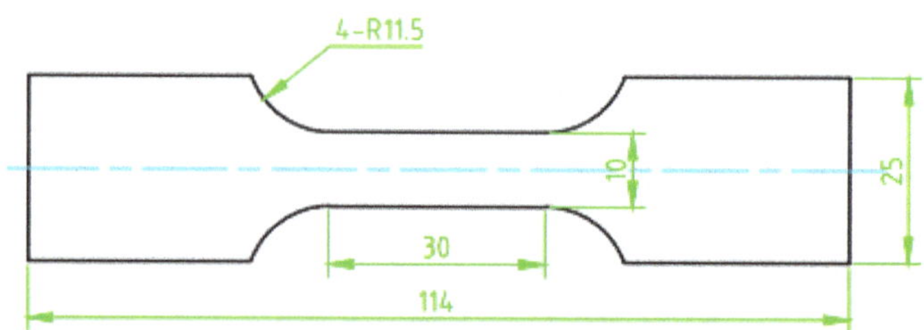

Fig. 4. Size of the tensile sample.

3. The Experimental Results

3.1. *Quenching rates*

Since the quench rate was not linear relative to the time, a fitting scheme was used to identify a representative quench rate. All representative quench rates in this work were determined by fitting a straight line onto the temperature-time curve of the blank in the range of 450 - 200 °C as shown in Fig. 5. The fit was performed using a linear least squares regression, with R2 values Ranging from 0.82 to 0.91. The slope of the line is assumed to be the representative quench rate at a given temperature. It shows the mold temperatures and equivalent quench rates used in this work in Table 2.

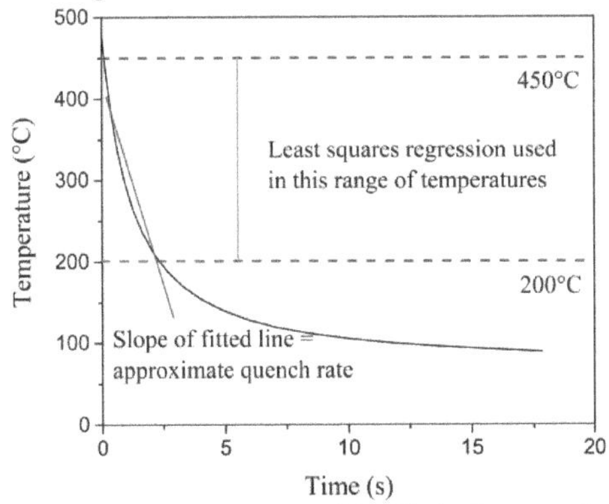

Fig. 5. The identification of the equivalent quench rate of 42°C/s in the range from 450 to 200 °C for a 7075 sample at a die temperature 100 °C.

Table 2. Die temperature and corresponding quench rates obtained using the least squares method in Fig.1.

Die Temperature (°C)	Equivalent Quench Rate (°C/S)
25	76
50	56
100	42
150	36
200	31

3.2. *Mechanical property*

The hardness test of the 7075 alloy before and after artificial aging was carried out to compare it with the 7075 aluminum alloy in the T6 state (hardness 188 HV1, maximum stress 604MPa). The results are shown in the Fig. 6 and Fig. 7 It can be seen that the hardening response of artificial aging strongly depends on the quenching rate.

Before artificial aging, the Vickers hardness values of 7075 alloy at different quenching rates are: which is 127 HV1 when it is at 36°C/s, which is 119 HV1 (31°C /s) when it is less than this rate, and which are 123 HV1, 120 HV1 and 120 HV1 respectively (42°C/s, 56°C/s and 76°C/s) when it is greater than this rate.

After being artificially aged for 24 h in a heating furnace at 120 °C, the 7075 alloy has the highest hardness of 193 HV1 when quenched at a rate of 42 °C / s. When the quenching rate is higher than 42°C/s (56°C/s and 76°C/s), the hardness of the 7075 alloy is 189 HV1 and 188HV1. The hardness of the alloy in the T6 state 7075 (188 HV1) is achieved. However, when the quenching rate is lower than 42°C/s (36°C/s and 31°C/s), the hardness value decreases significantly with the decrease of quenching rate, which is 182 HV1 and 171 HV1.

Tensile tests that conducted after three days of artificial aging displayed that the elongation after fracture of the alloy at different quenching rates. It can be seen that when the quenching rate is 42°C /s, the elongation after breaking reaches a maximum of 15.36%. The elongation at the remaining rate can also reach the elongation of the T6 state.

The maximum stress at different quench rates is also the maximum at a rate of 42°C/s, which is 616 MPa. the maximum stress is 610 MPa, at a rate of 76°C/s. the maximum stress values were 591MPa, 581 MPa and 573 MPa respectively at 31 °C/s, 36 °C/s and 56°C/s.

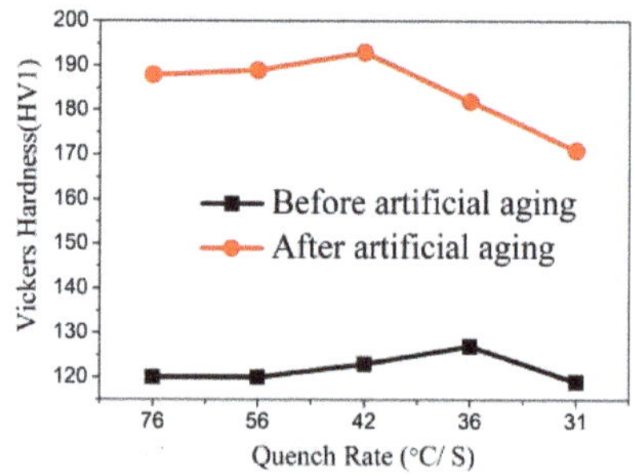

Fig. 6. Vickers hardness values obtained by different quenching rates before and after artificial aging, (Quenching rates of 76 °C/s , 56 °C/s, 42 °C/s, 36 °C/s and 31 °C/s, corresponding to mold temperature of 25 °C, 50 °C, 100 °C, 150 °C and 200 °C).

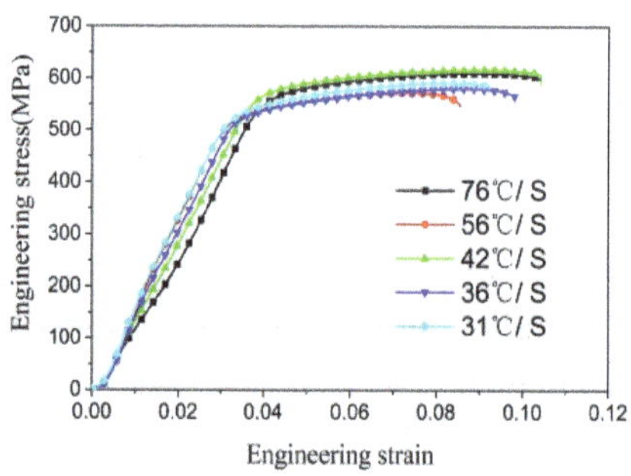

Fig. 7. Engineering stress-strain curves for different quench rate.

4. Conclusion

In the large-scale industrial production, due to the heat transfer between metals, the hot stamping die will gradually increase in temperature as the working time increases.

(1) Experiments conducted in this work show that the 7075 alloy is sensitive to the quench rate within the range considered in this study, as highlighted by the changes in the stress-strain curve in Fig. 7. Assuming that no significant deformation occurs during quenching, the maximum age hardening performance can be obtained after artificial aging when the alloy's quenching rate is about 42 °C/s.

(2) As is shown in Fig. 6, when the quenching rate is higher than 42 °C/s, the hardness of the alloy is 193 HV1, which is higher than 188 HV1 in T6 state. The maximum stress at this time is 616 MPa, which is higher than 604 MPa in T6 state. However, when the quenching rate is lower than 42 °C/s, the mechanical properties of the alloy can not reach the T6 state. The higher the quenching rate is, the faster the cooling rate of the part is.

Acknowledgments

This research work was financially supported by the National Natural Science Foundation of China (grant no. U1564203 and U1760205) and National Major Science Technology Project of China (grant no. 2018zx04023001).

References

1. K. Omer, A. Abolhasani, S. Kim, T. Nikdejad, C. Butcher, M. Wells, S. Esmaeili and M. Worswick, Process parameters for hot stamping of AA7075 and D-7xxx to achieve high performance aged products, *J. Mater. Process. Tech.* **257**, 170 (2018).
2. H. Wang, Y. Luo, P. Friedman, M. Chen and L. Gao, Warm forming behavior of High strength aluminum alloy AA7075, *Trans. Nonferrous Met. Soc. China* **22**, 1 (2012).
3. S. Liu, Q. Zhong, Y. Zhang, W. Liu, X. Zhang and Y. Deng, Investigation of quench sensitivity of high strength Al-Zn-Mg-Cu alloys by time-temperature-properties diagrams, *Mater. Des.* **31**, 3116 (2010).
4. H. G. Fabian and R. Wolter, Determination of the effective vacancy concentrations and the vacancy losses after the quench of Al–Zn–Mg alloys, *Cryst. Res. Technol.* **26**, 93 (1991).
5. M. J. Starink, B. Milkereit, Y. Zhang and P. A. Rometsch, Predicting the quench sensitivity of Al-Zn-Mg-Cu alloys: a model for linear cooling and strengthening, *Mater. Des.* **88**, 958 (2015).
6. Y. Zhang, B. Milkereit, O. Kessler, C. Schick and P. A. Rometsch, Development of continuous cooling precipitation diagrams for aluminum alloys AA7150 and AA7020, *J. Alloys Compd.* **584**, 581 (2014).
7. M. Rajamuthamilselvan and S. Ramanathan, Hot deformation behaviour of 7075 alloy, *J. Alloys Compd.* **509**, 948 (2011).

Tribology & Tools

High Performance Tool Steels for the Direct Hot Stamping Process of 22MnB5

M. Geile
voestalpine High Performance Metals Deutschland GmbH
E-mail: Marc.Geile@bohler.de

G. Jesner and C. Oberroither
voestalpine Böhler Edelstahl GmbH & Co KG, Kapfenberg, Austria

The demands for a stable hot stamping process are steadily increasing. Tool steel is still a major contributor when it comes to the issue of tool life improvement. In addition to thermal stresses, abrasion and adhesive wear are also crucial factors, which influence the material. Considering these factors, coefficient of friction tests were carried out using a kind of strip tensile test. The materials, which were used for these tests, were the sheet material 22MnB5 with varying surface coatings (AlSi / zinc) and various tool steels (including Böhler K353 and Böhler W360 ISOBLOC). Based on these experiments, the most suitable tool steel was determined. Furthermore, the topic of the hot- and cold punching of 22MnB5 was analysed. In addition to the determination of optimum punch geometry and cutting clearance, the abrasive wear behaviour of the punch was evaluated via a continuous stroke test. It could be proven again that the modern electro-slag remelted (ESR) tool steel Böhler K340 ISODUR or the powder metallurgical steel Böhler S390 MICROCELAN (S390MC) are suitable for this kind of application and can withstand its high mechanical loads. The implementation of laboratory results in practice is demonstrated by means of industrial application examples.

Keywords: Hot stamping; Tool steel; Friction; Wear behaviour.

1. Introduction

Hot stamped car body parts manufactured out of UHSS show an increase in quantity and variations of the component geometry and have proven to be a valuable asset in the field of crash safety and lightweight design. When it comes to process optimization in the field of direct hot stamping, the focus often is on tool steels. Based on the high thermal stress exposure of the tool, the standard solution for the selection of material is to use hot-work tool steels, such as 1.2343, 1.2344 or 1.2367. On the one hand, these materials show a limited hardness level of maximum 52 HRC due to their chemical composition. On the other hand, they can reach a high toughness, which guarantees an excellent resistance against fracture. The properties, which are required of tool steels, are:

- High compressive strength,
- Simple heat treatment (e.g. vacuum heat treatment),
- Sufficient wear resistance (abrasive/adhesive),
- Hardness of 42-60 HRC (depending on the hot stamping process),
- High thermal conductivity (cycle time reduction),
- Good weldability (in case of repairs or a geometry change).

Because of these qualifications, standard hot-work tool steels often reach their limits as hot stamping tools [1]. Especially the high hardness and abrasive wear resistance can be challenges and demand a tool steel with a higher carbide content such as cold-work steels and high-speed steels. However, based on their high alloy content, these steels possess a reduced toughness and thermal conductivity. In order to achieve the shortest possible cooling time, the sheet thickness, tool geometry and surface coating must be considered in addition to the thermal conductivity. In order to select the most suitable tool steel, its property profile must be determined especially for the hot stamping process. For this reason, two different experimental studies were carried out. Strip tensile tests were performed at "Steinhoff Engineering Consultants" to solve the issue of adhesive and abrasive wear resistance. The wear resistance of the tool steel can be derived from the experimentally obtained coefficient of friction μ. Furthermore, investigating the behaviour of the material during the cold and warm punch process was of interest.

The experimental investigations to detect the optimal punch geometry and cutting clearance were conducted at the Technical University of Munich. The data was verified by using a continuous stroke test. In the following pages, the results of these investigations are summarized and discussed.

2. Coefficient of Friction Investigation of Various Tool Steels and Sheet Materials Used in the Direct Hot Stamping Process

The common experiments used to investigate the abrasive wear behaviour of tool steels, such as the rubber wheel/dry sand wear test according to ASTM G65, are insufficient when it comes to the testing of materials used for hot stamping. They do not consider the thermal effect and do not deliver any results concerning the adhesive wear resistance. The strip tensile test was chosen as the main experimental method because of these special specifications. In this experiment, coated sheet metal strips (AlSi / zinc) were heated in a furnace up to the austenitizing temperature. Subsequently the sheet metal strips were clamped in between jaws made out of various tool steels and pulled through them. Due to the constant contact pressure of the jaws, the coefficient of friction was determined via the following equation:

$$\mu = \frac{Fz}{2Fd} \tag{1}$$

2.1. *Experimental setup of the coefficient of friction investigation*

Figure 1 shows the test setup used for the experiments. Samples made out of 22MnB5 with a thickness of 1.5 mm were tested by varying the typical process-parameters (austenitizing temperature, holding time, transfer time). The furnace atmosphere was air in the case of the coated blanks and inert gas in case of the uncoated blanks. The constant experimental parameters were the nominal surface pressure of 8 MPa and the pulling speed of 100 mm/s.

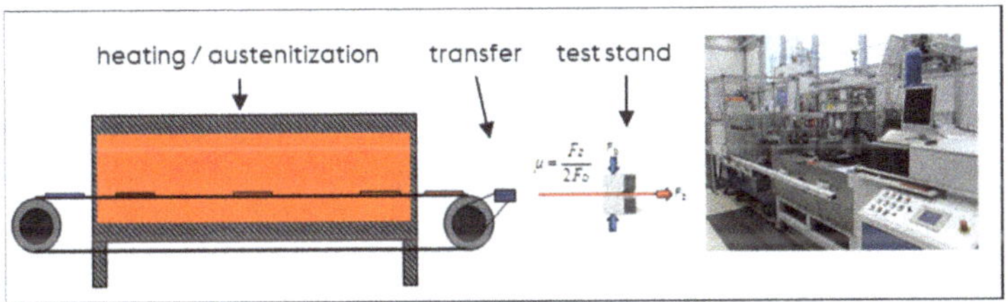

Fig. 1. Experimental setup of the friction coefficient investigation.

Böhler W300 (1.2343), Böhler W360 ISOBLOC (W360IB) and Böhler K353 were used as jaw material. Table 1 shows the chemical composition and the hardness level of the material samples.

Table 1. Chemical composition and hardness level of the samples [2].

Material	C	Si	Mn	Cr	Mo	V	Others	Hardness
Böhler W300	0.38	1.10	0.40	5.00	1.30	0.40	-	50 HRC
Böhler W360IB	0.50	0.20	0.25	4.50	3.00	0.60	-	56 HRC
Böhler K353	0.82	0.70	0.40	8.00	1.60	0.60	+ Al	58 HRC

The classic hot work tool steels e.g. Böhler W300, possess ideal toughness and red hardness. These standard tool steels are primarily used when the hot stamping process does not have any special demands concerning the abrasive wear behaviour and thermal conductivity. Böhler W360 ISOBLOC, which is a special developed grade, reaches a hardness level up to 56 HRC. This grade offers a convenient combination of hardness/abrasive wear behaviour and good toughness. Furthermore, this material shows an improved thermal conductivity. The Böhler W360 ISOBLOC can therefore be used in the direct and the indirect hot stamping. The cold work tool steel Böhler K353 with 8 % Cr content shows excellent abrasive wear resistance due to its high hardness up to 62HRC and his high carbide content. This material is a perfect tool steel solution for the direct hot stamping process, especially when the focus is on the abrasive wear resistance.

2.2. *Test results of the coefficient of friction study*

The coefficient of friction was determined through ten individual measurements.

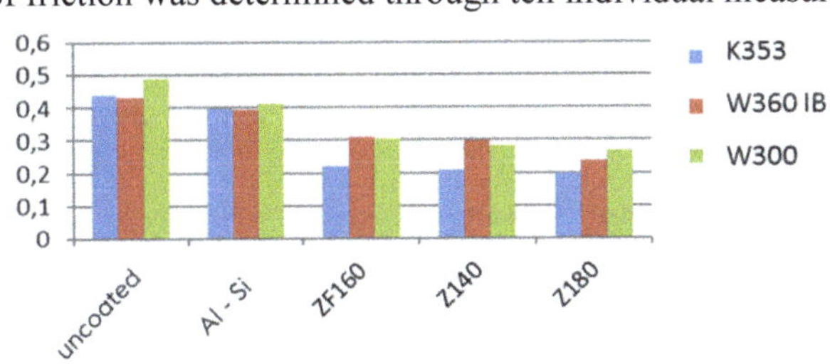

Fig. 2. Coefficient of friction in relation to the sheet strip surface and tool steel.

Figure 2 demonstrates the significant influence of the surface coating on the wear properties. The coefficient of friction of the uncoated samples reach high values of 0.42

to 0.48. The samples coated with Al-Si also reach high coefficient of friction regardless of the tool steel type. The zinc-coated samples show significant differences in their behavior [3]. The Böhler K353 with its high carbide content possesses the lowest coefficient of friction when compared to the hot work steels. Even the W360 ISOBLOC with its higher hardness shows a marginal improvement of the μ-values. The type of wear on the sample surface was determined by means of a white light interferometer [4]. Samples coated with Al-Si and zinc showed adhesive wear behaviour while uncoated samples demonstrated abrasive wear behaviour. This concludes that the Böhler K353 is the best choice when it comes to abrasive wear resistance. This cold work tool steel demonstrates a sufficient thermal stability due to the tempering temperature in the region of 530 °C

This experiment shows a clear trend which concludes, that steels with a high carbide content, such as the Böhler K353, have a clear advantage when it comes to abrasive wear in the direct hot stamping process. The influence on the tool lifetime of series tools cannot be detected by this experiment and requires testing on a tool as it is used in the industry. The Böhler K353 has already proven to be a reliable tool steel. One of its applications is e.g. as a tool in the production of B-pillars for the automotive industry. The Böhler K353 reaches twice the tool lifetime under abrasive wear conditions when compared to a 1.2367 type material.

3. Hot and Cold Punching of 22MnB5

The processing of hot stamped components via laser cutting and punching is state of the art. To improve productivity it is necessary to include the punch process into the forming process. The next question is, if the hole should be stamped in the hot or hardened and cold condition of the component. Cutting the product via warm cutting puts less stress onto the tool steel. In this case the martensitic transformation during cooling leads to a dimensional change which plays a major role and can therefore not be ignored. However, this solution could lead to high retraction forces. This aspect of the martensitic transformation can be ignored the cold component is punched. However, after hot stamping the component shows a strength of >1500 MPa which results in high cutting forces. Since only a few empirical values are known in this field, tests were conducted by means of a single-stroke experiment. A continuous-stroke experiment was performed for the cold punching process to determine the wear behaviour of the tools and to conclude the experiments.

3.1. *Experimental setup of the single stroke experiment for the warm and cold punching process*

The single stroke experiments were conducted on a hydraulic press. A modular punching tool was produced including a force / displacement measurement. The measurement technology was integrated and an arrangement of the measuring devices was established. The 22MnB5 strip samples used for the experiment possessed a thickness of 1.5 mm and a zinc coating. After the strips were heated up to austenitizing temperature, they were transferred into the compartment by using tongs. By closing the compartment, a constant transfer time of 16 s was set. After the closing process of the press, the steel strips were punched and removed after a defined holding time.

In addition to the force / displacement measurement, the surface of the cut surface was measured. The parameters rollover, burnish, fracture surface and burr height were determined. The optimal punch geometry and cutting clearance (10% or 15%) for the continuous-stroke experiment were supposed to be received from these experiments. For the single stroke test, the Böhler K340 ISODUR (K340ID) was used as punch material. The bulk material was Böhler K110 (1.2379). Table 2 shows the chemical analysis of the used materials.

Table 2. Chemical analysis of the active elements [2].

Material	C	Si	Mn	Cr	Mo	V	Others
Böhler K110	1.55	0.30	0.30	11.30	0.75	0.75	-
Böhler K340ID	1.10	0.90	0.40	8.30	2.10	0.50	+Al +Nb
Böhler S390MC	1.64	0.45	0.30	4.80	2.00	4.80	Co=8.00; W=10.40

Both, punch and die, were hardened and tempered to a hardness level of 58+2 HRC. The optimal cutting edge of 100 μm was adopted from a preliminary project, which dealt with shear cutting of 22MnB5.

3.2. *Results of the single stroke experiment for hot and cold punching*

Ten repeated experiments were performed in addition to the main experiments. The averaged results are presented in Figure 3.

Selection cutting clearance: average values - cold experiments

cutting clearance	cutting force [kN]	retraction force [kN]	rollover [μm]	burnish [μm]	fraction [μm]	burr [μm]
10%	77.49	-2.54	114	383	1002	21
15%	77.08	-1.97	174	339	980	20

Selection cutting clearance: average values - hot experiments

cutting clearance	cutting force [kN]	retraction force [kN]	rollover [μm]	burnish [μm]	fraction [μm]	burr [μm]
10%	23.72	-2.92	320	863	290	72
15%	21.45	-2.54	287	741	341	115

Selection punch geometry: average values – all experiments

punch geometry	cutting force [kN]	retraction force [kN]	rollover [μm]	burnish [μm]	fraction [μm]	burr [μm]
flat	42.46	-1.82	232	704	542	56
roof	40.95	-2.89	297	598	583	67
conical	39.40	-3.04	322	665	497	86

Fig. 3. Test results of the single stroke experiments.

The following aspects were observed during the punch experiments in cold condition:

- With 10% cutting clearance, the rollover is reduced, the burnish is increased.
- The fracture and burr height remain approximately the same.
- A cutting clearance of 15% is recommended to reduce the abrasive wear.

Following assumptions can be made when punching in the warm condition:

- With a cutting clearance of 10% the rollover, fracture and burr height are reduced while the burnish increases,
- This leads to better results for the cutting clearance of 10%,
- The burr height remains high.

The cutting force during hot punching is significantly lower than during cold punching, whereas the cutting clearance has a minor influence. The retraction force does not vary between the hot and cold punching process. No significant divergences were discovered when using different punch geometries (flat-, roof- or conical shaped punch). The cutting and retraction forces are hardly reduced. Thus, the flat shaped punch is sufficient, which ensures a cost-effective production of the punch.

3.3. *Experimental setup of the continuous stroke experiment for cold punching*

A stiff high-speed press was used for the continuous stroke experiments. The active elements were made out of Böhler K340 ISODUR or Böhler S390 MICROCLEAN. The punches were hardened and tempered to a hardness level of 62 HRC. The punch head was partially tempered to avoid punch head fraction. The rounding of the cutting edge was set to 100 µm and the cutting clearance to 15%. An additional trial was performed with a strip thickness of d=2 mm. The punch was repetitively measured after every 10000 strokes. After 50000 strokes, the experiment was stopped. At the end of the experiment, a metallurgical investigation was performed on the punches.

3.4. *Results of the continuous stroke experiment for cold punching*

The wear length of the punch and die was determined by measuring the cutting edge under an angle of 45°. Figure 4 shows the comparison between the wear length of the punches made of the materials Böhler K340 ISODUR and Böhler S390 MICROCLEAN.

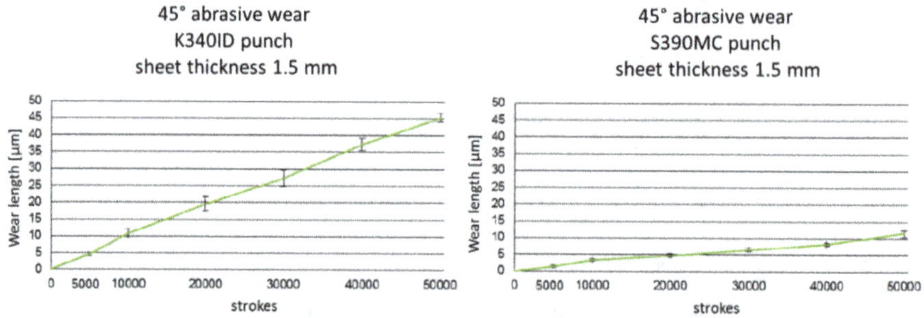

Fig. 4. Results of the continuous stroke experiment.

A linear development of the wear as a function of the stroke rate occurs in both of the tested materials. Furthermore, it becomes apparent that the higher carbide content of the Böhler S390 MICROCLEAN has a positive effect on the abrasive wear behavior. Its wear length after 50000 strokes is significantly lower than that of the Böhler K340 ISODUR. The dies show a similar behavior. In both cases, there was no fracture of the punch or die after 50000 strokes. A similar trend occurs regarding the sheet thickness of 2 mm. After the experimental trial was finished, the punches were tested visually and metallographically. Figure 5 shows the investigations of the punches after 50000 strokes. The lateral surface of the punch made out of Böhler S390 MICROCLEAN shows hardly any abrasive wear damage. Only after investigating the sample under the optical microscope, an influence on the lateral surface becomes evident. The Böhler K340 ISODUR shows a partial plastic deformation on its surface. This leads to the assumption that the stress during the stamping process is high enough to exceed the compressive strength of the material and thus creating plastic deformations. This effect is not as apparent in the case of the Böhler S390 MICROCLEAN. The compression strength is higher than that of the K340 ISODUR, due to its different alloy content. This confirms that the use of a high compressive strength tool steel makes sense for hot stamping and allows a long tool lifetime in continuous operations.

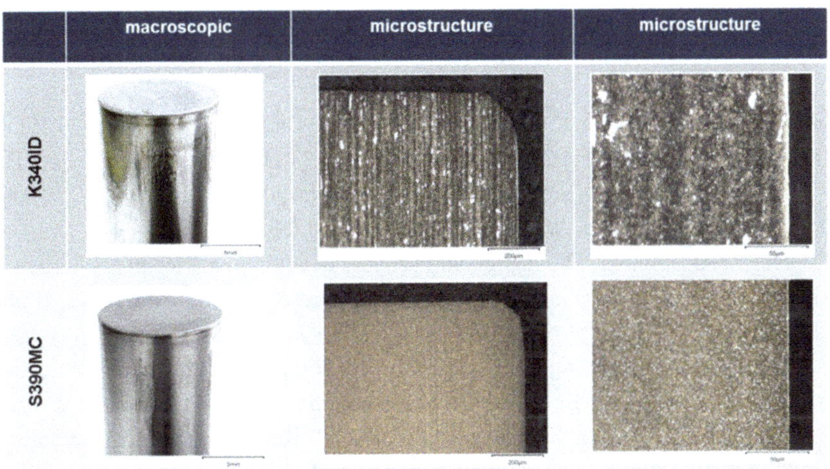

Fig. 5. Metallographic investigation of the punch after 50000 strokes.

The test series showed that the punching of hot stamped 22MnB5 sheets is a great challenge regarding the periphery (e.g. press) and tools. It is important that the construction is adjusted to the process. The cutting surface characteristics (indentation, burr height, burnish) are unaffected by the wear length of the active elements and remain almost the same [5].

4. Summary

Two different experiments were performed to optimize the characterisation of tool steels used for direct hot stamping. The coefficient of friction of various tool steels was determined based on a strip tensile test. In these tests, the cold-work steel Böhler K353 showed the most promising results. Furthermore, the tool life comparisons of tools in service confirm these results. The hot- and cold-punching processes were investigated. In the hot-punching process, it was observed that the proportion of the burnish is considerably higher than in the cold-punching process.

Likewise, the cutting forces are significantly lower. With regard to the retraction force, there was no difference between hot- and cold punching. The tool steels Böhler

K340 ISODUR and Böhler S390 MICROCLEAN showed no visible damage (e.g. chipping) after 50000 strokes in the continuous stroke experiments of the cold-punching process. The existing challenge is the process itself. The tool manufacturing process and the periphery has to be designed to withstand high stresses and loads. These investigations have shown that the existing alloy concepts of tool steels lead to good results concerning properties and tool life. It was proven that cold work steels can be used successful instead of hot work tool steels for high temperature applications such as the direct hot stamping process.

References

1. S. Gelder and G. Jesner, New high performance hot work tool steel with improved physical properties, in *9th International Tooling Conference,* (Leoben, Austria, 2012).
2. N. N., Datasheet Böhler K110 (K110 DE 10.98), Datasheet Böhler K340 ISODUR (K340 DE 02.2015), Datasheet Böhler K353 (K353 DE 01.2009), Datasheet Böhler S390 MICROCLEAN (S390 DE 04.2001), Datasheet Böhler W300 (W300 DE 04.02), Datasheet Böhler W360 ISOBLOC (W360 DE 05.2013).
3. R. Kelsch, A. Sommer, Die design and process developments for press-hardening of zinc coated steel, in *New developments in sheet metal forming conference,* (Fellbach, Germany, 2012).
4. R. Kelsch, Press hardening of galvanized steel components for BIW application, in *New developments in sheet metal forming conference,* (Fellbach, Germany, 2016).
5. S. Vogt, F. F. Neumayer, I. Serkyov, G. Jesner, R. Kelsch, M. Geile, A. Sommer, R. Golle and W. Volk, Experimental evaluation of tool wear throughout a continuous stroke blanking process of quenched 22MnB5 ultra-high-strength steel, in *The 36th International Deep Drawing Research Group Conference (36th IDDRG Conference)*, (Munich, Germany, 2017).

Innovative Die Technology — A Paradigm Shift for an Efficient Solution

M. M. Gharbi
Schuler Pressen GmbH, Schulerplatz 1, 73033 Goeppingen
E-mail: mohammad.gharbi@schulergroup.com

In this paper we consider the requirements on the die technology for a fast, stable and cost optimized hot stamping process. We presented an innovative technology to achieve high homogeneous contact pressure. Moreover, we use additive manufacturing for cooling channels, which allows us to have optimized and faster cooling which resulted in more efficient production. A new concept on die design to reduce the weight and increase the mechanical technological properties was put forward. Finally we show our concept to find the right material for the right hot stamping die based on wear resistance. No matter which method is used, the process to produce parts with tailored properties should be stable and reproducible. It is important that the overall process time stays almost the same in comparison with production of normal parts.

Keywords: Hot stamping; Pressure controlled hardening; Additive manufacturing; Die technology; Topology optimization; Wear resistance.

1. Introduction: Lightweight in the Automotive Industry

In the modern automotive industry, the need for lightweight design and construction is omnipresent. National and international regulations to reduce the CO_2 emission see a reduction of 25%-30% in the coming 5 to 10 years. This goal could just be realized while using lightweight concepts.

A reduction of car body weight by 100 kg, results a further weight reduction of other car components by adjusting and adapting them to car body up to 16 kg. This results fuel reduction of 0.3 - 0.6 lit/100km, depending on car category. The mentioned strict emission guidelines stand against the demand for durability, crashworthy qualities and passenger safety. Lightweight concept has the potential to fulfill all of these demands by three avenues: technology, form and material lightweight design. The use of lightweight technology has the most impact following by form and material.

In the most of the cases an intelligent combination of these factors is used and ensures the possibility to use the maximum potential of lightweight concepts. In the following, we consider on lightweight technologies, in some cases combined with lightweight materials, as a key solution to fulfill the functional requirements of the body in white (BIW). Generally, in BIW one could define three areas; exterior, safety cell, and front/rear area of deformation as shown in Fig. 1. The main challenge here is to use the right lightweight concepts to fulfill the desired characteristics, where the buckling resistance meets the strength.

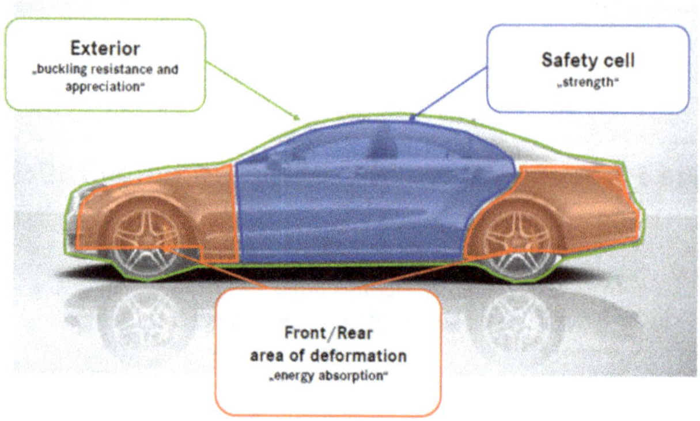

Fig. 1. Functional requirements in BIW.

2. Functional Requirements of the Body in White: Safety Cell

Considering safety cell, the strength and crash worthiness is the objective function. The need for an alternative material and new technology to form parts with higher strength is increased. On the other side the demand on part complexity and the design criteria are rises. In the following, the hot stamping technology to form parts which are mainly used for safety structure exhibiting high strength is presented.

3. Hot Stamping Technology

Due to the demand for reduced vehicle weight, improved safety, and crashworthiness qualities, the need to manufacture automobile structural components from ultrahigh strength steels is apparent (Åkerström, 2006). Hot stamping was developed and patented (GB1490535, 1977) by a Swedish company (Plannja), which used the process for saw blades and lawn mower blades. In 1984 Saab Automobile AB was the first vehicle manufacturer who adopted a hardened boron steel component for the Saab 9000 (Berglund, 2008). The number of produced parts increased from 3 million parts/year in 1987 to 8 million parts/year in 1997. Since the year 2000 more hot stamped parts have been used in the cars and the number of produced parts/year has gone up to approximately 700 million parts/year in 2017 (Aspacher, 2008). The applied hot stamped parts in the automotive industry are chassis components, like A-pillar, B-pillar, bumper, roof rail, rocker rail and tunnel which are shown in Fig. 2.

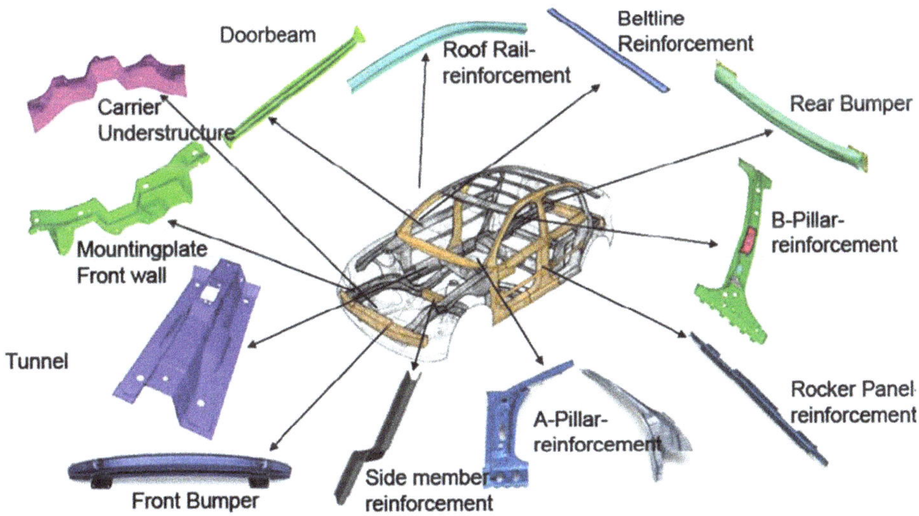

Fig. 2. Example of hot stamped parts.

The hot stamping process currently exists in two different main variants shown in Fig. 3. the direct and the indirect hot stamping method. In the direct hot stamping process, a blank is heated up in a furnace, transferred to the press and subsequently formed and quenched in the closed tool. The indirect hot stamping process is characterized by the use of a nearly complete cold pre-formed part which is subjected only to a quenching and calibration operation in the press after austenitization. (Merklein et al., 2008). Full martensite transformation in the material causes an increase of the tensile strength of up to 1500MPa.

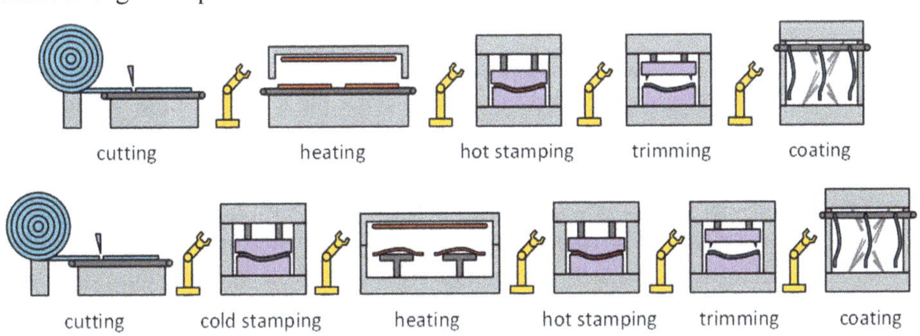

Fig. 3. Direct (top) and Indirect (bottom) hot stamping.

Investigation on alternative cold rolled medium Mn steel containing 9.76Mn-0.16C-l .37Al-0. 19Si (mass%) processed via a short time (3 min. soaking) continuous annealig-type treatment and then subjected to typical direct hot forming thermal cycles, with reheating at 650:800 °C could achieve an excellent mechanical properties (1330-1448 MPa tensile strength; 16.7-25.3% total elongation) of the hot formed-steel, were obtained with reheating at 700 °C. 22MnB5 steel grade is the most commonly used steel grade in hot stamping processes. The material exhibits a ferritic–pearlitic microstructure with a tensile strength of about 600 MPa

(Fig. 4). After the hot stamping process, the component finally has a martensitic microstructure with a total strength of about 1500 MPa. In order to achieve such a microstructure and hardness transformation, the blank has to be austenitized in a furnace for at least 5min at 950 °C. Afterwards, the blank is formed and quenched simultaneously by the water-cooled die for 5–10 s, depending on the part thickness. Due to the contact between the hot blank and the cold tool, the blank is quenched in the closed tool. If the cooling rate exceeds a minimum cooling rate of approximately 27 K/s, at a temperature of around 400 °C, a diffusionless martensitic transformation will be induced, which finally is responsible for the resulting high strength of the part. The martensite transformation begins at 425 °C (martensite start point Ms) and ends at 280 °C (martensite finish point Mf). The mechanical properties of steel after quenching change in dependence on its carbon content and consequently, the strength after quenching can be controlled by a proper adjustment of the carbon content (Somani et al., 2001).

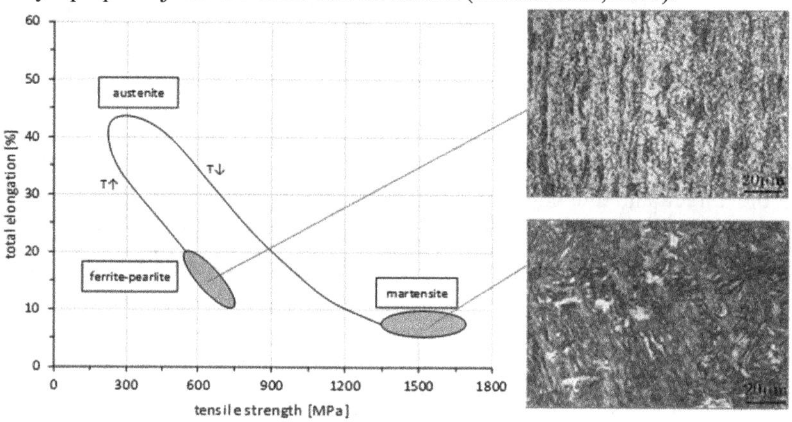

Fig. 4. Microstructure change during hot stamping process.

The desired phase transformation and hardenability is achieved by technically feasible cooling rates (Garcia Aranda et al., 2002). After the forming of the heated blank in the austenitic temperature range, the part is quenched in the closed tool until the entire martensite transformation of the part structure is completely. A cooling rate of more than 27 K/s is necessary for a full martensite microstructure of 22MnB5.

In hot forming process, the dimensional accuracy of produced components relies not only on an optimum friction level for the deformation of the workpiece, but it also gets affected by wear of the forming tools, which reduces the service life of the tool as well. It is desirable to enhance the durability of the tools by understanding the influence of contact conditions on tool wear.

3.1. *Cooling channels*

Figure 5 shows the heat transfer phenomena during hot stamping process. The quenching operation during the hot stamping process does not only influence the economy of the process but also the final properties of the component. The objective of the cooling ducts design is to quench the hot part effectively and to achieve a cooling rate of at least 27 K/s while martensite forms. The die cooling system is economical if a fluid coolant, such as water, is used, which flows through cooling ducts around the contours of the component. The heat flow in the formed component is dependent on the heat transfer from the component to the tool, the heat conductivity within the tool, and the heat transfer from the tool to the coolant (Karbasian and Tekkaya, 2010).

mechanism		equations
convection	free	$q_{conv} = h_{conv} \cdot A \cdot (T_1 - T_\infty)$
radiation		$q_{rad} = \varepsilon \cdot \sigma_{SB} \cdot A \cdot (T_1^4 - T_\infty^4)$
conduction	contact con-ductance	$q_{cont} = h_{cont} \cdot A \cdot (T_2 - T_1)$
	solid	$q_{cond} = \frac{k}{L} \cdot A \cdot (T_3 - T_2)$
convection	forced (fluid cooling)	$q_{conv} = h_{conv} \cdot A \cdot (T_3 - T_{fluid})$

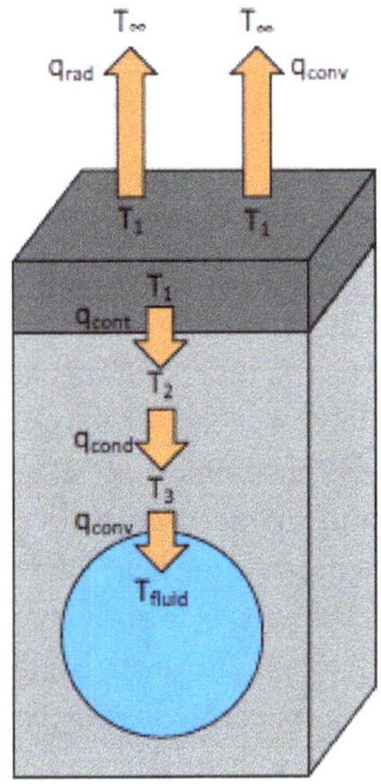

Fig. 5. Heat transfer phenomena during hot stamping process (Schuler Pressen GmbH).

For an optimum heat transfer between component and tool, the contact surface should not exhibit a scale or a gap. The heat conductivity within the tool can be considerably influenced by the choice of the tool material. Another important factor with respect to heat drain is the design of the cooling ducts, which is defined by the size, location, and distribution of the cooling ducts. The heat drain can be accelerated by using a coolant with a low temperature in order to increase the temperature difference between the coolant and the tool and therefore the resulting heat flux (Steinbeiss et al., 2007). The cooling holes can be drilled in the forming tool. For this method, the machining restrictions should be considered in the design of the hole position (Steinbeiss et al., 2007). Therefore, an optimal design of the cooling system as to heat transfer is not possible. Another alternative is to provide the cooling holes as pipes in the casting mold of the tool (Kuhn and Kolleck, 2006). The unrestricted design of the cooling system is the advantage of this method. Alternatively, the tools can be manufactured using the lasered blank segments, which are screwed and form the tool surface with integrated cooling holes (Freieck, 2007). This method is very cost-effective, but the lamellar design can have negative effects on the part surface and the heat transfer within the tool. Todays, additive manufacturing also assists design the cooling channels. Where the cooling channels cannot drill near to the active surface, one can substitute this segment with a 3D printed segment, which exhibits the same properties as conventional.

3.2. *Surface and contact pressure*

One of the most critical factors is the contact pressure between the blank or part and the die surface because it controls the heat transfer time very much and therefore the cycle time.

The experiments reveal that the interface heat transfer coefficient is proportional to contact pressure. It can be inferred that the contact condition at the interface between sheet metal and tool surfaces during the die quenching stage plays a very critical role in the cooling rate of the sheet metal (Hamasaiid et al., 2015).

Surface roughness and contact characteristics on the interface between blank and die, are critical. Surface irregularities promote the formation of gaps between the tool and blank, where air is retained. This lack of contact implies that heat transfer is not carried out completely by conduction phenomena, but through convection and radiation. Conduction is constrained only at some discrete points on the interface. The heat transfer through conduction and convection of the fluid in the cavities is negligible according to the reference because the heat transfer coefficient of the fluid is small. The radiant heat of the noncontact part is also very small (Abdulhay et al., 2001). Heat is primarily transferred through the heat conduction between for real contact spots between the asperities of the contacting bodies. Therefore, and as it is known, more surfaces roughness implies lower Heat transfer coefficient (HTC). In the case of hot stamping processes as a high pressure is applied, the contact spots should be increased by deforming or penetrating the asperities into the contacting surface. Consequently HTC increases when the forming pressure increases. A larger contact force will cause more deformation of the asperities, so the actual contact area increases, thus HTC also and the thermal contact resistance decreases. Higher pressure speeds up the blank cooling. At the beginning of contact, the temperature difference at the interface is large, so it is the thermal kinetic force, resulting in a rapid HTC increase. When blank temperature and die surface temperature approach, thermal kinetics becomes weaker and HTC decreases. Increasing the pressure entails higher values of HTC (Hamasaiid et al, 2015).

There are different possibilities to generate force to have high contact pressure during hot stamping process (Fig. 6). Conventional springs, nitrogen springs, hydraulic cylinders in the die are some of the possibilities to have higher contact pressure. Schuler Pressen GmbH new cushion technology, PCH© Flex (Pressured Control Hardening), is special developed system for press hardening with interacting individual cylinder. This system enables to reduce cooling time while generating high contact pressure which can be verify individually to the part. PCH© Flex made it possible to have high controlled pressure to compensate the material tolerances and faster cooling rate, respectively (Aspacher et al., 2015).

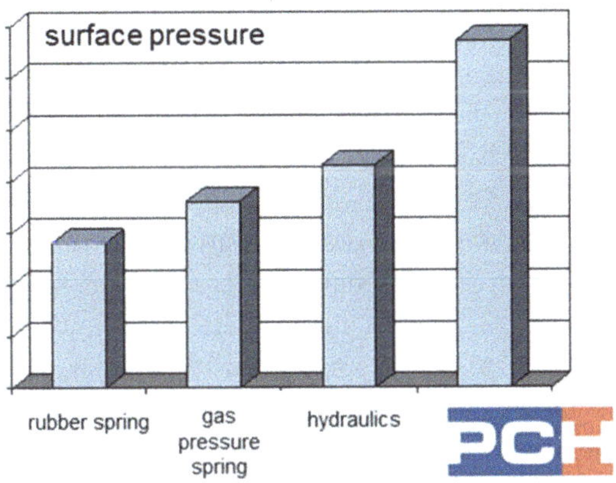

Fig. 6. Surface pressure affected by different strategy (Schuler Pressen GmbH).

3.3. *Wear*

The wear resistance of the tool has been measured by Dessain et al. (2008) using a strip drawing equipment which is adapted to high temperature tests. The resistance heating device allows heating up the Al–Si coated 22MnB5 strip. The heated strip slides over the die radius. For this test, the contact surface was made of abrasion and adhesion areas. The die abrasion was predominant and adhesion of Al–Si was observed at the end of the contact area with the blank. These built layers, formed very early at the beginning of the test, are known to form compacted layers on the tool surface during sliding of the strip (Hardell and Prakash, 2008) and exhibited low wear. Using the appropriate die material combined with an adjusted heat treatment to achieve the desired hardness and wear resistance is an important factor in hot stamping.

During hot stamping thermo-mechanical loads impact adhesive layer build-up on tools. Both, contact pressure and blank temperature influence the wear behavior of aluminum-silicon coated 22MnB5 only marginal. A trend of an overall reduction of wear height at high contact pressures could be stated whereas a reduction of wear height as a result of decreasing blank temperature was only seen for low contact pressures. Also forming induced loads affect wear behavior by breakout of particles out of the aluminum-silicon coating. This failure behavior occurs at a certain threshold temperature which shifts in dependency of compression strain rate. A reduction of real contact surface is the result, impacting adhesive layer build-up on hot stamping tools. A qualitative prediction of adhesion of the aluminum-silicon coating on hot stamping tools is possible by taking this failure behavior into account (Karbasian and Tekkaya, 2010).

3.4. *Tailored properties*

The fully martensitic transformation during hot stamping of the parts leads to a tensile strength of up to 1500MPa and a low elongation of about 5%. To improve and optimize the crash and deformation characteristics, press hardened components with partial adjusted mechanical properties - the so-called Tailored Properties - are realized, which are based on a local enhancement of the ductility. The main purposes are the well-directed deformation in the loading case, the enhancement of the maximum absorbable deformation energy and the prevention of the crack growth and spreading. The manufacturing of a single part with tailored properties can be carried out using different process control strategies or using tailor welded blanks. Alternatively, the thermal process can be influenced by reducing the cooling rate below the specific 27 K/s to avoid a fully martensitic microstructure in defined areas of the component during the forming process, or by reducing the annealing temperature below the material-specific Ac_3 temperature, which leads to an incomplete austenitization. Both methods implicate a lower strength and therefore a higher ductility. The other areas of the component are quenched, following the commonly known press hardening time–temperature profile (Stoehr et al., 2009).

Next to the well-known manufacturing strategies, the partial tempering of press hardened parts offers the opportunity of generating tailored Properties without regarding the preceded process chain and provides the potential of adjusting all requested mechanical properties, independent of the required geometry. Partial tempering of press hardened steel by direct flame impingement is introduced by (Zimmermann et al., 2015). A highly flexible manufacturing approach enables the generation of constantly reproducible mechanical properties in dependence of the maximum annealing temperature induced by partial flame impingement. The use of laser softening of press hardened steel in high volume production lines is discussed by (Schaefer et al., 2015), where the laser light is not only dedicated to cutting or welding but can also be used for softening local areas of hot stamped structural steel parts working like a single sided surface heat source.

In order to generate different cooling conditions, usually the tools are heated to decrease the cooling rate of the material. In contrast, the other possibility is based on a change of the heating process. In this case several process strategies are conceivable. First of all two different areas within a blank can be achieved if one area is completely austenitized and the second area is not austenitized due to a temperature level below the austenitisation start temperature. As the result of this approach, after press hardening, a part is provided with two areas exhibiting different mechanical properties (e.g. soft, hard). Another approach uses the possibility of partial austenitisation of the blanks before press hardening. Depending on the transformation grade of the microstructure during the heating of the material, different final mechanical properties can be obtained after quenching. Based on this approach studies on a new press hardenable steel grade were carried out. (Marten et al., 2015) use this method, where the results obtained are compared with another steel grade with similar tensile strengths after a conventional press hardening process. For the partial austenitisation process, material simulations were performed to provide information on the theoretical process windows. By means of basic material tests the simulations were evaluated afterwards. Finally, demonstration parts were manufactured by using the two different process routes.

Behrens et al. (2015) uses subsequent cooling outside the press for adjusting the tailored properties. With the designed quenching device cooling rates of part sections are varied flexibly by applying a water-air spray cooling. The high potential of this process is demonstrated on behalf of the mechanical properties which are compared for conventionally produced parts and parts featuring tailored properties induced by the modified process sequence. Last but not least, (Liu et al., 2015) presents a new hot stamping process to make tailored properties based on air cooling. In high strength region, microstructure of blanks with full austenite are quenched from austenite to martensite by tools with cooling water channel, and after air cooling microstructure evolution involving ferrite, bainite and martenstie have been fulfilled in absorption region. And tailored cooling rates are formed during transition region because of tailored contact pressures.

No matter which method is used, the process to produced part with tailored properties should be stable and reproducible. It is important that, the overall process time stay almost the same in comparison with production normal parts.

References

1. B. Abdulhay, B. Bourouga and C. Dessain, Experimental and theoretical study of thermal aspects of the hot stamping process, *Appl. Ther. Eng.* **31**, 674 (2001).
2. P. Åkerström, Modeling and simulation of hot stamping, Phd thesis, Luleå University of Technology (2006).
3. J. Aspacher, Forming hardening concepts, in 1st International Conference on Hot Sheet Metal Forming of High-Performance Steel, (Kassel, Germany, 2008).
4. J. Aspacher, D. Haller and P. Thom, From First Draft to Serial Production: Hot stamping Part design and Feasibility Study with Respect to Functionality and Optimization of Production Costs, in *Proceeding CHS2*, (Toronto, Canada, 2015).
5. B. Behrens, H. Maier, F. Nuenberger, J. Schroedler, L. Wolf and C. Gaebel, Hot Forming and Subsequent Cooling Outside the Press for Adjusted Tailored Properties of 22MnB5 Steel Sheets, in *Proceeding CHS2*, (Toronto, Canada, 2015).
6. G. Berglund, The history of hardening of boron steel in northern Sweden, in *1st International Conference on Hot Sheet Metal Forming of High-Performance Steel*, (Kassel, Germany, 2008).
7. U. Freieck, in zwei Wochen zum Serienwerkzeug, *Blech* **2**, 46 (2007).
8. L. Garcia Aranda, Y. Chastel, J. Fernández Pascual and T. Dal Negro, Experiments and simulation of hot stamping of quenchable steels, *Adv. Tech. Plasticity* **2**, 1135 (2002).
9. A. Hamasaiid, C. Casas and I. Valls, Role of Tool Material in the Interfacial Problematic of Tool/Blank in Press Hardening Process, in *Proceeding CHS2* (Toronto, Canada, 2015).
10. J. Hardell and B. Prakash, High Temperature Friction and Wear Behavior of Different Tool Steels During Sliding Against Al-Si-Coated High Strength Steel, *Tribol. Int.* **41**, 663 (2008).
11. H. Karbasian and E. Tekkaya, A Review on Hot Stamping, *J. Mater. Process. Tech.* **210**, 2103 (2010).

12. D. Kuhn and R. Kolleck, Warmumformung - den höheren Festigkeiten folgend, *MM das IndustrieMagazin* **17**, 86 (2006).
13. P. Liu, K. Wang, Z. Wang, B. Zhu and Y. Zhang, A New Hot Stamping Process to make Tailored Properties Based on Air Cooling, in *Proceeding CHS2,* (Toronto, Canada, 2015).
14. T. Marten, H. Block and T. Troester, Partial Hardening of New Press Hardenable Steels, in: *Proceeding CHS2,* (Toronto, Canada, 2015).
15. M. Merklein, J. Lechler and T. Stöhr, Characterization of tribological and thermal properties of metallic coating for hot stamping boron-manganese steels, in *Proceedings of the 7th International Conference Coatings in Manufacturing Engineering,* (Chalkidiki, Greece, 2008).
16. M. Schaefer, D. Schuoecker, O. Spitzer and T. Harrer, Laser Softening of Press Hardened Steel in High Volume Production Lines, in *Proceeding CHS2,* (Toronto, Canada, 2015).
17. M. C. Somani, L. P. Karjalainen, M. Eriksson and M. Oldenburg, Dimensional changes and microstructural evolution in a B-bearing steel in the simulated forming and quenching process, *ISIJ. Int.* **4**, 361 (2001).
18. T. Stoehr, J. Lechler and M. Merklein, Investigations on different strategies for influencing the microstructural properties with respect to partial hot stamping, in *2nd International Conference on Hot Sheet Metal Forming of High-Performance Steel*, (Luleå, Sweden, 2009).
19. H. Steinbeiss, H. So, T. Michelitsch and H. Hoffmann, Method for optimizing the cooling design of hot stamping tools, *Prod. Eng. Res. Dev.* **1**, 149 (2007).

Superior Tool Steels for Hot Stamping of AHSS

J. J. Wilzer
Special Steels, Dörrenberg Edelstahl GmbH, Engelskirchen, 51766, Germany
E-mail: jens.wilzer@doerrenberg.de

C. Escher
Quality Department, Dörrenberg Edelstahl GmbH Engelskirchen, 51766, Germany

Hot stamping of advanced high strength steels is a key technology in the automotive industry, so as to reduce the weight of the vehicle to save fuel consumption and therefore emission of carbon dioxide. However, tools to produce such parts are often exposed to high wear, thermal and mechanical loads. Standard hot work tool steels do not maintain a sufficient lifespan to ensure a high productivity and quality of the hot stamped parts they produced. Therefore, new tooling concepts and tool steel grades have been developed to improve the performance of hot stamping tools. In here, we study the development of a new hot stamping tool steel that not only has a higher thermal conductivity but also fulfills all other technical requirements. This paper gives an overview of the requirements of hot stamping tool steels and makes some recommendations of tool steels and heat treatment conditions. Finally, a short outlook is given into future research and development of the special steel CP2M® and sand-printing of casting molds, which could be combined using casted tools made from special tool steel GP2M.

Keywords: Tool steels; Hot stamping; Wear resistance; Thermal conductivity.

1. Introduction

The history of hot stamping started in 1977 with a Swedish patent that described a process for the production of hardened steel parts [1]. In the beginning, it wasn't very popular. The first hot stamping line came into operation in 1984 and three years later the world wide production of hot stamped car body panels reached approximately 3 million parts [2, 3]. Its popularity increased with the turn of the millennia and in 2011 the worldwide production of hot stamped parts reached more than 124 million [4]. Currently, additional hot stamping lines are installed in Europe, Asia, and in North America.

In contrast to classical cold forming processes, hot stamping tools get into contact with hot sheet metals which results in a warming of the tool surface. To avoid warming, the heat that was brought in needed to be removed quickly. Otherwise, wear of the tools surface increases and quality of the parts to be produced drops down [5-7].

In the past hot stamping tools were made from standard hot work tool steels like 1.2344 (X40CrMoV5-1) or 1.2367 (X38CrMoV5-3) because they had been used for years in other hot working applications like die forging, high pressure die casting, and aluminum extrusion. The development of new hot stamping tool steels was pushed by the automotive industry because they wanted to decrease cycle time and increase productivity of the hot stamping process. To achieve this, tool steels with higher thermal conductivity were needed. Therefore, first developments of hot stamping tool steels were focused on thermal conductivity [8-11]. However, these developments did not consider other technical requirements like mechanical properties, wear resistance, and heat treatment behavior. Some of these newly developed tool steels failed totally when they were tested under real conditions.

On the basis of these developments and practical experience, Dörrenberg Edelstahl started their own investigations and developed a new hot stamping tool steel not only has a higher thermal conductivity but also fulfills all other technical requirements [12-14].

The present report gives an overview of the requirements of hot stamping tool steels and makes some recommendations of tool steels and heat treatment conditions. Finally, a short outlook is given into future research and development done by Dörrenberg Edelstahl.

2. Requirements of Hot Stamping Tools

2.1. *Process description*

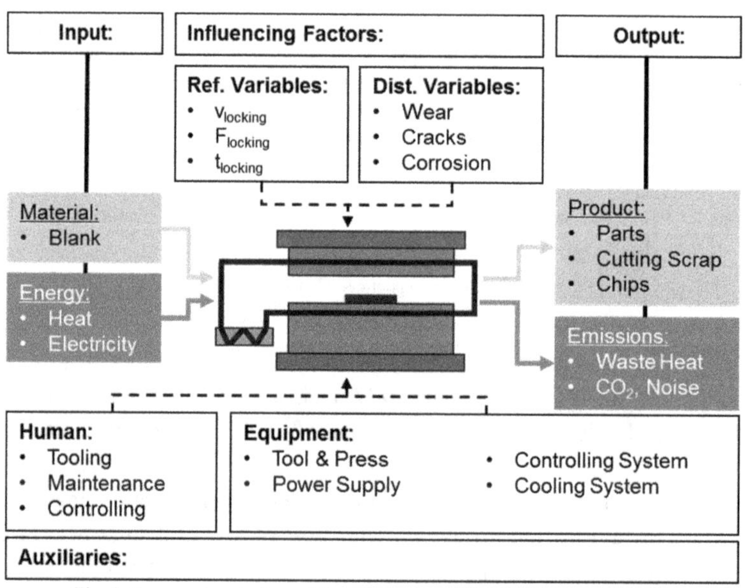

Fig. 1. Process description of the sub-process "forming and quenching".

Originally hot stamping was divided into the direct and indirect process. Nowadays the direct hot stamping process is widely used. Therefore, the present report focusses on the direct hot stamping process. For the hot stamping tool the sub-process "forming and quenching" is most important. This sub-process is schematically shown in Fig. 1.

On the input side there is the hot sheet metal blank and the electrical energy for the press. The hot sheet metal blank is transferred into the tool which then closes, forms, and quenches the blank. The reference variables are the locking speed, the locking force and the locking time. Auxiliaries can be human (tooling, maintenance, controlling) or equipment (tool & press, power supply, etc.). In ideal case, there are the hot stamped part and the waste heat as output of the hot stamping process. But in reality there are disturbance variables that have a significant influence on the hot stamping process. Due to friction between the sheet metal blank and the tool surface the drawing radii of the tool wear out and the tolerances of the parts to be produced can't be fulfilled anymore. Additionally, the contact pressure between the tool and the sheet metal drops down which lowering the heat transfer between both sides. And this may lower both productivity and product quality. Another disturbance variable is cracking of the tool. In some cases operators increase the overall contact pressure to enhance heat transfer in some areas. The problem is that in some other areas there will be a high mechanical load which may cause fatigue cracking after a short period of time. And if there is a corrosion attack inside the cooling channels due to contaminated cooling water cracking of the cooling channels will be promoted [15].

2.2. *Requirements on tool steels*

Wear resistance, toughness, and corrosion resistance are three properties that have to be considered with respect to hot stamping tool steels. In addition, a high thermal conductivity may increase heat transfer between the tool surface and the cooling channels which results in faster cooling and improved productivity. For the tooling shop machinability is more important. The heat treater instead looks for a material that can be heat treated easily without having a high distortion or crack risk. In some cases a welding operation may be needed on the tool. In this case the tool steels needs to be weldable. To sum up, a tool steel which is used for hot stamping tools needs to have a high wear resistance, thermal conductivity, and toughness. It should be less sensitive to corrosion, easy to machine, and easy to heat treat. And in case of emergency or for some maintenance work it should be able to weld.

3. Tool Steel Recommendations

Hot stamping tools can be classified into four different types according to Tab. 1. Prototype tools are not exposed to high thermal or mechanical loads. They are used to produce a few component examples. Due to engineering chance some tool segments need to be reworked or new segments have to be machined. Therefore, tool steels for prototype tools need to be low priced and they should have a good machinability. In this case, unalloyed cold work steel C45U (1.1730) features a low price whereas sulfur alloyed cold work steel 40CrMnMoS8-6 (1.2312) has an improved machinability. Small series tools should have a sufficient wear resistance. Besides tool steel 40CrMnMoS8-6, prehardened hot work steel 55NiCrMoV7 (1.2714) is the material of choice for such kind of tools. Tools for medium series should have a higher wear resistance. In this case hot work steels X40CrMoV5-1 (1.2344) and X38CrMoV5-3 (1.2367) prehardened to 45 HRC show a good performance. For mass production tools wear resistance, thermal conductivity and tempering resistance are the most important properties which have to be guaranteed by the tool steel. In the past, hot work tool steels like X40CrMoV5-1 and X38CrMoV5-3, quenched and tempered to 54 HRC, were used for such kind of tools. However, there was a lack of wear resistance for these tool steels which resulted in lots of maintenance and repair work. Nowadays, mass production tools are made of special tool steels like WP7V, DE-HS and CP2M®.

Table 1. Recommendations for tool steels used for hot stamping tools.

Type	Requirements	Tool Steel (Hardness)
Prototype	Low Costs, Easy Machining	C45U (soft annealed) 40CrMnMoS8-6 (~300 HB)
Small Series	Low Costs, Easy Machining, Sufficient Wear Resistance	40CrMnMoS8-6 (~300 HB) 55NiCrMoV7 (~380 HB)
Medium Series	Easy Machining, Good Wear Resistance, Good Tempering Resistance	55NiCrMoV7 (~380 HB) X40CrMoV5-1 (~45 HRC) X38CrMoV5-3 (~45 HRC)
Mass Production	Superior Wear Resistance, High Thermal Conductivity, Good Tempering Resistance	DE-HS (52-56 HRC) WP7V (54-58 HRC) CP2M® (56-60 HRC)

Special tool steel WP7V belongs to the 8%-chromium steels. Due to the medium carbon concentration of 0.50 wt.% and enhanced vanadium and molybdenum concentration it can be quenched and tempered to 58 HRC. Its big advantages over standard hot work steels are the better hardenability, higher hardness and improved wear resistance. Due to the slightly lower alloying concentration DE-HS achieves a higher toughness as compared to WP7V after quenching and tempering.

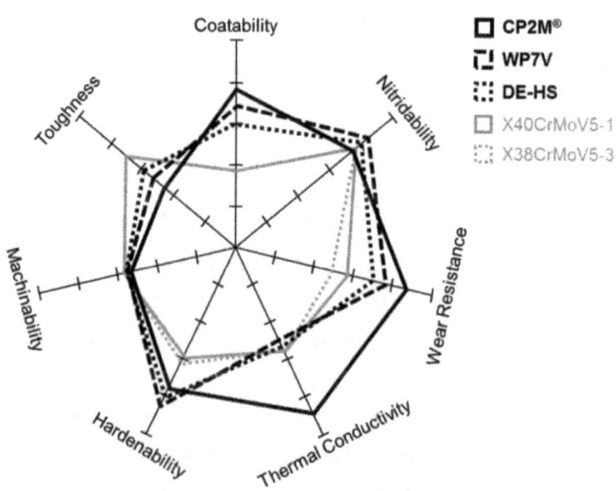

Fig. 2. Property profile of special steels CP2M®, WP7V, and DE-HS against hot work tool steels X40CrMoV5-1 (1.2344) and X38CrMoV5-3 (1.2367) (rating is based on measuring values for the quenched and tempered condition and on the experience of customers for hot stamping tool steels).

Special tool steel CP2M® was developed together with Ruhr-University Bochum in Bochum, Germany. The aim of the project was the development of a tool steel that combines all properties that are beneficial for hot stamping tools. To reach a hardness of up to 60 HRC after quenching and tempering CP2M® has a carbon concentration 0.65 wt.% and additions of carbide formers like vanadium, molybdenum, and tungsten. To have both a sufficient hardenability and high thermal conductivity, its chromium concentration was fixed to 2 wt.%. The big advantage of CP2M® over standard hot work steels is the combination of superior wear resistance and enhanced thermal conductivity. The relation-ship of hardness and thermal conductivity of standard hot work steels and special tool steels WP7V and CP2M® is shown in Fig. 3.

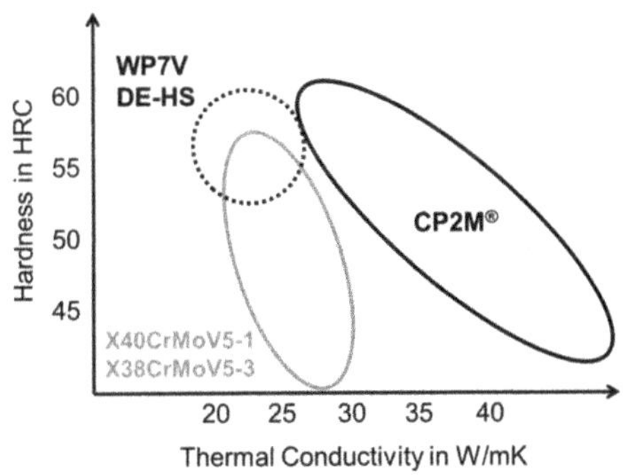

Fig. 3. Relationship of hardness and thermal conductivity of standard hot work steels X40CrMoV5-1 (1.2344) and X38CrMoV5-3 (1.2367) and of special tool steels WP7V, DE-HS, and CP2M®.

Due to the high chromium concentration, standard hot work steels and special tool steels WP7V and DE-HS do not achieve high thermal conductivities after quenching and tempering. The metallurgical background is described in other publications [8-14]. However, WP7V and DE-HS can be quenched and tempered to a higher hardness which then gives a better wear resistance than standard hot work tool steels. Special tool steel CP2M® instead reaches both a high hardness and high thermal conductivity due to the modified alloying composition. As a consequence it has a higher wear resistance in hot stamping applications because the surface temperature does not get too high during each stamping cycle. This relationship is illustrated in Fig. 4. Micrographs a) and b) were taken from a hot strip drawing die made of 5 wt% cold work steel that worked 1,600 strokes under test conditions. Due to the low thermal conductivity there is a serious tempering effect at the surface region that results in a stronger etching effect and lower micro hardness. At higher magnification a white etching zone is visible right on top of the surface. This zone looks as it suffered a lot of plastic deformation. However, it is not clear if this material layer belongs to the original surface or if it is an adhesive material transfer from the hot sheet metal blank.

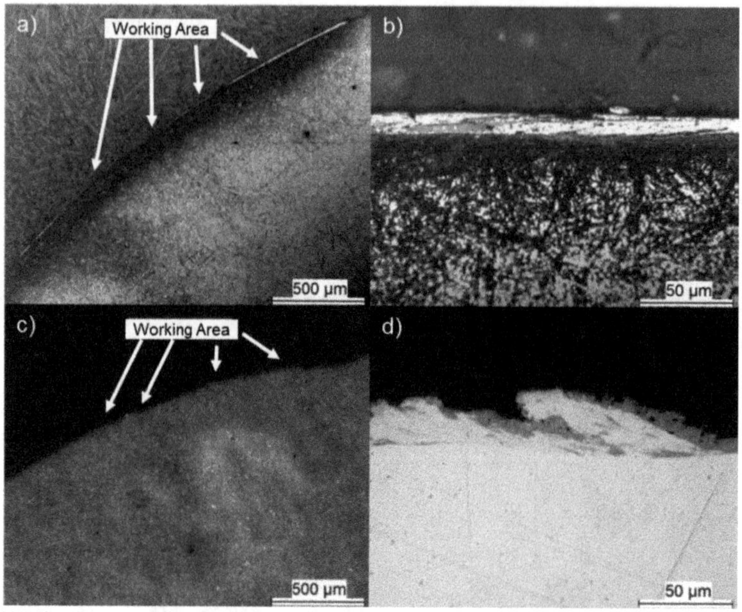

Fig. 4. a) + b) Microstructure of a cold work tool steel with 5 wt% chromium, quenched and tempered to 58-60 HRC, after 1,600 hot strip drawing tests. c) + d) Microstructure of special tool steel CP2M®, quenched and tempered to 58-60 HRC, after 65,000 strokes in a production tool.

In comparison to this Fig. 4 c) and d) show the results of a real hot stamping tool made of special steel CP2M® after 65,000 strokes. In this case, no tempering zone was detected on the surface. However, there was still plastic deformation on the surface. The question is if thermal conductivity not only influences heat removal rate inside the tool but also affects the wear resistance of the tool surface. In the current cases it looks like low thermal conductivity promotes tempering and softening of the surface region which then increases tool wear. This relationship needs to be investigated in the framework of future research. Within the next section we want to present practical experience our customers have won during testing of special steel CP2M®.

4. Practical Experience

The newly developed special tool steel CP2M® just has been tested by several partners under real condition. At this point we are able to present two examples to highlight the features of CP2M® over standard hot work steels in hot stamping applications.

One of the earliest practical test was done for a B-pillar tool that is shown in Fig. 5 a). The segments that were made of CP2M® are shown in Fig. 5 b). The segments were austenitized at 1050°C for 20 min, quenched with pressure gas (10 bar), and triple tempered at 550°C, 540°C, and 530°C for 3 hours each. The final hardness was 56-58 HRC. The four segments were tested in production, forming a B-pillar made of AlSi-coated 22MnB5 sheet metal blank. The benchmark for this tool was the standard hot work steel X38CrMoV5-3 (1.2367), quenched and tempered to 52-54 HRC. Tools made of this standard material lasted 200,000 to 250,000 strokes before being reworked or replaced. The segments made of CP2M® have been working for over 400,000 strokes and they are still running in production without any rework. In the end, CP2M® will have more than double the lifetime of standard hot work steels.

Fig. 5. a) B-pillar tool that was used for testing of special tool steel CP2M®, b) segments that were made of CP2M® and tested in the B-pillar tool.

A second tool that was used for tests of CP2M® was a bumper tool which is shown in Fig. 6 a). The segments that were made of special tool steel CP2M® are shown in Fig. 6 b). The segments were austenitized at 1070°C for 20 min, quenched with pressure gas (10 bar), and triple tempered at 550°C, 540°C, and 540°C for 2.5 hours each. The final hardness was 58-60 HRC. Additionally, the segments were plasma nitrided at 520°C for 18 hours. The benchmark for such hot stamping tools was standard hot work steel X38CrMoV5-3 (1.2367), quenched and tempered to 52-54 HRC with additional plasma nitriding. Lifetime of such tools made of X38CrMoV5-3 was 120,000 strokes. With special tool steel CP2M® the segments lasted 200,000 strokes which is an increase of 66%. This is a big benefit for the customer due to the fact that he saves time and money for rework or replacements of tools.

These two examples show that there is much room for improvement with respect to hot stamping tool steels. One way is to use special tool steels that have a property profile that fits well to the requirements of hot stamping tools. Especially special steel CP2M® fulfills all the requirements that are needed in hot stamping applications: high hardness and wear resistance, high thermal conductivity, good toughness, good machinability, good hardenability. However, there is further development work especially with respect to cooling strategies. One of these projects is production of cast hot stamping tools.

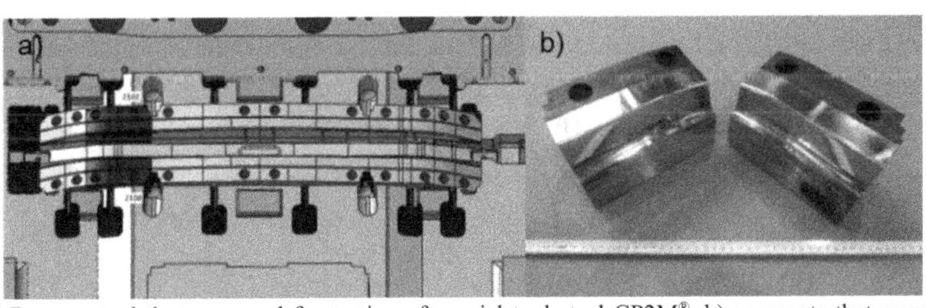

Fig. 6. a) Bumper tool that was used for testing of special tool steel CP2M®, b) segments that were made of CP2M® and tested in the B-pillar tool.

5. Cast Hot Stamping Tools

The first generation of cast hot stamping tools was produced using steel tubes for the cooling channels. As it is shown in Fig. 7, the first step was to produce a polystyrene model at which stainless steel tubes were mounted. The second step was the assembling of the casting mold followed by the casting process. Finally, the castings were fettled and heat treated. One of the biggest issues was the mounting of the steel tubes at the polystyrene model. If they were not mounted successfully they could change position during transportation or even during casting which means that the casted product was rejected. And especially transportation of the assembled model was not easy due to the weight of the steel tubes. These issues were the reasons for the development of a new process.

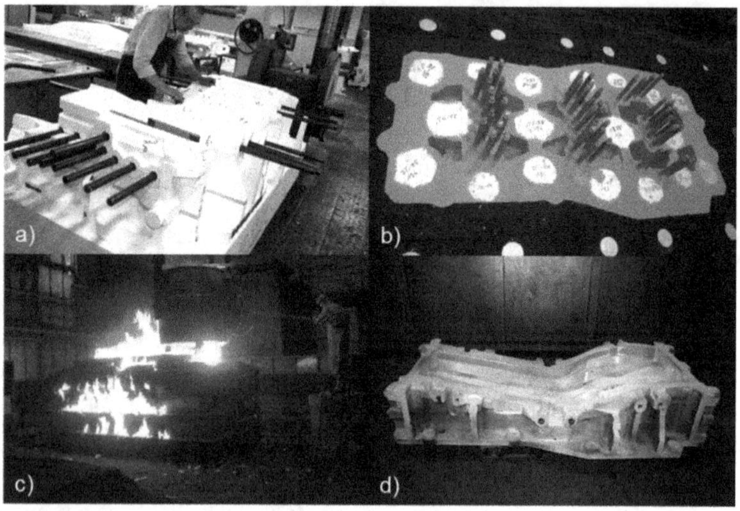

Fig. 7. Process flow for the classical production of casted hot stamping tools: a) assembling of the polystyrene model, b) assembling of the sand mold, c) casting, d) fettling of the casted tool.

The new generation of cast hot stamping tools is produced using the sandprinting process for the casting molds. As it is shown in Fig. 8 a), the first step during this production process is the engineering of the tool. This, of course, is an iterative process in which several constructive changes will happen from the initial tool geometry to the final geometry of the tool. The final CAD-file then needs to be translated into a compatible file for the sandprinter. The printing of the mold occurs in layers of 300 μm thickness. Each layer consists of sand and binder A. The printing head then sprays binder B on top of the area where the sand needs to get solid. Afterwards, the table moves down and the printing process starts from new. After 18 to 22 hours the whole printing process finishes and the printed sand molds can be taken out of the printer.

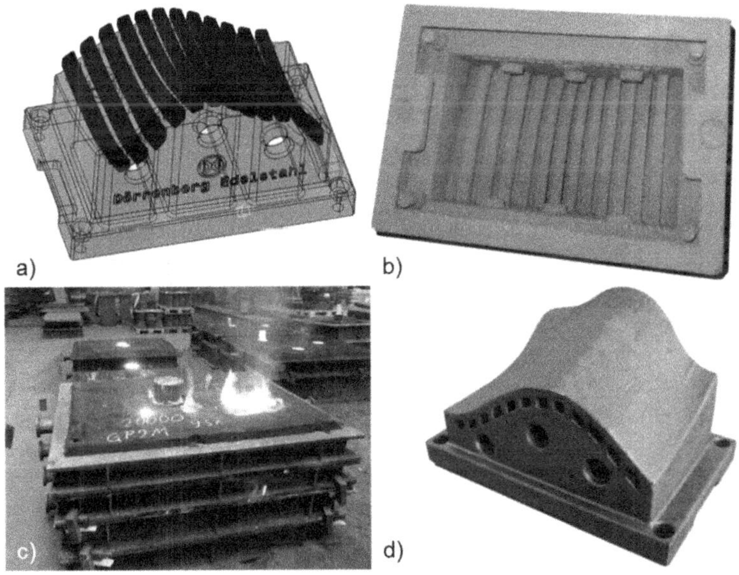

Fig. 8. Process flow for the casting process including sandprinting of the casting mold: a) construction and engineering process, b) 3D-printing of the casting molds, c) assembling of the molds and casting, d) fettling of the casted tools.

The advantages of the casting process including sandprinting of the casting molds over the classical casting process are as follows:

- Better cooling behavior due to improved geometry and trace of the cooling channels
- Better surface quality due to smoother surface of the printed sand molds
- Smaller dimensional and machining allowance are possible
- No flue gas of burning polystyrene, no air pollution, less material defects

The combination of both advantageous developments, special steel CP2M® and sand-printing of casting molds, can be combined using casted tools made of special tool steel GP2M®, the casted version of special tool steel CP2M®.

Acknowledgments

The authors gratefully thank the Chair of Materials Technology of the Ruhr-University Bochum for the project within special steel CP2M® was developed. Furthermore, we want to thank Mr. Matthias Kotzian from Volkswagen AG and Mr. Jörg Lerch from Magna International Stanztechnik GmbH for testing special steel CP2M® under real hot stamping conditions.

References

1. Norrbottens Järnverk A. B., Manufacturing a hardened steel article, Patent Specification 1490535 (November 1977).

2. G. Berglund, The history of hardening of boron steel in northern sweden, in Proceedings of the 1st International Conference on Hot Sheet Metal Forming of High-Performance Steel, (Kassel, Germany, 2008).
3. H. Karbasian and A. E. Tekkaya, A review on hot stamping, *J. Mater. Process. Tech.* **15**, 210 (2010).
4. R. Bartos, Heißes Eisen: warmumgeformte Stähle, *Stahl-Informations-Zentrum,* Medieninformation (2011).
5. M. Naderi and M. Ketabichi, Analysis of microstructure and mechanical properties of different boron and non-boron alloyed steels after being hot stamped, *Procedia Eng.* **7**, 460 (2011).
6. C. Boher and S. Le Roux, Experimental investigation of the tribological behavior and wear mechanism of tool steel grades in hot stamping of a high-strength boron steel, *Wear.* **3**, 286 (2012).
7. M. Merklein and T. Stoehr, Investigation on parameters influencing thermal and frictional properties within hot stamping, *HTM Journal of Heat Treatment and Materials* **5**, 391 (2013).
8. Valls and U. Paar, Capital role of tool materials in hot stamping of 22MnB5 sheet, in *Proceedings of the 8th International Tooling Conference,* (Aachen, Germany, 2009).
9. Valls and U. Paar, Benefits from using high thermal conductivity tool steels in the hot forming of steels, *La Metallurgia Italiana* **11**, 23 (2010).
10. S. Gelder and G. Jesner, New high performance hot work tool steel with improved physical properties, in *Proceedings of the 9th International Tooling Conference,* (Leoben, Austria, 2012).
11. E, Meurisse and C. Ernst, Improvement of thermal conductivity of hot-work tool steels by alloy design and heat treatment, in *Proceedings of the 9th International Tooling Conference,* (Leoben, Austria, 2012).
12. J. Wilzer, Wärmeleitfähigkeit martensitisch härtbarer stähle–physikalische zusammenhänge, Einflussfaktoren und technischer Nutzen, PhD thesis (Bochum, Germany, 2014).
13. J. Wilzer and S. Weber, Werkstofftechnische anforderungen an presshärtewerkzeuge am beispiel der werkzeugstähle X38CrMoV5-3, 30MoW33-7 und 60MoCrW28-8-4, *HTM Journal of Heat Treatment and Materials* **6**, 325 (2014).
14. J. Wilzer and Ch. Escher, Optimierte eigenschaften von werkzeugstahl für presshärtewerkzeuge, *HTM Journal of Heat Treatment and Materials* **1**, 35 (2016).
15. I. Schruff and P. Gümpel, Corrosion–An often neglected phenomenon in hot-work tools, in *Proceedings of the 10th International Tooling Conference,* (Bratislava, Slovakia, 2016).

Premium Tool Steels Solutions for Efficient Hot-stamping Applications

D. Wang[1,a,†] and I. Schruff[2,b]
[1]*Kind Special Alloys (Beijing) Trading Ltd., Chaoyang District, Beijing, 100122, China*
[2]*Kind & Co., Edelstahlwerk, GmbH & Co., KG, Wiehl, 51674, Germany*
E-mail: [a,†]David.Wang@Kind-special-alloys.hk, [b]Ingolf.Schruff@Kind-co.de

Based on a study on the loads of tools the paper describes requirements for tool steels for hot-stamping tools and presents two premium tool steel solutions specially developed for hot-stamping tools. For Tailored Tempering, a process that allows for local adjustment of the properties of the component, electrically heated tool segments are required in addition to cooled segments. The paper describes a special hot-work tool steel which provides the specific properties for this type of tool segments. Finally the paper offers recommendations for proper adjustment of hardness and successful avoidance of corrosion induced tool failure.

Keywords: Hot-stamping; Tool steel; Tailored Tempering; Hardness; Corrosion.

1. Introduction

Driven by the necessity to reduce the weight of vehicles hot-stamping has quickly developed to a leading technology for the production of light weight high-strength automotive components. Originally used to produce only few components for some passenger vehicles the variety of components as well as the total number of produced parts have been growing rapidly. Nowadays hot-stamped components are also used in the production of commercial vehicles as weight reduction is equally important for these vehicles. With the on-going development of e-mobility the importance of hot-stamping as a key production technology will be enforced, since weight reduction is supporting larger battery ranges and agility.

Hot-stamping is in a strong competition with technologies like pressure die casting of structural components – a technology which has also increasingly gained access to the car design. Efficient hot-stamping requires tools which allow a reliable production without premature tool failures or unexpected repair due to massive wear. In order to achieve short cycle times a quick transport of the heat out of the component and thus intense cooling is certainly desirable.

Some hot-stamped components require local variations of their mechanical properties. For crash safety reasons some sections of the B-pillar of passenger cars must have a drastically increased toughness in order to compensate crash energy by plastic deformation of the component. The tailored tempering technology using heated tool segments locally reduces the cooling rate of the sheet metal and this way avoids the martensitic hardening of the sheet metal, leaving this part of the component with more toughness.

The permanently on-going process development in the hot-stamping industry has led to various tool modifications. Shapes of components and tools have become more challenging, the distances between cooling channels and working surfaces have been reduced to a few millimetres to allow a faster heat transfer into the cooling channels and shorter production cycles. On the other hand, these developments have drastically increased thermal and mechanical stresses to which the tools are exposed during operation. e.g., due to intensive cooling corrosion induced cracks between cooling channels and working surface can increasingly be observed. Kind & Co. has investigated these phenomena and developed suitable recommendations and countermeasures to avoid these defects.

This paper describes the tool steel solutions recommended by Kind & Co. for the water-quenched as well as for the heated tool sections, advises on the appropriate hardness selection and offers help to avoid corrosion induced tool defects.

2. Stress Situation in Hot-stamping Dies and Required Tool - steel Properties

A detailed analysis of the stresses to which a hot-stamping die is exposed during operation is the best prerequisite for tool steel selection and heat treatment.

2.1. *Stress analysis*

During operation hot-stamping tools are exposed to a complex system of thermal and mechanical stresses. With each stroke the contact zone of the tool is heated up momentarily. According to own estimations [1] the peak temperature of the tools might rise up to approximately 490 °C within a working shift. Cyclic cooling and heating will result in thermal fatigue during the lifetime of the tools.

The stamping operation itself induces intensive compression stress in the dies and the relative movement of the sheet metal over edges causes intensive abrasive wear.

2.2. *Required tool steel properties*

The required properties of suitable tool steels for hot-stamping tools can be derived from the loads on the tools.

High high-temperature strength is required to withstand the mechanical loads during operation. This goes along a high thermal fatigue resistance to avoid undesired softening of the tools.

High hardness and tempering resistance are required to provide sufficient wear resistance. But wear resistance does not only depend on the steel's hardness but also on the type, size, and distribution carbides in the steel.

A fast heat transfer from the sheet into the cooling system requires a high thermal conductivity of the tool steel. Nevertheless, more relevant are design and distance of cooling channels from the working surface. While thermal conductivity brings a linear advantage to cooling efficiency, the distance of the cooling channels to the surface is of much higher order dependency.

Good toughness helps to avoid cracks between cooling channels and working surface of the tools.

3. Recommended Tool Steels

The recommendation of suitable tool steels for hot-stamping tools always depends on the combination of properties. It should never depend on only one single property such as thermal conductivity. It should always be kept in mind that material properties often interact so that high values of one property might lead to low values of another property.

3.1. *Tool steels for water cooled hot-stamping tools*

For water cooled tool segments Kind & Co. recommends the specially developed tool steel CR7V-L as I t combines the required properties in an excellent way. It has successfully proven its suitability in numerous hot-stamping tools. For applications which require an even higher abrasive wear resistance Kind & Co. has recently developed the grade HS 1. Table 1 gives a survey of their chemical compositions. The standardized hot-work tool steel 1.2344 (AISI H 13) is listed here only for reference purpose.

Table 1. Chemical compositions of the special steel grades CR7V-L and HS 1 (Alloy content in Mass-%).

Steel grade	C	Si	Mn	Cr	Mo	V	Recommended hardness in HRC
1.2344	0.40	1.00	0.40	5.20	1.30	1.00	---
CR7V-L	0.42	0.50	0.40	6.50	1.30	0.80	52 - 54
HS 1	0.50	0.90	0.80	8.00	1.50	1.70	54 - 58

As the tempering curves in Figure 1 clearly indicate, the special grades CR7V-L and HS 1 not only achieve higher hardness levels but also develop a better tempering resistance than the reference hot-work tool steel 1.2344. The high hardness contributes to the good abrasive wear resistance and the tempering resistance prohibits premature softening of the tool sets due to contact with the hot sheet metal.

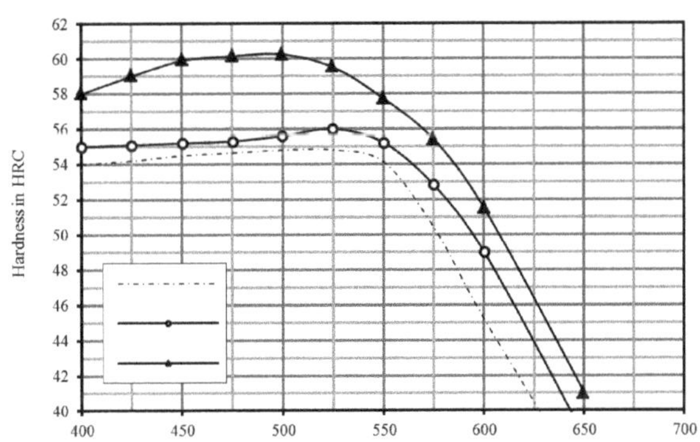

Fig. 1. Tempering curves of the steel grades 1.2344 (H 13), CR7V-L, and HS 1.

Comparing the recommended hardness values for these steels (Table 1) with their tempering curves shows that Kind & Co. strongly recommends not using the steels at their maximum hardness levels as the steels develop their lowest toughness at this hardness.

Abrasive wear resistance is a property which depends on hardness and carbide content, carbide types and carbide distribution. A schematic comparison of the three steels with respect to abrasive wear resistance is shown in Figure 2.

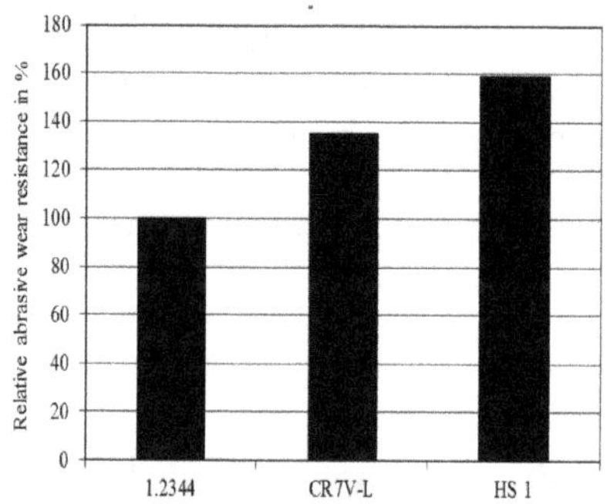

Fig. 2. Schematic comparison of the abrasive wear resistance of steel grades 1.2344, CR7V-L, and HS 1.

Thermal conductivity is responsible for the heat transfer from the sheet metal into the cooling channels. Shortening the cooling period has the best chance to reduce the cycle time of hot-stamping operations [2]. Nevertheless, in the process of steel selection for hot-stamping tools thermal conductivity should not be the only and decisive property as mechanical properties like tempering resistance, high-temperature strength, and wear resistance are opposing to thermal conductivity [1]. Table 2 underlines the good thermal conductivity of CRV-L and HS 1.

Table 2. Thermal conductivity λ in W/(m * K) of the steel grades 1.2344, CR7V-L, and HS 1.

Steel	Test temperature		
Grade	20 °C	200 °C	400 °C
1.2344	25.5	27.1	21.7
CR7V-L	26.7	29.8	30.8
HS 1	25.2	27.3	29.3

The high mechanical and thermal loads on hot-stamping tools not only require hardness and wear resistance but also toughness. As water-cooled hot-stamping tools operate at a high hardness level each contribution to improved toughness will directly influence the tool life. Apart from a correct hardness and heat treatment of the tool segments the metallurgical production process of the steels has an important influence on the toughness and tool life. The Electro-Slag-Remelting (ESR) process improves purity and homogeneity of the steels and this way contributes to an improved lifetime potential of the tools. While the grade HS 1 is produced exclusively via ESR the grade CR7V-L is available as air-melted steel as well as an ESR grade. For highly stressed tool segments Kind & Co. recommends the use of the ESR version of CR7V-L.

3.2. *Tool steels for tailored tempering*

Electrically heated to temperatures of 550 - 650 °C tool segments for tailored tempering prohibit the martensitic transformation of the steel sheets generating locally restricted zones of reduced strength and simultaneously increased plasticity in the components. Besides hardness and wear resistance steels for these tool segments must provide a high long-time tempering resistance in order to withstand the high tool temperatures for long times. As steel grades like 1.2344, CR7V-L, and HS 1 cannot maintain this requirement long enough Kind & Co. recommends the steel grade HMoD for heated tool steel segments in tailored tempering applications.

As Table 3 shows up this grade is highly alloyed with elements like molybdenum and cobalt which contribute to the steel's excellent high-temperature strength and tempering resistance.

Table 3. Chemical composition of the special steel grade HmoD (Alloy content in Mass-%).

Steel grade	C	Si	Mn	Cr	Mo	V	Co	Recommended hardness in HRC
HMoD	0.45	0.30	0.40	4.50	3.00	2.00	4.50	50 – 52

Due to the high alloy content Kind & Co produces the grade HMoD exclusively as ESR grade giving the steel improved purity, homogeneity, and toughness.

The tempering curve of HMoD (Figure 3) presents a much higher tempering resistance than CR7V-L and this way demonstrates the best suitability of HMoD for heated tool segments in tailored tempering applications.

The long-time tempering resistance of these two steel grades is expressed in Figure 4. Samples had been hardened and tempered originally to a hardness of 53 HRC and then exposed to a temperature of 550 °C for up to 100 hours. The curves show that over this long testing period the hardness HMoD declines only slightly. This demonstrates the enormous long-time tempering resistance of HMoD which makes this grade best suitable for heated tool inserts in tailored tempering applications.

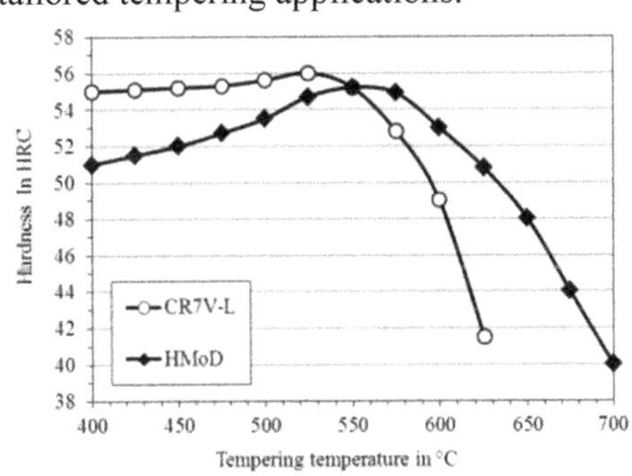

Fig. 3. Tempering behaviour of the steel grades CR7V-L and HMoD.

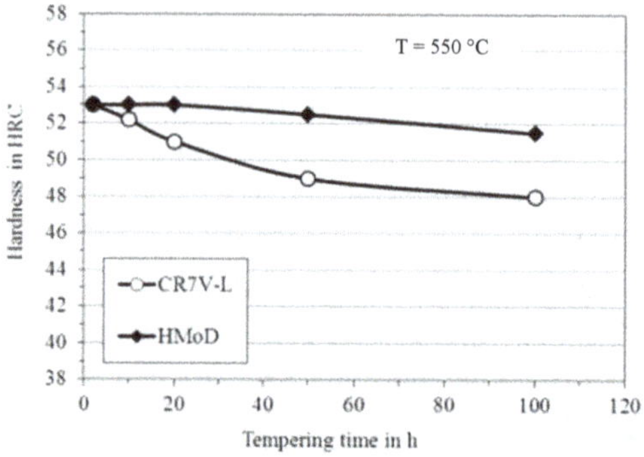

Fig. 4. Long-time tempering behaviour of the steel grades CR7V-L and HMoD at a temperature of 550 °C.

4. Recommendations for Improved Tool Life

Although being a fast growing technology hot-stamping is facing a technological and economic competition with technologies such as pressure-die casting of structural components. Economic hot-stamping production processes therefore require reliable tools which provide a long lifetime without interruptions or premature failure. Recent developments in the hot-stamping technology have led to many modifications of hot-stamping tools. Sizes and contours of the tools have changed significantly (Figure 5).

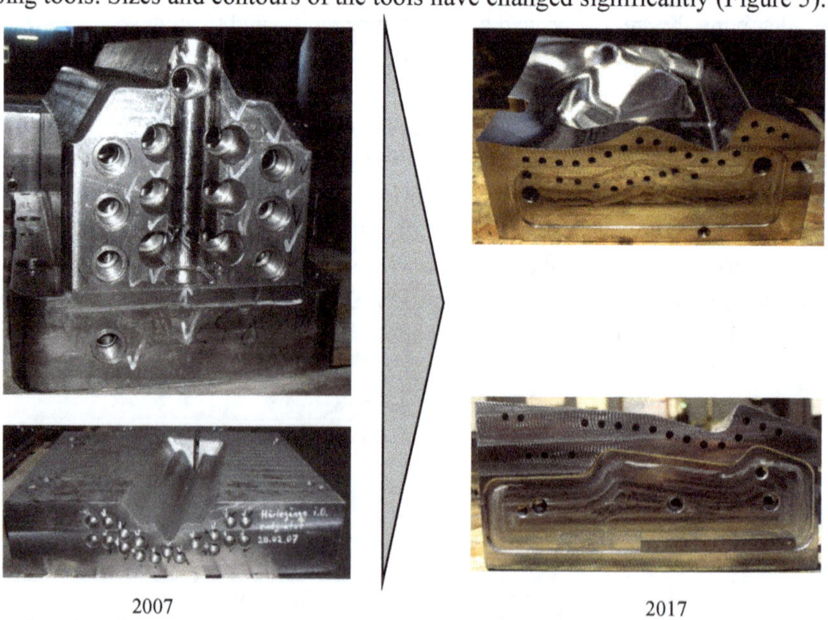

2007 2017

Fig. 5. Development of size, shape and cooling of hot-stamping tools between 2007 and 2017.

Hardness values of the tools had been increased in order to optimize wear resistance and to extend lifetime of the dies. In order to optimize the quenching of the sheet components the distances between cooling channels and working surface of the tools had been reduced significantly to only few millimeters. As a consequence of all these modifications the stress situation of the tools had been drastically intensified and tool defects like water leakage (Figure 6) had been observed frequently.

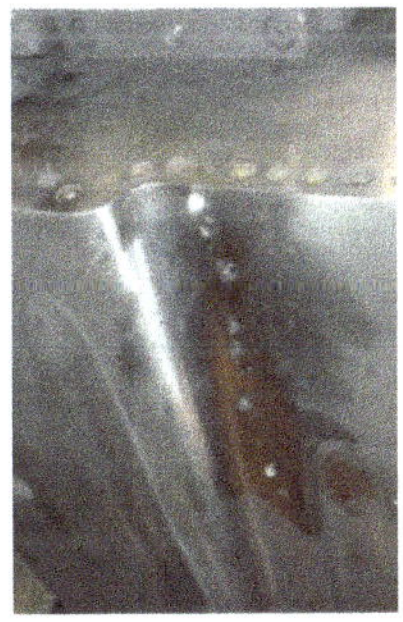

Fig. 6. Water leaking from a cooling channel in a hot-stamping tool.

Kind & Co. investigated numerous of these tool failures and developed strategies to avoid them. The photos in Figure 7 represent an investigated tool which failed prematurely due to corrosion induced cracks.

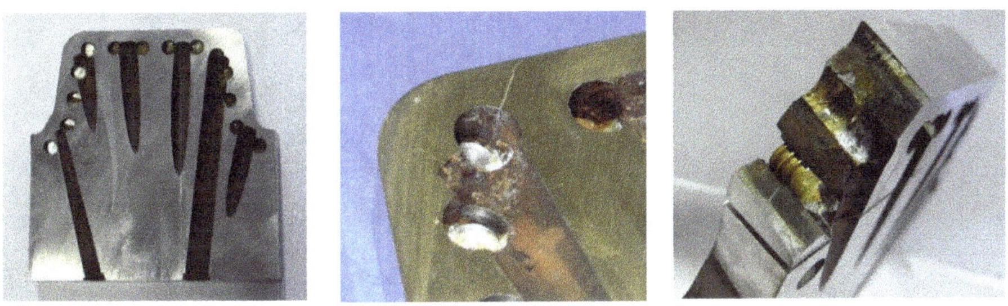

Fig. 7. Section through a defect hot-stamping tool: Left: Cross-section; Middle: Detail with crack from cooling channel to the working surface; Right: Corroded crack surfaces.

It can be seen in Figure 7 the traces of corrosion in the cooling channels are clearly visible. A crack had developed from the cooling channel and had propagated through the wall thickness of approx. 6 mm towards the working surface. The corroded surfaces of the crack demonstrate the lateral extension of the crack. The starting point of the crack was a corrosion pit within the wall of a cooling channel (Figure 8).

Fig. 8. Fragment from a hot-stamping die after corrosion induced crack (top; compare Figure 7 center); Corrosion pit as a starting point and rest lines of a dynamic crack (bottom).

These figures describe the dramatic effect of corrosion on tool life within only few months of tool operation.

Years ago, when the cooling channels were far apart (20 - 25 mm) from the contour of the tools corrosion induced tool failures were hardly observed. As only little information on corrosion of tool steels was available in the literature, Kind & Co. intensively investigated the corrosion behaviour of important tool steels. Apart from several steel grades the investigation included parameters such as metallurgical process route (ESR / Non-ESR), hardness of the samples, as well as water quality and test temperatures [3].

As expected none of the tool steels for hot-stamping tools is resistant against corrosion by cooling water. Three types of corrosion defects were observed:

- Rust,
- Pitting corrosion,
- Corrosion induced cracks.

Figure 9 illustrates the different types of corrosion induced defects.

Fig. 9. Photo micrographs of corrosion induced surface defects in hot-stamping tools: Left: Rust; Middle: Pitting corrosion; Right: Corrosion induced cracks.

Rust appears as a comparably harmless type of corrosion as it covers mainly the steel surface.

Pitting corrosion proved to be the most critical type of corrosion in hot-stamping tools. Hardly visible from the surface these pores expand into the steel. Under the influence of mechanical and/or thermal stresses these pores can grow rapidly (see Figure 9, right) and finally result in cracks. Pitting corrosion is caused by halogenides like chlorine dissolved in the cooling water.

Many factors influence the intensity of corrosion attack on hot work tool steels. Increasing water temperature and chlorine concentrations intensify the corrosive reactions and surface defects. Kind & Co. therefore urgently recommends controlling the quality of the cooling water quality regularly with special focus on chlorine concentration and pH-value.

Special precautions should be taken into account in order to avoid corrosion of the tools. A closed water circuit should be mandatory as the input of fresh water accelerates corrosion processes. The addition of corrosion inhibitors to the cooling water should also be considered. During the process of tool making only best and sharp drills should be used to manufacture the cooling channels. A rough surface of the channels offers more potential starting points for corrosion than a smooth surface.

The use of tool steel produced by the Electro-Slag-Remelting (ESR) process provides several advantages for hot-stamping tools. Due to the ESR process the steel has a higher purity than conventionally produced material. As corrosion preferably starts at non-metallic inclusions, ESR materials offer less potential starting points for corrosion than conventional steels. A further advantage is the higher toughness potential of ESR material which in some cases allows a slight increase of the hardness and a further improvement of the wear resistance. The most important advantage of improved toughness is that it can counteract to the growth of cracks through the steel between cooling channel and working surface.

The special steel grade CR7V-L is available as an ESR grade upon request. HS 1 is exclusively produced by ESR.

In order to prevent premature cracks of hot-stamping tools the hardness of the tool segments should not be adjusted to the maximum hardness. This advice is given because of two reasons. In case of maximum hardness the tool steels have their lowest toughness level-and insufficient toughness can favour cracks between cooling channel and working surface. The hardness of the steels can develop a significant influence on the corrosion behaviour as hardness is directly related to precipitation of carbides during tempering. The steels discussed in this paper have a significant chromium concentration and develop a secondary hardness maximum (see also Fig. 1). During tempering carbides are precipitated in the matrix of the steels. These carbides, originally fine and homogeneously dispersed in the matrix, can start to coagulate once the tools are tempered within the temperature range of the secondary hardness maximum. For this process of carbide growth the coagulated carbides consume the surrounding small chromium carbides. The consequence is the formation of a chromium depleted zone around the growing carbide. Within this zone the steel has a higher sensitivity against corrosion and local corrosive attack can easily start. This critical range of increased sensitivity extends to approx. 50 °C above the secondary hardness maximum. At higher tempering temperatures diffusion of alloying elements can compensate the chromium depletion and reduce the higher risk of corrosion [4].

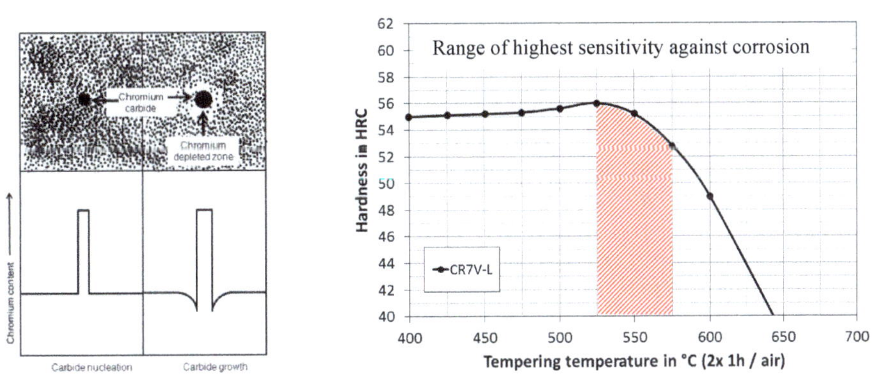

Fig. 10. Growth of chromium carbides and risk of corrosion during tempering at secondary hardness maximum.

Kind & Co. urgently recommends to follow this recommendation and to avoid maximum hardness of the tools, especially in case of tools with cooling channels only few mm below the working surface.

5. Conclusion

The fast growing market for hot-stamping applications requires highly developed tool steels solutions offering high hardness, high-temperature strength, wear resistance, enough toughness reserve for design optimization, and reasonable thermal conductivity. Fulfilling these requirements is an enormous challenge as these properties are opposing each other. With the two premium tool steel solutions CR7V-L and HS 1 Kind & Co. offers tailor-made steel grades for hot-stamping which successfully balance the required properties.

CR7V-L has proved its superior suitability in many industrial applications worldwide, not only for hot-stamping of steel sheet components but also for aluminium and nickel-base alloys.

For even higher abrasive wear resistance steel grade HS 1 can be recommended.

Recent developments which aimed at a reduced cycle time of hot-stamping operations have intensified the thermal and mechanical stresses in the tools and often resulted in tool defects like cooling water leakage. Own investigations demonstrated that corrosion and corrosion induced cracks are the most important failure reasons. Kind & Co. recommends balancing the properties in a way that provides sufficient toughness and avoids corrosion induced cracks. Lacks in thermal conductivity can be compensated by modern tool design.

Using the special steel grades CR7V-L as well as HS 1 and respecting the recommendations given in this paper will result in excellent tool life.

References

1. R. Rahn and I. Schruff, Modern tool steels for hot sheet metal forming applications, in *Proc. Hot Sheet Metal Forming of High-Performance Steel, 4th Int. Conference (chs² 2013)*, (Luleå, Sweden, 2013).
2. K. Steinhoff, N. Saba, M. Maikranz-Valentin and U. Weidig, Optimized processes and products in hot sheet metal forming, in *Proc. Hot Sheet Metal Forming of High-Performance Steel, 2nd Int. Conference (chs² 2009)*, (Luleå, Sweden, 2009).
3. P. Gümpel and I. Schruff, Corrosion - An Often Neglected Phenomenon in Hotwork Tools, in *Proc. 10th TOOL Conference 2016 (TOOL 2016)*, (Bratislava, Slovakia, 2016).
4. N. Arlt, H.-J. Fleischer et al., *Stand und Entwicklungstendenzen auf dem Gebiet der nichtrostenden Stähle*, 15th edn. (Thyssen Edelst, Techn. Ber, 1989).

Tool Steel Solutions for Tailored Property Parts Produced via Hot Stamping

C. Oberroither† and G. Jesner
voestalpine BÖHLER Edelstahl GmbH & Co KG,
Kapfenberg, 8600, Austria
†E-mail: christoph.oberroither@bohler-edelstahl.at

The demands on the hot stamping process, especially in the automotive sector, are steadily increasing. The constant change of complex lightweight constructions of sheet material puts high demands on the materials themselves and the production processes. Components with tailored properties are an elegant solution and state of the art. A customized selection of tool steels, depending on the hot stamping processes and the required mechanical and physical properties of the material, is therefore an essential part of optimizing the final component. Varieties of conventional, electro slag remelted and powder metallurgical steel grades are used to manufacture tools, which can realize the challenging standards, set by the automotive industry. This paper deals with the appropriate selection of modern high performance tool steels and high speed steels for the production of tailored property parts.

Keywords: Tool steel; Hot stamping; Tailored property parts; Tailored blanks; Heated segments; Differential cooling.

1. Introduction

An increasing number of car body parts are manufactured by the hot stamping (press hardening) process. The base material for this application is generally ultra-high strength steel sheet material. The most commonly used steel grades are boron alloyed steels (mostly 22MnB5, but also 27MnCrB5 and 37MnB4) [1]. In the beginning, the material possesses a ferritic & perlitic microstructure with tensile strength levels of 400-600 MPa [2]. After the hot stamping process, the steel sheet develops a fully martensitic microstructure and the strength levels reach 1500 - 2000 MPa [1-3]. Because of the high requirements of the automotive industry, it becomes more challenging to manufacture components that can achieve these high standards.

Tailored properties parts (TPP's) are a promising and elegant solution to construct automobiles with reduced weight and optimized component properties for ideal crash performance of the body in white (BIW) structure while reducing emissions.

To withstand the harsh wear conditions and high temperatures, which occur during the hot stamping process, high performance tool steels are needed which can reach the required high quality and property levels of the parts.

2. Hot Stamping Process

The aim of this process is to produce light and high strength components, which surpass conventional formed sheet material. There are mainly two assemblies to be distinguished, the indirect and the direct hot stamping process. In the indirect hot stamping process, the sheet is formed in its soft and cold condition. It is subsequently austenitized at temperatures of approximately 950 °C and then quenched between cooled tools. A final cutting procedure is not necessary in indirect hot stamping [1].

In the direct hot stamping process, the forming and quenching of the previously austenitized sheet occurs in one step. The forming operation during quenching takes place between 600 °C and 850 °C and not in the cold state as in the indirect process [2]. Then, the parts are hard-cut and receive their final shape.

3. Tailor Property Parts

The individual optimization of mechanical properties in specific regions of the component (e.g. B-pillar) can already start before the actual press hardening process. These tailored welded blanks (TWB) or tailored rolled blanks (TRB) can show a variation in thickness, properties of the sheet material and surface finish. Most importantly tailored blanks can lead to a weight reduction of the parts when compared to conventionally manufactured components. In general, there are four different types of tailored blanks [1].

Tailor welded blanks consist of at least two sheets which are welded together before being formed. Sheets with thicknesses of 0.8–2.8 mm can be joined together and therefore cannot only differ in their chemical composition and microstructure but also in their thickness.

Patchwork blanks are sheets, which are partially reinforced by spot welding with additional sheets of the same material. This method leads to lower production costs than conventional reinforcements in which the blank and the reinforcement are patched together after the forming process. Furthermore, it shows higher fitting accuracy between the sheets and the small size of the possible reinforcement [4].

Another way to manufacture blanks with varying sheet thickness is via a rolling process. Compared to the previously mentioned blank technologies the tailor rolled blank possesses better surface quality and less stresses due to a continuous transition between the thickness levels [4].

Tailor heat-treated blanks are a different technology than the mentioned methods above. Here, the various properties are not reached by structural alteration but rather by a customized heat treatment of the sheet material. There are different ways to heat treat blanks.

3.1. *Differential heating*

The first process route for tailored tempered properties in a hot stamped part is the differential heating of the sheet material. The entire sheet is heated up in a furnace.

Partial heating of the blank leads to the austenitization of certain regions of the blank, while the rest of the blank remains at a temperature below the austenitization temperature. The microstructure of these regions remains in the ferritic & perlitic state after the forming and quenching process while the austenitic regions transform into martensite. This process will not be discussed further in this paper.

3.2. *Differential cooling*

Another and more common way to produce components with soft and hard zone sectors is the differential cooling process. Here, the entire sheet is heated up in a furnace to austenitization temperature.

The hard zone parts, which should possess high hardness and strength, are quenched to reach a martensitic microstructure as it is common in the hardening process.

The soft zone areas are cooled down slower to reach a microstructure consisting of a ferritic and pearlitic microstructure. In this state, the material possesses lower hardness values and strengths. This is implemented through heated tooling segments, which slowly cool down the soft areas and leads to different microstructures in various areas of the hot stamped part. The advantage of this process is that the transition zone between the hard and the soft state of the blank is not as big as in differential heated blanks. These narrow transition zones are realized by placing segments which are heated up to temperatures between 500 and 650 °C right next to cooled segments. The exact temperature of the heated tool depends on the OEM's requirements regarding strength and toughness of the part.

To reach low hardness and strength values the temperature has to be increased and vice versa. This ensures that the 22MnB5 type steel sheet has no chance to become martensitic and transforms from austenite into a ferrite and perlite microstructure.

The heated segment technology brings some challenges regarding tool steels with it. With increasing application temperature, the hot hardness drops. This leads to a softening of the tool steel and to an increase of hot wear. The temperature of the heated segments is within or above the temperature range of the tempering process, which is the main issue for this technology.

Fig. 1 depicts a schematic common hardening - tempering curve of a tool steel. Such a curve displays the hardness value as a function of the tempering temperature at a specific austenitizing temperature. In general, high performance tool steels have a secondary hardness peak (S_H), which reaches its maximum hardness level at 500 °C - 540 °C. This depends on the alloying content and on the hardening temperature of the steel.

Typical tempering temperatures of such tool steels are at least 20 °C above their secondary hardness peak in order to achieve the optimal properties profile. Typical tempering temperatures for such tool steels are between 520 °C and 580 °C.

When the application temperature surpasses the peak of the secondary hardness maximum and reaches temperatures close to the tempering temperature, the hardness level drops immediately. Furthermore, the hot hardness also strongly depends on the thermal loading time of the segments.

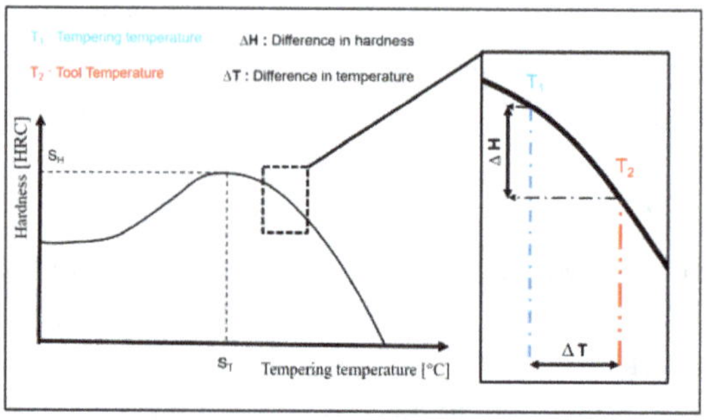

Fig. 1. Tempering temperature in comparison with the tool temperature.

The influence of the tempering temperature can be explained by an example (see Fig. 1). The tool steel material is tempered at T_1 = 550 °C. This means that the set hardness level stays constant up to application temperatures up to approximately 550 °C. However, if the tool is heated up to T_2 = 580 °C during the differential cooling process, the hardness drops by ΔH (as seen in Fig. 1). This leads to an immediate softening of the tool based on the temperature difference ΔT in accordance to T_1. To counteract this effect, the material is tempered at higher temperatures, for example up to 600 °C. This leads to a lower hardness level, but this hardness stays stable up to the tempering temperature of 600 °C, thus preventing a softening of the tool steel. However, this hardness level is generally too low for the hot stamping application and can lead to increased hot wear at elevated application temperatures.

The difficulty in this process is that common tool steels cannot be tempered up to such high temperatures while retaining a hardness level, which is acceptable for hot stamping applications.

By raising the hardening temperature, the maximum hardness level and the tempering temperature (and therefore the application temperature) can be moved to higher levels. However, when increasing the hardening temperature, the mechanical properties of the steel can change and therefore become unsuitable for the application. In order to achieve the perfect combination of hardness and temperature stability it is always advisable to consult the steel manufacturer.

4. Tool Steel Selection

A good and efficient way to optimize the hot stamping process is to make the right choices in the tool steel selection. The properties, which are required of tool steels in hot stamping, can be very different depending on the exact kind of process. Table 1 shows the demands of the hot stamping process on the tool steel.

In most cases, the optimum tool steel for hot stamping can only be a best compromise of all these requirements. Standard hot-work tool steels often reach their limits when faced with these demands.

Especially the high hardness and abrasive wear resistance demand tool steels with a high alloying and carbide content, such as cold-work tool steels and high-speed steels. However, based on their high alloy content, these steels possess a different properties profile and have therefore lower thermal conductivity and toughness levels than hot work tool steels.

Table 1. Tool steel properties demanded in direct hot stamping.

Common direct hot stamping	Direct hot stamping with heated segments
High compressive strength	High compressive strength
Simple heat treatment	Simple heat treatment
High thermal conductivity	High thermal conductivity
Good weldability	Good weldability
Sufficient wear resistance (abrasion/adhesion)	Sufficient wear resistance (abrasion/adhesion)
Hardness levels of 42-60 HRC	Hardness levels of 54-60 HRC
	High tempering resistance
	High hot / red - hardness & hot wear resistance

In order to achieve the shortest possible cooling time, the sheet thickness, tool geometry and surface coating of the sheet must be considered in addition to the thermal conductivity. To find the appropriate tool steel can be a challenge in itself when faced with these questions. This is why voestalpine Böhler Edelstahl recommends a range of special tool steels for hot stamping, to cover all of the demands and challenges that this procedure brings with it.

The Böhler W302 (H13 type) and W303 (1.2367 type) are conventionally melted hot work tool steels, which can reach hardness levels up to 52 HRC. Due du their chemical composition they have excellent hot strength, high toughness and very good machinability. These steel grades can be used for hot stamping applications without any specific requirements regarding thermal conductivity or abrasive wear resistance.

The Böhler K353 is a conventionally melted 8% Chromium cold work tool steel with a fine grain carbide structure. Due to its carbide content and chemical composition, it is able to reach high hardness levels of up to 62 HRC, which increases the abrasive wear resistance. It is the best solution when a long tool lifetime is required in direct hot stamping.

The Böhler W350 ISOBLOC is a specially developed ESR grade high performance tool steel. It possesses an excellent through hardenability up to 52 HRC and is therefore the perfect choice when it comes to big tools and segments with complex geometries. Compared to standard hot work tool steels the W350 ISOBLOC has an improved thermal conductivity as well as an improved toughness. It is therefore best suited as a tool steel for the direct hot stamping process when it comes to big parts and deep cavities.

The Böhler W360 ISOBLOC is also a specially developed ESR grade hot work tool steel. It reaches hardness levels up to 56 HRC. This grade offers an excellent combination of hardness, abrasive wear behaviour and good toughness. Furthermore, this material shows the highest thermal conductivity of all the previous mentioned grades. The Böhler W360 ISOBLOC is therefore perfectly suited for the direct hot stamping process. It is the best compromise regarding thermal conductivity and abrasive wear resistance. It is especially recommended for applications with a complex thermal and mechanical loading situation. Due to its high hot hardness level and thermal stability, the W360 ISOBLOC is also highly recommended as a soft zone tool steel material for the differential cooling process. Fig. 2 shows the Böhler tool steel selection in hot stamping when the focus is on the thermal conductivity and the abrasive wear resistance.

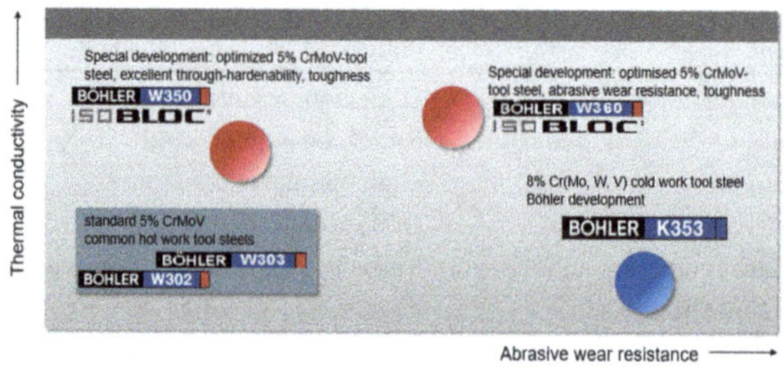

Fig. 2. Tool steel selection in hot stamping regarding thermal conductivity and abrasive wear resistance [5].

If the customers required strength levels of the sheet material are especially low, the application temperatures of the heated segments increase. Here, the required hot hardness values of cold work or hot work tool steels can exceed their capacities due to the softening behaviour of the material. In this case, the tool steel selection could be a high-speed tool steel like the Böhler S600 or the Böhler S790 MICROCLEAN. At first, this kind of tool steels seems like an unconventional choice for hot stamping applications, but these steels can withstand very high application temperatures occurring in the direct hot stamping forming process. Their microstructure possesses high carbide content and makes them a perfect choice against high abrasive wear at high working temperatures where other materials cannot compete. These high speed tool steels, originally designed for machining applications, can reach high hardness levels between 57 and 66 HRC. It is common to harden these types of steels at relatively high austenitizing temperatures in the range of 1200 °C. For hot stamping, it is crucial to avoid hardening the tools at these levels. Because of the larger dimensions in hot forming compared to machining, it can come to cracking of the material during the hardening process. It is therefore advisable to choose proper heat treatment parameters, to ensure that the tool steel possesses the ideal microstructure and properties for these tool dimensions.

5. Summary

Hot stamping of tailored property parts is a common process to achieve the challenging specifications required by the OEM's in the automotive industry. An efficient way to optimize the hot stamping process is the correct tool steel selection. In most cases, this is only a compromise between high thermal conductivity and high abrasive wear resistance. Cold work tool steels (Böhler K353) and hot work tool steels (Böhler W350 ISOBLOC & Böhler W360 ISOBLOC), but also high-speed steels (Böhler S600 & S790 MICROCLEAN) can be used for this application. Through differential heating and cooling of the sheet material, soft and hard zones can be created without structural alteration of the base material. Through heated segments in the hot stamping tool, the sheet metal can be cooled down to create regions with varying mechanical properties. The high process temperatures of the heated segments, which are accompanied by a softening of the tool, require a proper tool steel selection, which can withstand the hot wear and the softening behaviour. Furthermore, well adjusted heat treatment parameters can improve the temperature stability of the tool steel significantly.

References

1. H. Karbasian and A. E. Tekkaya, A review on hot stamping, *J. Mater. Process. Tech.* **15**, 2103 (2010).
2. M. Merklein and J. Lechler, Investigations on the thermal behaviour of ultra-high strength boron manganese steels within hot stamping, *Int. J. Mater. Form.* **2**, 259 (2009).
3. A. Keller (2018), https://www.voestalpine.com/thinkzinc/Tailored-property-parts-TPP.
4. M. Merklein, A review on tailored blanks - Production, applications and evaluation, *J. Mater. Process. Tech.* **2**, 151 (2014).
5. Data sheet Böhler, tool steel selection for hot stamping (BW 065DE05.2017)

A Tooling Heating Design for Tailored Die Quenching in the Hot Stamping Process

T. H. Hung[1,a], I. W. Chiu[1,b], S. W. Wang[2,c], C. H. Hung[3,d] and F. K. Chen[1,e,†]

[1]*Department of Mechanical Engineering, National Taiwan University, Taipei, 10617, Taiwan*
[2]*Product Application Research Section, Green Energy & System Integration R&D Department, China Steel Corporation, Kaohsiung, 81233, Taiwan*
[3]*Department of Mechanical Engineering, National Chiao Tung University, Hsinchu, 30010, Taiwan*
E-mail: [a]d99522038@ntu.edu.tw, [b]r05522528@ntu.edu.tw, [c]150169@mail.csc.com.tw [d]chhung@mail.nctu.edu.tw, [e,†]fkchen@ntu.edu.tw.

Tailored die quenching process is an advanced hot stamping technology that can be used to manufacture a partially high-strength and partially low-strength product by controlling the temperature of the tools used in hot stamping to obtain a product with the required properties. In this study, process parameters such as the heating temperature of the tools and die quenching time were investigated using a flat plate model through finite element analysis. The analysis results indicate that heating temperature is a crucial parameter that significantly affects the sheet product's hardness. The analysis results also reveal that the die quenching time slightly affects the sheet product's hardness when the die is heated to high temperatures. In this study, finite element model was employed to model heating tools. This study presents an initial tool design for the tailored die quenching process.

Keywords: Tailored die quenching; Heating temperature; Quenching time; Hot stamping.

1. Introduction

Tailor die quenching is one of the popular processes used to produce a hot stamped part with tailored properties. Through tailor die quenching, a single product that partially exhibits high-strength properties and partially exhibits low-strength properties can be realized by using a single sheet. However, for manufacturing a product with low-strength properties through tailor die quenching, the tools should be heated in the low-strength region of the product to reduce the sheet's cooling rate for generating a low-strength microstructure. Merklein et al. [1] discussed many kinds of methods in hot stamping process to obtain parts with tailored properties. George et al. [2] discussed the effects of the tool's heating temperature on sheet's hardness for the temperature range from 25 °C to 400 °C. In their study, it also revealed that if a tool is heated above 400 °C, then the sheet's hardness value could be effectively reduced. Omer et al. [3] applied a U-hat shaped tool to investigate the effects of the tool's temperature on the sheet's hardness. They designed tailor die quenching tools with grade soft zone which have different strength regions with a part. Wang et al. [4] investigated the effects of a tool's heating temperature on a sheet's hardness using a flat plate tool. In their study, the heating temperature range of the tool is from 25 °C to 400 °C. Tang et al. [5] conducted experiments and numerical simulations of tailor tempering, and the heating temperature range of the tool is 300 °C to 450 °C.

In this study, the effects of process parameters such as the heating temperature of a tool and die quenching time on a sheet's hardness were investigated to present the initial design for tailor die quenching tools using finite element analysis. A flat plate tool was realized using finite element model to investigate the crucial parameters that should be considered while heating the tailor die quenching tools.

2. Finite Element Analysis Model

2.1. *Flat plate tool model for tailor die quenching*

The effects of the process parameters on a forming sheet's hardness were investigated using a flat plate tool model by the finite element analysis, as presented in Fig. 1. Point A and point B are the measurement points of hardness and percent ratio of each microstructure phase in simulations. The hardness of blank and percent ratio of each microstructure phase was measured when the blank's temperature of heated zone was approximately at 100 °C. The analysis was conducted by using PAM-STAMP software [6], and the material used was 22MnB5. The flat plate tool contains two parts: a tool for the region that should be cooled and a tool for the region to be heated. To simplify the analysis model, the tools' temperatures were assumed to be constant during the process. The simulation contained three analysis processes: a transfer process, a die quenching process, and an air cooling process. The temperature of the tools and the die quenching time for each analysis case is listed in Table 1.

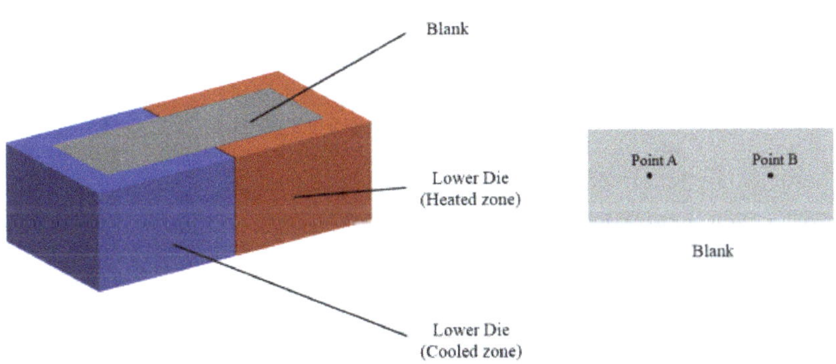

Fig. 1. Analysis model of the flat plate tool.

Table 1. Parameters for each analysis case.

Case number	Tool temperature in heat region (°C)	Tool temperature in cold region (°C)	Die quenching time (s)
1	450	25	5
2	450	25	10
3	450	25	15
4	450	25	20
5	550	25	5
6	550	25	10
7	550	25	15
8	550	25	20

2.2. *Heating model of the flat plate tool*

To heat this tool to a high temperature, heaters should be designed such that they can be set within the tool. Fig. 2 presents the finite element model for heating a flat plate tool. The cartridges were placed within the tool. The tool was heated at high temperatures, that is, from 25 °C to the 450 °C.

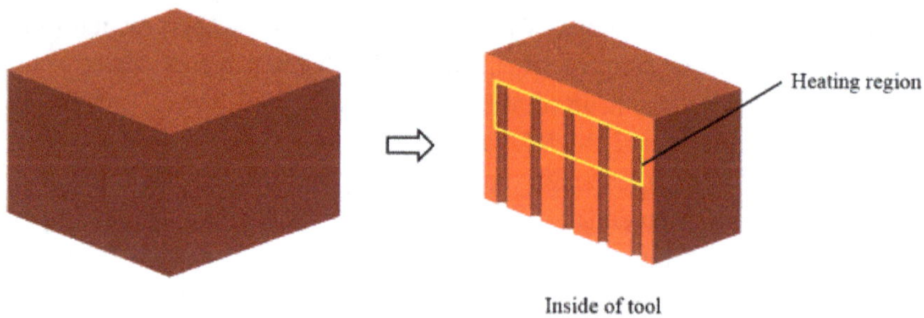

Fig. 2. Finite element model of heating a flat plate tool.

3. Results and Discussions

3.1. *Effects of the heating temperature of the modeled tool*

The results obtained after evaluating the effects of the tool's heating temperature on the sheet's hardness are presented in Fig. 3 and Fig. 4. The analysis results indicated that the tool's temperature is a crucial parameter that considerably affects the sheet's hardness. The microstructure distributions and sheet's hardness values for tools that are heated at 450 °C temperatures for a die quenching time of 10 s are presented in Fig. 5 and Fig. 6. The analysis results revealed that the tool's heating temperature affects the microstructure distribution of the sheet. This implies that the heating temperature affects the percent ratio of each phase of microstructure, and further affects the hardness of the sheet.

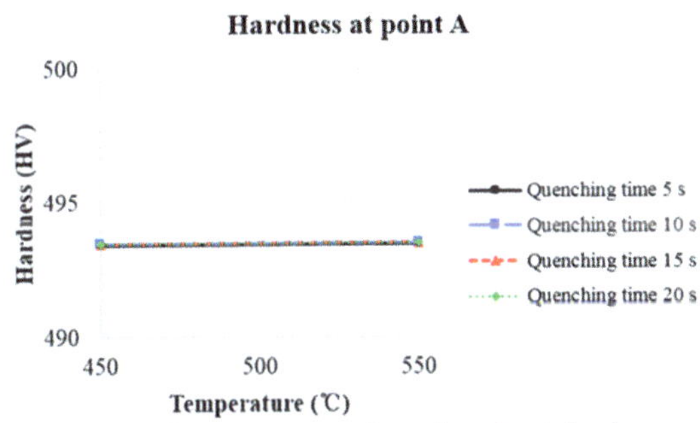

Fig. 3. Effects of heating temperature of a tool on sheet's hardness at point A.

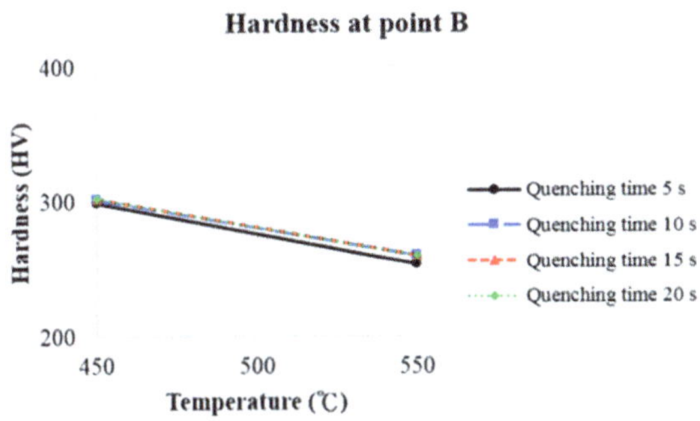

Fig. 4. Effects of heating temperature of a tool on sheet's hardness at point B.

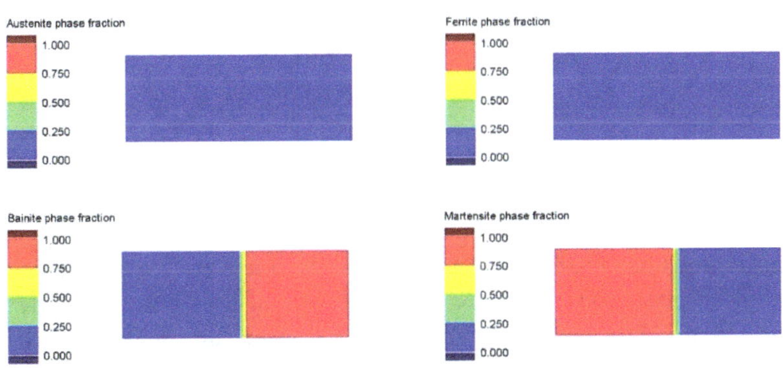

Fig. 5. Microstructure distribution of the sheet of analysis case 2.

Fig. 6. Hardness distribution of the sheet of analysis case 2.

3.2. *Effects of die quenching time*

The analysis results presented in Fig. 4 reveals that the die quenching time has a slightly effects at heated region when the heating temperature increases. This implies that die quenching time is a parameter that should be considered in relation to heating temperature of the tool while designing a tailor die quenching tool.

3.3. *Tool heating temperature and die quenching time for the initial tool design*

Fig. 7 presents the sheet's temperature histories of each analysis presented in Table 1. The analysis results reveal that when the heating temperature of tool increases, the blank's cooling rate decreases. From the material's continuous cooling transformation (CCT) diagram [7, 8], it can also find that the hardness would be affected as the cooling rate of material are different. This implies that the tool's heating temperature could be obtained through the CCT diagram or finite element analysis to realize the initial design of this tool.

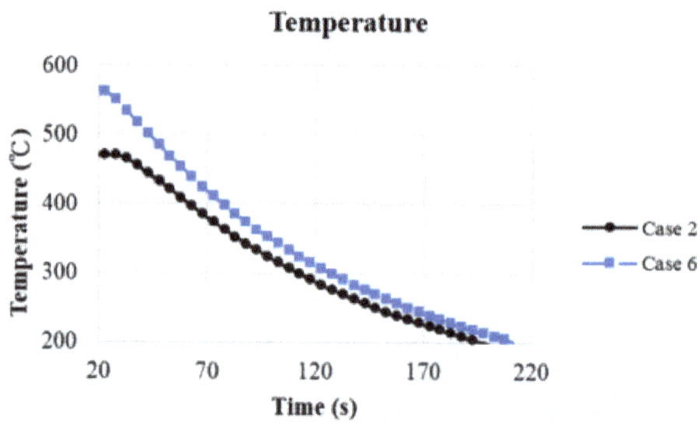

Fig. 7. Sheet's temperature histories.

3.4. *Heating performance of the flat plate tool model*

The effects of input power of cartridges and insulating a tool prevent heat loss were also analyzed by flat plate model presented in Fig. 2. The analysis results indicate that the input power of cartridges, the heating design of tools, and insulation layers all may affect the heating efficiency. This implies that these parameters should be considered designing tailor die quenching tools.

4. Conclusion

The effects of tool heating temperature and die quenching time on a sheet's hardness were investigated. The analysis results suggest that the tool's heating temperature is a crucial parameter that significantly affects the sheet's hardness. Moreover, the die quenching time is slightly affected when the tool's heating temperature increases. Either finite element analysis or the CCT diagram of the applied material can be used to prepare the initial design of tailor die quenching tools. An analysis of a heated tool revealed that the power of the cartridge, heating design of tools, and the insulation layer of the tool are the design parameters that should be considered when designing tailor die quenching tool.

Acknowledgments

The authors wish to thank the Ministry of Science and Technology of Taiwan for the grant and support under the project #MOST106-2622-8-006-001, which makes the research can be achieved. The authors also thank the ESI/France for providing the PAM-STAMP software to conduct the analysis.

References

1. M. Merklein, M. Wieland, M. Lechner, S. Bruschi and A. Ghiotti, Hot stamping of boron steel sheets with tailored properties: A review, *J. Mater. Process. Tech.* **228**, 11 (2016).
2. R. George, A. Bardelcik and M. J. Worswick, Hot forming of boron steels using heated and cooled tooling for tailored properties, *J. Mater. Process. Tech.* **212**, 2386 (2012).
3. K. Omer, R. George, A. Bardelcik, M. Worswick, S. Malcolm and D. Detwiler, Development of a hot stamped channel section with axially tailored properties – experiments and models, *Int. J. Mater. Form.* **11**, 149 (2018).
4. Z. Wang, P. Liu, Y. Xu, Y. Wang and Y. Zhang, Hot stamping of high strength steel with tailored properties by two methods, *Procedia Eng.* **81**, 1725 (2014).
5. B. T. Tang, S. Bruschi, A. Ghiotti and P. F. Bariani, Numerical modelling of the tailored tempering process applied to 22MnB5 sheets, *Finite Elem. Anal. Des.* **81**, 69 (2014).
6. PAM-STAMP software, ESI Group, Paris, France.
7. M. Nikravesh, M. Naderi, G. H. Akbari and W. Bleck, Phase transformations in a simulated hot stamping process of the boron bearing steel, *Mater. Design.* **84**, 18 (2015).
8. M. Naderi, M. Ketabchi, M. Abbasi and W. Bleck, Analysis of microstructure and mechanical properties of different high strength carbon steels after hot stamping, *J. Mater. Process. Tech.* **211**, 1117 (2011).

Prediction of Die Spotting Compensation Value of a Large Stamping Die Based on Virtual Clamping Technique

C. Y. Wang[†], P. F. Ye, J. H. Chen, Z. W. Xiang and J. Wu
School of Materials Science and Engineering, Hefei University of Technology, Hefei, 230001, Anhui, China
[†]E-mail: ahwangchengyong@163.com

Low die spotting efficiency of stamping die for automobile panels was focused. A fender panel die deformation was analyzed by the finite element method (FEM). The distributions of the defect region of die spotting were obtained. Modification of the die surface to compensate die deformation was proposed. The clearance value was predicted by the virtual clamping (VC) method. The discrepancy between virtual and the actual clamping clearance was analyzed. The feasibility of VC to optimize stamping offsets was confirmed.

Keyword: Die clearance; Stamping compensation; Virtual clamping; Die spotting efficiency.

1. Introduction

With the acceleration of automotive renewal, the traditional methods, repeated stamping and die-repair, are too inefficient to shorten the product development process [1]. And an efficient drawing die spotting for automobile panels helps to reduce the commissioning time. Considering the cumulative effects of calculation allowance, die casting distortion and machining accuracy error, the die spotting is inefficiency. Therefore, the clearance value distribution is inhomogeneous between the convex and concave dies. The assembly precision of stamping die and the forming precision of products are seriously affected. Besides, heavy repair workload prolongs the development process of the die.

Insufficient deformation occurs due to the complex surface of a large scale automotive panel [2]. Complex addendum surfaces must be set up, which reduce efficiency on the die spotting. It is difficult to judge the position where need to be spotted before the die commissioning. Therefore, it is vital to accurately predict the location of problem areas of the die spotting. And by using the corresponding methods to control the clamping clearance will shorten the development timescale of die, thus boost the work efficiency.

Die spotting commissioning is the most critical and time-consuming step in the die manufacturing process. The main factors on die spotting ratio are [3]: a) rationality of die structure design, b) quality of die casting, c) machining accuracy and d) processing methods, e) press precision and f) deformation. The fender forming die which is a typical auto-panel stamping tooling is studied in this article. The influence of die structure and press deformation on the die spotting is investigated. Then the improvement methods are determined. To reduce the costs of manufacturing and handling, traditional design usually cut down the die's weight as much as possible. But it will reduce the strength of die at the same time. The concave die of the vehicle fender shown in Fig. 1(a) is maximally carved out with thin walls. The elastic-plastic deformation will occur after the process of commissioning, which will reduce the die spotting ratio when the die is installed on the press for the first time.

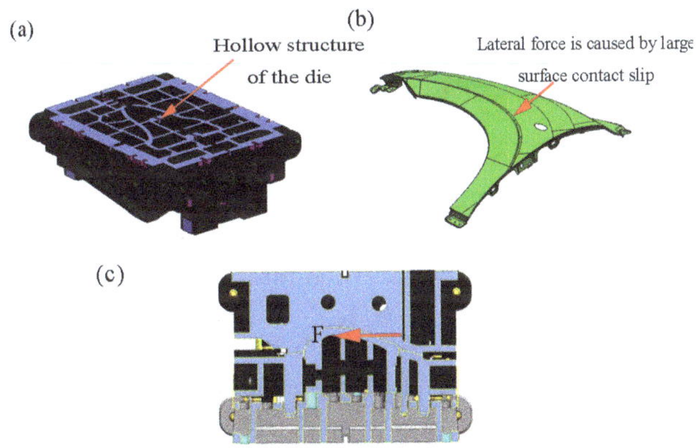

Fig. 1. (a) Hollow structure of upper die, (b) Shape of fender (c) Lateral force on dies

The structural design of forming dies for large-scale automotive panel infrequent take the differences of die structures caused by different parts under different stress [4]. The die generates a large lateral force due to the shape of the automobile panels' surface, which will cause lateral misalignment between the concave and convex dies. Fig. 1(b) shows a vehicle fender which has an asymmetric structure and a large forming area. The lateral movement of die is caused by small sliding along the die surface. The later force is schematically indicated in Fig. 1(c).

2. Material and Methodology

2.1. *Material*

The stamping process is modeled via the commercial FEA software Abaqus. The material parameters are shown in Table 1.

Table 1. Material properties.

Part	Mass density (kg/mm3)	Young's modulus (mN/mm^{2}kpa)	Poisson's ratio
Sliding block guide rail	7.85×10^{-6}	2.06×10^{8}	0.3
Die	7.3×10^{-6}	1.3×10^{8}	0.25

2.2. *Finite element modeling*

The die and were meshed with shell element (element type). Fig. 2 is an idealized model of a die without structures like small corners, small steps, small round holes and so on.

This idealized treatment neither affects the stamping deformation analysis, nor does it predict the distribution of defects in the die spotting. The small structures is neglected to speed up the simulation. Element size is 20 cm.

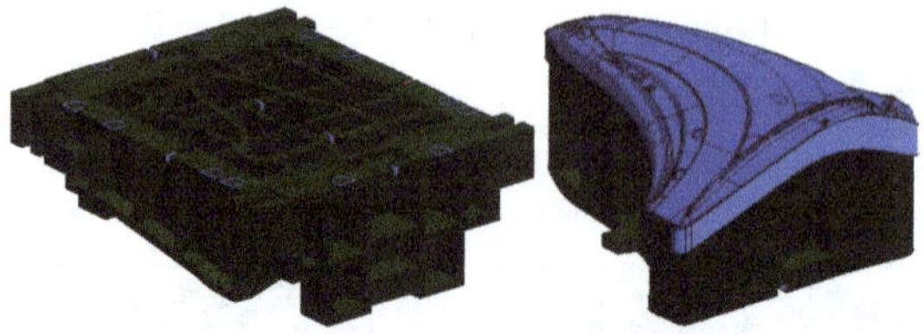

Fig. 2. Simplified model.

Fig. 3 shows the overall deformation of the die. When a vehicle fender was stamped, the total displacement of the machine tool reached 0.9mm (including the amount of deformation and the movement of machine tool due to the deformation of concave and convex dies).

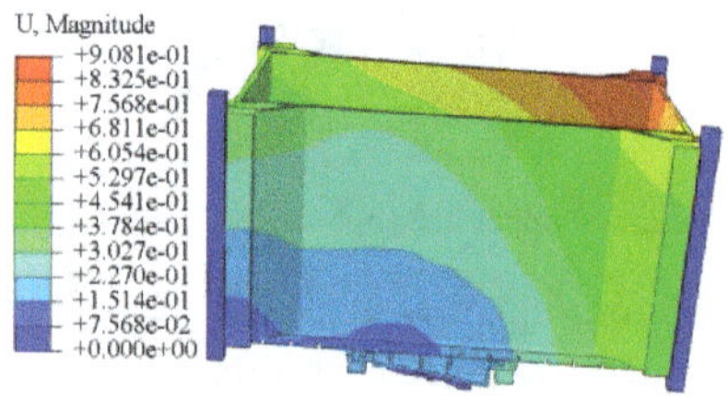

Fig. 3. The deformation of die.

The force of the concave die is analyzed to obtain the contact pressure, as shown in Fig. 4. It shows that the contact stress between the peripheral area and the main ridge line of the vehicle fender in the die cavity is the largest, and the maximum compressive stress is about 100 MPa. On the contrary, the contact stress in the large planar area of the vehicle fender center is small, which is an order of magnitude lower than the maximum compressive stress. The reason is that the forming force is transferred to the concave die body through the press. Because of the small deformation of the machine tool, the concave die appears concave deformation in the middle. Therefore, the key areas of die spotting are found in the large plane of vehicle fender's center, which is confirmed by the phenomenon of hanging in the middle of the die when the die coloring is applied.

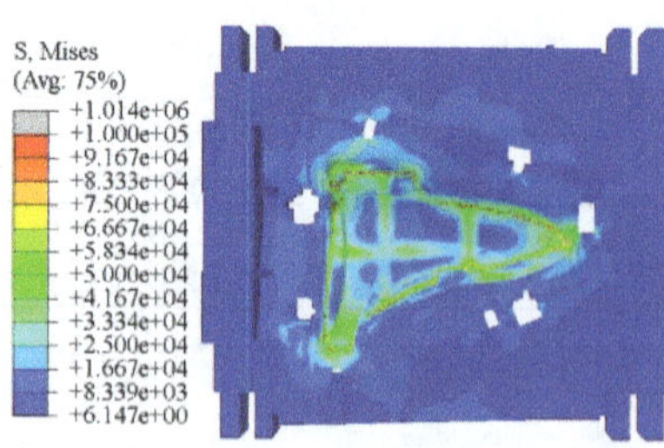

Fig. 4. Normal contact stress distribution graph of molding surface of die.

3. Die Surface Treatment Spotting

3.1. *Stamping deformation compensation*

According to the deformation amount of concave die surface shown in Fig. 10, the stamping deformation of die surface is predicted. The compensation value of concave die surface is determined by reverse compensation to the concave die when the die is spotting. It will avoid the phenomenon of surrounding dead and middle suspension in die spotting. The principle of stamping deformation compensation is shown in Fig. 5. The z-direction displacement is obtained by referring to the FEA results. Combined with production experience, the stamping compensation value is initially determined to 0.5 mm.

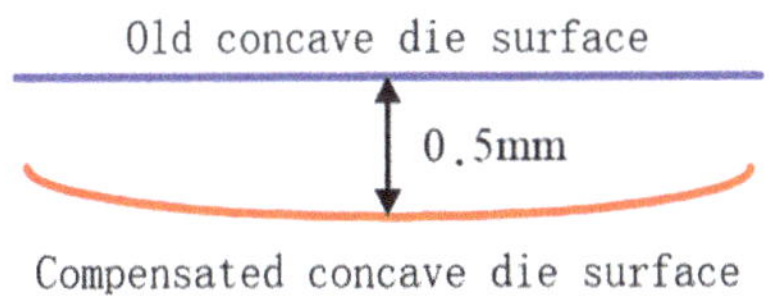

Fig. 5. Deformation compensation on stamping.

3.2. *The surface compression treatment*

The surface compression treatment is a measure to balance the local forming force of the sheet by adjusting die surface clearances. In other words, the local area within the non-machined reference surface along the normal direction to a slight value, which will reduce the local clearances and increase the local sheet plastic deformation [7]. In addition, the compression treatment helps to improve the die spotting rate and reduce the workload. Therefore, according to the design experience, the vehicle fenders were preset with 0.08mm of compression to change the die surface. The blank holder surface used a 0.1mm compression which is shown in Fig. 6.

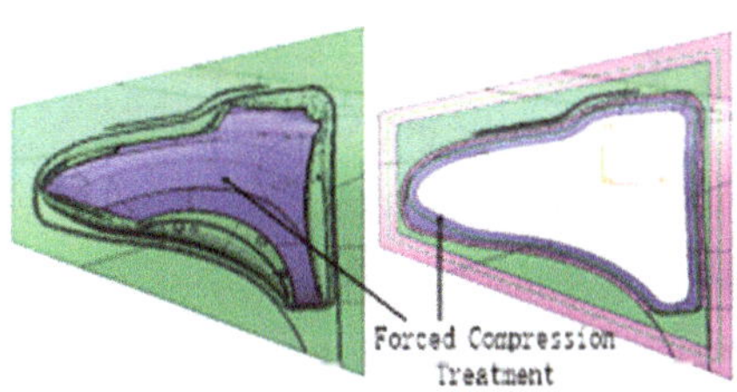

Fig. 6. Distribution of small mold clearance.

3.3. *Deviation machining of the addendum*

Deviation machining is to process the deviation of machining data based on the principle of convex die forming [8]. The areas after the deviation process can skip the die spotting, which will improve die spotting efficiency. In this article, the only the addendum area was deviation machined by 10mm away from the edge of the product (excluding the scrap cutters), which will increase the die spotting rate by reduce 30% of the die spotting area.

4. Experimental Results and Discussion

In conclusion, the final compensation value is 0.58mm which can greatly improve the die spotting. However, the above compensation value is based on experience setting. In the actual production, the die spotting rate is limited. Therefore, it is necessary to test and perform a verification optimization. The FEM is used to analyze the closed die, to determine the inaccuracy amount between the virtual the actual die clearance. Finally, the die spotting areas are checked by using the Pattern coloring technique, and the optimized compensation value is determined.

4.1. *Theoretical of the die surface clearance*

In the simulation of drawing process, contact offset settings are usually used between concave and convex dies to generate the clearance. [9-10]. Generally, the die clearance is 1.1 times than the sheet thickness. However, due to the influence of the working conditions such as the panel's precision, gravity, and the die's deformation, the actual die surface clearance changed while the die spotting rate dropped. In this paper, contact offset was completely inapplicable because the die surface was compensated, which means the physical offset should be used for simulation. After meshing, the initial positions of tools were all set to zero. By analyzing gravitational state of the tools, the theoretical clearances shown in Fig. 7 were obtained.

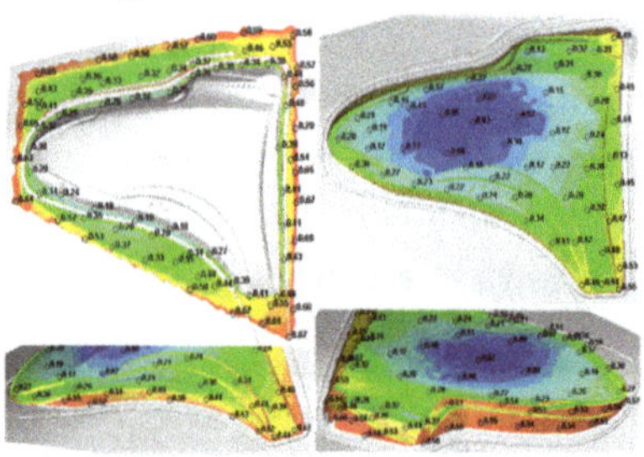

Fig. 7. The oretic clearance value when die closed.

It can be seen from Fig. 7, after the die surface compensation, the clearance value of the main surface is less than 0.1mm; the clearance distribution is relatively uniform between the blank holder and concave die, which is controlled at a range of 0.5 mm-0.7 mm. The lateral wall clearance is 0.5 mm.

4.2. *Frankincense test*

A qualified part is used as the benchmark for the precision test, the color of die surface can be used to judge whether the die surface meets the design requirements by observing the color status of each part. Figure 8 is a common frankincense test for the first time without optimization after stamping which has interference between the concave die surface and the blank holder surface. It proves that the die spotting is still a big problem by using the compensation of 0.58mm, and it needs to be further optimized.

After the die spotting is optimized, the coloring of surfaces shown in Fig. 9 is obtained, and the coloring is more uniform. The spotting rate between concave and convex dies was 90% and the blank holder was 95%. So it can be concluded that the die spotting was completed as required. In order to meet the above requirements, blank holder in the middle area completed the 0.3mm-0.4mm spotting and the 0.2mm spotting on both sides. This is consistent with the theoretical interference quantity which verifies the feasibility of the theoretical model and the accuracy of the prediction. The middle areas of the concave die are 0.25mm and gradually spread around to 0mm. The theoretical maximum static interference is 0.55mm.

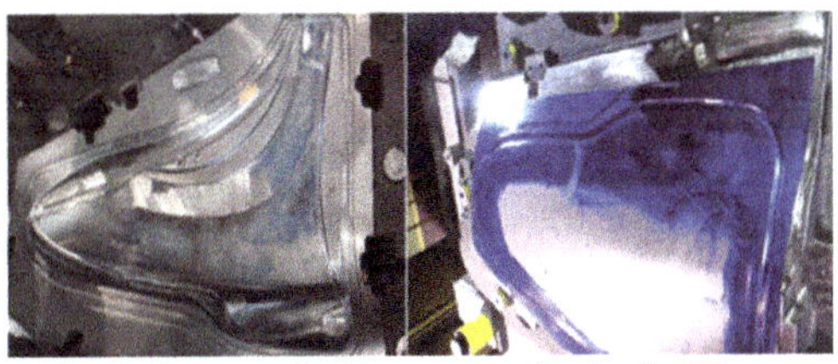

Fig. 8. Common frankincense test for the first time.

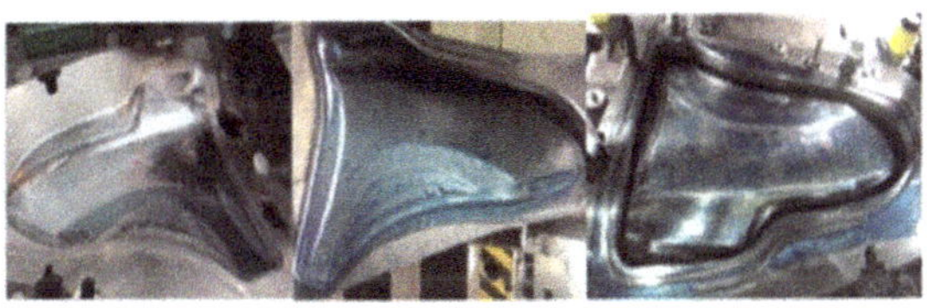

Fig. 9. Common frankincense test after die spotting.

The die surface clearance after the spotting is shown in Fig. 10. According to the clearance's value in the CAD drawing, the reasonable stamping compensation amount finally determined as 0.35mm. Although there is a slight deviation between the theoretical die surface clearance (0.58mm) predicted by the FEA and the actual clearance, but the inaccuracy is within an allowable range.

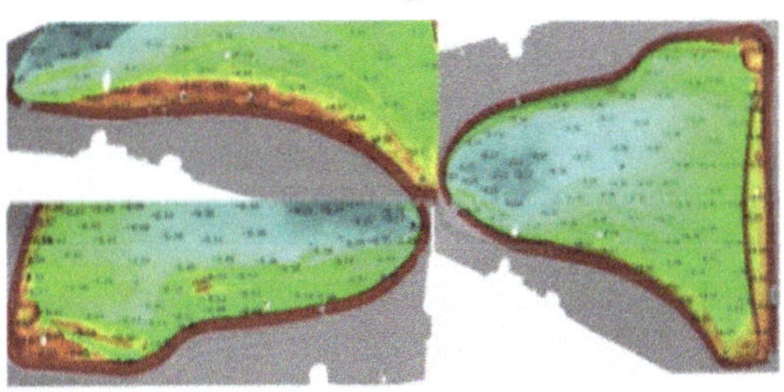

Fig. 10. Clamping clearance after die spotting.

5. Conclusion

(1) The value of die stamping deformation is numerically analyzed, and the distribution of defect areas is predicted.

(2) Comparing the virtual die surface clearance with the actual one, the maximum inaccuracy position and inaccuracy range are determined.
(3) Using virtual clamping technology, the compensation amount is 0.35 mm. The die spotting ratio is greatly improved and the die FEM commissioning time is reduced. And the forming precision of large-scale automobile panels is ensured.

References

1. D. D. Chen, B. S. Liu, Y. Liu and A. B. Wang, Researching on drawing process of a complex car panel component, *J. Plasticity Eng.* **15**, 13 (2008).
2. H. S. Park and T. T. Nguyen. Optimization of injection molding process for car fender in consideration of energy efficiency and product quality, *J. Comput. Des. Eng.* **1**, 256 (2014).
3. F. Sanling, Increase the Automobile Die Spotting Research, *J. Die Mould Manuf.* **8**, 29 (2014).
4. W. Li, X. S. Qin and F. Y. Hou, Improvement the Qualified Rate Measures Stamping Die Grind Rate, *J. Die Mould Manuf.* **9**, 4 (2014).
5. J. Z. Gao and W. L. Chen, *Plastic forming process and die design,* (China Machine Press, 2007).
6. W. L. Chen, *CAE analysis guide for sheet metal forming*, (China Machine Press, 2005).
7. Z. C. Tian, C. Y. Wang and W. Li, Redesign of plastic flange process for automotive sun-roof based on a compensation scheme of local redundancy, *J. Plasticity Eng.* **20**, 39 (2013).
8. L. J. Cui, *Stamping forming technology of automobile panel*, (China Machine Press, 2003).
9. S. P. Wang and E. Nakamachi, Nonlinear contact and friction modeling in dynamic explicit finite element analysis, in *Proceedings of the 3rd International Conference on Numerical Simulation of 3-D Sheet Metal Forming Processes Verification of Simulations with Experiments*, (Columbus, Ohio, USA,1996).
10. F. L. Wen, Y. Q. Liu and T. Du, Formability synchronous simulation system for automobile panel design based on CATIA user-defined features, *J. Plasticity Eng.* **20**, 14 (2013).

Investigation on Tribological Behavior of 7075 Aluminum Alloy at Elevated Temperatures

Y. P. Xu[1], Y. Liu[1], Z. J. Wang[1], L. F. Zhang[2], L. Li[2] and Y. L. Wang[1,†]
[1]*State Key Laboratory of Materials Processing and Die & Mould Technology, Huazhong University of Science & Technology, Wuhan, 430074, China*
[2]*Dongguan Vision Hot Stamping Technology Co., Ltd., Dongguan, 523475, China*
[†]*E-mail: wangyilin@hust.edu.cn*

In recent years, aluminum alloy hot forming-quenching (HFQ) has drawn more and more attention as a new technology for reducing the weight of automobiles. However, the friction behavior of aluminum alloy at elevated temperatures is a problem. This article focuses on the performance of the 7075 aluminum alloy at an elevated temperature of 450 °C, including the effect of sliding speed, normal force and graphite lubrication on the coefficient of friction (COF). Through the improved friction tester, COF at room temperature and 450 °C were measured first. Then the study investigated the effect of different sliding speeds (25mm/s, 50mm/s, 75mm/s and 100mm/s) and different normal forces (26.86 N, 47.53 N, 61.84 N and 82.52 N) respectively on COF of 7075 aluminum alloy at 450 °C with graphite lubrication. This study shows that COF of 7075 aluminum alloy is large and unstable at 450 °C, while graphite greatly improves its friction behavior. The sliding speed has little effect on COF of the 7075 aluminum alloy at 450 °C with lubrication, but the normal force has a great influence on it.

Keywords: Hot forming-quenching; Coefficient of friction; 7075 aluminum alloy; Graphite.

1. Introduction

Lightweight of automobiles has attracted more and more attention for the purpose of energy-saving and emission reduction. Compared with high-strength steels, aluminum alloys have more potential in the automotive lightweight. However, aluminum alloys cannot form complex parts because of its poor formability at room temperature. The hot forming–quenching integrated process has been developed for heat-treatable aluminum alloy forming, which was proposed firstly by Lin et al [1] to improve the formability. The friction behavior between plate and tools, which reduces product quality and tools life, is a key point in HFQ of aluminum alloy. Friction and wear in HFQ are known to be influenced by several factors, both physical, such as the tool geometry and surface conditions, and process-dependent, such as the sliding velocity, temperature and contact pressure [2]. Liu et al [3] studied the effect of friction coefficient of 6061 sheet at elevated temperatures by experiment and finite element analysis. Their results indicate that lubricants play an important role in the hot forming–quenching of B-pillars. Tian et al [4] studied the effect of friction coefficient of high strength steel at elevated temperatures. Their results show that temperature and drawing speed can considerably influence the friction and wear behavior. In the present work, COF of 7075 aluminum alloy at 450 °C were measured with different parameters by the improved friction tester.

Furthermore, in the absence of graphite lubrication, the adhesive points on the tools and scratches on the strip were observed.

2. Experimental

2.1. *Materials*

The H13 steel was selected for the friction pairs, and the 7075 aluminum alloy strip was selected as the friction sample. Table 1 is their alloy composition and room temperature hardness values. The size of friction pairs contacting aluminum alloy strip is 40 × 40mm, and 7075 friction strip sample size is 65 × 20 × 2mm.

Table 1. Main chemical compositions of H13 and 7075 aluminum alloy [wt. %].

Material	C	Mg	Mn	Si	Zn	Mo	V	Cu	Cr	Fe	Al	Hardness(HV)
H13	0.4	-	0.3	1.1	-	1.5	1.2	-	-	-	-	560
7075AL	-	3.16	0.18	0.03	5.9	-	-	1.64	0.21	0.14	Bal	201

2.2. *Testing machine*

Figure 1 shows the friction tester for experiment. The heating device selected a tubular resistance heating furnace, and the temperature of the heating furnace was controlled by a PID controller. The power transmission mode selected the AC servo motor - ball screw linear slide, which could change the sliding speed. The initial temperature of the strip entering the friction pairs was measured by an infrared thermometer and collected by the NI USB6008 DAQ card (sampling frequency 2000 Hz). The friction force was measured by a strain gauge and collected by the NI ENET-9237 DAQ card. Different mass blocks were placed to control the normal force. The formula for the friction coefficient $\mu = 0.5f / F_N$, where f is the friction force and F_N is the normal force.

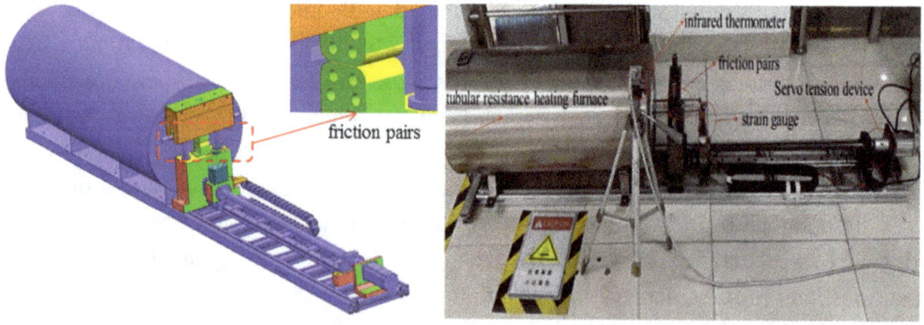

Fig. 1. 3D design and physical diagram of high temperature friction tester.

2.3. *Measurement procedures*

Accurate friction coefficient under certain conditions is critical to the accuracy of finite element simulation. Table 2 shows the parameters related to the measurement of the COF in this experiment. The specific procedures are as follows: (1) The friction pairs and sample strip polished with 1000-degree sandpaper and tested with a surface roughness tester (model SF200 Basic). The surface roughness of the sample strip was Ra= 0.169 um. A suitable amount of graphite was rubbed on the upper and lower friction pairs as lubrication; (2) Turning on the tube heater and adjusting the temperature to 450 °C. At the same time, turning on the computer that collects the experimental data; (3) When the temperature of the tube heater is stable, slowly putting the aluminum strip into the tube heater for heating 10 minutes; (4) Loading different normal forces on the friction pairs and pulling the aluminum strip at different speeds; (5) Collecting and recording experimental data by computer.

Table 2. Experimental parameters for measurement of COF.

Temperature (°C)	Sliding speed (mm/s)	Normal force (N)	Sliding speed (mm/s)	Normal force (N)
Room temperature	50	26.86	50	26.86
450	50	26.86	50	26.86
450	50	47.53	25	26.86
450	50	61.84	75	26.86
450	50	82.52	100	26.86

3. Results and Discussion

In order to study the high-temperature friction performance of aluminum strip, Experiments with 26.86 N normal force and 50mm/s sliding speed were performed by the friction tester on the condition of room temperature (28 °C), no lubrication at 450 °C, and graphite lubrication at 450 °C. Figure 2 shows the variation of COF over sliding time. It can be seen that the COF of aluminum alloy at room temperature is relatively small and stable with an average value of 0.173. The peak of curve is caused by the groove in the plane of the strip material shown in Figure 3(a).

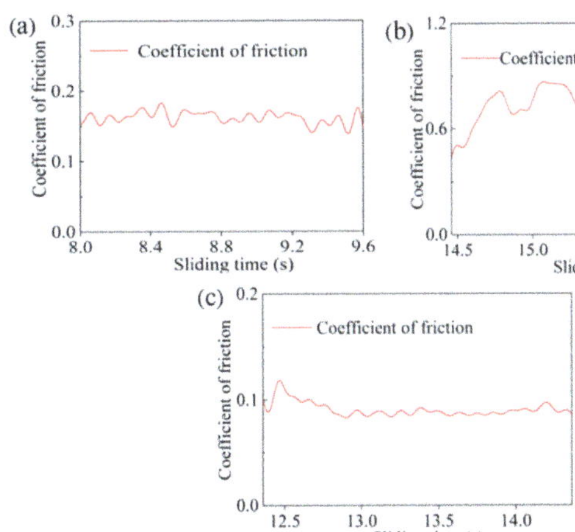

Fig. 2. The COF curves of different parameters: (a) RT, (b) 450°C, (c) 450 °C + graphite.

Under the condition of non-lubrication at 450 °C, like high strength steel [5], aluminum alloy experienced process can be roughly divided into three stages. The first stage can be considered as the frictional behavior of the strip and tools at the cold state. The second stage is the frictional behavior between the strip whose initial temperature change continuously and tools. It is also the most complicated stage. The third stage is the frictional behavior between the strip at the stable high temperature and tools. The third stage is the stable high temperature section that needs to be studied. Therefore, the last 2s of the experiment is taken as the research object. As shown in Figure 2(b), the COF became stable, with an average value of 0.652. From Figure 2(c), it can be seen that graphite can significantly improve the aluminum alloy friction behavior at 450 °C, with an average value of 0.091. Because of the poor frictional properties of aluminum alloys at elevated temperatures, the following experiments were performed under the condition of graphite lubrication.

Fig. 3. The defect left on the strip after friction: (a) RT, (b) 450 °C, (c) 450 °C+82.52 N.

3.1. *Effect of normal force on friction coefficient*

In order to study the effect of normal force on the COF of high-temperature aluminum alloys, 26.86 N, 47.53 N, 61.84 N and 82.52 N masses were selected, and high-temperature friction test with graphite lubrication were performed at sliding speed of 50 mm/s.

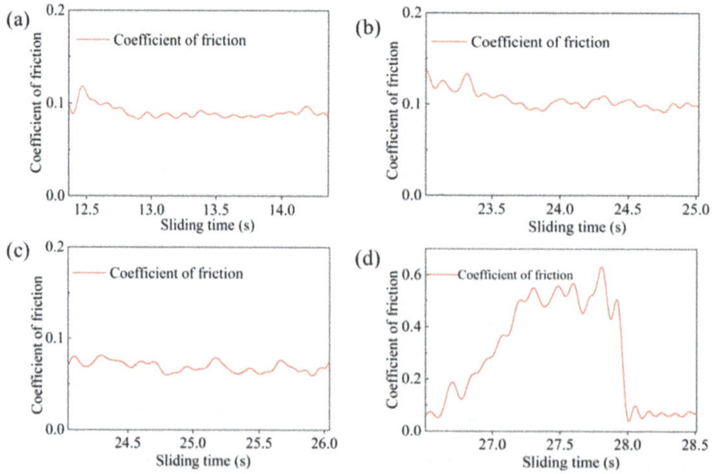

Fig. 4. The COF curves of different normal forces: (a) 26.86 N, (b) 47.53 N, (c) 61.84 N, (d) 82.52 N.

Figure 4 shows the variation of COF and temperature over sliding time under various normal forces condition. Figure 5(a) is their average COF value. It can be seen that COF changes little under the normal force of 26.86 N, 47.53 N and 61.84 N, but the COF value increases sharply when the normal force value is 82.52 N. Graphite has greatly improved the frictional performance by isolating the direct contact between the strip and the friction pairs. However, when the normal force is too large, graphite lubrication failure (because of the direct contact between the strip and the friction pairs) causes a large of scratches on the aluminum strip shown in Figure 3(c). Similar to the high-temperature friction experiment with non-lubrication, the COF at 82.52 N normal force becomes large and its oscillation is severe.

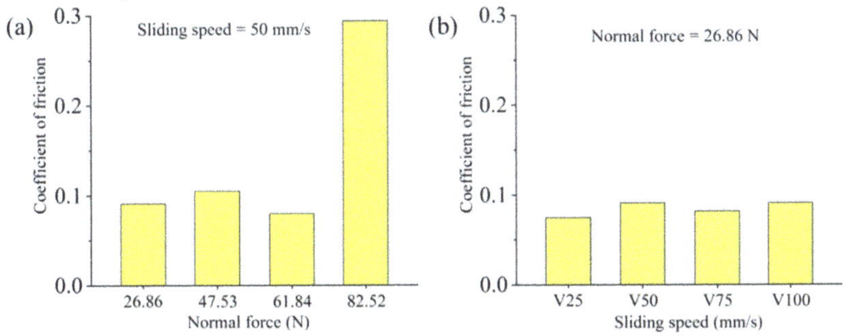

Fig. 5. Average COF value of different parameters at 450 °C.

3.2. *Effect of sliding speed on friction coefficient*

In order to study the effect of sliding speed on the COF of aluminum alloy at elevated temperature, sliding speed values of 25 mm/s, 50 mm/s, 75 mm/s and 100 mm/s were selected, and the friction experiment with graphite lubrication was performed at 26.86 N normal force. Figure 6 shows the variation of COF and temperature over sliding time under different speeds condition. Figure 5(b) is their average COF value. Figure 5(b) shows that as the sliding speed increases, there is no significant change for the COF.

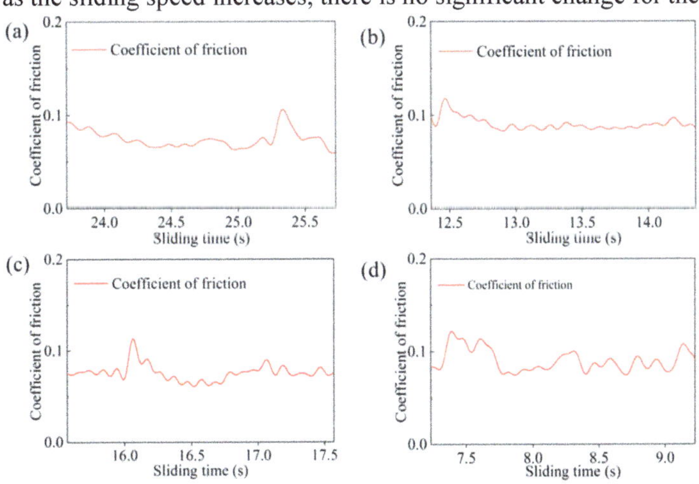

Fig. 6. The COF curves of different sliding speeds: (a) 25 mm/s, (b) 50 mm/s, (c) 75 mm/s, (d) 100 mm/s.

The change of the sliding speed affects the initial temperature of the friction material entering the friction pairs during the second stage of high-temperature friction and the friction time of the adhesive point. Due to the lubrication of graphite, there is no adhesive point in aluminum alloy friction at elevated temperatures, so the sliding speed has little effect on the coefficient of friction.

4. Conclusion

In this study, experimental studies about the tribological behavior of 7075 aluminum alloy have been made under different normal forces and sliding speeds. The obtained results showed that:

(1) Tribological behavior of 7075 aluminum alloy is very poor at 450 °C, while graphite greatly improves its friction behavior.

(2) Under the condition of graphite lubrication, the normal pressure within a certain range has little effect on COF of 7075 aluminum alloy at 450 °C. However, when it exceeds the limit value (this work is 82.52 N), graphite lubrication fails and COF becomes large and violent.

(3) Under the condition of graphite lubrication, the sliding speed has little effect on COF of 7075 aluminum alloys at 450 °C.

Acknowledgments

This research work was financially supported by the National Natural Science Foundation of China (grant no. U1564203 and U1760205) and National Major Science Technology Project of China (grant no. 2018zx04023001).

References

1. R. P. Garrett, J. Lin and T. A. Dean, Solution heat treatment and cold die quenching in forming AA6xxx sheet components: feasibility study, *Adv. Mater. Res.* **6,** 673 (2005).
2. H. Kim and N. Kardes, *Sheet Metal Forming-Fundamentals*, 1st edn. (ASM International, 2012).
3. Y. Liu, Z. J. Zhu, Z. J. Wang, B. Zhu, Y. L. Wang and Y. S. Zhang, Formability and lubrication of a B-pillar in hot stamping with 6061 and 7075 aluminum alloy sheets, *Procedia Eng.* **207**, 723 (2017).
4. X. W. Tian, Y. S. Zhang and J. Li, Investigation on Tribological Behavior of Advanced High Strength Steels: Influence of Hot Stamping Process Parameters, *Tribol. Lett.* **45**, 489 (2012).
5. Z. J. Wang and Y. S. Zhang, Investigation into Hot Stamping of High Strength Steel with Tailored Tempering Properties, Phd thesis, Huazhong University of Science and Technology (2016).

KDAHP1 Tool Steels for Hot Stamping Mold Dies

T. Tonomura[1] and P. Liu[2,†]
[1]*Nippon Koshuha Steel Co., Ltd., Japan*
[2]*Microfine Trading (Shanghai) Co., Ltd., China*
[†]*E-mail: ryu.pin@koshuha.jp*

While H13 hot work tool steel is the most commonly used steel for hot stamping molds, molds could suffer from a short service lifespan due to abrasions and abrasions and adhesion of the product materials. Moreover, as the hot stamping molding cycle is slower compared with cold press working, the need to develop molding could operate in a faster speed is a matter of priority. In this paper, we put forward KDAHP1 hot stamping steel as a concept to improve abrasion resistance and thermal conductivity. KDAHP1 is a self-lubricating steel, which forms a thin oxidized film on its surface when molding at a high temperature, such as when hot stamping is applied. This means significantly high abrasion resistance and longer lifespan for the die. Components of KDAHP1 are so designed that the hardness level improves to 55 HRC or above after quenching and tempering, permitting use with a higher level of hardness then H13. Furthermore, as far as the abrasion quantity is concerned, KDAHP1 is around one-third of H13. As a result, the higher hardness level permits a reduction in the mold die's plastic deformation during molding and higher adhesion between the base material and PVD surface treatment film. This means longer lifespan for die when PVD surface treatment is applied and opens the way for KDAHP1 to be used as a versatile steel.

Keywords: Hot stamping; Tool steel; Cooling efficiency; In-die trimming.

1. Introduction

In recent years, regulations on CO_2 emissions from automobiles have been tightened in response to strengthened international efforts to tackle global environmental issues. Even stricter regulations are to be imposed in the future. At the same time, high standards were set for collision safety as well, which resulted in growing safety-related needs. One method of complying with fuel-efficiency regulations is to reduce the weight of the vehicle body. Therefore, use of materials with high tensile strength (340 to 790 MPa) and those with super-high tensile strength (980 MPa of higher), which also fulfill the collision safety standards, has been increasing. Components made of these materials are mostly cold stamped. This has led to the development of tool steels for cold stamping mold dies and surface treatment films, which are compatible with the molding of high-tensile-strength materials. However, with regard to components made of steel sheets with tensile strength of 1,180 MPa or higher, demand for hot stamped ones is overwhelmingly high while some cold stamped components made of such steels have also been in practical use. Among tool steels for hot stamping mold dies, H13 hot work tool steels are used most widely. With these steels, however, die life may be shortened by abrasion, plastic deformation, and adhesion of product materials. Thus, extension of die life has been an issue for those steels. In addition, higher molding speed is also desired because molding cycle time is longer with hot stamping than with cold stamping. In response, we have developed KDAHP1 hot work tool steels. These steels are optimal for creating hot stamping mold dies which fulfill the above demand.

2. Properties Required of Materials for Hot Stamping Mold Dies

Steel sheets used for hot stamping mold dies include AlSi-plated ones, Zn-plated ones, and non-plated ones. These materials have different problems when used for mold dies. With AlSi-plated and Zn-plated steel sheets, oxidized AlSi or Zn adheres to the mold die, which results in progress of adhesive wear of the mold die. On the other hand, non-plated steel sheets are vulnerable to oxidized scale, which promotes abrasion of the mold die. In addition, they result in wrinkles developing on flange parts and parts in complicated structures. They cause localized stress to be applied to the mold die, resulting in plastic deformation. Because die life may be short even with surface treatment such as nitriding and PVD, there has been growing demand for higher abrasion resistance of steel materials and surface treatment films.

Further, tensile strength of a hot stamped steel sheet is 1,500 MPa to 1,800 MPa, which makes it difficult to trim or pierce it by means of cold stamping. Therefore, such a steel is usually cut with a laser. However, laser cutting has problems -- capital investment cost and long production cycles. This has led to the practical application and spread of a method, with which trimming, piercing, and molding are done in the same mold die. Use of H13 hot work tool steel for this mold die results in severe abrasion while use of a cold work tool steel, such as D2 and 8 % Cr steel, causes chipping, which is a problem. Table 1 shows problems and required properties of mold dies in hot stamping.

Table 1. Problems and required properties of mold dies in hot stamping.

	Mold die	Trimming die (in-die hot trimming)	
Tool steel used	Hot work tool steels, such as H13	Hot work tool steels, such as H13	Cold work tool steels, such as D2 and 8 % Cr steel
Problem	Abrasion/ adhesion Plastic deformation Breaking Slow molding speed	Abrasion	Chipping
Required property	Abrasion resistance Strength Surface treatment characteristics Toughness Thermal conductivity	Abrasion resistance	Toughness
Recommended typed of steel	KDAHP1(+surface treatment)	NOGA or KDAHP1(+surface treatment)	

3. Tool Steels for Hot Stamping

3.1. *Prolonging die life by increasing the abrasion resistance and strength*

Figure 1 shows the position of KDAHP1. In hot stamping, H13 is used with a targeted maximum hardness of around 52 HRC in many cases. This may still result in mold abrasion and a short die life. It is generally said that the harder a mold die, the more resistant to scratch abrasion and erosion abrasion it becomes. With regard to adhesive wear, the relationship between hardness and abrasion resistance is not necessarily clear. Generally, however, high hardness level is advantageous for resisting abrasion. Figure 2 shows the tempering curve of KDAHP1. Components of KDAHP1 are so designed that the hardness level becomes 55 HRC or above after quenching and tempering, permitting use with a higher level of hardness than H13. In addition, KDAHP1 is a self-lubricating steel, which forms a thin oxidized film on its surface when molding at a high temperature, such as hot stamping, is applied. This means significantly high abrasion resistance and enables an extension of die life. Figure 3 shows the abrasion resistance of KDAHP1. The abrasion quantity of KDAHP1 is around one-third compared to H13.

Moreover, the higher hardness level permits a reduction in the mold die's plastic deformation during molding and higher adhesion between the base material and PVD surface treatment film. This means that longer die life can be expected when PVD surface treatment is applied. Designed to permit heat treatment with H13, KDAHP1 is a versatile steel.

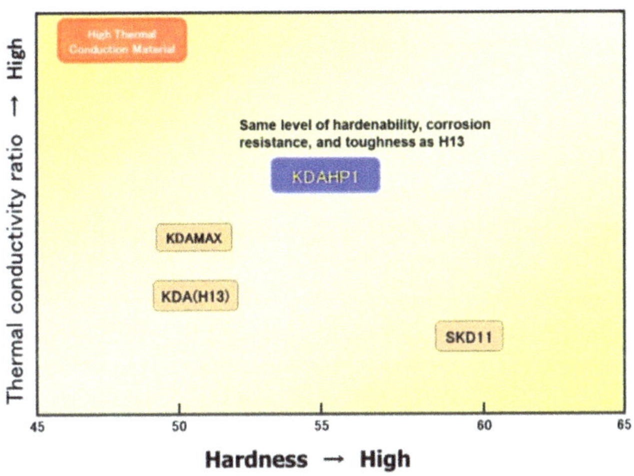

Fig.1. Positioning of KDAHP.

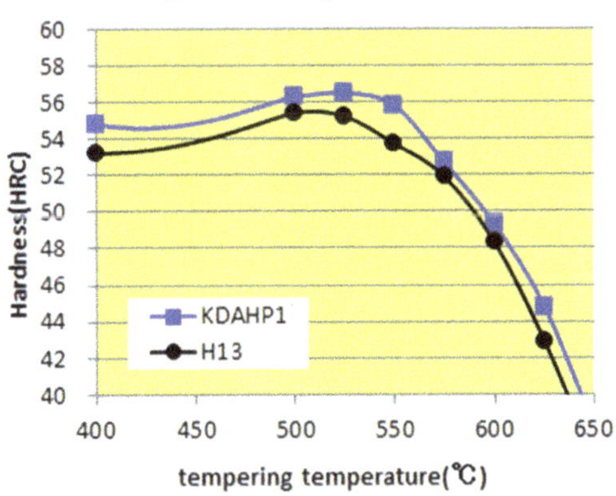

Fig. 2. Tempering diagram of KDAHP1 (Heat treatment condition: Gas cooling quenching: 1030 °C×90 min; Tempering: T °C×180min).

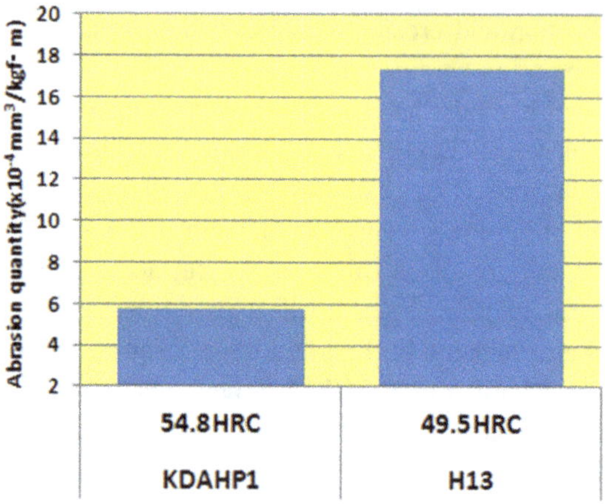

Fig. 3. Abrasion Resistance of KDAHP1 (Test conditions: Test machine: Ogoshi-typy; Target material: 590 MPa High-tensile steel; Lubrication oil: none; Friction distance: 100 m; Final lode: P6.3 kgf).

3.2. *Achieving higher cooling efficiency of steel sheets by improving thermal conductivity*

An effective method of improving the thermal conductivity of steel materials is to reduce Cr, Mn, and Si. However, reducing Cr or Mn lowers hardenability, which is a problem. Above all, many mold dies for hot stamping are large, which means hardenability is an important property. In addition, Cr contributes greatly to corrosion resistance. When Cr is reduced, corrosion resistance is significantly lowered. Corrosion resistance is also an important property of hot stamping mold dies, which need to be cooled with water from a water-cooling hole. It is generally known that reducing Si results in higher toughness as well as improved thermal conductivity. However, an extreme reduction in Si significantly lowers machinability, which is a shortcoming. In many cases, machining of mold dies for hot stamping is done by a department that also conducts machining of cold stamping mold dies. The machining load of hot stamping is high because it involves finish machining after quenching and tempering, which is usually not conducted for cold stamping mold dies. Accordingly, higher machineability is desired.

KDAHP1 is a steel material the thermal conductivity of which was maximized while maintaining the superior hardenability, corrosion resistance, and machinability of H13. Figure 4 shows the thermal conductivity of steel, measured by applying the laser flash method. The thermal conductivity of KDAHP1 is 30 % higher than that of H13, which means it is cooled more efficiently than H13 after hot stamping.

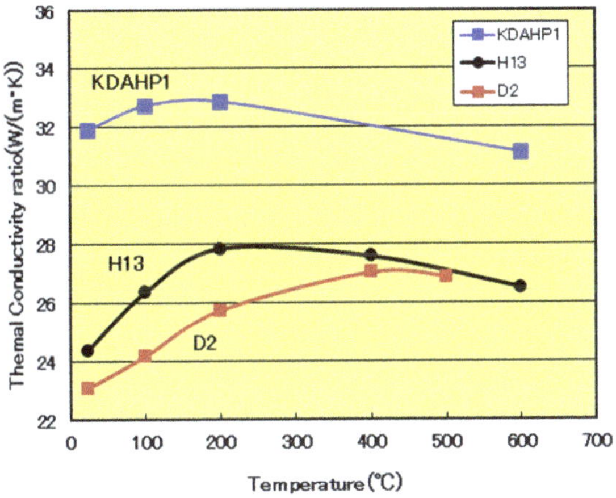

Fig. 4. Thermal conductivity ratio of KDAHP1 (Heat treatment condition: Gas cooling quenching: 1030 °C×90 min; Tempering: 530∼610 °C × 180 min 2 times).

The temperature of the commercially produced H13 mold die was measured, and it was found that the temperature of KDAHP1 is around 20 degrees Celsius lower. This result suggests that heat is removed from steel sheets more quickly when KDAHP1 is used.

Figure 5 shows the result of a machining test for comparing machinability. The test was carried out using a ball end mill that is used for finish machining after quenching and tempering, which features an especially high machining load. The machinability of KDAHP1 with the same hardness level as H13, namely 50 HRC, is equal to that of H13. With a hardness level of 55 HRC, the machinability of KDAHP1 is lower than that of H13. However, it is much higher than that of D2 (60 HRC). It can therefore be said that the machinability of KDAHP1 with 55 HRC is suitable for practical use.

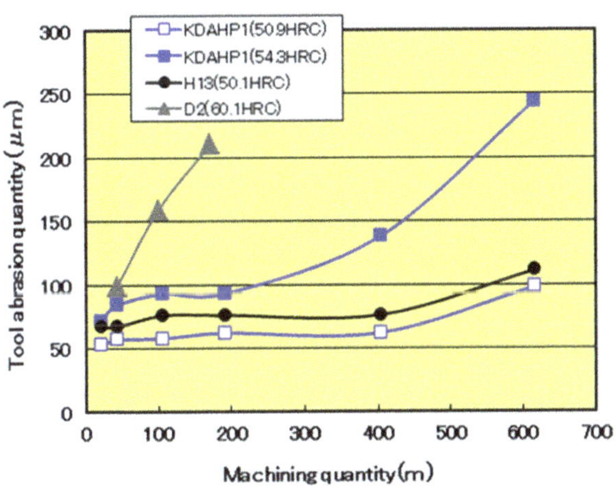

Fig. 5. Machinability of KDAHP1 (Test conditions: Machine: MC (BT40,5.5 kW); Tool: MITSUBISHI SRFH30S32M; Tip: MITSUBISHI SRFT30 MP8010; Revolution: 2650 min-1; Feed: 1660 mm/min; Depth of cut: Ad 0.3 mm, Rd 0.7 mm).

3.3. *Steel material for in-die trimming*

NOGA (+ surface treatment), a cold work tool steel from Nippon Koshuha Steel, has been used and highly acclaimed as a steel material for in-die trimming. While NOGA becomes as hard as 60 HRC, the same hardness level as D2, after quenching and tempering, it is more than twice as tough as D2. It therefore reduces the chipping of the cutting blade, which occurs when D2 is used, extending the die life. However, the use of KDAHP1 (+ surface treatment) is recommended when chipping occurs even with NOGA due to the shape or stamping method. KDAHP1 enables an extension of die life because it is superior to H13 in abrasion resistance and has a higher level of toughness than NOGA.

4. Summary

(1) The hardness level of KDAHP1 is as high as 55 HRC. It therefore improves the problems of die abrasion and plastic deformation that pertain to hot stamping.
(2) Because of its high thermal conductivity, KDAHP1 permits heat to be removed quickly after molding. It can therefore help shorten the cycle time.
(3) For in-die trimming, the use of NOGA featuring superior toughness reduces the chipping of blades and extends the die life. Where chipping occurs even with NOGA, it can be reduced using KDAHP1.

Equipment

High Performance Sheet Metal Hot Forming Technology and Service System

L. Wang† and Y. S. Zhang

State Key Laboratory of Materials Processing and Die and Mould Technology, Huazhong University of Science and Technology, Wuhan 430074, China
†E-mail: wangliang@hust.edu.cn

X. Y. An

Wuhan Zhongyu Dingli Intelligent Technology Co., Ltd., Wuhan 430074, China

Automobile parts manufacturers are cooperating with automotive companies in the early development of high performance metal sheet forming technology. For the early development of high strength steel / high strength aluminum material, especially the mechanical properties test of the material after forming, it needs a test line with large range of working parameters and flexible operation. In order to accurately obtain the physical parameters of the change of the material in the forming process, it is necessary to ensure that the control center needs the position of the slider movement of the press and the ability to perceive the exact data of the forming force of the material during the forming process. The process data acquisition and analysis of the production line has tended to the mode of big data to meet the needs of intelligent decision making. When the intelligent process system is connected to the platform of the industrial Internet, the production process of the hot forming production line in the production workshop of various places can be monitored at any time, and the early intelligent manufacturing workshop or factory can be realized.

Keywords: Intelligent manufacturing; Servo press; Intelligent engine; Industrial internet.

1. Introduction

Hot stamping technology of high-strength steel for automobile body-in-white has been widely used. With the continuous increasing requirements of the light-weighting, especially for the special needs of new energy vehicles, hot stamping technology of high-strength aluminum alloy sheet already appears. Auto parts suppliers are working with automotive companies to develop high-performance sheet metal forming technology. A small production line for hot stamping with a wide range of process parameters and flexible operating is required for the early development of high-strength steel/aluminum materials and especially for the mechanical properties testing after forming and strengthening. Along with increasing demand for hot stamping tooling development, the small production line for the purpose of tooling trial production has attracted much attention and been rapidly becoming important equipment for hot stamping parts suppliers and professional tooling suppliers.

This kind of small production line was developed by Huazhong University of Science and Technology (HUST) and WISCO Research Institution [1], shown in Fig. 1. The maximum stroke of the slider is 600 mm and the maximum forming load is 2000 kN.

The Fraunhofer Institute in Germany announced the Hot-stamping Laboratory based on Industry 4.0 in 2015 which is also a representative of the in-depth development of such kind of production lines. Nippon Steel and Sumitomo Metal also announced such test production lines in 2016.

In order to standardize and modularize the control system of each unit of the production line, the new generation intelligent production line has adopted a highly flexible unit control system and its process control integration technology with the industrial internet. In particular, emergency handling control of various event disturbances (equipment failures, accidents) that occur during the production process lays the foundation for the development of failure and safety control strategies [2].

Fig. 1. Hot forming test line of mechanical servo press developed by HUST-WISCO Joint Laboratory.

2. Mechanical Servo Press and Process Intelligent Engine

It can be seen from the hot stamping process that the influence of the holding pressure on the cooling rate of the blank is related to the material structure properties after forming. The gap between blank and tooling after die closing will affect the forming quality of the product (as shown in Fig. 2), which has received great attention during designing, manufacturing and production testing of tooling. In actual production, the static settings and parameters determined by trial production will change during the die-closing and cooling process. The mechanical servo press has certain abilities to sharply detect these changes and timely compensate by slider moving. This is the premise of realizing adaptive control process. That is to say, the intelligent process control requires precise dynamic control of die closing position and forming load.

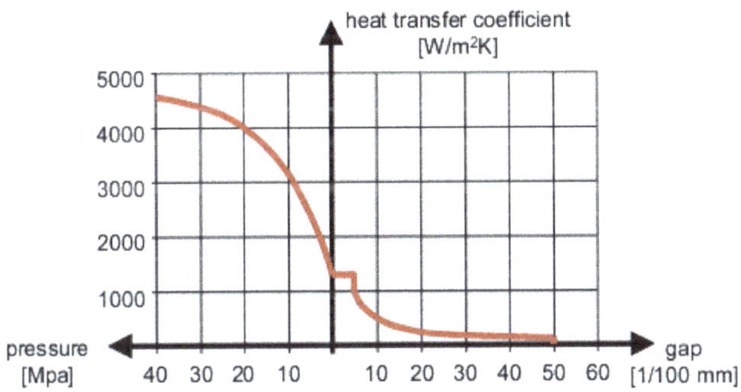

Fig. 2. The relationship between the gap between die and blank, positive pressure and thermal conductivity.

In order to accurately obtain the physical parameters of the material change in the forming process, it is necessary for the press to detect the accurate date of the position of the slider and the forming resistance during hot stamping process. Digital mechanical servo press is superior to conventional high speed hydraulic press. it is easily be seen from actual test data [3] that the mechanical servo press with a stroke of 800mm has a repeatability of 0.0135±0.005 mm at the bottom dead center position, as shown in Fig. 3. Due to the high precision of the bottom dead center position control, the mechanical servo press is able to detect the pressure change on the blank during its cooling and shrinking process, as shown in Fig. 4. The control system of servo press has the ability to perform automatically adjustment according to these changes. For example, a tiny movement of slider will be performed for gap compensation until the desired contact pressure between blank and tooling is reached and maintained.

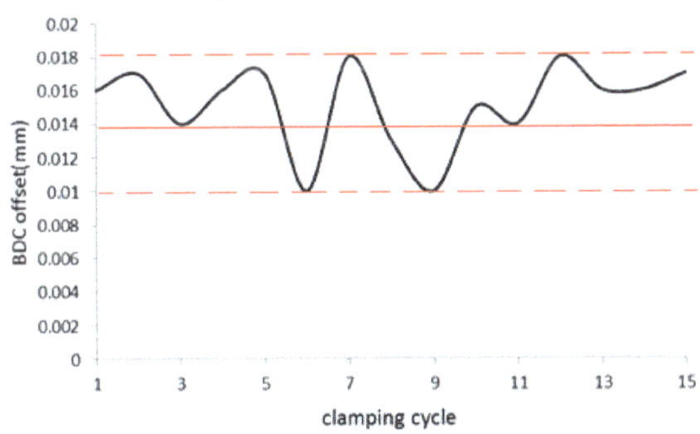

Fig. 3. BDC offset during slider reciprocating motion.

3. The Development of a New Generation of Intelligent Production Line

According to the on-site survey of the German automotive industry conference in September 2013, the future demand for hot forming production lines is shown in Table 1. It can be seen from the data: 1) requires strong adaptability of hot stamping for a variety of materials in the future. For example, the same production line should be able to perform production of high-strength steel and high-strength aluminum alloys. 2) Satisfies forming of the increasing complexity of parts, customized mechanical properties of parts, large-sized complex parts. For example, a direct hot stamping line for door ring. 3) Requires more advanced, low-energy heating and forming equipment, as well as smaller equipment footprint [4].

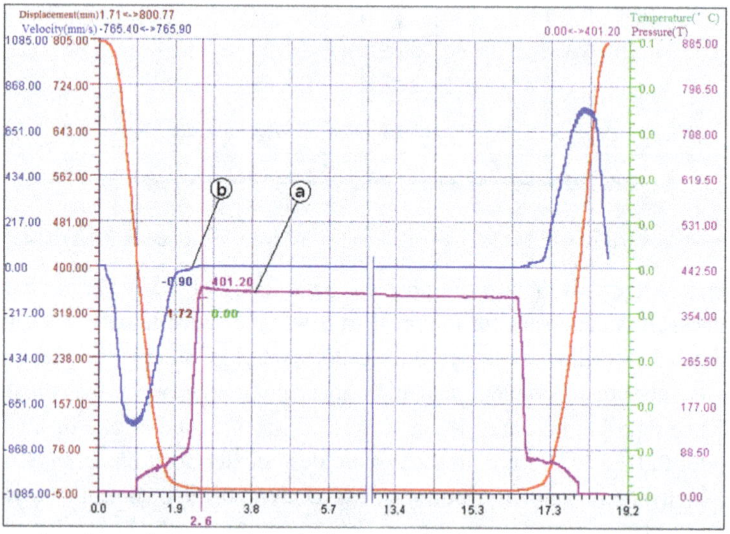

Fig. 4. Pressure data change during actual monitoring record forming process: (a) the change of pressure data during the holding process; (b) the change of press speed data during the closing process.

Table 1. Demand survey results of hot forming production line.

Demand	Percentage
Improve the production line beat (SPM)	10%
Adopt new materials	27%
Capable of producing more complex parts	24%
Higher flexibility / small batch manufacturing	32%
Other	8%

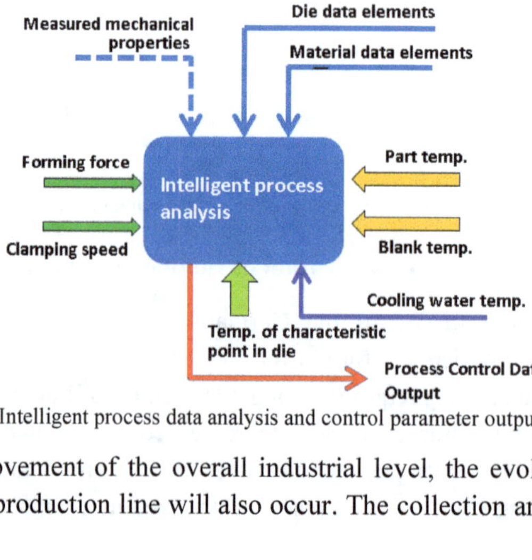

Fig. 5. Intelligent process data analysis and control parameter output system.

With the improvement of the overall industrial level, the evolution for automation and intelligence of production line will also occur. The collection and analysis of process data of production line approaches a model of big data to meet the needs of intelligent decision-making. Due to developing of material database of high-strength steel and high-strength aluminum alloy, and acquiring tooling data during forming process in real time, as long as the press has sufficient agility, the intelligent control system has the abilities to perform the best forming process according to the analysis result of the intelligent engine, shown in Fig. 5. By continuously adding new data from part performance testing into process database, the intelligent engine is assisted for learning and correction, so the forming process is continuously improving and the part quality and production efficiency are also improving.

4. Technical Service System

At present, the physical condition of the world's internet is that a complete high speed wireless communication network is being planned and tested. When the intelligent process system is connected to the industrial internet, the production of all the hot stamping lines in different workshops can be monitored at any time, which realizes the initial intelligent manufacturing workshop and factory.

An intelligent manufacturing technology system (as shown in Fig. 6) should be established to serve the needs of society. It is a service provider for manufacturing to connect the upstream and downstream of manufacturing, which is a hub to connect material suppliers and enterprise (mainly car and auto parts manufacturers here). This system provides fundamental informatization services on cloud platform and professional intelligent planning services for organizers and members of the intelligent equipment alliance. The construction of new manufacturing service-oriented enterprises and service platforms through technology development and capital investment is the key to the hot stamping intelligent manufacturing system.

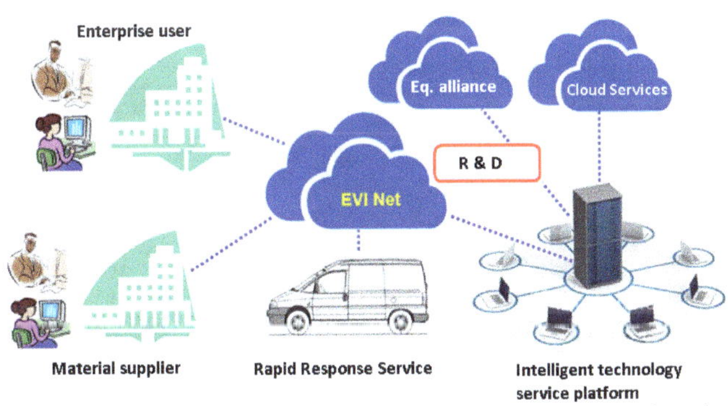

Fig. 6. The technical service system of hot stamping intelligent manufacturing.

5. Conclusion

With the increasing physical connectivity of the Industrial Internet, digital factory connections already have the necessary conditions. The advanced digital production line with servo press can quickly sense the changes of many parameters of the hot stamping process, thus realizing the real-time control of the process parameters, which helps to realize the intelligent process of hot stamping of high-strength steel/high-strength aluminum alloy sheets. The application of sensor networks, the completeness of advanced material data, and advances in non-destructive testing technology have made hot stamping intelligent production lines possible. An intelligent manufacturing technology system should be established to serve the needs of the society. It is a path to establish an intelligent manufacturing system for multi-species, customized and low-cost high-performance metal sheets.

Acknowledgments

This research work was financially supported by the National Major Science Technology Project of China (grant no. 2018zx04023001).

References

1. L. Wang, B. Zhu, Y. S. Zhang, Y. L. Wang, X. Y. An and Q. Wang, A smart process control strategy for press hardening production, in *6th International Conference Hot Sheet Metal Forming of High-performance Steel*, (Atlanta, USA, 2017).
2. L. Wang, B. Zhu and Y. S. Zhang, Production control and optimization of hot stamping line, in *5th International Conference on Hot Sheet Metal Forming of High-performance Steel*, (Toronto, Canada, 2015).
3. Q. Wang, L. Wang and Y. S. Zhang, Mechanical link servo press on hot stamping production line, in *Advanced High Strength Steel and Press Hardening: Proceedings of the 2nd International Conference*, (Changsha, China, 2015).
4. L. Wang, B. Zhu and Y. S. Zhang, A runtime intelligent iterative manufacturing system (Riims) for hot stamping, in *Advanced High Strength Steel and Press Hardening: Proceedings of the 2nd International Conference*, (Changsha, China, 2015).

Application of Mechanical Servo Press in Hot Stamping Production Line

X. Fang†, Z. M. Ke and L. Dou
Jiangsu Xuzhou Metalforming Machine Group Co., Ltd., Xuzhou, 221116, China
†E-mail: 727851852@qq.com

X. Y. An and L. Wang†
State Key Laboratory of Materials Processing and Die and Mould Technology, Huazhong University of Science and Technology, Wuhan, 430074, China
†E-mail: snigoal@qq.com

The article puts forward a new overall optimization solution for hot stamping production line, this scheme is based on the mechanical servo press which can be used in hot stamping production line of roller bottom furnace or multi-layer box heating furnace. The mechanical servo press can control the position, speed and load of the slider more accurately, and it also has higher movement agility. Compared with traditional high speed hydraulic press, it has the advantages of low energy consumption, simple structure, high digital level, high efficiency and simple maintenance, and has the potential to improve the technical level of hot stamping production line.

Keywords: Lightweight; Hot stamping; Mechanical servo press.

1. Introduction

The rapid development of automobile industry has resulted in three major problems of energy consumption, emission and safety, and automobile lightweight is the most direct and effective means of energy saving and emission reduction in automobile industry. The lightweight is particularly important for new energy vehicles due to the limitation of battery energy density and continuous driving distance. More use of high-strength steel and ultra-high-strength steel parts is an important way to realize vehicle lightweight and improve safety (Fig. 1). Hot stamping is a kind of advanced forming technology for producing ultra-high-strength automobile parts and effectively avoiding rebound, which can be widely applied to automobile safety structure parts such as front and rear bumper frame, A-pillar, B-pillar, etc. Under the premise of effectively reducing the deformation of the cockpit and protecting the safety of the drivers, the weight of the car body can be greatly reduced. Hot stamping technology is one of the most important advanced processing technologies to support automobile lightweight, which represents the development trend of automobile key parts manufacturing technology.

With the implementation of Chinese regulations on automobile energy consumption and light weight, Chinese fuel automobiles began to use high-strength steel plate hot stamping forming technology to manufacture body structures in a large scale. Up to now, more than 60 hot stamping production lines have been built in China. Judging by Chinese automobiles production and demand for hot stamping parts, it is expected that by 2020 China will need to build more than 50 new hot stamping production lines. The presses of the existing hot forming production lines are basically high-speed hydraulic presses. In the early stage, the hydraulic press is to use the energy accumulator to speed up the running speed of the slider and meet the needs of the hot stamping process. In recent years, the high speed hydraulic press adopts the proportional valve to directly control the movement of the oil cylinder, which improves the production rhythm and reduces the energy consumption. And the application of mechanical servo press driven directly by servo motor with low speed and high torque has just entered the field of hot stamping production.

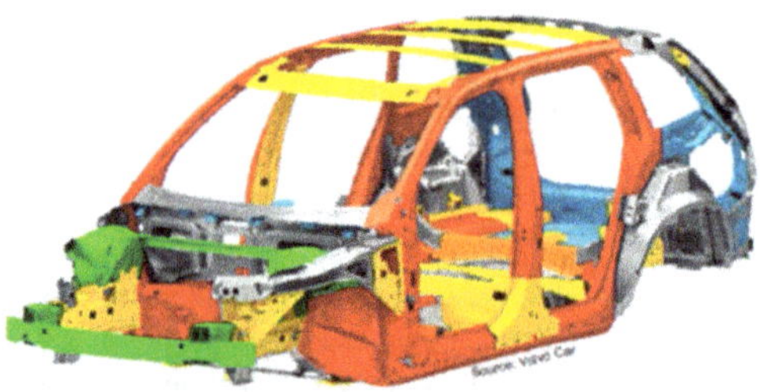

Fig. 1. Hot forming parts reach 38 % of the weight of a standard body in white (BIW).

2. High-strength Steel Sheet Hot Stamping Forming Technology

High-strength steel plate hot stamping technology is to heat special steel plate (boron alloy steel plate with tensile strength 400-600 MPa at room temperature) to austenitic temperature range, moving fast into the mould, stamping fast, and quenching and cooling by mould under the condition of press holding, finally obtain ultra-high-strength stamping parts (the tissue is martensite and its strength is about 1500 MPa or even higher).

The prototype of the high-strength steel plate hot stamping technology appeared in the 1980s, and it has been developed in Europe for a long time; in the 1990s, hot stamping forming equipment began to develop due to the expiration of hot stamping related patents, and a prototype of the high-strength steel hot stamping production line emerged. After 2000, the application began to expand and entered China around 2007.

3. Process for Hot Stamping

The process of hot stamping technology consists of blanking, heating, transfer, stamping and quenching, and subsequent processing. The core process is the heating of the steel plate, automatic transfer to the press, high-speed stamping, quenching and cooling by mould under the condition of press holding, as shown in Fig. 2.

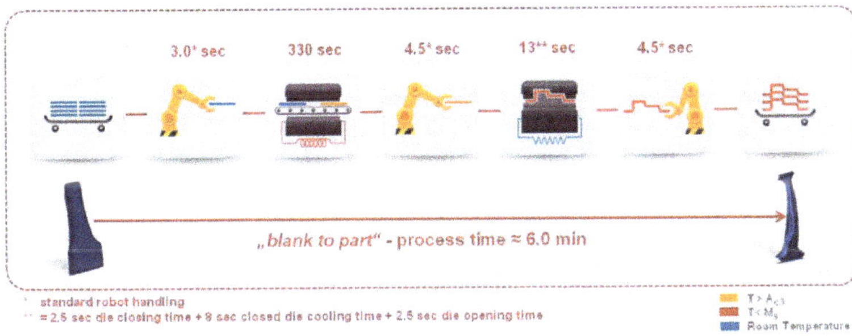

Fig. 2. Classic process for hot forming [1].

1) Blanking: It is the first step in the whole process, and punch the steel plate out of the desired profile.
2) Heating: The steel plate is heated to a suitable temperature, so that the steel plate is completely austenitic and has good plasticity.
3) Transfer: Remove the heated steel plate from the furnace and quickly put it into the mould. This process must be fast enough because on the one hand, it is to prevent the oxidation of the steel plate at high temperatures, and on the other hand, it is to ensure that the steel plate is still at a high temperature, to have good plasticity.
4) Stamping and quenching: After putting the hot steel plate into the mould, the hot steel plate should be stamped immediately so as not to affect the forming performance of the steel plate too much. After stamping, the mould must be closed under the condition of press holding for a period of time because on the one hand, in order to suppress the rebound and control the shape of the parts, and on the other hand, the cooling water channel set in the mold is used to quench the steel plate to form a uniform martensite structure and obtain good dimensional accuracy and mechanical properties. At present for the commonly used hot stamping steel, the minimum cooling rate for the transformation of austenitic to martensite is 27-30 degrees Celsius per second, so the cooling speed of the mould to the steel plate must be greater than this critical value.
5) Subsequent processing: After the forming parts are removed from the mould, some subsequent processing is required, such as the use of shot-blasting to remove the oxide skin on the surface of the parts and laser cutting and drilling of the parts.

4. The Optimization Solution for Hot Stamping Production Line

4.1. *Production line layout*

The layout of the hot stamping production line based on the mechanical servo press and the multi-layer heating furnace is that the mechanical servo press is centrally arranged, and two multi-layer heating furnaces are symmetrically arranged in front of the press. The T-shaped area between the press and the heating furnace is the motion space of the feeding manipulator, the front of the production line is equipped with unstacking truss manipulator, discharge manipulator is arranged behind the press, and Each furnace is equipped with a furnace door manipulator responsible for the opening and closing. This scheme covers a small area and has a compact structure.

4.2. *Mechanical servo press*

At present, hydraulic presses are used in nearly 100 hot stamping production lines worldwide. However, because the traditional hydraulic press does not have the function of rapid die closing and stamping, and the special developed rapid hydraulic press for hot stamping has high cost, high noise and high energy consumption. And the traditional mechanical press does not have the function of pressure maintenance and cannot meet the requirements of the hot stamping process.

This optimized solution proposes a press scheme for servo drive technology. In recent years, the high-speed development of mechanical servo presses has eliminated flywheel and clutch-brake, replaced traditional AC asynchronous motor with AC synchronous servo motor, and adopted digital control systems to adjust the speed of the slide at any position. The slide has high precision of motion and positioning, and has the characteristics of flexibility, high speed and high efficiency. The servo press in the scheme adopts a single low-speed and high-torque servo motor to drive the high-speed and heavy-duty sliding screw directly, and drives the slide movement through the elbow structure, as shown in Fig. 3. This scheme optimizes the structural design of servo press, reduces the transmission chain, and improves the overall transmission efficiency, control accuracy, and dynamic response of the press. The moment of inertia of the drive system is reduced by 25 % compared to the multi-motors drive, the drive efficiency is increased by 6.8 %, and the acceleration time is shortened from 0.6 s to 0.25 s. The design of high-speed and heavy-duty sliding screw and elbow structure improves the speed of motion (speed greater than 800 mm/s) when the slide is not loaded, and reduces the input torque of the servo motor. At the same time, this design also improves the repeated positioning accuracy (0.01 mm) of the bottom dead centre (B.D.C) and realizes the position following and torque compensation when pressure maintenance at the B.D.C. In the process of cooling shrinkage, according to the changes in pressure and position, the slide can dynamically follow the shrinkage of the parts, ensure the fitting state of the parts, improve the quality of the parts, and perfectly match the requirements of the hot stamping process, as shown in Fig. 4.

The electrical control system of the Mechanical servo press is shown in Fig. 5, using the professional industrial personal computer (IPC) as the controller and using the VARAN bus as the communication protocol of the master-slave IO station. Due to the addition of the single low-speed and high-torque servo motor, higher requirements are put forward for the servo drive. Innovative multi-drive unit parallel technology and synchronization technology are used in the scheme to overcome the problem of insufficient power of the single servo drive, and the system response of moving parts of press is further improved. The three servo drives composed of one master and two slaves are unified control through a main controller, ensuring three phase balanced outputs. Three servo drives share a dc bus, with the power output up to 1200 kw and the current up to 2200 A, as shown in Fig. 6. The motor uses a single winding interior permanent magnet synchronization technology that has twice the overload capability and can operate at high speed when the slide is not loaded. The controller is responsible for motion control processing, and the control information is sent to the servo drive through the Canopen bus to control the motor movement.

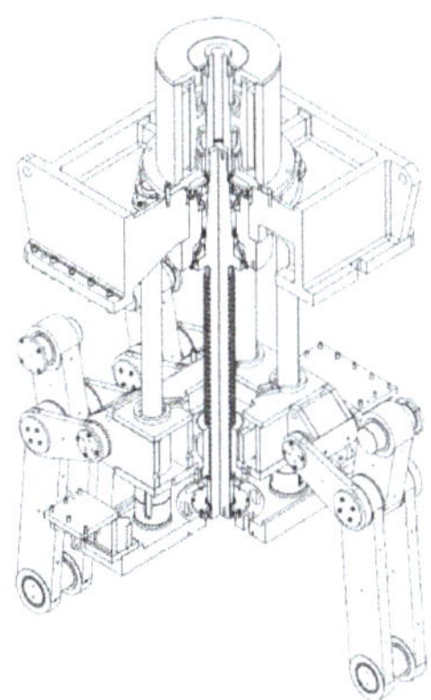

Fig. 3. Drive structure for servo motor-screw-elbow.

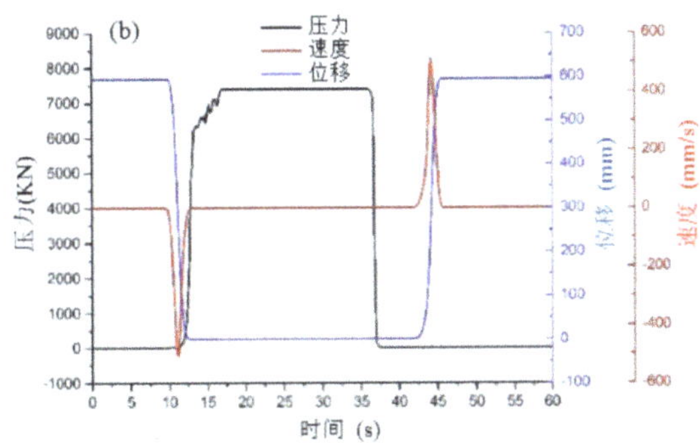

Fig. 4. Real-time state diagram of motion and load of slide.

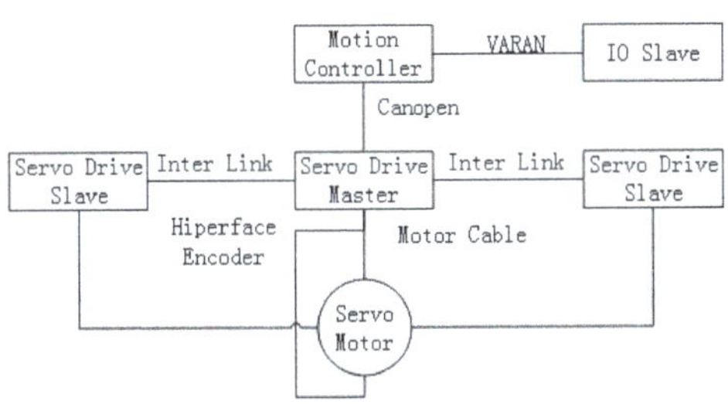

Fig. 5. Electrical control system structure drawing.

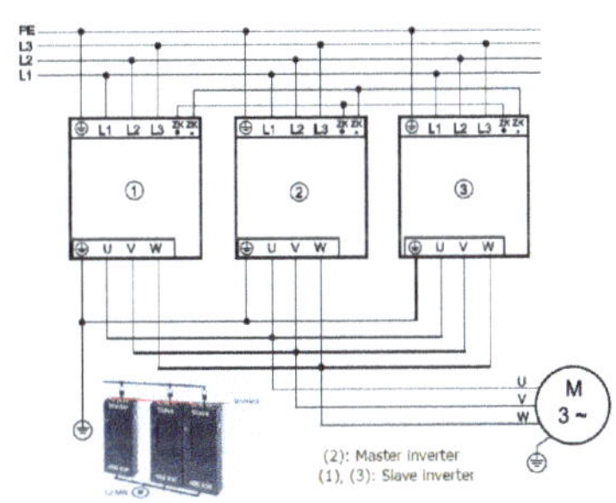

Fig. 6. Structure drawing for multi-drive unit in parallel.

4.2.1. *Motion curve control*

The user can set the speed of the slide according to the process requirements. It has the characteristics of high flexibility, intuition, etc. After the setting is completed, the set curve can be previewed, as shown in Fig. 7.

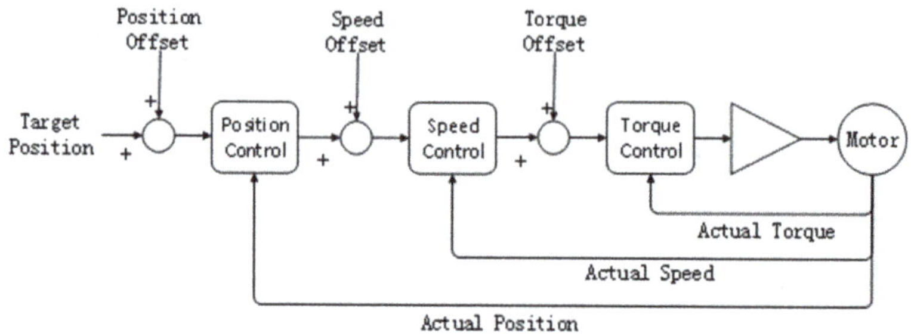

Fig. 7. Motion control structure drawing.

4.2.2. *Pressure maintenance control*

Due to the requirements of the hot stamping process, the pressure maintenance is an inevitable step. At this time, most of the current of the motor is used for heating, and the unique mechanical and electrical technology is used to reduce the current of pressure maintenance, thus reducing the motor heating situation and making it meet the actual work needs.

4.2.3. *Energy management system*

As shown in Fig. 8, the use of energy storage capacitors effectively reduces the peak current of external lines and reduces the impact on external power grids. At the same time, it can absorb and store the braking energy of drive system, improve the energy utilization efficiency, and save energy up to 30 %.

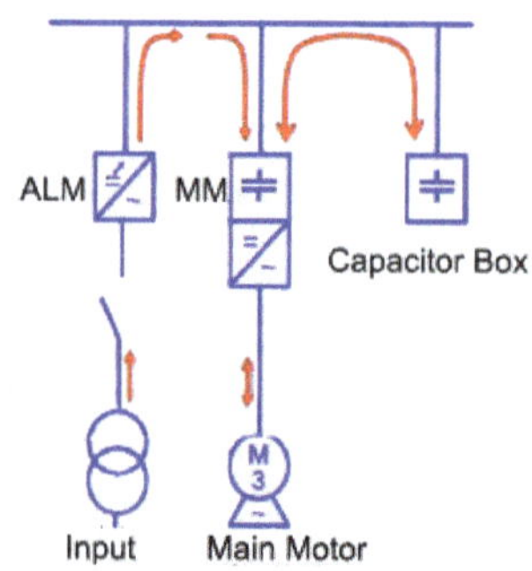

Fig. 8. Energy management system structure drawing.

4.2.4. *Auxiliary function*

Benefit from the development of automation technology, auxiliary functions such as automatic adjustment of the shut height and automatic change of the double moving table can greatly improve the mould exchange efficiency and reduce the auxiliary time.

The mechanical servo press is not only suitable for the stamping of high-strength steel plates, but also suitable for the stamping of aluminum alloys plates. With the use of multi-layer heating furnace, the production of steel and aluminum mixed wire is realized. It is the preferred equipment for the press in the hot stamping production line.

5. Conclusion

The hot stamping production line based on the mechanical servo press has improved the overall level of hot stamping production equipment, it can better meet the demand for high performance, high safety, and lightweight in the automobile and aviation industries, as shown in Fig. 9. Among them, the mechanical servo press has unique advantages in the forming of high-strength steel plates, aluminum alloy plates, and non-equal thickness welded steel plates. It is more than 50% energy-efficient than traditional hydraulic press, and the noise of stamping is greatly reduced. It provides important technical means for the application and popularization of new materials and the lightweight of automobile.

Fig. 9. Customer scene of hot stamping production line.

Acknowledgments

This research work was financially supported by the National Major Science Technology Project of China (grant no. 2018zx04023001).

References

1. K. Steinhoff, M. Schupfer, A. Ademaj and U. Weifig, All about press hardening- an overview on technology and markets, in *2th International Seminar CHS2*, (Hanover, Germany, 2012).
2. L. Wang, B. Zhu, Q. Wang and Y. S. Zhang, Modeling of hot stamping process procedure based on finite state machine (FSM), *Int. J. Adv. Manuf. Technol.* **89**, 857 (2017).
3. L. Wang, X. Y. An, B. Zhu, Y. L. Wang and Y. S. Zhang, Research on Intelligent Process and Production Method of Hot Stamping, in *International Conference on Advanced High Strength Steel and Press Hardening (ICHSU 2016)*, (Xian, China, 2016).
4. Q. Wang, L. Wang, Y. S. Zhang, L. M. Yao and Z. M. Ke, Mechanical Link Servo Press on Hot Stamping Line, in *International Conference on Advanced High Strength Steel and Press Hardening (ICHSU 2015)*, (Chongqing, China, 2015).

Enhance Reliability of Press Hardening Operations

D. R. Knecht
Quaker Chemical Co., Conshohocken PA, USA
Email: knecht@quakerchem.com

Presses for hot sheet metal forming are critical assets and significant investments for manufacturers engaged in press hardening operations. With metal sheets reaching temperatures of about 900 °C, a fire on the press where mineral oil based lubricants are involved can cause extreme hazards for operators, damages to or loss of the press itself, production losses due to suspended operations for extended periods of time, and damages to customer relationships resulting from a loss in trust in supply reliability. Ultimately, such losses could cost manufacturers millions. Protecting such assets by replacing mineral oil based lubricants with a reliable risk prevention product can reduce the risk of such losses and advance the safety efforts of the manufacturer. This paper describes a case where reliability of the production has increased considerably by choosing a fire resistant lubricant alternative for the hydraulic, greases, gear oils and slide way oil.

Keywords: Fire resistance; Lubricant.

1. Introduction

In the press hardening process red hot blanks come very near oil based lubricants every press cycle that lead to situations that can be classified as dangerous. Danger not only due to the plant's machinery, or the operations involving liquid or red hot steel, but also because of the escalated fire risk based on the lubricants being used.

In this paper we will focus on the potential risks associated with the use of mineral oil based lubricants like hydraulic fluids, gear oils and greases in for press hardening applications and what alternatives can be used without jeopardizing the performance or productivity of the press hardening production line.

2. The Fire Risks of Mineral Oil Based Lubricants

A fire is one of the events that, once experienced, leaves a huge impression on the people involved. In addition to the risk of personnel injuries, there is a likelihood of loss in both capital and production. These losses not only include damage to the building and equipment, but also encompass interruptions in production that can idle lines for days or even months.

One cause of fire at a press hardening line is the ignition of mineral oil based hydraulic fluids, gear oil or greases.

In the press hardening process blanks are being heated up to temperatures of ±950 °C (± 1742 °F). In most of these processes hydraulic units or gear boxes are used to operate the press, and in many cases a mineral oil based lubricant is chosen to fuel the hydraulic unit or the gear box. To lubricate the feeder system mineral oil based greases are widely used.

While mineral oil has the definite advantage of a good cost-performance ratio, it is a distillate from crude oil, and not always the safest choice, due to its tendency to catch fire easily and to continue burning.

Fire resulting from the ignition of a mineral oil based lubricant can happen in many different ways. Two that occur most often are when mineral oil based lubricants leak onto a very hot surface; or when hot metal land in a pool of mineral oil based lubricants.

The first instance becomes a larger issue when it happens with a liquid mineral oil based lubricants like a hydraulic fluid because the risk of larger spills or oil spray has a much wider reach. The second example is likely to happen with both liquid as well as greases like mineral oil based lubricants.

Fortunately, there are alternatives available to manage these risks and reduce the chance of an ignition.

3. Fire Resistant Lubricants

Several principles and test methods exist to quantify the level of fire resistance (or not). However, they are valid and designed for hydraulic fluids. The fundamental for the fire resistance of the base stock used is comparable for other lubricants as well.

3.1. *Types of fire resistant hydraulic fluids*

The standard hydraulic fluids used in steel production are mineral oil based. But an alternative to mineral oil hydraulic fluids are fire resistant hydraulic fluids, as described in Table 1 using the ISO 6743/4 classification.

Table 1. Description of Fire Resistant Hydraulic Fluids.

Water-based Fluids	Water-free Fluids
HFA-E: Oil in water emulsions water content > 80 % common use 1 to 5 %	**HFD-R:** Phosphate ester based. These products are less used because of CMR reputation
HFA-S: Synthetic aqueous solutions water content > 90 % common use 1 to 5 %	**HFD-U:** Based on other compounds, but mainly synthetic polyol ester and natural esters (renewable resources)
HFC: Water glycol solutions water content >35 %	

For each fluid type there are both pros and cons. Figure 1 shows a comparison of the performance properties for several hydraulic fluid types. The properties shown are considered important by both maintenance managers as well as purchasers.

Property	Mineral Oil	Phosphate Ester (HFDR)	Water Glycol (HFC)	Synthetic Polyol Ester (HFDU)
Fire resistance	– –	+ +	+ + +	+
Environmental performance	–	+ and –	+ and –	+ +
Thermal stability	+ +	+ +	–	+
Fluid maintenance	+	– –	– –	+
Component life/ System reliability	+	+ and –	– –	+
Price	+ +	– –	+ +	+ –
Total Cost of Operation	–	–	– –	+

Fig. 1. Hydraulic fluid comparison when used in fire hazardous situation.

The comparison table shows that mineral oil has good hydraulic fluid performance attributes, at a reasonable price. However, because mineral oil is not biodegradable, it is not environmentally friendly, and the rating for total cost of operation (TCO) is due to the aftermath experienced once a fire takes place.

Phosphate ester (HFDr) fluids are an older fluid technology, and are fire resistant by chemistry, but have a negative reputation. They are reported to be formulated with CMR (carcinogenic, mutagenic, reprotoxic) materials, and the combustion fumes they produce are reported to be neurotoxic. HFDr fluids can be 10 to 15 times more expensive than mineral oil and need to be carefully maintained as these products become aggressive acids as they age. These fluids are mainly used nowadays in power generation, although they are sometimes found in a steel plants as well.

HFC fluids, better known as water glycols, are widely used in steel plants as well as other industries, and represent about 50 % of the total fire resistant hydraulic fluids market. Their high water content makes them very good for fire resistance, and while they have a comparable price to mineral oil, they do not measure up in performance attributes. Additionally, compared to water free hydraulic fluids, the hydraulic units for HFC are more expensive to purchase, the service components have a shorter lifetime, more fluid management is needed, and energy consumption is 10 to 20 % higher compared to mineral oil or polyol ester based fire resistant hydraulic fluids.

Polyol ester based fluids (HFDu) are the best solution and alternative to mineral oil. Typically, no changes need to be made to the hydraulic unit when converting from a mineral oil or water glycol hydraulic fluid to a polyol ester fluid. They are more expensive than mineral oil (around 2 to 3 times more), but with the reduction in the risk of fire from the hydraulic fluid, the result is a lower total cost of operation and a much safer work environment. Additionally, nothing is sacrificed in terms of the fluid's performance, and the polyol ester based (HFDu) fluids have reduced environmental impact.

When the user knows the types of fire resistant hydraulic fluid available, a comparison between possible solutions can be made as shown in Table 2.

Table 2. Financial and technical evaluation of possible solutions to enhance safety.

Solution	Positive	Negative
Change design of the Unit to avoid mineral oil leakages close to the hot surface	Company can keep the same oil technology.	Does not avoid using a straight hydraulic oil (HLP) coming close to the hot surface. Does not avoid formation of pools.
Installation of a fire extinguisher system	Company can keep the same oil technology.	Very expensive and it does not catch the explosive ignition and fire balls.
Change hydraulic fluid to a Water Glycol HFC type	Offers a safe solution.	Reduction in lubrication performance; Expensive.
Change hydraulic fluid with HFD-U type	Hydraulic performance closest to Mineral oil based lubricants and general no investments needed on hydraulic systems.	Low risk of spreading fire (under control) but still possible on a 900 °C (1,650 °F) surface.

3.2. *How is fire resistance perceived*

The term "fire resistant" is often mistakenly understood to be the same as "fire retardant". It is not necessarily the same. Almost all fire resistant hydraulic fluids will burn under certain conditions.

- HFC fluids will ignite if a certain amount of water evaporates.
- Most HFDu fluids will burn, but will not give the vaporized kind of explosion mineral oil generates

The ignition-like explosion caused by the mineral oil is what leads to an uncontrollable situation. The only hydraulic fluids that can truly be considered fire retardant are the high water content (HFA) fluids.

Fluids can be tested to determine their fire resistance. The most common and generally accepted tests are those used by factory mutual (FM global), the testing and approval arm of a major industrial insurance underwriter (www.fmglobal.com). Using an FM global approved hydraulic fluid can reduce the premium a company needs to pay.

Additionally, beyond FM global, many other organizations and companies have developed fire resistance tests, usually to simulate a certain type of real-world accident.

The following video frame shots show the comparison between ignition of mineral oils and HFDu fluids.

3.3. *Comparison of mineral oil and hfdu polyol ester when poured on a 900 °C surface*

The still frames in Fig. 2 demonstrate the problem that typically occurs when a mineral oil based lubricant comes into contact with a hot surface. The mineral oil evaporates easily, and therefore, tends to build a vapour of oil droplets. Once ignition takes place, the oil droplets can catch fire and result in an explosion and/or fire ball. These two effects make the fire with a mineral oil dangerous and hard to control as the fire ball can go to the roof or to cables and can ignite that area. With the polyol ester based HFDu fluids this evaporation does not take place and thus no explosion or fire ball will be generated. The

HFDu fluid might burn as well, but there is no vapour or explosion and it is limited to the place it comes in contact with, so the situation remains under control.

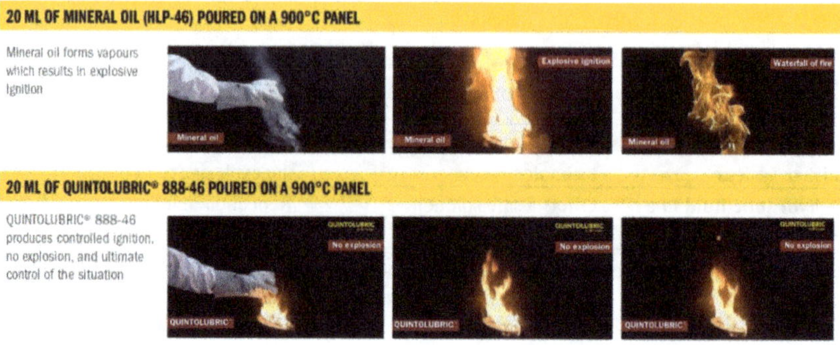

Fig. 2. Comparison of Mineral Oil and HFDu Polyol Ester [1].

3.4. *Comparison of a pool of mineral oil and HFDu polyol ester after a hot subject drops into it*

To explain this phenomenon some physical parameters in Table 3 should be understood.

Table 3. Physical Parameters of a Mineral Oil based lube compared to QUINTOLUBRIC® 888/HFDu.

Property	Mineral oil	QUINTOLUBRIC® 888 HFDU
Auto Ignition Point	300 °C 572 °F	460 °C 860 °F
Specific Heat	1.7-1.8 J/g.K	2.06 J/g.K
Heat of Combustion	43 KJ/g	38 KJ/g

A driving force in this process is the relative low auto ignition temperature of mineral oil in combination with a relative low specific heat and high heat of combustion.

- The auto ignition point of MO is much lower than that of QUINTOLUBRIC® 888.
- The specific heat expresses the energy needed to heat 1 g mineral oil 1 °C ≥10-15% less energy to heat Mineral oil based lubricants.
- The heat of combustion of a mineral oil based hydraulic fluid is typically about 43-44 kJ/g, whereas an HFDu, polyol ester fire-resistant hydraulic fluid has a heat of combustion of about 38 kJ/g. So chemically an HFDu fluid generates 10-15 % less heat during combustion.

So purely on physical facts it can be stated that less energy is needed to heat mineral oil based lubricants to reach the temperature needed to auto ignite (which is already relatively low). The relative high heat of combustion of mineral oil based lubricants acts as a catalysis in the process and creates the situation of the explosive ignition and propagation of the flames.

The mechanism described in this chapter is valid for the mineral oil based gear oil and mineral oil based grease as well.

(1) Case 1: The fire resistant hydraulic fluid.

In press hardening units numerous fires have taken place meanwhile causing in some occasions personal injuries but mainly causing financial damage as the presses needed repair and could not be ran for 1 or more days.

Replacing the standard mineral oil based hydraulic fluid by a HFDu type of fluid like QUINTOLUBRIC® 888 reduced the risk of an uncontrollable fire to an absolute minimum. Several practice incidents of hose rupture have proven that the incident remains limited to the hot blanc and the fire was not further extended. The situation remained under control.

(2) Case 2: The fire resistant grease.

The feeder systems are lubricated with a high grade but mineral oil based grease. A customer faced a situation where the grease on the feeder got ignited and the fire extended using the grease trail as fuse. It caused severe damage to the feeder system.

(3) Case 3: The fire resistant gear oil.

Mechanical or hybrid presses contain a gear box that operates the press. In many cases the chance that the oil from the gear box catches fire is not very large. However, in many press the gear oil is also used as a slide oil to lubricate the press slides. Here the gear oil/slide oil can easily get in contact with the hot blanc resulting in an uncontrollable fire.

The use of a fire resistant gear oil will eliminate this risk to a minimum.

4. Conclusion

Safety hazards are inherent at any press hardening line. In particular, fire hazards present a sizeable risk for line workers as the process requires significant heat to produce the desired finished products. Minimizing and controlling the hazards is very challenging and this paper explains how fire resistant fluids and greases can reduce some of those risks.

Because fluids and greases are typically mineral oil based, this becomes the fuel for a fire event. Using a synthetic ester oil based fluid and grease would reduce the likelihood of lubricants catching on fire. This ester technology is used in fire resistant hydraulic fluids for many years. This is further supported in that the ester technology requires a much higher energy level to cause and sustain combustion compared to mineral oil based lubricants.

References

1. Quaker Chemical, Fire-resistant hydraulic fluid test compilation, http://www.youtube.com/watch?v=bEtlikCMRWM

Design, Application and Challenges in Structural Ceramics for Press Quenching

T. L. Chen
Cencera Co., Haining City, Zhejiang 314400, China
E-mail: cencera@hotmail.com

R. X. Chai
College of Mechanical Engineering, Xi'an University of Science and Technology
Xi'an 710054, China
E-mail: joancra_030159@163.com

The heating system plays an important role in press quenching. Hydrogen embrittled parts under stress can fail suddenly without any warning, due to the introduction and subsequent diffusion of hydrogen into the metals. As temperatures increase, these individual hydrogen atoms within the metal gradually recombine to form hydrogen molecules, creating pressure from within the metal that weaken the ductility, toughness, and tensile strength, up to the point where cracks might develop (hydrogen-induced cracking or HIC). Hence, AHSS, hydrogen-induced cracking (HIC) is a crucial issue that we will be investigating in this paper.

In here, we explain the problems and provide solutions using an advanced ceramics CE-Pure, which is a fused silica matrix composite that could offer good thermal and energetic advantages for heating the blanks without hydrogen nor oxygen. The CE-Pure ceramic makes furnace design with good thermal and energetic advantages against traditional ones: it has dust-free chamber with three layers of CE-Pure lining; the heating element is mounted in the groove of the CE-Pure ceramic board; CONE-DUCT gas distribution system made of CE-Pure ceramic; shielding gas is injected into the furnace through well-designed small holes on the hollow CE-PURE roller; and over-lap sealing with CE-Pure tile. Results show that CE-Pure has good thermal shock resistance capability. When the blank enters the furnace, the temperature of the blank is about 25°C while the hearth has a maximum temperature of 1050°C. If the roller or support for the blanks has high thermal expansion coefficient, the roll or support will deform, or even break.

Keywords: Press quenching; Ceramic; Furnace; Advanced high strength steel.

1. Introduction

Press Quenching technology: to form lightweight body structures, in low cost, reduce the weight by 50 %, by 23–30 % reduction in fuel consumption. Press quenching process can work on different materials including the Advanced High-Strength Steels [1] (AHSS, the most recent and widely used one is 22MnB5), aluminum alloy, fibre, etc. By now press quenching is widely used with AHSS in automotive industry to produce the chassis components, A-pillar, B-pillar, bumper, roof rail, rocker rail, tunnel and door ring. The sheet metal or blank material is supplied in ferritic-pearlitic microstructure [2]. The blank is heated above Ac3 (Temperature at which austenite formation is completed during heating) till the whole blank is austenitized in the furnace. At last the blank is transferred into the mould, formed by pressing, and simultaneously quenched within the mould which has chilled water channels inside. The quenching time in the mould is depending on the tonnage of the press, the temperature of the mould, the thermal conductivity of the mould material, the quality of the blank material, etc.

The heating system plays key role in press quenching. Excellent press quenching results shall have homogenous microstructure, cleanest material surface, no material loss, and uniform mechanical properties But, for AHSS, hydrogen-induced cracking (HIC) is crucial issue which must be solved in priority [3, 4]. Hydrogen embrittled parts under stress can fail suddenly without any warning. Hydrogen is the most common element in the world and many acidic and oxidation reactions with steel will liberate hydrogen in various amounts depending on the specific chemical reaction.

Hydrogen embrittlement is the process by which hydride-forming metals such as titanium, vanadium, zirconium, tantalum, and niobium become brittle and fracture due to the introduction and subsequent diffusion of hydrogen into the metal [5-9]. Because the solubility of hydrogen increases at higher temperatures, raising the temperature can increase the diffusion of hydrogen. These individual hydrogen atoms within the metal gradually recombine to form hydrogen molecules, creating pressure from within the metal. This pressure can increase to levels where the metal has reduced ductility, toughness, and tensile strength, up to the point where it cracks open (hydrogen-induced cracking, or HIC).

Three major factors affecting delayed fracture susceptibility which are stress, environment such as density of hydrogen, and materials such as chemistry and microstructure. When the hydrogen diffused more in the blank, HIC has more possibility to happen. At the same time, the solubility of hydrogen increases at higher temperatures in the steel. But the heating is a must for press quenching. So, it is necessary to develop a system or solution for heating the blanks without hydrogen nor oxygen.

2. Heating Process and Potential Problems

The maximum temperature of the heating is about 980 °C in furnace for steel 22MnB5 which is most widely used material for press quenching in automotive industry. There are two kinds of heating systems available on the market which are roller hearth furnace and chamber furnace as shown in Fig. 1.

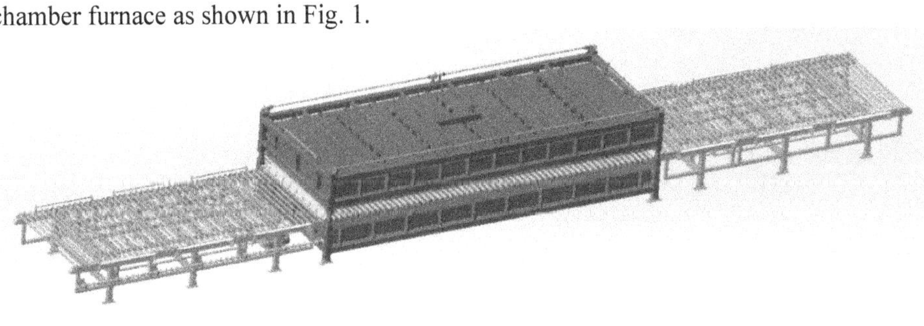

Fig. 1. Roller hearth furnace.

The roller hearth furnace is designed for mass production and a vast variety of blanks. The blanks can be heated in diverse atmospheres. Regardless of whether blanks are Al-Si or zinc coated or even uncoated blanks, the furnace provides just the right atmosphere. In addition to normal air, heat treatment can be performed under inert or reactive gases or in dried air. The supply of protective gas into the furnace is controlled automatically. Any gas escaping from the furnace is extracted at the furnace entry and exit ends after heat exchange system to save the energy. The furnace has dew point and oxygen monitoring and control system with which the atmosphere in the furnace can be perfect for processing blanks uncoated, or coated with zinc or AlSi, avoiding hydrogen-induced cracking (HIC). The atmosphere in roller hearth furnace is easily under control through design with at least three independent chambers at the entrance, soaking, and exit sections.

The chamber furnace can do test of new tools before delivery in mould factory, of new methods in hot stamping plant and research organization, of new materials in steel works. This kind of furnaces can make short production series, customer demonstrations, and prototype manufacturing. Multilayer feeder furnace can have one to sixteen chambers.

The door of chamber furnace opens and closes frequently, consequently the atmosphere and temperature distribution in the chamber changes rapidly. The change causes two problems, one is vapor comes into the chamber which causes HIC, another one is deformation of the steel structure [10]. Processing Al-Si or zinc coated or even uncoated blanks, the furnace must guarantee the workpieces are protected by shielding gases due to the water vapor and oxygen containing atmosphere. Dew point in the furnace normally is controlled under -30 °C. The chemical reaction of Al-Si coating at 480 °C as in Eq. (1)-Eq. (3).

$$2\mathrm{Al}+4\mathrm{H_2O}=2\mathrm{AlO(OH)}+3\mathrm{H_2} \tag{1}$$

$$2\mathrm{Al}+3\mathrm{H_2O}=\mathrm{Al_2O_3}+3\mathrm{H_2} \tag{2}$$

$$2\mathrm{Al}+4\mathrm{H_2O}=2\mathrm{AlO(OH)}+3\mathrm{H_2} \tag{3}$$

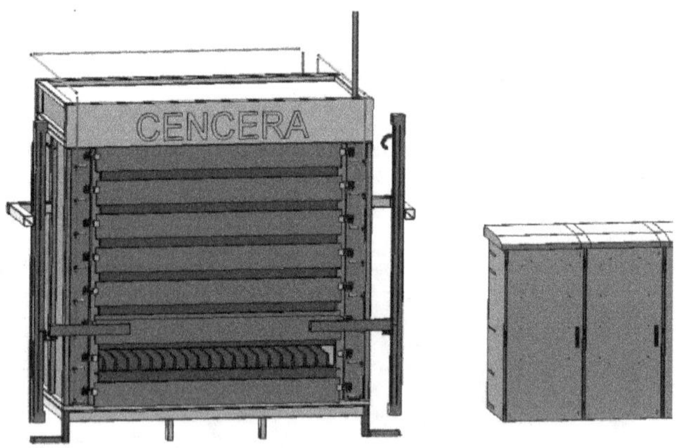

Fig. 2. Multilayer chamber furnace.

3. Advanced Ceramics Solutions

We developed advanced ceramics CE-Pure which is fused silica matrix composite. The following are typical compositions (Table 1), physical properties (Table 2) and thermal properties (Table 3) of CE-Pure ceramic material.

Table 1. Typical compositions of CE-Pure ceramic.

Typical compositions	Value
Silica SiO_2	≥99.5%
Iron Oxide Fe_2O_3	≤45ppm
Copper Oxide CuO	≤45ppm
Titania TiO_2	≤45ppm
Chromium Oxide Cr_2O_3	≤45ppm
Nickel Oxide NiO	≤45ppm
Magnesia MgO	≤40ppm
Sodium Oxide	≤40ppm
Potassium Oxide K_2O	≤40ppm

Table 2. The physical properties of CE-Pure ceramic.

Physical properties	Value
Density	1.90 – 2.00 g/cm^3
Apparent porosity	<10 - 11%
Compressive Strength	> 50Mpa
Modulus of Rupture	> 25MPa
Crystallinity	Crystobalite < 2% (wt)

Table 3. Typical compositions of CE-Pure ceramic.

Thermal properties		Value
Thermal Conductivity	20 °C	0.64 W/m°K
	500 °C	0.81 W/m°K
	1090 °C	0.88 W/m°K
Expansion Coefficient 20°C – 1000°C		0.54 x 10^{-6} /°C
Max Use Temperature (multiple thermal cycle)		1200 °C
Max Use Temperature (a single thermal cycle)		1650 °C
Melt point		1700 °C
Sodium Oxide		≤ 40 ppm
Potassium Oxide K_2O		≤ 40 ppm

From above Tables, CE-Pure has good thermal shock resistance capability. When the blank enters the furnace, the temperature of the blank is about 25 °C while the hearth has a temperature of max. 1050 °C. If the roller or support for the blanks have high thermal expansion coefficient, the roll or support will deform, or even break. The solution is to develop ceramic material having thermal expansion coefficient as low as possible, even as low to 'zero'.

4. UNIQUE Design Solutions

The CE-Pure ceramic makes furnace design good thermal and energetic advantages against traditional ones.

4.1. *Dust-free chamber with CE-Pure lining*

The wall of the 'dust-free' furnace has three layers, most outside is steel plate, middle fiber, and the inside lining is Ce-pure ceramic board. The fiber has crystallization trend. Crystallization starts at 800 °C. When the fire is crystalized consequently it breaks and the dusts appears which distributes everywhere in the furnace following strong convection of hot air. At the same time, the fibre material has much more shrinkage than Ce-pure ceramic. So, to prevent the dusts inside the furnace, Ce-pure ceramic lining insulates the wall chamber.

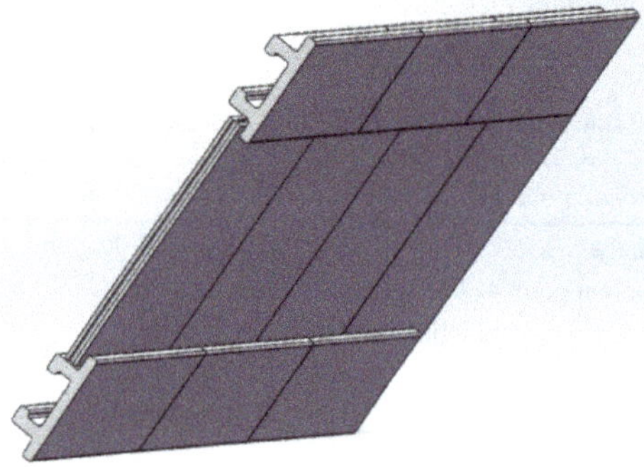

Fig. 3. Dust-free lining.

4.2. *Strongest reflection of heat radiation by CE-Pure ceramic*

The heating element is mounted in the groove of the CE-Pure ceramic board. CE-Pure material has strongest radiation refection property than other industrial materials, which makes the heating evenly.

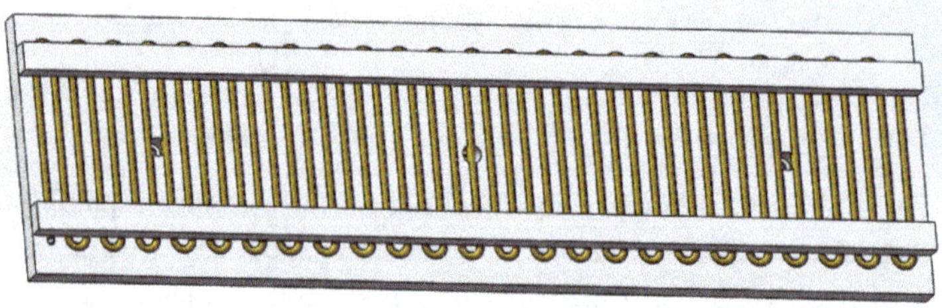

Fig. 4. Design of radiation reflection board.

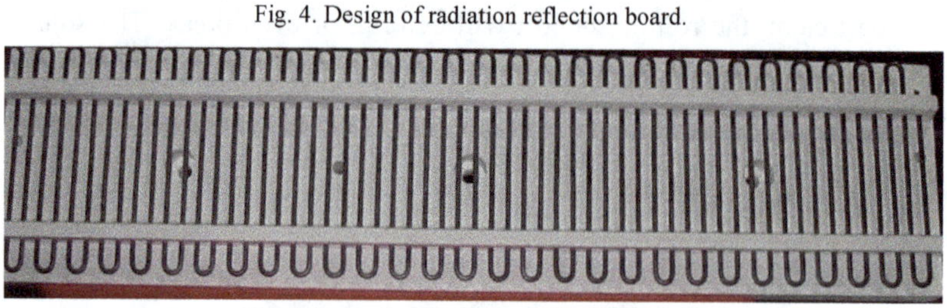

Fig. 5. Example of radiation reflection board.

4.3. *CONE-DUCT gas distribution system made of CE-Pure ceramic*

High purity nitrogen or other gas is injected in the hearth through CONE-DUCT gas distribution system which makes the nitrogen gas a curtain wall pushing out the oxygen and water vapor. CONE-DUCT design has stable and even pressure and volume supply of gas. This unique design removes the oxygen and hydrogen from the furnace atmosphere consequently the hydrogen embrittlement (Hydrogen Induced Cracks, HIC) can reduced greatly. The CONE-DUCT is made of CE-Pure material which has almost zero thermal expansion from 20 °C to 1000 °C which has very good thermal shock resistance property against rapid temperature changes.

The shielding gases prevent the uncoated blanks from oxidizing. The contents of shielding gases are hybrid of nitrogen and natural gas, or, nitrogen and methanol. The methanol is harmful to human body, so it is more common to use hybrid of nitrogen (95 % vol.) and natural gas (5 % vol.) as shielding gases.

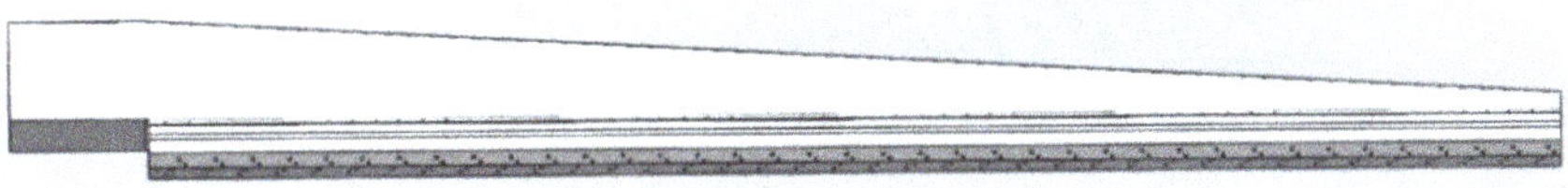

Fig. 6. CE-Pure CONE-DUCT gas distribution system.

4.4. *Air-cushion support reducing or removing friction*

The shielding gas is injected into the furnace through well-designed small holes on the hollow Ce-pure roller or flat board support. The gas will form 'air cushion' below the sheet metal to reduce the friction between the workpiece and the support, at the same time the protective gas around the workpiece is building an ultra-clean environment for each of workpieces.

Fig. 7. Example of air-cushion CE-pure plate.

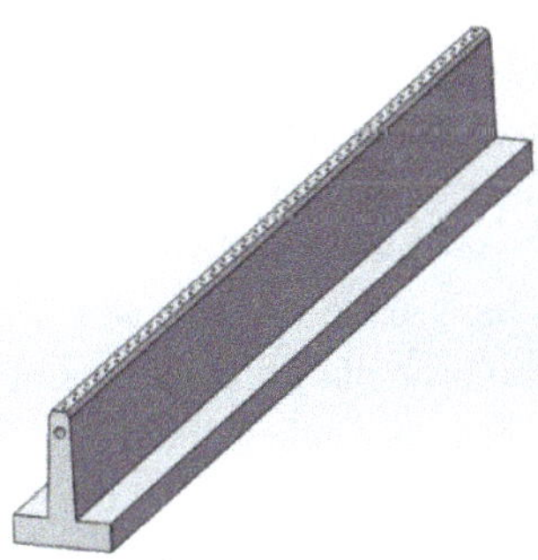

Fig. 8. Design of CE-pure Air-cushion support.

4.5. *Over-lap sealing with CE-pure tile*

In multilayer chamber furnace, each chamber could be independently operated. The insulation between each two layers must be tight. Two supports are mounted closely at room temperature as shown in Figure 9. When the furnace is heated up to around 1000 °C, the beam has thermal expansion 30 times more than Ce-pure ceramic, consequently each two supports have a gap in-between through which the heat flows to the upper layer from lower chamber. The sealing tile is designed as overlap to solve this problem as shown in Figure 10.

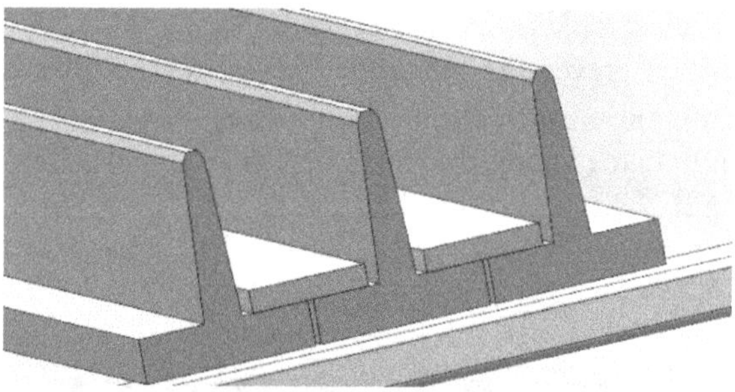

Fig. 9. No gap between two supports at room temperature.

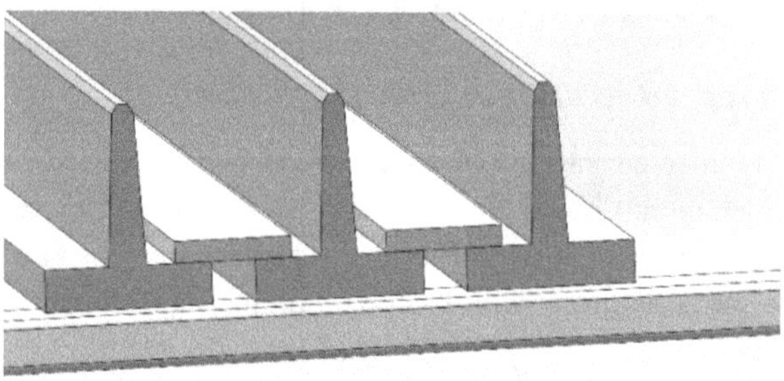

Fig. 10. Gap between two supports at 980 °C.

5. Conclusion

The control of heating and atmosphere is crucial for press quenching process. CE-Pure ceramic has advantages in this application for solving HIC, dusts, and energy-saving, as following:

(1) Dust-free chamber with CE-Pure lining has been designed. The wall of the 'dust-free' furnace has three layers, and the most outside is steel plate, middle is fiber, and the inside lining is CE-Pure ceramic board.

(2) CE-Pure material makes Cone-Duct gas distribution system possible even there is big temperature difference between the protective gas and the chamber in operation. The protective gas 'wall' can push out water vapor and oxygen completely.

(3) Over-lap design keeps each chamber independent.

(4) Air-cushion made of CE-Pure material reduces the friction to the workpiece. Air-cushion furnace can be next generation replacing some of roller hearth furnace.

Acknowledgments

The authors sincerely thank Prof. Yisheng Zhang of Huazhong University of Science & Technology for his instructive suggestions and help in writing this thesis.

References

1. M. M. Kuhlmann, I. O. Schwedler, M. N. Holtschke and I. S. Jüttner. Consideration of hydrogen transport in press-hardened 22MnB5, *Mater. Test.* **57**, 977 (2015).
2. W. S. Alvarez, M. J. Peet, E. J. Pickering, J. Jaiswal, A. Bevan and H. K. D. H. Bhadeshia, Synchrotron and neural network analysis of the influence of composition and heat treatment on the rolling contact fatigue of hypereutectoid pearlitic steels, *Mater. Sci. Eng. A* **707**, 259 (2017).
3. X. Zhu, W. Li, H. S. Zhao, L. Wang and X. J. Jin. Hydrogen trapping sites and hydrogen-induced cracking in high strength quenching & partitioning (Q&P) treated steel, *Int. J. Hydrogen Energ.* **39**, 13031 (2014).
4. J. Moon, J. M. Choi, S. K. Han, S. Huh, S. J. Kim, C. H. Lee and T. H. Lee, Influence of precipitation behavior on mechanical properties and hydrogen induced cracking during tempering of hot-rolled API steel for tubing, *Mater. Sci. Eng. A* **652**, 120 (2016).
5. X. F. Li, J. Zhang, Q. Q. Fu, X. L. Song, S .C. Shen and Q. Z. Li, A comparative study of hydrogen embrittlement of 20SiMn2CrNiMo, PSB1080 and PH13-8Mo high strength steels, *Mater. Sci. Eng. A* **724**, 518 (2018).
6. Liu, Q. J Zhou, M. X. Zhang and A. Atrens, Further study of the hydrogen embrittlement of martensitic advanced high strength steel in simulated auto service conditions, *Corros. sci.* **135**, 120 (2018).
7. W. J. Hui, Z. B. Xu, Y. J. Zhang, X. L. Zhao, C. E. Shao and Y. Q. Weng, Hydrogen embrittlement behavior of high strength rail steels: A comparison between pearlitic and bainitic microstructures, *Mater. Sci. Eng. A* **704**, 199 (2017).
8. F. J. Recio, M. C. Alonso, L. Gaillet and M. Sánchez. Hydrogen embrittlement risk of high strength galvanized steel in contact with alkaline media, *Corros. Sci.* **53**, 2853 (2011).
9. Q. L .Liu, Q. J. Zhou, J. Venezuela, M. X. Zhang and A. Atrens, The role of the microstructure on the influence of hydrogen on some advanced high-strength steels, *Mater. Sci. Eng. A* **715**, 370 (2018).
10. L. Vergani, C. Colombo, G. Gobbi, F. M. Bolzoni and G. Fumagalli, Hydrogen Effect on Fatigue Behavior of a Quenched & tempered Steel, *Procedia Eng.* **74**, 468 (2014).

Rolling Beam Furnace for Press Hardening

D. Angulo, J. Berasategui and A. Sanz†
GHI Smart Furnace, Galdakao, Bizkaia, 48960, Spain
†E-mail: asanz@ghihornos.com

This paper puts forward an innovative furnace for the press hardening process, to move blanks through the furnace, created by combining horizontal and vertical movements of the furnace beams. In this high-precision movement system, the blanks travel next to one another, maximizing floor space and ensuring high accuracy in the final position of parts.

The horizontal advancing mechanism is composed of metallic beams supported on rollers placed along the furnace, with a stroke that varies according to the size of the furnace, 2,900 mm being the highest achievable stroke, which allows for up to 8 blanks to be processed simultaneously in the press. The blanks travel with a minimum separation of 50 mm from one another and, depending on the production, the furnace is able to adjust both the beam stroke and the advancing speed. The vertical elevating mechanism is composed of several ceramic beams supported on steel frames and situated along the furnace, which enables the collection of blanks from the advancing beams in the upwards movement, and their subsequent placement further along on the advancing beams in the downwards movement.

This new technology avoids the contamination problems inherent to traditional roller-hearth furnaces, minimizing maintenance needs, boosting production rates and reaching very high OEE (Overall Equipment Effectiveness). Up to 8 blanks can be simultaneously extracted, centered and delivered to the press. The furnace is continuously supervised through Industry 4.0 analytics solutions to ensure the best real-time performance.

Keywords: Press hardening; Furnace; Rolling beams; GHI Smart furnaces; GHI hornos.

1. Introduction

1.1. *Press hardening process and furnaces*

Press hardening is a process increasingly used in the automotive industry that makes it possible to produce vehicle structure parts using ultra-high-strength hardened steel. These parts are mainly used in BIW (Body in White) and chassis components, that account for approximately 35% of total vehicle value [1]. This results in lower vehicle weight and better fuel efficiency; it also increases crashworthiness and therefore results in higher safety standards.

The demand for press hardened (or "hot stamped") parts in the automotive industry is increasing exponentially, driving the search of highly efficient equipment for this process. Several different furnace designs and technologies are available in the present day, each with their own strengths and weaknesses, all striving to become the most efficient and versatile, with the highest availability and the best process quality [2].

The traditional roller-hearth furnace has historically suffered of AlSi contamination that causes breakages during production and blank mislocation when moving through the furnace, as well as more frequent maintenance and new roller installation, which in turn decreases availability and increases exploitation costs.

GHI Smart Furnaces, a Europe-based company specialized in the tailor-made supply of industrial furnaces, has undertaken a great R&D effort geared towards developing an innovative furnace design that will face and overcome these challenges.

1.2. *Rolling beam furnace concept*

This new concept of furnace is based on an innovative technology for moving blanks through the furnace, created by combining horizontal and vertical movements of the furnace beams.

The horizontal advancing mechanism is composed of metallic beams supported on rollers placed along the furnace, with a stroke that varies according to the size of the furnace, 2,900 mm being the highest achievable stroke, which allows for up to 8 blanks to be processed simultaneously in the press. The blanks travel with a minimum separation of 50mm from one another and, depending on the production, the furnace is able to adjust both the beam stroke and the advancing speed.

The vertical elevating mechanism is composed of several ceramic beams supported on steel frames and situated along the furnace, which enables the collection of blanks from the advancing beams in the upwards movement, and their subsequent placement further along on the advancing beams in the downwards movement.

1.2.1. *Rolling beam movement sequence*

(1) A batch of blanks is placed on the elevating beams outside the furnace by a robot.
(2) The elevating beams descend and the batch is placed on the advancing beams.
(3) The advancing beams move forward in the required stroke of up to 2,900 mm, as determined by the production recipe.
(4) The elevating beams ascend and holds the batch above the advancing beams.
(5) The advancing beams move backwards in the same stroke described in step 3.
(6) A new batch is placed on the elevating beams outside the furnace by a robot.

2. Rolling Beam Characteristics

Blanks are transported along the furnace without relative movement to the transport system, which provides the following benefits:

2.1. *Blank crosswise traveling without relative movement*

As there is no relative movement of the blanks within the furnace, they can be as close to each other as 50 mm while traveling.

In addition, as the press-hardened blanks tend to be rectangular in shape, being longer than they are wide, the most suitable loading method is usually to arrange them crosswise, as seen on Fig. 1; because of the shape of the blanks, they will occupy less space in the flow direction.

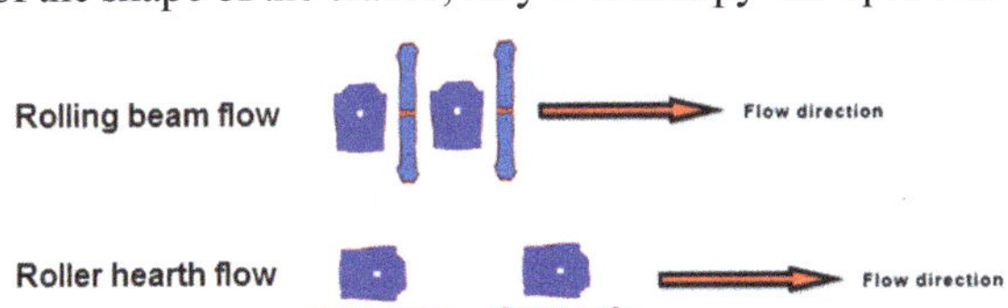

Fig. 1. Blanks being loaded on the elevating beams of the furnace by a robot.

2.2. *Precise location of each batch*

Given that blanks are traveling in batches that move step by step thanks to the rolling beam system, the blanks will only occupy a series of set positions, moving from one position to the next in steps and in batches. This means that blanks will only be in contact with specific parts of the furnace during the heating up of the blanks, thus reducing contamination of the furnace by the blanks.

2.3. *High speed heating of blanks*

The lack of ceramic rollers allows the operators to stablish a high temperature set point from the first control zone of the furnace, reducing the heating time of the blanks and shortening the required furnace length.

2.4. *Big data analysis*

Data acquisition equipment register large amount of information used to provide use improvements, assistance and to predict maintenance interventions.

3. Comparison of Rolling Beam Furnace and Traditional Roller-hearth Furnace

In this section, the advantages mentioned above are tested and compared to the traditional roller-hearth furnace in order to quantify the improvement achieved by the innovative rolling beam furnace.

3.1. *Furnace length comparison*

The space needed for producing some typical reference blanks is compared by processing them in a typical 40 m roller-hearth furnace and in the rolling beam furnace. The blanks are most often introduced in pairs, as most presses are fitted with two dies and are thus capable of processing two parts at a time, although GHI rolling beam furnaces can process up to eight blanks at a time. Blank thickness does not have an impact on the results. Table 1 shows the calculations regarding the furnace length required by the rolling beam system to maintain the same production, as well as the percentage of furnace length saved by loading the blanks on the rolling beam furnace instead of the traditional roller-hearth furnace. The innovative rolling beam design leads to a shorter modular furnace design, as a shorter length is needed to achieve the required production.

Table 1. Required rolling beam furnace length compared to the same production in a roller-hearth furnace.

Reference	Blank width	Blank length	Furnace length needed for same production	Length saving %
Blank 1	250 mm	1,000 mm	20 m	50%
Blank 2	250 mm	1,500 mm	14.3 m	62%
Blank 3	250 mm	2,000 mm	11.1 m	72%
Blank 4	500 mm	1,000 mm	36 m	10%
Blank 5	500 mm	1,500 mm	25.8 m	36%
Blank 6	500 mm	2,000 mm	20 m	50%

3.2. *Contamination comparison*

As explained above, the step-by-step movement of blanks reduces blank-furnace contact. In the roller-hearth, the blanks travel along all the rollers in the furnace in lengthwise direction, so even if two blanks are processed at the same time, only the longest blank will be taken in consideration for contamination calculation purposes.

The typical roller diameter is 60 mm, with a circumference of 188.5 mm. Dividing the length of a blank by said circumference will return the number of times the roller completes a full revolution, thus beginning a new contamination cycle. This means that, for a 1,500 mm length blank, the roller in the roller-hearth furnace will have turned 5.3 times, thus having 5.3 times more contact with the blank than in the rolling beam furnace. Table 2 shows a comparison for reference batches.

Table 2. Number of times a blank touches the same face of a roller/beam.

Reference	Blank width	Blank length	No. of times a blank touches a beam in the rolling beam furnace	No. of times a blank touches a beam in the roller-hearth furnace
Blank 1	250 mm	1,000 mm	1	5.3
Blank 2	500 mm	1,500 mm	1	7.9
Blank 3	250 mm	2,000 mm	1	10.6
Blank 4	500 mm	2,000 mm	1	10.6

As the contamination by AlSi is caused by the contact of the blanks with the furnace transport system, be it rollers or beams, the result is that the roller-hearth furnace will receive more contamination from the blanks than the rolling beam furnace, causing AlSi buildup which increases the roller diameter that gives the blanks different lineal speeds and causes them uncontrollable deviation, throwing them off-course.

4. Discussion

After comparing the transport systems, the following statements can be made.

4.1. *Space savings*

The length of the furnace needed for producing typical reference blanks in rolling beam systems is smaller than that of roller-hearth furnaces loading blanks lengthwise. This is true for the vast majority of blanks, the only exception being the shortest and widest blanks, which, due to their nearly square shape, will not yield this advantage.

4.2. *Transport-system contamination*

The contamination accumulated in the transport system will be less in the rolling beam furnace, as a result of reduced contact between the blanks and the transport system.

4.3. *No blank mislocation*

On the roller-hearth furnace, excessive AlSi buildup tends to deviate the blanks off-course, resulting in mislocations. In rolling beam furnaces, however, because of the aforementioned reduction in blank-furnace contact, AlSi buildup in the beams is curtailed; furthermore, due to the system of movement, this reduced buildup does not impact the final location of the blanks.

4.4. *Higher line availability*

As a result of the reduction in AlSi contamination, maintenance needs are heavily reduced. Furthermore, as there is no blank mislocation, there is no need to stop the line during production to correct mislocation issues. These two factors result higher availability of the furnace for production.

4.5. *Industry 4.0 services*

Big Data systems in operating furnaces reduce the risk of unexpected downtime and maximizes the potential of the intelligent furnaces, in addition to be validate the design, calculations and dimensioning of new and more effective equipment.

5. Conclusion

With the Onstamp rolling beam furnaces, GHI has developed its own innovative technology, providing a solution with a high availability and highly precise transport system that improves on the results obtained by traditional furnaces in press hardening.

References

1. M. Berret, Automotive metal components for car bodies and chassis-Global Market Study, *Roland Berger*, Munich, Germany (2017).
2. Industrial Furnaces and Ovens - Global Strategic Business Report, *Global Industry Analysts, Inc* (2018).

Advanced Design for Continuous Roller Furnace for Hot Forming Line

B. Dvorak[1,a,†], J. J. Tawk[2,b] and T. Vít[3,c]
[1]*BENTELER Mechanical Engineering GmbH, Automotive design, Liberec, Hodkovicka 981/42, 46006, Czech Republic*
[2]*BENTELER Maschinenbau GmbH, Frachtstraße 10-16, 33602 Bielefeld*
[3]*Technical University of Liberec, Department of energetical devices, Liberec, Studentska 2, 46117, Czech Republic*
E-mail: [a,†]*borek.dvorak@benteler.com,* [b]*jad.tawk@benteler.com,* [c]*tomas.vit@tul.cz*

Use of press hardened parts in BIW (Body in White) structures has evolved in recent years to encompass a wide range of part complexity, size and mechanical properties. In addition, the number of components per vehicle has also increased pushing demand for more capital investments. Suppliers of press hardened parts need to accommodate these changes while staying competitive. Advanced design of heat treatment furnace has to offer a unique furnace design that provides flexibility to handle future part sizes, minimizes down time to increase line utilization and offers a unique solution to produce tailor tempered parts for crash performance. This paper presents advanced innovative design of continuous roller furnace. These types of furnaces are generally used in hot forming lines. Design is focused on optimal heating layout, modern drives of rollers, new design and other items respecting the optimal technological and technical aspects. Technological functions like the dew point temperature regulation, oxygen rate regulation are considered. Based on the long-time experiences with manufacturing and development of the machinery for the automotive industry, new roller furnaces were designed using modern methods including the FEM analyses for numerical simulations of heating processes and heating power distribution. The numerical solution of many mathematical problems involves the combination of external and internal conditions and different technological processes.

Keywords: Hot Forming; Continuous roller furnace; Design characteristics.

1. Furnace Design Characteristics

An important parameter of the furnace design is its modularity, which allows an OEM to build the furnace according to user requirements (Figure 1). The design allows for comfortable access to the furnace for maintenance, safety features that meet updated international safety standards, and technological requirements such as the dew point regulation system and oxygen rate control.

There are many design consideration used for designing and manufacturing roller furnace for hot forming application. Due to new varieties of blanks (coated and uncoated) the hot forming furnace needs to be adaptive for future requirements. Therefore, the following key design characteristics are essential: Heating system and dew point regulation.

Rollers are driven by external bevel gear drive which is isolated from furnace heat (Fig. 2). No cooling system is required which reduces the complexity of the system and reduces required maintenance. But, the main advantage of this system is that each roller can be accessed and replaced individually from outside the furnace. Quick disconnect design separates the roller from the gear drive. Damaged roller can be replaced in less than 5 minutes without the need to stop production or purge the blanks from the furnace.

The gear drive also allows rotating the rollers in oscillation mode, i.e. rotating back and forth, during line stoppages. This feature maintains the flatness of the hot blanks in the furnace and prevents creep deflection of the rollers.

These features reduce down time of the furnace when maintenance or roller replacement is required. They also eliminate the need to purge and scrape blank from the furnace when short time stoppage is required and contributes to overall availability of the furnace for production.

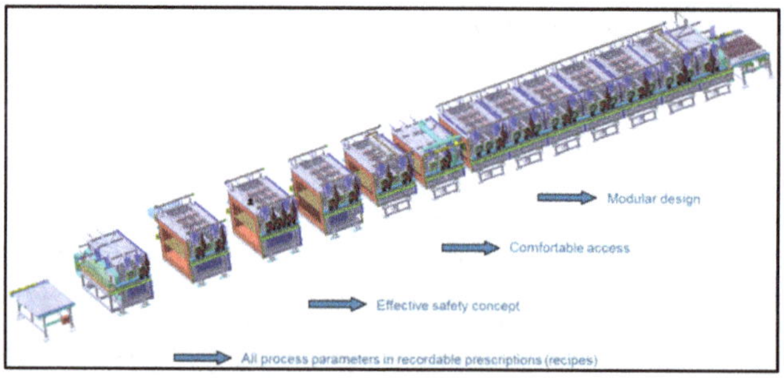

Fig. 1. Furnace modular concept.

Fig. 2. Blank transmission system through roller drivers.

2. Heating Concepts

2.1. *Heating transfer*

The heat transfer flux in the furnace and the homogeneous temperature distribution inside the furnace atmosphere are key factors to reach the required quality of the original equipment manufacturer (OEM). FEM analysis of heat transfer was used as the main tool for the design of advanced heating system. This calculation takes into account all mechanisms of heat transfer i.e. conduction, convection and radiation as shown in Fig. 3 [1].

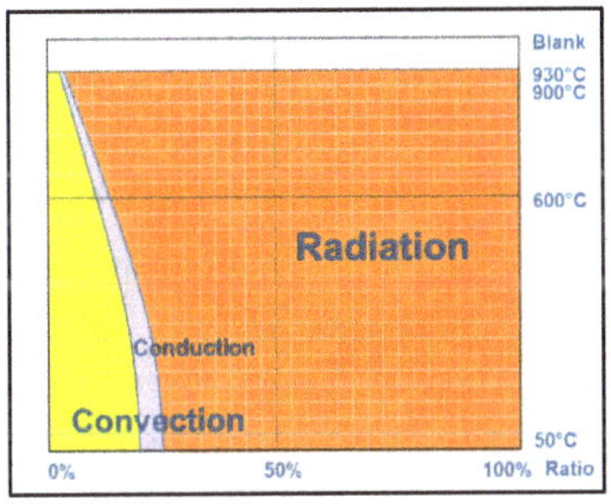

Fig. 3. Heating mechanism inside the hot forming furnace.

Heating concepts include basic models like the recuperative gas burners with the radiant tubes or electric resistance heaters [2].

The advanced system is based on efficient recuperative burners using the flameless oxidation (FLOX) technology equipped with the silicon reinforced silicon carbide radiant tubes produced by WS GmbH Company4. The main advantages of these burners are the higher efficiency calculated by using equation (1) and reduction of NOx-emissions. The efficiency is up 10% - 15% higher compared to common burners. Design also brings other benefits such as homogenous temperature distribution, reduction of the thermal stress in the burner, noise reduction and lower restrictions on fuels.

$$\eta=1-\frac{\text{exhaust gas losses}}{\text{fuel input}} \tag{1}$$

2.2. *Numerical simulation of the heat transfer*

FEM analysis of heat transfer was used as the main tool for the design of heating system. This calculation takes into account all mechanisms of heat transfer, i.e. conduction, convection and radiation. Table 1 shows an overview of different regimes of heat transfer for different temperature difference between the blank and the furnace. Radiation is the dominant heat transfer mechanism.

Table 1. Heat transfer inside the furnace.

Temperature of furnace [°C]	870	900	920	930.0
Temperature of blank [°C]	25	600	900	920.0
Emissivity	0.38	0.12	0.54	0.7
Convection coefficient h [W/m2K]	9.0	9.0	9.0	9.0
Heat flux radiation [kW/m2]	36.6	8.9	4.1	2.7
Heat flux convection [kW/m2]	7.6	2.7	0.18	0.09
Ratio between	4.8	13.3	22.8	29.4

It is possible to calculate radiation energy flux from so called “black” body e_b (W m^{-2}) using the famous Stefan-Boltzmann law as

$$e_b(T) = \sigma T^4 \tag{2}$$

Where σ=5.6707e-8 ($Wm^{-2}K^{-4}$) is Stefan-Boltzmann constant and T(K) is thermodynamic temperature. It is known, that non-black bodies absorb and emit less energy than black bodies, which are perfect emitters. It is possible to characterize the emissive power on a non-black body (or gray body) using a surface property called emittance (e). The equation (2) for non-black bodies should be written in the following form:

$$e(T) = \varepsilon\sigma T^4 \tag{3}$$

The emittance is a property of a surface, which depends on temperature and considered wavelength. For simplification, the concept of non-black body is established and supposes that emittance only depends on temperature.

When radiation is exchanged between bodies (1) and (2) that are non-black, the total net heat flux is calculated as:

$$Q_{net} = A_1 F_{1\text{-}2}\sigma(T_1^4 - T_2^4) \tag{4}$$

The transfer factor F1-2 depends on the emittances and geometrical view between bodies (1) and (2). That is also called the view factor. Transfer factor or view factor must be determined to calculate the radiative heat transfer. Generally, view factor can be calculated according to the equation:

$$F_{1\text{-}2} = \frac{1}{A_1}\int_{A_1}\int_{A_2}\frac{\cos\beta_1\cos\beta_2}{\pi s^2}dA_2 dA_1 \tag{5}$$

Where β_1 and β_2 are angles between surface dA_1 and dA_2 normal lines and a ray between dA_1 and dA_2, and s is distance between dA_1 and dA_2. Nevertheless, this calculation is impossible for complicated geometries. By FEM numerical simulations, the computational domain division into individual finite elements is advantageously used. The view factor is then numerically determined between each two elements in the system. Even here, the direct integration according the equation (5) would be very difficult. So it is necessary to use some appropriate statistical methods. Details about the radiation heat transfer could be found in [3].

3. Tailored Tempering

To improve energy absorption during crash or roll over, some BIW components like B-Pillar, A-Pillar and transmission tunnels are designed with lower strength areas within a high strength matrix. Several concepts are currently adopted at different OEMs. Earlier designs involved spot welding a conventionally stamped low strength component to a hot stamped main component. Others used tailor laser welded or tailor rolled part with thinner segments. Newer designs utilize a single blank with tailored mechanical properties.

BENTELER tailor tempering furnace consists of several standard furnace modules of which, the last 8 meters of the furnace chamber is divided into two longitudinal chambers. One chamber is maintained at 930 °C while the other is maintained at about 575 °C. Blanks are fully heated to Austenite (930 °C) before reaching the split zones.

Then, as the blanks move through the split zones, the portion of the blank required being at lower mechanical properties passes through the lower temperature chamber while the high strength portion is maintained at 930 °C in the high temperature chamber (Fig. 4).

The portion of the blank that passes through the low temperature chamber is cooled down under controlled conditions (rate) using dried air or Nitrogen showers transforming the material from Austenite to lower strength structures. The cooling rate is controlled and can be adjusted to influence final mechanical properties of the lower strength area. The blank is subsequently formed in a standard water cooled hot forming tool to get a press hardened part with lower strength at targeted area and high strength in the rest of the part (Fig. 5).

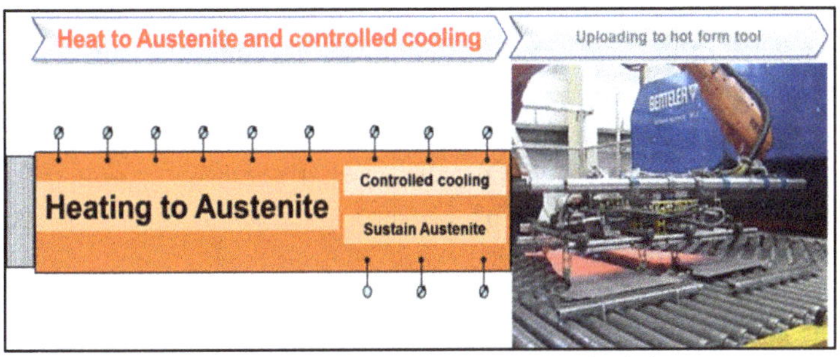

Fig. 4. Tailored tempering process - Patented solution.

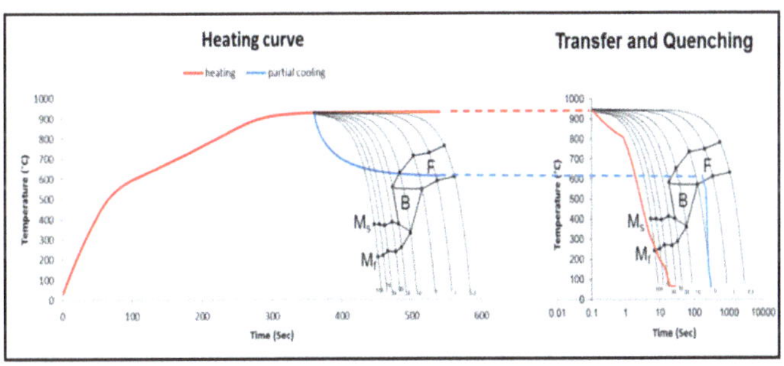

Fig. 5. Tailored tempering heating curve.

Both chambers of the split zones can be heated to 930 °C to run product with uniform mechanical properties without the need to do any physical changes to the furnace.

This solution offers the following advantages:

(1) One furnace can be used to produce tailor tempered or uniform strength parts without the need for additional setup of physical changes to the furnace.

(2) Use of one blank without the need for laser or spot welding.

a. Lower blank cost (no assembly required).

b. No investment required for assembly.

c. No performance issues related to laser or spot welding.

(3) This process can be used with tailor rolled or patched blanks.

(4) Standard hot form tools are used to form and quench the part. No need for heated tool details Therefore, dimensional quality of the part is equivalent to similar parts with uniform mechanical properties and better than that produced with heated dies.

(5) Overall heating time in the furnace is not impacted by the controlled cooling in the split zones. Therefore, standard cycle times can be achieved without penalties or any additional requirement on the tools or other equipment.

This process can be used for both Aluminized and uncoated steel. Al-Si coating is fully diffused in the entire blank and therefore, there will be no weld or paint issuesrelated to excessive Aluminum in the weld pool or paint tank.

4. Dew Point Regulation System

With the knowledge of the problem called the hydrogen embrittlement, a system for the dew point regulation has been developed and was patented in year 2011 [5]. The dew point is the temperature at which the water vapor in a sample of air at constant barometric pressure condenses into liquid water at the same rate at which it evaporates. Basically the dew point temperature is a function of the content of water vapor inside the furnace atmosphere.

Hydrogen embrittlement occurs when various metals, especially high-strength steel, become brittle and fracture following exposure to hydrogen. Hydrogen embrittlement is often the result of unintentional introduction of hydrogen into susceptible metals during heating, forming or finishing operations. The source of hydrogen is water inside the furnace. The chemical reaction during the embrittlement process is described by a basic equation:

$$3Fe + 4H_2O \rightarrow Fe_3O_4 + 8H \tag{6}$$

$$Fe + H_2O \rightarrow FeO + 2H \ \text{(Up 570 °C)} \tag{7}$$

This equation shows that it is quite important to control the mass of H2O inside the furnace to prevent the risk of hydrogen embrittlement. The risk is high especially for coated blanks. Basic chemical reactions for Al – Si coated blanks are:

$$2Al + 6H_2O \rightarrow 2Al(OH)_3 + 6H \tag{8}$$

$$2Al + 4H_2O \rightarrow 2AlO(OH) + 6H \tag{9}$$

$$2Al + 3H_2O \rightarrow Al_2O_3 + 6H \tag{10}$$

$$Si + 2H_2O \rightarrow SiO_2 + 4H \tag{11}$$

When the hydrogen (H) diffuses along the grain boundaries and combines with the carbon (C), which is alloyed with the iron inside the steel, it forms methane gas (CH_4). The methane is collected inside the small voids along the grain boundaries where it builds up enormous pressures that initiate crack.

This is quite important during the processing of Al-Si coated blanks; because the hydrogen can be easily collected and locked bellow the coating.

Possible sources of water (hydrogen) inside the furnace causing hydrogen embrittlement are:

a. Endo-Exo gas protective atmosphere inside the furnace (H_2, CO, N_2, CO_2, H_2O).
b. Nitrogen + natural gas mixture inside the furnace (CH_4, CO, N_2, CO_2, H_2O).
c. Open flame inside the furnace (N_2, CO_2, H_2O, CO).
d. Missing controlled atmosphere (N_2, CO_2, H_2O).

Nowadays, most OEMs only accept the parts produced under the controlled atmosphere with reduced dew point temperature to -10 °C and lower. The influence of hydrogen embrittlement can be decreased by using pure nitrogen (N_2) for uncoated blanks. This approach may reduce the dew point temperature and protect the surface against oxidation or using dried air (with dew point temperature -40 °C) atmosphere (CO_2, N_2, O_2) for blanks with Al-Si coating. This approach reduces the dew point temperature and supports the Al-Fe diffusion.

Cognizant of the problems that hydrogen embrittlement cause, a system for the dew point regulation was developed that saves energy and is based on the use of dried air or nitrogen [5]. The level of dew point is real time analyzed and regulated to reach permanently the required value.

Furnaces following the modern practices and latest trends used during the hot forming production are essential for an efficient press hardening process. Well-designed hot forming furnace contributes an improved OEE (overall equipment efficiency) of the hot forming line.

The mechanical design of the unloading area is based on similar configurations of the destacking area with different supplementary equipment and less complex design of the automation comparing to the destacking station. Nonetheless the cycle time in that area could be a challenging aspect when the availability of the space floor is limited and the required short cycle time of the customer has to be respected.

5. Conclusion

The latest trends and modern practices for manufacturing the hot forming parts has to be implemented by using process oriented equipment which enhances the productivity of the hot forming process to meet future evolution of press hardened parts at minimal capital Pinvestment. The hot forming furnace is also built to maximize line availability by minimizing down time, giving suppliers of press hardened parts higher throughput and more competitive costing. Many particular technologic drivers have been researched and developed in order to manufacture premium hot forming parts according to the OEMs required specifications and standards. Two main drivers contribute by respecting those required pressed hardened parts: The advanced utilized heating system based on the radiation heat transfer mechanism which allows the uniform heating transfer distribution along the hot forming parts and the active measurement of the dew point temperature through the automated dew point regulation system of the furnace atmosphere. Besides the advanced thermodynamic furnace design the cycle time for the material handling effectiveness of the complete hot forming line beginning with the stacking station through the hot forming furnace to the uploading station has to be mono-synchronized entire the complete line in order to respect the necessary heating treatment of the press hardened parts to ensure the required quality parameters and to avoid unnecessary down times along the complete line.

References

1. J. A. Wünning and J. G. Wünning, Flameless Oxidation to Reduce Thermal NO-Formation, *Prog. Energy Combust. Sci.* **23**, 81 (1997).
2. J. G. Wünning, Energy Saving Potentials for Gas Fired Industrial Furnaces, *in Thermoprocess Symposium*, (Düsseldorf, Germany, 2007).
3. H. C. Hottel and A. F. Saroffim, *Radiative Transfer* (McGraw-hill, 1967).
4. M. F. Cohen and D. P. Greenberg, The hemi-cube: a radiosity solution for complex environments, *ACM SIGGRAPH Computer Graphics* **19**, 31 (1985).
5. Patent, Patenschrift 10 2011 053 634.5, BENTELER Automobiltechnik GmbH, 33102, Paderborn, DE, 15.9.2011.
6. H. R. Hörmig, Metal und Wasser, *Vulkan Verlag Essen* (1978).
7. J. Petrovic and G. Thomas, Reaction of Aluminum with Water to Produce Hydrogen, *U.S. Department of Energy* (2008).

Optimization Strategy of Manipulator Structure for Multi-chamber Furnace in Hot Stamping Production Line

Z. T. Su[1,†], Y. M. Du[1], Y. Q. Li[1], Q. Wang[2], L. Wang[3] and Y. S. Zhang[3]
[1]*Qingdao Kingerobot Co., Ltd., Qingdao, 266000, China*
[2]*Jiangxi Haosite Automotive Components Co., Ltd., Xiushui, 332400, China*
[3]*State Key Laboratory of Material Forming and Mould Technology, Huazhong University of Science & Technology, Wuhan, 430074, China*
†*E-mail: suzt@kingerobot.com*

With the rapid development of hot stamping technology in China, the high speed conveying system has become an indispensable part of the hot stamping production line. In order to minimize the temperature loss during the hot billet transport, the smoothness and rapidity must be guaranteed in the hot billet conveying process. Firstly, taking a Cartesian coordinate manipulator as a kinematics model, the conveyor was optimized. At the same time, by optimizing the motion characteristics, the strategy of multiple robots was adapted to obtain the best conveyor program. Lastly, it took 7 seconds from the furnace to press mold. So the feasibility and efficiency of the optimization program is demonstrated.

Keywords: Cartesian coordinate robot; Optimization structure; Multiple robots work together; Hot stamping production line.

1. Introduction

Hot stamping production technology has become a research hotspot at home and abroad. The most important feature of thermoformed parts is high precision, high strength and light weight. Facing the energy crisis and environment pressure in the world, energy conservation and emission reduction has become an irresistible trend in the automotive industry. Under the high temperature, the plasticity of high strength steel is greatly improved, which can form up to 1500 MPa parts. After stamping and quenching of cold mould, the strength of parts can be further improved. The car parts which are manufactured by hot stamping are not only reduce car body weight but also improve the safety. It is also the best way to reduce car body weight and improve crash safety at the same time.

In order to ensure the quality of the hot stamping parts, the temperature of the steel sheet before forming must be guaranteed to be above about 780 °C, as shown in Fig. 1. It means that the time should be shorten in the air before heated steel plate stamped on the press. On the one hand, to shorten the oxidation time of heated steel plates by air. On the other hand, the production efficiency could be improved. The actual measurement of the material shows that when the plate material was cooled from 900 to 950 °C to about 780 °C, the time was only 7 to 8 seconds.

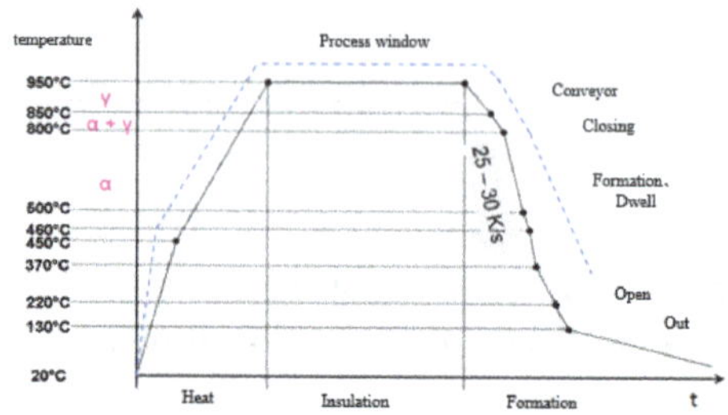

Fig. 1. High-strength steel plate hot stamping molding process.

In the conveying system design of thermoformed blanks, One solution is that multi-joint robot was used to load and unload of Heating Furnace and Press, and its advantage was that the routes planning was flexible, stable and reliable. The other solution is that Cartesian coordinate robot was used, and its advantage was that it had high precision and reliability in three-dimensional space, especially the high speed of its travelling. Choosing Cartesian coordinate robot was the best solution to realize hot forming line cold and hot billet conveying system, because the production line having a full route for straight line transportation.

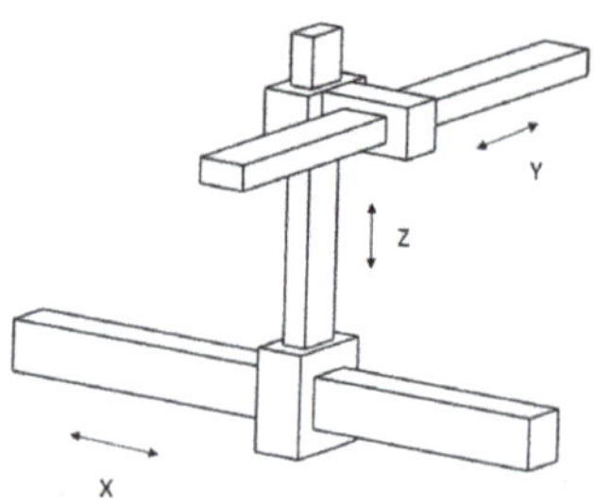

Fig. 2. Definition of the direction of Cartesian robots movement.

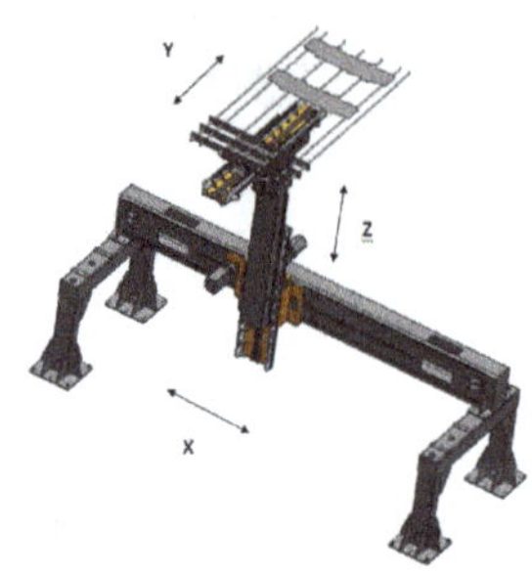

Fig. 3. The structure models of optimization Cartesian coordinate robot.

General Cartesian coordinate robot principle as shown in Fig. 2. Six degrees of freedom was had in the more complex and comprehensive Cartesian coordinate robot, included three rotational degrees of freedom. On the basis of general Cartesian coordinate robot, a group of Cartesian coordinate robot had been designed through structure optimization, part of structure models as shown in Fig. 3. General Cartesian coordinate robot had single stroke with single degree of freedom, but double stroke with single degree of freedom which is double speed were adopted in the direction of linear motion of optimization Cartesian coordinate robot. This research was mainly focused on the optimization of Cartesian coordinate manipulators for motion analysis and multi-robot cooperative work strategies.

2. The Structure and Motion Analysis of Cartesian Coordinate Robot

2.1. *Double speed principle*

Hot forming process was a forming control technology for hot stamping forming of high-strength steel sheet-constant pressure cooling phase transformation, and hot stamping lines are also more complex than conventional cold stamping lines. In the hot stamping product lines, because the sheet was thin and had a large surface area, the billet in conveying could be easily cooled.

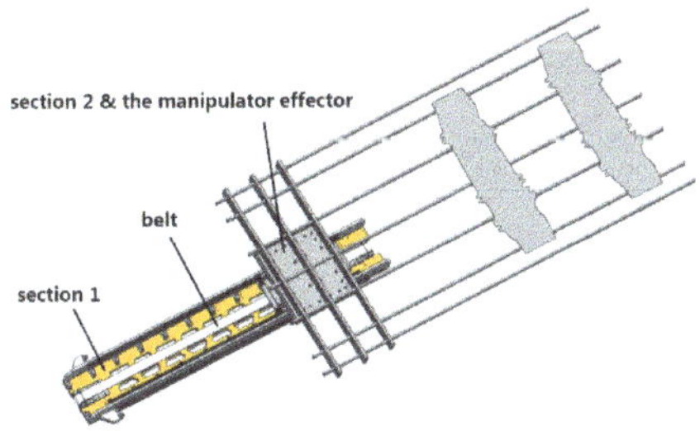

Fig. 4. The conveyor robot of hot stamping production line.

The initial forming temperature was requested strictly in hot forming process, so the rapidity and stability of conveyor robot was requested more strictly convey heated steel from Furnace to Press Mould. E.g. it was requested to finish 2-3 meters conveyor distance within 1-1.5 second, and that is to say the needed robot conveyor speed is 2 m/s. The optimized double speed robot principle model is shown in Fig. 4.

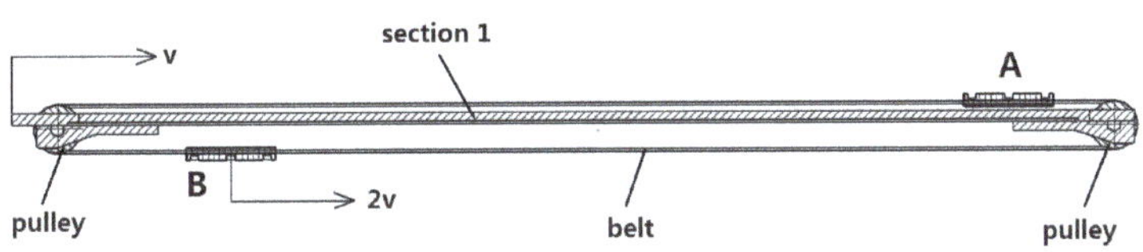

Fig. 5. The double speed structure movement principle.

The double speed structure is divided the motion section into two parts, including motion section and fixed section. The one section moved in a fixed direction drove by servo motor, at the same time the other section the same direction movement was performed at double speed. Its movement principle is shown in Fig. 5.

The end of the manipulator with the sports section 2 was fixed on the belt by point B, and point A on the belt was fixed to the outside world, at the same time the pulley was mounted on sports section1. When the sports section 1 moved along x-axis at velocity v on the geodetic coordinate system, point B moved at velocity 2v along x-axis direction because of point A fixed. So, the sports section 2 and the end of the manipulator were moved at velocity 2v.

The design of double-section double-speed structure not only shortened conveyor time by increasing sport speed of the end of the manipulator but also saved the space to make the layout of the production line more compact.At the same time, it improved the rigidity of the movement structure, and had a good application effect for light, high-speed conveying. In this research the distance of double-section double-speed sport structure is 2.4 meters, and its sports speed could reach 2 m/s.

2.2. *The structure optimization design*

In this research the motor is HG-SR152J with a rotated speed of 2000r per minute, its characteristic speed-torque characteristics as shown in Fig. 6.

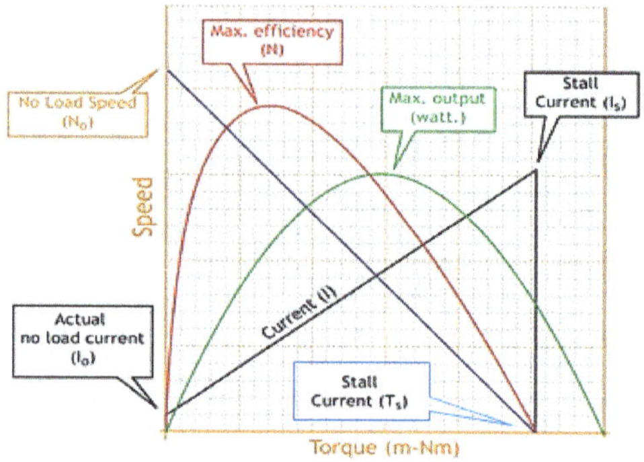

Fig. 6. Servo motor characteristic curve.

The X, Y, and Z directions of the manipulator have their own independent motor drive and transmission systems. The difference was that the main direction of motion (X-axis direction) used a double-section double-speed drive to achieve a higher movement rate. Motor-drive system principle as shown in Fig. 7.

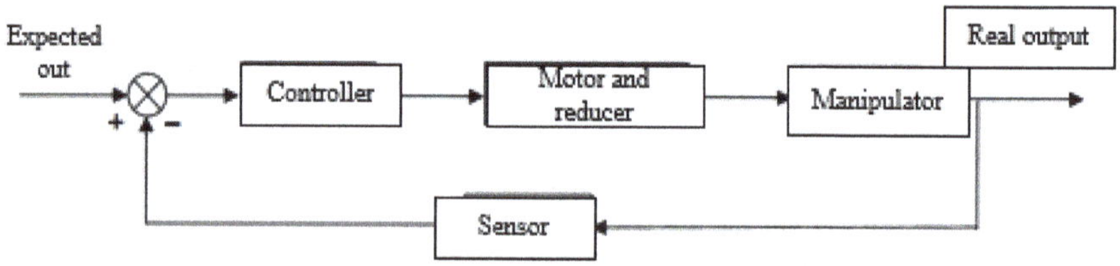

Fig. 7. Motor-drive system principle of conveyor robot.

According to classical mechanics theory, the general equation of dynamics of an object was following:

$$[M\{\ddot{x}\}]+[C]\{\dot{x}\}+[K]\{x\}=\{F(t)\} \tag{1}$$

In Eq. 1, [M] was the mass matrix, [C] was the damping matrix, [K] was the stiffness coefficient matrix, {x} was the displacement vector, and {F} was the force vector.

According to the general equation of dynamics and motor characteristic curve, in the case of a certain displacement x, without considering the effects of changing the damping and stiffness coefficients, if reduce the weight of the robot mass M could achieve faster speed and improve the robot handing speed. In this research, the weight of robot was minimized by material optimization and structure optimization.

As shown in Fig. 8, sports section 1 which was drove by motor moved along x-axis, sports section 2 and the end of the manipulator does the same movement drove by the structure design of double-section double-speed. The weight of sports section is about 140kg, and the weight of sports section B and the end of the manipulator is about 200kg.

Through the dynamic analysis of this manipulator, set the travelling distance of the end manipulator is 2.4 meters. The data were obtained as shown in Table 1.

According to dynamic analysis data, when the weight was 100 %, sports time was shortened 20 %, and the motor torque is increased 7.9 %. In the same sport time, after the weight of sport parts was lighted 30%, motor torque amplitude had been reduced 30%. Under this circumstance, the rapid technical indicators requested was easier to reach. At the same time, reduced motor torque will save energy, and it meets the global industrial trends.

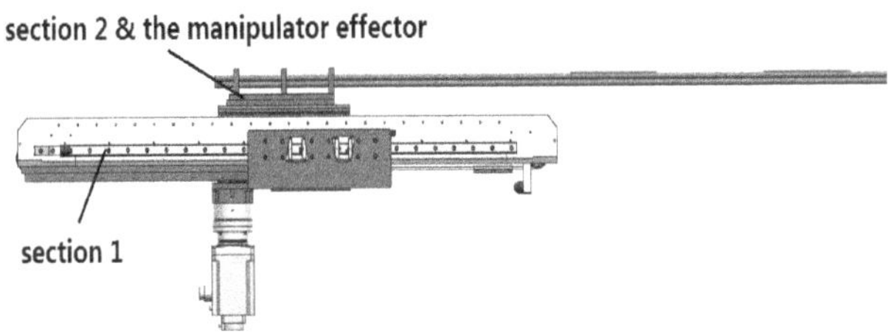

Fig. 8. The structure model of Cartesian coordinate robot.

Table 1. Dynamic analysis data.

	Time (s)	Speed amplitude (m/s)	Acceleration Amplitude (m/s²)	Motor torque amplitude (N·m)
100 %weight	1.5	1.6	3.2	1282
	1.2	2	5	1383
70 %weight	1.2	2	5	968

2.3. *The stability analysis*

When the conveyor robot suddenly start or stop, it would deviate from the normal movement trajectory due to the inertia of the movement, and the shifting and shaking would occur. The stability of the conveyor manipulator mainly refers to whether the manipulator could quickly return to the normal motion trajectory in case of suddenly start or stop, and ensured the accuracy of the motion trajectory without affecting the normal operation. In order to verify the movement rapidity and the stability, laser tracker was used to make dynamic test.

The optimization structure of load and unload manipulator for multi-chamber furnace was measured, and the stability accuracy of the entrance and exit direction of the five different working heights in the working space was shown in Table 2. When the manipulator was reached the height of the fifth floor as Z axis being raised, the vibrational frequency of the manipulator was 2.8~2.98 Hz, and vibration amplitude was ±0.336 mm. When the Z axis of the manipulator moved to lowest, the vibration frequency was 4.5Hz, and vibration amplitude was ±0.075 mm.

As shown in the test data, the stroke of x-axis was 2.4 meters, and maximum error of motion path was less than 0.03. It was demonstrated that the stability of the optimization structure conveyor manipulator in this study could satisfy the work requirements.

Table 2. Stability test data.

The floor of furnace		Vibration frequency	Vibration amplitude
first floor	into	4.55Hz	0.075mm
	out	5.43Hz	0.033mm
second floor	into	4.09Hz	0.156mm
	out	4.90Hz	0.107mm
third floor	into	3.86Hz	0.281mm
	out	4.37Hz	0.264mm
fourth floor	into	3.26Hz	0.563mm
	out	3.59Hz	0.679mm
fifth floor	into	2.77Hz	0.451mm
	out	2.98Hz	0.336mm

3. Conclusion

In the hot forming process, structural optimization and collaborative work strategy design were applied to the robot group to achieve rapid and stable conveying of the thermoformed sheet material and improve production efficiency. The work space of conveyor manipulator was saved by optimization structure of double-section double-speed, and the conveyor speed was guaranteed more than 2 meters per second stably. The weight of conveyor manipulator was reduced by optimization structure to reduce motor driving torque so that the conveyor manipulator could be easier to reach the highest speed. The multi-manipulators worked in tandem to make the manipulators work in parallel with each other. The space was more compact and did not interfere with each other. The production efficiency was increased by 143 %, and the completed transfer time of heat sheet is only 7 seconds. The time and space were fully used to complete the thermoforming production efficiently.

References

1. L. Wang, B. Zhu, Q. Wang and Y. S. Zhang, Modeling of hot stamping process procedure based on finite state machine (FSM), *Int. J. Adv. Manuf. Tech.* 1 (2016).
2. Z. Wang, Y. S. Zhang and H. Q. Liu, Optimization of industrial robots transminssion path in hot stamping production line, *Forging & Stamping Techn.* **37**, 31 (2012).
3. H. Q. Li, Y. S. Zhang, L. Wang, X. W. Tian, C. Wang and B. Zhu, Emissivity Calibration and Temperature Measurement of High Strength Steel Sheet in Hot Stamping Process, *Appl. Mech. Mater.* **148-149**, 1473 (2011).
4. W. Sun, Research on the Structure Design and Dynamic Performance Optimization of the High Speed Transportation Manipulators in Hot Stamping Production Line Candidate, PhD thesis, Huazhong University of Science &Technology (2016).

Optimization and Implementation of Efficiency-based Control Strategy: A Case Study of Hot Stamping Line

Q. Wang[1], L. Wang[2,†], B. Zhu[2], Y. Liu[2], Y. L. Wang[2], Y. S. Zhang[2] and X. Y. An[3]
[1]*Vision Hot Stamping Technology Co., Ltd.,Dongguan, 523460, China*
[2]*State Key Laboratory of Materials Processing and Die & Mould Technology*
Huazhong University of Science and technology, Wuhan, 430074, China
[3]*Wuhan Zhongyu Dingli Intelligent Technology Co., Ltd., Wuhan 430074, China*
[†]*E-mail: wangliang@hust.edu.cn*

A distributed modular production line is more powerful and flexible. The independent equipment module is controlled by a variety of different controllers. Due to the different implementation of control process for different controllers and the complexity of various control protocol conversions, it is not easy to achieve a unified and complete control system design and implementation at the production line level, often resulting in a control blind zone at the interactive boundary of multiple devices, resulting in functional defects at the entire line level. In order to solve this problem, this paper proposes an efficiency-based control strategy for heterogeneous distributed control industrial production lines. The model focuses on controller interaction boundaries, introduces two evaluation indicators: control instruction generation efficiency and data exchange efficiency, and form a method of efficiency function deployment. Finally, the efficiency-based control strategy was implemented on the hot stamping line which improves control efficiency and overall line operation reliability.

Keywords: Heterogeneous distributed system; Control architecture; Hot stamping.

1. Introduction

With the rise of intelligent manufacturing, the control system used in manufacturing is revitalized, and it is breaking the limitation of the original system [1], developing to the direction of intellectualization [2], complication and integration. Its development trend can be clearly observed from the dimensions of controller, system architecture and communication protocol.

The main developing trend of controllers is the integration of each other. The PLC has more and more strong logical operation ability, the IPC has the more and more rich industrial level communication interface. The development of control system architecture is from PLC, IPC and other single controller system, to homogeneous multiple controller system and then distributed heterogeneous control system. The distributed modular production line is a more powerful and flexible production line type. The independent device module is controlled by a number of different controllers. It is the industrial Internet which is one of the industrial communication protocols worth to be focused, and it promotes the evolution from the control system to the information system and provides the basis of communication and data for the intellectualization.

Hot stamping is a kind of advanced forming technology for forming and strengthening steel blank under high temperature [3]. The strength of hot stamping parts is obviously better than that of cold stamping parts, which is of great significance to improve the safety of automobiles and to promote energy saving and emission reduction of automobiles. In recent years, the proportion of hot stamping parts in automobile body-in-white has continued to increase. The hot stamping production is a large-scale batch manufacturing process under the condition of high temperature, which must be considered both in reliability and speediness. And its manufacturing process must be fully automated, because of the complex process it has a direct demand of the sensor network. The control system of traditional hot stamping production line adheres to the old control platform, its data exchange capability and the access ability is weak, it is difficult to integrate directly with ERP and MES system, and bury the data treasure underground.

The hot stamping production process requires the production line control system to have: reliability, to ensure stable and consistent parts performance; precision, continuous and efficient production of products; leading [4], equipment operation capability is significantly better than process requirements; traceability, conducive to discover the cause of equipment failures and parts defect.

2. Efficiency-based Control Strategy

In a distributed control system, an independent control system logically has a primary and secondary distinction due to different functions. Some control node are mainly for implementation, and there are relatively large executing structures, which have high requirements for the rapidity and accuracy of the execution process. Some control node are mainly for coordination and scheduling, with strong capabilities of data exchange and computing, and have high requirements for data exchange bandwidth and communication delay. The conventional control strategy design places all control logic and flow control on the scheduling node, and the execution node is responsible for instruction execution and a small amount of necessary logic operations, as shown in Fig. 1. The execution cycle t_{cc} of the control instruction is identified by the following formula.

$$t_{cc} = t_{dc} + t_p + t_{ic} + t_{ie} \tag{1}$$

In which, t_{dc} is the time of data exchange communication, determined by the amount of data exchanged and the communication bandwidth; t_p is the time of logical process; t_{ic} is the time of instruction exchange communication, mainly determined by the communication bandwidth due to little amount of data for several instructions; t_{ie} is the time of instruction execution, determined by the scan performance of the controller, which is usually 1~2 ms.

The control strategy is inefficient. Although it is a distributed control system, the control cycle t_{cc} contains multiple communication cycles, because the process control logic is concentrated in the Control Unit 1 (CU-1) in Fig. 1. CU-1 is likely to become an efficient bottleneck in the entire control system. One reason is that all related device state data must be transferred to CU-1 to begin executing process control logic. When node count exceeds a certain number, data exchange will be the main bottleneck. Another reason is that process control logic execution only utilizes the computing resources of a single node, and from the view point of the whole system, it wastes the computing power of other control nodes. When the logic of process control is complicated, it takes a long time to calculate the next control instruction in a single node.

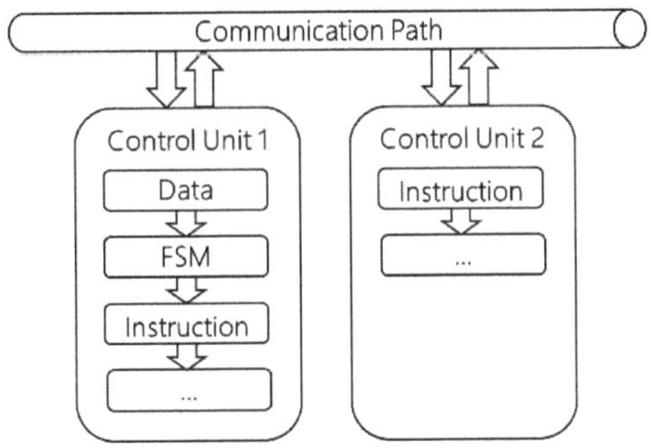

Fig. 1. Original inefficient control strategy.

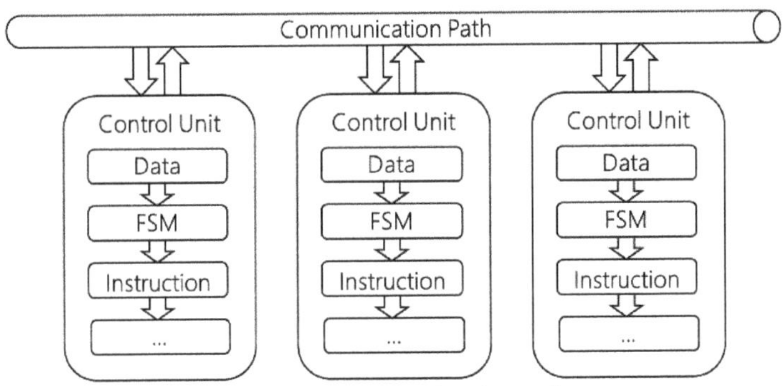

Fig. 2. Modeling of efficiency-based control strategy.

In order to solve this problem and improve the efficiency of the control system, an efficiency-based control strategy is presented in Fig. 2. This strategy fully utilizes distributed computing resources of distributed control system, decomposes process control logic into corresponding control units according to function relativity, balances compute load of each node, analyzes and optimizes the control instruction generation efficiency and data exchange efficiency, and determines the deployment position of function from the angle of execution efficiency, thereby improves the efficiency of the entire control system.

In this process, the control instruction generation efficiency refers to the time from the last control instruction in the independent control unit finished to the next control instruction to start to execute, which is used to evaluate the generation efficiency of the control instruction. The higher the control efficiency, the smaller the efficiency loss caused by the control logic. The data exchange efficiency refers to the time from the beginning of sending data of initiator to the end of data receiving, can be comprehensively evaluated the data processing ability of communication bandwidth and controller. Data exchange in the objective is uncertain. On the one hand, TCP/IP and other similar communication protocol are affected by a variety of factors interfere with communication delays. On the other hand, although with the same communication links and protocols, controllers with different calculate ability will be significantly different when receiving or sending data.

The control strategy unifies architectures of all control units, the basic logic is to obtain data, execute finite-state-machine-based control logic, calculate and obtain the control instruction, executes. The advantage of this strategy is that it reduces the requirement of communication delay, is applicable to all kinds of communication protocols, and is more advantageous to the integration of control system and information system.

The limitations of this strategy are as follows. The production beat should be far greater than the communication beat. The beat of the execution mechanism is far greater than that of the control logic execution. The real-time performance of the data and state of other units can not be mandatory.

3. Implementation and Analysis

The efficiency-based control strategy presented in this paper is a part of the control system of hot stamping production line with multi-layer box-typed furnace, which is developed by our research team.、The production line adopts heterogeneous distributed control system. For the line flow control unit, press, furnace and manipulator units, 4 kinds of control platforms such as Siemens, SigmaTek, BECKHOFF, Omron are selected according to the control complexity, control response requirement. Up to 7 kinds of control protocols are integrated based on industrial Ethernet. Finally, efficient production control is achieved. The hot stamping line is capable to produce many types of parts with the beat of 16s at the fastest.

The data exchange cycle of the control system is 10~100ms. This is mainly due to the use of 1000Mbps high-speed industrial Ethernet, to ensure the efficiency of data exchange. And at the same time it ensures that the industrial-computer-based line flow control system can be achieved 10~30 times monitoring and control every second.

In the production line, the automatic control system requires the highest speed. Its control instruction generation efficiency is 1~2 ms, which is the scanning period of Siemens motion controller. This is because the operating process of the production line is very suitable for the finite state machine. The data exchange period is completely masked by exchanging the data of the state of the device in the long period of the instruction execution (generally 1.5~5 s), so that the new control instruction can be calculated just by local relatively simple control logic operation, and the optimization of the control instruction generation efficiency is achieved.

Through the optimization of line layout [5] and automation control process, the operation of the line operating efficiency is successfully promoted from 24s to 16s, which is greatly improving the whole line of the operating cycle, and achieves an advanced level.

4. Conclusion

In this paper, a control strategy based on efficiency is proposed and applied to the design of the control system of the hot stamping production line. By adopting the control strategy, the unstable data exchange cycle is covered by the longer instruction execution cycle, which reduces the control instruction generation efficiency to 1~2ms. Then through optimization of the line layout and automation control process, hot stamping line operation beat is improved to 16s, which is an obvious progress. The control strategy also needs to test its versatility in other types of control systems through practice.

Acknowledgments

This research work was financially supported by the National Major Science Technology Project of China (grant no. 2018zx04023001).

References

1. A. Rizwan, M. S. I. Alvi, A. Sajid, A review of the evolution of microcontroller-based machine and process monitoring, *World Tribology Congress III* 869 (2005).
2. L. Wang, B. Zhu, Y. Zhang, Y. Wang, X. An and Q. Wang, A smart process control strategy for press hardening production, in *6th International Conference Hot Sheet Metal Forming of High-performance Steel*, (Atlanta, USA, 2017).
3. M. Ma, Y. Zhang, L. Song and Y. Wang, Research and progress of hot stamping in China, *Adv. Mater. Res.* **1063**, 151 (2015).
4. L. Wang, B. Zhu and Y. S. Zhang, A runtime intelligent iterative manufacturing system (RIIMS) for hot stamping, in *Advanced High Strength Steel and Press Hardening: Proceedings of the 2nd International Conference (ICHSU2015)*, (Changsha, China, 2015).
5. L. Wang, B. Zhu and Y. S. Zhang, Production control and optimization of hot stamping line, in *5th International Conference on Hot Sheet Metal Forming of High-performance Steel*, (Toronto, Canada, 2015).

The Application and Research of Lightweight Design of High-speed Manipulator in Thermoforming Lines

Z. T. Su†, M. S. Yu and X. B. Tao
Qingdao Kingerobot Co., Ltd., Qingdao, 266000, China
†E-mail: suzt@kingerobot.com

With the development of the demand for lightweight automobiles, in addition to improving the strength of car structure, how to increase the production efficiency of thermoforming lines has also become a focus of research. In the complex equipment and cumbersome production process of thermoforming process, the flexibility and speed of industrial manipulators had great influence on the compactness of the thermoforming line and the production cycle of the entire line, directly affecting the capacity and benefits of the entire thermoforming line. For the above problem, lightweight applications for manipulators in thermoforming lines have been put forward. A finite element analysis model with ANSYS workbench was built. The analysis and calculation of lightening for the Z-axis fixed joint of hearting furnace loading and unloading manipulator and the X-axis fixed joint of punch loading and unloading manipulator was done. The topology optimization for the Z-axis fixed joint was also done and the analysis and calculation for the optimized model was performed. The results show that when the performance and strength of the manipulator were satisfied, if the weight of the Z-axis fixed joint was reduced by 10.16 %, the transport time of the hot billet was shortened to 6.5~7 s and shortened by 18.75 %~22.22 %, which improved the speed and flexibility of the manipulator. And the lightweight manipulators have been put into production at a thermoforming company.

Keywords: Lightweight of car; Thermoforming; Industrial manipulators; Motion stability.

1. Introduction

With the demand for lightweight of automobiles, hot stamping forming technology has developed rapidly. Based on the hot stamping forming technology, the hot stamping forming line was a multi-manipulator collaborative working platform consisting of a conveying system, a heating system and a stamping forming system [1]. In hot stamping lines, the speed of high-speed manipulators to deliver hot billets has a significant effect on the stamping quality. At the same time, the complex equipment and cumbersome production processes of the thermoforming process put forward higher requirements for the hot stamping automated manipulator system: 1) The compact layout of the production line requires more flexibility for the manipulator; 2) The product performance was sensitive to the forming temperature. It was necessary to quickly transport the heated sheet material to the above of the mold, which imposes higher requirements on the speed of the manipulator; 3) The speed of the whole line's production tact directly affects the productivity and efficiency of the entire line and puts forward higher requirements for the collaborative work of the manipulators in the system [2, 3].

2. The Overview of High-speed Manipulators

In actual production, the hot stamping process was divided into a direct process and an indirect process. In the direct process, hot stamping production lines were divided into two types: roller hearth furnace and multi-layer box furnace, as shown in Fig. 1.

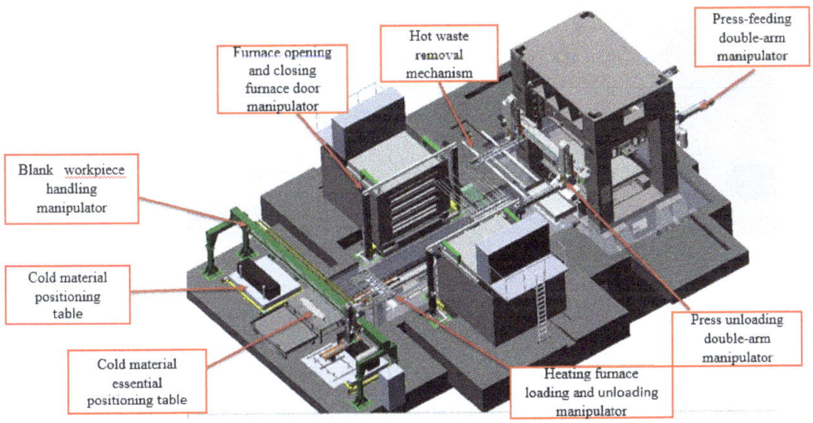

Fig. 1. The multi-chamber furnace.

The high-speed manipulators used in thermoforming lines include multi-station manipulators for breaking jam, high-speed manipulators for heating furnaces, operated manipulators for furnace door, and high-speed manipulators for press loading and unloading. Fig. 2 was the high-speed manipulators of the press loading and unloading. In the thermoforming production line, the high-speed conveying manipulator of the furnace and the high-speed manipulator loading and unloading of the press work together to complete the conveyance of the billet from the heating furnace to the press for pressing. The speed of the two manipulators determines the conveying speed of the hot billet in the production line. In order to ensure the stamping quality of hot stamping parts, the temperature of the steel plate before forming must be guaranteed at 780 °C or more. The actual measurement of the material shows that it takes 7~8 s for the billet to cool from 900~950 °C to 780 °C [4]. According to the actual test, from the hot billet out of the heating furnace to the press down, the hot billet exposure time in the air was 8~9 s, resulting in reduced stamping forming quality. Therefore, lightweight research was performed on the key moving parts of the high-speed conveying manipulators of the furnace and the high-speed manipulators for loading and unloading the press. And in the case of ensuring the strength and rigidity of moving parts, increase the movement speed of the manipulator and shorten the exposure time of hot billets in the air.

The walking part of the high-speed manipulators for loading and unloading of the press includes a Z-axis movable joint, a Y-axis rolling carriage, an X-axis fixed joint and an X-axis movable joint. During the movement of the press manipulator, the movement speed and stability in the X-axis direction were directly related to the feeding speed. In this paper, the Z-axis fixed joint of the heating furnace manipulators and the X-axis fixed joint of the press manipulators were mainly studied for lightening.

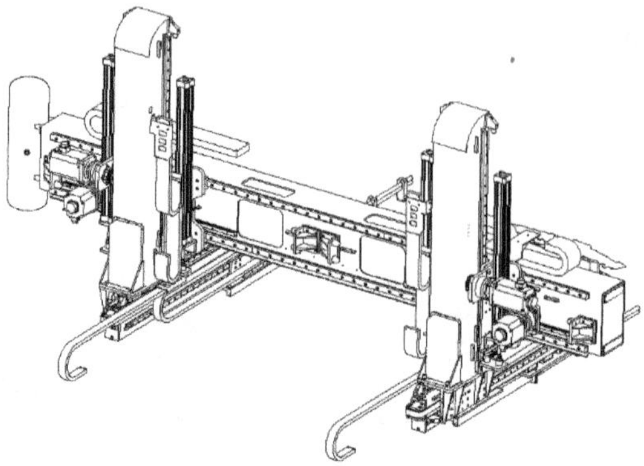

Fig. 2. High-speed manipulators for loading and unloading the press.

3. Major Headings

3.1. *Model establishment of key component*

Accurate and reasonable construction of the 3D model of the manipulator's key moving parts was the key to the finite element analysis and optimization design. The purpose of constructing a 3D model of a manipulator's key moving parts was to establish a mathematical model that can truly reflect the actual design prototype of a moving part [5]. In order to improve the calculation speed, the connection components (such as bolts and nuts), process holes (such as bolt holes, etc.) and transition fillets on the moving parts were ignored. These components, process holes, and transition fillets have little effect on the forces acting on the moving parts, but they increase the number of meshes when dividing the grid, reducing the computational speed of the computer. And for some of the larger holes will affect the strength of the moving parts, these holes should remain. Fig. 3 and Fig. 4 show the simplified model before and after the Z-axis fixed joint of the high-speed manipulator. The high-speed manipulator model was modeled using 3D software SolidWorks. Use SolidWorks to import the simplified 3D model into ANSYS work bench for finite element analysis.

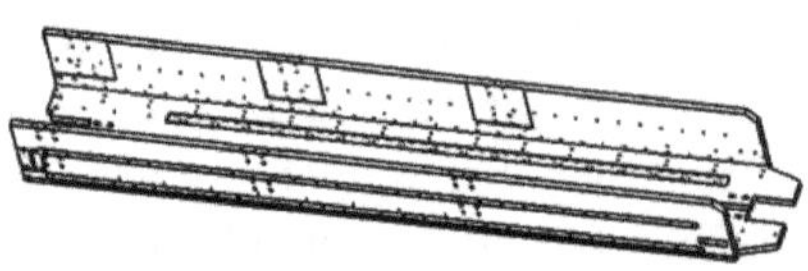

Fig. 3. Simplify before.

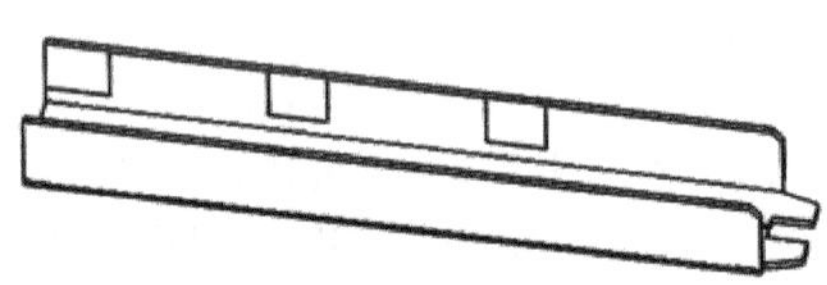

Fig. 4. Simplified after.

3.2. *Lightweight material analysis*

Lightweight materials include high-strength steels, lightweight alloys and composites. Aluminum alloy materials were widely used in light alloy materials. And compared with other alloy costs, aluminum alloy prices were lower. The magnesium alloy molding process was not mature, the price was 3 to 4 times of aluminum alloy. Titanium alloys have higher overall performance than aluminum alloys, but they were expensive [6].

Among composite materials, a company has successfully applied carbon fiber to the body of an electric vehicle. Its body weight has been reduced by 300 kg, and production models have been produced [7-9]. The parameters of the three materials were shown in Table 1.

Table 1. The parameters of steel, aluminum alloy and carbon fiber.

Material type	Elastic modulus (N/m^2)	Density (Kg/m^3)	Poisson ratio	Yield limit (MPa)
Steel	2.1E+11	7850	0.3	235
Aluminum alloy	6.9E+10	2700	0.33	275
Carbon fiber	2.31E+11	1780	0.307	4480

3.3. *Force analysis of moving parts*

3.3.1. *Force analysis of Z-axis fixing joint of high-speed manipulator of the heating furnace*

The Z-axis high-speed manipulator fixed joint of the heating furnace consists of one main board and two side boards, and the main board and the side board were connected by bolts. When the manipulator was working, the Z-axis structure was the most stressed, and the fixed joint in the Z-axis structure was the key moving part. The Z-axis fixing joint was equipped with rails on the side boards and the main board. The Z-axis fixing joint of the manipulator was fixed by the slider, and each of the rails uses three sliders. After the model was imported into the ANSYS workbench, the meshes were divided by the size-dividing mesh method, and the mesh size was set to 15 mm. After loading and constraint analysis and calculation, the analysis results were shown in Figure 5.

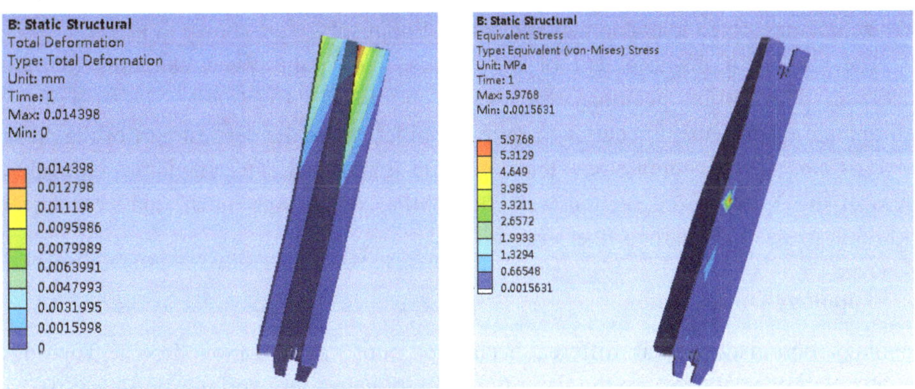

(a) Displacement deformation cloud picture. (b) Equivalent Stress Cloud picture.

Fig. 5. Z-axis fixing joint analysis results.

The Z-axis fixed joint was at the top of the two side plates. The maximum deformation displacement was 0.0144 mm, as shown in Fig. 5(a). The maximum stress position was at the position of the main board guide rail. The maximum equivalent stress was 5.98 MPa, as shown in Fig. 5(b). Since only one rail was mounted on the main board, the main board withstands stress concentration. However, the maximum equivalent stress was 5.98 MPa, which was far less than the allowable stress of aluminum alloy, so the structural strength can meet the requirements.

3.3.2. *Force analysis of X-axis fixed joint of press high-speed manipulator*

After the press feeding manipulator picks up the hot billet, the hot billet needs to be quickly transported to the press for stamping. Therefore, when the manipulator transfers the hot billet to the press, the hot billet needs to enter the press at a high speed and stable. When the press manipulator feeds, the speed and stability of the X-axis movement have a great influence on the stability of the hot billet into the speed of the press.

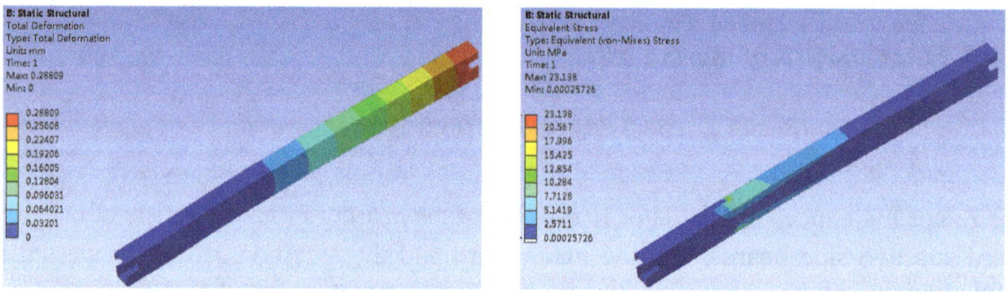

a. Displacement deformation cloud picture. b. Equivalent Stress Cloud picture.

Fig. 6. X-axis fixing joint analysis results.

The X-axis fixed joint's material was composite carbon fiber. Its density was 1.78g/cm^3, which was 0.66 times the density of aluminum alloy, 0.22 times of the steel density. The same volume was much smaller than other materials, which greatly improves the X-axis's running speed and manipulator flexibility. The X-axis fixed joint was subjected to force analysis and the vibration frequency of the fixed joint was analyzed. X-axis fixed joint analysis and calculation results were shown in Fig. 6.

The maximum displacement of the X-axis fixed joint was 0.288 mm, and the maximum deformation position was at one end of the fixed joint. The maximum deformation displacement occurs when the hot billet enters the deepest part of the press. At this time, the maximum stress that the fixed joint bears. The maximum equivalent stress of the X-axis fixed section was 23.14 MPa, which was much smaller than the allowable stress of the carbon fiber material.

4. Topology Optimization

Topology optimization was different from traditional optimization design. Topology optimization does not require the definition of parameters and optimization variables. It only needs to give material properties, models, loads, and optimization percentages. The objective function and design variables were predefined. Use ANSYS workbench to make the topology optimization for the Z-axis fixed joint, set the optimization percentage to 40 %, and obtain the material density cloud picture of the Z-axis fixed joint topology optimization results, as shown in Fig. 7. The dark part of the figure indicates that the material density was close to zero and can be removed in the structural design. And the light color part was the mass that needs to be retained. Taking into account the actual application requirements, the optimized model of the Z-axis fixed section structure was shown in Fig. 8.

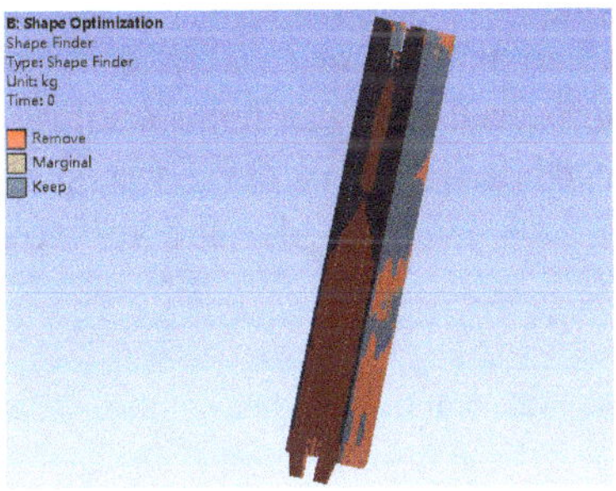

Fig. 7. Z axis fixed joint topology optimization cloud picture.

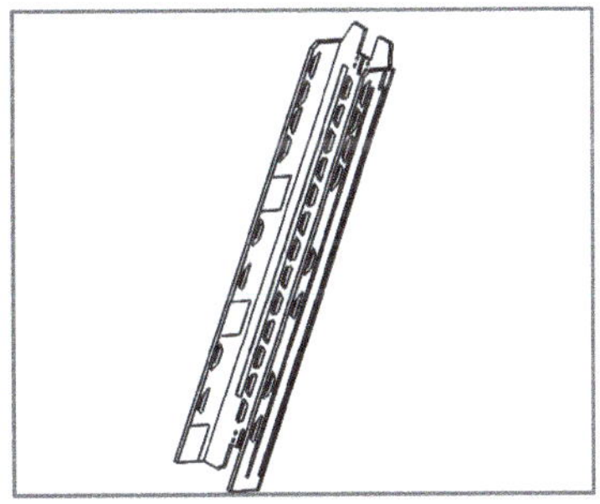

Fig. 8. Z axis fixed joint optimized model.

Static analysis was performed on the optimized model, and the maximum displacement of the model after optimization was 0.4185 mm, as shown in Fig. 9(a); the maximum equivalent stress was 9.48 MPa, as shown in Fig. 9(b). The results show that the displacement of the model after topology optimization was changed from 0.0144 mm to 0.4185 mm, which was still far smaller than the allowable displacement specified in the design requirements. Therefore, the stiffness of the model can meet the requirements after optimization. The maximum equivalent stress was changed from 5.98 MPa to 9.48 MPa. However, it was still far below the allowable stress of the material, so the strength of the model structure after the optimization can meet the requirements. At the same time, under the condition of satisfying the structural strength and stiffness, the weight of the Z-axis fixed joint was reduced by 10.16%, which reduced the weight of the manipulator, thereby improving the manipulator's movement speed and flexibility.

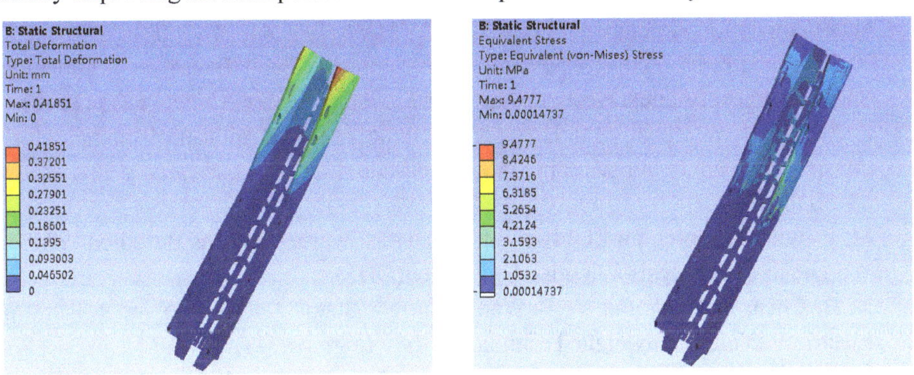

(a) Displacement deformation cloud picture. (b) Equivalent Stress Cloud picture.

Fig. 9. Results of optimized Z-axis fixed joint analysis.

5. Conclusion

Using the SolidWorks modeling, the model of the loading and unloading manipulators of the furnace and the manipulators for loading and unloading the press was established and analyzed and calculated by the ANSYS workbench. Through the use of aluminum alloy materials and topological optimization of the Z-axis fixed joint of the loading and unloading manipulators of the furnace, the weight was reduced by 10.16 % under the condition of ensuring the performance and strength of the manipulator. The X-axis fixed joint of the press feeding manipulator adopts carbon fiber material, and its stress was much lower than the allowable stress of the material and was more stable. After the lightweight design of the manipulator, the heat billet delivery cycle was shortened to 6.5~7 s, which was reduced by 18.75%~22.22%, and the demand to increase the manipulator's speed and flexibility was achieved.

Acknowledgments

This research work was financially supported by the National Major Science Technology Project of China (grant no. 2018zx04023001).

References

1. L. Wang, B. Zhu, Q. Wang and Y. S. Zhang, Modeling of hot stamping process procedure based on finite state machine (FSM), *Int. J. Adv. Manuf. Tech.* **89**, 857 (2016).
2. Y. Sun, L. Wang, B. Zhu, Y. L. Wang and Y.S. Zhang, The structure design and performance analysis of the transportation fork in hot stamping production line, in *5th International Conference on Advanced Design and Manufacturing Engineering*, (Shenzhen, China, 2015).
3. J. Q. Long, F. C. Lan and J. Q. Chen, New technology of lightweight and steel-aluminum hybrid structure car body, *Chin. J. Mech. Eng.* **44**, 27 (2008).
4. X. H. Fu and Z. Y. Yang, Analysis of Development Trend Based on Lightweight Special Purpose Vehicle Design, *Dual Use Technologies & Products* **4**, 21 (2016).
5. Z. Wang, Y. S. Zhang and H. Q. Liu, Optimization of industrial manipulator's transmission path in hot stamping production line. *Forging & Stamping Technology* **37**, 36 (2012).
6. M. L. Wan, Research on Lightweight Frame of Electric City Bus Based on ANSYS Workbench, PhD thesis, Qingdao University (2015).
7. X. B. Chen, P. Hang and Y. F. Wang, Current Status and Development Trend of Electric Vehicle Lightweight Technology, *Auto Engineer*, **11**, 23 (2015).
8. M. Anwar, I. C. Sukmaji, W. R. Wijang and K. Diharjo, Application of Carbon Fiber-Based Composite for Electric Vehicle, *Adv. Mater. Res.* **896**, 574 (2014).
9. N. Takahashi, Y. Kageyama and N. Kawamura, Research of Multi-Axial Carbon Fiber Prepreg Material for Vehicle Body, *SAE International in United States* **01**, 216 (2011).

Laser Softening of Press Hardened Steel for Novel Automotive Parts

T. Harrer† and M. Schäfer
TRUMPF Laser- und Systemtechnik GmbH,
Johann-Maus-Strasse 2, Ditzingen, 71254, Germany
†E-mail: thomas.harrer@trumpf.com

Press hardened Boron steel is still gaining importance for automotive body parts because of its advantageous material properties. For forming and hardening the flat metal sheets are heated up to Austenite temperature, transferred to the dye, high speed formed and hardened — while cooling down within the forming tool [1]. This process leads to high strength properties of about 1500 MPa and is used for weight reduction purposes in car design [2]. The combination of heat treatment and forming is also strongly influenced by the cooling gradients within the forming tool. High cooling gradients lead to bainitic and martensitic structure formation — structure parts which show high hardness [3]. Car manufacturers look into the application of flaps in the design of car body parts. Using flaps within the assembly process is helpful for part-positioning. In case of already press hardened Boron steel it is necessary to perform a heat treatment before bending to prevent crack formation in the bending zone. This paper shows how this process can be performed most efficiently within the laser cutting machine. Soft areas for rivet points to facilitate joining operations and ductile flange areas or complex 3D-contours to improve crash performance can also be realized by this technology. This paper shows a specially designed line optic for these applications.

Keywords: Laser; Laser softening; Hot forming; Press hardened steel.

1. Introduction

Laser cutting of press hardened structural steel components shows multiple advantages regarding production costs and flexibility. In that respect the absence of tool wear due to the contactless trimming as well as the high quality of the cutting edges compared to other cutting technologies has to be mentioned. Additionally there are almost no restrictions to the shape of the parts or the contours to be cut. The stiffness of the machine tool is responsible for the achievable part tolerances [1]. Manufacturing costs per part depend mainly on the laser-on-time respectively the ratio between cutting-time to part-changing-time. Therefore TRUMPF specifically developed a machine for hot-formed 3D structural components, the TruLaser Cell 8030 (Fig. 1).

But not only cutting is important for car manufacturers and the suppliers of structural parts. There is high interest on reducing the global hardness on areas of hot forming parts. Therefore latest process developments intend to fulfill that in a post-hardening process using laser beam as heat source on the surface. The intention of such a process step is to locally transform the martensitic structure to a phase with lower strength but higher ductility [4]. For Laser Annealing the following results has also been presented previously in [5] and [6].

This Laser annealing can be done with the same machine tool, the TruLaser Cell 8030 (Fig. 1) without any change of optics or fixture – compared to other methods no additional production equipment is needed. This approach allows the user to define the soft zones even in a very late project phase and gives almost unlimited flexibility of the size, location, shape and numbers of such soft zones at the part. In this respect, local distortion due to the heat input is important. Dependent on process parameters (feedrate and laser spot size) and conducted process strategy, maximum distortion less than 1.5mm is possible [3]. Furthermore, in a serial production, distortion coming from the heat input can be pre-compensated in the forming die – final distortion is 0mm.

Fig. 1. The TruLaser Cell 8030 (L60) is a specifically developed machine for hot-formed 3D structural components. However, not only cutting but also partial softening of these components is possible.

2. Experimental

2.1. *Flaps*

One technical demand for a post-hot-forming softening process is the bending of assembly assisting flaps as shown in Fig. 2. Dimensions of those areas are small and therefore the softening process is fast. For example the annealing of both flaps (left picture in Fig. 2, bended after softening) just takes 12 seconds.

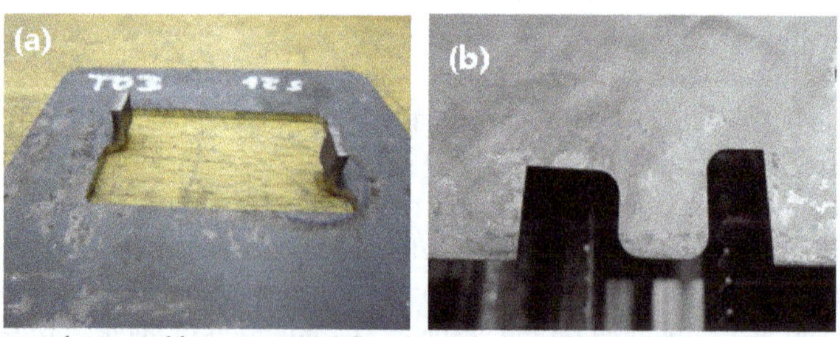

Fig. 2. Flaps used as assembly support: (a) 2 flaps bended after softening, (b) large flap before bending.

Flaps are dedicated to help positioning the part in the automotive assembly process. For this purpose the flaps mostly are part of the outer cutting contour. After laser cutting the flaps are bent into their final position. Most important, the flaps must not crack while bending. To prevent cracking of the flaps, hardness has to be reduced while ductility increases at the bending edge – softening the edge is the method of choice.

Softening small flaps is done by programming circles as NC-path in which the laser beam is guided to the work piece. The flaps are heated up to 850 °C, cooled down by shutting off the laser and post heated at approx. 600 °C. Alternatively the time for the cooling pause can be used for heating up a second closed-by flap (Fig. 2).

The resulting hardness profile in Vickers hardness shown in Fig. 3 (right picturc) and the original material hardness after hot-forming is displayed in the right graph of Fig. 3. Additionally different hardness levels of the transition curve are market with dashed squares and the corresponding microstructures are listed on the left side of Fig. 3.

Hardness decreases along the transition curve from measuring position “0” to higher values within less than 12 mm. While the hardness values on the left match the non-softened base material (approx. 480-500 HV1) the minimum hardness values of 220-230 HV1 in the softened zone correspond to those of the non-treated (not hot-formed) material. The initial microstructure at 220 HV1 exhibits a band-type ferritic - pearlitic microstructure (lower row, right picture). With increased hardness up to 250 HV1 the band-type microstructure is still visible. In contrast to this at 370 HV1 the microstructure changes to martensitic-bainitic, comparable to the microstructure at 480 HV1, where – corresponding to the hardness level – only tempered martensite exists.

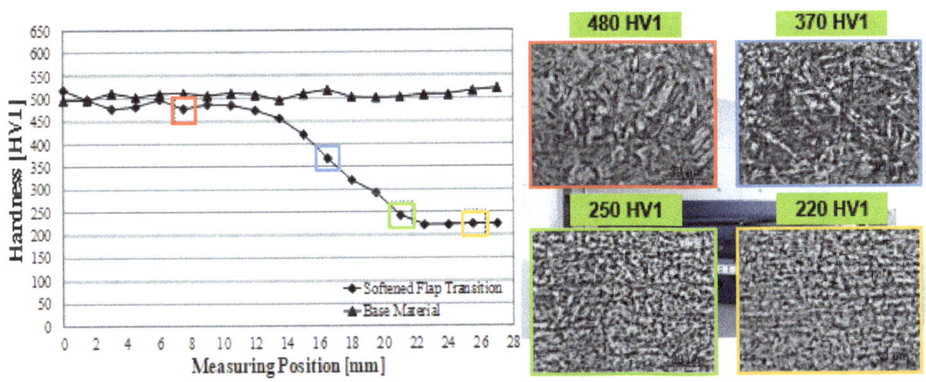

Fig. 3. Right: Hardness measurement at softened flap in the base material and in the transition zone between softened and non treated material; Left: Microstructure at different hardness levels market with rectangles in the chart.

2.2. *Partial softening using line optics*

In order to increase the annealed area per time step, the results of using specially designed process optics are presented. This optics forms a squared line of the originally round laser beam, where this line exhibits a width of 1 mm and the length can be modularly changed in steps of 5 mm between 10 mm and 25 mm. These optics are provided for the application to a 6-axis robot but as well for a TruLaserCell 8030, a machine originally dedicated for laser cutting of press hardened steel (depicted in Fig. 1). Both machine concepts provide the capability for 3D-processing.

In the experiments, the width of the line which was focused onto the work piece was l = 20 mm and the disk laser TruDisk 3001 provided the laser power of maximum 3 kW. The schematic optics, the beam profile of the laser line as well as the profile of the laser intensity in the focus which was measured with a Primes FocusMonitorTM, is depicted in Fig. 4. The annealing optics exhibits a long focal length of Zr = 13 mm which means that a variation of the distance between the optics and the work piece does not affect the annealing results. The optics can be easily integrated into the cutting machine due to the magnetic coupling and besides that, the programming of the 3D-beam path is the same as for cutting operations. Additionally, the right image in Fig. 5 shows a picture during annealing of an area located at the flange of a B-pillar.

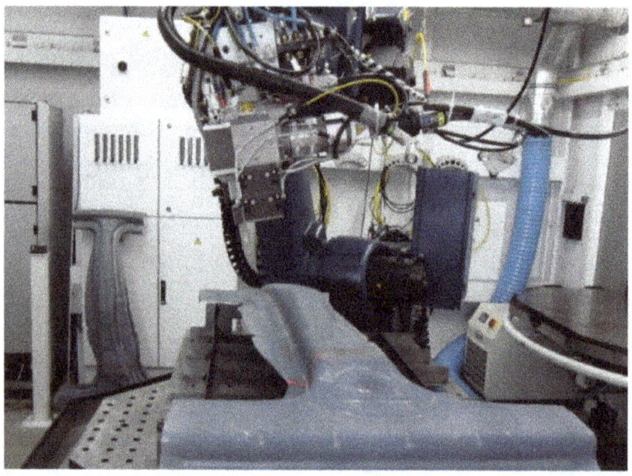

Fig. 4. The line optics dedicated for the annealing process can be applied to a 6-axis robot or the TruLaser Cell 8030 dedicated for laser cutting of press hardened steel.

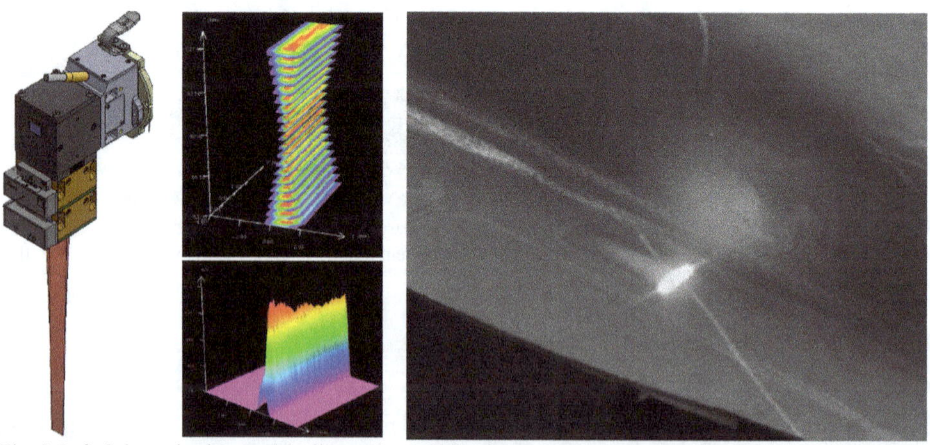

Fig. 5. Left: Schematic picture of the line optics. Middle: Beam profile of the laser line. Right: Snapshot of the annealing process using the line optics.

Compared to previous process developments [5], the feedrate for the annealing process could be increased significantly up to 1.2 m/min. To achieve a width of the annealed zone, two passes of the laser path were programmed. Due to the heat input of the first pass which is dissipated in the steel sheet, less power is necessary for the second pass. This reduction was determined empirically but could also be achieved by means of a closed-loop control in as much the temperature of the surface from the work piece would function as the input measure. Consequently, a productivity of vSoft = 1.1 cm^2/s can be achieved.

The resulting profile of the hardness distribution for two adjacent lines in the case of flat metal sheets of small thickness (1.2 mm) is shown in Fig. 6. As depicted, a very homogenous profile with an average hardness of less than 250 HV1 results in the annealed zone measured in the middle of the sheet thickness perpendicular to the feed direction. A small transition zone to the hardness of the untreated material (> 400 HV1) of less than 10 mm can be achieved. In the small inset, a staple of multiple annealed sheets is depicted, showing a very reproducible distortion related to the annealing. The mechanical properties of the annealed zones were tested by means of a tensile test where the yield strength of the annealed zone was measured to be $Rp_{0.2,mean}$ = 580 MPa with an ultimate strain of $A_{50,mean}$ = 10 %.

Additionally, in Fig. 6 (left) the homogeneous profile of the hardness for a higher sheet thickness (2.0 mm) are depicted. Depending on the process speed the impact of the heat capacity of the metal sheet can be observed: For 0.4 m/min the heat in the metal sheet keeps the material on a level of the temperature, where annealing takes place (about 600 °C) and the resulting hardness corresponds almost to the untreated base material (approximately 200 HV1). Compared to that, the heat dissipates faster for the feedrate of 0.8 m/min – resulting in a hardness level of about 300 HV1.

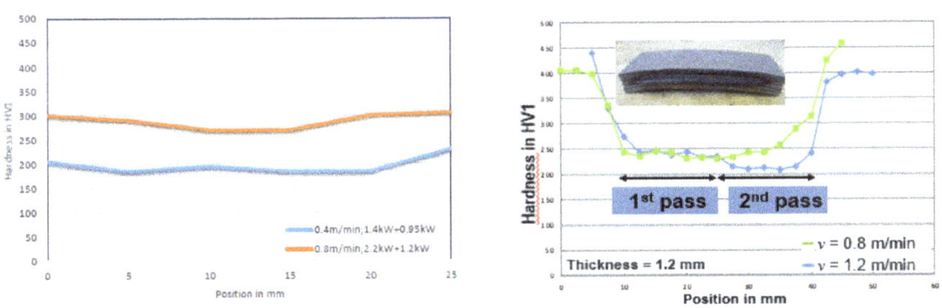

Fig. 6. Softened flanges relevant for crash situations (upper left). Soft zone within dashed lines (lower left). Hardness profile lateral and transversal to heat input (right picture).

3. Conclusion

Laser light is not only dedicated to cutting or welding but can also be used for softening local areas of press-hardened structural steel parts working like a single-sided surface heat source. Providing the right process parameters, a temperature level higher than the austenitization temperature (>860 °C) can be applied. To achieve homogeneous hardness profiles proper laser power, feedrate of the process optics and degree of defocusing have to be set.

As pointed out in the present paper a significant drop of hardness down to the values of the non treated, not hot-formed material can be achieved. The hardness drops evenly on the softened zones. It is shown, that the change of hardness correlates with the shape and distribution of microstructure phases. It is worth highlighting that the transition zone between hardness values of hot-formed and softened areas can be smaller than 10 mm.

By means of the method presented using laser as heat source it is possible to locally reduce hardness. Applications like softening of flanges to improve local ductility in crash-situations, softening of flaps to avoid cracks in the bending edges as well as softening of 3D-contours at complex automotive structural parts can be realized.

References

1. H. Karbasian and A. E. Tekkaya, A review on hot stamping, *J. Mater. Process. Tech.* **210**, 2103 (2010).
2. Salzgitter Flachstahl. Werkstoffblatt 22MnB5, Vergütungsstahl, borlegiert zum Kaltumformen für eine anschließende Warmumformung (presshärtbar) in kaltgewalzter Ausführung. Katalog, (2007), http://www.salzgitterflachstahl.de/fileadmin/mediadb/szfg/informationsmaterial/produktinformationen/warmgewalzte_produkte/deu/22mnb5.pdf.
3. G. Bergweiler, Lokale Wärmebehandlung mit Laserstrahlung zur Verbesserung der Umform- und Funktionseigenschaften von hochfesten Stählen, *Rwth Aachen*, (2013).
4. M. Sigvant, M. Fermer, O. Hedegdrd, R. Johansson and P. Nystrom, Improved Ductility and Spot Weld Strength by Local Annealing of Hot-Formed Steel, in Proc. of International Conference on Hot Sheet Metal Forming of High-Performance Steel, (Luleå, Sweden, 2013).
5. M. Schaefer, M. Fermér, O. Hedegärd, R. Johansson and P. Nyström, Laser softening of press hardend steel in high volume production lines, in *Proceedings of 5th International Conference on Hot Sheet Metal Forming of High-Performance Steel*, (Toronto, Canada, 2015).
6. M. Schaefer, P. Scheible, F. Spitz and T. Harrer, High Flexibility in Partial Laser Softening of Press-Hardened Steel, in *Proceedings of 6th International Conference on Hot Sheet Metal Forming of High-Performance Steel*, (Atlanta, USA, 2017).

Product Properties

Advanced Cleanliness Control in the 2nd Generation of Press Hardened Steels Produced by ArcelorMittal

C. Philippot[1,†], M. Beauvais[1], S. Sarkar[1], S. Cobo[1], E. Lucas[1], B. Thillou[2] and P. Drillet[1]

[1]ArcelorMittal Global R&D, Maizieres-les-Metz, F-57283, France
[2] ArcelorMittal Atlantique – Lorraine Grande Synthe, F-59381, France
[†]E-mail: clement.philippot@arcelormittal.com

The global reduction of CO_2 emission generated by cars is a major challenge of the 21st century. The lightweighting of automotive structures is mandatory to achieve the ambitious environmental targets. The use of high strength steels in the car design, and especially, Press Hardened Steels (PHS) with tensile strength higher than 1000MPa, is one the most cost-efficient way to help the car makers to keep excellent crash behavior on mass saving body-in-white structures. The 2nd generation of PHS recently developed at ArcelorMittal is being deployed worldwide with Usibor®2000 and Ductibor®1000. These new grades demonstrate that CMnB metallurgies, commonly used in hot stamping, must be optimized to offer steel solutions with suitable properties for anti-intrusion and energy absorption parts. As a key metallurgical feature of these martensitic steels, a good control of inclusions and precipitations is necessary to improve the crash ductility. Among different factors influencing the bending properties of the martensitic microstructure, the present paper focuses on the effect of cleanliness of advanced steels for press hardening. The characteristics of the particles formed in the earlier stage of the steelmaking production are linked to the final In-Use Properties of the PHS. The type of particles and its geometry, such as cubic TiN, spherical complex oxides or disk-like manganese sulfides plays a different role in the crack initiation and crack propagation mechanisms. A high density of large inclusions is to be avoided as they act as local heterogeneities in the microstructure which can be detrimental to achieve a good crash performance. Particles close to the surface are even more damaging than particles localized in mid thickness. As an indirect effect, grain size refinement induced by precipitation of micro-alloying elements (Ti, Nb, V) also contributes to the achievement of a fine martensite which is essential to obtain a PHS grade with superior crash ductility. All these aspects have been integrated in the metallurgical design of Usibor®2000 and Ductibor®1000 and contribute to the remarkable and robust performances of these new PHS products.

Keywords: Inclusions; Bendability; Press hardened steels; Precipitation; Crashworthiness.

1. Introduction

The metallurgical design of the Press Hardened Steels for automotive industry has significantly evolved since the first appearance of hot stamped die quenched parts in the 80's. The growing need for weight reduction of body-in-white led to a fast implementation of this steel solution but also generated needs for higher performance materials with additional functions (corrosion resistance, improved weldability, robust processability, superior productivit). The current challenge for steelmakers is to develop new PHS grades with tensile strength higher than 1000MPa and superior crash ductility to manufacture automotive parts which combine anti-intrusion and energy absorption ability [1].

The appropriate design of martensitic microstructures commonly used for press hardened parts is then a first requirement to achieve this challenging compromise. For a given level of mechanical properties, the improvement of bendability and crash performance can be achieved through different constituents:

(1) Adjustment of the chemical composition
(2) The initial microstructure prior to austenitization [2]
(3) Austenitic grain size refinement with microalloying elements such as Ti or Nb[3] or even with the use of adapted thermal cycles
(4) Surface decarburization or formation of a gradient material [4]
(5) Presence or absence of a second phase (residual austenite, bainite) [5, 6, 7]
(6) Tempering or auto-tempering [8]

In addition to these metallurgical factors, the presence of inclusions or precipitates may be extremely detrimental to the achievement of good crash performances for a given matrix. Through factual examples, the present paper intends to demonstrate that by mastering the population of these particles, ArcelorMittal managed to develop two new AlSi coated substrates with improved bendability and crash behavior in the name of Usibor® 2000 and Ductibor® 1000.

NB: In the present publication, the notion of "ductility" refers to the crash ductility measured by the fracture strain and not to the elongation achieved by the material.

2. Formation on Ti-based Particles in Press Hardened Martensitic Steels

Ti micro-alloying addition is a necessary protection which acts as nitrogen trapper to avoid boron-nitride formation. The chemistry of particles formed by Ti precipitation during the entire elaboration of Press hardened grades has been studied at laboratory scale on nine 22MnB5 chemical compositions (three Ti contents – 150, 400 and 1000ppm for three N contents - 30, 50 and 120 ppm). Automatic Scanning Electron Microscopy was used to quantify the surface fraction and surface density of Ti (C, N) particles. As a first observation, it can be seen on Fig. 1-A that the quantity of particles is mostly driven by the nominal amount of Ti. An increase of the Ti content leads to an increase of the surface fraction of Ti (C, N) particles. As an additional contribution to the formation of Ti (C, N) particles, the increase of nitrogen also leads to an increase of total amount. However, for high Ti content, this increase of nitrogen only contributes partially to the total amount of Ti-containing particles indicating that a significant part of the precipitation occurs under another form identified as TiC. A second aspect demonstrated by the Fig. 1-B is that the density of large particles increases linearly with the amount of Ti and N. These particles larger than 2 μm have been particularly selected for being a key point of the void nucleation as observed later on. This is the consequence of larger amount of Ti and N present during the solidification process and emphasized by the TiC formation taking place on TiN particles during the remaining production (cooling of the slab, hot rolling, aluminizin).

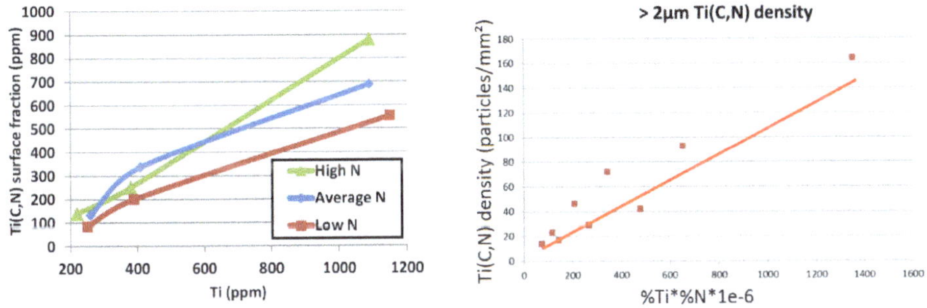

(a) (b)

Fig. 1. (a) Increase of the surface fraction of Ti (C, N) particles resulting of the increase of nominal Ti content. (b) Increase of the density of large TiN particles (> 2µm) linked to the nominal Ti and N content.

In details, the formation of Ti (C, N) or (Ti, Nb) (C, N) particles (when micro-alloyed with Nb) is far more complicated as many kinds of particles can be identified. An overview of the different particles formed is summarized in Table 1. TiN nucleation generally takes place at the solidification front due to the supersaturation occurring at austenite / liquid interface. The pr txesence of oxides may also be a preferential site for TiN formation as the energy required to nucleate on this kind of particles is low. TiN growth occurs with cubic or polygonal shape, either by Ti-N, Ti-N-C or Ti-C diffusion toward the nuclei depending on the temperature of precipitation and solubility product which is actually effective locally. This population of Ti-containing particles is extremely coherent with the characterization performed by Michelic et al. [9]. In a second stage, in presence of Nb, the growth of TiN may occur by NbC precipitation directly on those existing particles or new NbC nuclei may be formed elsewhere if the driving force for nucleation is sufficient.

Table 1. Identification of Ti-containing particles with and without Nb micro-alloying.

Ti precipitates	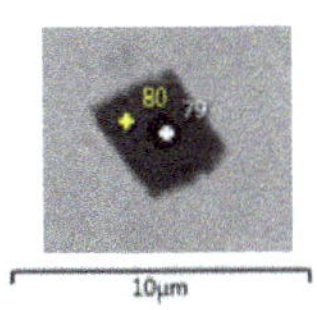 10µm TiN nucleated on oxide	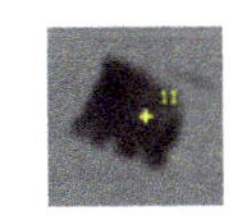 5µm	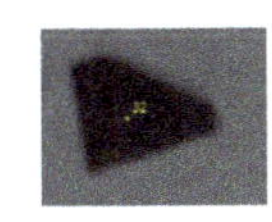 5µm TiN	 5µm (Ti,Nb)(C,N) 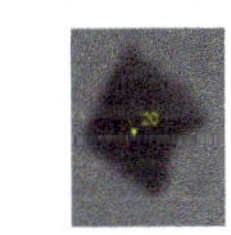 5µm (Ti,Nb)(C,N) 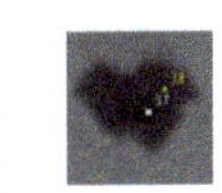 5µm (Ti,Nb)(C,N) nucleated on oxide
Nb precipitates	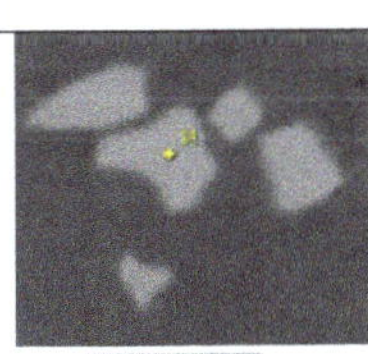 5µm Niobium carbides	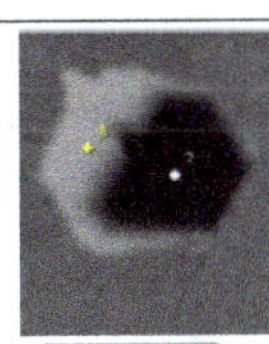 5µm Niobium carbide nucleated on oxide	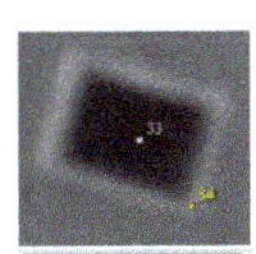 5µm Niobium carbide nucleated on (Ti,Nb)(C,N)	

3. Influence on Ti-based Particles on Bending Performance of Martensitic Steels

As demonstrated by Bardelcik *et al.* TiN particles have a significant contribution to the total void accumulation generated during mechanical testing either by decohesion or by cleavage [10]. Considering that martensite microstructures formed by hot stamping have a ductile fracture behavior with a very low amount of void formation as shown by Zhang *et al.* [11], only few particles localized in the strained area may influence the crack initiation and propagation during bending tests.

Typical examples of nucleation cases have been identified in Fig. 2. During VDA238-100 bending test and schematized in Fig. 3. For particles smaller than 2 μm (a), the main nucleation of void observed in the studied Press hardened steels is a decohesion taking place at the interface between the martensitic matrix and the TiN particle. The geometry of the particles also influences the localization of the nucleation as the angular part of TiN are even more prone to stress accumulation. For particles larger than 2 μm (b), cleavage is generally observed as a first stage of the void formation. The very low ductility of TiN particles and the stress accumulated inside particles and the surrounding matrix during bending test lead to a brittle failure. There is obviously a competition between stress accumulation in the TiN and deformation accommodation around the TiN which is size dependent. The most severe case is observed when large TiN is clustered in a same region (c). In this case, cleavage may take place inside the particles but the proximity between the particles can also lead to local decohesion. These clustered TiN are prone to initiate a high density of voids which can be detrimental to bending performance.

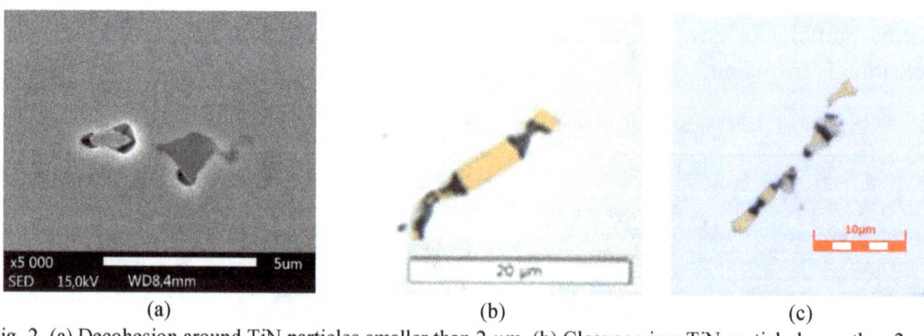

Fig. 2. (a) Decohesion around TiN particles smaller than 2 μm, (b) Cleavage in a TiN particle larger than 2 μm, (c) Mix between decohesion and cleavage in agglomerated TiN particles.

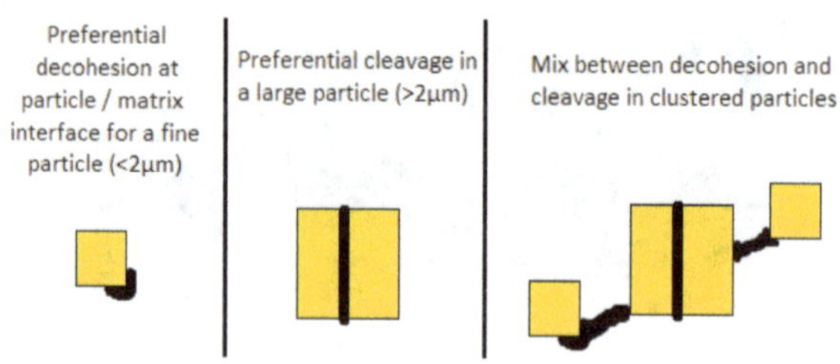

Fig. 3. Scenarios of void nucleation identified depending on the size and configuration of TiN particles.

Two bent samples with the same martensitic matrix are compared in Fig. 4 (a) and (b). Due to the high TiN content (with larger size particles and higher density), the void nucleation was denser than for the low TiN content sample. Despite the satisfactory achievement of a 60 ° bending angle after springback, the crack propagation is clearly driven by the presence of these voids. Indeed, it can be seen in Fig. 4 (c) that TiN particles are present all along the cracking path. These particles are typically larger than 2μm and have consequently undergone cleavage initiation mechanism. As a remaining proof of this cleavage, both sides of each particle can be found on the two surfaces of the crack.

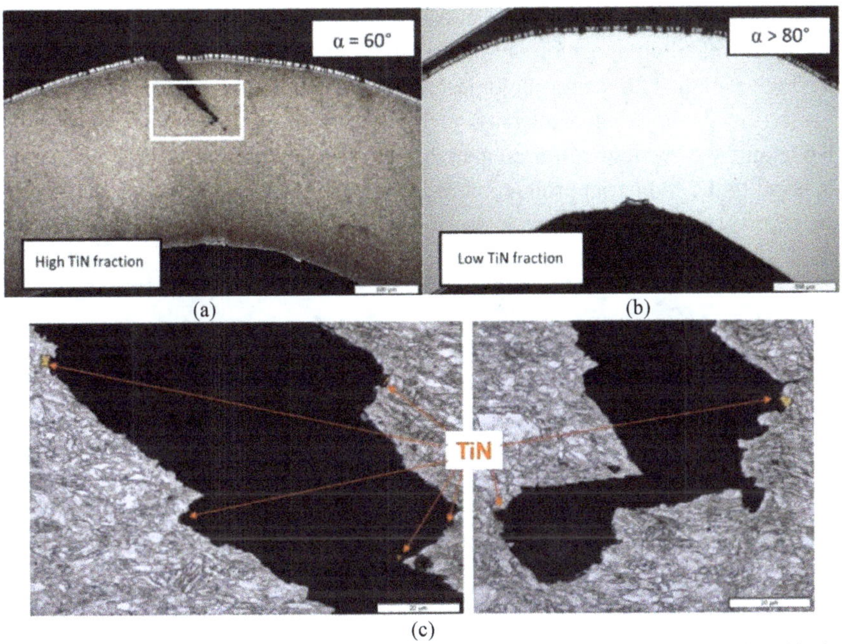

Fig. 4. (a) Crack propagation in high TiN fraction martensitic matrix, (b) superior ductility achieved in the same matrix through the production of low TiN fraction, (c) focus on the crack propagation of Fig. 4. (a).

The examples presented from the current lab studies are chosen cases to illustrate the influence of different TiN populations on final bending properties of press hardened steels. The PHS grades industrially produced by ArcelorMittal have a lower content of Ti and N content ensured by a strong quality control of these aspects. The mastering of advanced TiN cleanliness has been particularly important to achieve the outstanding bending properties of the 2nd generation of PHS with the Ductibor®1000 and Usibor®2000. A low density of fine Ti-containing particles is requested in these new advanced high strength steels to guarantee the superior crash performance and offer the high lightweighting potential of these new martensitic steels.

4. Bending Improvement in Transverse Direction Achieved with Low Sulfides Content

Sulfides and sulfur control is an attention point in the production of press hardened grades. The presence of elongated MnS particles in a martensitic phase such as 22MnB5 composition may create weak point in the microstructure especially when localized close to a surface. Decohesion with the matrix takes place creating a preferential propagation path for cracking. On Figs. 5 (a) and (b), it can be seen that the presence of MnS particles, identified before bending, can generate, during the bending test, secondary cracks perpendicular to the main propagation path. Fig. 5 (c) is a clear demonstration that bending performance can be improved in transverse direction by reducing the amount of sulfur in the nominal composition and consequently the total amount of MnS as well as their average length. Other morphologies of MnS particles obtained through different practices at the steelmaking plant may be less detrimental to bending properties. This knowledge on the control of sulfur content, already applied on the 1st generation of PHS, is transferred to the industrial production of 2nd generation of PHS to achieve a superior crash performance.

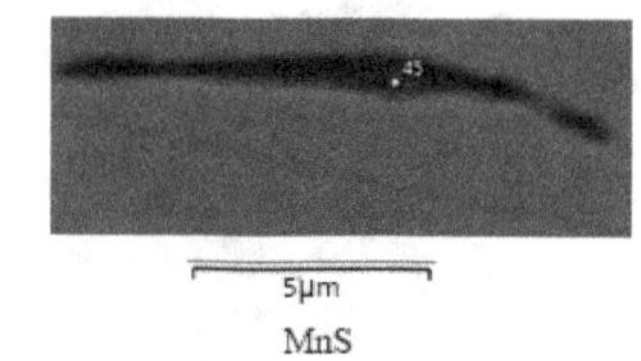

(a)

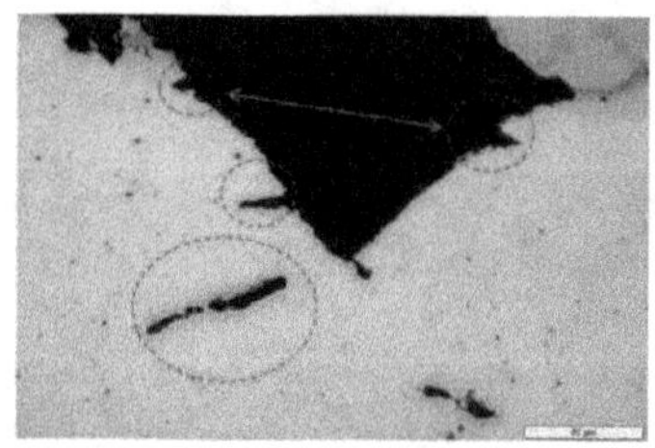

(b)

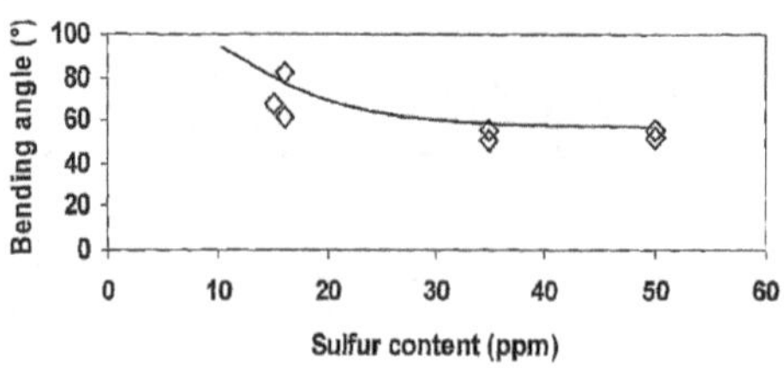

(c)

Fig. 5. (a) Elongated MnS particles in a press hardened steel, (b) example of oriented damage along a crack generated during bending test, (c) Improvement of the bending angle of 22MnB5 in transverse direction by reduction of sulfur content.

5. Case of Non-damaging Inclusions in a Ductile Martensite

The effect of large inclusions or precipitates, such as TiN or MnS, was demonstrated in the previous sections 3 and 4. However, the question of damage generated during bending is not only related to these particles but also to the surrounding martensitic matrix. Two lab materials were produced to achieve a ductile martensitic matrix like Ductibor®1000 but with very different inclusions populations. Large oxy-sulfides and complex oxides were voluntarily introduced during the elaboration process (Fig. 6 b).

The difference of inclusion size distribution can be seen in Fig. 6 (b). Particles larger than 10 μm could even be detected very close to the surface (Fig. 6 c).

Even with the presence of large oxides obtained in lab heats, when the material is processed according to the process window (see [12]), the high ductility martensitic matrix and the malleability of the particles of the present case do not generate damage during bending test. Fig. 7 shows bending angles achieved for the different steel cleanliness displayed Fig. 6 (a). The material with large particles is achieving the same level of bending performance than the typical base material without large particles (with a bending angle around 85 ° for 1.5mm thick material). This indicates the importance of the substrate design to ensure good crash performances specially to achieve high fracture strain that can be linked to good crash performance (in the present case, FS > 0.6). In particular, it shows that a ductile martensite can be produced with large particles without having consequences on bending properties. This observation may have been totally different for oxy-sulfides and complex oxides combined with another matrix with higher mechanical properties and lower intrinsic crash ductility.

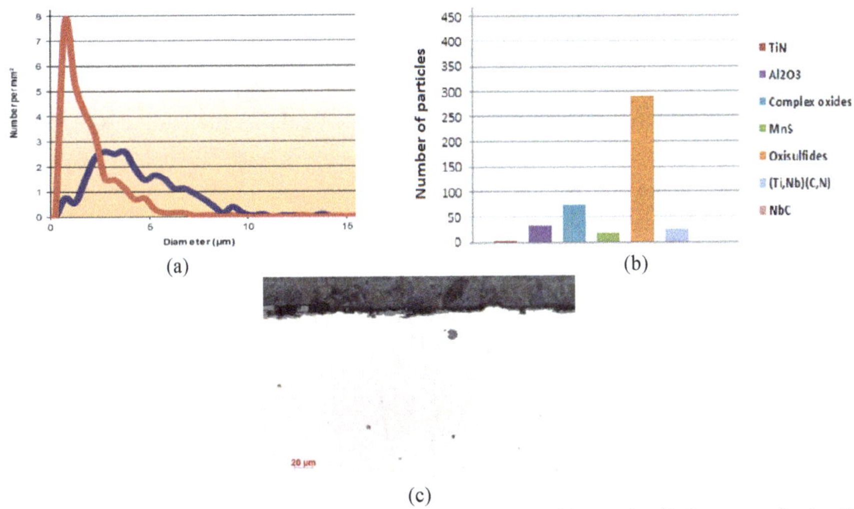

Fig. 6. (a) Modification of the inclusion population in a ductile martensitic matrix, (b) Presence of oxi-sulfides and complex oxides in the microstructure, (c) example of particle close to the surface.

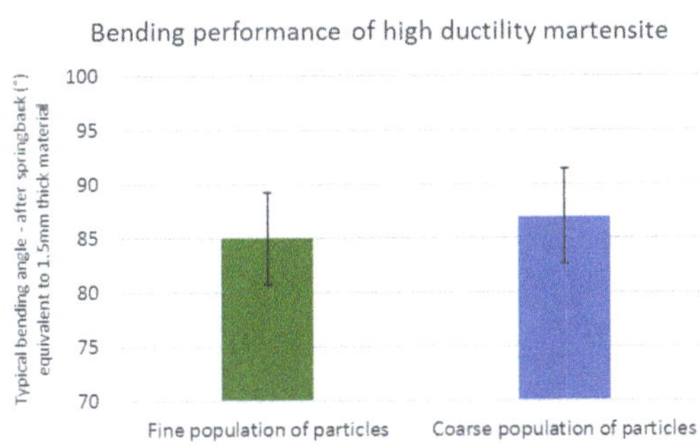

Fig. 7. Achievement of equivalent bending performance in press hardened high ductility martensitic steels despite the presence of very large inclusions (oxy-sulfides and complex oxides) in the microstructure.

6. Conclusion

The control of inclusions and precipitation in press hardened martensitic steels is of prime importance to achieve a good compromise between very high strength and good bending properties (which accounts for good crash performance). As key guidance for the design of the 2nd generation of PHS produced by ArcelorMittal, lab scale studies have demonstrated that:

(1) Ti and N contents should be strictly controlled during the entire process to avoid the formation of large Ti-containing particles which are detrimental to the bending performance of the martensitic steels when localized near the surface;

(2) Sulfur content should be very low in order to avoid the formation of large and elongated MnS particles that are weak points in the microstructure, for example, when bending in the transverse direction;

(3) As a general rule, for a given ductility of martensite, lowering the size and the density of large inclusions or precipitates leads to an improvement of the bendability and of the crash performance. Fig. 8 illustrates this tendency but also shows that for two different levels of martensite ductility, a high steel cleanliness can partially compensate for a poor microstructure performance such as martensite formed from a large austenitic grain size or with a high carbon content;

(4) The presence of large inclusions is even more critical that the hardness of martensite is high or its micro-constituents badly designed. An interesting case was found in which the presence of large inclusions is not detrimental to the bending performance of the material. The couple inclusion/microstructure should be considered in a larger extent.

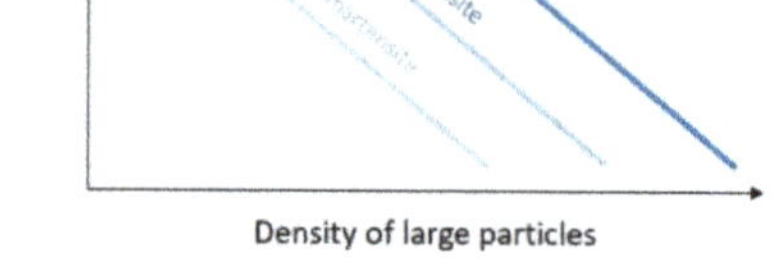

Fig. 8. Schematization of bending improvement obtained for different level of martensite ductility by lowering the density of large particles.

References

1. S. Sarkar, P. Drillet, M. Beauvais, N. Ramisetti and L. Dormegny, Ductibor® 1000 AlSi: A New PHS development for a crash ductility optimization, in *CHS2 conference proceeding* (Atlanta, USA, 2017).

2. H. Järvinen, M. Isakov, T. Nyyssönen, M. Järvenpää and P. Peura, The effect of initial microstructure on the final properties of hardened 22MnB5 steels, *Mater. Sci. Eng. A* **676**, 109 (2016).
3. J. Bian, W. Li, H. Mohrbacher, L. Hongzhou and W. Wenjun, Development of niobium alloyed press hardening steel with improved properties, *Adv. Mater. Res.* **1063**, 7 (2015).
4. J. Bian, H. Lu, W. Wang and A. Guo, Metallugical solutions to improve bending and crash performance of press hardening steels, in *CHS2 conference proceeding* (Atlanta, USA, 2017).
5. Y. Tomita, K. Okabayashi, Mechanical properties of 0.4%C-Ni-Cr-Mo high strength steel having a mixed structure of martensite and bainite, *Met. Trans.* **16**, 73 (1985).
6. H. Ohtani, S. Okagashi, Y. Fujishiro and Y. Ohmori, Morphology and properties of low carbon bainite, *Met. Trans.* **14**, 485 (1983).
7. J. Zhao, X. Zhao, C. Dong, X. Zhao and S. Kang, Effect of bainitic transformation combined with hot forming on the microstructure and mechanical properties of bainite-martensite multiphase steel, *Mater. Sci. Eng. A* **731**, 102 (2018).
8. Y. Tomita, K. Okabayashi, Heat treatment for improvement in lower temperature mechanical properties of 0.4%C-Ni-Cr-Mo ultra high strength steels, *Met. Trans.* **11**, 2387 (1983).
9. S. K. Michelic, D. Loder, T. Reip, A. A. Barani and C. Bernhard, Characterization of TiN, TiC and Ti(C, N) in Ti-alloyed ferritic chromium steels focusing on the significance of different particle morphologies, *Mater. Charact.* 61 (2015).
10. A. Bardelcik, C. J. Vowles and M. J. Worswick, A mechanical, microstructural, and damage study of various tailor hot stamped material conditions consisting of martensite, bainite, ferrite, and pearlite, *Metal. Mat. Trans.* **49**, 1102 (2018).
11. S. Zhang, Y. Huang, B. Sun, Q. Liao, H. Lu, B. Jian, H. Mohrbacher, A. Guo and Y. Zhang, Effect of Nb on hydrogen-induced delayed fracture in high strength ho stamping steels, *Mater. Sci. Eng. A* **626**, 136 (2015).
12. P. Drillet, Overview on coatings developed on press hardened steels for automotive applications, in *International Conference Hot Stamping UHSS (ICHSU)*, (Hefei, China, 2018).

The Effects of Strength for Hot Stamping Steel on Lightweight of Parts and Related Performance

M. T. Ma[1,†], X. J. Jin[2], Y. H. Guo[1] and Y. Zhao[1]

[1]*China Automotive Engineering Research Institute Co., Ltd., Chongqing, 401122, China*
[2]*Shanghai Jiaotong university, Shanghai, 200240, China*
†*E-mail: mingtuma@126.com*

In this paper, the change of thickness, stiffness, energy absorption of bending and buckling resistance of three point bending components for hot forming hardened steel with the different intensity (1500 MPa and 1800 MPa and 2000 MPa) are calculated by thumb rule and the related empirical equation. The impact energy absorption of parts during the trolley crash, intrusion speed and value, energy absorption during side impact of vehicle crash are simulated by computer. Results show that the tensile strength are raised from 1500 MPa to 1800 MPa, 2000 MPa respectively, the thickness of parts are decreased from 1.4 mm to 1.28 mm, 1.21 mm respectively. The rate of thinning is 8.5 %, 13.5 % respectively. The stiffness is decreased 23.6 % and 35.1 % respectively. The reduction of resistance of local instability during bending is similar to stiffness. The energy absorption during bending basically remains unchanged. The impact energy absorption of B pillar indicates is significantly decreased during trolley impact. The intrusion speed and value are increased during side impact of vehicle crash. These results show that when the ultra high strength steels are used in safety parts of automotive, it is necessary to comprehensively consider increasing strength, decreasing thickness, technical property, and performance over cost of the parts, we can pursue only high strength and lightweight effect.

Keywords: Hot stamping steel; B pillar; Stiffness; Thickness; Energy absorption of bending.

1. Introduction

Automotive lightweight is the important method and way which can solve fuel consumption and emissions produced from the automotive industry development [1]. The high strength steel and ultra-high strength steel have been widely applied due to increase lightweight and safety. Especially the application of hot stamping steel is the rapid developing in recent years [2]. Hot stamping steel has successively developed from 1500MPa to 1800 MPa and 2000 MPa which is a new ultra high strength steel. It makes the hot stamping parts further lightweight. The increasing strength of the ultra high strength martensite steel for hot stamping will make steel increase carbon and alloy elements, which affect the cost of car component. The increasing strength of material can make the material reduce thickness and lightweight of parts further increase, but this will influence on the technological property of the hot stamping process and the subsequent welding property of the parts. As automotive components, in addition to the strength requirement, stiffness is also required. The secure parts have more requirements such as impact absorbing energy and related connection technology, etc.. So people think that the strength increasing and thickness reduction should have a limit. The purpose in the paper is to use a series of empirical equation to calculate and analyze the relationship between material strength and the thickness reduction of parts, the relationship between stiffness and local buckling resistance in impact energy absorption of parts, and so on. Another purpose is to use computer simulation to evaluate relationship among the strength, thickness, lightweight, comprehensive performance and using function of parts. Based on the results, the suggestion to comprehensively consider the reasonable match of strength and lightweight of parts will be put out.

2. The Effect of Increasing Strength on the Lightweight and Related Properties

The parts with groove structure under three point bending stress, such as B pillar, its lightweight effect and related performance are calculated.

2.1. *Strength and lightweight effect*

Three strength of material (1500 MPa and 1800 MPa and 2000 MPa) are compared. Assuming the original thickness of material with1500 MPa is 1.4 mm, the lightweight thickness reduction is calculated using the thumb rule [3], the equation is used

$$1-\frac{t_2}{t_1}=1-\left(\frac{\sigma_1}{\sigma_2}\right)^{1/2} \tag{1}$$

t_1, t_2 is material thickness, σ_1, σ_2 is the using flow stress or tension stress of material corresponding t_1, t_2. According to the stress pattern and stress state of parts, yield strength, flow stress with certain deformation and tensile strength can be used. For convenient comparison, we will use tension strength. The results are shown in Table 1. It can be seen that strength increasing can effectively reduce thickness. The lightweight is realized, but the increased rate of strength is much greater than the reduction rate of thickness.

Table 1. The relationship between increasing strength and thickness reduction.

Strength of material/MPa	Strength increase rate/%	Thickness/mm	Reduction ratio /%
1500	——	1.40	——
1800	20.0	1.28	8.5
2000	33.3	1.21	13.5

2.2. *The effect of increasing strength and thickness reduction on the stiffness*

The relationship among the bending stress, thickness of plate, the modulus of elasticity which under bending stress is Eq. 2 [4]:

$$S_n^b=\frac{E_n}{E_0}\bullet\frac{t_n}{t_0}\bullet S_0^b \tag{2}$$

E_n, E_0 are modulus of elasticity, in the following calculation, the assumption is that $E_n = E_0$; t_0, t_n are thickness of different plate; S_n^b, S_0^b are the corresponding bending stiffness value.

In the calculation, the change of the stiffness is mainly decided by the thickness. The actual stiffness of the part is not only affected by the thickness of the plate, but also by the shape of the component. For the reduction in stiffness due to thickness reduction can be compensated by optimizing design of the shape for the parts. In the literature [5], it is proposed that the parts and material with rectangular section during three-point bending stress, its stiffness is calculated by Eq. 3.

$$S = \frac{E \bullet b \bullet t^3}{4l^3} \tag{3}$$

S is stiffness, *E* is modulus of elasticity, *b* is width, *l* is pivot distance at three-point bending, *t* is thickness.

The rate of changing in stiffness after the changing in thickness calculated by Eq. 2,3 is listed in Table 2.

Table 2. The thickness of the material and the rate of change in stiffness.

Strength of material /MPa	Thickness /mm	Rate of change of stiffness /% (Eq. 2)	Rate of change of stiffness /% (Eq. 3)
1500	1.40	—	—
1800	1.28	-8.6	-23.6
2000	1.21	-13.4	-35.1

Literature [4] has proposed calculation equation which is the relationship between the local buckling resistance and thickness, seen Eq. 4:

$$\frac{R_n}{R_0} = \frac{E_n}{E_0} \bullet \frac{1-\mu^2}{1-\mu_n^2} \bullet \left(\frac{t_n}{t_0}\right)^3 \tag{4}$$

R_0, R_n is local instability resistance or local buckling resistance; μ, μ_n is poisson ratio;

The change of local instability resistance calculated by Eq. 4 is consistent with the change of stiffness calculated by Eq. 3, as shown in Table 2.

As an automotive safety structure that is subjected to dynamic loads, crushing energy is an important parameter. It is proportional to the square of the material thickness and the 2n power of the tensile strength. (n=1/2~2/7), as shown in Eq. 5.

$$A_E = kt^2\sigma_b^{2n} \tag{5}$$

A_E is crushing energy absorption, J; *k* is proportional constant, m; *t* is thickness, mm; σ_b is tensile strength, MPa.

The changes of energy absorption which bent at three points caused by changes in strength and thickness of parts are shown in Table 3. In order to simplify the calculation, the n in Eq. 5 take 1/2.The data in the table indicate that to use the Eq. 5 and the thickness thinned value due to increasing strength of plate from 1500 MPa to 1800 MPa and 2000MPa calculated by the thumb rule, calculating the impact energy absorption value is basically unchanged under the comprehensive action of strength increasing and thickness reduction. i.e. It is about the same as the impact energy absorption value under the condition 1500 MPa tension strength and 1.4 mm thick. That is, the reasonable matching of the increasing strength and thinned thickness can keep the original energy absorption value unchanged.

Table 3. Energy absorption value of components with different strength and thickness.

Strength/MPa	Thickness/mm	Energy absorption /J	Rate of change of energy absorption/%
1500	1.40	2940.0k	——
1800	1.28	2949.0k	0.30
2000	1.21	2937.9k	-0.07

3. Computer Simulation

In order to further prove the change results of performance related to the strength improvement and thickness reduction obtained by the above empirical equation, a computer simulation of the B-pillar as a typical component was carried out.

3.1. *Three-point bending simulation with full constraint*

The three-point bending test with full constraint is adopted. The simulated model is shown in Fig. 1. In order to simplify the simulation calculation, the setting value of thickness is used. The simulation results of relevant performance are shown in Table 4. It can be seen that the thickness reduction will significantly increase the deformation of parts and reduce the stiffness of parts. The change trend is similar to the calculation results of Eq. 4.

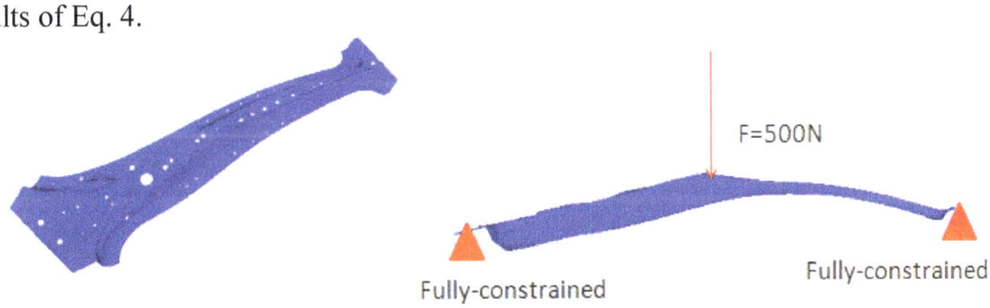

Fig. 1. Computer simulation model of B-pillar under three-point bending.

Table 4. Result of computer simulation for B-pillar under three-point bending.

Thickness/mm	Load/N	The biggest deformation/mm	Stiffness/N/mm	Rate of change of stiffness/%
1.4	500	0.2949	1695.5	——
1.2	500	0.3691	1354.6	-20.1
1.0	500	0.4825	1036.3	-38.9

3.2. *Trolley impact*

According to the deformation mode of B pillar in the side impact of vehicle, the design principle of the fixed tooling during side impact of the B-pillar part is as follows: the upper and lower ends of the B-pillar keep undeformed, and the intrusion deformation of the central region to the inside of the vehicle body is not affected during the impact.

The model of the B-pillar fixed and trolley impact is shown in Fig. 2. The model of trolley impact has 950 kg weight which is similar to the model used for side impact of the vehicle crash. Based on this model, impact simulation analysis is carried out.

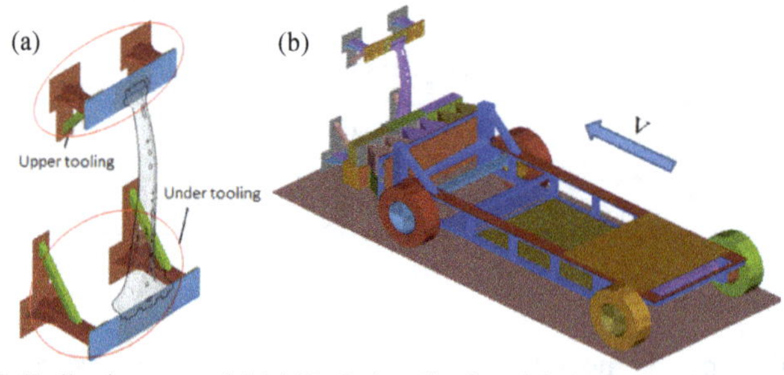

Fig. 2. Trolley impact model (a) Workpiece fixed model; (b) Trolley impact model.

During the side impact of vehicle crash, the maximum deformation of the B-pillar subjected to the impact area is between 80 mm and 100 mm. According to the simulation results of the deformation degree of B-pillar under different deformation speeds, the impact speed of the mobile trolley is determined to be 40 km/h. Using Ls-dyna software, the calculation results of energy absorption simulated by computer for B pillar during trolley impact are listed in Table 5. The data show that the increasing energy absorption value due to increasing strength cannot compensate for the reducing effect of the energy absorption value due to thickness reduction. In addition, the higher the strength is, the higher the energy absorption of elastic deformation when the part is deformed. The elastic recovery can occur after removing in the impact force. Therefore, the higher the strength is, the final energy absorption (plastic deformation energy) of the part after the impact may also decrease. This energy absorption value is significantly different from the conquassation energy of the part. The conquassation energy is the total energy absorbed by the quasi-static loading speed until unstable or broken of parts. In this time, it is not necessary to consider the influence of the strain rate on the hardening of the material.

Table 5. Calculation results of energy absorption of B-pillar by computer simulation.

Thickness/mm	Strength/MPa	Energy absorption of B pillar/KJ
1.4	1500	2.58
1.2	1800	1.77
1.0	2000	1.69

The impact absorbed energy during the trolley impact should include impact energy, strain rate energy, but it is not need to consider absorbed energy due to instable or fracture of the part, so the related results will be different from that of the quasi-static conquassation.

3.3. *Simulation of side impact for vehicle*

The simulation conditions and the side impact model of vehicle is different from trolley impact. At this time, the B-pillar and the body-related parts are connected together, and the parts connected to the B-pillar will be deformed when the side impact happen. The partial deformation energy of B-pillar can be absorbed or transferred by other connected parts. The speed of side impact for vehicle is 40 km/h. The weight of trolley impact is 950 kg. The Hyperwork software was used for pre- processing and post-processing during impact, and the Ls-dyna software was used for calculation. The intrusion speed and value of different positions of the B-pillars with different strengths and thicknesses were listed in Table 6, Table 7 respectively. The position on the B-pillar corresponding to each part of passenger is shown in Fig. 3.

Table 6. Intrusion speed of different positions for B-pillar during simulation of side impact.

Thickness/mm	Intrusion speed/m/s			
	Pelvis	Rib	Head	Abdomen
1.4 (1500 MPa)	5.42	5.80	3.91	5.65
1.2 (1800 MPa)	5.60	5.83	3.77	5.77
1.0 (2000 MPa)	6.12	6.26	3.61	6.32

Table 7. Intrusion value of different positions for B-pillar during simulation of side impact.

Thickness /mm	Intrusion value/mm			
	Pelvis	Rib	Head	Abdomen
1.4 (1500 MPa)	82.8	93.4	51.0	89.2
1.2 (1800 MPa)	89.3	98.9	52.4	95.4
1.0 (2000 MPa)	100.3	104.2	53.2	103.4

The data in Table 6 shows that the strength is increased, the thickness is thinned, and the speed of intrusion into the B-pillar during side impact is increased, except for a slight decrease in the head. The intrusion value increases with the increasing strength and thickness reduction. So the increasing strength can not compensate reducing safety indicator due to reduction of thickness during the impact. In general, the strength increase will help improve the safety indicators. However, the tensile strength of material is up to 2000 MPa, it can be lightweight for general safety parts such as B-pillars and maintain original safety for the material with strength 1500 MPa, thickness 1.4 mm based on the current simulation results. We need careful consideration whether the strength is further improved. It is necessary to know that further enhancement of strength and further reduction of plate thickness will affect process performance, weldability, alloy price, resisting sensitivity of hydrogen embrittlement of parts, etc. The stronger the B-pillar par is, the smaller the amount of deformation during the impact of the whole vehicle, the smaller the intrusion amount and the intrusion speed of the corresponding B-pillar, and the better the side impact performance of the whole vehicle. Therefore, in order to obtain a reasonable choice, the further improvement of the strength of ultra-high-strength steel and the application of lightweight components requires a various experimental work, comprehensive balance of advantages and disadvantages, and LCA evaluation.

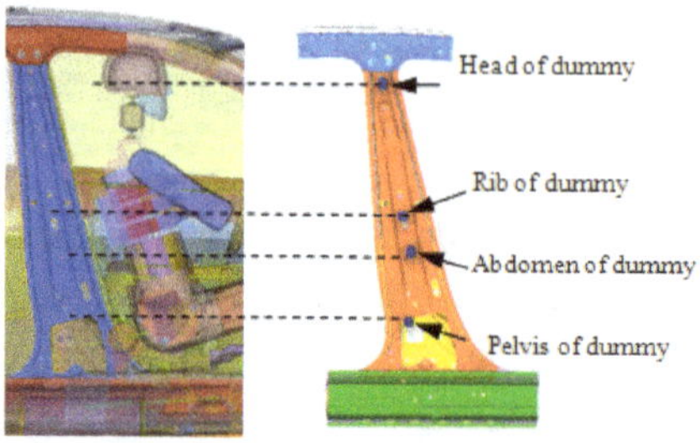

Fig. 3. The position in B pillar corresponding to some location in the dummy body.

4. Summary

The empirical equation is used to calculate the thickness reduction, stiffness, conquassation energy of the grooved component (different material strength) when subjecting to three-point bending stress, etc.. The energy absorption of B-pillar during the trolley impact and the intrusion amount and intrusion speed of different positions of the B-pillar during the side impact of vehicle were simulated by computer. The calculation results and simulation results were analyzed. The conclusions as follows:

(1) The strength increasing can make the thickness reduce. When the strength is greater than 1500 MPa, the strength increase rate is much larger than the thickness reduction rate.

(2) The conquassation energy absorption value under three point bending of material at the tensile strength of 2000 MPa basically remain unchanged during the strength increasing and thickness reduction, but the bending stiffness of groove component are significantly decreased.

(3) The simulation of the trolley impact of the B-pillar shows that the impact energy absorption value decreases significantly under the dynamic load state. That is, the increasing strength is difficult to balance the adverse effect of the thickness reduction on the impact energy absorption value.

(4) The results of side impact simulation of vehicle show that the intrusion speed and intrusion amount of various position in the B-pillar impact increase with the thickness reduction. That is, the strength is increased from 1500 MPa to 2000 MPa, the thickness is reduced from 1.4 to 1.0 mm, and it has a negative influence on the safety of side impact.

(5) In order to realize lightweight by further increasing strength and thickness reduction for the ultra high strength hot stamping component, it is necessary to comprehensively consider the influence of thickness reduction on the stiffness, energy absorption, the intrusion speed and the intrusion amount of side impact of vehicle as well as the comprehensive effect of strength improvement on the technological property, failure mode and cost performance of raw materials and component, etc.. It is not necessary to seek blindly for high indicators of strength and light weight.

Acknowledgments

Work supported by grant U1564203 of the National Science Foundation.

References

1. M. T. Ma, H. L. Yi and H. Z. Lu, On the automobile lightweight, *Eng. Sci.* **11**, 02 (2009).
2. M. T. Ma and Y. S. Zhang, *Research progress on hot stamping of ultra high strength steel. Trace research on advanced automobile manufacturing technology*, (Beijing institute of technology press, 2016).
3. M. T. Ma, *Advanced automotive steel*, (Chemical Industry Press, 2008).
4. A. S. Kasper, W. E. Swenson, S. Dinda and F. L. Cheng, Kinetic Modulus of Steel: a new automotive design parameter, *SAE paper* **88**, 1 (1979).
5. BOSCH, *Automotive handbook*, (Beijing institute of technology press, 1999).

Investigation on Mechanical Properties and Hydrogen Behavior Distribution of Nb and V Bearing 22MnB5 B-pillar

W. Ding[1,2,3], Y. Gong[1,2,3,a,†], W. Li[1,2,3], B. Zhu[4], Y. Zhao[5], X. J. Jin[1,2,3,b], Y. S. Zhang[4] and M. T. Ma[5]

[1]Shanghai Key Laboratory of Material Laser Processing and Modification, School of Materials Science and Engineering, Shanghai Jiaotong University, Shanghai 200240, China
[2]Institute of Advanced Steels and Materials, School of Materials Science and Engineering, Shanghai Jiao Tong University, Shanghai 200240, China
[3]Shanghai Innovation Institute for Materials, Shanghai, 200444, China
[4]State Key Laboratory of Materials Processing and Die & Mould Technology, Huazhong University of Science and Technology, Wuhan 430074, China
[5]Automotive lightweight Engineering Research Center, Chinese Automotive Engineering Research Institute Co., Ltd, Chongqing, 401122, China
E-mail: [a,†]yugong@sjtu.edu.cn, [b]jin@sjtu.edu.cn

It is assumed that the hot press deformation which will lead to various stress and strain states, as well as the various contact conditions at different parts of the hot stamping components can influence the subsequent microstructural evolution and the final mechanical properties. In the present project, the distribution of mechanical properties and hydrogen behavior of the industrially produced Nb and V bearing 22MnB5 B pillar was investigated. The additions of Nb and V to the traditional 22MnB5 steel were designed with the purpose of refining grain size, further strengthening the steel, as well as improving the resistance to hydrogen embrittlement. Our experimental results show that without hydrogen charging no obvious differences in the microstructure and mechanical properties can be found for the samples selected from 3 different typical positions of the B pillar. However, significant differences can be observed for these samples after hydrogen charging by the slow strain rate tensile tests and hydrogen permeation experiments. In the presence of hydrogen in the B pillar, sufficient attention should be paid to the inhomogeneity of the property distribution in order to avoid the potential danger of failure.

Keywords: Nb/V; Hot stamped B pillar; Hydrogen embrittlement; Hydrogen trapping.

1. Introduction

Advanced high-strength steels are widely used in the automobile industry in order to realize the lightweight of automobile. Hot stamping process is a main way to form ultra-high strength steel sheets into automobile parts, e.g. B pillar which is a typical hot stamping part [1-3]. Hot forming is a thermo-mechanical process so the final mechanical properties of the hot stamping parts are affected by both temperature and stress fields. According to the results of numerical simulation by Junying Min et al. [4], different positions have different strain fields during cold forming. Hyun-Ho Bok et al. [5] studied the temperature change during hot forming at different positions of B pillar by means of FEM simulation and found that the cooling rates at different positions were different due to the contact condition between the steel sheet and die. So for the newly designed niobium and vanadium alloyed steel which was designed based on the traditional 22MnB5 steel, it is important to study the distribution of the mechanical properties of the industrial produced B pillar. Hydrogen embrittlement of automobile parts has been paid more and more attention by researchers because it brings a delayed fracture. K. Mori et al. [3] indicated that hydrogen embrittlement is not investigated in detail and advised that more studies should be performed.

In this study, the microstructure and mechanical properties of the specimens from 4 typical positions of the industrial produced B pillar were compared to study the distribution of the mechanical properties. In addition, hydrogen embrittlement susceptibility and hydrogen permeation behavior were studied, which can evaluate the uniformity of microstructure from another aspect because hydrogen atoms are very sensitive to defects and stress states in the specimens.

2. Experimental Procedure

The niobium and vanadium alloyed B pillar used in the present study was supplied by Masteel. The composition, which is designed based on the traditional 22MnB5 steel, is shown in Table 1. According to Junying Min et al. (for strain distribution) and Hyun-Ho Bok et al. (for temperature distribution), four typical positions were selected, as shown in Fig. 1, and the samples from these four typical positions are named as 1#, 2#, 3# and 4# respectively.

Table 1. Chemical composition of the Nb & V alloyed B pillar (wt%).

C	Si	Mn	Cr	Nb	V	N	Ti	Al	B	P
0.24	0.23	1.24	0.16	0.04	0.042	0.0052	0.038	0.04	0.0028	0.013

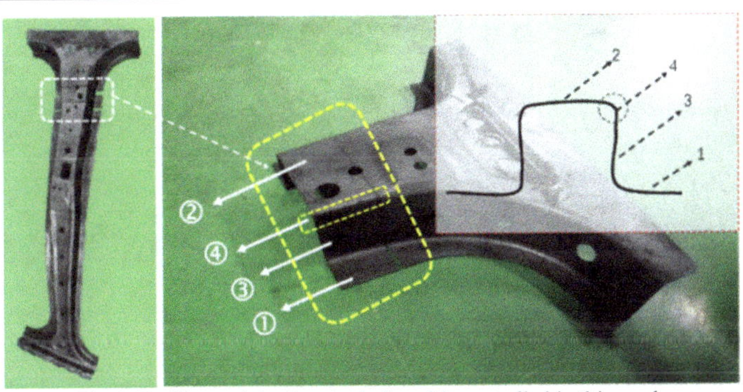

Fig. 1. B pillar and the four typical positions studied in this work.

The microstructure was characterized using scanning electron microscopy (SEM). Transmission electron microscopy (TEM) equipped with energy dispersive spectrometer (EDS) was used to study the size distribution of the precipitates. SEM samples were etched with 4 % nital solution. Carbon replica extraction samples were used for TEM observation, which was conducted in a JEOL JEM 2100F field emission transmission electron microscopy operated at 200 kV.

Tensile tests were carried out on samples with and without hydrogen charging. The hydrogen charging solution is composed of 30 g·L^{-1} NaCl and 3 g·L^{-1} thiocarbamide. Plate-type tensile samples with gage section 16 mm×5 mm×1.5 mm were used for tensile tests, which were conducted on a Zwick (BTC-T1-FR020 TN.ASO) universal testing machine at room temperature with a constant strain rate of 10^{-5} s^{-1}. Vickers hardness was measured by a hardness tester (1600–6406, Zwick/Roell, Germany) at a loading of 500 g and 1000 g.

The electrochemical hydrogen permeation experiment was operated using the modified Devanathan-Stachurski (D-S) cells [6-7], which consist of a cathodic cell and an anodic cell. The aqueous solution in the cathodic cell is 0.2 mol·L^{-1} NaOH + 3 g·L^{-1} NH_4SCN and in the anodic cell is 0.2 mol·L^{-1} NaOH solution.

3. Results and Discussion

3.1. *Microstructure*

3.1.1. *The microstructure of the specimens from different locations of the B pillar*

The microstructures of 1#-4# samples were observed by SEM in the center area of the observation plane which is far away from the surface region, as shown in Fig. 2. All of these four samples were found to have full martensite microstructure, and no obvious difference was found from the SEM microstructures of these four samples. At the surface region of the observation plane, the decarburization layer with thickness of around 20 μm was observed for all the samples, where the microstructure is different from that in the central region, as shown in Fig. 3.

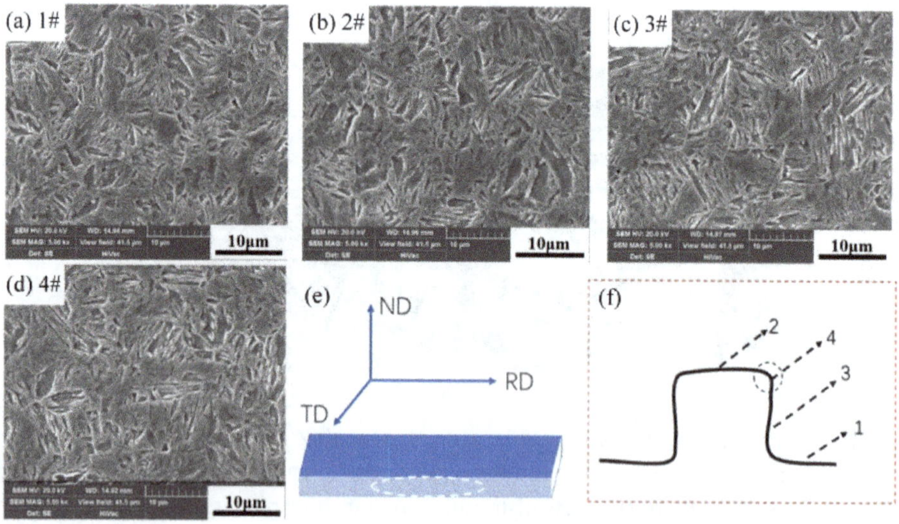

Fig. 2. SEM microstructures of the specimens (1#-4#) from 4 typical positions of the B pillar (the last two images (e-f) show the observation plane which is perpendicular to the transverse direction (TD) and the locations in the B pillar of these four specimens). ND=normal direction; RD=rolling direction.

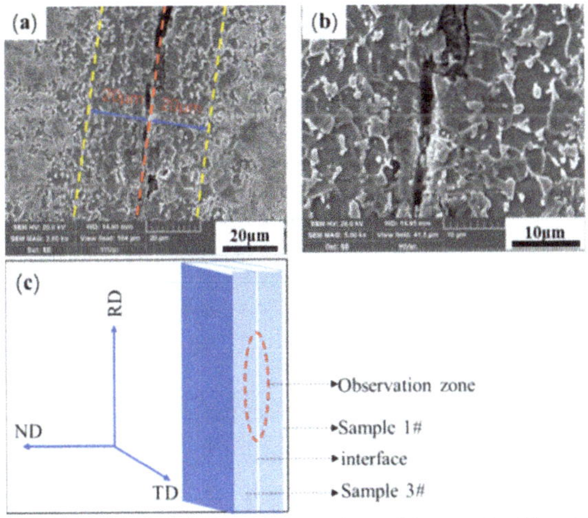

Fig. 3. The decarburization layers observed at the surface regions of the samples 1# and 3# which were mounted next to each other. (a) At magnification of 2000. (b) At magnification of 5000. (c) Schematic diagram of the observation zone marked by the red dashed circle.

3.1.2. *Morphology and distribution of niobium-vanadium carbides*

Figure 4 and Figure 5 show the niobium-vanadium precipitates observed in the 2# sample by TEM. The morphology of these precipitates can be seen clearly (as shown in Fig. 4), which is spherical, ellipsoidal or cubic in shape. The composition of the precipitate in the yellow circle in Fig. 5(a) was estimated by means of EDS analysis, which indicates that the precipitate is [(Nb_x, V_y, Ti_z) (C_i, N_j)], where x, y, z, i and j are the atomic ratios of these five elements. However, the accurate values of x, y, z, i and j can not be determined because EDS is just a kind of semi-quantitative analysis method, moreover, for the carbon replica extraction sample, the weight percent of C can be affected by the carbon film itself.

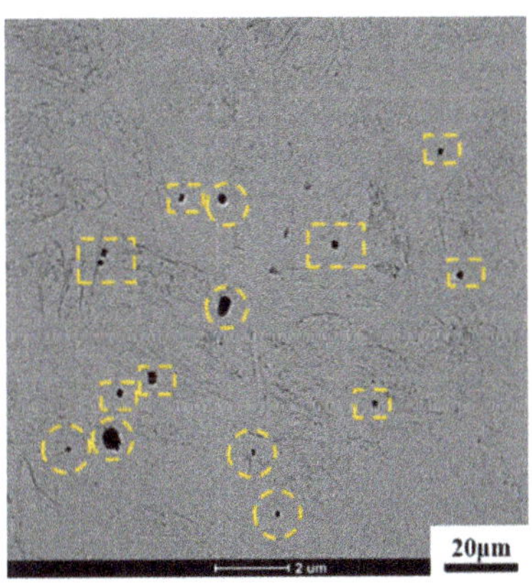

Fig. 4. TEM observation of the niobium-vanadium carbides in the carbon replica extraction sample 2#.

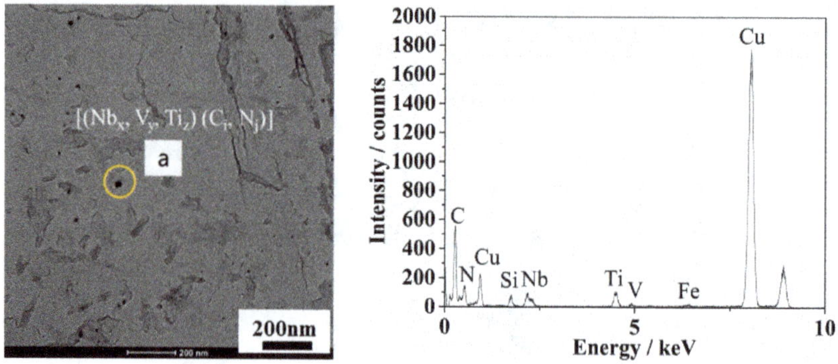

(a) (b)

Fig. 5. Morphology and composition of niobium-vanadium precipitates in 2# sample. (a) TEM images of niobium-vanadium precipitates. (b) EDS spectrum of the precipitate in the yellow circle.

From the above analysis in Fig. 4 and Fig. 5, niobium-vanadium carbides were observed in the B pillar. It is expected that precipitates in nano size can not only inhibit the grain coarsening but also play an important role in the precipitation strengthening [8-11]. The size distribution and morphology of the precipitates have great effect on the final strength of the steel.

The size distribution in 2# sample was studied by means of TEM. Large amount of precipitates in different zones of the sample were observed to improve the scientificity and credibility of the statistical data. Around 300 precipitates were studied and the size distribution is shown in Table 2 and Fig. 6. One can find that around 73% of the precipitates have a size less than 20 nm and 80 % of those precipitates have a size less than 50 nm. So, excellent precipitation strengthening effect can be expected. One can also find that there is a size distribution gap between 50 nm and 100 nm, which indicates that there exist two kinds of precipitates, the smaller type and the larger type in size, which may be precipitated at different heat treatment stages.

Table 2. Size distribution of niobium-vanadium precipitates in 2# sample.

Size interval/nm	0-10	10-20	20-30	30-40	40-50	100-200	200-300	300-400	400-500	500-600
Number	90	117	13	5	2	34	14	5	3	2
Number fraction	0.32	0.41	0.05	0.02	0.01	0.12	0.05	0.02	0.01	0.01
Average size/nm	7.34	14.26	23.4	32.01	45.46	157.08	238.15	353.48	445.89	564.57

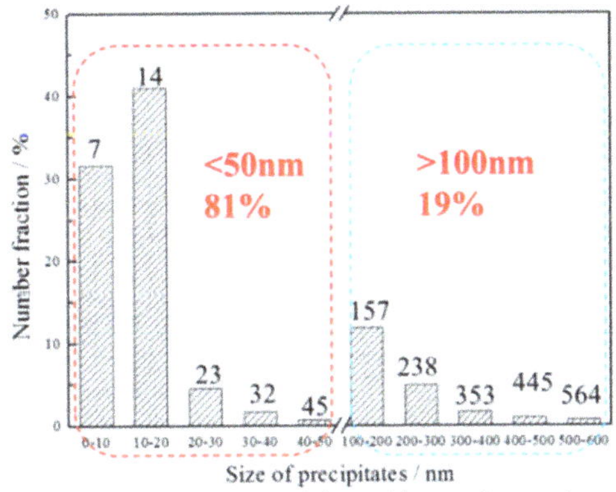

Fig. 6. Size distribution of the niobium-vanadium precipitates (the number on the top of each bar represents the average precipitate size of each interval).

3.2. *Mechanical properties*

The tensile test results were shown in Fig. 7 and Table 3. For samples without hydrogen charging, the ultimate tensile strengths and total elongations of 1#-3# samples are nearly the same, and the maximum differences of ultimate tensile strength and total elongation among the samples 1#-3# are 43 MPa and 2.1% respectively. Overall, the strength and elongation of the 1# sample were a little higher, and those of the 3# sample were a little lower.

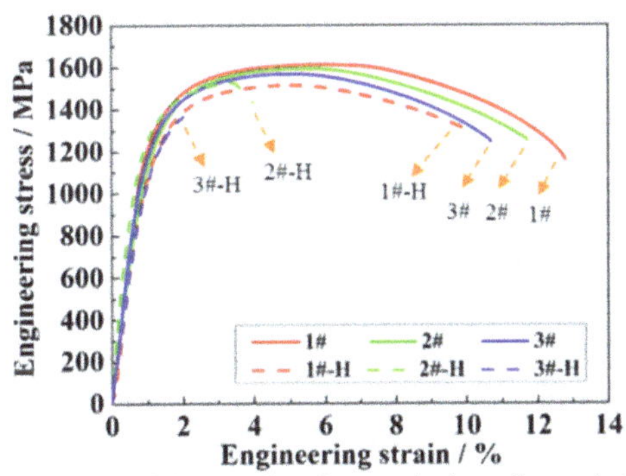

Fig. 7. Engineering stress-strain curves of the samples after standard tensile test (solid lines) and slow-strain rate pre-hydrogen charging tensile test (dashed lines). "H" after the number means that the sample was charged with hydrogen before tensile test.

However, for samples with hydrogen charging, there exist big differences in the mechanical properties for 1#-3# samples especially for the total elongation. The total elongation loss percentage calculated by Eq. 1 was used to evaluate quantitatively the hydrogen embrittlement susceptibility of the samples [12]. The calculated elongation loss percentages of all the investigated samples were presented in Table 4. One can find that the total elongation loss percentage of the sample in the 1# position is the smallest (21.1 %) and the total elongation loss percentage of the sample in the 3# position is the largest (80.4 %), which suggests that the sample in the 1# position has better hydrogen embrittlement resistance.

$$El_{loss}\,(\%) = (El_{uncharged} - El_{charged}) / El_{uncharged} \times 100 \quad (1)$$

where $El_{uncharged}$ is the total elongation of the sample before hydrogen charging, and $El_{charged}$ is the total elongation after hydrogen charging.

Table 3. Mechanical properties of the investigated samples with and without hydrogen charging.

	Yield strength (MPa)	Ultimate tensile strength (MPa)	Total elongation (%)
1#	1205	1610	12.8
2#	1200	1593	11.7
3#	1202	1567	10.7
1#-H	1058	1513	10.1
2#-H	972	1526	4.0
3#-H	987	1368	2.1

Table 4. Total elongation loss percentage of 1#-3# samples after hydrogen charging.

	$El_{\text{uncharged}}$ (%)	El_{charged} (%)	El_{loss} (%)
1#	12.8	10.1	21.1
2#	11.7	4.0	65.8
3#	10.7	2.1	80.4

The hardness of the B pillar at different positions was also measured by Vickers hardness test. The hardness values at the surfaces of the B pillar and at the inner regions were tested respectively, as shown in Table 5 and Table 6. It can be found that the hardness values at the surfaces of the 3 different positions are nearly the same, around 300 HV (500 g). Moreover, the hardness values at the inner regions of the B pillar at different positions are also very close to each other, which are around 200 HV higher than those at the surface regions. The lower surface hardness is due to decarburization layer on the surface, as can be seen in Fig. 3.

Table 5. Vickers hardness values tested on the normal planes of the investigated samples at a load of 500 g.

Hardness / HV (500 g)	1	2	3	4	5	6	Average	Standard deviation
1#	290.6	302.5	272.4	301.3	304.1	272.8	290.6	13.5
2#	301.9	294.7	285.1	284.9	284.1	311.8	293.8	9.7
3#	309	297.3	319.2	316.3	309.3	298.5	308.3	8.2

Table 6. Vickers hardness values measured from the plane perpendicular to the transverse direction at a load of 1000 g.

Hardness / HV (1 kg)	1	2	3	4	5	6	Average	Standard deviation
1#	501.5	482.9	520.5	512.5	502.9	512.1	505.4	11.8
2#	492.8	502.2	507.5	509.7	495.4	510.0	502.9	5.1
3#	498.0	484.4	520.0	510.5	516.9	509.2	506.5	11.5
4#	508.7	518.6	510.3	511.4	502.4	507.9	509.9	4.8

3.3. *Electrochemical hydrogen permeation test*

The electrochemical hydrogen permeation test device is shown in Fig. 8. The cell on the right is cathodic cell, where hydrogen ions get electrons to be reduced to hydrogen atoms and these hydrogen atoms will diffuse through the sample into the anodic cell on the left, in other words this cell provides hydrogen atoms for the permeation test so that the cell is also named hydrogen-charging cell. When these hydrogen atoms arrive at the anodic cell, they will be oxidized again to hydrogen ions and can be detected, thus the anodic cell is also named hydrogen-detecting cell. Then the permeation current curve (as shown in Fig. 9) can be obtained by electrochemical workstation.

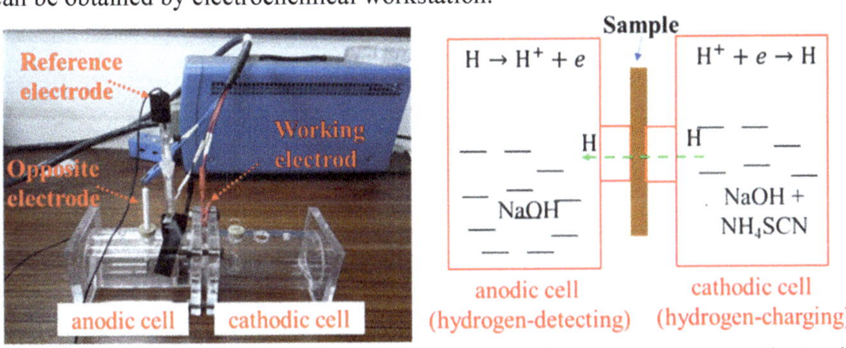

Fig. 8. Photograph (left) and schematic diagram (right) of the electrochemical hydrogen permeation test device.

Based on the analysis of permeation curve, the steady state permeation current I_∞ can be obtained directly from the curve, the apparent diffusion coefficient of hydrogen D_{app} can be calculated by Eq. 2 [13-14].

$$D_{app} = L^2 / 6t_L \tag{2}$$

where L (cm) is the thickness of the sample, and t_L(s) is the lag time where permeation current is 0.6 times of the steady state permeation current I_∞, and the hydrogen trapping sites density N_T can be determined by Eq. 3 [14-16].

$$N_T = N_L \times \left(\frac{D_L}{D_{app}} - 1 \right) \times \exp\left(-\frac{E_b}{RT} \right) \tag{3}$$

where N_T is the trapping sites density, N_L is the density of interstitial sites in steel, which is 5.1×10^{23} cm^3 suggested by Krom et al. [17]; D_L is the lattice diffusion coefficient of hydrogen in α-Fe, which is 1.28×10^{-4} cm² s⁻¹; D_{app} is the apparent diffusion coefficient; E_b is the hydrogen trap binding energy, 28905.6 J mol⁻¹; R is the universal gas constant and T is the absolute room temperature, 298 K.

The steady state permeation current I_∞ as well as the calculated apparent diffusion coefficient of hydrogen D_{app} and hydrogen trapping sites density N_T for all the samples were listed in Table 7. From Fig. 9 one can find that there exist significant differences among the three samples (1#-3#). Though all of these three samples have nearly the same I_∞ (around 6.7 μA), the sample from 2# position firstly reached the equilibrium current density, while 1# sample was the last one to reach its equilibrium current density. The calculated D_{app} values also provide the same trend. Sample 2# has the largest D_{app} value (2.09×10^{-7} cm²·s⁻¹) which is nearly two times of those of the other two samples. Samples 1# and 3# have nearly the same D_{app} value and the value of sample 1# is a little lower. The order of the magnitude of the hydrogen trapping sites density is: 1# (7.17×10^{21} cm⁻³), 3# (5.77×10^{21} cm⁻³), 2# (2.67×10^{21} cm⁻³). It can be easily understood that higher hydrogen trapping sites density can lead to slower diffusion of hydrogen atoms.

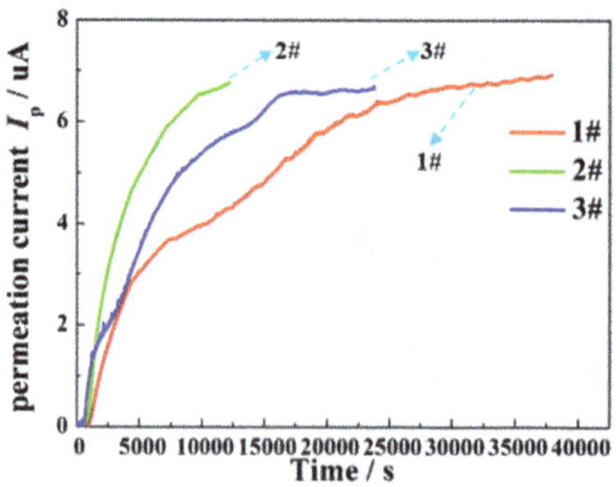

Fig. 9. Hydrogen permeation curves of the 3 investigated samples (1#-3#).

Table 7. Parameters of the samples (1#-3#) calculated from hydrogen permeation curves.

Sample	$t_{0.63}$ (s)	Thickness L (mm)	I_∞ (μA)	Apparent Diffusion coefficient D_{app} ($\times10^{-7}$ cm²·s⁻¹)	N_T ($\times10^{21}$ cm⁻³)
1#	13486	0.795	6.92	0.78	7.17
2#	3846	0.695	6.76	2.09	2.67
3#	6198	0.600	6.66	0.97	5.77

4. Conclusion

From this work, the following conclusions can be drawn:

(1) In the niobium and vanadium alloyed B pillar, nano-sized [(Nb_x, V_y, Ti_z) (C_i, N_j)] precipitates can be observed and 80 % of these precipitates have a size less than 50 nm.

(2) No obvious differences in the microstructure and mechanical properties can be found among the samples at the 3 different investigated positions without hydrogen charging.
(3) Significant differences can be found for the samples at the 3 different investigated positions of the B pillar after hydrogen charging, which were validated by the slow strain rate tensile tests and hydrogen permeation experiments.

Acknowledgments

The authors are grateful to the financial support of the National Key Research and Development Program of China (No. 2017YFB0304401), National Natural Science Foundation of China (U1564203, Nos. 51571141 and 51201105). The authors also gratefully acknowledge the support provided by Shanghai Key Laboratory of Materials Laser Processing and Modification, Shanghai Jiao Tong University.

References

1. M. Merklein, M. Wieland, M. Lechnera, S. Bruschi and Ghiotti A, Hot Stamping of Boron Steel Sheets with Tailored Properties: A Review, *J. Mater. Process. Tech.* **228**, 11 (2016).
2. H. Karbasian and A. E. Tekkaya. A review on hot stamping, *J. Mater. Process. Tech.* **210**, 2103 (2010).
3. K. I. Mori, T. Maeno, M. Sakagami, M. Ukai and Y. Agatsuma, Hot stamping of ultra-high strength steel parts, *CIRP Ann.-Manuf. Techn.* **207**, 681 (2017).
4. J. Y. Min, J. P. Lin, L. J. Xin and J. Y. Li, Indirect Hot Stamping of Boron Steel 22MnB5 for an Upper B-Pillar, *Adv. Mater. Res.* **314-316**, 703 (2011).
5. H. H. Bok, M. G. Lee, E. J. Pavlina, F. Barlat and HD Kim, Comparative study of the prediction of microstructure and mechanica properties for a hot-stamped B-pillar reinforcing part, *Int. J. Mech. Sci.* **53**, 744 (2011).
6. M. A. V. Devanathan and Z. Stachurski, The adsorption and diffusion of electrolytic hydrogen in palladium, in *Proceedings of the Royal Society of London Series A, Mathematical and Physical Sciences* **270**, 90 (1962).
7. P. W. Zhou, W. Li, Y. Li and X. J. Jin, The effect of MoS2 content on the protective performance of Ni-MoS2 composite coatings against hydrogen embrittlement in high strength steel, *J. Electrochem. Soc.* **164**, D23 (2017).
8. F. Perrard and C. ScottFabien, Vanadium precipitation during intercritical annealing in cold rolled TRIP steels, *ISIJ Int.* **47**, 1168 (2007).
9. Y. Q. Wang, S. J. Clark, V. Janik, R. K. Heenan, D. A. Venero, K. Yan, D. G. McCartney, S. Sridhar and P. D. Lee, Investigating nano-precipitation in a V-containing HSLA steel using small angle neutron scattering, *Acta Mater.* **145**, 84 (2018).

10. H. B. Wu, B. Ju, D. Tang, R. R. Hu, A. M. Guo, Q. Kang and D. Wang, Effect of Nb addition on the microstructure and mechanical properties of an 1800 MPa ultrahigh strength steel, *Mater. Sci. Eng. A* **622**, 61 (2015).
11. J. Takahashi, K. Kawakami, J. I. Hamada and Ken Kimura, Direct observation of niobium segregation to dislocations in steel
12. Research article, *Acta Mater.* **107**, 415 (2016).
13. J. H. Ryu, Hydrogen embrittlement in TRIP and TWIP steels, PhD thesis, Pohang University of Science and Technology (2012).
14. M. Devanathan and Z. Stachurski, The mechanism of hydrogen evolution on iron in acid solutions by determination of permeation rates, *J. Electrochem. Soc.* **111**, 619 (1964).
15. P. W. Zhou, W. Li, H. S. Zhao and X. J. Jin, Role of microstructure on electrochemical hydrogen permeation properties in advanced high strength steels, *Int. J. Hydrogen Energy.* 1 (2018).
16. C. Dong, Z. Liu, X. Li and Y. Cheng, Effects of hydrogen-charging on the susceptibility of X100 pipeline steel to hydrogen-induced cracking. *Int. J. Hydrogen Energ.* **34**, 9879 (2009).
17. D. F. Arau′jo, E. O. Vilar and J. P. Carrasco, A critical review of mathematical models used to determine the density of hydrogen trapping sites in steels and alloys, *Int. J. Hydrogen Energ.* **39**, 12194 (2014).
18. A. H. Krom and A. Bakker, Hydrogen trapping models in steel, *Metall. Mater. Trans. B* **31**, 1475 (2000).

Hot-stamping Application for Lightweight Torsion Beam of Passengers Vehicle

F. Xiong† and K. Yu
Qiantu Motors, Suzhou, 215163, China
†E-mail: jxiong001@126.com

Hot stamping parts have been widely used as re-enforcement for better crash safety and other attributes as well as in lightweight applications such as vehicle B-pillar, rocker and other structure parts. In this paper hot stamped torsion beam is studied as a function of product DVP and hot stamping process. Although hot-stamping process is fairly understood, the lightweight designed torsion beam's property as a function of various parameters is not well studied. In order to study the temperature field, stress field, strain field, thinning rate and phase distribution in the hot forming process of torsion beam, the finite element software was used in this paper. The study found the maximum principal strained area appeared in the highest temperature region due to the different regions of part contact with the tool in the different order while the maximum principal stress area is located in the lower temperature region. In the process of hot forming, the slow cooling rate of the part is 77 °C/s, which is greater than critical cooling rate of martensitic transformation 30 °C/s; With the increase of the initial temperature of sheet metal, the maximum content of austenite and thinning rate of the parts increase, and the forming quality of the parts decreases; with the increase of stamping speed, the thinning rate of parts is violent fluctuation. When the stamping speed is too large, parts of the temperature of parts is greater than the martensite transformation temperature in some regions at the end of forming; with the increase of friction coefficient, the thinning rate and the minimum content of martensite firstly increase and then decrease; the maximum content of austenite decreases first and then increases. Research shows that: when the initial temperature of blank is 850, stamping speed is 75 mm/s, coefficient of friction is 0.45, the quality of the parts is the best and thinning rate is 5.804 %. At the end of forming, the minimum temperature of parts is 285 °C; minimum martensite content is 75.63 % and the maximum content of austenite is 24.36 %.

Keywords: HPS; Lightweight; Thermal mechanical coupling; Hot forming; Cooling rate; Martensite; Austenite.

1. Introduction

Rear torsion beam is widely used in the A segment of passenger vehicle which part plays an important role for vehicle performance [1]. The strength, stiffness and fatigue strength are the key factors of the whole rear suspension system which are determined by the torsion beam's structure as well as the process. Torsion beam structure form not only has an important influence on the strength and stiffness of the parts, but also on the quality of the structure [2]. Four main types of torsion beam structures are current in application: V-shaped torsion beam, H-shaped torsion beam, semi-circular arc torsion beam and the pipe by mechanical pressing forming for closed section of the torsion beam. However, more hollowed variable shaped torsion beams are in use in order to meet the various strength, stiffness and modulus such as hydro formed parts.

This paper in combining with thermodynamic phase coupling numerical simulation, studied the hot forming process of temperature field, phase distribution, stress field and strain field distribution as a function of the initial billet temperature and punch velocity and friction coefficient on reduction rate of the thickness and the forming precision of torsion beam [3-7]. The results of the study have paved the way for process optimization and widen the application for hot stamping technique.

2. Materials and Parts

Commercial available hot stamping material of 22MnB5 is used for torsion beam. The initial material microstructure is ferrite and pearlite dual phase microstructure, yield strength and tensile strength were 370 Mpa and 510 Mpa, but after hot forming, the microstructure of part is lath martensite, which has greatly improved the mechanical properties, yield strength and tensile strength reached 1250 MPa and 1600 MPa respectively, and has good impact toughness.

In the process of hot forming, the temperature change is very violent which significantly influences the related properties. In this paper, Fig. 1 and Fig. 2 is the result of thermal physical properties of the material.

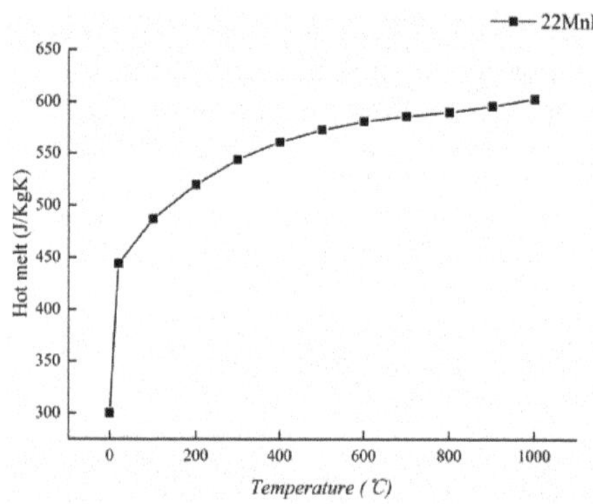

Fig. 1. Hot melt of materials.

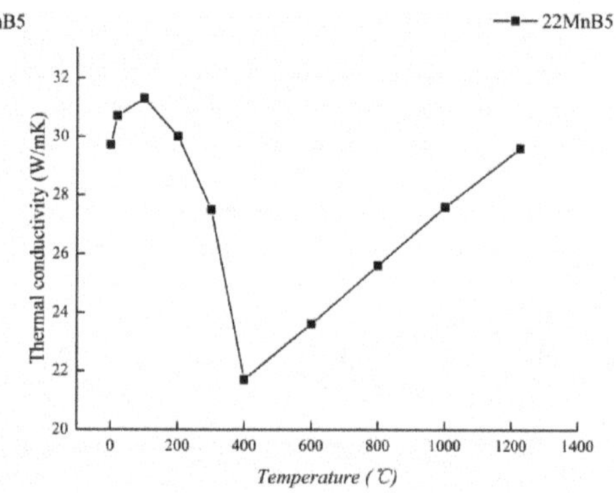

Fig. 2. Thermal conductivity of materials.

3. Hot Forming Process

The hot forming process is divided into direct and indirect hot forming and hot forming; these two kinds of hot forming have their advantages and disadvantages. Direct hot forming reduces the forming process and a compressor and process. So it reduces the production cost. This process is mainly used for simple shape and deformation of the small parts. Indirect heat forming process is before heating blank is preformed to 90 % of the parts of the final geometry by cold forming. This forming method is mainly used for the production of complex shapes, the billet deformation degree of parts [8-10]. Therefore, multiplex step forming is adopted.

In this paper, the advantages of direct and indirect heat forming are set up by using a multi - step process. The cost, the forming difficulty, the production efficiency has the irreplaceable advantage. The billet is heated to 1123 K in the heating furnace and the heat preservation time. Subsequently, the billet is transported to the forming machine, punch with 75 mm/s velocity 105 mm and ban. At this time, on both sides of the punch at the speed of 300 mm/s 150 mm forward motion, then still. At this time, the punch begins with the 75 m/s velocity 15 mm, and realizes the mold closing pressure quenching 3.4 s.

4. Results and Analysis

4.1. *Forming analysis*

Hot forming is a process that involves the coupling of thermal processes, and it is a dynamic process. In the numerical simulation process, due to the punch is applied the virtual punch velocity, in the forming process, if the ratio of the kinetic energy and internal energy of the billet occupies a large proportion., numerical simulation cannot reflect the real situation of the actual. Only when the ratio of the kinetic energy and internal energy is less than or equal to 5 %, the simulation results can reflect the real situation. Fig. 3 and Fig. 4 are the changes of kinetic energy, internal energy and total energy in the process of billet forming. From the graph, we can see that the kinetic energy is close to zero, and the ratio of total energy to internal energy is close to 1, which shows that the numerical simulation is credible.

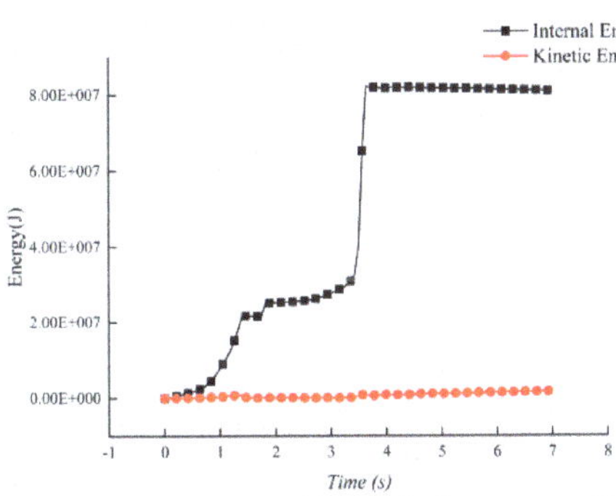

Fig. 3. Energy change with time in the process of hot forming.

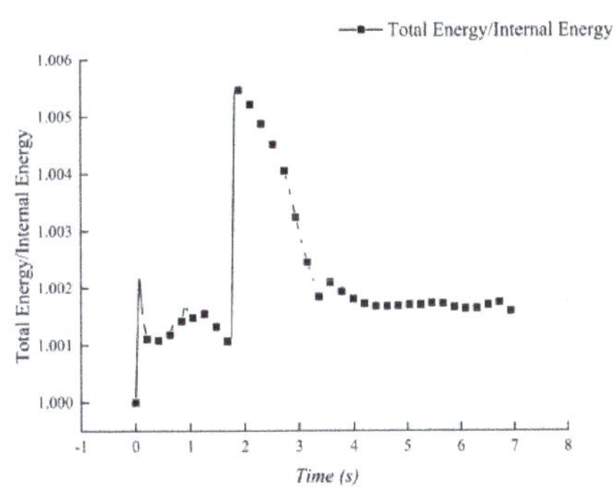

Fig. 4. Relationship between total energy and internal energy.

Fig. 5 shows the temperature field distribution in the tube blank forming process. Before left and right sides hold clamping, tube billet V-shaped section first contact with the mould with first quenching and low temperature. (Fig. 5a) After that, with the expansion of the amount of the increase, more area began to contact with mould, the temperature of these region began to decline rapidly. After the quenching, the maximum temperature is 284 °C. (Fig. 5b) The lowest temperature is in the V shaped corner. This is because from forming at the beginning, this region contact with mould tightly with a long mould quenching time and temperature can drop quickly. The highest temperature is in the lugs of V-shaped section. Before mould clamping, these has not contacted with mould with air cooling condition. After mould clamping, these region start quenching, so the quenching time is the shortest, the temperature drop is the slowest, so after the forming, the temperature is the highest.

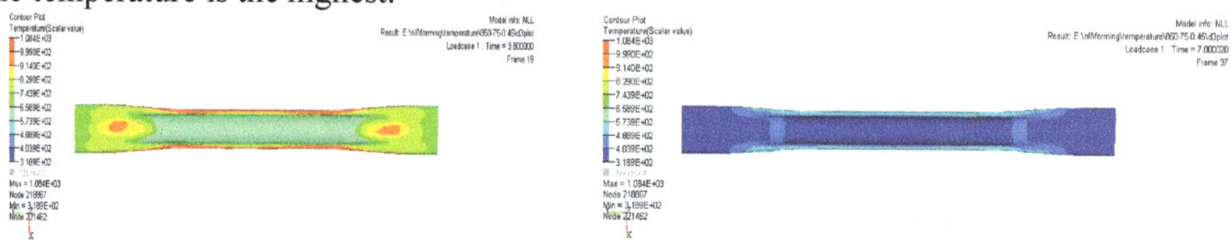

(a) End of forming-3.6 s (b) End of quench hardening-7 s

Fig. 5. Temperature field distribution during forming process.

The CCT curve of the hot forming material 22MnB5 is shown in Fig. 6. If the martensite transformation occurs during the hot forming process, the cooling rate curve of the billet is needed to avoid the curve of austenite, ferrite, bainite and pearlite. Therefore, if we want to get the complete martensite structure, the cooling rate of the billet needs to be greater than or equal to 30 °C/s.

After the end of the quenching and quenching, the maximum temperature of the part is 284 °C. By formula $Vt = \Delta T/\Delta t$. The ΔT is the change of the temperature value; the Δt value is the change of the time. So cooling rate is 73 °C/s which is greater than martensitic transformation critical speed of 30 °C/s.

After the quenching, the distribution of martensite is shown in Fig. 7. The B section is the transition region and martensite content was 77.46 %. This region is also the highest temperature at the end of the forming process. By CCT curve, before quenching, this area has undergone ferrite transformation, so quenching is inadequate. The red area martensite content is relatively high and after forming the temperature is the lowest. The quenching time of these regions is long and the highest content is 98.31 %.

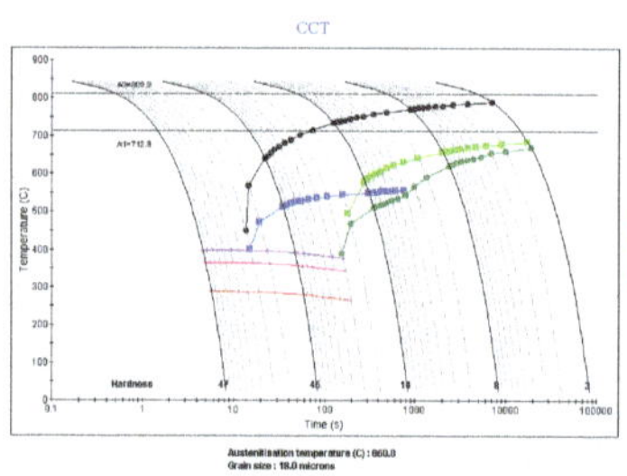

Fig. 6. 22MnB5-CCT curve.

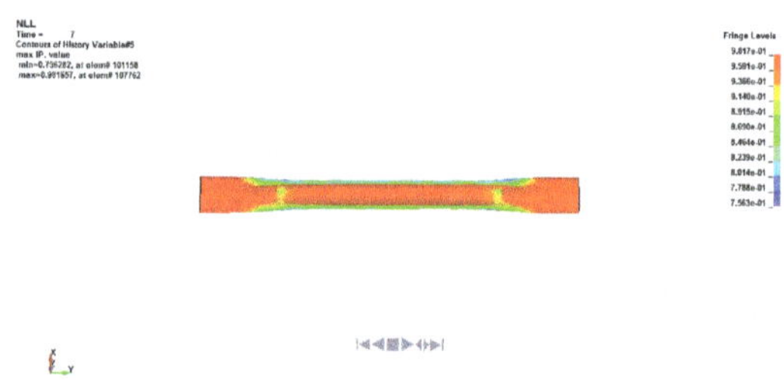

Fig. 7. Sketch map of martensite distribution.

4.2. *Parts forming results*

In the whole process of hot forming, there is a constant temperature field on the blank. Under the action of temperature field, the microstructure and mechanical properties of the billet are changed, which leads to the change of the stress field of the part, so the hot forming process is the coupling action of temperature field, stress field and phase change.

Plastic deformation occurs with high temperature and the deformation resistance is small. Hot forming can almost completely eliminate the springback, so we can obtain the higher geometric precision parts. Meanwhile, we can obtain martensite structure by quenching in the mould, so the performance of the parts has been greatly improved. So the martensite content is an important factor to test the forming quality of the hot forming.

The CCT curve of the hot forming material 22MnB5 is shown in Fig. 8. If the martensite transformation occurs during the hot forming process, the cooling rate curve of the billet is needed to avoid the curve of austenite, ferrite, bainite and pearlite. Therefore, if we want to get the complete martensite structure, the cooling rate of the billet needs to be greater than or equal to 30 °C/s.

We select A-E five points on parts Fig. 8(a), the relationship between the temperature and time changes as shown in Fig. 8(b). After the end of the forming, the temperature of five points is A > C > E > > B > D, where D is the lowest temperature, is 45 °C, A temperature is the highest, is 284 °C. Combined with CCT curve of 22MnB5, if we want to obtain fully martensitic structure, the cooling curve of the parts need to avoid the “nose” of the bainite and iron ferrite formation curve and the cooling rate should be greater than 30 °C/s. The temperature of bainite “nose” is 401 °C and ferrite “nose” is 450 °C.

From the temperature change curve in Fig. 8(b), we can get the temperature of A is 401 °C, and the forming time is 4.5 s. The initial temperature of the part is 800 °C. From formula $Vt = \Delta T/\Delta t$, ΔT is the temperature change value; Δt is the time change value. We can get cooling rate are 77 °C/s, which is larger than martensitic transformation critical speed 30 °C/s. So we can obtain complete martensitic structure. And also we can get the temperature of B is 117 °C after hot forming and the forming time is 4.5 s. The initial temperature of the part is 800 °C. We can get cooling rate are 151 °C/s.

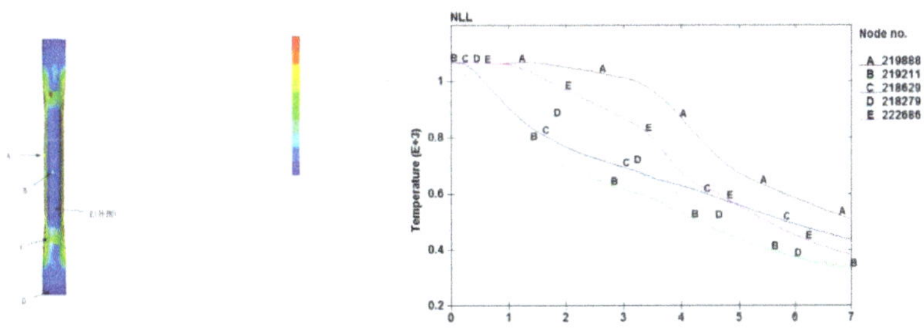

(a) Sketch map. (b) Temperature change curve with time.

Fig. 8. Relationship between temperature and time in different parts.

5. Conclusion

(1) After the hot forming, the highest temperature of the parts is 284 °C and the cooling rate is 77 °C/s which is greater than the martensite critical cooling rate 30 °C/s. So it can occur martensite transformation.

(2) The lowest temperature is in the V shaped corner. This is because from forming at the beginning, this region contact with mould tightly with a long quenching time and temperature can drop quickly. And after forming, the cooling rate of this area is 151 °C/s. So the martensite content of this area is the highest.

(3) The lowest temperature is in the B section. Before mould clamping, these has not contacted with mould with air cooling condition. After mould clamping, these region start quenching, so the quenching time is the shortest, the temperature drop is the slowest. And the cooling rate is 77 °C/s.

References

1. K. Snerider, H. Choi and J. Oh, High strength hydroformed automotive part developed by POSCO, in *hydroforming of sheet-tubes and profiles*, (Fellbach, Germany, 2008).
2. T. L. Kyung and J. B. Hong, Tube hydroforming process design of torsion beam type rear suspension considering durability, *Int. J. Mod. Phys. B* **31**, 6199 (2008).
3. C. Han and S. J. Yuan, Reduction of friction and calibration pressure by section preform during hydroforming of tubular automotive structural components, *Adv. Mate. Res*. 143 (2008).
4. J. Kiese, New lightweight stells for automobile applications potential, in *Int. conference on niobium microalloyed sheet steel for automotive applications*, (Araxa MG, Brazil, 2005).
5. J. E. Morral and J. B. Cameron, Boron hardenability mechanisms, in *the Metallurgical Society of the A. I. M. E,* (Warrendale, USA, 1980).
6. W. F. Jandeska and J. E. Morral, The distribution of boron in austenite, *Metall. T.* **11**, 2933 (1972).
7. D. A. Mortimer and M. G. Nicholas, Surface and grain-boundary energies of AISI 316 stainless steel in the presence of boron, *Metal Sci.* **10**, 326 (1976).
8. M. F. Li, T. S. Chiang, J. H. Tseng and C. N. Tsai, Hot stamping of door impact beam, in *11th Int. Conference on Tech. of Plasticity (ICTP)*, (Nagoya, Japan, 2014).
9. T. Altan, Hot-stamping boron-alloyed steels for automotive parts, part I: Process methods and uses. *Stamping J.* 4041 (2006).
10. J. T. Kim, B. M. Kim and C. G. Kang, Blank shape design process for a hot stamped front pillar and its experimental verification, *Trans. Mater. Process*. **3**, 186 (2012).

Research on the Influence of Advanced Hot Stamping Door Ring on Vehicle Performance

Y. C. Ling[1,2,†] and T. Q. Fan[1,2]

[1]*China Automotive Engineering Research Institute Co., Ltd.,Chongqing, 401122, China*
[2]*Chongqing Automotive Lightweight Engineering Research Center,Chongqing, 401122, China*
[†]*E-mail: lingyangchao@caeri.com.cn*

The structure design and optimization of the door ring are mainly based on the 25 % overlap frontal collision and side impact performance. The material thickness of each part of the door ring is rationally distributed by topology optimization, which makes the vehicle performance meet the design requirements. The analysis idea is to analyze the performance of the basic vehicle first and extract the corresponding performance parameters. Secondly, the structure of body door ring is built, and the simplified model of collision recognition is established. Thirdly, according to the simplified model displacement variation as an optimization constraint, an optimization model is established, and the thickness distribution of each part of the door ring is read to get the optimization plan. Finally, the collision performance is verified based on 25 % overlap frontal collision, and the collision safety performance meets the design requirements. The final hot stamping door ring reduces the number of parts around the door frame, simplifies the production and assembly process, makes the parts weight reduction 1.58 kg, and the weight reduction ratio of the components 10.5 %.

Keywords: Advanced hot stamping; Door ring; Passive safety performance; Lightweight structure.

1. Introduction

The hot stamping door ring can reduce the number of parts around the door frame, simplify the assembly process of the production machine, and optimize the safety performance of the car by reasonably arranging the thickness of the material in each area of the door ring. It has a certain light weight advantage, and has a broad application prospect and engineering value.

2. Performance Analysis of Basic Vehicle

The performance of the base vehicle is analyzed in 25 % overlapping frontal impact performance according to the US IIHS regulation. The IIHS 25 % bias vehicle structure rating scale includes four grades, and the limit of each grade corresponding to the entry point is shown in Figure 1. According to the requirements of the IIHS 25 % partial Collision Regulations, the measurement of the corresponding entry points of the foundation vehicle is measured, the measuring point on the body is shown in Figure 2.

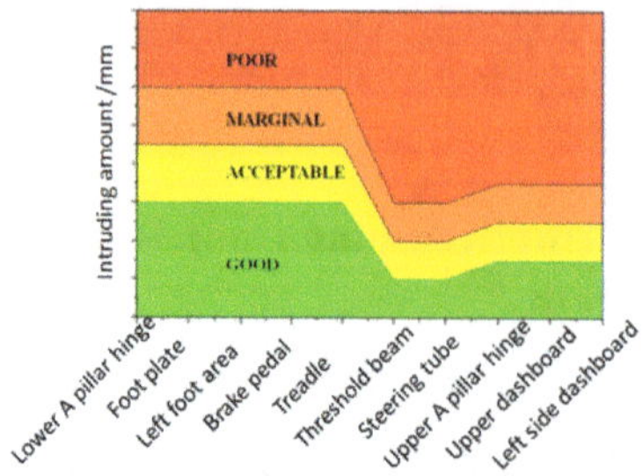

Fig. 1. Crash structure rating scale.

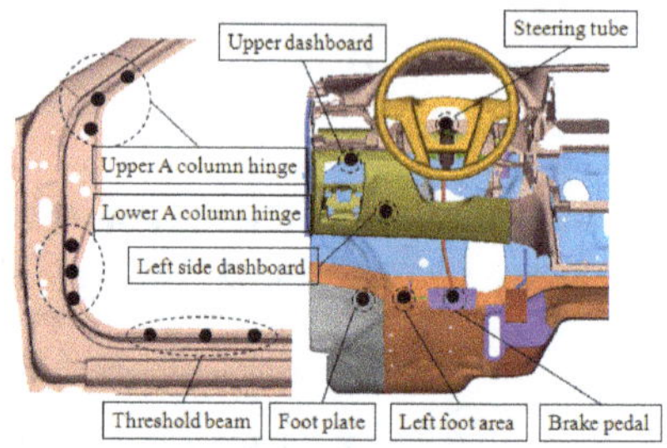

Fig. 2. Base car frontal impact test points.

The deformation of each point is shown in Table 1.

Table 1. Base car 25 % overlapped frontal impact performance.

Measurement term	Basic model	Measurement term	Basic model
Lower A pillar hinge(mm)	299.6	Steering tube(mm)	82.7
Foot plate(mm)	251.8	Upper A pillar hinge(mm)	155.4
Left foot area(mm)	151.2	Upper dashboard(mm)	188.6
Brake pedal(mm)	155.2	Left side dashboard(mm)	185.1
Threshold beam(mm)	147.2		

According to the IIHS 25 % overlap frontal impact scoring standard, the rating scale of the basic vehicle is shown in Figure 3. The 25 % bias grade of the base car is POOR.

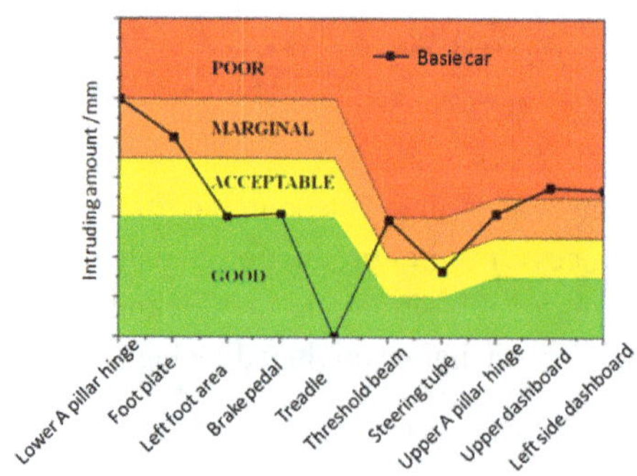

Fig. 3. Base car 25 % overlapped frontal collision results.

3. Analysis of Collision Mechanics Model

3.1. *Door ring design*

On the base of base car structure, the upper hinge mounting plate is removed, and the other parts around the front door integrate into one part, as shown in Figure 4.

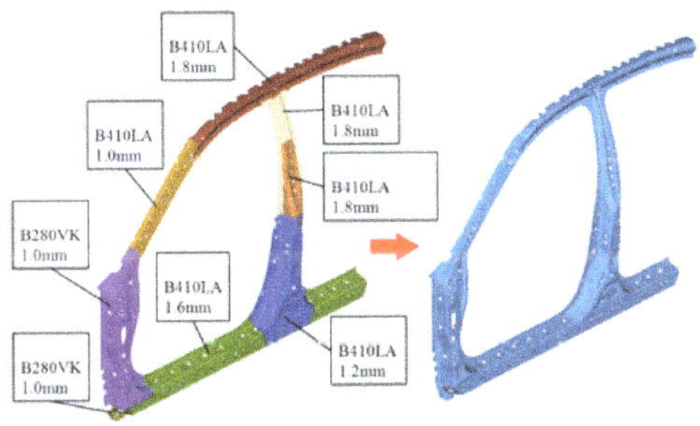

Fig. 4. Door ring design.

3.2. *Mechanical model of door ring*

A moment of dynamic mechanical model is transformed into a static mechanical model. For the 25 % overlapping frontal collision model, the degree of freedom of the upper end beam of the B column and threshold beam of can be restrained. The load of 2000 N along the collision direction is applied to the upper and lower hinge mounting holes respectively. For side collisions, the full freedom of the upper both sides of B column and the part of the threshold beam on both sides is restrained, The load on the B column corresponding to the chest of the dummy is applied along the side impact direction, the size is 2000 N. As shown in Figure 5.

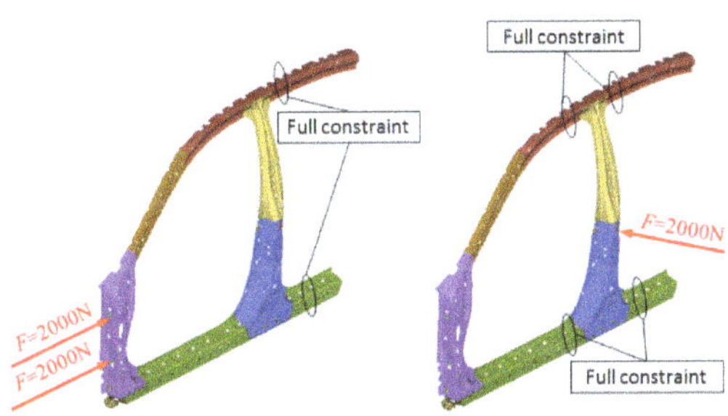

Fig. 5. A simplified model of collision mechanics for door rings.

The calculated results show that the X displacement values of the loading points of the upper and lower hinge mounting holes of the A column are 8.139 mm and 8.250 mm respectively, and the Y displacement values are -20.94 mm and -8.876 mm respectively. The Y displacement value of the loading point of the hinge mounting hole on the B column is 1.174 mm. The purpose of stress analysis of a simplified model of collision mechanics is to carry out topological optimization analysis of door ring.

4. The Optimum Analysis of the Thickness Distribution of the Door Ring

According to the simplified force model of the components around the door frame, the topological optimization model of the door ring is established, as shown in Figure 6.

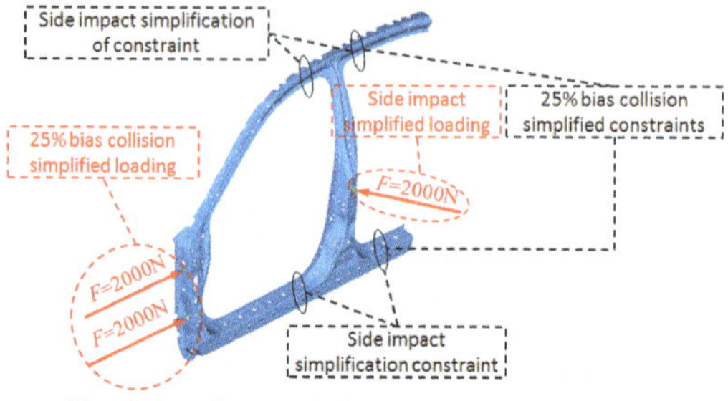

Fig. 6. Topology optimization model of door ring.

Using Optistruct to solve the optimization problem, we get the topology optimization analysis results of the door ring and the cloud map of material accumulation level. The thickness distribution of the door ring is designed according to the cloud map of material accumulation degree, as shown in Figure 7.

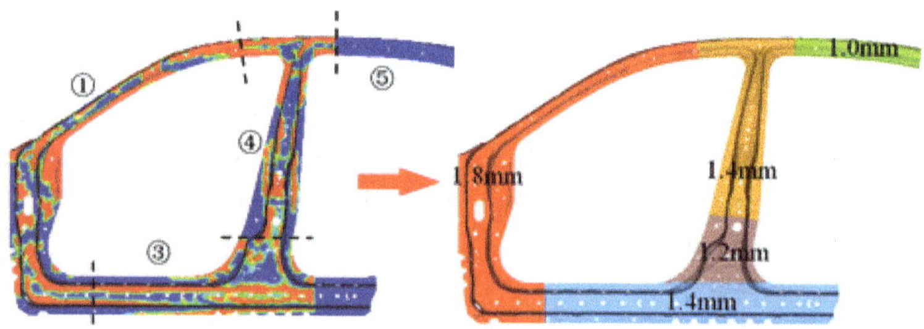

Fig. 7. An interpretation scheme for the thickness distribution of the material of the door ring.

5. Performance Verification

The thickness interpretation scheme is taken as the optimization scheme of the door ring into the whole vehicle, and the performance of the 25 % overlapped frontal collision is verified and analyzed. The performance indexes are compared to the base car changes as shown in Table 2.

Table 2. Optimization scheme 25 % overlap frontal crash performance index.

Measurement term	Basic model	Optimization scheme	Change quantity
The quality of the door frame (kg)	15.06	13.48	-1.58
Lower A column hinge (mm)	299.6	198.9	-100.7
Foot plate (mm)	251.8	209.7	-42.1
Left foot area (mm)	151.2	167.3	16.1
Brake pedal (mm)	155.2	131.2	-24
Threshold beam (mm)	147.2	90.8	-56.4
Steering tube (mm)	82.7	50.2	-32.5
Upper A column hinge (mm)	155.4	145.3	-10.1
Upper dashboard (mm)	188.6	114.7	-73.9
Left side dashboard (mm)	185.1	117.2	-67.9

The result shows that the optimization scheme is better than the base car. The 25% overlap frontal impact grade of the optimization scheme is ACCEPTABLE, which satisfies the target requirements. As shown in Figure 8.

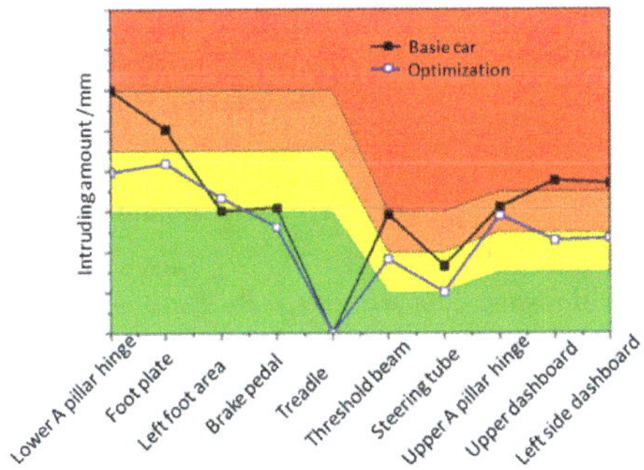

Fig. 8. Basic vehicle and optimization plan IIHS 25 % overlap frontal impact grade rating comparison.

6. Conclusion

The thickness distribution of each part of door ring is optimized by Optistruct, and the thickness distribution of door ring is obtained. The performance of 25 % overlapped frontal impact is verified. The analysis results show that the optimization scheme of door ring can meet the requirements of 25 % overlapping frontal crash and side collision.

The final door ring will integrate the upper plate of A column, the lower plate of A column, the upper beam, the upper plate of the B column, thelower plate of the B column, the threshold beam and the collector of the threshold beam and the A column into one part by stamping forming, which reduces the quantity of the development of the mould and reduces the cost of the process.

Statistical analysis results show that the design of the door parts reduces the weight of the parts by 1.58 kg, and the weight reduction rate of the parts is 10.5 %. The purpose of high-strength weight reduction is achieved through reasonable structural optimization.

Acknowledgments

This work is supported by Chongqing key industries common key technology innovation special project (No. cstc2015zdcy-ztzx60010).

References

1. L. Li, R. D. Liu, X. Xu, X. Wang, K. Q. Wang and J. Y. Guo, Research on non-isothermal warmed B pillar impact on the automotive side impact performance, *Shanghai Metal* **1**, 6 (2014).
2. F. Hou, Research on US side impact regulations based on computer simulation of the car door impact bars, 47 (2005).

The Analysis of Comprehensive Performance Evaluation of Press Hardened Steel by CSP Based on the Crashworthiness of B-pillar Reinforcement

Q. Yang[1,†], S. W. Jia[1], X. Wei[2], X. Pu[1], S. Lan[1], J. S. Zhang[1], Z. B. Wang[1] and Z. Y. Li[1]
[1]*Chongqing Changan Automobile Engineering Institute, Chongqing, China, 401122*
[2]*Wuhan Branch of Baosteel Central Reach Institute, Wuhan, China, 430080*
[†]*E-mail: yangqin1@changan.com.cn*

Exploration of press hardened steel by Compact Strip Production (CSP) is an important direction for the innovation of steel industry. It can lighten automobile body, guarantee crash safety, and reduce production cost at the same time. This paper presents the evaluated comparison of a press hardened B-pillar reinforcement and its raw material by conventional technique and CSP, which included chemical components, metallographic structure, mechanical properties, side impact simulation analysis, extreme tip cold bending angle test, three-point-bending and side impact test. The results show that, under the same condition, the performance of press hardened steel by CSP is within the allowable bearing range compared with conventional press hardened steel. Finally, this new technology used in the automobile industry has far-reaching significance, such as reducing the cost, saving energy and water, reducing emissions, and boosting the vehicle green manufacturing.

Keywords: B-pillar reinforcement; Press hardened steel; Three–point-bending; Side impact test.

1. Introduction

At present, the most challenging problems of the automobile industry are the improvement of safety and collision performance, fuel efficiency and environmental impact [1]. Due to the demand of reducing vehicle weight, improving safety and crashworthiness qualities, the need to manufacture automobile structural components by UHSS is desirable [2]. In recent years, press hardened steel has been increasingly used in car body structure for enhanced passive safety in the event of crash and reduce vehicle mass for achieving better fuel economy [3]. And use of press hardened steel with an ultimate tensile strength achieving 1500 MPa and above, has already become a common practice in designing component and sub-systems for anti-intrusion and crash energy management. So press hardened steel is a good solution to allow the automobile industry to significantly reduce the mass of vehicles to meet light-weighting requirements materially. And now we found that the exploration of press hardened steel by Compact Strip Production (CSP) has better effect than conventional press hardened steel in economy and environmentally friendly in full life circle of vehicles.

In the last 30 years, because of constant exploration and innovation, the Compact Strip Production is being continuously developed and improved, and has widely used in car body increasingly, which confirmed to be an efficient way to reduce the weight of car body and guarantees crash safety as well. At present, hot forming is an efficient way for automotive manufacturers to improve competence of their products.

In normal flow, a material needs to pass the following test verification before it is used on the collision safety parts, such as chemical components, metallographic structure, mechanical properties, side impact simulation analysis, extreme tip cold bending angle test, three-point-bending and side impact test and so on. Because, for the collision safety parts, it is necessary to ensure that the amount of intrusion is as small as possible and the energy is absorbed as much as possible when the collision occurs. Especially in the case of side impact, the cracking of safety parts is allowed only after a certain amount of energy is absorbed [4-5].

2. Material Performance Analysis

In order to verify the performance of press hardened steel by CSP, we found another conventional press hardened steels for comparative test. The measured material A (CSP-WHF1500H) is provided by Wuhan Branch of Baosteel Central Reach Institute. The material B with is provided by compared manufacturer, which is already widely used in various automobile enterprises. The thickness of both tested materials are 1.5 mm, And the chemical composition, mechanical properties and metallographic structure of these two materials are as presented in Table 1, Table 2, Fig. 1 and Fig. 2 respectively.

Table 1. Chemical composition of material A and B (wt%).

Grade	C	Si	Mn	P	S	Cr	B	Ti
A	0.21	0.32	1.35	0.014	0.013	0.32	0.0028	0.031
B	0.23	0.24	1.18	/	/	0.15	0.0025	0.0028

Table 2. Mechanical properties of static stretching of material A and B after tablet quenching.

Grade	$R_P0.2$ (MPa)	R_m (MPa)	A50 (%)
A	1030.5	1435.2	6
B	1150	1500	5

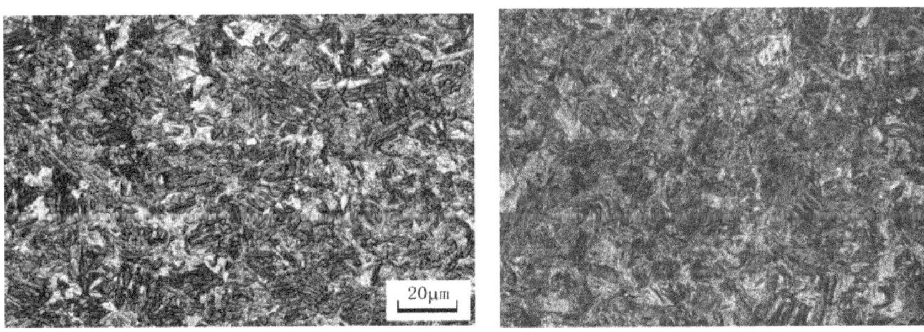

a) Material A a) Material B

Fig. 1. Metallographic structure of material A and B in after Tablet quenching.

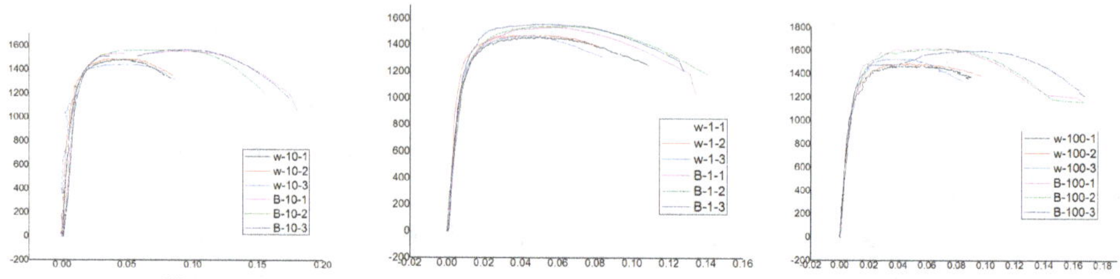

1/S 10/S 100/S

Fig. 2. Mechanical properties of High speed drawing of material A and B in after Tablet quenching.

The test results show that the chemical composition and mechanical properties of material A and material B both can meet the standard of press hardened steels after tablet quenching status. And the metallographic structure of both material A and B are martensite in strip shape and a small amount of ferrite, which is almost identical.

3. Component Performance Testing

3.1. *Performance testing for the component*

The locations of mechanical properties test samples of material A and material B are shown in Fig. 3, and the test results are shown in Table 3.

Sampling position of B-pillar reinforcement and the related analysis results are shown in Fig. 3 and Table 3 respectively. Side impact simulation analysis of B pillar reinforcement are shown in Fig. 4. What we can know from the curve in Fig. 4 is that there are some differences in the invasion velocity of material A and material B, which is still in an acceptable range.

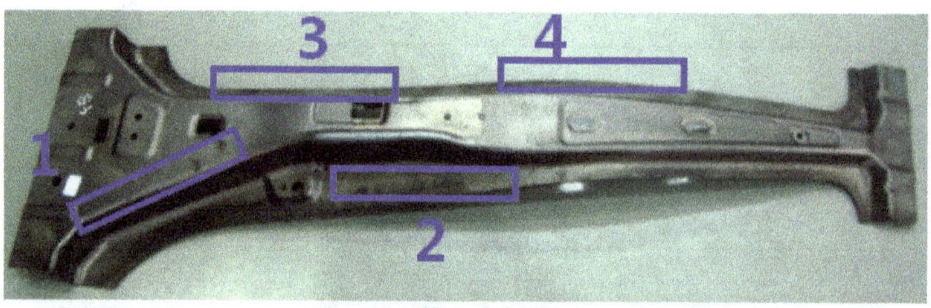

a) Material A

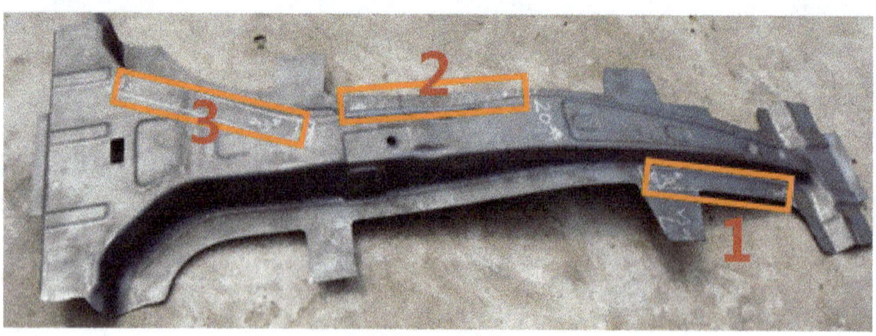

b) Material B

Fig. 3. B pillar reinforcement sample and sampling position for mechanical properties test.

Table 3. Mechanical properties of B-pillar reinforcement components.

Grade	Number	Rel/ MPa	Rm/ MPa	A50/%
A	1	1119	1466	5.8
	2	1050	1440	5.9
	3	1028	1460	5.9
B	1	1008	1493	7.3
	2	998	1481	6.4
	3	1043	1543	8.1
	4	1045	1524	6.8

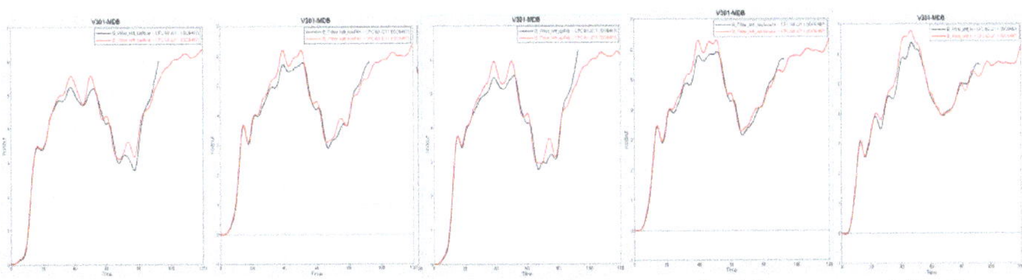

a) Safety belt b) Down Rib c) Up Rib d) Abdomen e) H-point

Fig. 4. Side impact simulation analysis of B pillar reinforcement.

As we can see from the above test results, the tensile strength of press hardened components is over 1500 MPa and the main microstructure is martensite.

3.2. *Component three point bending test*

Three-point-bending test is commonly used test method for mechanical properties, which can measure the bending strength of components stably and output the static pressure-displacement curve of test components. This test method is simple and the test results are intuitive, which can be used to evaluate the static anti-extrusion performance of a single component.

Three-point-bending is completed by a lightweight material and component testing system. As shown in Fig. 5, the loading force is 0-20 t, the displacement is 0-400 mm, the speed is 0.001-0.03 m/s, the data collection frequency is up to 10 kHz, and the fixed location of B pillar sample is also shown in the figure.

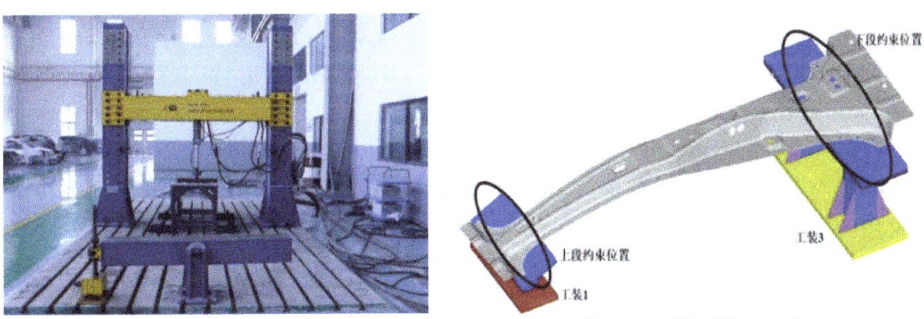

Fig. 5. Equipment and schematic diagram of three-point-bending test of B pillar reinforcement.

As the data presented in Table 4, peak static pressure of B-pillar reinforcement with material A and B are 21.86 kN, 21.24 kN, 23.70 kN and 21.94 kN respectively. Besides, the corresponding static pressure distances are 49 mm, 51 mm, 50 mm and 51 mm respectively, and the energy absorption are 1481.84 N•m, 1445.06 N•m, 1661.00 N•m and 1608.85 N•m. The results show that material A and B have similar static mechanical properties.

Table 4. Peak static pressure of B pillar reinforcement in three-point-bending test.

Grade	Number	Peak static pressure (KN)	Static pressure distance (mm)	Energy absorption (N•m)
A	1	21.86	49	1481.84
	2	21.24	51	1445.06
B	1	23.70	50	1661.00
	2	21.94	51	1608.85

3.3. *Extreme tip cold bending angle test*

Reference the German standard VDA238-100, which demand the extreme tip cold bending angle greater than 60°. As shown in Fig. 6, the extreme tip cold bending angle of material A is 65°±2°, which can meet the standard well.

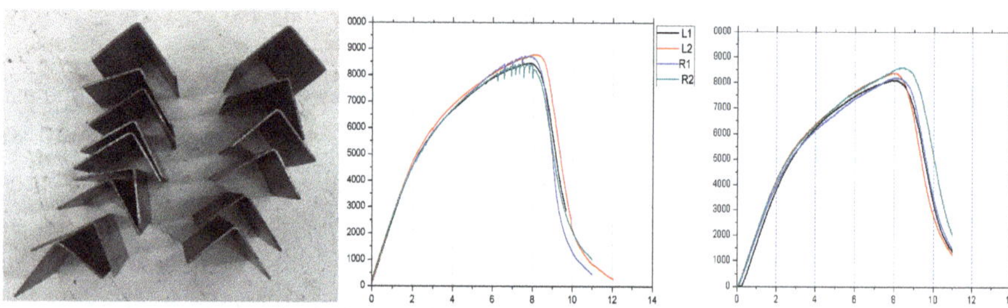

Fig. 6. Extreme tip cold bending angle test and curve.

3.4. *Crashworthiness testing*

The installation position of B pillar reinforcement in side impact test is shown in Fig. 7. B pillar sample is fixed on the rigid wall through the bolt, and the collision position of B pillar sample is determined by the normal loading position of vehicle. After welding the upper and lower ends with the mounting plate, it is fixed through the rigid wall of the bolt and collision experiment.

3 sensors are placed 006Fn the B pillar sample corresponding to the passenger's epipleural bone, abdomen and pelvis, where we can gain the intrusions at the same time. And in order to record the deformation process more clearly, two high-speed camera are placed in the left and right front of the tested B pillar sample respectively. The measured data are presented in Table 5.

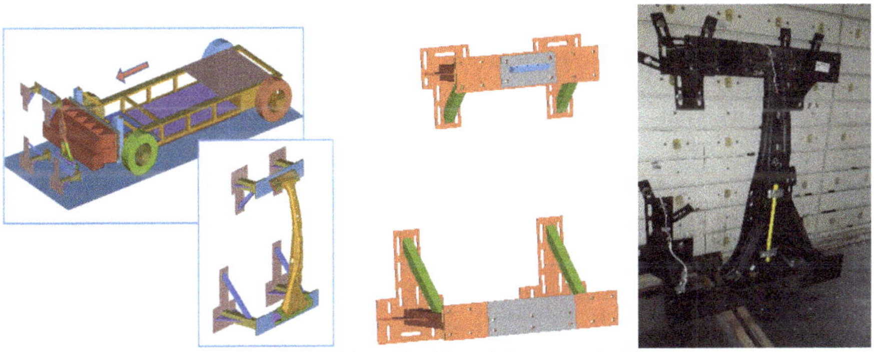

Fig. 7. The installation position of B pillar reinforcement in side impact test.

Table 5 depicts the testing conditions of side impact of B-pillar, weight and velocity of moving sled. Table 6 lists intrusion in different locations of B pillar reinforcement.

Table 5. Test conditions of side impact of B-pillar.

Test conditions	A	B	Standards
Weight of sled (kg)	950	950	950±20
Impact velocity (km/h)	11.25	11.27	11.0±0.5

Table 6. Intrusion in different locations of B pillar reinforcement.

Measured data	Epipleural bone (mm)	Abdomen (mm)	Pelvis (mm)	Impact velocity (km/h)
A	73	83	88	11.25
B	64	74	78	11.27

The results show that intrusion tested of material A is greater than material B, but still within the allowable bearing range, which means that material B has better crashworthiness than material A, and material A can also meet the safety and collision demands, and reduce the cost at the same time.

4. Summary

In terms of yield strength and tensile strength, material A have a better performance than material B, and in another term elongation, both of the two materials are similar. What we can get in this article is that after the chemical components, metallographic structure, mechanical properties, side impact simulation analysis, material A and material B hold the similar expression, which means that material A can instead of material B for commercial purposes in some condition. In conclusion, exploration of press hardened steel by Compact Strip Production (CSP) is really an important direction for the innovation of steel industry, which can meet the need of automobile industry well.

References

1. X. G. Shi, Y. Dong, B. Han, X. Xu and R. D. Liu, Influence of parameters of hot stamping technology on properties of AC1500HS steel sheet, *Angang Techol.* **383**, 5 (2013).
2. H. Karbasian and A. E. Tekkaya, A review on hot stamping, *J. Mater. Process. Tech.* **210**, 2103 (2010).
3. J. Bian and H. Mohrbacher, in *Proc. Of AIST International Symposium on New Developments in Advanced High Strength Sheet Steels*, (Colorado, USA, 2013).
4. M. Abspoel, B. M. Neelis and P. Liempt, Constitutive behavior under hot stamping conditions, *J. Mater. Process. Tech.* **228**, 34 (2016).
5. H. Z. Lu, W. J. Wang, A. M. Guo and J. Bian, Opimize the hot stamping technology base on vehicle safety, *Aut. Tech. Mater.* **10**, 8 (2013).

Technological Optimization of B-pillar Reinforced Panel Manufactured by a Press-hardening Steel Exceeding 1800 MPa

L. Huang, B. Liu†, X. D. Fang and X. Xu

OuShang Automobile Institute, Chongqing Changan Automobile Co., Ltd., Chongqing, 400000, China

†E-mail: liubo1@changan.com.cn

To validate the manufacturing feasibility of a novel Press Hardening Steel (PHS 1800), the B pillar reinforced panel was trail-produced by hot forming at temperature of 875 °C and 900°C. Besides, resistance spot welding was also carried out at welding current of 6-10 kA. Microstructure and mechanical properties of the B-pillar were examined using tensile test and scanning electron microscope (SEM). The results showed that the sample after hot formed at 900 °C exhibited the better mechanical properties, the tensile strength exceeded 1800 MPa; meanwhile, the elongation reached up to 8.6% which is comparative to conventional 1500 MPa counterparts. Under the fixed electron pressure and duration, the optimal welding current was determined to be 8.0-9.0 kA.

Keywords: Press Hardening Steel (PHS); Automobile; B Pillar reinforced panel; Mechanical Properties; Resistance spot welding.

1. Introduction

Light-weighting of the vehicles has been considered as a key technology for the global automobile manufacturers to achieve the national regulations on the energy-saving and environmental protection [1, 2]. To realize light-weighting, new materials and advanced technologies were developed and applied in the body-in-white (BIW), in which press-hardening steels (PHS) were the most popular candidate because of their cost performance [3-5]. Recently, press hardening steels has been widely used in the structural components which requires extra-high strength against the collision. At the same time, higher strength PHS was developed to reduce thickness and in turn realize the maximum weight loss. For example, a novel press hardening steel with tensile strength exceeding 1800 MPa has been developed by Benxi Steel to replace the conventional counterpart (22MnB5, 1500 MPa). The thickness of PHS 1800 sheet can achieve as low as 1.2 mm, which is usually 0.2 mm thinner than that of PHS 1500, i.e. nearly 14% weight reduction. Despite of the above advantages, there are still many forming and joining issues for the practical application of PHS materials. Therefore, it is of great importance to investigate of technology feasibility of PHS 1800.

In this paper, a B-pillar reinforced panel using PHS 1800 plate was trail-produced by hot forming at different temperatures. Subsequently, the mechanical properties and corresponding microstructure of the components after hot forming were investigated by uniaxial tension and electron scanning microscope (SEM), respectively. Additionally, the welding performance after resistance spot welding (RSW) were also carried out to find out the optimal welding parameters.

2. Optimization of Hot Stamping Temperature

In order to obtain appropriate stamping temperature, high temperature tensile test of PHS 1500 and PHS 1800 plates were performed at temperatures between 825 °C and 900 °C for a duration of 5 min. The strain rate was controlled to be 1.05 mm/min and 5.88 mm/min for the elastic and plastic stages, respectively. The tensile curves were plotted in Fig.1 and corresponding high temperature mechanical properties (yield strength, tensile strength and elongation) were measured and displayed in Table 1. The results shown that the tensile strength of both steels decreased with the increasing temperature, and negligible change for yield strength was observed. However, both steels exhibit distinct strain response at different temperature. PHS 1500 samples displayed a declining elongation with the rising temperature, for which tendency is contrary to that of PHS 1800 samples. At 825 °C and 850 °C, the elongation of PHS 1500 is apparently higher than that of PHS 1800. With increasing to 875 °C, the elongation almost equal to each other. Finally, the elongation reached up to 57% for the PHS 1800 at 900 °C which is twofold as compared to PHS 1500 counterpart. It is well-known that the PHS material needs good ductility during hot stamping. Therefore, the optimized stamping temperature can be recommended to be 875 °C and 900 °C.

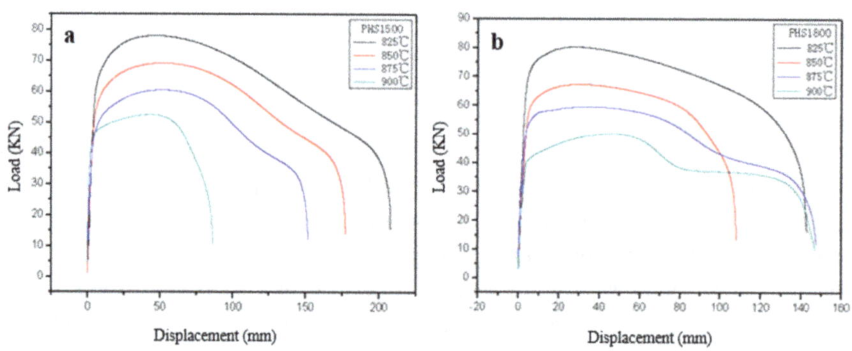

Fig. 1. Tensile curve for (a) PHS 1500 and (b) PHS 1800 plate at different temperatures.

Table 1. Tensile results of PHS 1500 and PHS 1800 at 825-900 °C.

Temperature(°C)	Steel grade	$Rp_{0.2}$(MPa)	Rm (MPa)	A (%)
825	PHS 1500	32.4	78	86
	PHS 1800	31.2	80.3	55
850	PHS 1500	28.9	69.2	64
	PHS 1800	32.4	67.4	39
875	PHS 1500	28.0	60.6	56
	PHS 1800	29.4	59.7	58
900	PHS 1500	32.8	52.6	30
	PHS 1800	25.8	50.2	57

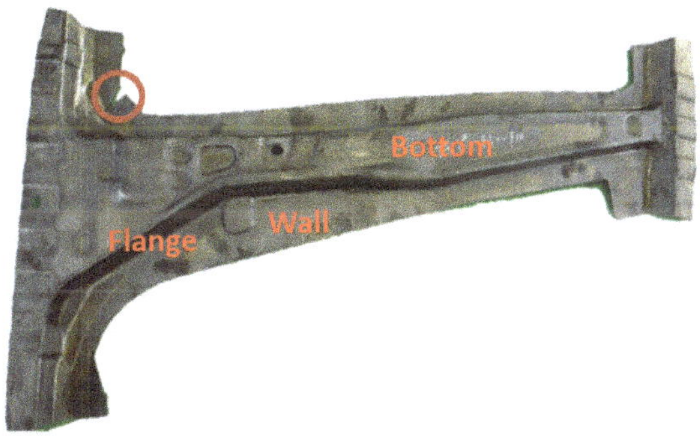

Fig. 2. Sampling location for the tensile test after hot stamping.

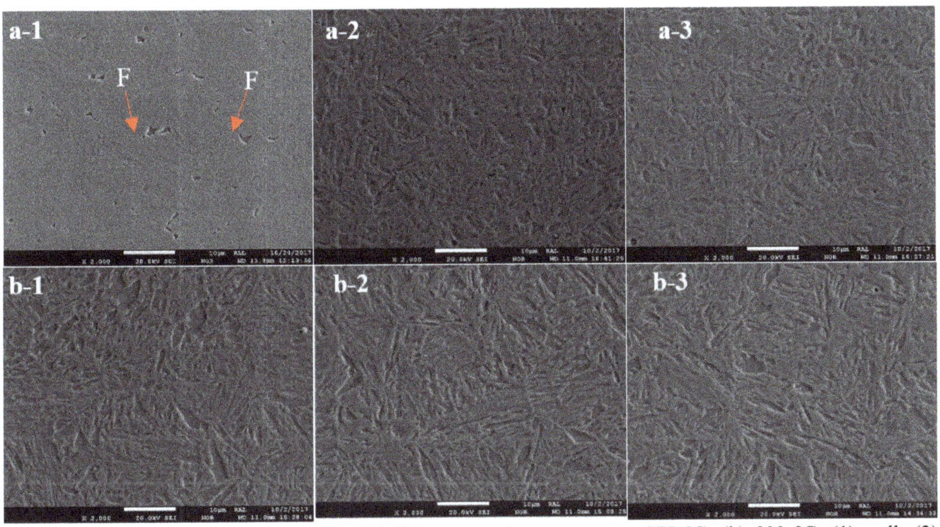

Fig. 3. SEM microstructure of B-pillar at different stamping temperature.875 °C; (b) 900 °C; (1) wall; (2) bottom; (3) flange.

The yield strength (YS), tensile strength (TS) and elongation of the samples on the B-pillar were measured by tensile tests, and the results were listed in Table 2. It can be found that the strength of "Wall" position was slightly lower than that of "Bottom" and "Flange", which is attributed to the formation of soft ferrite in this region. The yield strength of all samples exceeds 1526 MPa with a maximum of 1614 MPa, while the minimum tensile strength can reach up to 1872 MPa. In addition, an average elongation of 7.5% can be obtained. However, the average tensile strength of the samples at 900 °C is 71 MPa higher than that of the sample at 875 °C, which indicates that 900 °C is prove to the optimal hot stamping temperature for the PHS 1800.Conventional 22MnB5 (1500 MPa level) was also hot stamped at 900 °C and its mechanical property were measured as listed in Table 3. Compare with two steels, the tensile strength and elongation of PHS 1800 are higher than that of PHS 1500, which could be attributed to the combination of grain size and precipitation hardening [5].

To figure out the baking response of the PHS 1800, the samples was annealing at 170 °C for 20 min in the muffle furnace. The mechanical properties were also measured and listed in Fig. 2. After annealing at 170 °C, both the strength and ductility slightly increase, which means that baking is good for mechanical property.

Table 2. Mechanical properties of PHS 1800 B-pillar hot formed at 875 °C and 900 °C.

Location	Stamping temperature	170°C /20 min (Y / N)	YS $Rp_{0.2}$ (MPa)	TS Rm(MPa)	Elongation (%)
Wall	875	No	1526±43.0	1872±2.3	8.1±0.30
	900		1614±49.2	1940±96.8	5.6±1.41
Bottom	875		1570±0.24	1904±22	6.5±0.04
	900		1568±44.0	1961±34.4	7.9±0.96
Flange	875		1586±5.0	1913±15.1	8.6±0.6
	900		1654±17.2	2001±19.1	8.6±0.43
Wall	875	Yes	1591±17.2	1837±7.6	7.9±0.88
	900		1693±29.9	1918±27.4	5.8±0.63
Bottom	875		1635±15.2	1868±24.3	7.1±0.13
	900		1654±6.3	1909±6.8	8.0±0.62
Flange	875		1648±9.1	1882±2.9	8.8±0.05
	900		1707±21.1	1935±8.7	9.2±1.17

Table 3. Mechanical properties of 22MnB5 B-pillar hot formed at 900 °C.

Location	Stamping temperature	170°C /20 min (Y / N)	YS $Rp_{0.2}$ (MPa)	TS Rm(MPa)	Elongation (%)
Wall	900	No	1110±25.6	1600±32.1	5.8±0.23
Bottom	900		1100±12.9	1550±15.4	6.8±0.54
Flange	900		1080±8.5	1580±10.2	6.8±0.36

3. Optimization of Blanking Shape

From Fig. 2, it can been seen that one corner of B-pillar was cracked after hot forming (as marked by red circle). Similar failure phenomenon were also observed in the forming of PHS 1500 materials. Currently, to solve such problems, the most common approach is to adjust blanking size or shape on the crack region. In this case, the primary reason for the cracking could be resulted from the large friction stress around the corner. Therefore, the materials in the corner was cut off inward (as seen in Fig. 4) for the sake of reducing friction, and hence promoting the metal flow in this region. Finally, this issue was solved effectively by this means, and no cracking appeared in the subsequent B-pillar reinforced panel, as displayed in Fig. 5.

a 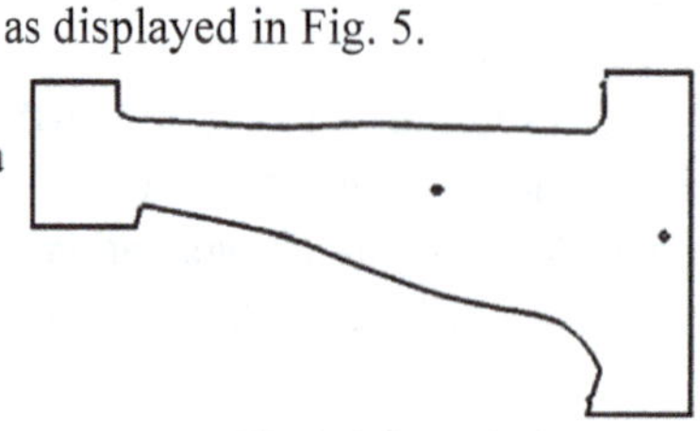

b 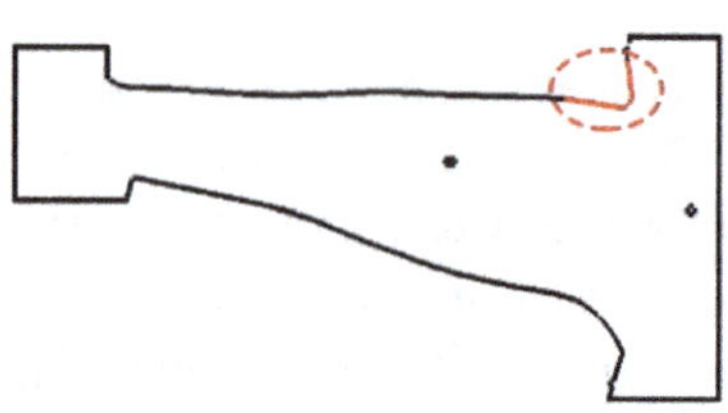

Fig. 4. Schematic diagram showing blanking shape optimization.

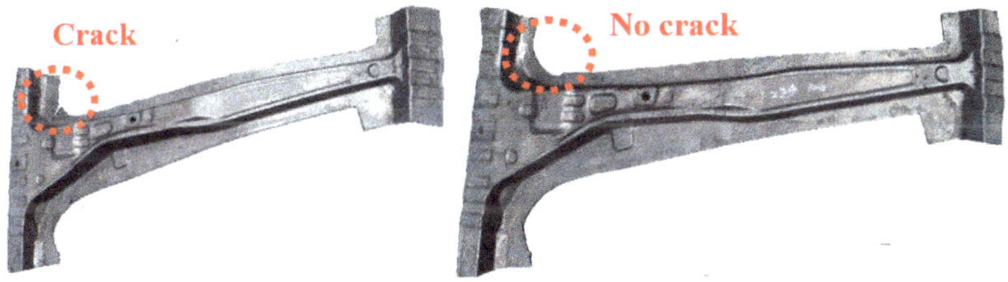

Fig. 5. No crack occurs after blanking shape optimization.

4. Optimization of Spot Welding Parameters

It is generally known that welding property is another key indicator for evaluating the industrial application of the press hardening steel. Hence, the influence of current on the welding performance and nugget diameter of resistance spot welding was investigated in the present section. According to the tensile strength of PHS 1800 in the previous section, the electron pressure (F) and welding duration (t) were determined to be F=10 kN and t=20 cyc, respectively. The shear fracture load and nugget diameter were measured in the welding current (C) range from 6 kA to 10 kA, and corresponding results were listed in Table. 4. It can be seen that the tendency of shear fracture load and nugget diameter is almost the same. Initially, an increasing current resulted in a raised shear fracture force and nugget diameter, reaching a peak of 24.44 kN and 7.47 mm when the current equals to 9 kA. Subsequently, both force and nugget diameter decreased with increasing welding current. Essentially, this demonstrated that the welding current is key factor to the nugget diameter, which in turn affected the shear strength. However, it does not mean that high welding current inevitably lead to large nugget diameter, because welding spatter may occur in the weld process when the current reach a certain value, such as C=9.5 in the present case. Based on the above results, the ideal welding current window for the PHS 1800 can be determined to be 8.0~9.0 kA under a fixed condition of the electron force was F=10 kN, welding duration t=20 cyc.

Table 4. Shear fracture force and nugget diameter of resistance spot welding at different current.

Number	Current (kA)	Shear fracture force (kN)	Nugget diameter (mm)	Welding condition	Fracture mode
1	6	8.96	4.08	Normal	Tearing at interface
2	6.5	14.08	5.24	Normal	Pulling out nugget
3	7	14.68	5.08	Normal	Pulling out nugget
4	7.5	16.78	6.91	Normal	Pulling out nugget
5	8	23.67	6.52	Normal	Pulling out nugget
6	8.5	22.56	7.23	Normal	Pulling out nugget
7	9	24.44	7.47	Normal	Pulling out nugget
8	9.5	23.68	5.76	Spatter	Pulling out nugget
9	10	18.36	7.31	Spatter	Pulling out nugget

5. Conclusion

In the present work, the B-pillar reinforced panel was successfully trail-produced using a novel PHS 1800 material. Hot forming temperature and blanking shape were designed to produce a quality B-pillar reinforced panel; in addition, the influence of welding current on the mechanical properties of the joint was investigated to acquire optimal welding windows. It can be draw some conclusions and listed as follows:

(1) The optimum hot stamping temperature of PHS 1800 was determined to be 900 °C, which resulted in 100% martensite after quenching.

(2) The PHS hot formed at 900 °C exhibited excellent mechanical properties with tensile strength exceeding 1800 MPa and an average elongation of 7.5%.

(3) Crack problem was solved effectively by means of optimizing the blanking shape which resulted in a reduced frication.

(4) The ideal welding current window for the PHS 1800 can be determined to be 8.0~9.0 kA.

Acknowledgments

This work is supported by a grant for Chongqing Innovation Program of the Key Industrial Generic Technology (No. cstc2017zdcy-zdzxX0006) and partially supported by Chongqing Research Program of Distinguished Youth Foundation (NO. cstc2014jcyjjq50002) and The National Key Research and Development Program of China (No.2016YFB0101600, No.2016YFB0101601).

References

1. M. T. Ma and H. L. Yi, Application of high strength steel to manufacturing auto, *Heat treatment* **26**, 9 (2011).
2. H. J. Kim, C. Mcmillan, G. A. Keoleian, and S. J. Skerlos, Greenhouse gas emissions payback forlight-weighted vehicles using aluminum and high-strength steel, *J. Ind. Ecol.* **14**, 929 (2010).
3. W. Y. Zhu, Development process on ultra-light steel and high strength steel sheet for automobile, *Shanghai Metal* **25**, 11 (2003).
4. Y. Y. Zhao, Study on hot forming process of car B-pillar reinforcement plate, Master thesis, Hefei University of Technology (2017).
5. M, Merklein and J. Lechler, Investigation of the thermo-mechanical properties of hot stamping steels, *J. Mater. Process. Tech.* **177**, 452 (2006).
6. H. L Yi, L. Sun and X. C. Xiong, Critical assessment: Challenges in the formability of the next generation of automotive steel sheets, *Mater. Sci. Technol.* **34**, (2018).

Simulation and Performance Analysis for Upper Legform to WAD 775 mm Test

S. J. Cui†, W. Xu, Q. J. Zhao, Q. Liu, K. Peng and G. Liu

State Key Laboratory of Vehicle NVH and Safety Technology, China Automotive Engineering Research Institute Co., Ltd., Chongqing, 4011222, China

†E-mail: cuishujuan@caeri.com.cn

This paper introduces the upper legform to WAD775mm test protocol, rating protocol and performance development strategies. Taking a real vehicle as an example, an upper legform to WAD775mm finite element model is established by Hyper-mesh. From the perspective of the component structure optimization, some suggestions are proposed to increase the absorption of impact energy. After analysis, the test score improves from 2.963 to 6.000 points.

Keywords: Pedestrian protection; Upper legform; WAD 775 mm.

1. Background and Current Status

China is a typical country where pedestrain and vehicles are mixed on road. According to the annual report of traffic accident data, from 2006 to 2013, among domestic road traffic accidents, the fatalities from pedestrian accidents accounted for about 25 % of the total number of deaths and increased year by year. In Europe, 23% of traffic fatalities are pedestrians and in the United States this data is 11 % [1].

Fig. 1 shows the distribution of human injuries in pedestrian accidents in China (taking the statistics of traffic accidents in Changsha city as an example) and abroad (taking Germany as an example) [2, 3]. Thus it can be seen that, in addition to the head and lower leg, pelvic injuries accounted for a higher proportion. It has great research significance.

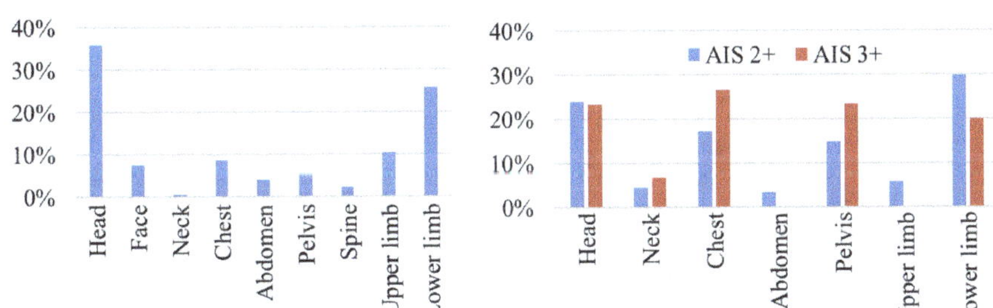

(a) Changsha, China (b) Germany

Fig. 1. Distribution of human injuries in pedestrian accidents in China and abroad.

The study on Pedestrian protection in China is relatively later than Europe and Japan. GB/T 24550-2009 Cars for pedestrian impact protection and C-NCAP (2018 version) only specify headform and legform tests. China insurance automotive safety index (C-IASI) released on July 20th 2017, taking the upper legform to WAD 775 mm test as a monitor item, which hope to carry out a study on the domestic vehicles in this test project and conducted a comprehensive assessment on the injuries suffered by pedestrians in traffic accidents.

This paper focuses on a upper legform to WAD 775 mm test of a vehicle, according to the requirements of C-IASI, and build the FE model. According to the test and simulation results, an optimal design of the vehicle is proposed.

2. Performance Analysis of Upper Legform to WAD 775 mm of a Vehicle

2.1. *Test and rating criteria of upper legform to WAD 775 mm*

In upper legform to WAD775mm test, use upper legform impactor instead of pedestrian thigh. The maximum bending moment M and the resultant force F are taken as the damage evaluation criteria [4], as shown in Table 1.

Table 1. Performing parameter of upper legform to WAD775 mm.

Criteria	Low value	Score	High value	Score
Bending moment / Nm	350	0.000	285	1.000
Resultant force / kN	6.0	point	5.0	point

Test point location aligned WAD 775 mm, and the impact angle of every test point α is shown in Fig. 2. During the impact process, the nominal impactor energy to be used in the test shall be calculated using the following formula, as in Eq. (1):

$$E_n = 0.5 \times m_n \times v_c^2 \tag{1}$$

Where, m_n=7.4 kg, v_c=v_0×cos (1.2α), v_0=11.11 m/s.

The test velocity v_t shall then be adjusted to meet the nominal energy by using the following formula [5, 6], as in Eq. (2).

$$v_t = \sqrt{\frac{2E_n}{10.5kg}} \tag{2}$$

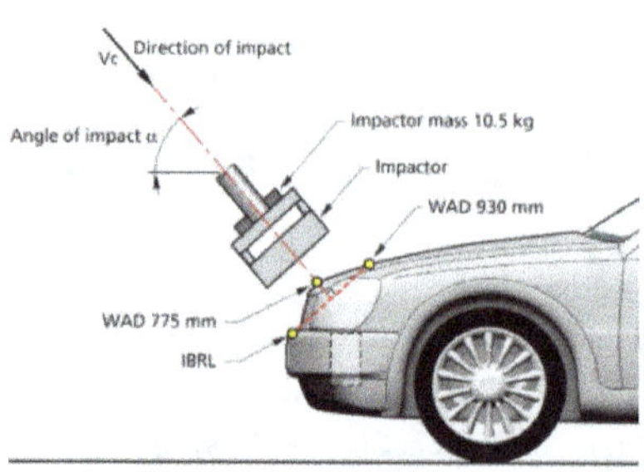

Fig. 2. Upper legform to WAD775mm test condition.

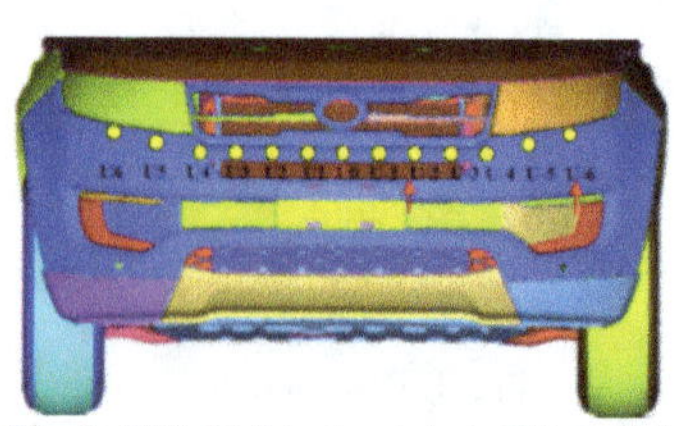

Fig. 3. U-2, U-6 test points in FE model.

2.2. *FE model*

FE model was established with Hyper-mesh software. Set the material and thickness of the parts according to the actual vehicle parameters, and establishing the connection between the various components by welding, bolts, hinges and etc. The analysis only cut off the front parts of the B-pillar of the vehicle, including the engine bonnet, the dash panel and other front parts, the components inside the engine compartment and the front suspension All the six degrees-of-freedom of the cross-section were constrained. Parameters of the impactor are set up in accordance with the protocol. The front bumper and its accessories are made of plastic, low strength and poor toughness, and the parts are easy to shear off in the crash process. In order to make the FE model closer to the reality, fracture failure model is set up for the key energy absorption components to improve the calculating accuracy.

Upper legform to WAD 775 mm test of U-2 and U-6 test points are simulated in FE model, as shown in Figure 3. Comparison of the simulation and test results are shown in Figure 4, the trend of the curves is similar and the peak value of the simulation and test curves are all within 10 %.

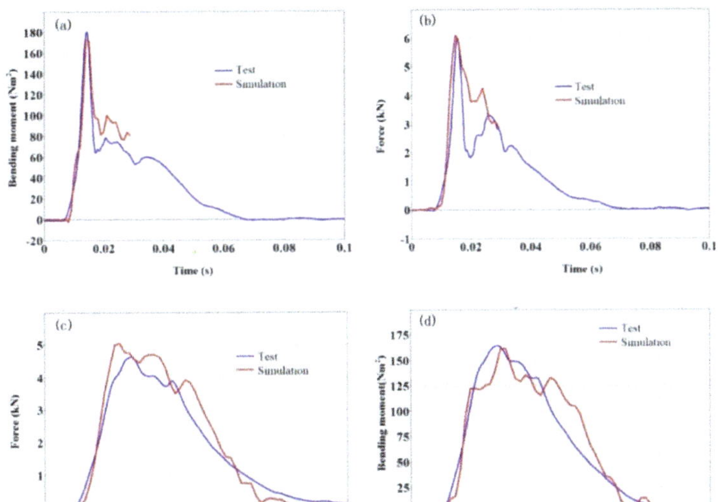

Fig. 4. Comparison of the test and simulation result curves: (a) Resultant force curve-U-2, (b) Max bending moment curve-U-2, (c) Resultant force curve-U-6, (d) Max bending moment curve-U-6.

2.3. *Performance development strategy of upper legform to WAD 775 mm*

2.3.1. *Automobile body shape-designing*

A test is not required if the calculated impact energy would be less than 160 J according to C-IASI protocol. The impactor energy only depends on the impact angle α,

$$0.5\times 7.4\times[11.11\times\cos(1.2\alpha)^2]<160 \tag{3}$$

After calculation, when $\alpha > 44.7$ °, the impact energy is less than 160 J and can reach the test exemption conditions, the test score can get the default green. The straight line passing through the internal bumper reference line and WAD 930 mm at the same lateral position from the vehicle centerline is more closer to horizontal line, the bigger the impact angel is. The flatter the front shape of the vehicle, the more it reduces the height of the WAD 930.So the impact angle α is related to the front of the vehicle shape, and the height of the bumper beam. Therefore, in the early stages of vehicle body shape-designing, setting the impact angle input as the goal of the vehicle body shape-designing, will effectively improve the vehicle crashworthiness [7]. As shown in Fig. 5, in E-NCAP upper legform to WAD 775 mm test, TOYOTA Prius has got the default green.

Fig. 5. The vehicle body shape.

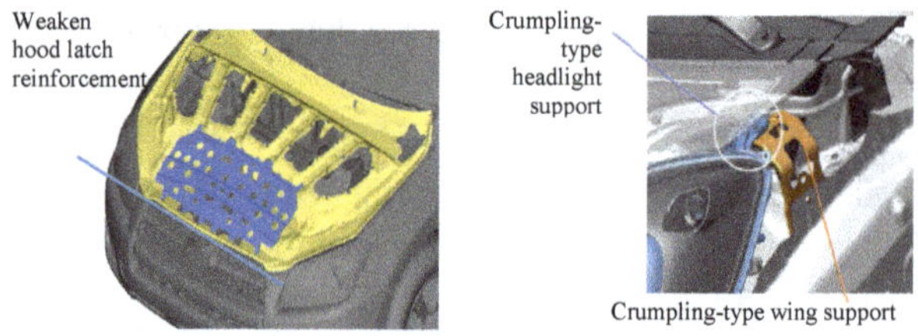

Fig. 6. Optimization of component structure.

2.3.2. *General layout*

According to the impact energy formula $E = F \times S$, when the impact energy is constant, the larger the impact energy absorption space S is, the smaller the impact force F is. Therefore, the components under the bonnet should have enough distance from the bonnet surface, so that the kinetic energy generated by the upper legform to WAD 775 mm test can be absorbed as much as possible [8]. Hard points and other dangerous features should be arranged as far as possible outside the test area.

2.3.3. *Component structure optimization*

The loss points of upper legform to WAD 775 mm test are usually concentrated in the hood latch, headlight, radiator upper mounting, and the support structure on the bumper and so on; these structures are usually maximum stiffness [9, 10]. Under the premise of outer shape and layout of the energy-absorbing space are curtained, only structural optimization of components is possible, weakening the strength of support components in the impact direction, fully deformed during the impact test. For example, some installing supports of the headlight using the overall collapse of the energy absorption structure to improve the local impact energy absorption effect, or using new materials to make it easy to shear off and deform. Hood latch reinforcement should be designed large surface, but have weak stiffness. In this condition, it can not only disperse the impact energy when close the bonnet, but also can reduce the damage to pedestrians, as shown in Fig. 6.

2.4. *Optimization of FE model*

As the development strategy mentioned before, and under the condition that the manufacturers do not want to change the outside shape of vehicle, we can only consider optimizing the energy absorption path and increase the deformation space.

Front bumper structure has small internal space, and the components layout is compact, there is no available deformation space. After analyzing and observing the simulation animation, we can find that in the impact process, the front bumper structure and its accessories are all rigid, inadequately deformed. Among them, the middle support structure of the front bumper is strong, and has little deformation. It is the biggest hard point in the process of upper legform deformation. In the test, this component also has varying degrees of cut off or shear off. Through rounds of simulation, the optimization of structure design is finally determined as shown in Table 2, to optimize the energy absorption and increase the deformation space.

(1) Opening poles on the front bumper symmetrically;
(2) Deleting the rib reinforcement of the trim in the middle of the front bumper, and opening poles too;
(3) Slotting both sides of the middle support structure of front bumper, opening poles and deleting the rib reinforcement, adding reinforcement.

Table 2. Optimization of structure design.

Before	After	Sample
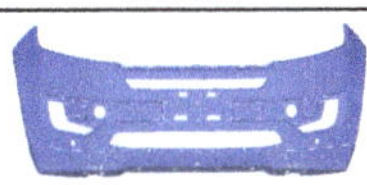	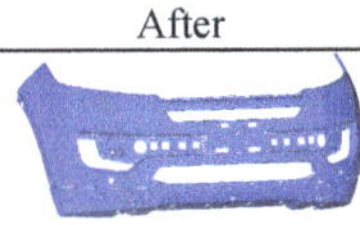	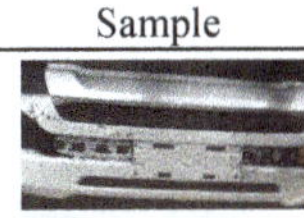
		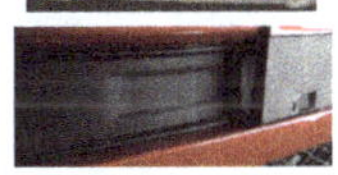

3. Performance Verification of Upper Legform to WAD 775 mm

3.1. *Test results before optimization*

The test zone is determined by the corner reference point, and a grid point is marked every 100 mm, a total of 13 grid points are marked on this vehicle, as shown in Fig. 7. From U0, six test points U5, U3, U0, U-2, U-4 and U-6 are selected for the test, see Fig. 8.

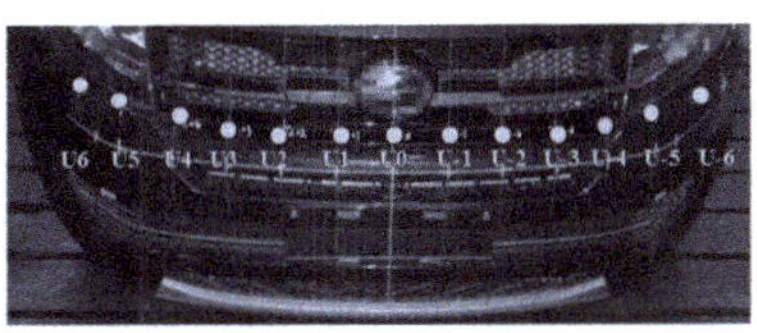

Fig. 7. Marking.

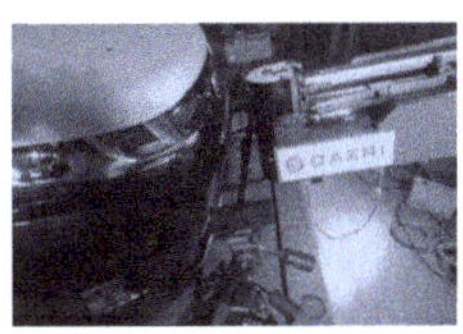

Fig. 8. Test.

Grid points that have not been tested will be awarded the worst result from one of the adjacent points. Symmetry will also be applied to all grid points on the opposite side of the vehicle. So the scores of each test point are: U6,1.000, U5,0.000, U4,0.750, U3,0.940, U2,0.010, U1,0.010, U0, 1.000, U-1,0.010, U-2,0.010, U-3,0.940, U-4,0.750, U-5,0.000, U-6,1.000. The upper legform to WAD775mm test score of the vehicle before optimization is 2.963 points. The results are shown in Table 3.

Table 3. Test results of each test points-before optimization.

	Upper bending Moment /Nm	Middle bending Moment /Nm	Lower bending Moment /Nm	Resultant Force /N	Score
U5	250	370	399	6.72	0.00
U3	169	256	271	5.06	0.94
U0	165	208	198	4.02	1.00
U-2	180	263	275	5.99	0.01
U-4	207	297	301	5.04	0.75
U-6	164	196	195	4.64	1.00

3.2. *Results of optimized model*

The optimized model is simulated and tested by U1, U3, U5, three impact points. The test result is shown in Table 4. After optimization; the resultant force and bending moment have a certain degree of reduction. Every grid point get 1.000 point, so the total score of upper legform is 6.000 points.

Table 4. Simulation and test results of optimized model.

	Resultant Force / kN		Upper bending moment /Nm		Middle bending moment /Nm		Lower bending moment /Nm	
Test point	Simulation	Test	Simulation	Test	Simulation	Test	Simulation	Test
U1	4.02	3.89	123.5	133.8	169.4	175.5	171.1	176.2
U3	4.59	4.79	140.7	142.0	192.5	195.5	195.4	201.8
U5	4.25	4.76	196.1	208.4	259.2	269.3	275.6	283.1

4. Conclusion

(1) According to the C-IASI Pedestrian Safety Index (2017 Version), this article first conducted a upper legform to WAD 775 mm test, getting a score of 2.963 points. The finite element model of this vehicle is established, and the results are compared with the test results. The accuracy of the model satisfied the requirements.

(2) The performance development strategy of the upper legform to WAD 775 mm test is described in detail. Based on performance development strategy and body structure, some suggestions are proposed to increase the absorption of impact energy. The results show that after applying the optimized design on the vehicle, the upper legform to WAD 775 mm test result increased to full score -6.000 points.

(3) At present, there are few researches on this project in our country. The development strategy introduced in this paper is forward-looking and it can be used for the development of subsequent models.

Acknowledgments

This work is supported by Special project of key technology innovation of key industries in Chongqing (CSTC2015ZDCY-ZTZX60010).

References

1. International traffic safety data and analysis group, *IRTAD Road Safety Annual Report*, 2009.
2. C. Y. Kong, J. K. Yang and L. Li, In-depth investigation of vehicle traffic injuries, in Changsha of China, in *Expert Symposium on Accident Research (ESAR)*, (Hannover, Germany, 2006).
3. O. Zander, D. U. Gehring and M. V. Ratingen, Beyond safety legislation: contribution of consumer information programmes to enhanced injury mitigation of pedestrians during accidents with motor vehicles, in *24th ESV conference proceedings*, (Washington DC, USA, 2015).
4. S. Mark, Pedestrian safety upper legform bumper impact simulation, in *The 17th International Technical Conference on the Enhanced Safety of Vehicles (ESV)*, (Amsterdam, Netherlands, 2001).
5. Pedestrian protection test protocol of pedestrian safety index. China insurance automotive safety index protocol (2017 version), China insurance automotive safety index administration center, (2017).
6. Pedestrian protection rating protocol of pedestrian safety index. China insurance automotive safety index protocol (2017 version), China insurance automotive safety index administration center, (2017).
7. Y. Okamoto, A. Akiyama, M. Okamoto and Y. Kikuchi, A study of the upper leg component tests compared with pedestrian dummy tests, in *The 17th Int. Tech. Conference on the Enhanced Safety of Vehicles (ESV)*, (Amsterdam, Netherlands, 2001).
8. J. Wang, D. H. Zhao, Z. Q. Lv, H. L. Yu, T. S. Cui and H. Zhao, Research on the performance of pedestrian upper leg impact bonnet leading edge, *Proceeding of China Society of Automotive Eng.*437 (2014).
9. B. B. Nie, Y. Xia, J. Huang, Q. Zhou, Q. Liu and H. B. Du, A study on vehicle front-end styling and structure concept design for pedestrian upper leg impact protection, *Automotive Eng.* **36**, 286 (2014).
10. B. B. Nie, J. Huang, Y. Xia and Q. Zhou, Influencing factors of contact force distribution in pedestrian upper legform impact with vehicle front-end, *SAE Int. J. Passenger Cars- Mechanical Systems.* **5**, 231 (2012).

Author Index

www.ingramcontent.com/pod-product-compliance
Lightning Source LLC
LaVergne TN
LVHW081351011026
849161LV00013BA/2789